THE CHILDREN OF ATOMIC BOMB SURVIVORS

A Genetic Study

edited by

James V. Neel and William J. Schull

NATIONAL ACADEMY PRESS
Washington, D.C. 1991

NATIONAL ACADEMY PRESS • 2101 Constitution Avenue, N.W. • Washington, DC 20418

The National Academy of Sciences is a private, nonprofit, self-perpetuating society of distinguished scholars engaged in scientific and engineering research, dedicated to the furtherance of science and technology and to their use for general welfare. Upon the authority of the charter granted to it by the Congress in 1863, the Academy has a mandate that requires it to advise the federal government on scientific and technical matters. Dr. Frank Press is president of the National Academy of Sciences.

The National Research Council was organized by the National Academy of Sciences in 1916 to associate the broad community of science and technology with the Academy's purposes of furthering knowledge and of advising the federal government. Functioning in accordance with general policies determined by the Academy, the Council has become the principal operating agency of both the National Academy of Sciences and the National Academy of Engineering in providing services to the government, the public, and the scientific and engineering communities. The Council is adminstered jointly by both Academies and the Institute of Medicine. Dr. Frank Press and Dr. Robert M. White are chairman and vice chairman, respectively, of the National Research Council.

Library of Congress Cataloging-in-Publication data

The children of atomic bomb survivors : a genetic study / edited by
 James V. Neel and William J. Schull.
 p. cm.
 Includes bibliographical references and index.
 ISBN 0-309-04488-X (hard); ISBN 0-309-04537-1 (soft)
 1. Children of atomic bomb victims —Health and hygiene—Japan—
Hiroshima-shi. 2. Children of atomic bomb victims—Health and
hygiene—Japan—Nagasaki-shi. 3. Radiation—Toxicology. 4. Genetic
toxicology. I. Neel, James V. (James Van Gundia), 1915–.
II. Schull, William J.
RA648.3.C49 1991 91-10042
614.5'909—dc20 CIP

Printed in the United States of America

Preface

The National Academy of Sciences–National Research Council, in concert with the appropriate Japanese agencies, has been engaged for some 45 years in an attempt to understand the late health effects on the survivors of the atomic bombings of Hiroshima and Nagasaki. From the outset of this long-term study, one of the major questions demanding consideration has been the possible genetic effects of this exposure. This issue became the first of many to be addressed on a comprehensive scale in the follow-up studies. Planning for the genetic program began in 1946, and data collection has been continuous since 1948. Over the years, the effort has been funded by the United States Atomic Energy Commission, the Energy Research and Development Administration, and the Department of Energy, and the Japanese Minstry of Health and Welfare.

For this book, Drs. Neel and Schull have selected for reprinting 13 publications which express their scientific views and those of their Japanese colleagues. Their purpose is to disseminate the results obtained in the genetic study, and they have added an analytical introduction and epilogue. This collection of reprints from scientific journals, and the authors' analysis of the possible course of future studies, should be of great interest to the scientific community and, ultimately, to a concerned public.

The NAS–NRC takes great pride in its involvement in this unique, long-term research program. The data obtained in the Atomic Bomb Casualty Commission–Radiation Effects Research Foundation (ABCC–RERF) studies, which are available to scientists throughout the world, have provided the basis for estimating genetic, as well as carcinogenic, risks resulting from radiation exposure. They have also been of international importance in the establishment of standards for protection of people in the work place and the general public.

FRANK PRESS
President
National Academy of Sciences

Contents

Orientation	1
Genetic Effects of the Atomic Bombs in Hiroshima and Nagasaki	9
The Effect of Exposure to the Atomic Bombs on Pregnancy Termination in Hiroshima and Nagasaki	13
Atomic Bomb Exposure and the Pregnancies of Biologically Related Parents	271
Some Further Observations on the Sex Ratio Among Infants Born to Survivors of the Atomic Bombings of Hiroshima and Nagasaki	280
A Cohort-Type Study of Survival in the Children of Parents Exposed to Atomic Bombings	291
Mortality in the Children of Atomic Bomb Survivors and Controls	327
Cytogenetic Study of the Offspring of Atomic Bomb Survivors, Hiroshima and Nagasaki	344
Search for Mutations Altering Protein Charge and/or Function in Children of Atomic Bomb Survivors: Final Report	363
Congenital Malformations, Stillbirths, and Early Mortality among the Children of Atomic Bomb Survivors: A Reanalysis	377

Malignant Tumors during the First 2 Decades of Life in the Offspring of Atomic Bomb Survivors	389
Mortality Among the Offspring (F_1) of Atomic Bomb Survivors, 1946-1985	401
The Children of Parents Exposed to Atomic Bombs: Estimates of the Genetic Doubling Dose of Radiation for Humans	431
The Comparative Radiation Genetics of Humans and Mice	451
The Future of These Studies	487
Biographical Sketches	495
Name Index	497
Subject Index	509

Orientation

The attempt to determine the nature and extent of the genetic effects of the atomic bombs detonated over Hiroshima and Nagasaki is at this writing by any one of several criteria far and away the most extensive exercise in genetic epidemiology ever undertaken. As the study has unfolded, it mirrors much of the amazing history of human (and general) genetics since World War II. This volume is designed to bring together a set of publications which will not only summarize the various investigations undertaken during the course of this program but also reflect the evolution of the program as it attempted to incorporate the new genetic technologies as they emerged.

The effort by geneticists to predict the genetic consequences for humans of exposure to ionizing radiation, of which this study is a principal component, has certainly been one of the several most serious (and controversial) social responsibilities they have faced in the past half century. But not only is this subject a major issue in its own right, it also serves as a prototype for the type of genetic epidemiology necessary to evaluate the impact on our genetic selves of other human perturbations of the environment in which our species now functions. It is a highly technical issue, ill served by overly simplistic approaches and hasty generalizations.

Between 1946 and 1974, the investigations reported herein were carried out as part of the activities of the Atomic Bomb Casualty Commission (ABCC), a program jointly operated by the U.S. National Academy of Sciences and the Japanese Institute of Health. In 1975, the ABCC was replaced by an independent binational foundation designated the Radiation Effects Research Foundation (RERF), with joint oversight by the U.S. National Academy of Sciences and the Japanese Ministry of Health and Welfare. Throughout the entire period, funding for the American contribution to the effort has been derived from the Atomic Energy Commission and its successor agencies, the Energy Research and

Development Administration and the Department of Energy; funding for the Japanese contribution has come from the Japanese Ministry of Health and Welfare. We, the Japanese and American authors of the papers which are reproduced in this volume, are grateful to the successive directors and the staff of the ABCC amd RERF, whose staunch support has so greatly facilitated the conduct of these studies.

BEGINNINGS

The early events in the history of the genetics program are recorded in Chapter 2. In 1946, knowledge both of the doses of radiation sustained by survivors and the sensitivity of the mammalian genome to radiation was far inferior to the present situation. Nevertheless, as the preliminary data on post-atomic bomb Hiroshima and Nagasaki emerged, it became very likely, given the number of survivors in the two cities and their probable gonadal doses, as well as the indicators one would be forced to pursue in any study, that even a very major effort would not yield a statistically significant difference between the children of survivors receiving increased radiation at the time of the bombings (ATB) and the children of suitable controls. Furthermore, it was clear that a valid study would be expensive and of considerable duration (but at the time, no one anticipated a program that may well extend into the twenty-first century). On the other hand, the exposure of so large a group of normal human beings to radiation, some up to the limits of human tolerance, was so singular in human history that there really was no question in the collective mind of the Committee convened by the National Academy of Sciences to consider the issues, but that a genetic study should be undertaken. Nor were there any doubts in the minds of our Japanese colleagues (see pp. 30-31 of Chapter 3).

There were other unique aspects to this nascent operation. The primary focus of the first round of studies was the occurrence of congenital defects in newborn infants. This is a sensitive subject to parents all over the civilized world, invoking the folklore stigma of some occult wrong doing. This was indeed true in Japan. The study, which would thus have been sensitive in any civilized country, clearly had to be as considerate as possible of Japanese mores and culture. There was from the outset a dynamic tension, between the need to respect the parents who would be involved and their cultural conditioning, and the need to determine pregnancy outcomes accurately, for the sake of these same worried survivors and a humanity just beginning to cope with the implications of the atomic era. In this regard, the Japanese survivors were, immediately after the war, being inundated by grossly exaggerated predictions concerning the fate of their children. Indeed, in those early years no less a figure in genetics than J.B.S. Haldane could suggest, in 1955, that the doubling dose of radiation for humans could be as low as 3 roentgens. In retrospect, we would like to believe that whatever offense our visiting medical teams occasionally rendered by intrusions on families in moments of sorrow is now more than offset by the dispersal of the ugly clouds which for a period were a serious impediment to the marriage of their children. By this statement we do not deny the genetic implications of these exposures but do strongly urge, as will become apparent, that they were far less than unbridled popular belief.

THE THREE PHASES OF THE STUDY

Phase I (1948-1954)

During this period, the program, in conjunction with the operation of a national rationing system, registered almost all pregnant women in Hiroshima and Nagasaki at the

completion of the fifth lunar month of pregnancy and then collected data following the birth of the child on viability, presence of congenital defect, neonatal death, and on sex of child and birthweight. Given the fact that during this period most births were attended by a midwife and occurred at home or in small facilities maintained by the midwives, collecting accurate data on the occurrence of congenital malformation with reference to all births in the two cities required a major effort directed at ensuring the examination of the child by a physician as soon as possible after birth. With respect to children who were stillborn or died during the neonatal period in Hiroshima, some 56% came to autopsy. Altogether, some 76,626 pregnancies were registered in Hiroshima and Nagasaki during this period; an additional 8,391 pregnancies were registered in Kure, which during 1948-1950 served as a "control" city. Approximately 30% of the surviving children in Hiroshima and Nagasaki were reexamined at age 8-10 months at the ABCC, the reexamination including physical measurements.

In view of recurrent conjectures concerning the possibility that critical data on the genetic effects of the bombs were missed because of the "delayed" initiation of the program (cf. Roberts, 1990), it seems worthwhile to review the time course of the early events rather carefully. The first children conceived after the bombings would (aside from premature terminations) have been born in May 1946. Planning for the genetic studies began in November, 1946; the registration of pregnancies in this program (at the completion of the fifth lunar month) was initiated in Hiroshima in February of 1948 and in Nagasaki in July of the same year, the midpoint of these two dates being approximately May 1948. The first pregnancies to be registered represented conceptions occurring in December, 1947. The "window" during which there was no direct coverage of newborn infants was approximately 18 months. However, the liveborn infants during this 18-month period were, on the basis of birth registrations, subsequently incorporated into the studies on F_1 mortality (Chapters 6, 7, and 12) and malignancies among the F_1 (Chapter 11), and also drawn upon for the cytogenetic (Chapter 8) and biochemical (Chapter 9) studies. The "window" is thus rather specific for congenital defect.

A preliminary analysis of the accumulated data in 1953 revealed no significant difference between children according to the exposure category of their parents. The experience between 1948 and 1953 with the numbers of children born to parents thought to have received significant amounts of radiation at the time of the bombings made it possible to predict probable future trends with respect to the number of offspring of such parents. There was a precipitous drop in the Japanese birth rate in the early 1950s, due to the introduction of liberal abortion laws designed to help Japan cope with the rapidly expanding population created by the repatriation of Japanese from China, Korea, Formosa, and Manchuria following World War II. This fact, plus the attrition inevitable in a study of this nature, suggested that with respect to congenital malformation and stillbirth, the additional numbers to be obtained in the future were very unlikely to alter the small differences between exposure subgroups from statistically non-significant to significant.

At our request, in July of 1953, a Committee was convened by the National Academy of Sciences to examine both the analysis and the projection. The Committee concurred with the analysis, and voted to recommend termination of the clinical aspects of the program in early 1954. It was agreed, however, that the registration of births and the study of the sex-ratio and survival of the offspring should be continued, not only because of the intrinsic interest of these indicators but also with the thought that future genetic developments might bring new approaches to the cohorts which were being identified.

We then undertook the definitive analysis of these data which constitutes Chapter 3.

The results of the preliminary analysis were confirmed and extended. Further details concerning the pattern of congenital malformation in Japanese, as revealed by the at-birth and nine-months examinations, and the autopsy program in Hiroshima, are to be found in Neel (1958). A particular focus of the monograph which constitutes Chapter 3 was the exclusionary power of the data, i.e., the magnitude of the effect which could exist and go undetected, at various probability levels. Although on the basis of a burgeoning experimental literature it could not be doubted that the survivors of the bombings had sustained some genetic damage, our philosophy during this period was that it would be inappropriate to speculate on the magnitude of that damage in the absence of statistical significance. This decision was fortified by the relatively poor understanding at that time of the genetic (mutational) component in the various indicators under scrutiny.

Because marriages between cousins were relatively common among registrants in the program (6% in Hiroshima, 8% in Nagasaki), and because the individuals engaging in cousin marriages might not be randomly distributed in the two cities with reference to the hypocenter (a possibility which would introduce a source of bias into the data), we felt it best to exclude children resulting from such marriages from the analysis described in Chapter 3. With the main analysis completed, however, we then undertook a separate analysis of the much smaller sample of children born to cousins (Chapter 4), which analysis, within the limits of sample size, failed to suggest that these children exhibited any greater sensitivity to the genetic effects of radiation than the children of unrelated parents. These children of consanguinity, plus suitable controls, later became the subjects of a major study (Schull and Neel, 1965).

Periodically we encounter questions concerning (unspecified) pressures brought to bear on the study by the Atomic Energy Commission during these early days. Let us say very simply that we were never, either in Washington (where the pressure would be on the National Academy of Sciences) or in Japan, aware of any political pressure meant to influence the organization of the study or the way the data were analyzed. Problems there were, in the conduct of the study and the analysis of the data, but this was not one of them.

Phase II (1955-1968)

During this period, data collection was essentially limited to the sex-ratio of newborn infants in the two cities, and the survival of liveborn infants. The sex-ratio studies were thought to be important because of earlier, equivocal findings that the ratio had indeed been altered by parental exposure, whereas the continuing registration of liveborn infants (now through civil procedures rather than a pregnancy registry) ensured a proper population for study, should future events dictate a resumption of a hands-on program. In addition, an effort was initiated to determine if there was a relation between parental radiation histories and a simple battery of anthropometric measurements of Hiroshima schoolchildren obtained in 1965 in the course of an annual school examination program.

Among births occurring in Hiroshima and Nagasaki during this period, only some 15% involved parents who were thought to have incurred significant radiation exposures ATB. Since the periodic updating of the survival status of each liveborn child was time consuming, and there was a great excess of births to parents receiving no radiation ATB in the city samples, the study of survival was reorganized in 1958. Three cohorts of children were established at this time, namely, one comprised of all children born to survivors one or both of whom were "significantly" exposed (within 2,000 meters of the hypocenter ATB, the "proximally exposed") and two age-sex-city matched control cohorts, one of children born to parents not in the city ATB, and the other to parents one or both of whom

were "distally exposed" (i.e., in either city but beyond 2,500 meters from the hypocenter ATB). Chapters 5, 6, and 7 present analyses based on these cohorts describing the findings on sex ratio and survival of liveborn infants through 1969. Chapter 6 is particularly important because it describes and compares in considerable detail these three cohorts, from which the subjects for studies yet to come were derived.

A word must be said about the assignment of radiation doses during this period. The analyses described in Chapters 3-6 are based on radiation categories established by us on the basis of distance from the hypocenter, shielding, and symptoms ATB. A potential weakness in the data was that the mother supplied the father's radiation history at the time she registered the pregnancy. It was not until the early 1960s that the Dosimetry Section of the ABCC had in hand detailed radiation histories on all significantly exposed survivors residing in these cities at the time of the 1950 Japanese national census, at which point a direct assessment of fathers' exposures became available. A codified approach to calculating doses for survivors, known as the Tentative 1965 Dosimetry System (T65D), was introduced in 1965 and was the basis for the analysis of Chapter 7. This dose schedule was replaced in 1986 by a system of assigning doses which is presumably definitive, the Dosimetry System, 1986 (DS86). It is noteworthy that the results of the analyses of phase III, employing the DS86 schedule, are in substantial agreement with the results of the analyses during phases I and II, employing other dose systems, although there was a substantial gain in statistical power in the transition from the early analyses based on group doses to the later analyses based on individual doses.

Phase III (1969-1990)

This phase of the study recognized both the emergence of important new technologies for the study of these cohorts of children and the growing insights into the genetic contribution of mutation in the parental generation to congenital defect, malignancy, and early death. The first of the studies to be initiated during this period, in 1968, dealt with the cytogenetic findings in a subset of the F_1 Mortality Cohorts. Inasmuch as for practical reasons the necessary venous blood samples were not obtained until the child had attained 13 years of age, the findings are considered valid with respect to sex-chromosome aneuploidy and balanced chromosomal rearrangements, conditions in which survival is essentially normal, but not with respect to unbalanced chromosomal rearrangements or autosomal aneuploidy (Chapter 8). In 1975, after a three-year pilot study, a search was initiated for mutations resulting in electrophoretic variants of a series of 30 proteins of serum and erythrocytes, and in loss of activity in a series of 11 erythrocyte enzymes, again in a subset of the F_1 Mortality Cohort sampled at age 13 or above (Chapter 9).

As knowledge of the genetic basis of the childhood malignancies improved, it was felt worthwhile to make the incidence of these malignancies in the cohorts being followed for survival the object of a special study (Chapter 11). Finally, the early data on untoward pregnancy outcomes were reanalyzed (Chapter 10), and the data on survival were brought up to date through 1985 (Chapter 12). The DS86 dose system mentioned earlier was available for the latter three analyses but not for the first two (the cytogenetic and biochemical studies). However, in the paper (next section) which explores the implications of the findings for the genetic doubling dose, the data of Chapters 8 and 9 have also been put in the context of DS86 doses.

AN ESTIMATE OF THE GENETIC DOUBLING DOSE OF RADIATION

The complex of data assembled in the course of these studies can first of all be viewed as an empirical exercise in evaluating the tangible risks to children of parental radiation exposure. In particular, we note that within the limits of the statistical error inherent in estimations of this type, they refute many of the unbridled and ill-advised speculations appearing in the past in the public presses of the world, speculations whose impact on the morale of exposed survivors the reader can well imagine. It would in our opinion, however, have been a breach of public trust to terminate the analysis at this point, especially in light of the explosive increase in recent years in knowledge concerning the mutational component of the various indicators employed in the study. In this connection, we note that the control data collected during phase III provide major input to our knowledge of spontaneous mutation rates with respect to certain types of chromosomal and protein variants.

None of the analyses of the data sets just mentioned yielded results suggestive of a significant difference between the children of exposed and of controls. The average combined gonadal doses of the exposed parents, however, approximated what on the basis of experiments with mice have been extrapolated to be the gametic doubling dose of acute radiation for humans. As these analyses unfolded, we began to sense a possible discrepancy between these findings on the children of survivors and the expectations based on projections from the mouse paradigm which had dominated the risk assessments of various national and international committees. There might be a message in the consistent lack of any notable difference between the cohorts under study.

On the grounds that we could accept the hypothesis that ionizing radiation produced mutations in humans, and that the corpus of data from Japan was the most substantial to be available in this century, we accordingly undertook a cautious exploration of the implications of the data for the human genetic doubling dose of ionizing radiation (Chapter 13). The latter is defined as the amount of acute or chronic (including low-level intermittent) radiation that will produce the same mutational impact on a population as occurs spontaneously each generation. We prefer this metric to any absolute calculation of the genetic impact of radiation because of the perspective which a doubling dose estimate supplies. In point of fact, doubling doses undoubtedly differ according to the genetic phenomenon under consideration; the estimate which was derived was an average. Undertaking to estimate a doubling dose represents a major evolution in our thinking concerning the most appropriate way to address these data.

The application of this approach to these data has not been easy. The procedures we have employed have no precedent. In the absence of the statistical difference between study groups, which is the traditional point of departure for most scientific propositions, we could just be manipulating the noise in the system. Furthermore, to develop the argument, we must make assumptions concerning the contributions which spontaneous mutation in the parental generation made under the conditions existing in post-war Japan to the complex of congenital defect, stillbirth, and death during the neonatal period, as well as to death during infancy and childhood. We needed also to devise the most appropriate way to combine results across studies, and to decide on the appropriate dose rate factor, in extrapolating from the results of acute radiation to what the results of chronic radiation might be.

The resulting estimate of the gametic doubling dose, of approximately 2 Sv for the acute radiation exposures experienced by the survivors of the bombings, and, employing a dose rate factor of 2, of 4 Sv for chronic or intermittent, low-LET radiation, is substantially

higher than several national and international committees have projected for humans on the basis of their evaluation of the experimental murine data. Inasmuch as for technical reasons discussed in Chapter 13 the estimate could not incorporate the data on physical development, sex ratio, and balanced chromosomal rearrangements, none of which indicators were suggestive of a radiation effect, the estimate is thought to be conservative. This apparent discrepancy between the human and mouse data led to an effort to reanalyze the existing published data on murine radiation genetics, together with the addition of some previously unpublished data, the better to understand the reason(s) for the apparent difference. In fact (Chapter 14), in our hands much of the discrepancy disappeared: the best estimate of the gametic doubling dose for the levels of acute radiation employed in the experiments on mice is approximately 1.4 Sv, and for chronic or intermittent, low-LET radiation, employing a dose rate factor of 3 for the extrapolation, in the neighborhood of 4.0 Sv. (The higher dose rate factor for the murine data is necessitated by the substantially larger gonadal doses employed in the experiments with mice.) For reasons that become apparent in Chapters 13 and 14, we find it impossible to assign the usual statistical limits to these two estimates, which in fact are not based on homologous material. We accept, however, that the error to be attached to both these estimates is wide, and both estimates may be subjected to considerable revision, which, however, seems unlikely to drop either estimate below 2 Sv for chronic, low-LET radiation.

It is hoped this anthology will mirror to a considerable degree the increasing sophistication of the study of human genetics during this 40-year period. Had the techniques that became available in the 60s and 70s been available in 1946, the study would have followed a very different course. We will return in the final chapter to what this continuing evolution in genetic technology suggests for future studies.

Inasmuch as even today the issues are occasionally confused, a clear distinction must be drawn between these studies and studies on the children *in utero* at the time of the bombings. The extensively documented microcephaly, mental retardation, and other abnormalities of these children (reviewed in Miller, 1956; Schull et al., 1989) reflect the well-known direct somatic (teratogenic) effects of radiation but not maternal genetic damage.

Even as these studies were unfolding, remarkable insights into the ways the genetic material protects itself against damage were developing. At the outset of these studies, virtually nothing was known about genetic repair mechanisms, nor about the array of endogenous and exogenous mutagenic influences to which DNA is "normally" exposed. Radiation appeared then as a unique insult; now it can be seen as one of many insults for which evolution has prepared DNA. We suggest that radiation—whose adverse somatic and genetic effects are of course clear—can now be regarded as only one of many disruptive forces with which DNA must cope and that the higher doubling doses which we are now postulating for mice and humans should be viewed as encouraging evidence of the robustness of the genetic material. This statement is not meant to encourage complacency in the face of the many biological challenges facing humans and the ecosystems in which they function but rather to contribute to the growth of perspective in setting ecosystem and exposure priorities.

There has been a keenly felt responsibility associated with the conduct of these studies. It is the essence of science that one investigator's results can be duplicated by others. But the bombings were a unique situation, and, because of both the post-war circumstances and the necessary scale of any program which purported to meet the demands of the situation, only a single comprehensive study has been undertaken. Furthermore, certain aspects of the study have involved carefully negotiated access to confidential vital statistics, with

great care to preserve personal anonymity. In principle, much of the data reported herein on morbidity and mortality could be reconstructed from government vital statistics, given access to the exposure data of the RERF, and the cytogenetic and biochemical observations could be independently repeated. In the real world, this is unlikely to happen. On the other hand, there is a precedent for the RERF making its data available to qualified investigators not associated with the organization, a precedent which the Foundation is of course prepared to meet with respect to the genetic data.

Finally, we would be remiss if we failed to acknowledge the remarkable cooperation that this study has enjoyed throughout its history. No study undertaken under these postwar circumstances could expect to escape criticism, both individual and organized, and the programs of the ABCC and RERF, including this one, are not exempt. In the final analysis, however, the ultimate test of the public's attitude toward a program is its cooperation. We note that the cytogenetic and biochemical studies, which, long after the Occupation had been terminated, involved obtaining venous blood samples from individuals 13 years or older (i.e., "free agents"), enjoyed 90% cooperation from the "children" and those of their parents still living in the cities at the time they were selected for study. Those familiar with such studies will realize what a remarkable record this is.

REFERENCES

Haldane, J.B.S. 1955. Genetical effects of radiation from products of nuclear explosions. *Nature* 176:115.

Miller, R.W. 1956. Delayed effects occurring within the first decade after exposure of young individuals to the Hiroshima atomic bomb. *Pediatrics* 18:1-18.

Neel, James V. 1958. A study of major congenital defects in Japanese infants. *Amer. J. Hum. Genet.* 10:398-445.

Roberts, L. 1990. British radiation study throws experts into tizzy. *Science* 248:24-25.

Schull, W.J., and Neel, J.V. 1965. *The Effects of Inbreeding on Japanese Children*. New York: Harper and Row. pp. *xii* and 419.

Schull, W.J., Otake, M., and Yoshimaru, H. 1989. Radiation-related damage to the developing human brain. In: Baverstock, K.F., and Stather, J.W. (eds.), *Low Dose Radiation: Biological Bases of Risk Assessment*. London: Taylor and Francis. pp. 28-41.

Genetic Effects of the Atomic Bombs in Hiroshima and Nagasaki

Genetics Conference, Committee on Atomic Casualties, National Research Council

The Atomic Energy Commission recently formally signified its intention of supporting long-range medical studies of the survivors of the atomic bombings in Japan, to be conducted by the Committee on Atomic Casualties of the National Research Council. One aspect of these studies will concern the much-discussed potential genetic effects of the bombs. The background of this program begins shortly after Japan's surrender, when a Joint Army-Navy Commission made extensive observations in Hiroshima and Nagasaki on the survivors of the bombings. At the conclusion of the Commission's work its chairman, Col. A. W. Oughterson, M.C., AUS, recommended to the Surgeon General of the Army that the Council be requested to undertake a long-range study of the medical and biological effects of the atomic bomb and this recommendation was transmitted by Surgeon Gen. Norman T. Kirk to Lewis H. Weed, chairman of the Division of Medical Sciences. As a result, in June 1946 a conference group was convened by the Council, and in November, following its recommendation, a five-man commission composed of representatives of the Council, the Army, and the Navy left for Japan for the purpose of determining the current status of Japanese work on atomic bomb casualties, evaluating the feasibility of American participation in continued research on these casualties, and indicating the lines along which such studies should proceed. This commission, known as the Atomic Bomb Casualty Commission and composed of Austin Brues, Paul S. Henshaw, Lt. Melvin Block, M.C., AUS, Lt. James V. Neel, M.C., AUS, and Lt. (j.g.) Frederick Ullrich, (MC) USNR, submitted a report of its findings to the Council in January 1947.

The June 1946 conference group had recommended that appropriate action be taken to obtain a Presidential Directive authorizing the National Research Council to initiate a long range study of the atomic bomb effects. This Directive was issued at the request of the Secretary of the Navy, James T. Forrestal, in November 1946, and on its authority the Council, in January 1947, established a Committee on Atomic Casualties, composed of Thomas M. Rivers (chairman), George W. Beadle, Detlev W. Bronk, Austin Brues, George M. Lyon, C. P. Rhoads, Shields Warren, Stafford L. Warren, George H. Whipple, and Raymond E. Zirkle.

The potential genetic effects of the atomic bomb were apparent to all interested students from the day the first bomb was dropped—in fact, to some, well before that time. A consideration of genetic studies was one facet of the work of the Atomic Bomb Casualty Commission, and a section of its January 1947 report was devoted to this subject. This phase of the work was to a large extent the responsibility of Lt. James V. Neel.

On June 24, 1947, the Committee on Atomic Casualties arranged a conference on the potential genetic effects of the atomic bombs. At this meeting, which was attended by George W. Beadle (chairman), Donald R. Charles, Charles H. Danforth, Herman J. Muller, Laurence H. Snyder, and Lt. Neel, the latter submitted a report of preliminary genetic studies, based on his observations in Japan during the preceding six months. Following a thorough appraisal of the problem, the conference voted to recommend to the Committee on Atomic Casualties that a program be undertaken in Japan along the lines sketched out in the Neel report. This recommendation was accepted at a meeting of the Committee on June 26, 1947. The conference also recommended that a statement be prepared, briefly summarizing the current status of the problem. This statement follows.

WHETHER THE ATOMIC BOMBS dropped on Hiroshima and Nagasaki will have detectable genetic effects on the Japanese is a question of widespread interest. The purpose of the present note is to show briefly that (1) many difficulties beset any attempt to obtain a valid answer to this question and (2) even after a long-term study, such as that outlined below, it still may not be possible to determine just how much genetic damage was done at Hiroshima and Nagasaki.

This memorandum is essentially a partial summary of the material presented by Lt. James V. Neel at the meeting of the Conference on Genetics convened by the Committee on Atomic Casualties of the National Research Council on June 24, 1947, but with certain additional considerations which grew out of the deliberations of the Conference.

It must first be recognized that, inasmuch as the majority of mutations occurring in animals are recessive, only the relatively small proportion of mutations which are dominants may be expected to show effects in the first postbomb generation. The potential range in their effects is very wide. Dominant mutations with large, clear-cut manifestations can be expected to be much rarer than those with smaller, but possibly quite significant, effects on bodily dimensions, life span, etc. But the detection of these latter is a matter of great difficulty with present techniques. For practical considerations investigation will have to be concentrated chiefly on the class with such large effects as may lead to stillbirths, to live births with gross external abnormality, or to internal defects causing death or serious illness in infancy.

Since there is no general agreement as to what proportion of cases of abnormal fetal development is genetically determined, and what proportion is due to nongenetic factors, an increased incidence of morphologically abnormal fetuses following irradiation may not be used as an index of the frequency of genetic change until the non-

SCIENCE, October 10, 1947

Reprinted, with permission, from *Science*, vol. 106, no. 2754, October 10, 1947.

genetic effects of this irradiation on the reproductive history of the mother have been determined. This point will be very difficult to evaluate.

It is obvious that in this case the approach to the problem of genetic effects is the statistical one. It is unlikely that any individual and specific pathology in a post-irradiation generation can ever be attributed with certainty to the effects of the bomb, but if there is a definite increase in the occurrence of abortions, miscarriages, stillbirths, and abnormal products of conception, one may surmise that this is related to the bombing—although some of the effects need not necessarily be genetic. Appropriate control studies in other Japanese cities are therefore of the utmost importance.

The survivors of the bombing received amounts of irradiation ranging from negligible to just short of lethal. It is impossible to say on a priori grounds whether an adequate number of people absorbed sufficient radiation to produce enough dominant mutations to result in detectable effects. However, comparison of this situation with the experimental data on infrahuman material suggests that these effects, if detectable, will be small. The median lethal dose for whole body irradiation in man is probably in the neighborhood of 500 Roentgen units (Shields Warren. *Physiol. Rev.*, 1944, **24**, 225–238). It is likely that many individuals in Hiroshima and Nagasaki who received high but sublethal doses of irradiation sustained other injuries which, in combination with the radiation damage, resulted in death. Thus, the mean radiation dose received by the surviving population will be lower than expected from a consideration of the median lethal dose alone. In terms of radiation genetics this is a small exposure, expected from data on animals to produce a frequency of dominant mutations which would increase the normal frequency of abnormal offspring by so small a proportion as to be difficult to demonstrate. Large-scale studies, utilizing all available material, plus accurate vital statistics, are thus necessary if data of value are to be obtained. In this connection it is important to bear in mind that the available children born to parents who received significant amounts of irradiation will probably not exceed 12,000 or 13,000 within the next 10 years.

The Japanese had recognized the importance of genetic studies and, under great difficulties, were organizing a program in Hiroshima when the Atomic Bomb Casualty Commission arrived. They had planned to compare the present and future frequency of abnormal births in Hiroshima with the frequencies reported in their medical literature and vital statistics during the prewar years. But it is by no means sure either that the prewar figures were sufficiently accurate or that the present reporting of vital statistics would be wholly effective in detecting rare effects of the atomic bomb radiations. It cannot be too strongly emphasized that there is at present absolutely no reliable evidence on which to base any opinion concerning the absolute or relative frequency of congenital abnormalities among children being born in Hiroshima and Nagasaki. Unfortunately, a good deal of misinformation is currently in circulation.

Two independent programs, one Japanese and one American, would involve needless duplication of effort. A joint undertaking is therefore indicated. In view of the fact that the Japanese are actively attempting to initiate genetic studies, it seems that any American efforts which may materialize should contemplate cooperation with the Japanese in an attempt to ensure an efficient and satisfactory program. The Neel report contains the following seven specific recommendations as to the organization of a program:

(1) Organize, in Hiroshima, Nagasaki, and a control area or areas, a modified system of pregnancy registration, this to include the irradiation history of the parents.

(2) Obtain as complete information as possible on the outcome of each registered pregnancy.

(3) Follow up each report of an abnormal termination of pregnancy or a congenital malformation with detailed family studies.

(4) Develop a system of checking on the completeness and accuracy of registration of births and deaths, such as requiring at intervals dual registration by both the family and the obstetrician or midwife.

(5) Conduct these studies on a sufficiently large scale that the results will have statistical significance.

(6) Integrate this program with a system of periodic examination of the offspring of irradiated persons and with careful death certification, so that genetic effects not apparent at birth but detected subsequently may be recorded. In particular, causes of infant mortality should be accurately recorded.

(7) Place this program in competent Japanese hands, through the Japanese Government, with only enough American supervision and cooperation, including supplies, to facilitate a successful program.

This program must extend over a period of 10–20 years before a significant amount of data can be accumulated, and quite possibly an even longer period of study, extending to the second and subsequent generations, will be indicated.

Certain practical limitations of the program may be considered at this point. The most difficult problem will be to obtain the necessary completeness of reporting. This will require constant effort, a wide educational program, and frequent cross-checks. Congenital malformations occurring within Japanese families may sometimes not be reported. This is perhaps more likely to occur in Japan than in this country, because probably less than 10 per cent of Japanese births occur in hospitals as these are defined in the United States. To what extent stillbirths and malformations occurring outside a hospital will be recorded depends on the vigor with which the problem is pursued. It will be difficult to get evenly matched teams of investigators for bombed and control areas. Furthermore, once people living in Hiroshima and

Nagasaki learn that stillbirths and malformations may possibly be attributed to the effects of the bomb, they will probably lose some of their reluctance to report such matters, whereas this will not be the case in a control area.

Japan is now a defeated and occupied country, under severe postwar stress, whose people have a very different psychology from our own. A program such as that under consideration will proceed much more slowly there than it would in this country.

In order to reduce the possibility that a negative result of the investigation on Japanese material be interpreted by the medical and lay public as meaning that important genetic effects were not produced, it is essential that a comparable effort be expended in experimentation on other mammalian material, in which genetic effects of different kinds can much more readily be brought to light. In this way it should be possible to throw light upon the proportion of the total genetic effects produced by the radiation that would have been detectable by the methods used in the investigation on the human material, and the serious danger of misinterpretation of the latter results would be minimized.

Recognizing the difficulties briefly touched upon in the foregoing paragraphs, the Conference on Genetics voted unanimously to record the following expression of its attitude toward the genetic program: "Although there is every reason to infer that genetic effects can be produced and have been produced in man by atomic radiation, nevertheless the conference wishes to make it clear that it cannot guarantee significant results from this or any other study on the Japanese material. In contrast to laboratory data, this material is too much influenced by extraneous variables and too little adapted to disclosing genetic effects. In spite of these facts, the conference feels that this unique possibility for demonstrating genetic effects caused by atomic radiation should not be lost."

THE EFFECT OF EXPOSURE TO THE ATOMIC BOMBS ON PREGNANCY TERMINATION IN HIROSHIMA AND NAGASAKI

By

J. V. NEEL AND W. J. SCHULL

DEPARTMENT OF HUMAN GENETICS
UNIVERSITY OF MICHIGAN

in collaboration with

R. C. ANDERSON	M. KODANI	W. J. WEDEMEYER
W. H. BORGES	D. J. MCDONALD	J. W. WOOD
R. C. BREWER	N. E. MORTON	S. W. WRIGHT
S. KITAMURA	M. SUZUKI	J. N. YAMAZAKI
	K. TAKESHIMA	

ATOMIC BOMB CASUALTY COMMISSION
HIROSHIMA, JAPAN

Publication No. 461

NATIONAL ACADEMY OF SCIENCES—NATIONAL RESEARCH COUNCIL
WASHINGTON 25, D. C.
1956

Library of Congress Card Catalogue No. 56-60060

TABLE OF CONTENTS

		PAGE
I.	BACKGROUND	1
1.1	THE GENERAL ADMINISTRATIVE FRAMEWORK OF THE STUDY	1
1.2	THE BEGINNINGS OF THE GENETICS PROGRAM	2
1.3	THE EARLY JAPANESE EFFORTS ALONG THESE LINES	2
1.4	SCIENTIFIC AND ADMINISTRATIVE CONSIDERATIONS SHAPING THE NATURE OF THE PROGRAM IN JAPAN	3
1.5	LOCAL CONSIDERATIONS SHAPING THE NATURE OF THE PROGRAM IN JAPAN	3
II.	THE PLAN	5
2.1	PREGNANCY REGISTRATION	5
2.2	SPECIAL STUDIES ON ABNORMAL TERMINATIONS	9
2.3	THE "10-PER CENT SAMPLE"	9
2.4	THE ACCURACY AND REPRODUCIBILITY OF THE ANAMNESTIC DATA OBTAINED ON THE GENETICS SHORT FORM AND THE GENETICS LONG FORM	9
2.5	THE AUTOPSY PROGRAM	14
2.6	THE COLLECTION OF DATA ON INFANTS AGED 9 MONTHS	14
2.7	THE PROCESSING OF THE DATA	18
2.8	THE STUDY OF SPONTANEOUS ABORTIONS	18
2.9	CYTOGENETIC EFFECTS OF THE ATOMIC BOMBS	18
2.10	THE DECISION TO DISCONTINUE WORK IN KURE	18
2.11	THE TERMINATION OF THE PROGRAM IN JANUARY, 1954	19
2.12	ACKNOWLEDGMENTS	19
III.	A COMPARISON OF HIROSHIMA AND NAGASAKI	21
3.1	THE PEOPLING OF JAPAN; POSSIBLE DIFFERENCES BETWEEN THE INHABITANTS OF HONSHU AND KYUSHU	21
3.2	NON-JAPANESE ELEMENTS IN THE TWO CITIES	21
3.2.1	EARLY NAGASAKI CONTACTS WITH THE WEST	21
3.2.2	THE DUTCH ON DESHIMA	22
3.2.3	FROM THE REOPENING OF JAPAN TO WORLD WAR II	22
3.3	THE BIOLOGICAL INFLUENCE OF "FOREIGNERS" ON NAGASAKI AND HIROSHIMA	23
3.4	THE DIFFERENT IMPACTS OF THE ATOMIC BOMBS ON THE TWO CITIES	28
3.4.1	TYPES OF BOMBS	28
3.4.2	EFFECTS OF THE BOMBS ON THE TWO CITIES	28
3.5	THE DEVELOPMENT OF THE ABCC PROGRAM IN THE TWO CITIES	29
IV.	THE CRITERIA OF RADIATION EMPLOYED IN THE STUDY	33
4.1	THE COMPLICATED NATURE OF THE INJURIES SUSTAINED BY SOME SURVIVORS; "DISASTER EFFECT" VS. "RADIATION EFFECT"	33
4.2	THE QUESTION OF RESIDUAL RADIATION FOLLOWING AN ATOMIC BOMBING	33

Table of Contents

		PAGE
4.3	Factors Determining the Nature of the Radiation Data Collected in This Study	34
	4.3.1 The Syndrome of Radiation Sickness Due to Whole-Body Irradiation	34
	4.3.2 The Relation between Distance from the Hypocenter and Radiation Dosage	36
	4.3.3 The Role of Shielding in Determining Radiation Dose	36
4.4	The Type of Radiation Data Collected in This Study	38
4.5	The Relation between Distance, Shielding, and Symptoms in These Data	41
4.6	Factors Contributing to the Validity of the Radiation Histories	44
4.7	Definition of Radiation Categories	44
4.8	Considerations in the Estimation of the Average Amount of Radiation Received by Persons in Each of the Five Radiation Categories	45
	4.8.1 The Estimated Distance-Dosage Curve	46
	4.8.2 The Observations of the Joint Commission Regarding Leucopenia	46
	4.8.3 The Proportion of Individuals Protected by Various Types of Shielding	50
4.9	Estimates of the Average Amount of Irradiation Received by Individuals in the Various Exposure Categories	50

V. THE COMPARABILITY OF IRRADIATION SUBCLASSES 53
 5.1 Consanguinity 53
 5.2 Age and Parity 55
 5.3 Economic Status 59
 5.4 Frequency of Positive Serological Test for Syphilis 61
 5.5 Frequency of Induced Abortions and of Dilatation and Curettage of the Uterus (D and C) 61
 5.6 The Frequency of Repeat Registrations 63
 5.7 Parental Cooperation 63
 5.8 Late Sequelae of Exposure to the Bombs 69
 5.9 The Changing Proportion of Control and Irradiated from Year to Year 69
 5.10 The Background of Group 1 Individuals 71
 5.11 Summary 71

VI. STATISTICAL METHODS 72
 6.1 The Problem and the General Plan 72
 6.2 Indicators of Radiation Damage and the Problem of Non-Overlapping Measurements 72
 6.3 Concomitant Variation 73
 6.4 Rejected Observations 77
 6.5 The Analysis of the Attribute Data 78
 6.6 The Analysis of the Measurement Data 82
 6.7 Some Further Problems 85
 6.8 The Use of Exposed Persons as Controls 86
 6.9 Presentation of Material 87

VII. ANALYSIS OF THE SEX RATIO DATA 88
 7.1 The Trait 88
 7.2 The Genetic Argument for Radiation-Induced Changes in the Sex Ratio . 88

Table of Contents v

		PAGE
	7.3 Concomitant Variation Influencing the Indicator	89
	7.4 The Data	89
	7.5 Summary	96

VIII. ANALYSIS OF THE MALFORMATION DATA 99
 8.1 The Trait 99
 8.2 Reliability of Diagnosis 99
 8.3 The Genetic Argument for Radiation-Induced Changes . . 104
 8.4 Concomitant Variation Influencing the Indicator . . 105
 8.5 The "At-Birth" Data 110
 8.6 The "9-Months" Data 115
 8.7 Analysis by Specific Malformation Type 117
 8.8 Summary 117

IX. ANALYSIS OF THE STILLBIRTH DATA 118
 9.1 The Trait 118
 9.2 The Genetic Argument for Radiation-Induced Changes . . 118
 9.3 Concomitant Variation Known to Affect the Stillbirth Rate . . 118
 9.4 The Data 124
 9.5 Summary 129

X. THE ANALYSIS OF THE BIRTHWEIGHT DATA 131
 10.1 The Trait 131
 10.2 The Genetic Argument for Irradiation Effects 131
 10.3 Concomitant Variables Known to Affect Birthweight . . . 131
 10.4 The Data and Their Analysis 132
 10.5 Summary 150

XI. ANALYSIS OF THE DATA CONCERNING DEATH DURING THE NINE-MONTH PERIOD FOLLOWING DELIVERY 151
 11.1 The Trait 151
 11.2 The Genetic Argument for Radiation-Induced Changes in the Neonatal Death Rate 151
 11.3 Concomitant Variables Known to Influence the Occurrence of a Neonatal Death 152
 11.4 The Data 157
 11.5 Summary 162

XII. THE ANALYSIS OF THE ANTHROPOMETRIC DATA 164
 12.1 The Measurements Obtained at Nine Months 164
 12.2 The Genetic Argument for Irradiation Effects 164
 12.3 Concomitant Variables Known to Affect Growth and Development During the First Year of Life 164
 12.4 The Data 165
 12.4.1 The Multivariate Means 168
 12.4.2 The Equality of the Generalized Variances . . . 175
 12.4.3 Within-Cell Heterogeneity 179
 12.5 Summary 179

XIII. THE AUTOPSY FINDINGS 184
 13.1 The Randomness of the Hiroshima Autopsies 184
 13.2 The Data 187
 13.3 Summary 191

THE CHILDREN OF ATOMIC BOMB SURVIVORS

Table of Contents

		PAGE
XIV.	RECAPITULATION	192
	14.1 The Findings	192
	14.2 The Question of Evaluating the Over-all Direction of the Indicators	196
	14.3 The Confidence Limits Determined by These Observations	196
	14.4 A Résumé of Work on Mammalian Material Pertinent to the Interpretation of These Findings	199
	14.4.1 Effects on Sex Ratio	200
	14.4.2 Effects on Malformation Frequency	201
	14.4.3 Effects on Stillbirth Frequency	202
	14.4.4 Effects on Birthweight	203
	14.4.5 Effects on Neonatal Death Rates	203
	14.4.6 Effects on Growth and Development	203
	14.5 Interpretation of the Findings	203
XV.	PERMISSIBLE INFERENCES	205
	15.1 Basic Data Necessary to Reaching Permissible Inferences	205
	15.2 The Spontaneous Mutation Rate in Man	205
	15.3 The Radiation-Induced Mutation Rate in Man	210
	15.4 Estimates of the Number of Genes in Man	212
	15.5 The "Accumulation Factor"	212
	15.6 The Nature of Natural Selection	213
	15.7 Concluding Remarks	216
APPENDIX		218
	Procedure for Coding Genetics Short-Form Questionnaires	218
	Coding Instructions for Genetics Follow-up Questionnaires	221
	Coding Instructions for Pediatrics Follow-up Questionnaire	225
REFERENCES		230

LIST OF TABLES

Chapter II PAGE

TABLE 2.1 Per Cent of All Registrations (Including Those Involving Parental Consanguinity) with At Least One Parent Heavily Exposed (Radiation Categories 4, 5) 19

Chapter III

TABLE 3.1 Population Figures by Nationality for Foreigners Resident in Nagasaki City between 1864 and 1870 23

TABLE 3.2 The "Foreign" and Total Population of Nagasaki City, and the "Foreign" Population of Nagasaki Prefecture, 1897–1923 24

TABLE 3.3 The Ethnic Composition of the Foreign Component of Nagasaki City, for the Years 1910 and 1930 25

TABLE 3.4 The Age Composition and Ethnic Status of the Total Foreign Population of Nagasaki in 1920, as well as of the Four Principal Ethnic Groups in This Population 26

Chapter IV

TABLE 4.1 Frequency of Occurrence of Certain Symptoms in Persons Alive 20 or More Days Following the Atomic Bombings, as Related to Distance from the Hypocenter 35

TABLE 4.2 The Effectiveness of Shielding in Protecting against Radiation Sickness in Hiroshima 37

TABLE 4.3 Distribution by Distance and Shielding of *Husbands* of Wives Registering Pregnancies with the Genetics Program: Hiroshima 38

TABLE 4.4 Distribution by Distance and Shielding of *Wives* Registering Pregnancies with the Genetics Program: Hiroshima 39

TABLE 4.5 Distribution by Distance and Shielding of *Husbands* of Wives Registering Pregnancies with the Genetics Program: Nagasaki 40

TABLE 4.6 Distribution by Distance and Shielding of *Wives* Registering Pregnancies with the Genetics Program: Nagasaki 41

TABLE 4.7 The Definition of "Exposure Categories" to be Employed in This Analysis. 44

TABLE 4.8 Distribution of Registered Births by Parental Exposure 45

TABLE 4.9 The Findings of the Joint Commission in Hiroshima with Regard to the Occurrence of Epilation, Petechiae, and Leucopenia in Persons Falling into Various Exposure Categories 47

TABLE 4.10 The Exposure Categories Defined by the Joint Commission, to be Applied to the Interpretation of Table 4.9 47

TABLE 4.11 The Distribution of Leucocyte Values in Hiroshima Japanese Who Failed to Develop Epilation, Petechiae, or Gingivitis Following the Bombing, in Relation to Distance from Hypocenter and Type of Shielding . . . 48

TABLE 4.12 Proportions of Parents Exposed in the 1,800–2,500 Meter Ring Who Reported Various Types of Shielding 50

Chapter V

TABLE 5.1 Frequency of Consanguineous Marriages (First Cousins, First Cousins Once Removed, Second Cousins) by City and Parental Exposure . . . 54

TABLE 5.2 Chi-Square Analysis of the Frequency of Consanguineous Marriages by City and Parental Exposure 55

vii

viii *List of Tables*

		PAGE
TABLE 5.3	Mean Maternal Age by City and Parental Exposure	56
TABLE 5.4	Analysis of Variance: Mother's Age by City and Parental Exposure	57
TABLE 5.5	The Distribution of Mean Squares for Maternal Age by City, Sex of Infant, and Parental Exposure	57
TABLE 5.6	Mean Parity by City and Parental Exposure	58
TABLE 5.7	Analysis of Variance: Parity by City and Parental Exposure	59
TABLE 5.8	The Distribution of Mean Squares for Parity by City, Sex of Infant, and Parental Exposure	59
TABLE 5.9	Economic Status by City and Parental Exposure	60
TABLE 5.10	Chi-Square Analysis of the Distribution of Economic Statuses by City and Parental Exposure	61
TABLE 5.11	Frequency of Positive Serology by Parental Exposure, City, and Group: "Zero" Parents Only	62
TABLE 5.12	Chi-Square Analysis of the Frequency of Positive Serologies by City and Parental Exposure	63
TABLE 5.13	Frequency of Mothers Reporting One or More Induced Abortions by Parental Exposure, City, and Time	64
TABLE 5.14	Chi-Square Analysis of the Frequency of Mothers Reporting One or More Induced Abortions by Parental Exposure, City, and Time	66
TABLE 5.15	Frequency of "Dilatation and Curettage" by Parental Exposure and City: Zero Terminations	67
TABLE 5.16	Chi-Square Analysis of the Frequency of "Dilatation and Curettage" by Parental Exposure and City: Zero Terminations	68
TABLE 5.17	Mean Number of Registered Pregnancies per Mother by Parental Exposure and City	68
TABLE 5.18	Incidence of Leukemia in the Hiroshima Survivors of the Atomic Bombing as Related to Distance from the Hypocenter and the Presence of Severe Radiation Complaints (After Moloney and Kastenbaum, 1955)	70
TABLE 5.19	The Frequency of Malformations by Year among the Offspring of Parents Neither of Whom Was Exposed to the Atomic Bombs	70
TABLE 5.20	The Frequency of Stillbirths by Year among the Offspring of Parents Neither of Whom Was Exposed to the Atomic Bombs	71

Chapter VI

TABLE 6.1	The Number of Infants Rejected from the Study, Tabulated by Reason for Rejection	77
TABLE 6.2	An Accounting of the Number of Observations Considered at Representative Stages in the Analysis of the "At-Birth" Data, and the Number of Rejected Observations with the Cause of Rejection	79
TABLE 6.3	An Accounting of the Number of Observations Considered at Representative Stages in the Analysis of the "9-Months" Data and the Number of Rejected Observations with the Cause of Rejection	79

Chapter VII

TABLE 7.1	The Frequency of Male Births by Parental Exposure and City	90
TABLE 7.2	Chi-Square Analysis of the Frequency of Male Births by City and Parental Exposure	91
TABLE 7.3	Selected Comparisons Regarding the Effect of Irradiation on Sex Ratio	92
TABLE 7.4	National Statistics of Livebirths, 1935–1952	94
TABLE 7.5	The Frequency of Male Births among Infants Born After April, 1946 but Prior to June, 1948, by Parental Exposure	95
TABLE 7.6	Chi-Square Analysis of the Frequency of Male Births among Infants Born After April, 1946 but Prior to June, 1948	95

THE CHILDREN OF ATOMIC BOMB SURVIVORS

List of Tables

		PAGE
TABLE 7.7	The Frequency of Male Births by Parental Exposure and City, 1954–1955	97
TABLE 7.8	Chi-square Analysis of the Frequency of Male Births during the Years 1948–1955 by Time of Birth, City, and Parental Exposure	98

Chapter VIII

TABLE 8.1	An Alphabetical Listing of Those Malformations Observed in This Study which Occurring Alone or in Combination with One Another Were Graded as Major Congenital Defect	100
TABLE 8.2	The Types and Frequency of Various Major Congenital Malformations Observed at the Tokyo Red Cross Maternity Hospital during the Years 1922 through 1940	101
TABLE 8.3	A Comparison, for Hiroshima and Nagasaki, of the Findings as Regards Major Congenital Malformation in Infants Examined at Approximately Age 9 Months, and in Those Same Infants When Examined Shortly After Birth	103
TABLE 8.4	The Effect of Maternal Age at Fixed Parity on the Frequency of Malformed Infants	106
TABLE 8.5	The Effect of Maternal Parity at Fixed Age on the Frequency of Malformed Infants	108
TABLE 8.6	The Frequency of Malformed Infants by Parental Exposure, Sex of Infant, and City	111
TABLE 8.7	The Frequency of Malformed Infants by Parental Exposure, City, and Maternal Age	113
TABLE 8.8	Chi-Square Analysis of the Frequency of Congenitally Malformed Infants by Sex, City, and Parental Exposure	114
TABLE 8.9	Chi-Square Analysis of the Frequency of Congenitally Malformed Infants by City, Maternal Age, and Parental Exposure	114
TABLE 8.10	The Distribution of Frequency of Malformed Infants Classified by Mother's Age and Exposure *Only*	115
TABLE 8.11	Chi-Square Analysis of the Effect of Mother's Exposure on the Frequency of Malformed Infants at Each of Five Different Age Levels	115
TABLE 8.12	The Distribution by Maternal Exposure and Parity of Malformed Infants Born to Mothers of Ages 15–20	115
TABLE 8.13	The Distribution of Frequency of Malformed Infants Classified by Mother's Age and Father's Exposure *Only*	115
TABLE 8.14	The Frequency of Malformed Infants among All Infants Re-examined at 9 Months of Age, by City and Parental Exposure	116
TABLE 8.15	Chi-Square Analysis of the Frequency of Malformed Infants at 9 Months of Age by City and Parental Exposure	116
TABLE 8.16	The Distribution by Maternal Exposure of the Seven Most Common Major Congenital Malformations in the Japanese, Exclusive of Congenital Heart Disease	116

Chapter IX

TABLE 9.1	Congenital Syphilis among Living Infants Born in Nagasaki in 1951: Incidence and Relation to Maternal Age (After Wright, S. W. et al., 1952)	119
TABLE 9.2	The Effect of Maternal Age at Fixed Parity on the Frequency of Stillborn Infants	120
TABLE 9.3	The Effect of Maternal Parity at Fixed Age on the Frequency of Stillborn Infants	122
TABLE 9.4	Frequency of Stillbirths by Sex, Parental Exposure and City	125

THE CHILDREN OF ATOMIC BOMB SURVIVORS

List of Tables

		PAGE
TABLE 9.5	Chi-Square Analysis of the Frequency of Stillbirths by Sex, Parental Exposure and City	127
TABLE 9.6	The Frequency of Stillbirths by Parental Exposure, City and Parity	128
TABLE 9.7	Chi-Square Analysis of the Frequency of Stillbirths by Parental Exposure, City and Parity	129
TABLE 9.8	The Frequency of Stillbirths among Firstborn Infants by City and Paternal Exposure	129
TABLE 9.9	The Frequency of Stillbirths among Firstborn Infants by Paternal Exposure and Maternal Age	130

Chapter X

TABLE 10.1	Mean Birthweight by Parental Exposure, Sex and City	134
TABLE 10.2	Analysis of Variance of Birthweight by Parental Exposure and City	135
TABLE 10.3	The Distribution by Parental Exposure of the Weighted Mean Squares of Deviations	136
TABLE 10.4	Analysis of Covariance of Birthweights: Males, Hiroshima	136
TABLE 10.5	Tests of the Significance and Homogeneity of the Regressions of Birthweight on Maternal Age and Parity: Males, Hiroshima	137
TABLE 10.6	Analysis of Variance on the Adjusted Birthweight Means: Males, Hiroshima	138
TABLE 10.7	The Adjusted Birthweight Means: Males, Hiroshima	138
TABLE 10.8	The Residual Mean Squares from the Individual Cell Regressions: Males, Hiroshima	139
TABLE 10.9	Analysis of Covariance of Birthweights: Females, Hiroshima	139
TABLE 10.10	Tests of the Significance and Homogeneity of the Regressions of Birthweight on Maternal Age and Parity: Females, Hiroshima	140
TABLE 10.11	Analysis of Variance on the Adjusted Birthweight Means: Females, Hiroshima	140
TABLE 10.12	The Adjusted Birthweight Means: Females, Hiroshima	141
TABLE 10.13	The Residual Mean Squares from the Individual Cell Regressions: Females, Hiroshima	141
TABLE 10.14	Analysis of Covariance of Birthweights: Males, Nagasaki	142
TABLE 10.15	Tests of the Significance and Homogeneity of the Regressions of Birthweight on Maternal Age and Parity: Males, Nagasaki	142
TABLE 10.16	Analysis of Variance on the Adjusted Birthweight Means: Males, Nagasaki	143
TABLE 10.17	The Adjusted Birthweight Means: Males, Nagasaki	143
TABLE 10.18	The Residual Mean Squares from the Individual Cell Regressions: Males, Nagasaki	144
TABLE 10.19	Analysis of Covariance of Birthweights: Females, Nagasaki	144
TABLE 10.20	Tests of the Significance and Homogeneity of the Regressions of Birthweight on Maternal Age and Parity: Females, Nagasaki	145
TABLE 10.21	Analysis of Variance on the Adjusted Birthweight Means: Females, Nagasaki	145
TABLE 10.22	The Adjusted Birthweight Means: Females, Nagasaki	146
TABLE 10.23	The Residual Mean Squares from the Individual Cell Regressions: Females, Nagasaki	146
TABLE 10.24	A Summary of the Salient Findings of the Covariance Analysis	147
TABLE 10.25	The Distribution by Parental Exposure of the Residual Mean Squares After Removal of Variation Due to Year of Birth of the Infant: Males, Hiroshima	149
TABLE 10.26	The Distribution by Parental Exposure of the Residual Mean Squares After Removal of Variation Due to Year of Birth of the Infant: Females, Hiroshima	149

THE CHILDREN OF ATOMIC BOMB SURVIVORS

List of Tables

		PAGE
TABLE 10.27	The Distribution by Parental Exposure of the Residual Mean Squares After Removal of Variation Due to Year of Birth of the Infant: Males, Nagasaki	149
TABLE 10.28	The Distribution by Parental Exposure of the Residual Mean Squares After Removal of Variation Due to Year of Birth of the Infant: Females, Nagasaki	149

Chapter XI

TABLE 11.1	The Effect of Maternal Age at Fixed Parity on the Frequency of Neonatal Deaths	154
TABLE 11.2	The Effect of Parity at Fixed Maternal Age on the Frequency of Neonatal Deaths	156
TABLE 11.3	The Frequency of Neonatal Deaths by Parental Exposure, City and Sex of Infant	158
TABLE 11.4	The Frequency of Neonatal Deaths by Parental Exposure, City and Parity	160
TABLE 11.5	Chi-Square Analysis of the Frequency of Neonatal Deaths by Parental Exposure, City and Sex	161
TABLE 11.6	Chi-Square Analysis of the Frequency of Neonatal Deaths by Parental Exposure, City and Parity	162
TABLE 11.7	The Frequency of Deaths between Birth and Nine Months of Age by Parental Exposure and City	163
TABLE 11.8	Analysis of the Frequency of Deaths between Birth and Nine Months of Age, by Parental Exposure and City	163

Chapter XII

TABLE 12.1	Distribution of Mean Weight in Decagrams at 9 Months of Age by City, Sex and Parental Exposure	165
TABLE 12.2	Distribution of Mean Height in Millimeters at 9 Months of Age by City, Sex and Parental Exposure	166
TABLE 12.3	Distribution of Mean Head Girth in Millimeters at 9 Months of Age by City, Sex and Parental Exposure	167
TABLE 12.4	Distribution of Mean Chest Girth in Millimeters at 9 Months of Age by City, Sex and Parental Exposure	168
TABLE 12.5	Analysis of Dispersion (All Exposure Cells). (a) Sums of Squares and Cross Products of Deviations for the Two-Factor Interactions. (b) Mean Squares for Individual Analyses of Variance	169
TABLE 12.6	Analysis of Dispersion (All Exposure Cells). (a) Sums of Squares and Cross Products of Deviations for Main Effects and Additivity. (b) Mean Squares for Individual Analyses of Variance (on Main Effects and Additivity). (c) Analysis of Dispersion Test, Wilks' (Using Bartlett's Approximation)	170
TABLE 12.7	Estimates of Constants and Their Variances for Tests of Equality	171
TABLE 12.8	A Summary of the Significance of Tests Comparing All Possible Pairs of Exposure for Each Parent with Respect to the Variables w, x, y, and z	172
TABLE 12.9	Analysis of Dispersion (Only Those Cells Where Both Parents Were Exposed). (a) Sums of Squares and Cross Products of Deviations for the Two-Factor Interactions. (b) Mean Squares for Individual Analyses of Variance (on Two-Factor Interactions)	173
TABLE 12.10	Analysis of Dispersion (Only Those Cells Where Both Parents Were Exposed). (a) Sums of Squares and Products of Deviations for Main Effects and Additivity. (b) Mean Squares for Individual Analyses of Variance for Main Effects and Additivity	174

THE CHILDREN OF ATOMIC BOMB SURVIVORS

		PAGE
TABLE 12.11	Estimates of Constants and Their Variances for Test of Equality	175
TABLE 12.12	Mean Product Matrices: Males, Hiroshima	176
TABLE 12.13	Mean Product Matrices: Females, Hiroshima	176
TABLE 12.14	Mean Product Matrices: Males, Nagasaki	177
TABLE 12.15	Mean Product Matrices: Females, Nagasaki	177
TABLE 12.16	Tests of the Homogeneity of the Variances of the Anthropometric Measurements Over All Exposure Cells, and Subdivisions Thereof, for Specified Sex and City	178
TABLE 12.17	Tests of the Homogeneity of the Variances of the Anthropometric Measurements Over All Exposure Cells, and Subdivisions Thereof, for Specified Sex and City	178
TABLE 12.18	Test of the Generalized Variances on the Anthropometric Measurements by Specified Sex and City	179
TABLE 12.19	Analysis of Dispersion, with Age at Examination Included as a Way of Classification (All Exposure Cells). (a) Sums of Squares and Cross Products of Deviations for the Two-Factor Interactions. (b) Mean Squares for Individual Analyses of Variance	180
TABLE 12.20	Analysis of Dispersion with Age at Examination Included as a Way of Classification. (a) Sums of Squares and Products of Deviations for Main Effects and Additivity. (b) Mean Squares for Individual Analyses of Variance for Main Effects and Additivity	181
TABLE 12.21	Estimates of Constants and Their Variances for Tests of Equality	182
TABLE 12.22	A Summary of the Significance Tests Comparing All Possible Pairs of Exposure for Each Parent with Respect to the Variables x, y, w, and z, after Allowance Is Made for Age at Examination	183

Chapter XIII

TABLE 13.1	A. The Randomness of the Distribution of Autopsied Infants by Sex of Infant and Parental Exposure. B. Chi-Square Analysis of the Frequency of Autopsy by Sex of Infant and Parental Exposure	185
TABLE 13.2	Mean Maternal Age of Infants Stillborn or Dying during the First Six Days of Life by Sex of Infant, Parental Exposure and the Occurrence of Autopsy	186
TABLE 13.3	The Distribution of Economic Status among Autopsied and Non-Autopsied Infants	186
TABLE 13.4	The Distribution of Positive Serologies by Parental Exposure among Autopsied and Non-Autopsied Infants	187
TABLE 13.5	The Distribution by Parental Exposure of Infants Born in Hiroshima and Found to be Grossly Abnormal at Autopsy	188
TABLE 13.6	The Distribution by Exposure Class of the Exposed Parents Given in Table 13.5	188
TABLE 13.7	Hayashi's Data on Congenital Abnormalities in Relation to Exposure of Parents (After Sevitt, 1955)	189
TABLE 13.8	A Comparison of the Exposure Distribution of Hayashi's Data and That of the Atomic Bomb Casualty Commission Collected in Nagasaki	190
TABLE 13.9	Distribution by Exposure Class of Exposed Parents Whose Infants Came to Autopsy at ABCC in Nagasaki	190

Chapter XIV

TABLE 14.1	A Summarization of the Comparisons of the Various Indicators with Parental Exposure When (a) All Exposure Cells Are Considered, and (b) Only Those Cells Where Both Parents Were Exposed Are Considered	195

THE CHILDREN OF ATOMIC BOMB SURVIVORS

List of Tables

		PAGE
TABLE 14.2	The Effect of Irradiation of Male Mice on the Sex Ratio of Offspring Conceived during the Post-Sterile Period (After Hertwig, 1938)	200
TABLE 14.3	The Sex Ratio among Livebirths and Fetal Deaths in the Study of Macht and Lawrence (1955)	201
TABLE 14.4	The Effect of Irradiation of Male Mice on the Frequency of Stillbirths among Offspring Conceived during the Post-Sterile Period (After Hertwig, 1938)	202

Chapter XV

TABLE 15.1	Frequency of Occurrence of Spontaneous "Visible" Mutations in Drosophila and in the House Mouse	208
TABLE 15.2	Estimated Average Mutation Rates per Lethal-Producing Locus in Several Drosophila Species (After Dobzhansky, Spassky, and Spassky, 1952)	209
TABLE 15.3	Frequency of Occurrence of Nine Different Dominant or Sex-Linked Mutations in Man	210

LIST OF FIGURES

		PAGE
Chapter II		
FIGURE 2.1	English Translation of the Japanese-Language Genetics "Short Form"	6
FIGURE 2.2	English Translation of the Japanese-Language Pamphlet Distributed at the Time of Pregnancy Registration	8
FIGURE 2.3	The Genetics "Long Form"	10
FIGURE 2.4	The Form Used for Recording Pediatric Information at Age 9 Months	16
Chapter III		
FIGURE 3.1	The Topography of the Hiroshima City Region	30
FIGURE 3.2	The Topography of the Nagasaki City Region	31
Chapter IV		
FIGURE 4.1	Symptom Ratio in Relation to Distance from Hypocenter for Hiroshima (Husbands and Wives Combined)	42
FIGURE 4.2	Symptom Ratio in Relation to Distance from Hypocenter for Nagasaki (Husbands and Wives Combined)	43
FIGURE 4.3	Total Dosage of Initial Gamma Radiation as a Function of Distance from the Hypocenter of the Explosion of a "Nominal" Atomic Bomb, from "The Effects of Atomic Weapons"	46
FIGURE 4.4	Fast and Slow Neutrons Delivered per Square Centimeter as a Function of Distance from the Hypocenter of the Explosion of a "Nominal" Atomic Bomb, from "The Effects of Atomic Weapons"	46
FIGURE 4.5	Neutron and Gamma Radiation Distance-Dosage Curves for the Atomic Bomb Explosions at Hiroshima and Nagasaki	51
Chapter VI		
FIGURE 6.1	A Schematic Representation of the Method of Sorting the Data to Obtain Non-Overlapping Indicators	73
Chapter VII		
FIGURE 7.1	The Distribution of the Frequency of Male Births by Age of Mother at the Birth of the Infant with Parity Ignored	89
FIGURE 7.2	The Distribution of the Frequency of Male Births by Parity with Maternal Age Ignored	89
Chapter VIII		
FIGURE 8.1	The Distribution of the Frequency of Infants with Major Malformation by Maternal Age for Specified Parities	107
FIGURE 8.2	The Distribution of the Frequency of Grossly Malformed Infants by Parity for Specified Maternal Ages	108
Chapter IX		
FIGURE 9.1	The Distribution of the Frequency of Stillborn Infants by Age of Mother for Specified Parities	123
FIGURE 9.2	The Distribution of the Frequency of Stillborn Infants by Parity for Specified Maternal Ages	124
Chapter X		
FIGURE 10.1	The Distribution of Mean Birthweight in Decagrams by Parity for All Maternal Ages, and for Maternal Age 30 Only	133

List of Figures

		PAGE
FIGURE 10.2	The Distribution of Mean Birthweight in Decagrams by Maternal Age for Fixed Parities (Parities 1 and 4)	133

Chapter XI

FIGURE 11.1	The Distribution of the Frequency of Neonatal Deaths by Age of Mother for Specified Parities	152
FIGURE 11.2	The Distribution of the Frequency of Neonatal Deaths by Parity for Specified Maternal Ages	153

Chapter XIII

FIGURE 13.1	The Distribution of the Frequency of Autopsied Infants with Major Malformation, Relative to All Autopsied Infants, by Age of Mother with Parity Ignored	185

Chapter XIV

FIGURE 14.1	A Graphical Representation of the Effect of Parental Exposure on: (1) the Sex Ratio; (2) the Frequencies of Malformed Infants, Stillborn Infants, and Infants Dying in the Neonatal Period; (3) Birthweight Means; and (4) Birthweight Variances	193
FIGURE 14.2	A Graphical Representation of the Effect of Parental Exposure on: (1) the Frequency of Malformed Infants Alive at Age Nine Months; (2) the Frequency of Death in the First Nine Months of Life; and (3) the Anthropometric Measurements of Weight, Height, Head Circumference and Chest Circumference	194
FIGURE 14.3	The Adequacy of the Data with Regard to Sex Ratio as Indicated by the Operating Characteristic (OC) Curves for Analyses Based on Sample Sizes of 5,629 Mothers and 2,453 Fathers, Where the True Proportion of Successes is Assumed to be 0.5198, the Value Observed in the Control Population	198
FIGURE 14.4	The Adequacy of the Data with Regard to Malformations, Stillbirths, and Neonatal Deaths as Indicated by the Operating Characteristic (OC) Curves for Analyses Based on Samples of Size 1,097 Parents, Where the True Proportion of Successes are Assumed to be 0.0090 for Malformations and 0.0142 for Stillbirths and Neonatal Deaths, the Control Values	199

Chapter XV

FIGURE 15.1	A Schematic Representation of Two Different "Mutation Spectra" with Reference to Degree of Viability, Both Compatible with the Existing Data	207

Chapter I

BACKGROUND

The present monograph is designed as a detailed report on certain efforts made during the period 1946–1955 to provide answers to the following two questions:

1. Can there be observed, during the first year of life, any differences between the children born to parents, one or both of whom were exposed to the effects of the atomic bombings of Hiroshima and Nagasaki, and the children born to suitable control parents, and
2. If differences do exist, how are these to be interpreted?

1.1 *The general administrative framework of the study.* — The study to be described was undertaken as one facet of a comprehensive attempt to obtain detailed information concerning the various possible late or delayed biological effects of exposure to an atomic bombing. Certain information pertinent to the development of the over-all program is essential to an understanding of the manner in which the effort to provide answers to the two questions listed above evolved.

The background of this program begins shortly after Japan's surrender, when a Joint Army-Navy Commission made extensive observations in Hiroshima and Nagasaki on the survivors of the bombings. At the conclusion of the Commission's work its chairman, Col. A. W. Oughterson, M.C., AUS, recommended to the Surgeon General of the Army that the National Academy of Sciences — National Research Council be requested to undertake a long-range study of the medical and biological effects of the atomic bomb. This recommendation was transmitted by Surgeon Gen. Norman T. Kirk to Lewis H. Weed, chairman of the Division of Medical Sciences of the National Research Council. As a result, in June of 1946, a conference group was convened by the Council, and in November, following its recommendation, a five-man commission composed of representatives of the Council, the Army, and the Navy left for Japan for the purposes of (1) determining the current status of Japanese work on atomic bomb casualties, (2) evaluating the feasibility of American participation in continued research on these casualties, and (3) indicating the lines along which such studies should proceed. This commission, composed of Austin Brues, Paul S. Henshaw, Lt. Melvin Block, M.C., AUS, Lt. James V. Neel, M.C., AUS, and Lt. (j.g.) Frederick Ullrich, (MC) USNR, submitted a report of its findings to the Council in January, 1947.

The June, 1946 conference group had recommended that appropriate action be taken to obtain a Presidential Directive authorizing the National Research Council to initiate a long-range study of atomic bomb effects. This Directive was issued at the request of the Secretary of the Navy, James T. Forrestal, in November, 1946, and on its authority the Council, in January, 1947, established a Committee on Atomic Casualties, composed of Thomas M. Rivers (chairman), George W. Beadle, Detlev W. Bronk, Austin Brues, George M. Lyon, C. P. Rhoads, Shields Warren, Stafford L. Warren, George H. Whipple, and Raymond E. Zirkle.

At its first meeting, on March 25, 1947, the Committee on Atomic Casualties went on record to the effect that a large-scale program should be organized towards the end of learning as much as possible of medical significance from the Japanese experience. Financial support for the program was sought from the U.S. Atomic Energy Commission, which, during the fall of 1947, formally signified its intention of financing the program in Japan.

Ever since November of 1946, i.e., beginning with the visit of the five-man commission referred to earlier, there has been resident in Japan a group of investigators and supporting

personnel concerned with various phases of the over-all program. This group has been known as the Atomic Bomb Casualty Commission (ABCC). Starting with a scant handful, it grew until at its peak strength it consisted of 143 "allied" personnel (Americans, English, Australians, etc.) and 918 Japanese Nationals.

At the time that the ABCC began its operations in Japan, that country was of course under U.S. Military Occupation. It was Occupation policy that groups such as the Atomic Bomb Casualty Commission, having to do primarily with the Japanese people, should work with and through some existing Japanese agency. In early 1947, the Japanese Ministry of Welfare, at the instigation of the Public Health and Welfare Section of GHQ, SCAP, established a National Institute of Health, designed to occupy roughly the place in Japanese medicine which the National Institutes of Health then occupied in the United States. Several months later, in June of 1947, Brig. Gen. C. F. Sams, then Chief of the Public Health and Welfare Section, suggested that it would be appropriate for the ABCC to develop its program in cooperation with the Japanese National Institute of Health. Eventually there was established an Atomic Bomb Casualty Section of the Japanese National Institute of Health. The personnel of this Section have been closely integrated with ABCC personnel, in such a manner as to make the program truly a combined Japanese-American effort.

1.2 *The beginnings of the Genetics Program.* — So well known are the genetic effects of the irradiation of a variety of plant and animal material that inevitably one of the foremost questions in the minds of those considering the possible late consequences of the atomic bombings had to do with the characteristics of the children of exposed parents. As a member of the five-man survey commission which went to Japan in the late fall of 1946, the senior author was especially responsible for an evaluation of the lines along which the study of this question could best proceed. The results of this evaluation were laid before an ad hoc Genetics Conference convened by the National Research Council in the summer of 1947, composed of G. W. Beadle (chairman), D. R. Charles, C. H. Danforth, H. J. Muller, L. H. Snyder, and Lt. J. V. Neel. It was clearly recognized by the members of this Conference that the demonstration of the potential genetic effects of the irradiation of the human species presented many difficulties not encountered in laboratory material. It was further recognized that post-war conditions in Japan were by no means the most favorable for a study of this type.

In finally recommending that a rather large-scale effort be undertaken in Japan, the Conference felt constrained to make the following statement:

"Although there is every reason to infer that genetic effects can be produced and have been produced in man by atomic radiation, nevertheless the conference wishes to make it clear that it cannot guarantee significant results from this or any other study on the Japanese material. In contrast to laboratory data, this material is too much influenced by extraneous variables and too little adapted to disclosing genetic effects. In spite of these facts, the conference feels that this unique possibility for demonstrating genetic effects caused by atomic radiation should not be lost." (Genetics Conference, 1947.)

Although the study which was undertaken will for purposes of convenience be termed the "Genetics Program," because of its obvious implications, it must be emphasized that those concerned in its organization and conduct have always regarded it first and foremost as an effort to collect data on the characteristics of the children born to the irradiated survivors of the atomic bombings, data which, if indicating that the bombs had had some effect, were then subject to several possible interpretations, genetic and otherwise.

1.3 *The early Japanese efforts along these lines.* — The U.S. Army-Navy Joint Commission which studied the medical effects of the atomic bombs in Japan as soon as possible after the surrender worked in close cooperation with the Medical Section of a special Committee for the Investigation of the Effects of the Atomic Bombs, appointed by the Japanese National Research Council. The members of this group also recognized clearly the desirability of long-range studies on the medical effects of the bombs, but in the organization of such studies were greatly handicapped by post-war conditions in Japan. The administrative reorganization of the Japanese National Research Council and other aspects of Japanese science which the Occupation sponsored also unavoidably created uncertainties which delayed the work of this group.

Despite these many difficulties, at the time that the U.S. National Research Council's survey group visited Hiroshima in December of 1946 they found the Japanese planning to obtain data on the characteristics of the children then being born. This program, under the immediate supervision of Dr. I. Matsubayashi, was carefully reviewed. It was felt to be inadequate in several respects. Nevertheless, it is certainly a favorable commentary on Japanese interest in the problem and desire to carry on research that plans for even a small program had been set up under the circumstances then prevailing. This program was in effect very briefly and then merged with the program of the ABCC; Dr. Matsubayashi became a member of the ABCC staff.

1.4 *Scientific and administrative considerations shaping the nature of the program in Japan.* — The lines along which the Genetics Program in Japan was laid out in 1946–1947 were dictated not only by purely scientific considerations but also by certain practicalities of the situation. The more important scientific considerations were as follows:

1.4.1 The possible observable genetic effects of irradiation upon the first generation born after an atomic bombing are many and varied. These include changes in the sex ratio, an increase in the frequency of stillbirths, an increase in the frequency of congenital malformation, an increase in infant mortality, etc. *Each of these possible indicators of genetic damage is also influenced by a number of other factors; there are no known unique yardsticks of genetic damage.* Under these circumstances, the crux of any program of study was the feasibility of establishing control material which insofar as possible differed from the irradiated only with respect to the radiation factor.

1.4.2 At the time this program was organized, although there was available a mass of data concerned with spontaneous mutation rates and the genetic effects of irradiation on Drosophila, little was known concerning spontaneous mutation rates in mammals, including man, and, with the exception of the work of Charles (1950), still less concerning the mutagenic effects of the irradiation of mammalian material. Largely by extrapolation from the Drosophila material, it could be anticipated that in the light of the probable irradiation dosages sustained by the survivors of Hiroshima and Nagasaki, only very slight genetic effects should be detected in the first generation. In particular, it is worth pointing out that although the atomic bombs were dropped in August of 1945, because of the various time-consuming administrative developments recounted above, involving policy decisions and their means of implementation for the National Research Council, the U.S. Atomic Energy Commission, the U.S. Army of Occupation in Japan, and the Japanese Government, it was March of 1948 before the Genetics Program as it will be described presently was actually in action. The first children to come under the scrutiny of the program were conceived in October of 1947. There was thus a loss of information for the two years following the bombing. For a variety of reasons, some to be discussed below, it was felt that no attempt to reconstruct the frequency of malformations, stillbirths, or neonatal death during these first two post-bomb years could succeed. The practical corollary of these considerations was the necessity, once the program got under way, of a large-scale effort which would utilize as much of the available material as possible.

1.4.3 It was apparent from the outset that the Genetics Program could not and would not operate independently of the various other facets of the activities of the ABCC, but as one of a collection of integrated units. However, the organization of the Genetics Program proceeded somewhat more rapidly than the organization of other segments of the ABCC. Furthermore, there was for some time uncertainty as to the scale of operations in Japan. Under these circumstances, it was necessary to design a program with a basic, irreducible framework to which additions could be made later if circumstances permitted, but the additions had to be of such a nature that they would not invalidate a comparison of the early and later data. It was further necessary that the observations selected as possible indicators of genetic damage be relatively simple, as devoid as possible of the subjective element, and capable of being carried out under the conditions in Japan to be described below.

1.5 *Local considerations shaping the nature of the program in Japan.* — Among the conditions in Japan which were determining elements in the program, the following deserve special mention:

1.5.1 In Japan, the practice of obstetrics

is still largely in the hands of midwives. They attend approximately 96 per cent of the deliveries. Thus, in 1948, out of a total of 2,681,624 deliveries in Japan, 2,468,821 were attended by midwives, 102,627, by physicians, and 110,176 by unlicensed attendants. The deliveries for which midwives and unlicensed attendants are responsible usually occur at home.

The prerequisites for licensure as a midwife were revised upwards during the Occupation. However, the great majority of midwives now in practice were of course licensed under the previous system. This required that a candidate for admission to a school of midwifery had completed a kōtōshōgakkō education (about eight years, corresponding to grammar school). Although not required, some candidates had also completed chūgaku (middle school, about four years, corresponding to high school). The course in midwifery covered two years. One could either spend all two years in a school, or spend one year in school and the last year obtaining experience either in the obstetrics-gynecology department of a hospital or in association with a senior midwife. One had then to pass a prefectural examination for licensure. The legal minimum age was 20 years. Not uncommonly a licensed midwife married soon after completing school and did not return to the practice of midwifery for some years.

It is apparent that whatever the midwife's qualifications for the practice of obstetrics may be, her training had not prepared her for the detection and careful description of congenital malformations. This made it necessary with respect to the Genetics Program to develop a system whereby each newborn infant was seen by a physician as soon as possible after birth.

1.5.2 The Japanese economy during the immediate post-war years, when this study was instituted, was marginal. There was strict rationing of food and articles of clothing. Since official rations were inadequate to meet caloric needs, there was a widespread "black market" in food. One of the features of the ration system was special provision for pregnant women. Such women (or their designated representative) upon certification of pregnancy by a midwife or physician could register at any time after the fifth lunar month of pregnancy and thereby obtain access to certain items of value to them and their future children. Registration was revealed by a preliminary study to be in excess of 95 per cent complete. There was thus available a system whereby very nearly all pregnant women could be contacted at mid-pregnancy.

1.5.3 Finally, mention must be made of the psychological and sociological pitfalls involved in a study of this type. This is scarcely the place for a detailed analysis of the psychological problems inherent in the operations of the ABCC in Hiroshima and Nagasaki. Some of the inevitable problems in Japanese-American relationships are apparent. Others were unexpected (Matsumoto, 1954). Each step in the program had to be planned and taken cautiously, after every effort had been made to explore the possible repercussions. In retrospect, the first year of activity in Japan may be characterized as an apparently interminable series of conferences punctuated by weekly crises which, although often inconsequential in retrospect, at the time threatened to stall the entire operation until met. Lacking any semblance of authority in Japan, the ABCC was wholly dependent upon the voluntary cooperation of the Japanese people. The good will of the city officials, local physicians, and — above all — of the midwives, was, if not indispensable, highly desirable. It should particularly be mentioned that in Japan the social stigma attached to the birth of a malformed child is rather considerable. Every effort had to be made to develop a program which would not antagonize the mothers of malformed children by exposing them to what they considered undue publicity. In this effort, the midwives were the key.

Although the destruction and desolation which were the aftermath of the atomic bombings have been many times described, neither photographs nor words are adequate to the occasion. In addition to the post-war stresses and strains to which all the inhabitants of Japan were subject, the citizens of Hiroshima and Nagasaki were confronted with a more formidable job of reconstruction than the inhabitants of most Japanese cities, badly bombed though these had been. These are not the circumstances in which research or even cooperation in the research of others flourishes.

Chapter II

THE PLAN

WITH the considerations outlined in Chapter I in mind, a plan of attack on the problem was developed which will now be described.

2.1 *Pregnancy registration.* — As mentioned earlier, in the post-war years the ration system in Japan was such that pregnant women upon registration of their pregnancy following the completion of the fifth month of gestation could thereby acquire access to certain rationed items. With the cooperation of the city administrators of Hiroshima, Nagasaki, and Kure (a control city), a system was instituted whereby at the time of her registration at the city hall (or district office) for ration purposes, each pregnant woman or her representative in these three cities completed the first two-thirds of a questionnaire which included such items as identifying information, a brief radiation history of the husband and wife, a short summary of the past reproductive performance, and pertinent details concerning the present pregnancy. In the majority of instances the informant for this questionnaire was the pregnant woman herself. Only rarely were both parents available. Accordingly the radiation information on this questionnaire tended to be more reliable for the wife than the husband (cf. Sec. 4.6).

Figure 2.1 is an English translation of this questionnaire, which will be referred to as the "Genetics Short Form." This questionnaire was administered by trained clerks who occupied a special office in the city hall (or district office). The actual collection of data began in Hiroshima in February, 1948; in Kure in March, 1948; and in Nagasaki in July, 1948. The questionnaire was filled out in duplicate; the original was then given the registrant, while the Commission retained the copy. At the time of termination of the pregnancy, the midwife or physician in attendance completed the questionnaire by answering certain questions pertaining to the characteristics of the child and delivery.

More specifically, information was requested on the following *possible* indicators of a genetic difference between the children of control and irradiated parents: sex, birthweight, stillbirth, and presence of malformation. In case there was an abnormal termination (e.g., stillbirth, malformation), the midwife informed the Commission by telephone as soon as possible. If the outcome appeared normal, the questionnaire was held by the midwife until collected by a Commission clerk. Such collections were at first at weekly and later twice weekly intervals. Regardless of the type of termination, a Japanese physician in the employ of the Commission or the Japanese National Institute of Health called to examine the child — at once, if there was a report of an abnormal termination, or on a somewhat more leisurely schedule if the termination was reported as normal. Midwives received a small fee for each questionnaire they completed. Because of this lag between birth, questionnaire collection, and examination by a physician, it was also possible to obtain rather complete information concerning death during the first 7 days post partum, hereafter referred to as neonatal death.

The cooperation both of the mother and of the midwife in this study was of course entirely voluntary. An attempt was made at the time of pregnancy registration to explain the rudiments of the program to the registrants, and, in addition, each mother was given a brief printed description of the program. Figure 2.2 is a translation of this description. Numerous meetings were held with the Midwives' Associations of the cities to explain the program and answer questions. Attendance at the meetings was usually excellent, although, the Japanese birth rate being what it was, apt to diminish appreciably between the opening and closing of any given meeting.

Approximately a year and a half after the

5

INVESTIGATION SHEET FOR BIRTHS AFTER ATOM BOMB
PRINTING OF JULY, 1949

		Husband	Wife
1.	Name of city and investigation sheet number		
2.	Day, month, and year of registration _____ 3. Expected date of birth _____		
4.	Name (Maiden name in case of wife)		
5.	Birth dates of husband and wife		
6.	Age (exact number of years and months)		
7.	Present in Hiroshima or Nagasaki at time of bombing		
8.	Location at time of bombing (street and number)		
9.	Distance from hypocenter		
10.	Indoors		
11.	Type of building		
12.	Did you have or not have subcutaneous bleeding		
13.	Did you have or not have gingivitis		
14.	Did you have or not have bloody diarrhea		
15.	Did you have or not have epilation		
16.	Did you have or not have fever		
17.	Did you have or not have burns		
18.	Did you have or not have external injuries		
19.	Date, month, and year of beginning cohabitation		
20.	Number of months interruption of cohabitation		
21.	Total number of months cohabitation		
22.	Number of months cohabitation before August, 1945		
23.	Number of pregnancies before August, 1945		
24.	Number of spontaneous stillbirths before August, 1945		
25.	Number of therapeutic abortions before August, 1945		
26.	Number of months cohabitation after August, 1945		
27.	Number of pregnancies after August, 1945 (including present)		
28.	Number of spontaneous stillbirths after August, 1945		
29.	Number of therapeutic abortions after August, 1945		
30.	Total number of pregnancies		
31.	Total number of spontaneous stillbirths		
32.	Total number of therapeutic abortions		
33.	Marriage of blood relations (first cousin, one and one-half cousins, second cousins, etc.)		
34.	Present address and occupation of husband		
35.	Day, month, and year of beginning of last menstrual period of wife		
36.	Day, month, and year of birth expectation (according to calculation sheet)		
37.	Present month of pregnancy		
38.	Day, month, and year of termination of birth		
39.	Number of weeks of pregnancy		
40.	Course of labor: Spontaneous _____ Induced _____ Duration _____ Use of instruments _____		
41.	Condition of newborn: Live birth after 38 or more weeks _____ Premature birth under and including 38th week _____ Miscarriage 20 weeks or under _____ Stillbirth 21-29 weeks _____ Stillbirth 30-38 weeks _____ Stillbirth after 38 weeks _____		

THE CHILDREN OF ATOMIC BOMB SURVIVORS

42. Multiple birth (2, 3, etc.)_____ Order of birth_____
43. Sex of newborn_____ 44. Weight (grms.)_____
45. Presence or absence of malformation_____
46. Type of malformation (give details)_____

47. Date of death of newborn child_____
48. Date of termination of any pregnancies after January, 1948_____
49. Remarks_____
50. Name and address of attendant at birth_____

FIGURE 2.1 — English translation of the Japanese-language "Genetics Short Form."

program went into operation, an attempt was made to ascertain what proportion of births in Hiroshima were to mothers registered in the Genetics Program, i.e., to determine the completeness of our pregnancy registration. Inasmuch as studies by the Public Health and Welfare Section, GHQ, SCAP (Mr. L. V. Phelps) had revealed that the official Japanese registration of births by place of occurrence during the period of this study was quite complete,[1] it was felt that the most valid (as well as convenient) approach to this evaluation would be to compare the number of terminations recorded by the Genetics Program each month with the number officially registered in the city. The comparison follows:

1948	City	ABCC	%
August	713	653	91.6
September	723	674	93.2
October	602	614	102.0
November	625	596	95.4
December	555	577	104.0
1949			
January	1,001	914	91.3
February	759	703	92.6
March	740	675	91.2
April	719	630	87.6
May	707	634	89.7
June	728	711	97.7
July	826	759	91.9
August	762	703	92.3
September	740	685	92.6
	10,200	9,528	93.4

Percentages greater than 100 are probably best explained by changing city boundaries and monthly carry-overs, especially in the case of the December, 1948 figure, because of the Japanese custom of delaying registration on a child born in late December. Assuming that terminations known to the ABCC were also known to the city, this indicates that approximately 7 per cent of the births occurring in the city were to mothers not registered with the Genetics Program.

A large proportion of these births to parents not registered with the Genetics Program of the ABCC subsequently came to the attention of the Program in the following two ways: (1) At the time of registration of the birth with the city, the city clerks routinely inquired if the pregnancy was known to the ABCC, and if not, directed the registrant to the Genetics Program offices in the city hall. (2) Private doctors and midwives frequently reported the names of all women whom they attended who were not registered with the Genetics Program. Births coming to the attention of the Program through these channels were placed in an "Unregistered Series." From time to time, in an effort to improve the coverage of the Program, the reasons were investigated for the failure of the pregnancies resulting in these births to be included in the routine registration. One of the more complete investigations of this type was carried out in Hiroshima in October of 1952 and dealt with 147 unregistered births occurring in the period of June through September of that year. The reasons for not registering given by the mothers of these 147 children were as follows:

Reason for failing to register	No.
1. Arrived in city just prior to birth of child..	20
2. Registered pregnancy at city hall, but failed to visit ABCC office there.............	37
3. Failed to register either with city for ration purposes or with ABCC, although in town during pregnancy:	
Too busy	23
Forgot	19

[1] The actual figures for registration of live births during the years 1948, 1949, and 1950 are 98.1%, 98.8%, and 98.9% respectively, while for stillbirths the corresponding figures were 98.7%, 99.4%, and 99.4% (cf. Public Health & Welfare in Japan, 1948, 1949, 1950, by Public Health & Welfare Section, GHQ, SCAP).

First seen by midwife late in pregnancy; felt registration wasn't worth bothering with	5
Requested someone else to register; that person failed to do so	17
Too sick to register	4
Didn't know about registration	12
4. Miscellaneous causes	10
	147

differences in city administration practices, the percentage of unregistered births was always less in Nagasaki than in Hiroshima.

The question naturally arose as to whether the information collected concerning registered births should also be obtained for these unregistered births, following which the two sets of data would, for analytical purposes, be com-

TRANSLATION OF PAMPHLET DISTRIBUTED AT THE TIME OF PREGNANCY REGISTRATION

To All Prospective Mothers:

The ABCC has a request to make of all prospective mothers. I presume you are familiar with the research project of the ABCC. The American and Japanese doctors work together in giving physical examinations to all babies born in Hiroshima, Kure, and Nagasaki, and cooperate in carrying on other scientific researches. We hope that those who have encountered the bomb and also those who have not will respond to our program. A comparative study of the physical conditions of the persons who experienced the bomb and persons who did not, insures the scientific accuracy of our studies.

When you register your pregnancy at the city office will you kindly fill out a questionnaire for ABCC? Our representative will ask you questions and fill out the questionnaire. You need not be worried about the questionnaire for it contains only questions concerning your expected baby, name of parents, birthdates, date of marriage and history of exposure to the atomic bomb. If you have any questions to ask, our representative will be happy to assist you. The form, on a white sheet, which you will be asked to take home with you, will be filled out by your attending midwife or doctor and returned to ABCC.

Within a month after delivery an ABCC doctor will call on you and make a physical examination of your baby. (If unfortunately your pregnancy terminates in abortion, stillbirth, or any other abnormality, an ABCC doctor will also call on you.) By this examination you will be able to know your baby's true physical condition and at the same time you will be making an important contribution to medical science.

In the ABCC clinic the most modern X-ray and other medical equipment are available if the necessity of a thorough examination of your baby is found necessary. In such cases the ABCC will call for you and bring you and your baby to the clinic. The results of the examinations are strictly confidential but if you desire they may be available to your family doctor. All services are performed free of charge.

NOTE:

1. Regardless of whether or not you experienced the atomic bomb, whether your pregnancy terminates in abortion or stillbirth, your cooperation will be appreciated.
2. Please preserve the questionnaire that you were asked to take home and have the attending midwife or doctor fill it out after delivery. If you lose the questionnaire it will be issued to you again, if you will present your Expectant Mother's Notebook at the city office.
3. Please inform the ABCC if you change address at any time.

Your kind cooperation is requested in this medical research program.

FIGURE 2.2 — English translation of the Japanese-language pamphlet distributed at the time of pregnancy registration.

Reason (1) appears to represent an unavoidable loophole in the program, but inasmuch as mothers falling into this category as a rule were unexposed, this does not represent a serious loss of data. Reason (3) also represents an unavoidable loophole in a program geared as this was to a civic function. Reason (2), on the other hand, represented an unnecessary loss of data (which could be and was rectified), due for the most part to the fact that the city clerk failed to direct the registrant to the ABCC office, either through forgetfulness or, in the case of new clerks, not knowing that this should be done. Because of

bined. In view of the extent to which recent arrivals to the city enter into the composition of the unregistered group (20/147 in the preceding analysis), it was felt that it would be better not to combine the two series, even though this involved the loss of a certain amount of badly needed data.

As mentioned earlier, it was not necessary that the pregnant woman register in person, it being possible for her designated representative to register for her. In approximately 4 per cent of all registrations, neither the prospective mother nor the father appeared at the city hall,

the registration being carried out by some third party. Such individuals of course were unable to answer many of the items on the questionnaire. In these instances, a clerk was sent to the home to obtain the necessary data.

2.2 *Special studies on abnormal terminations.* — In the event that a pregnancy terminated abnormally, as in a stillbirth or a child with a congenital malformation, a supplementary questionnaire was completed in the patient's home by a doctor in the employ of the ABCC. This questionnaire covered in some detail gynecologic history, maternal disease during pregnancy, past reproductive performance, and economic status. The actual questionnaire is reproduced in Figure 2.3 and will be referred to as the "Genetics Long Form." In addition, in the case of abnormal terminations, blood was obtained from the mother for a serological test for the presence of syphilis.

If the physician charged with completing the supplemental questionnaire encountered an unusual abnormality, or one concerning whose nature he was unsure, he could, if the parents were willing, arrange for the child to be seen by an American-national pediatrician at the ABCC headquarters, for such diagnostic studies as seemed necessary, as well as photographic documentation of the case.

The Japanese physicians concerned with the home examination of these infants were for the most part recent medical school graduates. Many of them were employed on a half-time basis, the remainder of their time being devoted to hospital duties. Each of these men, before he was sent out on home calls, received instruction in the systematic examination of the newborn infant. In addition, from time to time lectures were given on the recognition of the more common congenital abnormalities. Between 30 and 40 full or part-time physicians were employed in this capacity at any one time. For a variety of reasons, personnel turn-over was at first relatively high; this necessitated a constant training program.

2.3 *The "10-percent sample."*—Each woman as she registered received a registration number for her pregnancy, these being assigned in sequence and without respect to radiation history. For every tenth registration, which is to say, all registrations for which the terminal digit in the registration number was zero, the same supplemental questionnaire just described in connection with abnormal terminations was completed in the home. A serological test for syphilis was also carried out on the mother of every tenth termination. In this way a 10 per cent sample was obtained with which to supplement the information obtained on the original questionnaire concerning the comparability of control and irradiated parents. In addition, information was obtained of value in analyzing the causation of abnormal pregnancy terminations.

2.4 *The accuracy and reproducibility of the anamnestic data obtained on the Genetics Short Form and the Genetics Long Form.* — The information accumulated in the course of this study is of two types: anamnestic, and observational. The accuracy of both types of information was of course a matter of vital concern. During the period covered by this investigation, many women in the study cities had several pregnancies. Each pregnancy of a given woman was registered independently of any others. Discrepancies in the answers to particular questions in the course of multiple registrations by a given woman provide some insight into the reliability of the material. In addition, because of overlaps in the various segments of the ABCC program, the same information might be obtained independently in different studies.

The information obtained on the Genetics Short Form is of three main types: (1) radiation history, (2) history of past reproductive performance, and (3) observations by midwives and physicians on the outcome of the current pregnancy. The reliability of information of types (1) and (3) will be examined in detail in Chapters IV, VIII, IX, and X. The attempt to obtain information of type (2) was motivated by the possibility of detecting an increase in the proportion of abortions and miscarriages among conceptions occurring in the period immediately following the bombing. However, a preliminary analysis of repeat registrations by the same mother, in May of 1951, suggested that the error in the reporting of abortions and miscarriages was such that any attempt to utilize these data would be ill-advised. This study thus utilizes only direct observations collected under supervision during the years 1948–1954.

The material sought on the Genetics Long Form was collected with three purposes in mind: (1) the description in some detail of congenital defect, (2) the recording of certain

SPECIAL BIRTH QUESTIONNAIRE

Date of investigation_____ Name of investigating Doctor

1. Sex and full name of baby: M F
 (*Romaji, Kanji, Furigana*)
2. Name of place of birth :
 Home Hospital
4. Date of birth :

3. Questionnaire Number

5. Attendant at birth: Dr. Mw. Name:
 Address:

6. Diagnosis of defects of child and details:
 Blood taken Y N Birth weight_____Gms. Birth length_____In.
 Seen by ABCC doctor Y N Mm. Sun.
 Specimen taken Y N Circumference: Head_____In. Chest_____In.
 Photo or X-ray taken Y N

Diagnosis : _____

7. Place of death: Age: NEONATAL
 Home Hospital
8. Attendant at death: Dr. Mw. Name:
 Address:
9. Autopsy: Yes No Unknown Date:
 Place of autopsy and Doctor's name:

Page 1 of 2 pages ABCC Form No. GEN-2 (OCT 50 R3)
 BIRTH QUESTIONNAIRE

FIGURE 2.3 — The Genetics "Long Form." Explanation in text.

THE CHILDREN OF ATOMIC BOMB SURVIVORS

10. Name of father:
 (*Romaji, Kanji, Furigana*)
11. Address:
12. Date of birth:
13. Age: yrs. mons.

14. Does the father have the same defect as this child? Yes No Unknown
 Details:

 Does the father have any other defect? Yes No Unknown
 Details:

 Does any of the father's relatives have the same defect as this child? Yes No Unknown
 Details:

 Does any of the father's relatives have any other defect? Yes No Unknown
 Details:

15. Blood relationship: Father and Mother Maternal Grandparents Paternal Grandparents
 Yes No Unknown Yes No Unknown Yes No Unknown
 Details:

16. Economic status of family: Very poor Poor Average Well-to-do Very rich

17. Name of mother:
 (*Romaji, Kanji, Furigana*)
18. Address:
19. Date of birth:
20. Age: yrs. mons.

21. Does the mother have the same defect as this child? Yes No Unknown
 Details:

 Does the mother have any other defect? Yes No Unknown
 Details:

 Does any of the mother's relatives have the same defect as this child? Yes No Unknown
 Details:

 Does any of the mother's relatives have any other defect? Yes No Unknown
 Details:

22. Menstruation: Age of onset Duration in days Interval in days

23. Irregularities menstruation: Yes No Unknown
 Details:

ABCC Form No. GEN-2 (OCT 50 R3)
BIRTH QUESTIONNAIRE

FIGURE 2.3—*Continued*

24. Pelvic operations, diseases, or treatments prior to present pregnancy: (*Particularly D & C*) Yes No Unknown
 Name of hosp. & Dr.: Date:
 Address of hosp. & Dr.:
 Details:

RECORD OF THIS PREGNANCY

25. General diseases and surgical operations
 (*particularly of a febrile nature*): Yes No Unknown Date:
 Details:

 Fever: Onset Duration

 Rash: Yes No Unknown Details:

26. Uterine bleeding in the first six months of pregnancy: Yes No Unknown

 Onset Hours of duration:

27. Presentation of fetus at delivery: Occiput Face Breech Foot Arm Caesarean Unknown

SEROLOGY

28. Mother:

TEST	RESULT	PLACE	DATE	TEST	RESULT	PLACE	DATE
		ABCC		WASSERMANN			
		ABCC		MURATA			
		ABCC		IDE			
		ABCC		HOKKEN			
		ABCC		MEINICKE			

INDICATE : POSITIVE as +
 DOUBTFUL as ±
 NEGATIVE as −

29. Infant:

TEST	RESULT	PLACE	DATE	TEST	RESULT	PLACE	DATE
		ABCC		WASSERMANN			
		ABCC		MURATA			
		ABCC		IDE			
		ABCC		HOKKEN			
		ABCC		MEINICKE			

30. Remarks:

Page 2 of 2 pages

ABCC Form No. GEN-2 (OCT 50 R3)
BIRTH QUESTIONNAIRE

FIGURE 2.3—*Continued*

FIGURE 2.3—Concluded

supplementary technical observations of possible pertinency to pregnancy outcome such as economic status and maternal luetic infection (as revealed by serological testing), and (3) the recording of considerable background information of possible relevancy to the manner in which a given pregnancy terminated, with particular reference to the problem of congenital defect. This included a brief family history of both father and mother, a detailed reproductive history of the mother, disease during pregnancy, etc. The reliability of the material collected under headings (1) and (2) will be discussed in Chapters V and VIII. The reliability of anamnestic data mentioned under heading (3) will be briefly considered here. This material is in the strict sense not relevant to the problem of radiation effects, the questions having been introduced because of the opportunity of adding to existing knowledge concerning the etiology of congenital defect. Unfortunately, particularly with reference to the occurrence of congenital defect in other members of the family, several analyses have raised serious doubts as to whether these histories are of sufficient reliability for research purposes. There is in general less major congenital defect reported in family histories than would be expected on the basis of the observations made during the course of this investigation. This may in part be due to lack of information and forgetfulness, but is probably in no small measure due to the social stigma connected with congenital defect, and the consequent efforts to conceal it.

2.5 *The autopsy program.* — As the over-all program of the ABCC developed, and particularly after a Department of Pathology was established, it became possible to initiate the performance of autopsies on deceased or stillborn infants falling within the scope of the program. Because of the greater concentration of Commission personnel in Hiroshima than in Nagasaki, infant autopsies were begun in Hiroshima in December, 1948, but not in a methodical fashion in Nagasaki until several years later.

A request to perform an autopsy was routinely made of the parents of each stillborn infant and of each infant dying during the neonatal period coming to the attention of the Commission. If permission was granted, the contactor brought the cadaver to the ABCC pathology laboratory. A complete autopsy was performed in all cases save where maceration made this impossible. Photographs and X-rays were taken when indicated. A copy of each autopsy protocol was ultimately filed with the appropriate Genetics Long Form. The cadaver was then carried to a crematorium in an ABCC vehicle. Ashes were returned to the family in a suitable receptacle on request. Cost of the coffin and cremation fee were borne by the ABCC.

When the autopsy program was well under way in the two cities, some 50 per cent of the total available material came to autopsy. For instance, an analysis of 300 consecutive registrations terminating in stillbirths or neonatal deaths, beginning January, 1950 in Hiroshima, revealed that out of the total of 311 infants involved, 158 (50.8%) came to autopsy. Of the remaining 153, 46 (14.8%) were seen by an ABCC physician but did not come to autopsy. There were 107 (34.4%) who were not seen by an ABCC physician, and concerning whose appearance at birth there is only the statement of the mother and attending midwife. This represents the most serious potential loss of information in the entire program, since a certain amount of congenital malformation could well go unrecognized or concealed. However, these unexamined infants can introduce bias into the findings only if there is a difference in reporting on the part of exposed and unexposed parents. We shall examine the latter possibility in Chapter XIII.

2.6. *The collection of data on infants aged 9 months.* — Japanese homes are not always well lighted and, because of their construction and the fuel shortage, are decidedly cold in winter. These conditions are not favorable for an adequate physical examination of a newborn child. The possibility had to be recognized that for these reasons, as well as diagnostic oversights, some malformations were not being observed at birth. Furthermore, certain defects, such as congenital dislocation of the hip, spastic paraplegia, deafness, blindness, congenital heart disease, or mental defect are not always readily diagnosable at birth. Accordingly, in January of 1950, the earliest date at which the over-all development of the ABCC permitted such action, a program was inaugurated to bring into the central clinical facility at age 9 months as many of the children examined shortly after birth as possible, both as a check on diagnostic

oversights at the time of the first examination and on supplementary diagnoses. Certain anthropometric measurements were also obtained as an index of general physical development. Furthermore, because of the possible relationship between genetic damage and death during the neonatal period of infancy, information as to time and manner of death was obtained for deceased children. The age for this study was set at 9 months rather than, e.g., one year, because of the relative ease with which a 9-month-old infant may be examined. The form on which the data were recorded is reproduced as Figure 2.4, and will be referred to as the Genetics 9-Months Follow-up.

Since it was not possible to conduct studies of all the babies examined at birth who were still alive 9 months later, a system of obtaining a random sample had to be set up. This was done by the simple expedient of calling in babies for examination according to the terminal registration digit of the pregnancy. Each month all babies with certain specified terminal registration digits were seen, the number of children seen being regulated by the other demands on the clinical facility and the personnel available. In this fashion, a random 10, 20, 30, etc. per cent of the original group could be called in for examination. Parental cooperation was usually excellent. Where a child who was included in the sample could not be examined, an attempt was made to establish why, in an effort to detect possible sources of bias.

A comparison of the recorded findings on a series of infants examined shortly after birth and again at 9 months provides a check on both the clerical and medical efficiency of the program. Such a comparison was carried out in 1952, based on 4,578 pregnancy terminations studied at birth and followed up again at age 9 months in Hiroshima in 1951. The comparison involved the findings coded on the IBM cards used for the final analysis (see below) rather than on a matching of the two original forms, since the IBM cards are of course the basis for the final tabulations. Particular attention was directed towards the occurrence of gross malformation in the two series.

A total of 38 clerical errors came to light. Approximately half of these arose in the following manner: a tentative diagnosis of major malformation made when a child was seen in the home was not confirmed when the child was brought into the central facility for examination a short time thereafter. When this same child was seen at age 9 months and the defect again not observed, this was erroneously coded as a refutation of an earlier diagnosis, when in fact no actual diagnosis had been reached. This error did not affect the actual analysis of the data. The other 20 clerical errors were of a more serious nature, involving for the most part an error in the use of the 6-digit code developed in connection with this study for the classification of congenital malformation.

From the medical standpoint, the comparison of the two sets of records brought out the unreliability, under the conditions of this study, of the diagnosis shortly after birth of congenital heart disease and congenital torticollis. This finding led to omitting these two diagnoses from the "at birth" data, although both were included in the "9 months" data.[2] Excluding these two diagnoses, there were 48 instances of major defect listed among the 4,578 terminations studied shortly after birth. The follow-up examination at age 9 months increased the number of diagnosed major defects (exclusive of congenital heart disease) to 122. On the face of it, this amounts to a 154 per cent increase. However, analysis of the data revealed that three diagnoses contributed disproportionately to the 9-months total, as follows:

Diagnosis	No. of times made on first examination	No. of times made on second examination
Dysplasia of acetabulum	3	26
Pilonidal sinus	5	22
Inguinal hernia (females only)	5	19
	13	67

Exclusive of these three diagnoses, there were 35 diagnoses of major defect at the time of the first examination (all confirmed later), as contrasted to 55 at the second examination, an increase of 57 per cent. It was apparent from this that the 9-months examination not only served as a valuable check on the "at birth" program but significantly increased the amount of congenital disease recognized in these children, although the bulk of this contribution was due

[2] The diagnosis of congenital heart disease was based on one or more of the following criteria: persistent cyanosis, a grade III or IV apical systolic murmur, a precordial thrill, or cardiomegaly in the absence of another adequate explanation.

FIGURE 2.4 — The form used for recording pediatric information at age 9 months. Explanation in text.

The Plan

ATOMIC BOMB CASUALTY COMMISSION

PEDIATRIC FOLLOW-UP (Interval History)

(Romaji) Family name	Given name	(Kanji and Furigana)	Master file no.

Location: H ☐ K ☐ N ☐ Exposed ☐ Control ☐ Sex: M ☐ F ☐ Day Month Year

Birth Date: Day Month Year (Era Year) Age (mos.) Dept. No.

Interval History: Bowel movements–
　Feedings per day–
　Vomiting or regurgitation–
　Sleep–

Measurements:

Weight	lbs.	oz.	Abdomen (cm)	
Height (cm)			Pelvis (cm)	
Health (cm)				
Chest (cm)				

Physical Examination:　　　Temp: (O) (R)_____　Pulse:_____

Teeth:　Present (X)　Erupting (O)

Right						Left					
VI	V	III	II	I	I	I	II	III	III	V	VI

Motor Development: (good, fair, poor)
Muscles: (good, fair, poor)
Subcutaneous tissues: (good, fair, poor)

Laboratory:　(X) Examination Completed; (A) Abnormal Finding

Hematology	
Urine	
Serology	
Stool	
X-ray (hand)	
X-ray (chest)	

Remarks:

Examining Physician
ABCC Form No. PED-6 (FEB 51)
PED FOLLOW-UP INT.

FIGURE 2.4—*Concluded*

to a relatively few diagnoses. The question of whether these three common diagnoses "unbalance" the 9-months data will be examined later. On the basis of these findings, the decision was made to conduct separate analyses on the malformation findings "at birth" and the *additional* findings at the later examination. It should perhaps be emphasized at this point that although we have presented the above analysis in terms of specific defects, the unit in the statistical analysis to be presented later of the relationship between radiation history and congenital defect was the malformed child rather than specific defects, i.e., for analytic purposes each child is scored only once regardless of the number of major defects present.

2.7 *The processing of the data.* — All of the questionnaires employed in this study were checked for completeness by trained clerks. In the event of an omission or an obvious error, the individual concerned was queried either by mail or by a "contactor." Where a discrepancy appeared between the information obtained on a first registration and on a subsequent registration by the same couple, an attempt was made to determine which of the two statements was correct and the reason for the discrepancy. When the questionnaires were completed, the data necessary to the projected analyses were coded, and the coded results transferred to standard 80-column machine tabulation cards. The codes used for the Genetics Short Form, the Genetics Long Form, and the Genetics 9-Months Follow-up are reproduced in Appendix 1.

2.8 *The study of spontaneous abortions.* — Early in the planning of the study, the possibility presented itself that a significant fraction of the detectable genetic effects of the atomic bombs on the first post-bomb generation would be in the form of dominant lethals which would find expression during the early stages of pregnancy. Although some of these dominant lethals might manifest themselves prior to implantation, others might not be effective until a month or two after implantation. Theoretically, the occurrence of dominant lethals in any number might be detectable through an increase in early spontaneous abortions.

In the fall of 1949 an attempt was initiated in Hiroshima to obtain as many records as possible of pregnancies terminating spontaneously prior to the twentieth week of gestation. Data were collected through the personal contact of one physician with the practicing obstetricians and gynecologists of the city. Radiation histories were obtained on each couple involved, with the intent of determining whether these histories differed significantly from those obtained from parents when the pregnancy was of twenty or more weeks' duration.

In the fall of 1950, the physician originally responsible for the collection of these data severed his association with the ABCC; the collection of the data was continued by two younger men. The collection of data was discontinued in early 1952. Data on 1,053 early spontaneous terminations were obtained during the first portion of this study, and information on 638 during the latter portion.

Many problems were anticipated in the collection of data on early spontaneous terminations; most of these anticipations were realized. During the spring of 1952, a preliminary analysis of these data was carried out. This analysis clearly revealed heterogeneity within the data, in terms of differences between the findings of the first and second portions of the program. Because of the impossibility of determining the source of these differences, as well as continuing difficulty in the collection of such data, this aspect of the program was abandoned, and will not be referred to again.

2.9 *Cytogenetic effects of the atomic bombs.* — As one facet of the over-all Genetics Program, the possibility has been explored of detecting by cytological methods chromosomal damage among the survivors of the bombings. Testicular material obtained either at the time of surgery, by biopsy, or at autopsy was examined, using appropriate methods. These studies, which will be described in detail elsewhere by Dr. M. Kodani, revealed an unsuspected amount of chromosomal polymorphism among normal Japanese, a finding which has served as a serious drawback in the use of this material as originally envisioned.

2.10 *The decision to discontinue work in Kure.* — At the outset of the study, when satisfactory information was not available concerning the composition in terms of irradiation experience of the population of the cities of Hiroshima and Nagasaki, an effort was made to collect control data in the city of Kure, located some 18 miles from Hiroshima. Preliminary analyses of the year-to-year data quickly revealed

that a considerable proportion of the inhabitants of the cities of Hiroshima and Nagasaki during the years 1948 through 1953 had not been there at the time of the atomic bomb explosions. Thus there existed in these two cities a source of internal controls which appeared to obviate the necessity for a separate control city. Accordingly, in September, 1950, after data had been collected on some 8,391 pregnancy terminations, work was discontinued in Kure.

2.11 *The termination of the program in January, 1954.* — At the time that the program which has just been described was initiated, Japanese birth rates were at record levels. However, in 1948 and 1949 the Japanese government, as one of a series of measures designed to reduce the disparity between population number and available food resources, liberalized the legal indications for the performance of "therapeutic" abortions by physicians. Although exact figures are impossible to obtain, in 1951 approximately 300 pregnancies were being interrupted each month in Hiroshima alone. In consequence of this and possibly other measures, the birth rate in Japan, including, of course, Hiroshima and Nagasaki, underwent one of the most spectacular declines ever recorded in a civilized country (Koya, 1953, 1954; Population Reference Bureau, 1953). The decline in actual number of births registered with the Genetics Program in the two study cities is shown in Table 2.1. In addition, during the first five years of the study, there was not only an absolute decline but in Hiroshima, the largest source of data, there was also a tendency toward a relative decrease in the numbers of infants born to more heavily irradiated parents. The reasons for this are not entirely clear. In part the finding is undoubtedly due to emigration from the city and completion of reproductive span on the part of older exposed individuals, without proportional replacement by younger age groups, but other factors may also be involved.

In the original planning of the program, the anticipated duration had been approximately ten years. By 1952 the annual decline in the amount of data becoming available had reached the point where it was apparent that a serious reconsideration of the duration of the program was indicated. Accordingly, on July 10-11, 1953, a Second Genetics Conference met to consider the results of a preliminary analysis of the data, and to reach recommendations concerning the future conduct of the work. The members of this conference were Dr. G. W. Beadle, Dr. D. R. Charles, Dr. C. C. Craig, Dr. L. H. Snyder, and Dr. Curt Stern (chairman) with Drs. W. J. Schull and J. V. Neel functioning ex officio. In view of the relatively small expected return, in terms of reducing the sampling variances of possible differences, from observations extending over an additional four years, and in the light of the very real problems and the expense involved in maintaining the program at a satisfactory level of efficiency, it was the unanimous recommendation of the Conference that the program be terminated in the near future. This recommendation was accepted

TABLE 2.1 PER CENT OF ALL REGISTRATIONS (INCLUDING THOSE INVOLVING PARENTAL CONSANGUINITY) WITH AT LEAST ONE PARENT HEAVILY EXPOSED (RADIATION CATEGORIES 4, 5)

	Nagasaki			Hiroshima		
Year	M or F, 4 or 5	Total	%	M or F, 4 or 5	Total	%
1	31	778	4.0	356	3,805	9.4
2	269	8,736	3.1	593	8,064	7.4
3	259	7,621	3.4	488	6,878	7.1
4	226	7,093	3.2	393	6,064	6.5
5	222	6,664	3.3	360	5,239	6.9
6	158	5,348	3.0	317	4,723	6.7
	1,165	36,240	3.22	2,507	34,773	7.21

by the Committee on Atomic Casualties; actual data collection was suspended in February of 1954.

2.12 *Acknowledgments.* — It is obvious that a program of this complexity and magnitude depends for its success on the efforts of many people in addition to the authors of this monograph. In particular, our heartfelt appreciation is due Lt. Col. Carl F. Tessmer, Dr. Grant Taylor, Dr. John Morton, and Dr. F. H. Connell, who, as successive directors of the Commission faced with the difficult task of balancing many demands on limited facilities and personnel, were always most generous in their treatment of the Genetics Program. Without the staunch, long-time support of the Committee on Atomic Casualties of the National Research Council and the Division of Biology and Medicine of the U.S. Atomic Energy Commission, this program could not have been undertaken nor continued. In particular, we are indebted to Drs. Max Zelle, Harold Plough, and Earl

Green, who, during their tenure as geneticists with the Division of Biology and Medicine of the Atomic Energy Commission, gave freely of their time and advice, and to Dr. A. E. Brandt, who, as biometrician to the Health and Safety Division of the New York Operations Office of the U.S. Atomic Energy Commission, offered a number of helpful suggestions. Turning now to Japan, it is a pleasure to mention the support received from various members of the Public Health and Welfare Section, GHQ, SCAP, and in particular from Brigadier General C. F. Sams, Chief of the Section, Col. Harry Johnson, Chief, Medical Service Division, and Mr. L. V. Phelps, Chief, Health Statistics Division. Dr. Harry C. Kelley, Deputy Chief, Scientific and Technical Division, Economic and Scientific Section, GHQ, SCAP, was an important link with Japanese science. Our special thanks are due Dr. M. Tsuzuki, who, as Chairman of the Medical Section of the Special Committee for the Investigation of the Effects of the Atomic Bombs of the Japanese National Research Council, assisted the work in many ways during the early days of the study. Later, when the Japanese National Institute of Health entered into the picture, Dr. R. Kobayashi, Director of the Institute and Dr. I. Nagai, Chief of the Atomic Bomb Section, were most helpful. Dr. H. Maki, as Director of the National Institute of Health staff in Hiroshima and Nagasaki, was an unfailing source of counsel in meeting local problems as they arose. Dr. Taku Komai served as a frequent and greatly appreciated source of contact with Japanese geneticists. The statistical analysis profited greatly from discussions and correspondence with Dr. C. R. Rao, Dr. H. L. Lucas, Dr. Robert Krooth, Dr. Marvin Kastenbaum, and Mr. Donald E. Lamphiear. Mrs. Betty Hsiao provided invaluable assistance in the computations. In addition the following persons have been so kind as to read and criticize all or part of the manuscript: Dr. H. Fairfield Smith, Dr. I. Olkin, Dr. C. C. Craig, Dr. P. S. Dwyer, Dr. Joseph Ullman, Dr. Curt Stern, Dr. L. S. Penrose, Dr. Bradford Hill, and Dr. P. Armitage. The burden of preparing the manuscript has been substantially lightened by the conscientious assistance of Miss Frances Davidson, Mrs. Jane Schneidewind, Mrs. Barbara Seijas, and Miss Grace Yesley. We are grateful to the Rockefeller Foundation for defraying publication expenses through a grant to the National Academy of Sciences. Finally, to the literally hundreds of other people, American and Japanese employees of the Commission and local Japanese, who have contributed so much to whatever success this program may have enjoyed, our sincerest thanks.

Chapter III

A COMPARISON OF HIROSHIMA AND NAGASAKI

HIROSHIMA, situated facing the Inland Sea on the southern coast of the Japanese island of Honshu, was at the time of the atomic bombing a city of approximately 350,000 inhabitants (including military personnel). Nagasaki, located on the western side of the Japanese island of Kyushu, was at that time a city of approximately 250,000 persons. In addition to the obvious differences in size and location, there are a number of other respects, pertinent to this study, in which the two cities are not comparable.

3.1 *The peopling of Japan; possible differences between the inhabitants of Honshu and Kyushu.* — The origin of the present-day inhabitants of Japan, like the origins of so many of the peoples of the earth, is a tantalizing riddle. Most standard reference works on the subject recognize the possibility of three distinct prehistoric streams of immigration into the Japanese islands, one entering Kyushu from the south by way of Formosa and the Ryukyu Islands, and ultimately derived from southern China,[1] a second entering northern Kyushu and southern Honshu from Korea, and ultimately derived from Manchuria, and a third stream, represented by the contemporary Ainu and having affinities with the present-day inhabitants of Siberia, northern Russia, Finland, and northern Sweden, entering from the north (Munro, 1908; Brinkley, 1915; Murdoch, 1926; Sansom, 1943; Beardsley, 1955). But while Japanese mythology, the earliest written records, and the archeological findings all supply reasonably good evidence for such waves of immigration towards the end of the Stone Age, it is not at all clear whether these immigrants found Japan already inhabited and, if so, the provenance of these very earliest inhabitants (cf. Kiyono, 1949). Almost equally uncertain is the relative timing of these waves of immigration, and the proportions in which these waves, together with the possible even earlier inhabitants, blended to form the modern Japanese type. Suffice it for our purposes to recognize the possibility that some thousands of years ago there existed significant anthropological differences between the inhabitants of the vicinity of Nagasaki in southern Japan and of Hiroshima in central Japan, and the further possibility that today, despite the many historical developments which would tend to obliterate such differences, some vestige still remains.

3.2 *Non-Japanese elements in the two cities.* — The present-day inhabitants of Nagasaki may differ genetically from those of Hiroshima for reasons other than just outlined. Historically, Nagasaki is pre-eminent among all Japanese cities as a point of contact with Western culture. The problem to which we must now address ourselves briefly is the question of the extent to which these contacts have been accompanied by intermarriages and arrangements of convenience which have left a lasting imprint on the biotype of the inhabitants of this area.

3.2.1 *Early Nagasaki contacts with the West.* — From our standpoint, the history of these contacts is best divided into three periods. The first of these begins in 1542 or 1543, when three Portuguese traders who had taken passage in a Chinese junk for Liampo were driven north by a typhoon and landed on a small island off the coast of southern Japan. Within a few years they were followed by Portuguese trading ships, which also brought Jesuit priests from the missions at Macao and Goa. The next 100 years were characterized by a considerable Japanese trade with the West, much of it funneling

[1] Some have suggested Indonesia, Malaysia, or Polynesia.

through Nagasaki. This trade was at first dominated by the Portuguese, but later shared in by Spanish, Dutch, and English ships. Concurrently, Portuguese Jesuit and Spanish Franciscan missionaries were busy. The activities of these missionaries, at first readily tolerated, at length reached the point where, both in terms of numbers of converts and political overtones, they were felt by the Tokugawa shogunate to pose a threat to the stability of Japan. In 1612, an earlier ban against Christianity was for the first time rigorously enforced. The Christian converts who refused to renounce their faith — and there were many — were vigorously persecuted. At the same time, the entrance of foreigners into Japan, as well as their movement about the country, was increasingly restricted. In 1636, Japanese ships and Japanese individuals were forbidden to go abroad. By 1639, all foreigners are reported to have been expelled from Japan, and the country had embarked on an era of self-imposed seclusion.

3.2.2 *The Dutch on Deshima.* — This severance of ties with the West was not quite complete. From 1640 until 1853, when Commodore Perry was successful in the first steps at re-establishing intercourse with the West, the Dutch, presumably because of the non-political and non-religious nature of their prior activities, were permitted to maintain a small trading station on Deshima in Nagasaki. This span of 213 years is the second of the three periods we must recognize. During this period, Chinese were also permitted to trade at Nagasaki and in much greater numbers than the mere handful of Dutch. Thus, Kaempfer (1728) describes a Chinese section of Nagasaki with upwards of 1,000 inhabitants, and further estimates, on the basis of the number of junks coming to Nagasaki and their size, that in the years 1683 and 1684 (which may or may not be representative of previous years) there were "for each year not less than 20,000 Chinese visitors." A year later trade with China was, at least officially, much more restricted, to 70 junks per annum with crews of not more than 30. Throughout the next century and a half the Dutch continued as in the past to send on the average one or two ships a year to Nagasaki, while Chinese activities were still further restricted, only a dozen junks a year being permitted to visit the port by 1820 (Murdoch, 1926).

3.2.3 *From the reopening of Japan to World War II.* — The third period may be dated from 1853 to the outbreak of World War II. Perry in his negotiations of 1853 and 1854 for port facilities declined the Japanese offers of Nagasaki, apparently feeling that its past would be more hindrance than help in establishing his new era, but under an agreement negotiated in 1857 by Harris, the first American Consul-General to Japan, Nagasaki became one of three treaty ports into which American ships could enter freely. However, the Dutch, from their beachhead at Deshima, had already profited from Perry's visit. In 1853, immediately following Perry's visit, the Japanese entered into negotiations with the Dutch for the purchase of men-of-war. In 1855, the Dutch presented the Japanese with the *Soembing,* the first unit of Western construction to be acquired by the Japanese Navy. In that same year, the Japanese established a navigation school and ship-building yard in Nagasaki, instruction being furnished by 22 Dutchmen. In 1857 another Dutchman, Dr. Pompe van Meerdervoort, assumed charge of a newly established school of medicine.

During the first decade following Perry's visit, while Japanese relations with foreigners were most unsettled, the number of Europeans in Nagasaki remained quite small, but beginning with the mid-1860's, and particularly after the initiation of the pro-foreign Meiji era in 1868, there arose a sizeable "foreign colony" in Nagasaki, largely concentrated on land on the eastern side of the harbor specifically set aside for this purpose. We have found it difficult to locate any exact data concerning the "foreign colony" between the reopening of Japan and 1897, with the exception of some statistics for 1864–1870, 1882, and 1889. Concerning the situation after 1897 there appears to be considerable information, but unfortunately sometimes conflicting in nature. This conflict is not so great as to invalidate an approximate evaluation of certain matters pertinent to this study.

Table 3.1 summarizes the earliest complete data on this period which we have been able to locate, made available through the courtesy of the Nagasaki Prefectural Library. Between 1864 and 1870 there were on the average 150–200 Occidentals in the city, as well as a rapidly increasing number of Chinese, the number of the latter growing from 141 in 1864 to 366 in

1870. The biological significance of this number of persons can only be evaluated in terms of the city's total population, which in 1870 was given as 29,127 (Nagasaki since the Restoration, 1925). Occidentals thus accounted for approximately 0.6 per cent of the population at this time.

Data concerning only two years during the interval 1871–1896 have come to our attention. In 1882, when the population of Nagasaki was 39,963, the total foreign population had risen to 829. Of these, six hundred and some were Chinese and the rest Occidentals, including approximately 100 English, 30 French, 30 Americans, and some Russians, Austrians, Dutch, and Danes (Nagasaki since the Restoration, 1925). By 1889, when the city population was 54,502, there were 354 Occidentals and 701 Chinese in residence (Nagasaki since the Restoration, 1925).

Beginning with 1897, more complete data became available. Table 3.2 summarizes city and prefectural census reports from 1897 to 1923.[2] The data are drawn from different sources, the city data from a book issued by the municipal government in 1925 (Nagasaki since the Restoration), the prefectural data from the Japanese Empire Statistical Annual (Nihon Teikoku Tōkei Nenkan). It is apparent that most of the foreigners residing in Nagasaki prefecture were concentrated in the city proper, making it possible in an approximate treatment such as this to substitute prefecture for city figures where the latter are lacking. That one or the other or both sets of data are not completely accurate is suggested by the fact that for several of the years, city figures exceed those for the prefecture, a manifest impossibility unless "residence" is defined differently in the two sets of data, a point not entirely clear. City figures are not available after 1923, but those for the prefecture indicate a slow, continual increase in the number of Occidentals residing in the city, as illustrated by the figures for 1930 given in Table 3.3. Throughout the first 40 years of this century, something like 0.2 per cent of the population of the city was Occidental, and an additional 0.7 per cent, Chinese.

The ethnic breakdown of the figures for two representative years, 1910 and 1930, is indicated in Table 3.3. These simple totals fail to provide a true insight into the "dynamics" of the situation. In Table 3.4 figures based on the 1920 census report are given concerning the age composition and the marital status of four of the principal ethnic groups, as well as for the total foreign population. Attention is directed towards the high proportion of unmarried males in the 20–39 age interval. Furthermore, in evaluating the significance of the number of married women, it should be borne in mind that in cases of mixed marriages, the wife and children assumed the citizenship of the husband (Izumi, 1921; Sasano, 1921).

3.3 *The biological influence of "foreigners" on Nagasaki and Hiroshima.* — It is a manifest absurdity to attempt to quantitate in any way the extent to which foreign contacts during these three periods left a biological imprint on the face of Nagasaki. However, one is perhaps permitted certain impressions. It seems unlikely for at least two reasons that the foreigners who visited Japan during the first of the three periods defined above contributed in any significant way to the genetic constitution of the present-day inhabitants of this area. For one thing, the

TABLE 3.1 POPULATION FIGURES BY NATIONALITY FOR FOREIGNERS RESIDENT IN NAGASAKI CITY BETWEEN 1864 AND 1870

(Abstracted from "Records of the Investigation by Nationality of Foreigners Resident in Nagasaki City — 1864 through 1879," from the official files of the Nagasaki magistrate's office)

Year	Britain	U.S.A.	Germany	France	Russia	Portugal	Holland	Others	Chinese	Total
1864	49	37	10	10	—	1	24	—	141	272
1865	66	33	10	11	2	3	26	—	246	397
1866	66	36	15	16	—	5	38	1	224	401
1867	66	35	19	14	—	5	30	4	305	478
1868	81	39	20	15	—	8	30	6	375	574
1869	79	23	21	15	1	5	23	2	333	502
1870	89	29	25	14	1	6	20	3	366	553

[2] The prefecture is a geographical unit roughly corresponding to the state of the U.S.A.

systematic suppression of Christianity, thought to involve the death of at least 20,000 Japanese converts, and perhaps 100,000 or even more (Kaempfer, 1728; Murdoch, 1926; Sansom, 1943), may have decimated the very group in which the offspring of Caucasian-Japanese unions were most apt to be found.[3] For another thing, the Japanese, during the period ending in 1639 when they were ridding themselves of foreign influences, were systematic in their uprooting of all traces of the intruders, to the extent that, among other actions, it is recorded that they exiled to Macao in 1636 some 287 women and children known to be related to the Portuguese by marriage or birth (Kaempfer, 1728; Woolley, 1881).

It is characteristic of the thoroughness of the Japanese in ridding themselves of foreign influences that during the second of the three periods we have defined, the Dutch were forced to live on a small, artificial island in the Nagasaki harbor, termed Deshima, measuring some 600 × 240 feet. The number of Dutch in residence was severely limited, usually to about 10 to 20, and the movements of these carefully restricted (cf. Kaempfer, 1728). But the Japanese are above all else realists. Alone among the Japanese people, prostitutes were permitted to visit Deshima, and periodically such of the Dutch as desired — their numbers perhaps swelled by the arrival of a ship — were permitted to visit Maruyama, then (and still) the brothel district of Nagasaki. These activities, like everything else the Dutch did, were carefully noted. Thus it is a matter of record that in one year (the 7th year of Kyōho, 1722) there were 270 Dutch visits to Maruyama — during that same year there were 20,738 Chinese visits (Boxer, 1950). In view of the practice of abortion and infanticide during the Tokugawa Era and the official attitude toward foreigners, it would be strange if any considerable number of children

TABLE 3.2 THE "FOREIGN" AND TOTAL POPULATION OF NAGASAKI CITY, AND THE "FOREIGN" POPULATION OF NAGASAKI PREFECTURE, 1897–1923

Year	Households	Males	Females	Total	Foreign population, prefecture	Total population, city
1897	290	851	271	1,122	—	73,974
1898	561	1,218	342	1,560	—	113,307
1899	731	1,345	357	1,702	1,743	120,865
1900	662	1,442	476	1,918	1,983	129,597
1901	526	—	—	2,104	2,037	142,811
1902	542	1,304	355	1,659	1,725	148,882
1903	640	1,334	409	1,743	1,698	154,727
1904	538	1,170	367	1,537	1,579	159,041
1905	487	1,121	365	1,486	1,535	163,324
1906	430	1,057	448	1,505	1,553	168,436
1907	416	1,061	402	1,463	1,523	173,118
1908	380	867	395	1,262	1,282	175,936
1909	321	800	419	1,219	1,290	176,970
1910	326	756	389	1,145	1,186	178,074
1911	274	639	329	968	1,127	179,257
1912	276	657	370	1,027	—	154,351
1913	318	754	396	1,150	1,189	160,450
1914	328	800	405	1,205	—	164,272
1915	304	763	392	1,155	1,200	174,077
1916	314	772	403	1,175	1,189	182,695
1917	316	772	405	1,177	1,173	188,006
1918	345	804	393	1,197	1,261	197,500
1919	347	810	397	1,207	1,311	205,958
1920	351	810	403	1,213	1,342	233,813
1921	355	806	406	1,212	1,217	245,954
1922	362	822	417	1,239	1,261	256,316
1923	370	832	423	1,255	1,303	264,669

[3] Exact data on this point are of course unobtainable. From our standpoint it is important to recognize that the permanent flight of Christians from Nagasaki to more inaccessible regions to escape persecution may, from the genetic standpoint, have done as much to obliterate any effects of interbreeding in that city as the actual death of Christians.

from such relationships reached maturity. The few of whom there is any record are cited in reference works primarily as "curiosities" (e.g., Thunberg, 1795, 1796).

It is more difficult to evaluate the extent to which racial admixture occurred during the period ushered in by Commodore Perry's visit. The fraction of one per cent of the Nagasaki population which has been Occidental has been a very mixed group — diplomats, missionaries, teachers, and commercial persons — many of whom, of course, did not intermarry or otherwise contribute to the Nagasaki gene pool. However, in addition to these permanent residents, there were relatively many transient seamen. The Russian fleet was stationed in Nagasaki during the winter months prior to the Russo-Japanese War in 1904. The intellectual climate of Japan during the late nineteenth and early twentieth centuries was characterized by the enthusiastic acceptance in some quarters of many aspects of Western civilization — there was apparently no particular opprobrium at-tached to mixed marriages or even temporary arrangements of convenience, as witnessed by the well-known story of Madame Butterfly, the locale for which was Nagasaki. One can state with considerable assurance that limited opportunities for racial admixture existed in Nagasaki between 1870 and 1940 [4] — the data do not permit one to go much further.

These are the bare historical facts. To what extent the present-day inhabitants of Nagasaki differ from those of Hiroshima because of racial admixture can only be a matter for conjecture. On the face of the evidence, it seems very unlikely that *at most* more than a few per cent of the corporate genetic constitution of present-day Nagasaki is non-Japanese in origin. This conclusion is borne out by the fact that the published A-B-O blood group frequencies of

TABLE 3.3 THE ETHNIC COMPOSITION OF THE FOREIGN COMPONENT OF NAGASAKI CITY, FOR THE YEARS 1910 AND 1930

[The figures for 1910 are based on the city alone (Kitano, 1911), while those for 1930 on prefectural census reports.]

Nationality	1910 Males	1910 Females	1910 Total	1930 Males	1930 Females	1930 Total
English	52	43	95	129	57	186
American	31	48	79	84	35	119
French	30	12	42	14	13	27
Russian	17	21	38	9	9	18
Danish	7	9	16	14	11	25
German	17	11	28	2	4	6
Portuguese	3	6	9	2	3	5
Italian	9	4	13	3	5	8
Austrian	2	4	6	—	—	—
Turk	2	1	3	1	—	1
Rumanian	2	5	7	—	—	—
Norwegian	6	4	10	30	1	31
Bulgarian	—	1	1	—	—	—
Dutch	—	—	—	31	5	36
Polish	—	—	—	8	1	9
Belgian	—	—	—	1	—	1
Swedish	—	—	—	2	4	6
Swiss	—	—	—	2	2	4
Canadian	—	—	—	2	—	2
Subtotal	178	169	347	334	150	484
Chinese	578	220	798	1,883	441	2,324
Other	—	—	—	1	—	1
Totals	756	389	1,145	2,218	591	2,809

[4] The opportunities which arose during the Occupation, starting with the arrival of the First Marine Division in Nagasaki, are scarcely pertinent to the problem of the ancestry of those individuals forming the parentage of the children under study.

THE CHILDREN OF ATOMIC BOMB SURVIVORS

TABLE 3.4 THE AGE COMPOSITION AND ETHNIC STATUS OF THE TOTAL FOREIGN POPULATION OF NAGASAKI IN 1920, AS WELL AS OF THE FOUR PRINCIPAL ETHNIC GROUPS IN THIS POPULATION

(From Naikaku tōkei-kyoku [Cabinet statistical bureau], Census Report, 1930, V. 4, 4, p. 40.) Abbreviations: T, total; U, unmarried; M, married; and WD, widowed or divorced.

Age	All foreigners Total	All foreigners ♂	All foreigners ♀	Chinese Total	Chinese ♂	Chinese ♀	American Total	American ♂	American ♀	English Total	English ♂	English ♀	Russian Total	Russian ♂	Russian ♀
All ages															
T	2,504	1,925	579	1,363	1,001	362	444	391	53	346	315	31	244	162	82
U	1,511	1,181	330	760	539	221	327	303	24	237	222	15	112	78	34
M	924	714	210	563	441	122	114	87	27	105	93	12	119	78	41
WD	69	30	39	40	21	19	3	1	2	4	—	4	13	6	7
0–4															
T	192	101	91	152	82	70	11	6	5	5	4	1	16	5	11
U	192	101	91	152	82	70	11	6	5	5	4	1	16	5	11
M	—	—	—	—	—	—	—	—	—	—	—	—	—	—	—
WD	—	—	—	—	—	—	—	—	—	—	—	—	—	—	—
5–9															
T	192	109	83	158	88	70	10	8	2	6	4	2	12	6	6
U	192	109	83	158	88	70	10	8	2	6	4	2	12	6	6
M	—	—	—	—	—	—	—	—	—	—	—	—	—	—	—
WD	—	—	—	—	—	—	—	—	—	—	—	—	—	—	—
10–14															
T	165	91	74	127	76	51	7	4	3	6	2	4	10	6	4
U	165	91	74	127	76	51	7	4	3	6	2	4	10	6	4
M	—	—	—	—	—	—	—	—	—	—	—	—	—	—	—
WD	—	—	—	—	—	—	—	—	—	—	—	—	—	—	—
15–19															
T	221	180	41	140	112	28	12	10	2	43	42	1	19	12	7
U	199	173	26	122	106	16	11	9	2	43	42	1	16	12	4
M	22	7	15	18	6	12	1	1	—	—	—	—	3	—	3
WD	—	—	—	—	—	—	—	—	—	—	—	—	—	—	—
20–24															
T	442	395	47	150	125	25	155	148	7	92	91	1	35	26	9
U	352	338	14	94	88	6	140	138	2	83	83	—	26	24	2
M	88	56	32	54	36	18	15	10	5	9	8	1	9	2	7
WD	2	1	1	2	1	1	—	—	—	—	—	—	—	—	—

A Comparison of Hiroshima and Nagasaki

persons living in Nagasaki do not differ strikingly from those of their neighbors, although one wonders about the selection for typing studies of "pure" Japanese (summary in Boyd, 1939). It is unfortunate that studies on the Rh gene frequencies are not available inasmuch as these, because of the difference between Caucasian and Oriental populations (summary in Mourant, 1954), would be expected to be especially revealing.

That there has been some admixture can scarcely be challenged. One who visits the three cemeteries where foreigners were customarily buried is impressed by the frequency with which there appear on the tombstones of the past three-quarters of a century Japanese female given names in combination with non-Japanese surnames. Unfortunately, the local church records, which might have been of real value in this connection, fared less well than the tombstones in the atomic holocaust. One of these cemeteries is the large and picturesque, semi-official "Foreign National" cemetery, conveniently subdivided into Russian, Dutch, English, etc., sections. There is a marked preponderance of males buried here. It would be passing strange if, during the 70 years preceding World War II, these men, even more than the casual sailors from so many ports, failed to leave a genetic heritage paralleling their socio-economic stamp.

Finally, some of the authors have the distinct impression of encountering from time to time in Nagasaki, individuals who, because of hair or eye color or facial conformation, strongly suggested Caucasian ancestry. Such persons are a very small minority but, in view of the general dominance in mixed Japanese-Caucasian marriages of the straight, black hair, the dark eyes, and the facial appearance of the Japanese, cannot be easily disregarded.

In striking contrast to Nagasaki, the Hiroshima area, although it has supplied relatively many emigrants to Hawaii and the U.S.A., has itself been characterized by very limited contacts with the West, even down to the time of World War II. It would seem that the possibility of a Caucasian element in this population may safely be ignored.

In addition to the possible role of historical (and pre-historical) factors in creating biological differences between the inhabitants of Hiroshima and Nagasaki, certain obvious present-day differences should be mentioned. We are indebted to Mr. Fu, Chinese Consul in Nagasaki in 1952, for the information that in February of that year there were 600 Chinese citizens in the city. These were not all "pure" Chinese; on the other hand, there were known to be many persons in Nagasaki whose ancestry was in part Chinese who no longer claimed Chinese citizenship. In Hiroshima there was no significant number of Chinese, but, by contrast, a relatively large "Korean colony," numbering, according to data supplied by the Hiroshima Municipal Government, some 5,000 persons in January of 1952. There is reason to suspect that because of illegal entry, the number was actually somewhat larger. There were relatively few Koreans in Nagasaki.

3.4 The different impacts of the atomic bombs on the two cities. — There are important differences between Hiroshima and Nagasaki in respect to their experience with the atomic bombs.

3.4.1 Types of bombs. — Different kinds of bombs were used on the two cities, that dropped over Hiroshima being a uranium-235 bomb, whereas the one used against Nagasaki was composed of plutonium-239. As will be brought out in the next chapter, the radiation spectrum of these two bombs differed.

3.4.2 Effects of the bombs on the two cities. — The over-all effects of the atomic bombs on Hiroshima and Nagasaki have been adequately described elsewhere (British Mission, 1946; United States Strategic Bombing Survey, 1946; Los Alamos Scientific Laboratory, 1950; Oughterson et al., 1951). Suffice it to say here that both the mortality and the morbidity from the bombs differed markedly in the two cities. Because of the deterioration in Japanese vital statistics during the war and the destruction of records in consequence of the bombings, exact casualty figures will never be available. However, it is usually stated that in Hiroshima approximately 60,000 *inhabitants* were killed immediately or died within a few weeks of the effects of the explosion, and an additional 70,000 sustained overt injury. This figure may be a very conservative estimate of the total casualties for two reasons. As the headquarters of the Second Grand Army, the chief concentration of military power in Central

Japan, Hiroshima was a major "staging area" for the South Pacific theater of war. The elaborate facilities of the Second Army were almost completely destroyed. Because of wartime secrecy plus the deliberate destruction of surviving military records, the military casualties will never be known, but they number well into the thousands. In addition, on the day of the bombing there were a number of work parties from neighboring towns in the area. The total number of persons killed may exceed 100,000. With respect to Nagasaki, it is usually stated that there were approximately 33,000 civilian deaths, and 25,000 surviving injured. Nagasaki contained no military installations of any significance, so that the problem of accounting for military personnel does not exist for this city. Since many more records survived here than in Hiroshima, it is felt that the figures for Nagasaki are reasonably accurate.

The plutonium-type bomb detonated over Nagasaki actually had a greater explosive power than that used on Hiroshima. The reason for the greater number of casualties in the latter city is to be sought in large part in differences in the physical features of the two cities. Hiroshima is built on the triangular delta of the river Ota (Fig. 3.1). Only one small "mountain" (Hijiyama, height 69 meters) breaks the flatness of the terrain occupied by the great majority of the city. As indicated in Fig. 3.1, the bomb was detonated not far from the "center" of this delta. The topography of Nagasaki is very different (Fig. 3.2). The city lies at the head of a long, narrow bay, running up from which there is a "mountain," with a valley on either side. The city extends along both sides of the bay and up into the two valleys, thus roughly resembling in its outlines the letter "X." As indicated on the map, the bomb was detonated over one of the valleys, in which there was a heavy concentration of war industry (and, incidentally, the largest Christian colony and church in Japan, and the Nagasaki Medical School and its hospital). The serious effects of the bomb were largely confined to this one valley.

The official statistics concerning the effects of the bombs are paralleled by the experience of the ABCC in the two cities. For instance, in consequence of a Radiation Census carried out in 1949, together with certain later supplementary data, it can be estimated that in 1949 there were some 31,000 inhabitants of Hiroshima who had been within 2,000 meters of the hypocenter at the time of the explosion, whereas the corresponding figure for Nagasaki was 9,850. It can be further estimated that approximately 6,000 persons then resident in Hiroshima, and 2,000 in Nagasaki, had shown such symptoms of relatively heavy irradiation as epilation, purpura, and/or oropharyngeal lesions following the bombings (ABCC Semi-Annual Report, January-June, 1954). As can be seen from Table 2.1, among the parents of children falling within the scope of the Genetics Program, there were roughly twice as many relatively heavily irradiated in Hiroshima as in Nagasaki.

3.5 *The development of the ABCC program in the two cities.* — Despite the number of persons on the ABCC roster (p. 2), there was, considering the magnitude of the total problem to be attacked, a chronic shortage of trained personnel, this imposed in part by budgetary considerations and in part by recruitment difficulties. The original plan had been that the ABCC would develop programs in Hiroshima and Nagasaki which would be quite comparable in size. From the first, however, the concentration of personnel and facilities in Hiroshima far outstripped that in Nagasaki. The reasons were chiefly two: (1) Given the personnel shortages alluded to above, and the greater number of relatively heavily irradiated survivors in Hiroshima, it was obviously more economical of available personnel to concentrate them in Hiroshima. (2) For a number of reasons which need not be entered into here, logistical problems, including the matter of housing, were less serious in Hiroshima. To those of your authors who found themselves curiously stirred by the colorful and dramatic history of Nagasaki — a history whose shadows confronted one at many turns — it has always seemed regrettable that practical considerations dictated putting so much more effort into Hiroshima.

Because of the clear need from the outset for *all* the "genetic" data which could be collected from *both* cities, the Genetics Program came closer to an equality of effort in the two cities than did any other facet of the ABCC's activities. Every possible attempt was made to ensure the comparability of the genetics programs in

FIGURE 3.1.— The topography of the Hiroshima City region, with particular reference to distance from the hypocenter of the atomic bomb explosion. Although the grid is laid out in 1,000-yard intervals, the concentric rings indicate distance from the hypocenter in 1,000-meter intervals.

FIGURE 3.2 — The topography of the Nagasaki City region. Explanation as for Figure 3.1.

Hiroshima and Nagasaki. The two chief factors in this effort were the formulation of a rather rigid set of procedures to be adhered to in the two cities, and frequent exchanges of personnel. It is felt that in the main, this effort was successful. On the other hand, as will become evident in Chapter V, some of the reported differences in pregnancy termination between the two cities may not actually reflect true biological differences.

Chapter IV

THE CRITERIA OF RADIATION EMPLOYED IN THE STUDY

THE present chapter will be devoted to an examination of the available and practical criteria for evaluating the amount of radiation received by individuals exposed in Hiroshima and Nagasaki to the effects of the atomic bombs.

4.1 *The complicated nature of the injuries sustained by some survivors; "disaster effect" vs. "radiation effect."* — The detonation of an atomic bomb may be harmful to persons in its vicinity for a variety of reasons. The air blast may inflict injury directly or secondarily, from flying debris and falling walls and roofs. The thermal radiation produces severe burns, and, in addition, further burns may be sustained as a result of fires kindled by the explosion. Under the conditions prevailing in Hiroshima and Nagasaki, serious secondary infection of these burns was the rule rather than the exception. Finally, the explosion is accompanied by the release of a variety of deleterious high energy radiations, particularly X-rays, gamma rays, and fast and slow neutrons. The complex manner in which these various noxious agents affect persons exposed to the explosion of an atomic bomb is well depicted in the Report of the Joint Commission for the Investigation of the Effects of the Atomic Bomb in Japan (Oughterson et al., 1951).

Moreover, in the wake of such a cataclysmic event as the detonation of an atomic bomb over a city, come certain well recognized accompaniments of any large-scale disaster, notably, disturbed nutrition and increased morbidity from disease. These "disaster effects" are here so complexly intertwined with the primary event that it will be difficult, if not in fact impossible, in any analysis of atomic bomb sequelae, to effect a realistic and completely satisfactory separation of the relative importance of what transpired at the time of the bombing and what transpired in the next several months.

In this particular instance, the general problem of disentangling primary and secondary effects upon the manner in which pregnancies are terminating is still further complicated by the fact that certain delayed somatic effects of irradiation have been shown to occur in the population under study. The "late" effects include leukemia (Folley, Borges, and Yamawaki, 1952; Lange, Moloney, and Yamawaki, 1954; Moloney and Lange, 1954) and cataracts (Cogan, Martin, and Kimura, 1949; Sinsky, 1955). Although both of these conditions affect only small numbers of the more heavily irradiated survivors, they do occur in the group most critical to this study. It seems entirely possible that there are other late sequelae, still undefined, which may influence reproductive behavior.

In subsequent chapters an attempt will be made to correlate exposure history with a variety of possible indicators of genetic damage. *However, it must be clearly understood that if a significant association can be shown to exist, this demonstration of itself does not enable one to conclude that exposure to the bomb has genetic consequences; the latter conclusion is justified only if an effect remains after the various other factors associated with the irradiation have been assigned their proper role.*

4.2 *The question of residual radiation following an atomic bombing.* — The ionizing radiation of an atomic bomb explosion is released over a relatively brief period, of approximately 100 seconds. However, as a result of the contamination of the area due to the fall-out of radioactive by-products of the explosion, as well as secondarily induced radioactivity, there is a certain amount of residual radioactivity in an area over which a bomb has been detonated.

The estimation of the amount of residual radiation at Hiroshima and Nagasaki presents many difficulties. The distribution of residual radiation around the hypocenter in the two cities was asymmetric, the exact pattern depending on local meterological details (Warren, 1945; Warren, 1946). Based on the observed residual radioactivity at intervals following the bombing, Warren (1945) has estimated that the maximum irradiation due to residual radioactivity was in Hiroshima during the first 60 days following the bombing the equivalent of 4.2r, and in Nagasaki during the first 47 days, the equivalent of 14.2r. Wilson (1951), using similar data, has suggested that in Hiroshima the *maximum* cumulative dose of residual radiation at distances greater than 1,000 feet (300 meters) was not more than 10r, with somewhat higher doses prevailing in Nagasaki, especially in a fall-out area extending eastward from the hypocenter, where the cumulative dose from radioactivity might have amounted to 100r. Irradiation of this latter degree would only be experienced by individuals establishing continuous residence near ground zero immediately after the bombing, a situation which rarely, if ever, obtained, although persons coming into the area later or intermittently might of course receive lesser amounts. On the other hand, there are persistent Japanese reports of members of rescue parties and others not actually in the two cities at the time of the bombing later developing symptoms of radiation sickness, or leucopenia (Japan Science Council, 1951; Appendix 18, ABCC Semi-Annual Report, 1 January 1952–30 June 1952); this would suggest a dose of 200r (cf. Sec. 4.8). This apparent conflict of evidence remains unresolved at present; we shall adhere for the time being to the more conservative view concerning residual radioactivity. By and large, individuals exposed to the effects of the atomic bombs tended to leave the area as rapidly as possible. Accordingly, for the purposes of this study it has been felt that although exposed individuals and also those entering the city immediately after the bombings may have been subjected to some residual radiation, by comparison with the amount absorbed by those present at the moment of the explosion, this was on the average small and could be disregarded. Moreover, it might be pointed out that in the case of an individual engaged, e.g., in rescue work in Hiroshima or Nagasaki during the first few days following the bombing, it appears impossible to reconstruct his behavior during those several days in a fashion which would permit a realistic estimation of the total radiation dose which he absorbed during this period.

4.3 *Factors determining the nature of the radiation data collected in this study.* — At the time this study was being planned, in 1947, no data of any sort concerning the types of radiation emitted by an atomic bomb and the distance-dosage relationship were available to any of those closely connected with the program. In retrospect, in view of the many revisions the physical estimates have undergone, this is perhaps fortunate. Be that as it may, under the circumstances extensive recourse was had to the findings of the Joint Commission, then available in preliminary form. These findings, as later published (1951), established the following facts of importance in the design of the present observations:

4.3.1 *The syndrome of radiation sickness due to whole-body irradiation.* — Although the therapeutic uses of radiation had long since laid the groundwork for an appreciation of the signs and symptoms of "radiation sickness," it remained for the observations of the Joint Commission to provide the definitive description of this entity. The findings, of course, vary according to the radiation dose. We are not here concerned with the early effects of lethal doses of whole-body irradiation, since persons receiving such doses can scarcely provide the parentage for a study of this nature. The Joint Commission observed that individuals receiving relatively large but yet sublethal doses of irradiation (as judged by proximity to the hypocenter in the absence of marked shielding) may or may not develop nausea and vomiting within a few hours of their exposure. There follows a relatively asymptomatic period of approximately two weeks, after which characteristic signs and symptoms appear. Some of these, such as fever, malaise, anorexia, nausea, and vomiting, have a low degree of specificity, being present in a wide variety of diseases. Other findings were much more specific. The chief among these are summarized in Table 4.1. These are figures for all survivors, irrespective of the amount of shielding which protected them. Let us restrict ourselves, for the moment, to a consideration of the findings in persons reporting themselves

TABLE 4.1 FREQUENCY OF OCCURRENCE OF CERTAIN SYMPTOMS IN PERSONS ALIVE 20 OR MORE DAYS FOLLOWING THE ATOMIC BOMBINGS, AS RELATED TO DISTANCE FROM THE HYPOCENTER

(Data of Joint Commission)

Hiroshima

Ring	Distance (meters)	Total no. of people	Epilation No.	Epilation %	Purpura No.	Purpura %	Oropharyngeal lesions[a] No.	Oropharyngeal lesions[a] %	Necrotic gingivitis No.	Necrotic gingivitis %	Diarrhea[b] No.	Diarrhea[b] %	Bloody diarrhea No.	Bloody diarrhea %	Other hemorrhage No.	Other hemorrhage %
1	0–1,000	749	520	69.4	366	48.9	458	61.1	77	10.3	375	50.1	80	10.7	285	38.1
2	1,100–1,500	1,125	341	30.3	241	21.4	381	33.9	43	3.8	476	42.3	75	6.7	197	17.5
3	1,600–2,000	1,824	151	8.3	78	4.3	286	15.7	12	0.7	666	36.5	99	5.4	112	6.1
4	2,100–2,500	1,450	69	4.8	27	1.9	230	15.9	16	1.1	519	35.8	95	6.6	94	6.5
5	2,600–3,000	700	16	2.3	13	1.9	105	15.0	1	0.1	260	37.1	34	4.9	41	5.9
6	3,100–4,000	576	7	1.2	4	1.2	39	6.8	0	0	134	23.3	13	2.3	22	3.8
7	4,100–5,000	239	0	0	4	1.7	21	8.8	0	0	59	24.7	2	0.8	5	2.1
Total	0–5,000	6,663	1,104	16.6	736	11.0	1,520	22.8	149	2.2	2,489	37.4	398	6.0	756	11.3
9[c]	Over 5,000	219	0	0	1	0.5	16	7.3	1	0.5	67	30.6	2	0.9	8	3.7
Grand total	—	6,882	1,104	16.0	737	10.7	1,536	22.3	150	2.2	2,556	37.1	400	5.8	764	11.1

Nagasaki

Ring	Distance (meters)	Total no. of people	Epilation No.	Epilation %	Purpura No.	Purpura %	Oropharyngeal lesions[a] No.	Oropharyngeal lesions[a] %	Necrotic gingivitis No.	Necrotic gingivitis %	Diarrhea[b] No.	Diarrhea[b] %	Bloody diarrhea No.	Bloody diarrhea %	Other hemorrhage No.	Other hemorrhage %
1	0–1,000	789	249	31.6	168	21.3	275	34.9	20	2.5	337	42.7	54	6.8	110	13.9
2	1,100–1,500	1,882	460	24.4	324	17.2	607	32.3	33	1.8	736	39.1	86	4.6	233	12.4
3	1,600–2,000	1,034	128	12.4	92	8.9	200	19.3	16	1.5	358	34.6	56	5.4	78	7.5
4	2,100–2,500	672	41	6.1	24	3.6	101	15.0	4	0.6	224	33.3	23	3.4	36	5.4
5	2,600–3,000	644	16	2.5	7	1.1	84	13.0	6	0.9	195	30.3	28	4.3	18	2.8
6	3,100–4,000	1,141	16	1.4	17	1.5	121	10.6	1	0.1	268	23.5	21	1.8	28	2.5
7	4,100–5,000	265	1	0.4	1	0.4	19	7.2	0	0	23	8.7	3	1.1	5	1.9
Total	0–5,000	6,427	911	14.2	633	9.8	1,407	21.9	80	1.2	2,141	33.3	271	4.2	508	7.9
9[c]	Over 5,000	194	0	0	0	0	5	2.6	0	0	23	11.9	1	0.5	2	1.0
Grand total	—	6,621	911	13.8	633	9.6	1,412	21.3	80	1.2	2,164	32.7	272	4.1	510	7.7

[a] Includes necrotic gingivitis.
[b] Includes bloody diarrhea.
[c] There is no ring 8 in the Joint Commission system of designating distance.

within 1,000 meters of the hypocenter. From the standpoint of distinctiveness, two symptoms — epilation and purpura — are outstanding. Oropharyngeal lesions also were frequently encountered. Diarrhea, while reported by large numbers of persons, could, at 20 days after the bombing, in any individual case be as well attributed to the poor hygienic conditions following the bombing as to the bomb itself, and does not seem to be a satisfactory symptom on which to base a study. Bloody diarrhea is a somewhat more reliable symptom, but still, from the findings of the Joint Commission, less useful than the other three symptoms already discussed.

Further evidence as to the relative validity of the findings listed in Table 4.1 as indicators of radiation exposure comes from an analysis of the manner in which the various findings tended to be associated. Epilation and petechiae were more highly associated than any other pair of symptoms in both cities. The next highest associations observed were, in both cities, epilation with oropharyngeal lesions, and petechiae with oropharyngeal lesions.

On the basis of these findings, it was felt that the appearance of epilation, petechiae, or oropharyngeal lesions, singly or in combination, within three months of the bombing, provided a relatively specific and objective yardstick of the absorption of a certain amount of radiation, such as could be employed in the present study (see also, inter alia, Warren and Bowers, 1950; Los Alamos Scientific Laboratory, 1950; vor der Bruegge, 1952; Hempelmann, Lisco, and Hoffman, 1952). Although exact data are lacking, it seems likely, on the basis of animal data, that there is considerable individual variation in susceptibility to these symptoms, a point which must of course be borne in mind when we come later to the problem of estimating radiation dosages. It will be noted that there are significant differences between Hiroshima and Nagasaki in the percentages of persons exhibiting certain symptoms subsequent to exposure within 1,500 meters of the hypocenter. These differences are customarily attributed primarily to local differences in shielding factors (see below), but may also be due in part to actual dose differences.

4.3.2 *The relation between distance from the hypocenter and radiation dosage.* — Although the intensity of radiation obviously must decrease with distance from the hypocenter, the exact form of the distance-dosage relationship in Hiroshima and Nagasaki remains to this day uncertain. In part this is due to inadequate information concerning certain physical properties of these atomic explosions, in part to lack of detailed information concerning atmospheric moisture content at the time of the explosion. Furthermore, there is the possibility that the distance-dosage relationship is not the same in all radii from the hypocenter. Be this as it may, at the time this study was planned it was clear from such data as are given in Table 4.1, concerning the relationship between the occurrence of certain of the symptoms described above and distance from the hypocenter, that it would be important to record position at the time of the bombing as accurately as possible.

4.3.3 *The role of shielding in determining radiation dose.* — The third type of information thought to be especially valuable in estimating radiation dosage concerned the amount of shielding protecting the individual from the effects of the explosion. Table 4.2 illustrates some of the findings of the Joint Commission in this respect. For the sake of brevity, only the Hiroshima data are reproduced; the findings in Nagasaki were essentially similar. An obvious question which had to be disposed of early in the design of the program concerned the elaborateness of the shielding data to be collected. Obtaining a complete shielding history can be quite time consuming. Theoretically, knowing the distance-dosage relationship and the precise shielding, one can arrive at a relatively accurate estimate of the amount of irradiation received by an individual. We have already mentioned the uncertainties surrounding the distance-dosage relationship. These uncertainties are particularly acute as regards the neutron component of the radiation spectrum, a component which in animal experiments has been shown to have a high relative biological effectiveness with respect to genetic effects. Furthermore, in the authors' opinion there is no realistic approach to the problem of radiation scatter and secondary radiation as a factor in the exact dosage received by any individual. Finally, there is room for reasonable doubt as to most individuals' abilities to reconstruct their exact positions at the time of so traumatic an event as this, and yet nothing less than such a reconstruction will suffice for a precise evaluation of shielding. With all these considerations in mind, it was

The Criteria of Radiation Employed in the Study

TABLE 4.2 THE EFFECTIVENESS OF SHIELDING IN PROTECTING AGAINST RADIATION SICKNESS IN HIROSHIMA
(Data of Joint Commission)

Outdoors or in Japanese-type building[c]

Ring	Distance (meters)	Total no. of people	Epilation No.	Epilation %	Purpura No.	Purpura %	Oropharyngeal lesions[a] No.	Oropharyngeal lesions[a] %	Necrotic gingivitis No.	Necrotic gingivitis %	Diarrhea[b] No.	Diarrhea[b] %	Bloody diarrhea No.	Bloody diarrhea %	Other hemorrhage No.	Other hemorrhage %
1	0–1,000	570	434	76.1	310	54.4	356	62.5	67	11.8	301	52.8	66	11.6	240	42.1
2	1,100–1,500	960	315	32.8	209	21.8	305	31.8	37	3.9	405	42.2	66	6.9	179	18.6
3	1,600–2,000	1,633	145	8.9	73	4.5	225	13.8	10	0.6	613	37.5	90	5.5	100	6.1
4	2,100–2,500	1,415	68	4.8	26	1.8	190	13.4	16	1.1	508	35.9	93	6.6	94	6.6
5	2,600–3,000	674	16	2.4	12	1.8	82	12.2	1	0.1	251	37.2	32	4.7	39	5.8
6	3,100–4,000	548	7	1.3	7	1.3	26	4.7	0	0	124	22.6	13	2.4	22	4.0
7	4,100–5,000	202	0	0	3	1.5	12	5.9	0	0	50	24.8	1	0.5	4	2.0
Total	0–5,000	6,002	985	16.4	640	10.7	1,196	19.9	131	2.2	2,252	37.5	361	6.0	678	11.3
9[d]	Over 5,000	46	0	0	1	2.2	8	17.4	0	0	18	39.1	0	0	4	8.7
Grand total	—	6,048	985	16.3	641	10.6	1,204	19.9	131	2.2	2,270	37.5	361	6.0	682	11.3

Indoors, in heavy building

Ring	Distance (meters)	Total no. of people	Epilation No.	Epilation %	Purpura No.	Purpura %	Oropharyngeal lesions[a] No.	Oropharyngeal lesions[a] %	Necrotic gingivitis No.	Necrotic gingivitis %	Diarrhea[b] No.	Diarrhea[b] %	Bloody diarrhea No.	Bloody diarrhea %	Other hemorrhage No.	Other hemorrhage %
1	0–1,000	113	60	53.1	38	33.6	55	48.7	5	4.4	46	40.7	8	7.1	30	26.5
2	1,100–1,500	118	15	12.7	21	17.8	26	22.0	3	2.5	58	49.2	8	6.8	13	11.0
3	1,600–2,000	94	2	2.1	3	3.2	12	12.8	1	1.1	35	37.2	5	5.3	8	8.5
4	2,100–2,500	12	1	8.3	1	8.3	1	8.3	0	0	7	58.3	2	16.7	0	0
5	2,600–3,000	14	0	0	1	7.1	3	21.4	0	0	2	14.3	1	7.1	2	14.3
6	3,100–4,000	13	0	0	0	0	2	15.4	0	0	5	38.5	0	0	0	0
7	4,100–5,000	27	0	0	1	3.7	1	3.7	0	0	7	25.9	1	3.7	1	3.7
Total	0–5,000	391	78	19.9	65	16.6	100	25.6	9	2.3	160	40.9	25	6.4	54	13.8
9[d]	Over 5,000	2	0	0	0	0	0	0	0	0	1	50.0	0	0	0	0
Grand total	—	393	78	19.8	65	16.5	100	25.4	9	2.3	161	41.0	25	6.4	54	13.7

[a] Includes necrotic gingivitis (within 39 days).
[b] Includes bloody diarrhea.
[c] Includes people indoors, type of building unknown.
[d] There is no ring 8 in the Joint Commission system of designating distance.

concluded that while the experience of the Joint Commission left no doubt as to the desirability of collecting shielding data, this should be simple in nature.

4.4 *The type of radiation data collected in this study.* — In addition to the biological considerations which we have just discussed, certain "practical" considerations entered into the determination of the type of data to be collected. The histories were to be obtained by clerks with the equivalent of a high school education. These clerks, although they could be carefully drilled in their duties, could not be expected to exercise a great deal of judgment if the approach to the problem involved the compilation of an elaborate history. It could be anticipated that the turn-over among clerks would be rather high.

Finally, since the history taking was to extend over an indefinite number of years, much was to be said for concentrating on certain salient features of the individual's experience, which could be as readily recalled relatively late as relatively early.

With all these considerations in mind, the following information was obtained concerning each pregnant woman and her spouse included in this study:

1. Presence in Hiroshima or Nagasaki at the time of the bombing.
2. Location in city at time of bombing.
3. Distance from the hypocenter [calculated from (2)].
4. Indoors or outdoors.
5. Type of building.

TABLE 4.3 DISTRIBUTION BY DISTANCE AND SHIELDING OF *HUSBANDS* OF WIVES REGISTERING PREGNANCIES WITH THE GENETICS PROGRAM (HIROSHIMA)

(For a definition of the classes see Sec. 4.5.)

Distance (hundreds of meters)	No. persons symptomatic	P[a]	G[a]	E[a]	PG	PE	GE	PGE	No. persons asymptomatic	Total no. persons exposed (a)	Total no. symptoms experienced (b)	Ratio (b/a)
In open												
00–04	1	—	—	—	—	—	—	1	2	3	3	—
05–09	15	1	—	6	1	—	2	5	12	27	28	1.04
10–14	53	4	3	22	—	3	9	12	102	155	89	.57
15–19	44	5	6	23	2	3	3	2	287	331	56	.17
20–24	29	4	6	12	1	2	4	—	409	438	36	.08
25–29	11	4	—	4	—	1	2	—	213	224	14	.06
30+	17	1	7	5	—	2	2	—	868	885	21	.02
									1,893	2,063		
In Japanese building												
00–04	2	—	—	—	—	—	—	2	1	3	6	—
05–09	39	1	1	8	—	4	9	16	13	52	84	1.62
10–14	127	11	8	29	10	16	14	39	387	514	245	.48
15–19	54	20	11	10	3	4	4	2	695	749	69	.09
20–24	27	9	6	5	—	2	3	2	634	661	36	.05
25–29	11	2	1	2	1	2	2	1	444	455	18	.04
30+	24	5	3	10	—	1	4	1	1,636	1,660	31	.02
									3,810	4,094		
In other shelter												
00–04	—	—	—	—	—	—	—	—	2	2	—	—
05–09	14	2	—	5	1	3	—	3	17	31	24	.77
10–14	17	6	4	1	1	1	2	2	33	50	25	.50
15–19	6	1	3	—	2	—	—	—	61	67	8	.12
20–24	2	1	1	—	—	—	—	—	46	48	2	.04
25–29	2	—	1	—	—	—	—	1	51	53	4	.08
30+	2	1	—	1	—	—	—	—	200	202	2	.01
									410	453		

[a] P = Petechiae; G = Gingivitis; E = Epilation.

6. Occurrence of subcutaneous bleeding (petechiae).
7. Occurrence of gingivitis.
8. Occurrence of bloody diarrhea.
9. Occurrence of epilation (partial or complete).
10. Occurrence of burns.
11. Occurrence of external injuries.

of calculation, meters are rounded to decameters, and decameters further rounded to hectometers, so that the coded interval in hectometers denotes $(x-55)$ to $(x+44)$ meters.

Items 4 and 5 on the above list represent an attempt, however rough, to accumulate certain data relative to the evaluation of shielding. Item 4 was to be answered simply as indoors or outdoors. Item 5 pursued the point somewhat further, admitting of ten alternatives, namely, if indoors, then

1. Inside concrete building,
2. Inside brick building,
3. Inside wooden Japanese or other type building,
4. Inside cave or "bunker,"

TABLE 4.4 DISTRIBUTION BY DISTANCE AND SHIELDING OF *WIVES* REGISTERING PREGNANCIES WITH THE GENETICS PROGRAM (HIROSHIMA)

Distance (hundreds of meters)	No. persons symptomatic	P[a]	G[a]	E[a]	PG	PE	GE	PGE	No. persons asymptomatic	Total no. persons exposed (a)	Total no. symptoms experienced (b)	Ratio (b/a)
In open												
00–04	—	—	—	—	—	—	—	—	1	1	—	—
05–09	14	—	—	4	1	—	6	3	8	22	27	1.23
10–14	75	9	2	24	1	8	13	18	156	231	133	.58
15–19	85	10	9	50	3	4	3	6	489	574	107	.19
20–24	53	7	6	26	1	—	8	5	560	613	72	.12
25–29	18	7	3	4	—	—	2	2	311	329	24	.07
30+	24	2	6	11	—	1	4	—	1,020	1,044	29	.03
									2,545	2,814		
In Japanese building												
00–04	2	—	—	—	—	—	—	2	1	3	6	—
05–09	52	1	1	7	2	15	6	20	23	75	115	1.53
10–14	254	41	17	66	8	38	23	61	881	1,135	445	.39
15–19	116	36	24	36	1	8	4	7	1,668	1,784	143	.08
20–24	65	18	13	22	1	6	1	4	1,398	1,463	81	.06
25–29	28	13	4	6	2	2	1	—	1,205	1,233	33	.03
30+	31	9	11	8	—	2	1	—	3,369	3,400	34	.01
									8,545	9,093		
In other shelter												
00–04	3	1	—	—	—	2	—	—	—	3	5	—
05–09	35	3	—	5	1	8	6	12	75	110	74	.67
10–14	13	3	2	1	2	1	2	2	102	115	22	.19
15–19	11	3	—	6	—	1	1	—	181	192	13	.07
20–24	8	5	1	2	—	—	—	—	117	125	8	.06
25–29	1	—	1	—	—	—	—	—	148	149	1	.01
30+	3	1	—	1	—	1	—	—	188	191	4	.02
									811	885		

[a] P = Petechiae; G = Gingivitis; E = Epilation.

In locating the position of an individual at the time of the bombing an effort was made to pinpoint position as accurately as possible. The distance from the hypocenter was then determined by actual measurement on a large-scale map of the city. The method of measurement has varied from time to time, involving variously the use of plastic and metal tapes, coordinates, and concentric circles. In the process

and if outdoors, then

5. In open,
6. Behind (within two meters) wall,
7. In trench,
8. Behind post or tree,
9. In tram, train, or car, and
10. Under eaves of house (i.e., shielded by house).

In the original design of the study, no symptom among those listed was to be recognized unless it developed prior to December 15, 1945. Some instances have come to light where the alleged onset of symptoms after that date have been recorded. This cannot be attributed to exposure to the bomb. There is a good reason to believe that the erroneous inclusion of such re-

TABLE 4.5 DISTRIBUTION BY DISTANCE AND SHIELDING OF *HUSBANDS* OF WIVES REGISTERING PREGNANCIES WITH THE GENETICS PROGRAM (NAGASAKI)

Distance (hundreds of meters)	No. persons symptomatic	P[a]	G[a]	E[a]	PG	PE	GE	PGE	No. persons asymptomatic	Total no. persons exposed (a)	Total no. symptoms experienced (b)	Ratio (b/a)
In open												
00–04	—	—	—	—	—	—	—	—	1	1	—	—
05–09	5	—	—	1	1	1	2	—	6	11	9	.82
10–14	20	3	2	10	—	3	—	2	51	71	27	.38
15–19	11	2	—	5	1	1	—	2	72	83	17	.20
20–24	4	1	1	2	—	—	—	—	94	98	4	.04
25–29	7	2	1	4	—	—	—	—	165	172	7	.04
30+	8	3	—	4	—	—	—	1	1,151	1,159	10	.01
									1,540	1,595		
In Japanese building												
00–04	—	—	—	—	—	—	—	—	1	1	—	—
05–09	6	—	—	1	—	1	2	2	10	16	13	.81
10–14	45	6	6	9	3	7	2	12	186	231	81	.35
15–19	23	2	5	9	1	2	2	2	216	239	32	.13
20–24	2	—	2	—	—	—	—	—	220	222	2	.01
25–29	2	1	—	1	—	—	—	—	387	389	2	.005
30+	11	4	2	3	—	—	2	—	2,714	2,725	13	.005
									3,734	3,823		
In other shelter												
00–04	—	—	—	—	—	—	—	—	1	1	—	—
05–09	10	1	1	3	—	2	—	3	34	44	18	.41
10–14	43	5	4	10	5	4	6	9	188	231	76	.33
15–19	14	3	—	4	1	2	1	3	171	185	24	.13
20–24	2	—	1	1	—	—	—	—	103	105	2	.02
25–29	2	—	1	—	—	—	1	—	101	103	3	.03
30+	4	2	1	1	—	—	—	—	1,560	1,564	4	.003
									2,158	2,233		

[a] P = Petechiae; G = Gingivitis; E = Epilation.

Items 6–9, bearing on radiation sickness, were to be answered simply as "yes" or "no," after a brief explanation of the symptom by the clerk, if necessary. Limiting answers to "yes" or "no" obviously required the clerk to exercise her judgment in some cases. For the more vague term "oropharyngeal lesions" we have substituted gingivitis, since a necrotic-type of gingivitis with gingival bleeding was one of the more prominent and specific of the oropharyngeal lesions.

ports of "delayed effects" occurred very rarely.

Items 10 and 11 were included as one of several approaches to the problem of disentangling the possible radiation effects of the bombs from other effects and also the general disaster effect. Thus, it was anticipated that in the event of a positive correlation between some aspect of pregnancy termination and radiation history, an attempt could be made to determine if the correlation still held up after the elimina-

tion of those recording burns or external injuries of any type.

Those familiar with the problem of evaluating radiation exposure will be quick to recognize the very simple nature of the data collected in this study. The adequacy of these data as a basis for quantitative estimates of irradiation has been the subject of considerable discussion.

4.6 summarize the findings of the present study as regards the relationship between certain types of shielding, certain symptoms, and distance from the hypocenter. Based on items of the questionnaire (see preceding section), two degrees of shielding have been recognized, namely, that provided by being inside a Japanese-type home at the time of the explosion, and that

TABLE 4.6 DISTRIBUTION BY DISTANCE AND SHIELDING OF *WIVES* REGISTERING PREGNANCIES WITH THE GENETICS PROGRAM (NAGASAKI)

Distance (hundreds of meters)	No. persons symptomatic	P[a]	G[a]	E[a]	PG	PE	GE	PGE	No. persons asymptomatic	Total no. persons exposed (a)	Total no. symptoms experienced (b)	Ratio (b/a)
In open												
00–04	—	—	—	—	—	—	—	—	1	—	—	—
05–09	5	—	—	2	—	—	2	1	7	12	9	.75
10–14	24	3	2	7	1	2	3	6	60	84	42	.50
15–19	18	2	2	9	—	2	1	2	108	126	25	.20
20–24	7	—	—	5	—	—	2	—	173	180	9	.05
25–29	9	1	1	5	—	2	—	—	274	283	11	.04
30+	10	2	3	4	—	—	1	—	1,674	1,684	11	.01
									2,297	2,369		
In Japanese building												
00–04	—	—	—	—	—	—	—	—	—	—	—	—
05–09	4	—	—	1	—	1	—	2	7	11	9	.82
10–14	129	11	7	35	7	15	22	32	292	421	237	.56
15–19	55	12	6	19	1	6	6	5	459	514	78	.15
20–24	26	5	9	9	1	1	1	—	842	868	29	.03
25–29	6	2	—	3	—	—	1	—	1,106	1,112	7	.006
30+	19	7	5	6	—	—	1	—	7,162	7,181	20	.003
									9,868	10,107		
In other shelter												
00–04	—	—	—	—	—	—	—	—	5	—	—	—
05–09	19	1	2	9	1	1	4	1	74	93	27	.29
10–14	56	3	6	24	4	5	7	7	185	241	86	.36
15–19	25	6	1	11	—	—	1	6	109	134	38	.28
20–24	9	2	3	3	—	—	—	1	145	154	11	.07
25–29	5	1	2	1	—	—	1	—	154	159	6	.04
30+	7	1	3	2	1	—	—	—	2,005	2,012	8	.004
									2,677	2,793		

[a] P = Petechiae; G = Gingivitis; E = Epilation.

Specifically, the possible desirability of a much more elaborate history was repeatedly discussed. It is our contention, admittedly subjective but documented in part by the internal consistency of the data presented in the following sections, the recall factor in these situations being what it is, and in view of the radiation variables, that this represents the most practical approach to this problem which could be adopted.

4.5 *The relation between distance, shielding, and symptoms in these data.* — Tables 4.3 to

provided by being *inside* any other type of structure — concrete or brick building, cave or bunker-type air raid shelter. Figures 4.1 and 4.2 present these same findings in graphic form. For ease of presentation, a "symptom ratio" has been used, consisting of the total number of symptoms (epilation, petechiae, or gingivitis) recorded by the survivors at a particular distance, divided by the number of survivors giving information. Data are not plotted for the 00–04 distances because of the few individuals in-

volved. Several important conclusions emerge from a consideration of these data:

4.5.1 For all three groups, the symptom ratio falls off rapidly with distance. In the light of these findings, as well as the distance-dosage relationship estimated on physical grounds (see below), it seems unlikely that there was significant radiation beyond 3,000 meters from the hypocenter. The few individuals reporting epilation, petechiae, and/or gingivitis beyond that distance are, with possible rare exceptions, almost certainly in error in attributing their symptoms to irradiation. Taking this 1–2 per cent as a baseline for "false reporting," it follows that the symptomatic group within 3,000 meters may also be somewhat "diluted" by false reporting.

4.5.2 There is evidence that shielding in "other style" buildings substantially reduced the proportion of persons with symptoms at distances within 1,500 meters.

FIGURE 4.1 — Symptom ratio in relation to distance from hypocenter for Hiroshima (husbands and wives combined). Explanation in text.

4.5.3 In what appears at first sight to be a paradox, in the 500–900 meter ring a substantially higher proportion of persons in Japanese homes developed symptoms than did persons in the open. The reason for this seems clear. At this distance, persons reporting themselves in the open must actually have been shielded by one or more buildings from the bomb burst, otherwise they almost certainly

would have sustained fatal burns. The biological evidence suggests that the total *average* shielding of such persons actually exceeded that of persons in Japanese homes.

4.5.4 In the 1,000–1,900 meter ring, there is a suggestion that in Hiroshima, although not so clearly in Nagasaki, presence in a Japanese-style home conferred some protection against the development of symptoms, the basis for comparison being persons reporting themselves

and the consequent variation in the angle of incidence of the radiation. However, for radiation incident at right angles, the average tile roof has an absorptive value in the neighborhood of 45 mm. of aluminum. Beyond 1,000 meters, persons in Japanese-type buildings in Hiroshima at the time of the explosion probably averaged several such walls or roofs between themselves and the exploding bomb. This is a significant degree of shielding. In very round

FIGURE 4.2 — Symptom ratio in relation to distance from hypocenter for Nagasaki (husbands and wives combined). Explanation in text.

"outdoors." The average amount of shielding by which a person "in the open" was protected is very difficult to estimate. It is somewhat easier to quantitate the shielding of the persons in Japanese-style buildings. Woodbury (1953 and unpublished) reports that X-ray photographs taken at 62-kv indicate that the absorption of the bamboo-lattice and mud walls of the average Japanese house is equal to about 33 mm. of aluminum in the densest portion, and 15 mm. in the lightest. The absorptive value of the tile roofs which are standard construction in Japanese houses is undoubtedly more variable, depending on the distance from the hypocenter

terms, such shielding would screen out in the neighborhood of 50 per cent of the dosage at a level of 100r of the high-energy X-ray and gamma radiation released by the bomb.

It seems worth pointing out that the amount of shielding implied in the statement, "in a Japanese-style home," varies directly with distance from hypocenter, since at greater distances, because of the angle of incidence of the radiation, more walls and roofs would usually intervene between the source of radiation than at lesser distances. Persons in Nagasaki in Japanese homes at the time of the explosion on the average were probably less shielded than in

Hiroshima, because of the difference in terrain and the extension of homes up the sides of the valley.

4.6 *Factors contributing to the validity of the radiation histories.* — The value of several different persons obtaining a history from a given individual is well known in clinical medicine. The changes in emphasis and the new material introduced, as the patient passes from one interviewer to the next, are sometimes remarkable. From the standpoint of verifying information, the Genetics Program has been fortunate for the following reasons:

4.6.1 Over the 6-year span covered by this program, many women registered two or more times (cf. Sec. 5.6). At each registration, a history was obtained without reference to any previous history.

4.6.2 In 1950 a Radiation Census was carried out in both Hiroshima and Nagasaki. This census obtained brief data on each survivor of the bombings then living in Hiroshima and Nagasaki, including position at the time of the bombing.

4.6.3 Some of the parents falling within the scope of the Genetics Program — especially the more heavily irradiated — also came under scrutiny in connection with other special studies of the ABCC, such as the Adult Medical Program, the Ophthalmology Program, etc. Each of these latter programs obtained a very detailed radiation history.

All of the information available on any given person was correlated through a Master File. Particular attention was directed towards a comparison of the radiation histories obtained from a given person on different occasions. In the event of a discrepancy, an effort was made to query the individual concerned, in an attempt to resolve it. On the basis of the information obtained on the query, a single uniform entry was made on all the forms involved. As might be expected, changes on the Genetics Short Form were not infrequent, although the data have not been kept in such a fashion as to permit a precise analysis. However, it is our impression that important changes were not common.

4.7 *Definition of radiation categories.* — We are now in a position to define certain "radiation categories" which will be basic to the analysis which follows. This is done in Table 4.7. "Heavy" shielding denotes presence in concrete or brick building or air raid shelter at the time of the explosion. "Moderate" shielding includes being within a street car, train, or car, behind a wall or under the eaves of a house, i.e., on the side of the house *away* from the hypocenter. Finally, "light" shielding includes those giving their location as in a Japanese-style building or in a trench or behind a post or tree. Trench-type shielding is classified as "light" because of the difficulty in establishing how well an individual was protected by a trench.

TABLE 4.7 THE DEFINITION OF "EXPOSURE CATEGORIES" TO BE EMPLOYED IN THIS ANALYSIS

	Groups				
	1	2	3	4	5
Unexposed	+	—	—	—	—
Over 3,000 meters, with or without alleged symptoms	—	+	—	—	—
2,000–2,900 meters					
Heavy shielding	—	+	—	—	—
Moderate shielding	—	+	—	—	—
Light shielding	—	+	—	—	—
No shielding	—	—	+	—	—
1,500–1,900 meters					
Heavy shielding	—	+	—	—	—
Moderate shielding	—	+	—	—	—
Light shielding	—	—	+	—	—
No shielding	—	—	—	+	—
1,000–1,400 meters					
Heavy shielding	—	+	—	—	—
Moderate shielding	—	—	+	—	—
Light shielding	—	—	+	—	—
No shielding	—	—	—	+	—
Under 1,000 meters					
Heavy shielding	—	+	—	—	—
Moderate shielding	—	—	+	—	—
Light shielding	—	—	—	+	—
No shielding	—	—	—	+	—
Symptoms (epilation, petechiae, gingivitis), under 3,000 meters	—	—	—	—	+

Inspection of Table 4.7 in the light of the preceding discussion suggests that this is a somewhat "conservative" classification, in terms of correction for shielding. Thus, individuals reporting "heavy" shielding but asymptomatic have all been relegated to group 2, despite the fact that the symptom ratio in heavily shielded persons within 1,000 meters is approximately 0.7. Since relatively few persons were "heavily" shielded (see below), the net effect of this will be to increase slightly the average radiation dosage of group 2. On the other hand, this removes from categories 3 and 4, which received intermediate amounts of radiation, a group of

individuals whose average radiation dose would be extremely difficult to evaluate.

The distribution of all the parents of registered terminations included in this study in terms of their radiation category is given in Table 4.8. Approximately 78.3 per cent of all the fathers who were registered, and 57.0 per cent of all the mothers, were not in the cities at all at the time of the explosion. Of those present in the cities, a substantial majority was beyond 3,000 meters from the hypocenter. Only 1.2 per cent of fathers, and 2.4 per cent of mothers, reported symptoms characteristic of radiation sickness.

In a preliminary report on this study (Neel et al., 1953), a system of radiation categories was employed which failed to take into account shielding. Thus, in this treatment group 4 was defined simply as: "In one of the two cities, and less than 1,845 meters from the hypocenter, but asymptomatic. Most of these individuals were shielded to a greater or lesser extent from the full effects of the bombs." In this preliminary treatment, for some purposes the offspring of parents both of whom were group 4 or 5 were considered. In the case of the malformation analysis, for example, there were available 596 infants born to parents meeting these restrictions. But with the radiation categories which have just been defined, there are in the malformation analysis (Table 8.4) only 145 infants born to parents both in categories 4 or 5. The difference in numbers is due almost entirely to the attempt to correct for shielding.

4.8 *Considerations in the estimation of the average amount of radiation received by persons in each of the five radiation categories.* — We come now to an attempt to estimate the average amount of radiation represented by each of the five radiation categories. It should at once be made clear that with the possible exceptions of categories 1 and 5, any estimate is at best an exceedingly rough approximation. For categories 2, 3, and 4, we will do well if we establish the range within which the true average is likely to be found. A particular difficulty in estimating radiation dosage, already alluded to, is the fact that a wide spectrum of radiant energy is involved, with the consequent complications in expressing tissue dosage in simple terms, a difficulty only partially circumvented by expressing dosage in terms of the "roentgen equivalent physical" (*rep*) unit. This latter may be defined as that amount of any ionizing radiation (beta rays, neutrons, etc.) which dissipates the same amount of energy per gram of water as one roentgen of X-rays or gamma rays. All estimates refer to surface dose. The actual amount of irradiation delivered to the gonad will be less, although because of the high energy content of atomic bomb irradiation, the gonad dose will actually be rather close to the skin dose.

Category 1 individuals, outside the cities at the time of the bombing, of course have received no radiation from the atomic bomb (ignoring, of course, the possibility of residual radiation). Category 5 individuals, symptomatic and within 3,000 meters, are considered to have received as a minimum the equivalent of 200 roentgens of gamma rays (Los Alamos Scientific Laboratory, 1950). The maximum dose received by a category 5 individual is in the neighborhood of the equivalent of 600 roentgens of gamma rays. The latter figure is based on the premise that the LD_{50} in man for whole-body radiation is about 400 roentgens of gamma rays, but that a few individuals may survive doses amounting to 600*r* of gamma rays (Hempelmann, Lisco, and Hoffman, 1952). The average dose received by category 5 individuals can probably be placed at the equivalent of 300–400 roentgens of

TABLE 4.8 DISTRIBUTION OF REGISTERED BIRTHS BY PARENTAL EXPOSURE

		Hiroshima						Nagasaki					
		Mother's exposure						Mother's exposure					
		1	2	3	4	5	Total	1	2	3	4	5	Total
Father's exposure	1	18,723	5,721	2,320	417	791	27,972	16,338	10,141	823	116	480	27,898
	2	1,611	1,993	451	90	127	4,272	2,420	4,483	298	38	91	7,330
	3	648	416	545	49	67	1,725	258	301	109	14	21	703
	4	147	119	81	28	26	401	42	55	14	5	1	117
	5	275	145	80	19	54	573	107	133	25	3	21	289
Total		21,404	8,394	3,477	603	1,065	34,943	19,165	15,113	1,269	176	614	36,337

gamma rays. We will return shortly to the important question of the magnitude of the neutron component of this estimated dosage. Suffice it to say at this point that category 5 persons as a group received a relatively larger amount of neutron radiation than any other group, and because of the relatively greater biological effectiveness of neutrons as contrasted to gamma rays, generally speaking, the dose in *rep* units received by category 5 persons may be substantially below the dose as expressed in roentgens.

pressed distance. Figure 4.3, if taken at face value, indicates that beyond approximately 1,600 meters the amount of gamma radiation received was less than 100*r*. Our own observations, involving the reported appearance of symptoms in, e.g., significant numbers of persons in the 2,000–2,400 meter ring, lead us to feel this estimate is somewhat conservative. Also, unpublished calculations of Wilson (1951) suggest that, at least for Hiroshima, the neutron curve given in Figure 4.4 is on the conservative

FIGURE 4.3 — Total dosage of initial gamma radiation as a function of distance from the hypocenter of the explosion of a "nominal" atomic bomb, from "The Effects of Atomic Weapons."

FIGURE 4.4 — Fast and slow neutrons delivered per square centimeter as a function of distance from the hypocenter of the explosion of a "nominal" atomic bomb, from "The Effects of Atomic Weapons."

In arriving at our estimate of the amount of radiation received by category 2, 3, and 4 individuals, three considerations are outstanding:

4.8.1 *The estimated distance-dosage curve.* — Figures 4.3 and 4.4 are the distance-dosage curves in terms of neutron and gamma radiation for the explosion of a "nominal" atomic bomb, roughly comparable to what was detonated over Hiroshima and Nagasaki, as published in "The Effects of Atomic Weapons." In the absence of other unclassified data, this must of necessity serve as the only guide available. Note that the abscissa in both figures is scaled in terms of *feet* rather than the *meters* in which we have ex-

side. On the other hand, both from these two curves and on the basis of our own observations regarding the shape of the distance-dosage curve, it seems likely that persons at distances in excess of 3,000 meters received little if any radiation.

4.8.2 *The observations of the Joint Commission regarding leucopenia.* — One of the well recognized effects of whole-body irradiation in amounts compatible with survival is a temporary leucopenia. Information concerning the proportion of individuals in the radiation categories defined above who developed leucopenia following their exposure would therefore be of value in any attempt to estimate radiation

dosage. Certain of the published observations of the Joint Commission are quite useful in this respect. The more pertinent portion of their published data is summarized in Table 4.9 (see also LeRoy, 1950). Leucopenia is here defined as a total leucocyte count per mm.³ of less than 3,000 at some time during the second through the fifth weeks following the bombing. Some of the individuals on whom the percentages in Table 4.9 are based had repeated leucocyte determinations during a hospitalization, others had but a single count, at any time

TABLE 4.9 THE FINDINGS OF THE JOINT COMMISSION IN HIROSHIMA WITH REGARD TO THE OCCURRENCE OF EPILATION, PETECHIAE, AND LEUCOPENIA IN PERSONS FALLING INTO VARIOUS EXPOSURE CATEGORIES

(Further explanation in text.)

	Exposure category			
	A	B	C	D
Epilation				
No. examined	570	1,119	1,817	1,604
No. epilated	434	400	172	74
% epilated	76.1	35.7	9.5	4.6
Petechiae				
No. examined	570	1,119	1,817	1,604
No. with petechiae	310	262	108	31
% with petechiae	54.4	23.4	5.9	1.9
Leucopenia				
No. examined	195	211	156	111
No. with leucopenia	146	121	31	10
% with leucopenia	74.9	57.3	19.9	9.0

between the second and fifth weeks. The figures are thus at best estimates of the percentage which on careful studies would be found to develop leucopenia. The exposure categories utilized by the Joint Commission and reproduced in Table 4.9 are somewhat different from our own, the four most severe, the ones with which we are concerned, being defined as shown in Table 4.10.

It will be noted that the figures regarding leucopenia are based on fewer individuals than those with respect to epilation and petechiae. This is understandable in view of the difficulties in obtaining leucocyte determinations imposed by the conditions prevailing following the bombing. As the Commission report points out, these figures must be accepted with some reservations in view of the unfavorable circumstances under which they were obtained; on the other hand, the manner in which the findings regarding leucopenia parallel those on petechiae and epilation gives indirect confirmation of their validity. Since the Commission worked with a selected sample, composed largely of injured and, in the majority of cases, hospitalized survivors, caution must be exercised in extrapolating to our own, *total* sample. It will be noted, however, that with the exception of exposure category A, more of the persons examined de-

TABLE 4.10 THE EXPOSURE CATEGORIES DEFINED BY THE JOINT COMMISSION, TO BE APPLIED TO THE INTERPRETATION OF TABLE 4.9

Description	Exposure group					
	A	B	C	D	E	F
0–1,000 meters						
Outdoors or in Japanese buildings	+	—	—	—	—	—
In heavy buildings	—	+	—	—	—	—
In bomb shelters	—	—	+	—	—	—
1,100–1,500 meters						
Outdoors or in Japanese buildings	—	+	—	—	—	—
In heavy buildings	—	—	+	—	—	—
In bomb shelters	—	—	—	+	—	—
1,600–2,000 meters						
Outdoors or in Japanese buildings	—	—	+	—	—	—
In heavy buildings	—	—	—	+	—	—
In bomb shelters	—	—	—	—	+	—
2,100–2,500 meters						
Outdoors or in Japanese buildings	—	—	—	+	—	—
In heavy buildings	—	—	—	—	+	—
In bomb shelters	—	—	—	—	—	+
2,600–3,000 meters						
Outdoors or in Japanese buildings	—	—	—	—	+	—
In heavy buildings	—	—	—	—	—	+
In bomb shelters	—	—	—	—	—	—

veloped leucopenia than either epilation or petechiae.

Through the courtesy of The Armed Forces Institute of Pathology, it has been possible to obtain certain very useful additional tabulations of the material collected by the Joint Commission. Among exposed Japanese in Hiroshima who received one or more leucocyte determinations sometime between the second and fourteenth weeks, who *failed* to develop epilation, petechiae, or gingivitis, and *who are not known to have died within the fourteen weeks following the bombings,* the distribution of *lowest observed* leucocyte counts by distance is as shown in Table 4.11. More intensive studies would have shown a higher proportion with leucopenia.

TABLE 4.11 THE DISTRIBUTION OF LEUCOCYTE VALUES IN HIROSHIMA JAPANESE WHO FAILED TO DEVELOP EPILATION, PETECHIAE, OR GINGIVITIS FOLLOWING THE BOMBING, IN RELATION TO DISTANCE FROM HYPOCENTER AND TYPE OF SHIELDING

(Data of Joint Commission, as tabulated by the Armed Forces Institute of Pathology)

Distance from hypocenter less than 1,500 meters

Leucocytes/mm.³	Outdoors, unshielded No.	Cum. %	Outdoors, shielded No.	Cum. %	Indoors, Jap. bldg. No.	Cum. %	Indoors, heavy bldg. shelter No.	Cum. %
0– 500	1	1.4	0	0	0	0	0	0
600–1,000	1	2.9	1	2.7	1	.5	0	0
1,100–1,500	1	4.3	1	5.4	2	1.6	1	1.0
1,600–2,000	0	4.3	1	8.1	6	4.9	3	4.1
2,100–2,500	1	5.7	2	13.5	6	8.2	1	5.2
2,600–3,000	1	7.1	1	16.2	9	13.0	1	6.2
3,100–3,500	3	11.4	1	18.9	16	21.7	3	9.3
3,600–4,000	5	18.6	2	24.3	22	33.7	7	16.5
4,100–4,500	7	28.6	6	40.5	11	39.7	0	16.5
4,600–5,000	5	35.7	3	48.6	14	47.3	1	17.5
5,100–5,500	3	40.0	3	56.8	16	56.0	18	36.1
5,600–6,000	6	48.6	9	81.1	9	60.9	7	43.3
6,100–6,500	8	60.0	3	89.2	16	69.6	11	54.6
6,600–7,000	7	70.0	0	89.2	15	77.7	8	62.9
7,100–7,500	2	72.9	0	89.2	8	82.1	4	67.0
7,600–8,000	4	78.6	2	94.6	7	85.9	8	75.3
8,100–8,500	2	81.4	0	94.6	9	90.8	3	78.4
8,600–9,000	5	88.6	1	97.3	3	92.4	7	85.6
9,100–9,500	1	90.0	1	100.0	2	93.5	2	87.6
9,600–	7	100.0	0	100.0	12	100.0	12	100.0
	70		37		184		97	

Distance from hypocenter 1,600–2,000 meters

Leucocytes/mm.³	Outdoors, unshielded No.	Cum. %	Outdoors, shielded No.	Cum. %	Indoors, Jap. bldg. No.	Cum. %	Indoors, heavy bldg. shelter No.	Cum. %
0– 500	0	0	0	0	1	.4	0	0
600–1,000	0	0	0	0	0	.4	0	0
1,100–1,500	3	1.1	1	1.7	0	.4	1	1.9
1,600–2,000	1	1.5	0	1.7	2	1.1	0	1.9
2,100–2,500	3	2.6	0	1.7	5	3.0	1	3.7
2,600–3,000	5	4.5	1	3.4	4	4.5	0	3.7
3,100–3,500	11	8.6	4	10.2	10	8.2	1	5.6
3,600–4,000	16	14.7	2	13.6	16	14.1	2	9.3
4,100–4,500	17	21.1	6	23.7	26	23.8	6	20.4
4,600–5,000	10	24.8	1	25.4	17	30.1	1	22.2
5,100–5,500	30	36.1	8	39.0	23	38.7	6	33.3
5,600–6,000	18	42.9	11	57.6	27	48.7	8	48.1
6,100–6,500	22	51.1	5	66.1	17	55.0	4	55.6
6,600–7,000	26	60.9	2	69.5	27	65.1	2	59.3
7,100–7,500	14	66.2	3	74.6	21	72.9	3	64.8
7,600–8,000	21	74.1	3	79.7	14	78.1	3	70.4
8,100–8,500	9	77.4	3	84.7	16	84.0	4	77.8
8,600–9,000	21	85.3	3	89.8	11	88.1	8	92.6
9,100–9,500	9	88.7	2	93.2	8	91.1	1	94.4
9,600–	30	100.0	4	100.0	24	100.0	3	100.0
	266		59		269		54	

THE CHILDREN OF ATOMIC BOMB SURVIVORS

TABLE 4.11—Continued
Distance from hypocenter 2,100–3,000 meters

Leucocytes/mm.³	Outdoors, unshielded No.	Cum. %	Outdoors, shielded No.	Cum. %	Indoors, Jap. bldg. No.	Cum. %	Indoors, heavy bldg. shelter No.	Cum. %
0– 500	0	0	0	0	0	0	0	0
600–1,000	0	0	0	0	0	0	0	0
1,100–1,500	0	0	0	0	1	.2	0	0
1,600–2,000	0	0	0	0	2	.6	0	0
2,100–2,500	5	1.9	0	0	3	1.3	0	0
2,600–3,000	3	3.1	1	2.6	9	3.2	1	9.1
3,100–3,500	8	6.1	1	5.3	16	6.6	0	9.1
3,600–4,000	12	10.7	1	7.9	16	10.0	0	9.1
4,100–4,500	14	16.0	0	7.9	28	16.0	2	27.3
4,600–5,000	20	23.7	1	10.5	29	22.2	0	27.3
5,100–5,500	19	30.9	5	23.7	50	32.9	3	54.5
5,600–6,000	27	41.2	2	28.9	43	42.1	0	54.5
6,100–6,500	27	51.5	4	39.5	47	52.1	1	63.6
6,600–7,000	23	60.3	4	50.0	34	59.4	1	72.7
7,100–7,500	15	66.0	2	55.3	34	66.7	0	72.7
7,600–8,000	14	71.4	2	60.5	25	72.0	1	81.8
8,100–8,500	18	78.2	2	65.8	34	79.3	0	81.8
8,600–9,000	17	84.7	4	76.3	26	84.8	1	90.9
9,100–9,500	6	87.0	1	78.9	10	87.0	0	90.9
9,600–	34	100.0	8	100.0	61	100.0	1	100.0
	262		38		468		11	

The concentration of values in the "9,600 and above" class would seem to indicate a grouping of values which would vitiate attempts to calculate a mean. However, generally speaking, these observations suggest that a significant number of persons in our radiation categories 3 and 4 probably developed leucopenia as that term is here defined. An exact estimate of the proportion is difficult because of the difference between the way the Joint Commission's and our own data are broken down, but in round figures, based in part on the data in Table 4.11, it can be estimated that some 5–10 per cent of group 4 parents and 2–4 per cent of group 3 parents developed leucopenia.

The average amount of whole-body irradiation which, when delivered over a period of a day or two, will produce in man leucopenia of this degree is not known with certainty. For obvious reasons, there are not many pertinent observations. Thus, studies of patients with leukemia or multiple myeloma (e.g., Collins and Loeffler, 1956), while valuable for therapeutic reasons, obviously do not permit extrapolation to normal persons. Of the more recent investigations, the most valuable for present purposes — although still, because of the nature of the subjects, to be interpreted with caution — are those of Nickson (1951), who found that in three patients with carcinoma with metastases, 120r of whole-body irradiation with 400-kv X-rays produced no significant leucopenia. Two of these patients had a moderate leucocytosis at the time of treatment, amounting to 14,700/mm.³ and 13,900/mm.³, respectively. A fourth patient, also with carcinomatosis, who was treated with 107r of 200-kv whole-body X-rays on two successive days likewise failed to develop significant leucopenia. The impropriety of extrapolating from such patients to normal individuals is apparent. Furthermore, it does not appear that these patients were followed long enough for the detection of late leucopenic effects. The studies of Hempelmann, Lisco, and Hoffman (1952) on the few persons who have been involved in industrial radiation accidents are also especially pertinent. In one case, an individual was exposed to an amount of irradiation estimated as the equivalent of 186 roentgens of 80-kv X-rays and 10.7 roentgens of γ-rays, and in another case, to the estimated equivalent of 140r of 80-kv X-rays and 8.7r of γ-rays. In neither case did leucopenia as here defined develop.

In March of 1954 certain natives of the Marshall Islands and a lesser number of Ameri-

can military personnel were accidentally exposed to the fall-out of radioactive materials subsequent to the explosion of an experimental thermonuclear device. Detailed hematological studies were carried out on these individuals (Cronkite, Bond, and Dunham, in manuscript). Through the courtesy of Dr. E. P. Cronkite and Mr. Hyman Hechter, the original data on the Marshall Islanders regarding total leucocyte counts during a 10 to 70-day period following the exposure have been made available to us. There were two groups of Marshall Islanders involved, one group of 64 estimated to have received an average dose of 175r of whole-body, gamma radiation as measured in air, and a second group of 18 estimated to have received an average dose of 69r of gamma radiation as measured in air. The actual dose received by the Marshallese has been the subject of lively discussion. The observed maximum depression in the total leucocyte count occurred between days 39 and 51 following the exposure (counts were made every four days). On day 39, 3.1 per cent of the Group I Marshallese were found to have a leucopenia of 3,000 WBC/mm.3 or less; on day 43, 6.5 per cent; on day 47, 3.1 per cent; and on day 51, 0.0 per cent. None of the Group II natives showed a leucopenia of this degree during this period.

A comparison of these figures with the Japanese figures cited above is biased by the fact that the Marshall Island figures refer to the period of peak depression, whereas the Japanese data include a longer period of time. Nevertheless, the important fact emerges that in a group of healthy persons known with some accuracy to have received a dose of whole-body gamma irradiation of approximately 175r, the frequency of leucopenia was probably no greater than (if as great as) that occurring in the category 4 parents of this study. The obvious inference is that the average radiation dose experienced by category 4 parents is at least the equivalent of 200r of whole-body gamma radiation, with category 3 parents receiving a smaller average dose, but very likely one in the neighborhood of the equivalent of 100r of whole-body gamma radiation.

4.8.3 *The proportion of individuals protected by various types of shielding.* — The final consideration to be introduced into this attempt to approximate the average radiation exposure in the various categories is the proportion of individuals experiencing various types of shielding. Data are not available for the whole sample. However, in a special study carried out in 1953 on all parents falling in the 1,800–2,500 meter ring, a group especially critical to this study, the breakdown shown in Table 4.12 was observed for 4,515 Hiroshima parents and 2,083 Nagasaki parents. The important point which emerges is the relatively small proportion of persons classified as receiving heavy or moderate shielding. It will be recalled that in the radiation classification adopted here, the report of "heavy" or "moderate" shielding resulted in an individual being placed in a lower radiation category. The contribution of such individuals — whose shielding is especially difficult to evaluate — to the composition of any radiation category is small.

TABLE 4.12 PROPORTIONS OF PARENTS EXPOSED IN THE 1,800–2,500 METER RING WHO REPORTED VARIOUS TYPES OF SHIELDING

Degree of shielding	Hiroshima	Nagasaki
Heavy	4.9	15.9
Moderate	2.5	3.0
Light	68.0	67.0
None	24.6	14.1
	100.0	100.0

4.9 *Estimates of the average amount of irradiation received by individuals in the various exposure categories.* — We are now in a position to approximate the average amount of whole-body irradiation received by individuals in categories 2 through 5. Category 2 individuals, on the basis of the distance-dosage curve, probably received on the average no more than 5–10 roentgens equivalent physical (*rep*'s). Category 3 survivors, on the same basis, as well as the known incidence of leucopenia and the shielding data, may be estimated on the average to have received 50–100 *rep*'s. The estimation of the dosage in *rep*'s for groups 4 and 5 is more difficult because, in contrast to groups 2 and 3, there is, on the basis of the distance-dosage curves and the occurrence of radiation cataracts in appreciable numbers of these survivors, a significant neutron component in the irradiation experienced by these persons. To make approximate allowance for this, we will express "the equivalent of 300–400 roentgens of gamma rays" suggested earlier as the average dose in category 5, as 200–300 *rep*'s, and on the same grounds estimate the average dosage for cate-

gory 4 individuals as 100–150 *rep*'s. The arbitrary nature of these estimates will be apparent to all students of radiobiology. Any one might well be off by a factor of 2. We feel that any attempt to be more specific cannot be justified. In arriving at what some will term overly conservative estimates, we have been influenced by the fact that traumatic and thermal injuries may in a significant proportion of persons have

neutrons in inducing genetic change (summary in Symposium: Some Biological Effects of Radiation from Nuclear Detonation, 1954). However, it is important to recognize that the average dosages estimated above include, in the case of category 4 and 5 persons, a component of high effectiveness in producing genetic change.

Even should the upper rather than the lower estimates of the average amount of radiation

FIGURE 4.5 — Neutron and gamma radiation distance-dosage curves for the atomic bomb explosions at Hiroshima and Nagasaki.

combined with ordinarily sublethal radiation effects in causing death, thus lowering the mean radiation dose expected in the various categories on radiological grounds alone. On the other hand, any allowance for residual irradiation would revise the dose upward.

There are at present no published data which permit making a realistic estimate of the contribution of fast and slow neutrons to the total radiation spectrum. This is unfortunate in view of the relatively high biological efficiency of

received by survivors prove correct, the fact is clear that in the terms of the radiation geneticist, these are small doses indeed. This was recognized from the outset. But despite the improbability of being able to detect induced genetic effects in this material, there was no doubt in the minds of those who considered the problem initially, at all echelons, that "this unique possibility for demonstrating genetic effects caused by atomic radiation should not be lost" (cf. p. 2).

Note added in proof. — At the time this chapter was written, only the distance-dosage curves for a "nominal" atomic bomb had been published. However, since this manuscript went to press, distance-dosage curves, in terms of neutron and gamma radiation, have been declassified for both Hiroshima and Nagasaki. These are presented in Fig. 4.5. The gamma curves for the two cities are quite comparable, and correspond satisfactorily with those published earlier for the explosion of a "nominal" atomic bomb. However, it will be noted that the neutron curves for the two cities appear to differ significantly, with, at a given distance, rather more radiation of this type in Hiroshima than in Nagasaki. The curve for the latter city approximates that published earlier for the explosion of a "nominal" bomb. In view of the many uncertainties that enter into assigning mean dosages to our radiation exposure categories, it is not felt that these new data warrant a revision of the estimates given in this chapter. On the other hand, it is clear that there is a very strong neutron component in the radiation experienced by category 4 and 5 parents in Hiroshima, and in this connection, it will be recalled that approximately 68 per cent of all category 4 or 5 parents were from Hiroshima. In view of the high relative effectiveness of neutrons in inducing genetic change, the publication of these new curves has the effect of increasing the genetic significance to be attached to the observations herein to be reported.

Chapter V

THE COMPARABILITY OF IRRADIATION SUBCLASSES

The infants examined in Hiroshima and Nagasaki during the course of this study may be apportioned among 25 subclasses on the basis of the radiation exposure categories of their fathers and mothers, as these categories were defined in the preceding chapter. The crux of our problem is a comparison of the characteristics of the infants comprising these subclasses. As has been emphasized several times already, all of the indicators of possible genetic damage utilized in this investigation may also be influenced by a variety of other factors. It will be the purpose of this chapter to undertake a detailed comparison of the parents of the infants comprising these subclasses with respect to certain possible differences which might influence the outcome of pregnancy. Of the many possible differences which could be explored, we shall restrict ourselves to those which in our opinion are most pertinent to the problem at hand and for which we have more nearly satisfactory data.

In the comparisons which will be presented, usually no allowance has been made for the fact that because of repeated pregnancies the same couple may be represented more than once among the parentage of a subclass. The tests of significance that follow assume no duplicate registration and thus yield underestimates of the errors of differences. It might be pointed out that, in the main, we should underestimate variances in the parental population more than in the offspring population.

5.1 *Consanguinity.* — It is generally recognized that the offspring of consanguineous marriages more often exhibit the consequences of genetic homozygosity than do the offspring of non-consanguineous unions. To the extent that there is a recessive, incompletely recessive, or semi-dominant genetic component in the etiology of congenital defect, stillbirth, or neonatal death, such homozygosity might be expected to alter the frequency of these events sufficiently to obscure real irradiation differences or to create spurious ones. The occurrence within the populations under study of a non-random distribution of consanguineous marriages thus might introduce a source of bias into the findings. This, then, was the theoretical consideration which led to the inclusion of an item in the Genetics Short Form regarding parental consanguinity. The problem seemed of special importance because of the relatively high frequency of consanguineous unions in Japan (Neel, Kodani, Brewer, and Anderson, 1949).

In Tables 5.1 and 5.2 the present data are examined as regards a relationship between exposure subclass and the frequency of consanguinity.[1] Although all known degrees of consanguinity were recorded by the clerks, in the analysis attention has been restricted to marriages of first cousins, first cousins once removed, and second cousins. A few uncle-niece unions, and some involving more remote degrees of consanguinity than second cousin marriages, have been recorded during the study; all such unions have been excluded from these considerations because of the uncertainty regarding the exhaustiveness of the ascertainment.

There is significantly more consanguineous marriage in Nagasaki than in Hiroshima, this presumably a reflection of social and cultural differences between the two localities. Moreover, within the cities there is heterogeneity among exposure subclasses. In general this consists in less consanguineous marriage among the "exposed" (categories 2, 3, 4, 5) than among the

[1] The differing forms utilized in the presentation of the results of the analysis of the various tables will be evident following the reading of Chapter VI.

TABLE 5.1 CONSANGUINITY (FIRST COUSIN, FIRST COUSIN ONCE REMOVED, SECOND COUSIN) BY CITY AND PARENTAL EXPOSURE

Hiroshima

			Fathers[a]			
		1	2	3	4–5	Total
1	n	18,723	1,611	648	422	21,404
	r	1,242	97	42	23	1,404
	P	.0663	.0602	.0648	.0545	.0656
2	n	5,721	1,993	416	264	8,394
	r	311	117	27	16	471
	P	.0544	.0587	.0649	.0606	.0561
3	n	2,320	451	545	161	3,477
	r	97	23	16	4	140
	P	.0418	.0510	.0294	.0248	.0403
4–5	n	1,208	217	116	127	1,668
	r	77	9	6	6	98
	P	.0637	.0415	.0517	.0472	.0588
Total	n	27,972	4,272	1,725	974	34,943
	r	1,727	246	91	49	2,113
	P	.0617	.0576	.0528	.0503	.0605

(Mothers[a] along left axis)

Nagasaki

			Fathers[a]			
		1	2	3	4–5	Total
1	n	16,338	2,420	258	149	19,165
	r	1,541	211	13	8	1,773
	P	.0943	.0872	.0504	.0537	.0925
2	n	10,141	4,483	301	188	15,113
	r	683	293	22	10	1,008
	P	.0674	.0654	.0731	.0532	.0667
3	n	823	298	109	39	1,269
	r	69	14	7	4	94
	P	.0838	.0470	.0642	.1026	.0741
4–5	n	596	129	35	30	790
	r	33	10	—	2	45
	P	.0554	.0775	—	.0667	.0570
Total	n	27,898	7,330	703	406	36,337
	r	2,326	528	42	24	2,920
	P	.0834	.0720	.0597	.0591	.0804

[a] In this and subsequent tables, the term "Fathers" will be used as an abbreviation for Father's Exposure Category, and the term "Mothers" similarly.

unexposed (category 1), although the tendency is statistically significant only with respect to the mothers. There is also a significant city-mother interaction. Category 1 contains a high proportion of persons who did not legally reside in either of the two cities at the time of the bombings. It may be surmised that a substantial proportion of these persons have rural antecedents; the frequency of consanguinity is known to be higher in rural communities (Neel et al., 1949).

ables, parental age[2] and parity,[3] and most of the indicators of possible genetic damage herein studied is attested to by a voluminous literature (e.g., Ciocco, 1938; Yerushalmy, 1938; Penrose, 1939; Murphy, 1947; Landtman, 1948; Record and McKeown, 1949; Sutherland, 1949; Carter, 1950; Lowe and McKeown, 1950; Worcester, Stevenson, and Rice, 1950; Hegnauer, 1951; Karn and Penrose, 1951; Novitski, 1953; Salber and Bradshaw, 1953; MacMahon and Gordon, 1953; Myers, 1954). Our own

TABLE 5.2 CHI-SQUARE ANALYSIS OF THE FREQUENCY OF CONSANGUINEOUS MARRIAGES BY CITY AND PARENTAL EXPOSURE

Source	DF	χ^2	P
Total	31	261.453	< .001
Interactions, first order			
CM	3	16.913	< .001
CF	3	1.610	.50–.70
MF	9	6.317	.70–.80
Main Effects			
City [a]			
Mother's exposure — 1	1	101.499	< .001
Mother's exposure — 2	1	10.257	.001–.01
Mother's exposure — 3	1	22.672	< .001
Mother's exposure — 4, 5	1	0.031	.80–.90
Mother [b]			
Hiroshima (H)	3	37.775	< .001
Nagasaki (N)	3	83.002	< .001
Father [c]			
(H) Mother's exposure — 1	3	1.783	.50–.70
Mother's exposure — 2	3	1.292	.70–.80
Mother's exposure — 3	3	4.156	.20–.30
Mother's exposure — 4, 5	3	2.123	.50–.70
(N) Mother's exposure — 1	3	9.581	.02–.05
Mother's exposure — 2	3	0.947	.80–.90
Mother's exposure — 3	3	4.950	.10–.20
Mother's exposure — 4, 5	3	3.210	.30–.50
Sum	24	28.042	.20–.30

[a] Adjusted for mothers. (The meaning of the term "adjusted" is explained in Sec. 6.5.)
[b] Adjusted for cities.
[c] Adjusted for cities and mothers.

In Nagasaki, the bomb was detonated over that area of the city in which there was the highest concentration of Christians; Schull (1953) has shown that there is less consanguineous marriage among Christian (Catholic) Japanese in Nagasaki than among non-Christians. Thus, it may be surmised that among the non-Christian Japanese, the differences in frequency of consanguineous unions between Hiroshima and Nagasaki are even more striking than shown in Table 5.1.

5.2 *Age and parity.* — A probable or certain relationship between the correlated variables, parental age and parity, and most of the indicators of possible genetic damage herein analysis of these relationships in the present data will be published elsewhere. Suffice it to say at this point that significant associations can be readily demonstrated.

The findings in these data with respect to mean maternal age and parity differences among the mothers of the children falling into the various subclasses are shown in Tables 5.3

[2] The terms "parental age," "maternal age," or "paternal age" as here used refer simply to age of parent(s) at birth of child.
[3] The term "parity" refers simply to total number of conceptions, including the present.

TABLE 5.3 MEAN MATERNAL AGE BY CITY AND PARENTAL EXPOSURE
(Based on unrelated parents *only*)

Hiroshima

		Fathers					
Mothers		1	2	3	4	5	Total
1	n	17,481	1,514	606	136	263	20,000
	\bar{x}	27.0469	27.3250	27.5165	27.4853	28.2700	27.1013
2	n	5,410	1,876	389	108	140	7,923
	\bar{x}	26.8366	30.1311	29.2956	28.0926	28.7500	27.7883
3	n	2,223	428	529	81	76	3,337
	\bar{x}	26.9523	29.1402	30.9981	30.0123	29.0395	27.9961
4	n	388	88	45	24	19	564
	\bar{x}	27.7861	28.3636	29.6667	31.1667	29.6842	28.2340
5	n	743	120	65	26	52	1,006
	\bar{x}	27.1252	29.3667	29.3385	28.5385	31.4808	27.7972
Total	n	26,245	4,026	1,634	375	550	32,830
	\bar{x}	27.0087	28.9091	29.1989	28.5147	28.8509	27.3988

Nagasaki

		Fathers					
Mothers		1	2	3	4	5	Total
1	n	14,797	2,209	245	41	100	17,392
	\bar{x}	28.5427	29.1421	29.5265	28.4878	30.2800	28.6425
2	n	9,458	4,190	279	51	127	14,105
	\bar{x}	27.6087	30.8477	30.3118	30.1176	29.8976	28.6540
3	n	754	284	102	12	23	1,175
	\bar{x}	27.6578	29.8415	31.5294	31.5833	30.6522	28.6204
4	n	105	35	14	4	3	161
	\bar{x}	29.9333	30.4857	34.6429	32.0000	34.6667	30.6025
5	n	458	84	21	1	20	584
	\bar{x}	26.7445	28.4762	29.1905	13.0000	29.4000	27.1490
Total	n	25,572	6,802	661	109	273	33,417
	\bar{x}	28.1447	30.2207	30.2648	29.5780	30.1172	28.6299

through 5.8. It will be noted that the comparison is restricted to *unrelated* parents. In the following chapter we shall advance reasons for excluding children born to *related* parents from the analysis. Because of the potential great importance of age-parity differences in this analysis, it has seemed wise to anticipate this decision with respect to this tabulation.

Not only do the cities differ but there is also significant heterogeneity within cities among exposure subclasses as regards both mean maternal age and mean parity. In Hiroshima there is a clear tendency for maternal age to be greater when the mother or father is exposed (category 2, 3, 4, or 5) than when non-exposed (category 1). This appears to be true with reference to fathers' but not mothers' exposure in Nagasaki. An obvious explanation of the findings with regard to the relation between paternal exposure and mean maternal age is the fact that the younger males tended to be in military service while the older males remained at home (and so were exposed), and mother's age is correlated with father's. Heterogeneity between the subclasses is also indicated by the fact that most of the interaction terms are also significant.

No less variable than the mean maternal ages and parities are the mean squares (variance esti-

TABLE 5.4 ANALYSIS OF VARIANCE: MOTHER'S AGE BY CITY AND PARENTAL EXPOSURE
(Unrelated parents)

(In this and subsequent tables, one asterisk will be used to denote a value significant at the 5 per cent level, and two asterisks, a value significant at the 1 per cent level.)

Source	Sums of squares of deviations	DF	Mean square	F
Main effects				
City (C)	21,983.321551	1	21,983.3216	830.86**
Father (F)	43,401.030323	4	10,850.2576	410.08**
Mother (M)	947.687594	4	236.9219	8.95**
Interactions				
First order				
CF	74.432938	4	18.6082	1.42
CM	3,107.667017	4	776.9168	29.36**
MF	18,682.813301	16	1,167.6758	44.13**
Higher orders	871.870114	16	54.4919	2.06**
Between classes	94,513.849388	49	1,928.8541	72.90**
Within classes	1,751,484.087605	66,197	26.4587	—
Total	1,845,997.936993	66,246		

TABLE 5.5 THE DISTRIBUTION OF MEAN SQUARES FOR MATERNAL AGE BY CITY, SEX OF INFANT, AND PARENTAL EXPOSURE (Unrelated parents)

(The sample numbers are the same as in Table 5.3. The form of the analysis will be discussed in Chapter VI.)

Hiroshima

		Fathers			
	Sex	1	2	3	4–5
Mothers 1	♂	22.160	27.336	29.653	25.529
	♀	21.628	24.205	29.271	23.310
Mothers 2	♂	21.999	28.964	33.690	28.247
	♀	22.327	29.351	40.519	34.122
Mothers 3	♂	24.216	29.592	32.618	25.009
	♀	23.381	29.429	31.245	36.640
Mothers 4–5	♂	26.112	28.495	47.317	35.253
	♀	24.226	36.713	27.148	36.738

$\chi^2_\male = 121.021^{**}$ DF = 15
$\chi^2_\female = 144.251^{**}$ DF = 15

Nagasaki

		Fathers			
	Sex	1	2	3	4–5
Mothers 1	♂	25.376	36.213	40.472	38.920
	♀	25.822	33.257	32.491	32.455
Mothers 2	♂	28.335	37.926	43.489	31.397
	♀	26.361	38.289	42.377	40.381
Mothers 3	♂	30.266	49.061	42.831	18.423
	♀	29.384	39.629	37.398	31.729
Mothers 4–5	♂	27.956	28.292	32.515	28.743
	♀	27.720	48.456	39.000	55.269

$\chi^2_\male = 222.236^{**}$ DF = 15
$\chi^2_\female = 192.154^{**}$ DF = 15

TABLE 5.6 MEAN PARITY BY CITY AND PARENTAL EXPOSURE
(Based on unrelated parents *only*)

Hiroshima

Mothers		Fathers 1	2	3	4	5	Total
1	n	17,481	1,514	606	136	263	20,000
	\bar{x}	2.4228	2.4108	2.3762	2.3676	2.4677	2.4207
2	n	5,410	1,876	389	108	140	7,923
	\bar{x}	2.4222	3.6151	3.2339	2.7963	2.9429	2.7588
3	n	2,223	428	529	81	76	3,337
	\bar{x}	2.4368	3.2897	3.9187	3.9136	3.3158	2.8370
4	n	388	88	45	24	19	564
	\bar{x}	2.6031	3.2273	3.4889	3.8750	3.3684	2.8511
5	n	743	120	65	26	52	1,006
	\bar{x}	2.4078	3.5333	3.0615	3.4231	3.5192	2.6680
Total	n	26,245	4,026	1,634	375	550	32,830
	\bar{x}	2.4261	3.1167	3.1377	2.9947	2.8364	2.5596

Nagasaki

Mothers		Fathers 1	2	3	4	5	Total
1	n	14,797	2,209	245	41	100	17,392
	\bar{x}	2.9008	3.0946	3.1551	2.4878	3.1500	2.9295
2	n	9,458	4,190	279	51	127	14,105
	\bar{x}	2.7294	3.9754	3.6810	3.2941	3.6457	3.1287
3	n	754	284	102	12	23	1,175
	\bar{x}	2.8594	3.7254	4.0000	5.0000	3.6957	3.2060
4	n	105	35	14	4	3	161
	\bar{x}	3.4762	3.8000	5.3571	4.7500	6.3333	3.7950
5	n	458	84	21	1	20	584
	\bar{x}	2.6201	3.1667	3.6190	1.0000	3.7500	2.7705
Total	n	25,572	6,802	661	109	273	33,417
	\bar{x}	2.8335	3.6680	3.5688	3.2110	3.5055	3.0247

mates) for these two variables.[4] From Tables 5.5 and 5.8 we note a significant heterogeneity among exposure classes, within each sex and city, in the variance estimates. It can also be shown that the variance estimates differ significantly between cities. In general, the variance in post-war Japan, those persons closer in having suffered in material ways to a greater extent than those farther out. Economic status appears to be related to some of the indicators of radiation effect here under consideration, notably stillbirth frequency, birthweight, and neonatal

TABLE 5.7 ANALYSIS OF VARIANCE: PARITY BY CITY AND PARENTAL EXPOSURE
(Unrelated parents)

Source	Sums of squares of deviations	DF	Mean square	F
Main effects				
City (C)	1,663.729494	1	1,663.7295	531.52**
Father (F)	6,053.779421	4	1,513.4449	483.51**
Mother (M)	852.862735	4	213.2157	68.12**
Interactions				
First order				
CF	181.504471	4	45.3761	14.50**
CM	62.510339	4	15.6276	4.99**
MF	2,637.923744	16	164.8702	52.67**
Higher orders	77.508327	16	4.8443	1.55
Between classes	13,593.325165	49	2,774.1480	886.27**
Within classes	207,206.531342	66,197	3.1302	—
Total	220,799.856507	66,246		

TABLE 5.8 THE DISTRIBUTION OF MEAN SQUARES FOR PARITY BY CITY, SEX OF INFANT, AND PARENTAL EXPOSURE (Unrelated parents)

(The sample numbers are the same as in Table 5.6.)

Hiroshima

	Sex	Fathers 1	2	3	4–5
Mothers 1	♂	2.255	2.156	2.360	2.067
	♀	2.076	2.311	1.691	2.217
Mothers 2	♂	2.242	3.438	4.023	2.286
	♀	2.232	3.840	3.659	2.729
Mothers 3	♂	2.361	3.610	4.460	4.286
	♀	2.134	3.274	3.785	4.251
Mothers 4–5	♂	2.239	4.171	4.999	5.328
	♀	2.206	4.526	3.884	3.441

$\chi^2_{\male} = 281.717^{**}$ DF = 15
$\chi^2_{\female} = 324.417^{**}$ DF = 15

Nagasaki

	Sex	Fathers 1	2	3	4–5
Mothers 1	♂	2.867	4.111	4.717	3.266
	♀	2.899	4.036	3.297	2.328
Mothers 2	♂	3.022	4.933	4.652	3.984
	♀	2.795	5.183	4.284	3.992
Mothers 3	♂	3.059	4.849	4.306	4.474
	♀	3.624	4.647	3.350	5.862
Mothers 4–5	♂	3.258	2.967	4.434	6.495
	♀	3.434	4.878	4.529	5.359

$\chi^2_{\male} = 337.847^{**}$ DF = 15
$\chi^2_{\female} = 397.544^{**}$ DF = 15

estimates increase with increasing conjoint parental exposure.

5.3 *Economic status.*—On a priori grounds it seemed quite possible that there might be a correlation between distance from the hypocenter at the time of the explosion and economic status death (e.g., Ebbs et al., 1942a, b; Balfour, 1944; Antonov, 1947; Smith, 1947; Burke et al., 1949; Dean, 1950; Nixon, 1950). It therefore seemed advisable to make some attempt to evaluate this possibility.

The economic status of the home into which the child was born was estimated for all pregnancies terminating abnormally, as well as every pregnancy whose registration number ended in

[4] The effect of heterogeneity of the variances on the tests given in Tables 5.5 and 5.7 will be discussed in Section 6.6.

TABLE 5.9 ECONOMIC STATUS BY CITY AND PARENTAL EXPOSURE
(The figure given for economic status is proportion of homes falling into the "poor" and "very poor" categories.)

Hiroshima

Mothers			Fathers 1	2	3	4–5	Total
1		n	1,770	143	65	46	2,024
		r	144	4	6	—	154
		p	.0814	.0280	.0923	—	.0761
2		n	569	171	47	28	815
		r	49	24	4	2	79
		p	.0861	.1404	.0851	.0714	.0969
3		n	218	26	56	19	319
		r	14	5	6	1	26
		p	.0642	.1923	.1071	.0526	.0815
4–5		n	103	21	10	11	145
		r	9	1	1	1	12
		p	.0874	.0476	.1000	.0909	.0828
Total		n	2,660	361	178	104	3,303
		r	216	34	17	4	271
		p	.0812	.0942	.0955	.0385	.0820

Nagasaki

Mothers			Fathers 1	2	3	4–5	Total
1		n	1,459	227	27	14	1,727
		r	226	33	8	—	267
		p	.1549	.1454	.2963	—	.1546
2		n	894	387	33	14	1,328
		r	125	74	5	1	205
		p	.1398	.1912	.1515	.0714	.1544
3		n	61	29	10	6	106
		r	7	5	2	2	16
		p	.1148	.1724	.2000	.3333	.1509
4–5		n	56	12	4	3	75
		r	9	2	1	—	12
		p	.1607	.1667	.2500	—	.1600
Total		n	2,470	655	74	37	3,236
		r	367	114	16	3	500
		p	.1486	.1740	.2162	.0811	.1545

zero, in connection with the use of the Genetics Long Form (Secs. 2.2 and 2.3). This evaluation was made according to a very rough scoring system admitting of five economic classifications: very poor, poor, average, well-to-do, and rich. The evaluation was carried out by the Japanese physician at the time of the home visit. The physicians were instructed that it was expected that most homes would be graded as average, i.e., that existing rather than pre-bombing standards should serve as the yardstick of comparison. Inasmuch as in the course of time a physician employed in the program found himself in all sections of the city, it was felt that this crude system of rating would pick up any marked differences between exposure categories.

The findings are given in Tables 5.9 and 5.10. The data are presented in simplified form, in terms of proportion of all homes graded as "poor" or "very poor." In view of the subjective nature of the ratings, it is difficult to attach significance to the apparent difference between the two cities. There is no apparent relation between economic status and mother's exposure category, nor is there a relation between father's category and economic status. At first glance this observation would seem to be counter to common sense, since it would seem obvious that persons close to the hypocenter would have suffered greater material losses than those more distant, losses which would reflect themselves in economic level. However, the effect of the atomic bomb was but one of a number of sources of severe economic dislocation in post-war Hiroshima and Nagasaki. The obvious conclusion from the findings summarized in this section would be that these other causes of economic readjustment tended to complicate and even to nullify the effects of the bombs.

5.4 *Frequency of positive serological test for syphilis.*—In view of the well-known relationship between syphilis and stillbirth and neonatal death, a serological test for syphilis was carried out in the ABCC laboratories on each woman whose registration number terminated in "0," as well as for any woman whose pregnancy terminated abnormally. In about 3 per cent of the terminations no test was performed, usually because of the disinclination of the mother to submit to venipuncture. The tests used were the cardiolipin microflocculation and the Kline. Tables 5.11 and 5.12 present the findings.

There is a significant difference between cities, but no indication of heterogeneity between parental exposure classes.

5.5 *Frequency of induced abortions and of dilatation and curettage of the uterus (D and C).*—During the post-war years, in an effort to control birth rates, the Japanese government relaxed the indications for legal "therapeutic" abortion. Exact figures on the frequency with which abortions were performed are impossible to obtain for a variety of reasons. However, because of the possibility that the late complications of uterine infection following an induced

TABLE 5.10 CHI-SQUARE ANALYSIS OF THE DISTRIBUTION OF ECONOMIC STATUSES BY CITY AND PARENTAL EXPOSURE
(Unrelated parents)

Source	DF	χ^2	P
Total	31	120.378	< 0.001
Interactions, first order			
CM	3	2.268	.50–.70
CF	3	0.365	.90–.95
MF	9	17.683	.02–.05
Main effects			
Cities (C)	1	82.529	< 0.001
Fathers (F)[a]			
Hiroshima	3	3.782	.20–.30
Nagasaki	3	6.262	.05–.10
Sum	6	10.044	.10–.20
Mothers (M)[a]			
Hiroshima	3	3.354	.30–.50
Nagasaki	3	0.028	$> .99$
Sum	6	3.382	.70–.80

[a] Adjusted for cities.

abortion might influence pregnancy termination, an attempt was made to obtain some information on this point through the use of the Genetics Short Form. Each registrant was questioned concerning the occurrence of abortions. The abortions which were reported were divided into two categories, depending on occurrence before or after the date of the atomic bombing, and further subdivided as to whether they were spontaneous or induced. The findings concerning induced abortions are shown in Tables 5.13 and 5.14. The figures refer to the proportion of women reporting one or more induced abortions. Each mother is scored only once, on the basis of the first registered pregnancy. Induced abortions are reported with a significantly greater frequency in Hiroshima than in Nagasaki and, in both cities, appear to be more

TABLE 5.11 FREQUENCY OF POSITIVE SEROLOGY BY PARENTAL EXPOSURE, CITY, AND GROUP
(Unrelated "zero" parents only)

Hiroshima

Mothers		Fathers 1	2	3	4-5	Total
1	n	1,610	131	60	41	1,842
	r	61	3	5	2	71
	p	.0379	.0229	.0833	.0488	.0385
2	n	509	145	43	26	723
	r	12	10	4	2	28
	p	.0236	.0690	.0930	.0769	.0387
3	n	197	25	51	18	291
	r	5	1	1	1	8
	p	.0254	.0400	.0196	.0556	.0275
4-5	n	84	18	9	11	122
	r	5	—	1	1	7
	p	.0595	—	.1111	.0909	.0574
Total	n	2,400	319	163	96	2,978
	r	83	14	11	6	114
	p	.0346	.0439	.0675	.0625	.0383

Nagasaki

Mothers		Fathers 1	2	3	4-5	Total
1	n	1,401	214	26	14	1,655
	r	70	11	4	—	85
	p	.0500	.0514	.1538	—	.0514
2	n	853	363	32	11	1,259
	r	42	23	2	—	67
	p	.0492	.0634	.0625	—	.0532
3	n	62	25	10	6	103
	r	11	—	—	—	11
	p	.1774	—	—	—	.1068
4-5	n	55	12	4	1	72
	r	3	—	—	—	3
	p	.0545	—	—	—	.0417
Total	n	2,371	614	72	32	3,089
	r	126	34	6	—	166
	p	.0531	.0554	.0833	—	.0537

common following the bombing than before. However, there is no apparent, consistent relation to exposure history. Three of the interaction terms are significant.

In view of the possibility that it might be easier to obtain a history of a "therapeutic" D and C than of an induced abortion, a question concerning D and C was included in the Genetics Long Form. The findings in those terminations where the registration number ended in "0" are shown in Tables 5.15 and 5.16. The figures refer to the proportion of women reporting one or more D and C's, with no reference as to whether the event preceded or followed the bombings. There is a striking difference between the cities, similar in direction to that reported with respect to induced abortions, but much greater in magnitude. However, no difference with respect to exposure groups is apparent, nor are any of the interaction terms significant.

5.6 *The frequency of repeat registrations.* — Depending upon the number of pregnancies which she experienced during the period covered by this study, a woman residing in Hiroshima or Nagasaki could bear one, two, three, four or even more infants whose examination fell within the scope of the program. Table 5.17 summarizes the mean number of infants per mother in relation to the exposure categories of mother and father, by city. There is a tendency towards a higher average number of pregnancies in Nagasaki than in Hiroshima. Furthermore, particularly with respect to mothers' exposure in Nagasaki, there is apparent a tendency for the more heavily irradiated mothers to have borne more children.

5.7 *Parental cooperation.* — If for any reason one group of parents was more cooperative than another in permitting an examination of their newborn child, herein lies a source of bias. Fortunately, parental cooperation was excellent throughout the course of this study. Refusals to permit the ABCC physician to examine a child were infrequent, so infrequent that no analysis of the phenomenon has been carried out. It should in this connection be pointed out that such refusals could influence only the data on malformation, since sex, birthweight, and the occurrence of stillbirth were reported by the midwife, and since, further, it would be impractical to attempt to conceal a neonatal death.

In Japan as in the United States, the birth of a malformed child tends to stigmatize the parents in the lay mind. In a country where marriages are still often arranged to a large extent by the families concerned, and where the *koseki* (census register) is freely consulted when a marriage is under consideration, there are at work social factors encouraging the concealment of congenital abnormality. In this connection, the possibility has to be recognized of subtle influences tending to make one of the groups under study more cooperative than another group. More specifically, one had to recognize the possibility that some of the "exposed" group, learning from the newspapers that congenital abnormalities were a possible aftermath of exposure to an atomic bomb, would cooperate more completely than the controls, and thus introduce a bias.

The chief opportunity for concealing major congenital defect lay in children who were stillborn or died shortly following birth, and whose bodies were disposed of without being seen by an ABCC physician. As mentioned earlier (Sec. 2.5), an attempt was made to obtain autopsies on as many children who were stillborn or died during the neonatal period as possible. In 1950 and 1951, when the autopsy program was in full stride in Hiroshima, autopsies were obtained on approximately 50 per cent of all children who were stillborn or died during the neonatal period. Another 15 per cent of all children who were stillborn or died during the neonatal

TABLE 5.12 CHI-SQUARE ANALYSIS OF THE FREQUENCY OF POSITIVE SEROLOGIES BY CITY AND PARENTAL EXPOSURE
(Unrelated parents)

Source	DF	χ^2	P
Total	31	60.330	.001–.01
Interactions, first order			
CM	3	6.645	.05–.10
CF	3	3.522	.30–.50
MF	9	12.876	.10–.20
Main effects			
Cities (C)	1	8.231	.001–.01
Fathers (F)[a]			
Hiroshima	3	6.469	.05–.10
Nagasaki	3	3.106	.30–.50
Sum	6	9.575	.10–.20
Mothers (M)[a]			
Hiroshima	3	2.136	.50–.70
Nagasaki	3	6.099	.10–.20
Sum	6	8.235	.20–.30

[a] Adjusted for cities.

TABLE 5.13 FREQUENCY OF MOTHERS REPORTING ONE OR MORE INDUCED ABORTIONS BY PARENTAL EXPOSURE, CITY, AND TIME
(Unrelated parents)

Hiroshima — before Aug. 6, 1945

		Fathers				
		1	2	3	4–5	Total
Mothers	1 n	4,738	310	115	79	5,242
	r	207	16	4	4	231
	p	.0437	.0516	.0348	.0506	.0441
	2 n	1,037	902	137	78	2,154
	r	42	28	6	1	77
	p	.0405	.0310	.0438	.0128	.0357
	3 n	437	165	281	83	966
	r	19	3	6	1	29
	p	.0435	.0182	.0214	.0120	.0300
	4–5 n	186	73	62	66	387
	r	2	7	4	—	13
	p	.0108	.0959	.0645	—	.0336
	Total n	6,398	1,450	595	306	8,749
	r	270	54	20	6	350
	p	.0422	.0372	.0336	.0196	.0400

Hiroshima — after Aug. 6, 1945

		Fathers				
		1	2	3	4–5	Total
Mothers	1 n	10,599	799	426	223	12,047
	r	581	50	20	19	670
	p	.0548	.0626	.0469	.0852	.0556
	2 n	2,860	402	128	85	3,475
	r	187	18	9	4	218
	p	.0654	.0448	.0703	.0471	.0627
	3 n	1,185	127	99	42	1,453
	r	77	8	7	—	92
	p	.0650	.0630	.0707	—	.0633
	4–5 n	599	51	39	19	708
	r	45	4	7	—	56
	p	.0751	.0784	.1795	—	.0791
	Total n	15,243	1,379	692	369	17,683
	r	890	80	43	23	1,036
	p	.0584	.0580	.0621	.0623	.0586

THE CHILDREN OF ATOMIC BOMB SURVIVORS

TABLE 5.13—*Continued*

Nagasaki — before Aug. 9, 1945

			Fathers				
			1	2	3	4–5	Total
Mothers	1	n	5,616	706	65	34	6,421
		r	121	17	—	—	138
		p	.0215	.0241	—	—	.0215
	2	n	1,918	2,322	128	80	4,448
		r	27	27	3	1	58
		p	.0141	.0116	.0234	.0125	.0130
	3	n	171	120	58	29	378
		r	—	1	—	1	2
		p	—	.0083	—	.0345	.0053
	4–5	n	87	48	21	10	166
		r	—	—	—	1	1
		p	—	—	—	.1000	.0060
	Total	n	7,792	3,196	272	153	11,413
		r	148	45	3	3	199
		p	.0190	.0141	.0110	.0196	.0174

Nagasaki — after Aug. 9, 1945

			Fathers				
			1	2	3	4–5	Total
Mothers	1	n	7,953	1,073	129	85	9,240
		r	182	23	2	—	207
		p	.0229	.0214	.0155	—	.0224
	2	n	4,692	1,028	90	55	5,865
		r	98	20	3	—	121
		p	.0209	.0195	.0333	—	.0206
	3	n	361	120	25	3	509
		r	4	4	—	—	8
		p	.0111	.0333	—	—	.0157
	4–5	n	291	56	13	8	368
		r	5	1	—	—	6
		p	.0172	.0179	—	—	.0163
	Total	n	13,297	2,277	257	151	15,982
		r	289	48	5	—	342
		p	.0217	.0211	.0195	—	.0214

TABLE 5.14 CHI-SQUARE ANALYSIS OF THE FREQUENCY OF MOTHERS REPORTING ONE OR MORE INDUCED ABORTIONS BY PARENTAL EXPOSURE, CITY, AND TIME (Unrelated parents)

Source	DF	χ^2	P
Total	63	574.384	< .001
Interactions, first order			
CT	1	3.057	.05 –.10
TM	3	16.265	.001–.01
TF	3	3.725	.20 –.30
CM	3	10.087	.01 –.02
CF	3	2.616	.30 –.50
MF	9	30.308	< .001
Main effects			
Cities (C)			
Before bombing	1	95.230	< .001
After bombing	1	270.261	< .001
Sum	2	365.491	< .001
Time (T)			
Hiroshima (H)	1	40.677	< .001
Nagasaki (N)	1	5.401	.02 –.05
Sum	2	46.078	< .001
Mothers (M)			
(H) Before bombing	3	6.191	.10 –.20
After bombing	3	9.001	.02 –.05
(N) Before bombing	3	15.700	.001–.01
After bombing	3	0.459	.90 –.95
Fathers (F)			
(H) Before bombing	3	5.039	.10 –.20
After bombing	3	3.127	.30 –.50
(N) Before bombing	3	3.898	.20 –.30
After bombing	3	0.851	.80 –.90
Sum	12	12.915	30. –.50

TABLE 5.15 FREQUENCY OF DILATATION AND CURETTAGE BY PARENTAL EXPOSURE AND CITY: ZERO TERMINATIONS

Hiroshima

Mothers			Fathers 1	2	3	4–5	Total
1	{	n	1,770	143	65	46	2,024
		r	266	19	10	5	300
		p	.1503	.1329	.1538	.1087	.1482
2	{	n	569	171	47	28	815
		r	75	16	5	6	102
		p	.1318	.0936	.1064	.2143	.1252
3	{	n	218	26	56	19	319
		r	34	1	11	1	47
		p	.1560	.0385	.1964	.0526	.1473
4–5	{	n	103	21	10	11	145
		r	17	4	2	1	24
		p	.1650	.1905	.2000	.0909	.1655
Total	{	n	2,660	361	178	104	3,303
		r	392	40	28	13	473
		p	.1474	.1108	.1573	.1250	.1432

Nagasaki

Mothers			Fathers 1	2	3	4–5	Total
1	{	n	1,459	227	27	14	1,727
		r	25	3	1	—	29
		p	.0171	.0132	.0370	—	.0168
2	{	n	894	387	33	14	1,328
		r	12	12	1	1	26
		p	.0134	.0310	.0303	.0714	.0196
3	{	n	61	29	10	6	106
		r	—	1	1	—	2
		p	—	.0345	.1000	—	.0189
4–5	{	n	56	12	4	3	75
		r	—	—	—	—	—
		p	—	—	—	—	—
Total	{	n	2,470	655	74	37	3,236
		r	37	16	3	1	57
		p	.0150	.0244	.0405	.0270	.0176

TABLE 5.16 CHI-SQUARE ANALYSIS OF THE FREQUENCY OF DILATATION AND CURETTAGE BY PARENTAL EXPOSURE AND CITY: ZERO TERMINATIONS (Unrelated parents)

Source	DF	χ^2	P
Total	31	372.060	<.001
Interactions, first order			
CM	3	3.099	.30–.50
CF	3	7.426	.05–.10
MF	9	10.799	.20–.30
Main effects			
Cities (C)	1	346.140	<.001
Mothers (M)[a]			
Hiroshima	3	3.212	.30–.50
Nagasaki	3	1.718	.50–.70
Sum	6	4.930	.50–.70
Fathers (F)[a]			
Hiroshima	3	4.034	.20–.30
Nagasaki	3	5.185	.10–.20
Sum	6	9.219	.10–.20

[a] Adjusted for cities.

TABLE 5.17 MEAN NUMBER OF REGISTERED PREGNANCIES PER MOTHER BY PARENTAL EXPOSURE AND CITY

Hiroshima

		Fathers				
		1	2	3	4–5	Total
Mothers	1 n	13,577	1,155	490	301	15,523
	\bar{x}	1.288	1.311	1.237	1.326	1.288
	2 n	3,964	1,472	308	189	5,933
	\bar{x}	1.365	1.274	1.263	1.312	1.335
	3 n	1,653	318	420	121	2,512
	\bar{x}	1.345	1.346	1.260	1.298	1.328
	4–5 n	821	156	82	99	1,158
	\bar{x}	1.378	1.333	1.341	1.222	1.356
	Total n	20,015	3,101	1,300	710	25,126
	\bar{x}	1.311	1.298	1.257	1.303	1.307

Nagasaki

		Fathers				
		1	2	3	4–5	Total
Mothers	1 n	10,581	1,582	166	97	12,426
	\bar{x}	1.398	1.396	1.476	1.454	1.400
	2 n	6,246	2,971	188	122	9,527
	\bar{x}	1.514	1.410	1.484	1.459	1.481
	3 n	476	196	75	20	767
	\bar{x}	1.632	1.449	1.360	1.750	1.532
	4–5 n	345	81	24	21	471
	\bar{x}	1.632	1.469	1.458	1.333	1.582
	Total n	17,648	4,830	453	260	23,191
	\bar{x}	1.449	1.408	1.459	1.469	1.441

period were seen by a physician in the employ of the ABCC but did not come to autopsy. It follows that opportunities for the concealment of defect existed with respect to only 35 per cent of all stillbirths and neonatal deaths. In 1952 and 1953 the efficiency of the autopsy program increased, to the point where some 60–70 per cent of the possible material was being autopsied, with a corresponding decrease in opportunities for concealment. Tables 13.1 through 13.4 analyze the material coming to autopsy with reference to its randomness. It can be shown that there is no detectable bias as regards the exposure history of the parents. Since this autopsy material includes the majority of the stillbirths and neonatal deaths, it seems unlikely that there exists the possibility of a serious bias as regards parental exposure among the stillbirths and neonatal deaths who were not seen. If, for instance, unirradiated mothers of stillborn (malformed) children were especially prone to dispose of these infants without their coming to the attention of the ABCC, then, since approximately 10 per cent of stillborn children are malformed, this should depress the representation of unexposed mothers in the parents of the autopsy material; such a disproportion was not observed. On the other hand, the data are of course not sufficiently extensive to exclude small biases in this direction. In this connection, however, it must be borne in mind that all of the infants falling in this 25 per cent were seen by a Japanese midwife or obstetrician, which provides a partial safeguard against the concealment of defect, even though there were occasionally encountered striking shortcomings in midwife reporting.

5.8 *Late sequelae of exposure to the bombs.* — Reference has already been made (Sec. 4.1) to the occurrence among the parents of the infants under study of certain late sequelae of exposure to the atomic bombs. The best documented of these late effects are cataracts (Cogan, Martin, and Kimura, 1949; Sinsky, 1955) and leukemia (Folley, Borges, and Yamawaki, 1952; Lange, Moloney, and Yamawaki, 1954; Moloney and Lange, 1954; Moloney and Kastenbaum, 1955). Refractory anemia may also be a delayed manifestation of radiation injury (Lange, Wright, Tomonaga, Kurasaki, Matsuoke, and Matsunaga, 1955). These events occur with a frequency which cannot be ignored in a study of this type. Thus, the findings with respect to leukemia are given in Table 5.18. To date, the over-all frequency of this disease in individuals who at one time displayed one or more of the three radiation symptoms, epilation, petechiae, or gingivitis, is 0.5 per cent.

While the occurrence of cataracts would not be expected to influence pregnancy outcome, there is little doubt that a disease with the systemic manifestations of leukemia has profound effects. Whether in addition to leukemia there are other serious sequelae which have so far escaped detection cannot be said. Fillmore (1952) was not able to detect any significant sequelae in a general medical examination of 78 persons who had received sufficient radiation at the time of the bombings to develop, later, radiation cataracts. On the other hand, Lorenz and his collaborators (1954) have demonstrated a shortened life span in a variety of animals exposed to chronic irradiation, associated with the development of such conditions as pancytopenia and lymphoid, pulmonary, ovarian, and mammary tumors. Whether there are comparable effects in man is not yet known.

Whereas techniques exist for coping with the age-parity and consanguinity differences between the groups of parents with whom we are concerned, it is much more difficult in any plan of analysis to make allowance for the somatic effects of irradiation which might influence the outcome of a pregnancy. These somatic effects should be exerted largely through the mother. Thus, "effects" apparently consequent upon maternal radiation which are not confirmed by a corresponding analysis with regard to paternal radiation history must be viewed with reservations.

5.9 *The changing proportion of control and irradiated from year to year.* — Table 2.1 presented a summary of the number and proportion of registrations with at least one parent falling into categories 4 or 5 for the years 1948 through 1953. There was apparent a recent decrease in both the absolute and the relative representation of the more heavily irradiated. If there were any marked tendency for the level of diagnostic accuracy or parental cooperation with respect to congenital malformation to change during the course of the study, or if because of post-war improvements in medical and economic levels the stillbirth or the neonatal death rate fell appreciably or the birthweight increased, then here again are factors capable of introducing a source of spurious conclusions. Tables 5.19 and 5.20 summarize our annual

TABLE 5.18 INCIDENCE OF LEUKEMIA IN THE HIROSHIMA SURVIVORS OF THE ATOMIC BOMBING AS RELATED TO DISTANCE FROM THE HYPOCENTER AND THE PRESENCE OF SEVERE RADIATION COMPLAINTS
(After Moloney and Kastenbaum, 1955)

Distance from hypocenter (meters)	Population[a] SRC[b]	NRC[c]	Total
0– 999	750	450	1,200
1,000–1,499	2,250	8,250	10,500
1,500–1,999	1,750	16,950	18,700
2,000–2,499	950	16,250	17,200
2,500 and over	850	49,650	50,500
Total	6,550	91,550	98,100

Cases of leukemia			Incidence		
SRC	NRC	Total	SRC	NRC	Total
14	1	15	1: 53	1: 450	1: 80
15	9	24	1:150	1: 917	1: 438
3	2	5	1:583	1: 8,475	1: 3,740
1	1	2	1:950	1:16,250	1: 8,600
0	4	4	—	1:12,412	1:12,625
33	17	50	1:198	1: 5,385	1: 1,962

[a] Population estimated and rounded off to the nearest 50 persons. These population figures were based on the Commission's 1949 radiation census and the Japanese national census (1950). Numbers of survivors with severe radiation complaints were estimated from an analysis of the pregnancy registration data.
[b] SRC: severe radiation complaints (heavily irradiated).
[c] NRC: no radiation complaints (lightly irradiated).

TABLE 5.19 THE FREQUENCY OF MALFORMATIONS BY YEAR AMONG THE OFFSPRING OF PARENTS NEITHER OF WHOM WAS EXPOSED TO THE ATOMIC BOMBS

Hiroshima

	1948	1949	1950	1951	1952	Total
Total births	1,756	4,005	3,602	3,324	3,084	15,771
Malformations	14	34	34	36	35	153
Percentage	0.80	0.85	0.94	1.08	1.13	0.97

$\chi^2 = 2.496$ DF = 4 $0.70 > P > 0.50$

Nagasaki

	1948–49	1950	1951	1952	Total
Total births	3,934	3,243	3,189	3,123	13,489
Malformations	31	36	45	32	144
Percentage	0.79	1.11	1.41	1.02	1.07

$\chi^2 = 6.584$ DF = 3 $0.10 > P > 0.05$

THE CHILDREN OF ATOMIC BOMB SURVIVORS

The Comparability of Irradiation Subclasses

TABLE 5.20 THE FREQUENCY OF STILLBIRTHS BY YEAR AMONG THE OFFSPRING OF PARENTS NEITHER OF WHOM WAS EXPOSED TO THE ATOMIC BOMBS

Hiroshima

	1948	1949	1950	1951	1952	Total
Total births	1,742	3,971	3,564	3,287	3,037	15,601
Stillbirths	51	87	64	58	61	321
Percentage	2.93	2.19	1.80	1.76	2.01	2.06

$\chi^2 = 9.544$ \quad DF $= 4$ \quad $0.05 > P > 0.02$

Nagasaki

	1948–49	1950	1951	1952	Total
Total births	3,902	3,200	3,134	3,084	13,320
Stillbirths	63	61	49	67	240
Percentage	1.61	1.91	1.56	2.17	1.80

$\chi^2 = 4.372$ \quad DF $= 3$ \quad $0.30 > P > 0.20$

figures for the frequency of gross malformations and stillbirths for the first five years of the study among the offspring of parents neither of whom was exposed to the atomic bombs. The frequency of malformations and stillbirths appears to remain sufficiently constant from year to year, that time trends should not complicate the analysis of these two indicators. On the other hand, we shall in a subsequent section (Sec. 10.4) present evidence that time trends may be of importance in the birthweight analysis.

5.10 *The background of group 1 individuals.* — There remains one final problem for discussion. There are important differences in background between the parents in radiation category 1 and those in categories 2 through 5. The latter have all been urban dwellers since 1945 or before. The former parents, although in part composed of residents of Hiroshima and Nagasaki who were away at the time of the bombings, and also of urbanites who have come to Hiroshima and Nagasaki from the other cities of Japan, in addition include a high proportion of emigrants from rural areas and repatriates from Manchuria, Korea, and Formosa. In addition, among the men there are many with extensive overseas military service. Although there are well-known urban-rural differences with respect to the indicators of possible genetic effect here considered, it is by no means clear to what extent these are apparent and to what extent real, and if the latter, whether they are socio-economic or biological in origin. Finally, if biological, there are no data as to how long they persist after urbanization. In the face of this situation, one can only conclude that the significance of any finding based solely on a difference between category 1 and categories 2 through 5 collectively must be viewed with reservations.

5.11 *Summary.* — Of the differences between the individuals in the various exposure subcategories which have come to light in the foregoing analysis, some appear to be inconsequential, of such a nature that they can be ignored with safety. Others of the differences appear to represent sources of potentially significant bias. These latter differences fall into a natural dichotomy, depending on whether or not allowance can be made for them in an analysis. Thus, one way or another, differences in the frequency of consanguinity, age, and parity can be circumvented. This is not so either with respect to the occurrence of late sequelae of the bombing among the more heavily irradiated, the possible implications of the progressive decline during the six years covered by this study in the proportion of category 4's and 5's among the registrants, or the differences in background among persons falling into category 1 as opposed to categories 2 through 5. In the following chapter, the steps taken to meet this situation will be described.

Chapter VI

STATISTICAL METHODS

THE problems posed when one attempts to employ survey data in an analytical fashion are legion. As a consequence, it is doubtful whether, given a body of survey data, any two competent statisticians would evolve essentially the same approach. Accordingly, it seems important that an attempt be made to establish the line of reasoning which led to the final form which the analysis of the Japanese data assumed. In this chapter, therefore, we shall set out in some detail the analytical plan, and consider briefly the tests of significance to be employed.

6.1 *The problem and the general plan.* — It has been previously stated that the purposes of this study were to answer the questions, "Is there a difference between the progeny of irradiated and non-irradiated parents?", and "If a difference exists, how is it to be explained?" The latter question is, of course, outside the purview of statistics, but the former may be paraphrased in terms of the general statistical problem posed by this study. We may state the problem as follows: What is the significance of the association of parental exposure with those variables indicative of genetic damage due to irradiation? A meaningful answer to this question requires that no differences with respect to any of the indicators exist between the exposure groups being compared save those differences which have been taken into account in the course of the analysis or which arise from exposure experience itself. Available to us are the following two collections of data:

1. Observations on some 75,000 pregnancies terminating sometime after 20 weeks of gestation. These observations are distributed over two cities and 25 parental exposure combinations.
2. Clinical and anthropometric examinations at 9 months of age of some 21,000 infants randomly selected from pregnancies comprising (1). These observations are also distributed over two cities and 25 parental exposure combinations.

The fact that observations have been obtained in two cities permits us to view these data as constituting two approximate replications of one basic experiment.

While the basic problem is readily formulated statistically, its solution is complicated by two factors, namely, extraneous (concomitant) variation and differing numbers of observations in the various exposure cells. Before we consider the impact of these factors on the form of the analysis, let us examine the indicators of radiation damage, that is, the measurements by which we shall attempt to determine whether radiation has or has not resulted in a measurable effect upon the outcome of pregnancy terminations.

6.2 *Indicators of radiation damage and the problem of non-overlapping measurements.* — Among the numerous measurements or attributes by which a newborn infant or a 9-months-old child may be classified there exist at least six which *a priori* may be expected to reflect genetic changes due to irradiation. These six indicators of genetic damage are (1) the sex ratio, (2) birthweights, (3) measurements of bodily development, and the frequencies of (4) stillbirths, (5) neonatal deaths, and (6) gross malformations. As has already been made clear, none of these measurements is a unique yardstick of radiation damage; this is an inherent difficulty in the problem. Moreover, it is apparent that these indicators are not all mutually independent measurements of radiation damage. Some are correlated, and many would measure, to some degree, the same genetic damage.

Extrapolation from experiments involving the irradiation of laboratory animals at the relatively low levels obtaining in Hiroshima and Nagasaki suggested that the effects appearing in the human populations in question would undoubtedly be small, small enough that such effects would be demonstrable only with a very large sample (review of literature in Chapter XV). This in

conjunction with the lack of unique indices of radiation damage led us to believe that it was advisable to attempt to develop a method whereby information gained with respect to the various indicators of radiation damage would be additive, e.g., a method permitting combining information from tests of significance through the chi-square transformation of probabilities. This required that the tests of significance performed on various segments of the data be independent of one another. It was clear, however, that insofar as children who were grossly malformed were apt to be stillborn, or stillborn infants were apt to be small, tests on malformation and stillbirths, or stillbirths and birthweight, would not be wholly independent measures. We chose to remove effects for example, amounted to less than 3 per cent of the total observations.

Presumably the device just described would not be necessary were we able to combine non-independent tests of significance. However, as Wallis (1951) has pointed out, to combine dependent tests of significance requires that we be able to specify the n-dimensional surface formed by groups of probabilities of n events which are not all necessarily equally probable. To specify this surface, we must know the exact kind of dependence which is present, and we obviously do not know this.

6.3 *Concomitant variation.* — As stated earlier, the analysis of these data is complicated not only by extraneous variation, but also by disproportion in the number of observations

FIGURE 6.1 — A schematic representation of the method of sorting the data to obtain non-overlapping indicators.

such as those just indicated by a pyramidal handling of the data. Under the scheme employed, the first attribute to be measured was the sex ratio. This was followed by the frequency of malformation. In this and all subsequent partitions sex was taken into account. All grossly malformed infants were then excluded, and the frequency of stillbirths obtained. The stillborn infants were discarded in turn, and birthweights distributed on the remainder. Thus the frequencies of stillbirths in the various exposure categories were based only on those infants with no clinically obvious malformation. Similarly, birthweights were based only on those liveborn infants without clinically recognizable gross abnormalities. The order of the testing is indicated in Figure 6.1. This approach must lead to the loss of some data. However, the advantages to be gained by having measurements which were essentially non-overlapping seemed to outweigh the loss in data, particularly since the "loss" with respect to birthweight, within the exposure cells. The problems posed by the latter we shall treat of presently; for the moment let us concern ourselves only with extraneous variation.

In Chapter V attention has been called to a rather large number of variables in which the exposure subpopulations differ significantly within or between cities. Thus, the reader has been apprised of differences in (1) the frequency of consanguineous marriages, (2) mean maternal age at birth of a registered infant, (3) mean parity, (4) the frequency of "D and C," (5) the frequency of a positive serology, (6) the frequency of induced abortions, (7) the frequency with which repeat registrations occurred, and (8), possibly, the economic status of the parents. A number of these concomitant variables are known to influence the outcome of pregnancy terminations, for example, the frequency of malformation increases with increasing age, and birthweight increases with parity. We might rightly ask, however, whether the

effects of these variables are of sufficient consequence to lead to misinterpretation of the radiation effects, or to the obfuscating of a real effect if one should exist. It is obviously impossible to answer this question categorically with regard to all of the concomitant variables. It is possible, however, to make a general appraisal of the effect of these variables, and then to make a value judgment with regard to whether or not the effect of a given concomitant is of a magnitude large enough to warrant consideration in the analysis of a given indicator. The methods of appraisal are not fully rigorous but are descriptive and convenient to use. Since the use of these methods is not widespread and since concomitant variation is a major problem in the analysis of the data to be presented in subsequent chapters, we shall present the methods which we have employed in some detail.

In the case of a continuously distributed indicator we shall employ the ratio of two residual mean squares obtained under different assumptions regarding the population sampled, to estimate the amount of variation in the indicator ascribable to a concomitant variable and to variations in the relation between the indicator and a concomitant variable. The basis for using the ratio in this way was first pointed out to us by Dr. H. L. Lucas, Jr. whose argument, which is unpublished, we have been kindly permitted to reproduce here. The argument is as follows:

Consider the variable y, which is known or thought to be influenced by the concomitant variable, x. A simple model of this relationship would be

$$y_{ij} = m_i + \beta_i x_{ij} + \epsilon_{ij}$$

where

$i = 1, 2, \ldots, p$, designates the group,
$j = 1, 2, \ldots, n_i$, designates the individual within a group $(\sum_i n_i = n)$,
$m_i =$ the intercept of the i^{th} group,
$\beta_i =$ the regression coefficient of the i^{th} group,
and $\epsilon_{ij} =$ random, independent errors with variance σ^2.

Within this model, we may work out the expectations of the following quantities:

(1) The mean square for y after correction for the group means but not for regression on x. This quantity would be

$$s_o^2 = \frac{\sum_i \sum_j (y_{ij} - \bar{y}_i)^2}{n - p}$$

(2) The mean square for y after correction for (a) the group means and (b) the common regression on x. This would be computed as

$$s_c^2 = \frac{1}{n-p-1}\left[\sum_i \sum_j (y_{ij}-\bar{y}_i)^2 - \frac{\left[\sum_i \sum_j (y_{ij}-\bar{y}_i)x_{ij}\right]^2}{\sum_i \sum_j (x_{ij}-\bar{x}_i)^2}\right]$$

(3) The mean square for y after correction for (a) the group means and (b) separate regressions on x in each group. This residual mean square we compute as

$$s^2 = \frac{1}{n-2p}\left[\sum_i \sum_j (y_{ij}-\bar{y}_i)^2 - \sum_i \left\{\frac{\left[\sum_j (y_{ij}-\bar{y}_i)x_{ij}\right]^2}{\sum_j (x_{ij}-\bar{x}_i)^2}\right\}\right]$$

If now we let

$$f_o = \sum_i f_i = \sum_i (n_i - 1) = n - p$$

and

$$\bar{\beta} = \frac{\sum_i \beta_i \sum_j (x_{ij}-\bar{x}_i)^2}{\sum_i \sum_j (x_{ij}-\bar{x}_i)^2}$$

We find the expectations for a given set of x_{ij} to be

$$E(s_o^2) = \sigma^2 + \frac{1}{f_o}\sum_i (\beta_i - \bar{\beta})^2 \sum_j (x_{ij}-\bar{x}_i)^2$$
$$+ \frac{1}{f_o}\bar{\beta}^2 \sum_i \sum_j (x_{ij}-\bar{x}_i)^2$$

$$E(s_c^2) = \sigma^2 + \frac{1}{f_o - 1}\sum_i (\beta_i - \bar{\beta})^2 \sum_j (x_{ij}-\bar{x}_i)^2$$

$$E(s^2) = \sigma^2.$$

Now, if we define the second and third terms of $E(s_o^2)$ as σ_d^2 and σ_c^2, we can write

$$E(s_o^2) = \sigma^2 + \sigma_d^2 + \sigma_c^2$$

$$E(s_c^2) = \sigma^2 + \frac{f_o}{f_o - 1}\sigma_d^2$$

$$E(s^2) = \sigma^2.$$

Or, since $f_o/(f_o - 1)$ is essentially unity, we could in the above have written with little error

$$E(s_c^2) = \sigma^2 + \sigma_d^2.$$

An objection could be made, of course, that the meanings of σ_d^2 and σ_c^2 depend upon the sample pattern of the x_{ij}. Let us assume, however, (a) that the sample of size $n = \sum_i n_i$ was drawn from a population of size N which divides into p groups with N_i individuals in a

group, (b) that the within-group variance of x is σ_x^2 in all groups, (c) that the sample is random and large with the restriction (d) that

$$\frac{f_i}{f_o} = \frac{N_i}{N}.$$

In this event, stable meaning can be given to σ_d^2 and σ_c^2. Define

$$\sigma_\beta^2 = \frac{\sum_i (\beta_i - \beta_o)^2 f_i}{f_o} = \frac{\sum_i (\beta_i - \beta_o)^2 N_i}{N}$$

and

$$\beta_o = \frac{\sum_i \beta_i f_i}{f_o} = \frac{\sum_i \beta_i N_i}{N}$$

Clearly under the assumptions and restrictions just given, $\bar{\beta} = \beta_o$ to a close approximation and

$$E\left[\frac{\sum_i \sum_j (x_{ij} - \bar{x}_i)^2}{f_o}\right] = \sigma_x^2$$

$$E\left[\frac{\sum_j (x_{ij} - \bar{x}_i)^2}{f_i}\right] = \sigma_x^2.$$

Then we see that to a close approximation

$$\sigma_d^2 = \sigma_\beta^2 \sigma_x^2$$
$$\sigma_c^2 = \beta_o^2 \sigma_x^2$$

These are definitions in terms of population values. In practice, of course, instead of large random samples, arbitrarily chosen samples of any size may be used that satisfy the restrictions

$$\frac{f_i}{f_o} = \frac{N_i}{N_o}$$

and

$$\frac{\sum_j (x_{ij} - \bar{x}_i)^2}{f_i} = \frac{\sum_i \sum_j (x_{ij} - \bar{x}_i)^2}{f_o} = \sigma_x^2$$

to a close approximation.

To estimate, now, the percentage of variation in y ascribable to the common relation to x, we form the ratio

$$L_1 = \frac{s_o^2}{s_c^2}$$

The percentage ascribable to between-group variation in the relation to x is obtained by computing

$$L_2 = \frac{s_c^2}{s^2}.$$

Now

$$\frac{100(L_1 - 1)}{L_1} = \text{per cent of the variation remaining after correction for group means that is ascribable to the common regression, and}$$

$$\frac{100(L_2 - 1)}{L_2} = \text{per cent of the variation remaining after correction for the group means and common regression that is ascribable to the different regressions.}$$

When L_1 and L_2 are near unity, as in our data, $100(L_1 - 1)$ and $100(L_2 - 1)$ differ but negligibly from $100(L_1 - 1)/L_1$ and $100(L_2 - 1)/L_2$, and we shall employ the former. If L_1 and L_2 differ materially from unity, then $100(L_1 - 1)/L_1$ and $100(L_2 - 1)/L_2$ are appropriate when one wishes to speak of the per cent of the total variation due to a specified source. Even under the latter circumstance, however, $100(L_1 - 1)$ and $100(L_2 - 1)$ would be appropriate if one should wish to speak of the per cent increase in error variance which would result from a failure to remove a given variance source.

It seems appropriate at this point to comment briefly on the simplifying assumptions which we have used in the foregoing presentation. In the main these assumptions are not particularly restrictive when one notes that the purpose of this procedure is not a precise estimate of the variation contributed by a particular variance source, but rather to determine the order of magnitude of the contribution of this source.

In the case of a discrete indicator, we shall employ a method devised by Krooth (1955) for the analysis of the "importance" of an effect of maternal age on the presence or absence of some character among the mother's offspring. A slight modification of Krooth's method has been necessary to permit estimating, independently, the "importance" of parity in addition to maternal age. This modification, which merely involves holding one concomitant constant while measuring the other, will be apparent from a study of the tables dealing with maternal age and parity in Chapters VIII, IX, and XI.

Generally, problems posed by concomitant variation are met by one or another of the following three techniques: balanced sampling, covariance analysis, or the addition of another way of classification to the analysis wherein this

classification corresponds to intervals in the distribution of the concomitant variable. We will have occasion in analyzing these data to employ all of these alternatives save the technique of balanced sampling. The latter was not employed because in those instances where it was applicable, exact balancing led to a very large loss of data, and balancing in terms of intervals in the distribution of the concomitant variable does not generally relieve one of the responsibility of a covariance analysis. We shall now turn to a brief consideration of the courses of action adopted in each of the concomitant variables.

(1) *Consanguinity.* — Among these data over 90 per cent of the observations represent observations on pregnancies occurring to unrelated parents. Thus, if warranted, all pregnancies occurring to consanguineous unions could be excluded without an exorbitant loss of data. The justification for and the decision to exclude these terminations rests primarily on two factors. Firstly, Schull (in manuscript) and Morton (in manuscript) have shown fairly general, if not large, effects of consanguinity on pregnancy outcome as here measured. In the main, these effects consist of an increase in frequency of malformation with increasing consanguinity, an increase in infant mortality with increasing consanguinity, and a decrease in birthweight. Secondly, the distribution of consanguineous marriages by parental exposure is such as to introduce a bias (tending to minimize exposure differences if such exist).

(2) *Maternal age and parity.* — Adjustment for one or both of these variables has been undertaken for all indicators save the anthropometric measurements obtained at 9 months of age. For the analysis of the malformation data, stillbirth data, and infant mortality data, compensation for these variables took the form of adding to the analysis another level of classification. For the birthweight data, compensation for these variables took the form of an analysis of covariance.

(3) *Economic status.* — In respect to only one variable, birthweight, has an attempt been made to determine the effect of economic status. This stems from three considerations. Firstly, it was possible to obtain information on the economic status on only 10 per cent of the infants, so that adjustment for this variable, in the total data, is impossible. Secondly, economic status is probably of importance only insofar as it is a measure of nutritional standards. Thirdly, the only differences with regard to economic status in these data which are demonstrable are between cities and not parental exposure. It seemed dubious, therefore, whether any form of adjustment which could be entertained would justify the effort.

(4) *Dilatation and curettage.* — The more or less standard procedure in Japan for interrupting a pregnancy or treating a woman following a spontaneous abortion consists of dilating the cervix and curetting the uterus. It seems logical to suppose that repeated performance of this routine can lead to the formation of sufficient scar tissue in the uterus to pose an obstacle to the successful implantation and development of subsequent concepti. On this thesis the frequency of D and C was investigated with the full knowledge that possibly no adequate adjustment could be determined if exposure or city differences obtained. In Chapter V, we have indicated that while city differences obtain, no exposure differences are demonstrable. We are inclined to view the recorded city differences as being largely a reflection of differences between the cities in the enthusiasm with which this question was approached by the examining physicians. No attempt has been made to take into account this variable.

(5) *Positive serology.* — Congenital syphilis markedly affects the frequency of stillbirths, and the neonatal mortality rate. Again, information on maternal serology was limited to but 10 per cent of the total sample. This sample, however, revealed that, within cities, no significant differences exist among exposure groups (see Sec. 5.4). The paucity of data precluded any attempt to take into account this variable in the analysis.

(6) *Induced abortions.* — The liberalization of the Japanese abortion law has resulted in a large-scale interruption of pregnancies. This could obviously pose a serious bias if in some parental exposure categories more interruptions occurred than in others. Moreover, undetected interruptions could play havoc with an attempt to assess the "importance" of parity on a given indicator. The data reveal no consistent exposure differences although the city rates are significantly different (see Sec. 5.5). This variable has been ignored in the analysis of the indicators.

(7) *Repeat registrations.* — We have stated that in the course of this study some mothers registered more than one pregnancy. Since the indicators being used are in part genetically determined, there will be a non-zero sib-sib correlation for many, if not all, of them. Therefore when repeated births involving the same parent or parents are entered into the radiation subclasses, these entries will be non-independent. The tests of significance which follow assume that each entry is independent. Consequently, the standard errors, or their equivalents in more complex tests, may in general be slightly too low, since there is actually rather less information than the test assumes. The data suggest that the frequency of repeat registrations is not the same over all exposure cells (see Sec. 5.6). The bias which this might introduce into the estimates of within-cell variances would probably be small in view of the number of repeat registrations. We have, as a consequence, ignored the fact that repeat registrations occur with different frequencies in the various exposure cells.

In general, in the succeeding chapters we shall have occasion to present analyses of the data where, on the one hand, the effect of concomitant variation is ignored, and, on the other hand, some adjustment has been made for one or more of the above-mentioned concomitants. The primary purpose of presenting the analysis in this extended fashion is to enable the reader to judge the necessity of correcting for concomitant variation, and to ascertain the effects of such corrections on the data.

6.4 *Rejected observations.* — As would be surmised in a study of this kind, there arise instances in which observations of dubious validity occur, and instances where the information relative to a particular variable is incomplete. In Table 6.1 are presented the number of infants who were excluded from the final analysis of the "at-birth" data along with the reasons for exclusion. Several entries in this table require comment. Firstly, it will be noted that the two largest numbers of rejections, in each city, occur by virtue of the fact that the pregnancy was unregistered (and these pregnancies are known to be biased exposure-wise and in the frequency with which abnormal terminations occur), or the pregnancy occurred to parents related as first cousins, first cousins once removed, second cousins, or occasionally closer or slightly more remote relationships. Secondly, a fairly large number of rejections occurred where the infant was described as representing an "induced termination where the birthweight was less than 2,500 grams." The argument for rejecting these infants hinges primarily on the word "induced." In Japan, it is customary to view any termination in which medicinal or mechanical assistance was given to the laboring mother as an induced termination.[1] This definition, while a patently plausible one, is much broader than occurs elsewhere. It was generally agreed that any pregnancy which was terminated before the natural occurrence of labor could not be scored in this study. The reason for this is two-fold, namely

TABLE 6.1 THE NUMBER OF INFANTS REJECTED FROM THE STUDY, TABULATED BY REASON FOR REJECTION

	Hiroshima	Nagasaki
Unregistered births	2,372	892
Registered births		
Consanguinity	2,184	2,979
Induced terminations where birthweight less than 2,500 gms.	379	338
Unknown birthweight	274	209
Unknown parity	—	2
Gestation less than 21 weeks, or unknown and infant less than 2,500 gms.	149	44
Unknown sex	4	8
Unknown maternal age	—	—
Unknown distance	177	164
Exposed in one city, now residing in other city	52	152
Total	5,591	4,788

(1) the possibility that such terminations would be non-randomly distributed with respect to parental exposure, and (2) the high probability that an induced termination will result in a stillborn infant, or one dying during the neonatal period and wherein the cause of death is directly or largely attributable to the inducing agent. Our problem, therefore, was that of sorting out of all terminations loosely labelled "induced" those in which there was probably a true induction of labor. The only reasonably reliable standard for which data existed appeared to be birthweight. It seemed highly probable that if an induced termination gave rise to an infant weighing less than 2,500 grams then

[1] This use of the word "induced" did not become known to us until a large number of terminations had been scored as induced in the Japanese sense.

induction was an overt attempt to end the pregnancy rather than an attempt to assist the parturient mother. Lastly, a comment is in order regarding the category "gestation less than 21 weeks or unknown and the infant less than 2,500 grams." It has been mentioned that the rationing system obtaining in these cities presented possibilities for error and that occasionally a pregnant female registered prior to the time when the law permitted. This would present no problem if the pregnancy continued past the time officially designated for registration. If, however, the pregnancy terminated, naturally or artificially, prior to the 21st week, then the registration represented a type of pregnancy not normally coming to our attention. Since such registrations could not be viewed as necessarily representative of pregnancies terminating before 21 weeks of gestation and because of a possible exposure bias, it seemed advisable to reject them. In some instances, mothers would not or could not provide information which would permit an estimation of the length of gestation. Accordingly, if the duration of gestation could not be estimated and the infant was apparently premature (as judged by birthweight), the termination was excluded.

To appraise the effect on the data available for analysis of the most frequent cause of exclusion of *registered* pregnancies, consanguinity, the reader's attention is directed to Table 5.1 wherein the distribution by exposure of the consanguineous unions is given. It will be noted from Table 5.1 that exclusion of the consanquineous marriages is more at the expense of the unexposed and lightly exposed parents than the heavily exposed parents.

The reader will find in Tables 6.2 and 6.3 an accounting, at representative stages in the analysis, of all observations which were rejected from the "at-birth" or "9-months" data. An explanation will be found in the tables for those rejections which have not been accounted for in the previous paragraphs of this section, in Section 6.2, or in Section 6.3(1).

6.5 The analysis of the attribute data. — The analysis of attribute data presents a number of formidable problems not the least of which is the appropriate specification of the hypotheses of interest. More exactly, difficulties arise in the formulation of hypotheses regarding "main effects" and "interactions." Specification of these hypotheses becomes increasingly difficult as the number of ways of classification of the observations increases. The statistical literature outlining tests of significance in multi-way classifications of attribute data is surprisingly scanty, when one ignores that portion of the literature devoted to transformations necessary to fit attribute data into one of the conventional methods of handling measurement data. Until recently, the one and only case to be considered in any detail was the 2 x 2 x 2 system of classification (Bartlett, 1935). A generalization of Bartlett's approach is to be found in Roy and Kastenbaum (1956), on which we shall draw freely. The latter authors succeed in more sharply defining the parallelism between the analysis of variance for continuous data and the analysis of attribute data.

The method of analysis which we shall outline in the succeeding paragraphs is complex. Lest the reader doubt the necessity of so complex an analytical form it is worth pointing out that our problems stem largely from the numerous ways in which these data are partitioned. The approach could certainly be simplified by ignoring some of the ways in which we have elected to partition the data such as, say, city of birth, sex, and the concomitant variation. It is our contention that such an omission is unjustifiable. If one accepts this point of view, then there is no alternative known to us other than a multi-way analysis. In attribute data, this poses problems often more complex than those which arise in the analysis of continuously distributed data. Our approach is essentially one of pooling information from different ways of classification *but only after such pooling can be shown to be justified.* When pooling cannot be justified, alternative statistical procedures, to be explained later, will be adopted. Few, if any, instances in the statistical literature exist wherein attribute data have been employed in the fashion required here. This is a commentary, in part, on the difficulties which arise when multi-way classification of attributes occurs, and on the biological complexity of the indicators of irradiation damage.

Before we discuss some of the particulars in the analysis of the Japanese data, we shall consider in some detail the basic arguments underlying the tests of "main effects" and "interactions." For illustrative purposes, let us examine a simple problem. We shall assume that we are

Statistical Methods

TABLE 6.2 AN ACCOUNTING OF THE NUMBER OF OBSERVATIONS CONSIDERED AT REPRESENTATIVE STAGES IN THE ANALYSIS OF THE "AT-BIRTH" DATA AND THE NUMBER OF REJECTED OBSERVATIONS WITH THE CAUSE OF REJECTION

	Available observations			Rejected observations		
	Hiroshima	Nagasaki	Total	Hiroshima	Nagasaki	Total
Total infants seen	38,421	38,205	76,626	—	—	—
Rejected because the pregnancy was unregistered, parental exposure was unspecifiable, consanguinity or other observations were incomplete (see Table 6.3)	—	—	—	3,478	1,868	5,346
Considered for consanguinity	34,943	36,337	71,280	—	—	—
Rejected consanguinity	—	—	—	2,113	2,920	5,033
Considered for maternal age	32,830	33,417	66,247	—	—	—
Rejected multiple births	—	—	—	365	451	816[a]
Considered for sex ratio	32,465	32,966	65,431	—	—	—
Considered for malformations	32,465	32,966	65,431	—	—	—
Rejected malformations	—	—	—	313	281	594
Rejected congenital heart disease	—	—	—	44	53	97
Total	—	—	—	357	334	691
Considered for stillbirths	32,108	32,632	64,740	—	—	—
Rejected stillbirths	—	—	—	472	482	954
Considered for neonatal deaths	31,636	32,150	63,786	—	—	—
Rejected neonatal deaths	—	—	—	414	480	894
Considered for birthweights	31,222	31,670	62,892	—	—	—

[a] In Hiroshima one set of registered triplets and 181 sets of registered twins occurred; in Nagasaki there were one set of registered triplets and 224 sets of registered twins.

TABLE 6.3 AN ACCOUNTING OF THE NUMBER OF OBSERVATIONS CONSIDERED AT REPRESENTATIVE STAGES IN THE ANALYSIS OF THE "9-MONTHS" DATA AND THE NUMBER OF REJECTED OBSERVATIONS WITH THE CAUSE OF REJECTION

	Available observations			Rejected observations		
	Hiroshima	Nagasaki	Total	Hiroshima	Nagasaki	Total
Total infants on whom there exists some follow-up study	14,768	12,324	27,092	—	—	—
Rejected inadequate exposure history, infant not part of 9-months program, etc.	—	—	—	3,422	1,882	5,304[a]
Total infants considered under the 9-months program	11,346	10,442	21,788	—	—	—
Rejected consanguinity	—	—	—	694	828	1,522
Rejected incomplete measurements	—	—	—	140	308	448
Considered for neonatal death	10,512	9,306	19,818	—	—	—
Rejected neonatal deaths	—	—	—	484	458	942
Considered for malformation	10,028	8,848	18,876	—	—	—
Rejected malformations	—	—	—	183	195	378
Considered for anthropometrics	9,845	8,653	18,498	—	—	—

[a] A word about this total is in order since it may appear to the reader as an inordinate loss of information. This total includes 5,089 infants who were seen at some age other than 9 months (in fact, 8–10 months). Many of these entries represent visits to the newborn malformation verification clinics. The latter infants are, of course, scored under the "at-birth" program. A second major contributor, particularly in Hiroshima, to this total of 5,089 infants stems from the initial indecision as to the "best age" at which to examine the infants. Some of the first infants seen under what was subsequently called the 9-months program were a year and a half old. These children have been rejected here in order to minimize differences between cities and, within cities, between exposure categories in the age at examination. The importance of standardizing, insofar as possible, the age at examination need hardly be labored.

dealing with a variate of the presence-absence variety, say, presence or absence of malformation, classified by sex and city. Our data, then, form an array of eight cells, and we shall denote by n_{ijk} the observed number of individuals and by p_{ijk} the true proportion, under any given hypothesis, of observations in the (ijk)-th cell where $i=1, 2$; $j=1, 2$; and $k=1, 2$, and where i, j, and k are, respectively, the classifications with regard to sex, city, and the variate under study. In such a table, there are only seven comparisons with regard to k which can be made, namely, (1) the influence of sex and city on the variate (ij with k), (2) the influence of sex on the effect of city on the variate (j on k among i), (3) the influence of city on the effect of sex on the variate (i on k among j), (4) the influence of sex on the variate for *each* city (i with k for *each* j), (5) the influence of city on the variate for *each* sex (j with k for *each* i), (6) the influence of sex on the variate when cities are pooled (i with k ignoring j), and (7) the influence of city on the variate when sexes are pooled (j with k ignoring i). Let us now examine each of these comparisons in terms of their meaning and the tests which they afford.

(1) *The influence of sex and city on the variate.* — Under this comparison we are concerned with whether the variate is independent of the sex-city cell, or, alternatively stated, whether the variate is distributed homogeneously over the sex-city cells. The null hypothesis in this instance is

$$H_o: p_{ijk} = p_{ij.}p_{..k}$$

and asserts that the variate has a uniform distribution over the sex-city cells, that is, that the ratio of individuals in category $k=1$ to the individuals in category $k=2$ is the same in each sex-city cell. This hypothesis affords the basis for a test which might be termed "total χ^2," and which is, in effect, an omnibus test of the effect of city *and* sex including, of course, interaction. Non-significance implies no effect of city or sex. The degrees of freedom associated with this test are 3. In general, if $i=1, \ldots r$; $j=1, \ldots s$, and $k=1, 2$, then the degrees of freedom are $(rs-1)$.

(2) and (3) *The influence of sex on the effect of city on the variate, or the influence of city on the effect of sex on the variate.* — The null hypothesis is now

$$H_o: \frac{p_{222}p_{112}}{p_{212}p_{122}} = \frac{p_{221}p_{111}}{p_{211}p_{121}}$$

and asserts that the variate has the same distribution with respect to cities for *all* sexes (or sexes for *all* cities). This hypothesis affords the basis for a test which we shall term a test of "interaction," or more specifically, the "interaction of sex with city (or city with sex)." Non-significance at this level does not imply no effect of sex or city, but merely that the effect of city is the same over all sexes (or sex over all cities). The degrees of freedom associated with this test are 1, or for the general case $(r-1)(s-1)$. This use of the term interaction appears to be due to Bartlett (1935).

(4) and (5) *The influence of sex on the variate for each city, or the influence of city on the variate for each sex.* — Under this comparison we are concerned with whether the variate is homogeneously distributed over sexes for *each* city (or cities for *each* sex). It is important to note that the null hypotheses, which are

$$H_o: p_{i1k} = p_{i1.}p_{.1k}$$

and

$$p_{i2k} = p_{i2.}p_{.2k}$$

or

$$p_{1jk} = p_{1j.}p_{.1k}$$

and

$$p_{2jk} = p_{2j.}p_{.2k},$$

where now

$$\sum_{ik} p_{i1k} = \sum_{ik} p_{i2k} = 1 \text{ (or } \sum_{jk} p_{1jk} = \sum_{jk} p_{2jk} = 1\text{)},$$

do not specify that the variate must have the same distribution over sexes for *all* cities (or cities for *all* sexes). This hypothesis affords the basis for a test of the effect of city (or sex) on the variate for each sex (or city). There can be as many such tests as there are sexes (or cities). The degrees of freedom associated with each test are 1, or in general (r_1-1), (r_2-1), etc., where $r_1 = r_2 = \ldots = r$. The χ^2's associated with these individual tests are additive; the sum, however, confounds "main effects" and "interaction."

(6) and (7) *The variate with sex, or the variate with city.* — Under this comparison we are concerned with whether the variate is distributed homogeneously over *all* sexes neglecting cities (or *all* cities neglecting sexes). The null hypothesis which is

$$H_o: p_{i.k} = p_{i..}p_{..k}$$

or

$$p_{.jk} = p_{.j.}p_{..k}$$

asserts that the variate has the same distribution over the sexes neglecting the cities (or over

cities neglecting sexes). The hypothesis affords a test of the effect of sex (or city) on the variate *assuming* (1) no main effect of city (or sex), and (2) no interaction between sex and city. This we shall term a "main effects" test. The degrees of freedom associated with this test are 1, or in general $(r-1)$ or $(s-1)$.

We shall now turn our attention to the appropriateness of these tests as illustrated by the analyses of the Japanese data. Essentially we are concerned with testing (1) whether sex has an effect on our variate independent of cities, (2) whether city has an effect independent of sex, and (3) whether the effects of sexes and cities can be considered independently. Or, to use the language of the analysis of variance, we are concerned to measure (1) the main effect due to sexes, (2) the main effect due to cities, and (3) the interaction of sex with city. The procedure for testing, which we shall outline as it would occur in the three-way table we have used for illustrative purposes, is quite general and can be extended to an n-way table. The procedure is as follows:

1. Test the hypothesis of "no interaction" of sex and city with the variate.

2. If the hypothesis of "no interaction" is accepted, then

(a) Test the hypothesis of "no main city effect," ignoring sex.

(b) If the hypothesis of "no main city effect" is accepted, then test the hypothesis of "no main sex effect" ignoring city.

(c) If this hypothesis is also accepted, then we may accept the general hypothesis of no main effects and no interaction.

(d) If the hypothesis of "no main city effect" (ignoring sex) is rejected, then a test of the sex effect may be influenced by the city differences. We are free to test the hypothesis of "no main sex effect" ignoring city *only if cities are equally or proportionally represented among the sexes*. If this does not obtain, that is, if the cities are disproportionately represented among the sexes, then any differences in sex when city is ignored are apt to be attributable to the differences between cities. Unless these city differences are taken into account, a test of the effect of sex confounds the effect of city. One possible way of getting around this problem is to consider the hypothesis of "no main sex effect" at each city level. The test of this hypothesis is a χ^2 which is the sum of the χ^2 tests at each city level, with appropriate degrees of freedom (the sum of the individual tests). We shall term this test the "sum test" of sexes. This test will answer the question "Is there a main effect of sex on the variate, assuming no interaction of sex with city but a possible contribution of city?" By this procedure we may pick out the levels of city which contribute most heavily to the total χ^2. We will refer to this test as a test of the sex effect adjusted for cities, the "adjustment" being merely a consideration of the sex effect at all possible levels of cities. In the absence of an interaction, this will be our best test of the sex effect.

(e) If the hypothesis of "no sex effect" ignoring cities is rejected, we follow the same procedure as outlined in (d).

(f) If both hypotheses, namely, "no main city effect" and "no main sex effect," are rejected, we follow the procedure outlined in (d) for both effects. This would yield two sets of tests. It is important to note that the χ^2 and degrees of freedom *are additive within sets but not between sets.*

3. If the hypothesis of "no interaction" is rejected, then the tests of sex ignoring city, and of city ignoring sex may be biased. Accordingly, our procedure will be as follows:

The effect of sex will be evaluated at each level of city, and city at each level of sex. Here, however, the "sum test" obtained by the addition of the two χ^2 tests of sex (one for each city) or the two χ^2 tests of city (one for each sex) is not a meaningful test of the main effect due to sex or city. This stems from the fact that the presence of an interaction reveals a significant inconsistency in the direction of the effect of the ways of classification on the variable. Otherwise stated, the "sum test" as a test of main effect is not meaningful because it confounds interaction.

In Chapter V, and in the chapters to follow, we have adopted the convention of indicating (1) each of the individual tests whenever adjustment is necessary, and (2) the "sum test" only when the "no interaction" hypothesis is accepted.

The adjusted tests in 2(d), (e), and (f) above are somewhat analogous to adjusted tests in the analysis of variance in the sense that

though the degrees of freedom are additive, the χ^2's are not. Thus in (d) we will have

Source	D F
city (unadjusted)	$(s-1)$
sex (adjusted)	$s(r-1)$
Total	$(rs-1)$

and for (e) we will have

Source	D F
sex (unadjusted)	$(r-1)$
city (adjusted)	$r(s-1)$
Total	$(rs-1)$

In the n-way classification, the interaction hypotheses are more numerous. For example, in a four-way table with dimensions h, i, j, and k, we have

"First order interactions"

(1) h on k among i
(2) h on k among j
(3) i on k among j

"Second order interactions"

(1) h on k among i among j.

While the procedure for testing outlined in the previous pages can be so extended that our initial test is a test of "no second (or higher) order interaction," we shall in the analysis to follow assume, in general, that all interactions higher than the first are not significant. The validity of this assumption can, of course, be questioned. In the event that serious ambiguity in the interpretation of "main effects" or "first order interactions" might arise through ignoring the higher order interactions, then the second order interactions will be explored. In view of the great amount of labor involved in the calculation of the first order interactions, involving in this case simultaneous cubic equations with one unknown for each degree of freedom, the Michigan Digital Automatic Computer has been utilized.

The above outlined procedures are, obviously, not the only possible approaches to these data. However, the logical basis for some of the alternative, simpler methods, such as Brandt's factorial chi-square, have not been set out in detail in the statistical literature. Other alternatives which will find favor in some quarters are (1) to transform the attribute data and employ an analysis of variance on the transformed variate (see Eisenhart, 1947, or Rao, 1952), or (2) to attempt a regression form of analysis of the indicator on dose of irradiation. With regard to the latter, we believe this approach is fraught with danger for at least two reasons. Firstly, the estimates of average dose in each of the five categories of parental exposure are most tenuous, and secondly, even if these estimates are reasonably reliable the distribution of doses within a given category of exposure is unknown. In the latter connection, it seems most probable that in many, if not all, exposed cells the median dose will be less than the mean dose (judging from the distance distribution of survivors). Be that as it may, for the data to follow on sex ratio, malformation, stillbirth, and neonatal death, one or more of these alternative methods of analysis was routinely performed. Since these alternatives did not give rise to results differing substantially from those obtained by the method of Roy and Kastenbaum, the results of the alternative analyses will not be presented.

The use of chi-square as a test of significance in the procedure outlined here requires certain assumptions regarding the distribution of χ^2 where

$$\chi^2 = \sum_i \frac{(x_i - m_i)^2}{m_i}$$

and where x_i and m_i are respectively the observed number in a cell and the expected number based on some null hypothesis. Cochran (1952) has discussed these assumptions in considerable detail, and has formulated a number of operating rules regarding the minimum expectation in a cell. One of these rules, on which we shall draw heavily, is concerned with tables with more than 1 degree of freedom and some cells with expectations greater than 5. Cochran asserts that χ^2, without correction for continuity, is a satisfactory approximation in this instance. In instances where the expectation in a cell was less than two, the effect of this cell on the total chi-square was carefully noted. When the total chi-square was significant and due in large measure to a single cell with an unusually small expectation, an alternate scheme of classification was employed to increase the expectation in the various cells.

6.6 The analysis of the measurement data. — In general, in the analysis of the measurement data, that is, the data with respect to birthweight and the anthropometric measurements obtained at 9 months of age, we have had occasion to

employ three common statistical procedures, namely, the analysis of variance, the analysis of covariance, and the analysis of dispersion. As is frequently true when one passes from a theoretical consideration of a test of significance to the application of such a test to a body of data, certain of the assumptions underlying the test cannot be met in the strict sense. It seems appropriate, therefore, that we consider the assumptions underlying the tests here employed, indicating where the data do not or may not satisfy the assumptions, and then to discuss briefly the variations of the basic tests necessary to meet certain problems posed by these data.

Firstly, let us consider the analysis of variance when there exist multiple ways of classification. In the classical test of the significance of the differences in a set of k means associated with a main effect, we make four basic assumptions in order to test the null hypothesis that $m_1 = m_2 = \ldots = m_k$, namely,

1. that the observations in a cell (for all cells) are values of random variables distributed about a true mean which is a fixed constant;
2. that the true cell means are simple additive functions of the corresponding marginal means and the general mean;
3. that the observations are uncorrelated, and have equal variances; and
4. that the observations are jointly distributed in a multivariate normal distribution.

For a detailed consideration of these assumptions the reader is referred to Eisenhart (1947); we shall concern ourselves merely with the validity and importance of these assumptions as they bear on the Japanese data. Assumption (1) needs no comment since it is basic to any statistical analysis, in a sense, and is merely an assertion that we are dealing with random variables. Assumption (4), which to some extent impinges on assumption (1), is probably not strictly satisfied in the Japanese data. In general, it has been found that variables such as height, weight, etc., are non-normal; however, the departure from normality is generally not sufficient to jeopardize seriously the validity of the test. Moreover, the analysis of variance is known to be very insensitive to non-normality (see Box, 1953). From the purely practical standpoint, assumptions (2) and (3) are the most troublesome. Assumption (2), the assumption of additivity, disallows the possibility of interactions. Alternatively stated, if additivity does not prevail then we assert that there are interactions between the ways of classification; however, when additivity does not prevail, we can still obtain a test of the main effects. The principal effect of non-additivity rests in the alteration of the model from which the effects of classification are estimated, and the generalizations of which the new tests will admit. As an illustration, suppose we have a variable, x_{ij}, which we shall assume is normally distributed. Suppose, moreover, that a given observation can be classified with respect to properties A, and properties B. The additivity assumption asserts then that the expected value of x in the $(ij)^{\text{th}}$ cell is

$$E(x_{ij}) = m + A_i + B_j,$$

that is, that the expected value is a linear function of the true general mean, the effect due to A, and the effect due to B. Alternatively, if additivity does not obtain, then the expectation in the (ij) cell is

$$E(x_{ij}) = m + A_i + B_j + (AB)_{ij},$$

that is, the expected value is a function of the general mean, of A, of B, and a function of A and B taken conjointly. In the orthogonal case of the analysis of variance, whichever of these hypotheses obtains, the computation of the sums of squares due to interaction and to main effects remains the same. The difference between the models enters the picture only in the formation of the appropriate variance ratio, and its interpretation. If additivity prevails, then our test of the main effect due to A, say, is the ratio of the mean square due to A to the mean square within cells (error mean square). If additivity does not obtain, then we may make the comparison just stated or we may compare the mean square due to A with the mean square interaction. The former test would permit us to draw inferences with respect to A only over the circumstances which obtain with respect to A in this experiment. The latter ratio would permit us to make broader statements regarding the effect of A. For example, if the effect of mother's exposure was judged by the ratio of the mean square due to mothers to the mean square within cells and if an interaction involving mothers and, say, cities obtained, then we could make statements regarding mother's exposure only with respect to the situation obtaining in Hiroshima and Nagasaki. On the

other hand, if our contrast involved mother's exposure and the interaction of mothers and cities, then our statements with respect to the effect of mother's exposure would apply to other cities subjected to the same exposure conditions experienced by these two cities and where these other cities may be assumed to fulfill the remaining experimental conditions. While both of these comparisons have meaning, in general we shall be concerned with the broadest possible statement regarding mother's effect. It might be noted that the use of the interaction to test main effects is not good when the cell numbers are unequal (either proportionate or disproportionate). For a more complete discussion of this aspect of the analysis of variance the reader is referred to Fisher (1949).

Assumption (3) may be violated because the observations within a cell are correlated, or the variances are unequal, or both. If the variances are unequal and if we are contrasting but two means, we face the classical Fisher-Behrens problem (Fisher, 1939). For our purposes two comments here seem sufficient. Firstly, it is not inconsistent to test the same body of data under the hypothesis that the means are equal and the variances are equal, and under the hypothesis that the means are equal but the variances unequal (the Fisher-Behrens problem). Secondly, the work of Box (1953, 1954 a and b) suggests that the test of the equality of a set of means is not seriously affected if there exists only a moderate inequality of the variances and if the cell numbers are equal ("moderate" envisages an inequality of the variances wherein the larger is three times the smaller). Much larger discrepancies, however, arise if the same moderate inequality exists, and if the cell numbers are markedly unequal. The inequality of the cell entries becomes less important as the differences in the variances diminish. When an inequality in the variances exists in these data, this inequality is small and can be ignored without seriously jeopardizing the inferences which may be drawn from the tests on the means.

Thus far we have considered only tests on the means; needless to say, we shall also be interested in testing the equality of the variances. The assumptions for a valid test of the variances are less numerous. We merely assume that we are dealing with values of a random variable which are normally distributed and uncorrelated. Comparison of the variances of a series of exposure cells has meaning, however, only if all extraneous sources of variation which may be dissimilarly distributed between the exposure cells are removed. To see that this is true will be a matter of prime concern in the succeeding chapters.

The assumptions for a valid test of the equality of variances and means set out in the preceding paragraphs have been phrased in a manner appropriate to the univariate case. By a slight extension, these assumptions are equally valid for the multivariate case wherein we analyze the dispersion of a set of observations. Specifically, in the multivariate case we shall be concerned with testing two hypotheses, namely,

1. The equality of the dispersion matrices of k p-variate normal populations

$$H_o: \Sigma_1 = \Sigma_2 = \ldots = \Sigma_k$$

and where Σ_i is the variance-covariance matrix in the i^{th} class.

2. The equality of k means for each of p variates for k p-variate normal populations with the same covariance matrix.

$$H_o: \xi_1 = \xi_2 = \ldots = \xi_k.$$

The ξ_i are now vectors of means.

The approach to these data which we have outlined in this and the preceding section calls, in essence, for the use of so-called "omnibus" or "portmanteau" tests.[2] Not all readers will subscribe to this since omnibus tests tend to be less sensitive with respect to a particular comparison than a more specialized test. The primary justification for the omnibus test is, in our minds, the fact that such a test does not require the measure of specification of the alternatives to the null hypothesis required by a more specialized statistical tool. It is our opinion that, with the possible exception of the sex ratio, our knowledge with regard to the types of changes which may arise in human beings consequent to parental irradiation is so poorly understood as to make any real attempt to specify direction of change specious. To some our attitude will seem much too conservative, and for those readers we would point out that the more specialized tool is quite useless under the wrong conditions (Pearson, 1936). Furthermore, since the problem of radiation-induced genetic change in human beings may

[2] An omnibus test is generally defined as one that has good discriminating power with regard to a large variety of alternatives to the null hypothesis.

well constitute the most important problem in human biology in our generation, we believe quite strongly that at this stage our approach must be an open-minded one which does not draw too heavily in any particulars upon infra-human data, the more so because of the great gaps which exist at present in comparable observations on laboratory material. The prior specification of a subset of alternatives to the null hypothesis required by a specialized test would imply a greater knowledge of the genetic effects of irradiation on the indicators here studied than we are willing to assume.

6.7 Some further problems. — There remain two more subjects for our consideration regarding the analysis of the measurement data, namely, within-cell heterogeneity and unequal numbers of observations within cells. Firstly, a brief description of what we have elected to term *within-cell heterogeneity*. It is patent that so long as the observations within a cell represent values of a random variable the observations will be heterogeneous in the sense that they will not all be like-values. The heterogeneity with which we are concerned is not of this variety, but rather the heterogeneity which arises if the observations within a given cell are drawn at random from not one but *several* normal parent populations. We are concerned with such heterogeneity since it represents a violation of assumption (3) in Section 6.6. Let us consider what may happen when this circumstance prevails. There are two chief aspects of the problem:

1. The parent populations may differ with respect to the mean, with respect to the variance, or with respect to both the mean and the variance.

2. The parent populations may or may not be represented with the same relative frequency in each of the several exposure classes. If the relative frequencies are identical among the several exposure cells, we shall say the heterogeneity is "uniform."

What now are the consequences of within-cell heterogeneity? Let us tabulate the case:

1. Within-cell heterogeneity with respect to the mean alone.

 (a) *Uniform consequences*

 (1) Inflation of within-cell sum of squares, thus reducing the sensitivity of the test.

 (2) Usually a departure of the within-cell distribution from normality with possible plurimodality.[3]

 (b) *Non-uniform consequences*

 (1) Those listed for the uniform case.

 (2) In the event that the within-cell heterogeneity arises from concomitant variation, having nothing to do with irradiation, a spurious heterogeneity of cell means may be observed, *or* a true heterogeneity of cell means may be concealed.

2. Within-cell heterogeneity due to variances alone.

 (a) *Uniform consequences*

 (1) Departure from normality (persistence of higher cumulants but no plurimodality).

 (2) Inflation of within-cell sums of squares, leading, as before, to a reduction in the sensitivity of tests.

 (b) *Non-uniform consequences*

 (1) Those listed for the uniform case.

 (2) In the event that within-cell heterogeneity arises from concomitant variation having nothing to do with irradiation, a spurious heterogeneity of cell variances may be observed, *or* a true heterogeneity of cell variances may be concealed.

3. Within-cell heterogeneity with respect to both means and variances.

 Consequences

 Any or all of those listed above may prevail.

It is thus clear that within-cell heterogeneity may lead to any or all of the following:

1. Departures from normality.

2. Inflation of the within-cell sums of squares with consequent reduction in the sensitivity of statistical tests.

3. Detection of spurious statistical effect — or concealment of true ones, *provided*

 (a) the within-cell heterogeneity does not reflect an effect of irradiation itself, but

[3] A discussion of the circumstances under which the combination of Gaussian distributions leads to plurimodality can be found in Harris and Smith (1947). These authors consider the case of but two parent distributions.

rather the operation of some purely concomitant factor or factors, *and*

(b) the relative frequencies with which the several parent populations are represented in each cell are not uniform over all cells.

Clearly, the possibility of within-cell heterogeneity has to be explored in all tests, and wherever important concomitant factors are discovered they must at least be shown to exert a uniform effect or else be incorporated into the analysis. To ignore a concomitant variable uniformly distributed among the exposure groups assumes, of course, that the uniformly distributed variable does not interact with a variable which may be non-uniformly distributed, and the inflation of the error sums of squares is negligible. This topic will be discussed again in connection with the analysis of birthweight and anthropometric data.

The analysis of variance (or dispersion) is, as a general rule, computationally simple and interpretively straightforward when the number of observations within a cell is the same for all cells, or when the number within the cell is proportional to the marginal totals. This situation is often referred to as the orthogonal case of the analysis of variance. Not infrequently, the numbers within a cell do not satisfy this stricture of proportionality. When this occurs, the addition theorem for sums of squares fails and the usual computational procedures for the analysis of variance do not yield valid tests of main effects or interactions. There exist, however, a number of techniques which are appropriate to this situation, among them being the method of expected subclass numbers (Snedecor and Cox, 1935; see Snedecor, 1946), the method of weighted means (Yates, 1934), and the method of "fitting constants" (inter alia Wilks, 1938). In our analysis, we shall have frequent occasion to employ the method of fitting constants as described by Wilks, and logical extensions of this method appropriate to the multivariate analysis of dispersion. In the analysis of the anthropometric data we shall employ a procedure devised by Rao (1955) which has the added advantage of permitting one to find the standard errors of differences in the estimated constants. While this technique allows for the fact that the cell numbers are disproportionate it does not create additivity among the tests of main effects or interactions. There is a valid test for any set of main effects which can be used irrespective of the presence of interaction but such tests would confound interaction if present. For a more complete discussion of the problem of unequal cell numbers the reader is referred to Kendall (1946).

In the normal procedure in the analysis of variance for the non-orthogonal case, one would begin by inquiring into the presence or absence of interactions and, frequently, at the same time estimating main effects under the additive assumption. Main effects so estimated provide, as we have mentioned, only approximate tests if an interaction is present. If an interaction is present and a more accurate test is desired, the main effects must be estimated anew from a model in which the interaction is now accounted for. Often, however, approximate tests of the main effects may suffice if the primary concern is to establish heterogeneity between cells and not to inquire exhaustively into the main effects. We have, in Chapter V, frequently settled for approximate tests on the main effects because the presence of an interaction no less than main effects differences reveals heterogeneity between exposure cells.

6.8 *The use of exposed persons as controls.* — We have indicated that in the analysis of the data with respect to the various indices of radiation damage a variety of tests will be presented. Specifically, we have stated that analysis will be presented in which either (1) concomitant variation is ignored, or (2) major sources of concomitant variation are accounted for. To this list we now shall add a third comparison and indicate its purpose.

It has been stated that each parent of a registered infant has been placed into one of five exposure categories depending upon his or her position relative to ground zero, to the amount of shielding between the parent and the explosion, and to the array of symptoms experienced or not experienced following the bombing. When both parents are considered, then a given registered infant can be assigned to one and only one of twenty-five exposure cells. Unfortunately, the numbers of terminations to parents one or both of whom were in exposure categories 4 or 5 are so small as to necessitate pooling of these exposure categories. Accordingly, in the analysis to follow, an infant will have been assigned to one and only one of sixteen exposure cells wherein the appropriate cell was determined by whether the mother was

in category 1, 2, 3, or 4–5, and similarly for the father.

It will be recalled that exposure category 1 includes those individuals who were not present in Hiroshima or Nagasaki at the time of the atomic bombings. In Chapter V we have advanced reasons for doubting whether category 1 parents afford an entirely valid comparison with those parents who experienced some measure of exposure to the bombs. If exposure 1 parents are not an "adequate control," then the only meaningful comparison which can be made to determine the effects of irradiation on our indices of genetic damage would be a comparison involving only those infants where both parents were present in the city at the time of the detonation of the bomb. Moreover, even if one accepts the validity of the comparison utilizing category 1 parents, a real irradiation effect would also lead to differences among the terminations to 2, 3, and 4–5 parents. For these two reasons, in the analysis of the indicators there will be presented two analyses, one wherein the parents one or both of whom are in category 1 are excluded, and one where they are included. It should be pointed out at this juncture that the differences which we can detect by statistical procedures are largely a function of sample number. Accordingly, it may be that differences demonstrable in the 4 x 4 comparison including category 1 parents will not be demonstrable in the 3 x 3 comparison excluding these parents solely because of the curtailment of sample size occasioned by the exclusion of the category 1 parents. The differences among the remaining exposure cells brought about by exposure, while no longer significant, should, of course, persist even following exclusion of the category 1 parents.

6.9 *Presentation of material.* — It would be highly desirable in a problem of this nature to present in detail the tabulations on which the various analyses are based. However, these tabulations are extremely bulky, requiring, even for presentation in a somewhat condensed form, an estimated 1,000 pages. Moreover, because of differences in statistical approach, many investigators might wish for tabulations other than those presented. Under the circumstances, it would seem that the matter of making available the raw material of this study is best met by the following procedure: the investigator who desires to verify some of the calculations presented in the following chapters, or to explore other lines of analysis, can apply to the Division of Biology and Medicine, U.S. Atomic Energy Commission, or the Committee on Atomic Casualties, National Research Council, for a duplicate set of the IBM cards on which this analysis is based. The investigator must be prepared to meet the costs of duplicating the cards and all shipping charges.

Chapter VII

ANALYSIS OF THE SEX RATIO DATA

The characteristic for which the most data are available is the sex of the offspring.

7.1 The trait. — Among the various characteristics of the infants examined in connection with this study, sex is undoubtedly that attribute in whose determination there is the smallest margin of error. Although the sexing of prematurely borne infants presents certain well-recognized problems (cf. Tietze, 1948), by the twentieth week of gestation, the earliest at which pregnant women were permitted to register, there is seldom any room for doubt concerning the sex of a prematurely born infant.[1]

7.2 The genetic argument for radiation-induced changes in the sex ratio. — Lethal mutations, i.e., mutations resulting in the death of the organism sometime prior to reproduction, are an important type of mutation arising in consequence of irradiation of experimental animals. Induced lethal mutations may be dominant or recessive to their normal alleles. Such mutants are of concern to us in this chapter only insofar as they affect the sex ratio. The heterogametic nature of the human male results in (a) a differential distribution to his offspring of his sex chromosomes, and (b) a difference in the dominance relationships between the two sexes. Thus lethal mutations borne on the X- or Y-chromosomes afford ample opportunity for an alteration of the sex ratio. Accordingly, our discussion shall concern itself with this class of mutants, i.e., the X- and Y-borne lethals. Let us examine separately the possible effects on the sex ratio of maternal irradiation, paternal irradiation, and conjoint parental exposure.

Maternal irradiation would be tantamount to the irradiation of X-chromosomes alone. Since there is no differential distribution of the maternally derived X-chromosomes among the offspring, induced sex-linked dominant lethals would be expected to result in comparable reductions in the frequency of male and female births. The same would hold true for all partially sex-linked lethals if such exist. Induced sex-linked recessive lethals would find expression only in the hemizygous male or the homozygous female. The much more frequent occurrence of the former would lead to a greater diminution in male than in female births with a subsequent decrease in the sex ratio. Partially sex-linked lethal genes would result in distortion of the sex ratio within specific families, but at the population level, the sexes would be equally affected. Thus the net effect of maternal irradiation insofar as it affects the sex ratio would appear to be a reduction in the frequency of male births.

In the event of paternal exposure, the situation would be appreciably altered. Induced sex-linked lethal mutants, be they recessive or dominant, would be distributed only to the female. Thus sex-linked lethals arising in the differential segment of the paternal X-chromosome would lead to a reduction in the frequency of female births, or an augmentation of the sex ratio. Similarly, Y-borne lethals would be distributed only to sons of irradiated fathers; however, the evidence for the existence of any Y-borne genes in man is of the poorest order. The net effect of paternal exposure, all types of lethal mutants considered, would be a reduction in the frequency of female births.

If both parents were exposed, the net effect on the sex ratio would presumably be the sum of the parental effects considered separately. Since the effect on the sex ratio is different and in opposite directions for the two parents, it

[1] Because of unavoidable uncertainties in determining the duration of gestation, a few women may have registered prior to the twentieth week of gestation. However, the contribution of infants of less than twenty weeks of gestation in age to the sex ratio may be regarded as negligible. Sex ratio is here defined as the proportion of male births among all births.

would appear that if both parents were exposed there would be either no appreciable effect on the sex ratio, or if an effect existed it would be at the expense of the males. The latter is conceivable because sex-linked recessive lethals are more frequent in occurrence than sex-linked dominant lethals.

So far as irradiation effects are concerned, the sex ratio is unique among the indicators of genetic damage. Its uniqueness arises from the fact that, as has been indicated, maternal exposure would be expected to produce an effect different from paternal exposure.

7.3 Concomitant variation influencing the indicator. — Before we turn to a consideration of the data regarding the effect of parental exposure on the sex ratio, a brief consideration of the sex ratio as a variable is indicated. While it is true that in some respects the sex ratio may be the most specific of the indicators of genetic damage herein studied, in other respects it is a less satisfactory variable. A variety of factors have been alleged to influence the outcome, with regard to sex, of a pregnancy. Among these factors are maternal age (Lowe and McKeown, 1950), paternal age (Novitski, 1953), birth order (Ciocco, 1938), race (Strandskov, 1945), urban versus rural origin of the parents (Ciocco, 1938), and social strata (Bernstein, 1948). In the main, the effects of these variables are small. For example, Ciocco (1938) has shown that the difference in the sex ratio between first and fifth or higher born children is the difference between 0.5153 and 0.5124. The relationship in these data between maternal age and parity, on the one hand, and sex ratio, on the other hand, is illustrated in Figures 7.1 and 7.2. The fact that these concomitants exert such small effects and the difficulty in classifying some of them, has led us not to attempt any adjustment for maternal age, paternal age, birth order, social strata, or origin of the parents in the Japanese data.

7.4 The data. — In Table 7.1 are given the frequencies of male births by parental exposure and cities. Inspection of this table reveals no striking differences in the sex ratio among the classes of parental exposure. The analysis of these data is given in Tables 7.2a and 7.2b. Analysis fails to reveal a significant effect of cities, mother's exposure, or father's exposure, and there exists no evidence from the interactions of significant heterogeneity among these data.

As we have previously indicated, the sex ratio is the one indicator of irradiation damage wherein genetic theory specifies, with the greatest precision, the direction of change anticipated if irradiation exerts an effect on the sex ratio. Accordingly, we may ask what these data would show under a more restrictive set of alternatives to the null hypothesis, that is to say, under a one-tailed significance test. This alternative

FIGURE 7.1 — The distribution of the frequency of male births by age of mother at the birth of the infant with parity ignored.

FIGURE 7.2 — The distribution of the frequency of male births by parity with maternal age ignored.

analysis could proceed as follows: it may be argued that (1) the genetic effect of irradiation mediated through exposed mothers would be greater than that mediated through exposed fathers and hence the status of the father's exposure may be disregarded without undue loss of information. It may be further argued that (2) the absence of evidence for heterogeneity between cities permits pooling the observations from the two cities. Finally, (3) we have advanced reasons for believing that parents who were unexposed may not be *in pari materia*

TABLE 7.1 THE FREQUENCY OF MALE BIRTHS BY PARENTAL EXPOSURE AND CITY
(Unrelated parents)

Hiroshima

			Fathers				
Mothers			1	2	3	4–5	Total
	1	n	17,294	1,500	596	395	19,785
		♂	9,005	772	313	209	10,299
		P	.5207	.5147	.5252	.5291	.5205
	2	n	5,368	1,850	385	244	7,847
		♂	2,832	941	197	122	4,092
		P	.5276	.5086	.5117	.5000	.5215
	3	n	2,185	424	521	157	3,287
		♂	1,114	220	268	80	1,682
		P	.5098	.5189	.5144	.5096	.5117
	4–5	n	1,117	202	110	117	1,546
		♂	571	109	54	56	790
		P	.5112	.5396	.4909	.4786	.5110
	Total	n	25,964	3,976	1,612	913	32,465
		♂	13,522	2,042	832	467	16,863
		P	.5208	.5136	.5161	.5115	.5194

Nagasaki

			Fathers				
Mothers			1	2	3	4–5	Total
	1	n	14,610	2,170	243	139	17,162
		♂	7,608	1,120	129	75	8,932
		P	.5207	.5161	.5309	.5396	.5205
	2	n	9,316	4,144	273	178	13,911
		♂	4,849	2,112	140	103	7,204
		P	.5205	.5097	.5128	.5787	.5179
	3	n	747	279	94	35	1,155
		♂	360	134	51	14	559
		P	.4819	.4803	.5426	.4000	.4840
	4–5	n	559	116	35	28	738
		♂	279	56	18	15	368
		P	.4991	.4828	.5143	.5357	.4986
	Total	n	25,232	6,709	645	380	32,966
		♂	13,096	3,422	338	207	17,063
		P	.5190	.5101	.5240	.5447	.5176

THE CHILDREN OF ATOMIC BOMB SURVIVORS

with parents who were exposed, and therefore terminations occurring to mothers who were unexposed should be rejected. If now we proceed on these three assumptions we may test whether the sex ratio is lower among the offspring of all mothers in exposure classes 3, 4, and 5 than it is among the offspring of all mothers in exposure class 2. The pooling of exposure class 3 with 4 and 5 serves to increase the number of observations which may be brought to bear on the question of a maternal exposure effect, and may be interpreted as a conservative step since

TABLE 7.2 CHI-SQUARE ANALYSIS OF THE FREQUENCY OF MALE BIRTHS BY CITY AND PARENTAL EXPOSURE (Unrelated parents)

a. All exposure cells (4 × 4)

Source	DF	χ^2	P
Total	31	20.118	.90–.95
Interactions, first order			
CF	3	1.498	.50–.70
CM	3	2.582	.30–.50
MF	9	3.606	.90–.95
Main effects			
City (C)	1	0.219	.50–.70
Mother (M)	3	5.327	.10–.20
Father (F)	3	2.630	.30–.50

b. Excluding parents with exposure 1 (3 × 3)

Total	17	8.485	.95–.98
Interactions, first order			
CF	2	1.940	.30–.50
CM	2	1.099	.50–.70
MF	4	1.675	.70–.80
Main effects			
City (C)	1	0.002	.95–.98
Mother (M)	2	0.082	.95–.98
Father (F)	2	0.127	.90–.95

it lowers the mean dose in the "heavily" exposed group. The results of such a comparison are presented in Table 7.3 (Comparison 1) from which we note a $\chi^2 = 3.925$ which on a one-tailed test of significance corresponds to a probability level of 0.02. There may, then, be an effect of maternal exposure on the sex ratio. We could now ask whether there exists ancillary evidence which may shed light on the reality of this observed difference.

Let us consider first what may be termed the "internal" evidence, that is to say, evidence which can be obtained from further analysis of the data given in Table 7.1. To this end four other comparisons are given in Table 7.3. These comparisons are as follows (the numbers correspond to the numbers given in Table 7.3): a comparison of the sex ratio among terminations occurring to (2) mothers in class 2 with mothers in classes 3, 4, and 5 *but* eliminating all those observations where the father was unexposed, (3) mothers in class 2 whose spouses were unexposed with mothers in classes 3, 4, and 5 whose spouses were also unexposed, (4) mothers in class 2 whose spouses were unexposed with mothers in class 2 whose spouses were exposed, and lastly (5) mothers in class 2 with mothers in classes 3, 4, and 5 *but* eliminating the single cell where mothers were in class 2 and had unexposed spouses, i.e. the M_2F_1 cell. The arguments underlying these comparisons and the results obtained are as follows:

Comparison 2: The exclusion of mothers in class 1 is based on the assumption that this group is not comparable to groups 2 to 5 with regard to a number of concomitants, including age of mother, parity, and origin. From Tables 5.3 and 5.6 it should be noted that exclusion of mothers in class 1 alone leaves a series of cells, namely, mothers 2, 3, 4, and 5 and fathers 1, in which the mothers are clearly younger and have had fewer pregnancies, on the average, than is true of the mothers in the remaining exposure cells. While the effect of parity and maternal age on the sex ratio is certainly not striking, what effect there is suggests that younger mothers have a higher proportion of male births, and that first parities more frequently terminate in male infants (Ciocco, 1938). In this connection, it is interesting to note that in both cities one of the highest sex ratios among the numerically large cells occurs in that cell where the mothers are the youngest and have had the fewest pregnancies (the mothers 2-fathers 1 cell). It would be informative, then, to know whether a significant difference would obtain if the differences in maternal age and parity were further reduced by eliminating those cells in which the fathers were unexposed. Comparison 2 in Table 7.3 affords information on this point. We find not only no significant difference when unexposed fathers are rejected but also that the absolute value of the difference is smaller than that obtained when unexposed fathers were included in the comparison. The direction of the difference is, however, still consistent with the genetic argument. It may, of course, be argued that the lack of a

significant effect in this comparison is not unexpected because of the depletion of the sample, and the fact that the absolute difference is smaller could merely reflect an effect of fathers.

Comparison 3: The comparison which would maximize an effect due to mother's exposure if

TABLE 7.3 SELECTED COMPARISONS REGARDING THE EFFECT OF IRRADIATION ON SEX-RATIO

(See text for assumptions underlying the selection of comparisons.)

(1) Comparison: M_2 with M_{3-5} (summing over cities and fathers)

	Sex of infant ♂	♀	Total	Number of males per 100 females
Mother's exposure M_2	11,296 (0.5192)	10,462	21,758	107.99
$M_{(3-5)}$	3,399 (0.5054)	3,327	6,726	102.18
Total	14,695	13,789	28,484	—

$\chi^2_1 = 3.925*$

(2) Comparison: M_2 with M_{3-5} (summing over cities and fathers but eliminating fathers 1)

	Sex of infant ♂	♀	Total	Number of males per 100 females
Mother's exposure M_2	3,615 (0.5110)	3,459	7,074	104.50
$M_{(3-5)}$	1,075 (0.5076)	1,043	2,118	103.09
Total	4,690	4,502	9,192	—

$\chi^2_1 = 0.786$

(3) Comparison: M_2F_1 with $M_{3-5}F_1$ (summing over cities)

	Sex of infant ♂	♀	Total	Number of males per 100 females
Parental exposure M_2F_1	7,681 (0.5231)	7,003	14,684	109.69
$M_{3-5}F_1$	2,324 (0.5043)	2,284	4,608	101.73
Total	10,005	9,287	19,292	—

$\chi^2_1 = 4.937*$

(4) Comparison: M_2F_1 with M_2F_{2-5} (summing over cities)

	Sex of infant ♂	♀	Total	Number of males per 100 females
Parental exposure M_2F_1	7,681 (0.5231)	7,003	14,684	109.69
M_2F_{2-5}	3,615 (0.5110)	3,459	7,074	104.50
Total	11,296	10,462	21,758	—

$\chi^2_1 = 2.782$

TABLE 7.3—*Continued*

(5) Comparison: M_2 with M_{3-5} (summing over cities, fathers, but excluding fathers 1, mothers 2 cell)

	Sex of infant ♂	♀	Total	Number of males per 100 females
Mother's exposure M_2	3,615 (0.5110)	3,459	7,074	104.50
M_{3-5}	3,399 (0.5054)	3,327	6,726	102.18
Total	7,014	6,786	13,800	—

$\chi^2_1 = 0.444$

the effect of father's exposure was not negligible would be a comparison of the sex ratio among the children of mothers in class 2 whose husbands were unexposed with the ratio among the children of mothers in classes 3 to 5 whose husbands were also unexposed. Moreover, from Tables 5.3 and 5.6 we note that these cells are not too dissimilar with regard to mean maternal age and parity. This comparison results in a $\chi^2 = 4.937$, which is clearly significant. The occurrence of a more striking difference here than in comparison 1 could be interpreted as indicating that the effect of father's exposure is not negligible.

Comparison 4: If the effect of father's exposure is not negligible, as the preceding comparison might suggest, we could appraise this effect by holding mother's exposure constant (class 2) and then contrasting the offspring of unexposed fathers with those of exposed fathers. This should give rise, on genetic theory, to a higher sex ratio among the children of exposed fathers than among unexposed fathers. From comparison 4, however, we note that the direction of change is opposite to hypothesis, but this difference is not significant. The comparison does, however, raise questions as to the validity of using as a "control" the mothers 2-fathers 1 data. The latter is further borne out by a comparison of M_2F_1 with M_2F_2 wherein we would expect no differences, whereas we find a striking though not significantly increased sex ratio in the M_2F_1 cell ($\chi^2 = 3.221$).

Comparison 5: Comparison 4 leads one to conclude that it would be informative to contrast the sex ratio when mothers are in class 2 with mothers in classes 3 to 5 *but* eliminating the single cell in which the mothers were 2 and the fathers were unexposed. When this is done, a non-significant difference results. This further suggests the critical role played by the rejected

cell in the tests of significance. At this point it seems appropriate to introduce evidence on the sex ratio available from the Japanese vital statistics which further imply the occurrence of an inexplicable inflation of the sex ratio in the mothers 2-fathers 1 cell. In Table 7.4 are given the national average sex ratio and the average sex ratio for Hiroshima and Nagasaki for the period 1935–1952. In terms of numbers of males per 100 females we note that at no time during the period 1935–1952 do the combined city averages or the national average equal, much less exceed, the 109.7 found in the M_2F_1 cell.[2] The sex ratio obtaining in Kure during the years 1948–1950 was 0.5133 or 105.4 males per 100 females, a figure also lower than that observed in the M_2F_1 cell. It should be noted, however, that the data from "heavily" exposed mothers deviate from the national average as well as the city averages in the manner anticipated by genetic theory.

Let us turn now to the "external" evidence bearing on the difference we have been discussing, namely, the evidence afforded by comparisons other than those arising from Table 7.1. Several sources of evidence are pertinent here, namely, the stillbirth data, the early termination data, and the sex ratio among births occurring prior to the period covered by the Genetics Program.

If the depression of the sex ratio observed among terminations to "heavily" exposed mothers is due to induced sex-linked lethal genes, then we might logically expect to find evidence of increased male mortality among stillborn infants or among pregnancies terminating before 21 weeks of gestation. In this connection, it is interesting to note that the sex-mother interaction in Table 9.5 (the stillbirth data) reveals no evidence of heterogeneity among the sexes with regard to mother's exposure. We have previously stated (Sec. 2.8) that an attempt was made to obtain data on pregnancies terminating before 21 weeks of gestation. We have further stated that because these data are deficient in a number of respects they will not be presented in this report. However, for what it is worth, a preliminary analysis of these data failed to reveal evidence that "heavily" exposed mothers more frequently abort or miscarry male infants than do unexposed or "lightly" exposed mothers. If, then, one accepts the observed "exposure effect" as real, it is necessary to postulate that the induced sex-linked lethals are either gametic lethals or act so early in embryogenesis that the affected feti are resorbed or ejected at a time when the sex of the abortus cannot be readily determined.

Finally, when in a preliminary analysis of the data in 1953 it seemed likely that a significant relationship between maternal exposure and the sex ratio might exist, an effort was made to collect and scrutinize other data which would bear on this relationship. To this end, data were gleaned from the records of the Genetics Program as well as other ABCC records pertinent to the sex ratio among births occurring in Hiroshima and Nagasaki following the atomic bombings but prior to the inception of the Genetics Program.

The data to be presented in this section were collected from the histories of reproductive performance prior to approximately June, 1948 obtained on:

1. Those women who had been delivered of a registered or unregistered infant in the years 1948–1952 and known to ABCC through the Genetics Program. This information was available on only those parents who had produced a child on whom a "Genetics Long Form" was available (see Sec. 2.2).

2. All adults seen by the Commission under Project ME-55 (a random sample of adults exposed at distances of less than 2,000 meters and their controls).

3. Those women who had reported a spontaneous abortion and were seen under the study of early terminations mentioned in Section 2.8.

These three sources of data were cross-referenced to prevent duplication. In all, information was obtained on 8,824 pregnancies terminating between April, 1946 and approximately June, 1948. The former date was selected to avoid inclusion of pregnancies wherein the fetus might have been exposed *in utero*, and the latter date was so chosen as to prevent overlapping of these data with the Genetics Program. The sex of these infants is given in Table

[2] It should be noted that the data in Table 7.4 represent only liveborn infants. Since the data presented in Table 7.1 are for all births, and since there is a relative excess of males among stillbirths, the sex ratios given in Table 7.4 are not strictly comparable with those in Table 7.1, being biased in a downward direction. It seems unlikely that allowance for this bias would bring the M_2F_1 cell findings in line.

TABLE 7.4 NATIONAL STATISTICS OF LIVEBIRTHS, 1935–1952

Year	All Japan Total births	All Japan Males	All Japan Sex-ratio	Hiroshima City Total births	Hiroshima City Males	Hiroshima City Sex-ratio	Nagasaki City Total births	Nagasaki City Males	Nagasaki City Sex-ratio	Hiroshima and Nagasaki combined Total births	Males	Sex-ratio
1935	2,190,704	1,122,867	105.2	7,261	3,724	105.3	5,503	2,869	108.9	12,764	6,593	106.8
1936	2,101,969	1,076,197	104.9	6,831	3,502	105.2	5,499	2,790	103.0	12,330	6,292	104.2
1937	2,180,734	1,116,154	104.8	7,066	3,681	108.7	5,821	2,989	105.5	12,887	6,670	107.3
1938	1,928,321	990,888	105.7	6,077	3,128	106.1	5,671	2,880	103.2	11,748	6,008	104.7
1939	1,901,573	973,744	104.9	6,292	3,225	105.1	6,515	3,381	107.9	12,807	6,606	106.5
1940	2,115,867	1,084,282	105.1	7,466	3,823	104.9	6,871	3,506	104.2	14,337	7,329	104.6
1941	2,277,283	1,165,437	104.8	9,039	4,666	106.7	8,262	4,213	104.0	17,301	8,879	105.4
1942	2,233,660	1,145,068	105.2	8,720	4,548	109.0	7,775	3,955	103.5	16,495	8,503	106.4
1943	2,267,292	1,163,000	105.3	9,111	4,618	102.8	7,735	3,930	103.3	16,846	8,548	103.0
1944	2,171,621	1,112,778	105.1	—	—	—	—	—	—	—	—	—
1945	1,677,620[a]	—	—	—	—	—	—	—	—	—	—	—
1946	1,922,383	993,515	107.0	—	—	—	—	—	—	—	—	—
1947	2,678,792	1,376,986	105.8	6,696	3,548	112.7	4,841	2,461	103.4	11,537	6,009	108.7
1948	2,681,624	1,378,564	105.8	—	—	—	—	—	—	—	—	—
1949	2,696,638	1,380,008	104.8	7,941	4,154	109.7	9,137	4,716	106.7	17,078	8,870	108.1
1950	2,337,507	1,203,111	106.0	7,032	3,620	106.1	8,088	4,227	109.5	15,120	7,847	107.9
1951	2,137,689	1,094,641	104.9	6,409	3,349	109.4	7,856	4,095	108.9	14,265	7,444	109.1
1952	2,002,254	1,026,611	105.2	5,974	3,073	105.9	7,551	3,859	104.5	13,525	6,932	105.1
Total	37,825,911	19,403,851	105.3	101,915	52,659	106.9	97,125	49,871	105.5	199,040	102,530	106.2

[a] Not included in total.

7.5 where the terminations are distributed by city and parental exposure.

The exposure categories given in Table 7.5 are different from those presented elsewhere in this report, being the categories in use during the preliminary analysis. No attempt has been been made to take into account shielding due to slight differences in the way in which the three different segments of the data were collected. The radiation categories are here defined as follows:

(A) Not exposed, i.e., not present in the city at the time of bombing.
(B) Exposed, no symptoms, over 2,545 meters from the hypocenter.

of city, but no effect of maternal or paternal exposure.

The actual worth of these data is conjectural. All of the information is anamnestic, and the misreporting of the sex of an infant, particularly those who were stillborn or who died at an early age, is a very real possibility.

The preceding paragraphs serve to point up two items of general interest. Firstly, they indicate the interpretive difficulties which arise when one begins to select specific cells or groups of cells on which to base comparisons. Secondly, they serve to indicate that while an elegant genetic argument can be advanced for expecting changes in the sex ratio consequent to parental

TABLE 7.5 THE FREQUENCY OF MALE BIRTHS AMONG INFANTS BORN AFTER APRIL, 1946 BUT PRIOR TO JUNE, 1948, BY PARENTAL EXPOSURE

Mothers		Fathers Hiroshima				Fathers Nagasaki			
		A	B, C	D, E	Total	A	B, C	D, E	Total
A	n	2,380	214	81	2,675	2,002	292	53	2,347
	m	1,239	112	41	1,392	1,037	142	29	1,208
	p	.5206	.5234	.5062	.5204	.5180	.4863	.5472	.5147
B, C	n	738	403	69	1,210	1,072	682	87	1,841
	m	377	221	30	628	531	363	42	936
	p	.5108	.5484	.4348	.5190	.4953	.5323	.4828	.5084
D, E	n	345	73	127	545	125	33	48	206
	m	201	37	58	296	75	22	22	119
	p	.5826	.5068	.4567	.5431	.6000	.6667	.4583	.5777
Total	n	3,463	690	277	4,430	3,199	1,007	188	4,394
	m	1,817	370	129	2,316	1,643	527	93	2,263
	p	.5247	.5362	.4657	.5228	.5136	.5233	.4947	.5150

(C) Exposed, no symptoms, 1,845–2,544 meters from the hypocenter.
(D) Exposed, no symptoms, 1,844 or less meters from the hypocenter.
(E) Exposed, symptomatic, i.e., reporting epilation and/or petechiae and/or gingivitis.

It should be noted that the average exposure in categories A, B, and E will be essentially unchanged from those in categories 1, 2, and 5. But the average exposure in categories C and D will be less than those in 3 and 4.

Inspection of these data, to the extent that trends are apparent, reveals exposure differences which are diametrically opposed to those seen in Table 7.1. The changes observed with respect to both maternal and paternal exposure are contrary to those predicted by the genetic hypothesis. The analysis of these data (see Table 7.6) reveals no significant interactions, an effect

TABLE 7.6 CHI-SQUARE ANALYSIS OF THE FREQUENCY OF MALE BIRTHS AMONG INFANTS BORN AFTER APRIL, 1946 BUT PRIOR TO JUNE, 1948

Source	DF	χ^2	P
Total	17	228.999	<.001
Interactions, first order			
CM	2	1.034	.50–.70
CF	2	0.700	.70–.80
MF	4	8.848	.05–.10
Main effects			
Cities	1	5.347	.02–.05
Mothers			
Hiroshima	2	1.035	.50–.70
Nagasaki	2	3.559	.10–.20
Sum	4	4.594	.30–.50
Fathers			
Hiroshima	2	4.168	.10–.20
Nagasaki	2	0.616	.70–.80
Sum	4	4.784	.30–.50

exposure, the sex ratio, as a variable, leaves much to be desired. A voluminous literature purports to show any number of factors which can alter the sex ratio; adequate explanation for the peculiar variations which occur in the sex ratio due to those factors have not been advanced.

In view of the unsatisfactory state in which we find ourselves with regard to the effect of parental irradiation on the sex ratio, data continue to be collected in Japan which are pertinent to this problem. The present findings fail to confirm unequivocally the apparently significant effect of exposure on sex ratio reported in a preliminary note (Neel et al., 1953), although the direction of the observed effect remains the same. Among possible reasons for the disparity the following should be considered: (1) the different (improved) classification of parental exposure employed in the present analysis, (2) the superior statistical techniques which have become available since that preliminary report, and (3) the accumulation of additional data.

It is worth noting that the direction of deviation among the cells is generally consistent with that to be expected from genetic theory, namely, a reduction in the sex ratio with increasing maternal exposure, and an augmentation of the sex ratio with increasing paternal exposure.

7.5 Summary. — No significant association between sex ratio and parental exposure is demonstrable in either the comparison involving non-exposed parents or in the comparison among differing classes of parental exposure. The direction of deviation is, in general, consistent with the genetic hypothesis.

Note added in proof.—Since the previous portion of this chapter was written it has been possible to bring certain additional evidence to bear on the question of an altered sex ratio. Briefly, the source of this evidence is as follows: When in 1953 it was decided that the Genetics Program should be terminated for the reasons advanced in Section 2.11, there existed some question as to whether or not the sex ratio had been significantly altered among the pregnancies occurring to exposed parents. It seemed desirable that additional observations be made with regard to the sex of infants being born to exposed and non-exposed parents in Hiroshima and Nagasaki. Through the cooperation of the municipal authorities, during 1954 and 1955 access was obtained to the birth records on both live and stillborn infants. From these records the requisite information to identify the infant as well as his or her parents was copied along with the weight of the child, the sex, and place in sibship. Parental exposure was secured from (1) the master file on exposed individuals maintained by the ABCC, or (2) a home visit if one or both parents were not listed in the master file. In the years 1954 and 1955 there were 22,710 births concerning which, during 1954, 1955, and the early part of 1956, it was possible to obtain the information listed above. These data, which are presented in Table 7.7, were analyzed during the summer of 1956. The series is continuous with that previously reported; no infant is recorded in both sets of data.

In Table 7.8 are presented the results of an analysis of the combined data, that is, of the data available for 1948–1955. From the latter table we note that when all exposure cells are considered there is no evidence of heterogeneity between the two sets of data nor evidence for a parental exposure effect. When those infants one or both of whose parents were unexposed are excluded, a somewhat different situation obtains. We note a significant difference in the sex ratio between the two sets of data with male births being somewhat more common in 1954–1955. Since none of the interactions involving time is significant, there is no evidence that this increase is at the expense of a particular city or parental exposure. The "sum" test is, therefore, a valid measure of the effect of parental exposure on the sex ratio. Neither father's nor mother's exposure can be shown to exert a significant effect.

It is particularly interesting to examine the combined data under those assumptions given in Section 7.4, Comparisons 1 and 3, which gave rise to a significant effect of mother's exposure on the sex ratio. If in the combined data father's exposure is ignored, mother's exposure classes 3, 4, and 5 are pooled, and those infants whose mothers are in exposure class 1 are rejected, no significant effect of mother's exposure can be demonstrated ($\chi^2 = 1.634$, DF $= 1$, and the probability on a one-tailed test is greater than 0.10). This is, of course, at variance with the findings for the 1948–1953 data given in Section 7.4. Stated in terms of regression analysis, if sex ratio is regressed on "mean" maternal exposure using as the estimated doses 8, 60,

TABLE 7.7 THE FREQUENCY OF MALE BIRTHS BY PARENTAL EXPOSURE AND CITY 1954–1955: UNRELATED PARENTS

Hiroshima

Mothers		Fathers 1	2	3	4–5	Total
1	n	6,200	744	275	150	7,369
	♂	3,254	386	156	78	3,874
	P	.5248	.5188	.5673	.5200	.5257
2	n	1,790	435	120	94	2,439
	♂	933	242	61	54	1,290
	P	.5212	.5563	.5083	.5745	.5289
3	n	657	121	77	50	905
	♂	358	65	40	29	492
	P	.5449	.5372	.5195	.5800	.5436
4–5	n	374	60	34	23	491
	♂	188	35	12	14	249
	P	.5027	.5833	.3529	.6087	.5071
Total	n	9,021	1,360	506	317	11,204
	♂	4,733	728	269	175	5,905
	P	.5247	.5353	.5316	.5521	.5270

Nagasaki

Mothers		Fathers 1	2	3	4–5	Total
1	n	5,440	754	112	69	6,375
	♂	2,813	388	55	35	3,291
	P	.5171	.5146	.4911	.5072	.5162
2	n	3,136	1,039	100	80	4,355
	♂	1,579	564	68	47	2,258
	P	.5035	.5428	.6800	.5875	.5185
3	n	369	91	30	16	506
	♂	204	46	13	6	269
	P	.5528	.5055	.4333	.3750	.5316
4–5	n	218	29	13	10	270
	♂	123	13	8	4	148
	P	.5642	.4483	.6154	.4000	.5481
Total	n	9,163	1,913	255	175	11,506
	♂	4,719	1,011	144	92	5,966
	P	.5150	.5285	.5647	.5257	.5185

and 200 *reps* for exposure categories 2, 3, and 4–5 respectively, one obtains a regression coefficient equal to $-0.55\%/100$ *reps* for the combined data and $-0.81\%/100$ *reps* for the 1948–1953 data alone. The former is not significantly different from zero even on a one-tailed test whereas the latter is significant at the 5 per cent level if a one-tailed test is used. The net effect of the 1954–1955 data is, then, to render more questionable the occurrence of a significant effect of parental exposure on the sex ratio in the Hiroshima-Nagasaki data.

TABLE 7.8 CHI-SQUARE ANALYSIS OF THE FREQUENCY OF MALE BIRTHS DURING THE YEARS 1948–1955 BY TIME OF BIRTH, CITY, AND PARENTAL EXPOSURE: UNRELATED PARENTS. TWO INTERVALS OF TIME ARE RECOGNIZED, NAMELY, 1948–1953 AND 1954–1955.

All exposure cells

Source	DF	χ^2	P
Total	63	60.900	>.50
Interactions, First Order			
TC	1	0.760	.30–.50
TM	3	4.758	.10–.20
TF	3	4.845	.10–.20
CM	3	1.453	.50–.70
CF	3	1.589	.50–.70
MF	9	3.310	.95–.98
Main Effects			
Time (T)	1	1.203	.20–.30
City (C)	1	1.109	.20–.30
Mother (M)	3	2.291	.50–.70
Father (F)	3	1.431	.50–.70

TABLE 7.8—*Continued*

Excluding Parents with Exposure 1

Source	DF	χ^2	P
Total	35	39.421	>.25
Interactions, First Order			
TC	1	0.009	.90–.95
TM	2	1.851	.30–.50
TF	2	0.273	.80–.90
CM	2	2.330	.30–.50
CF	2	3.546	.10–.20
MF	4	4.285	.30–.50
Main Effects			
Time (T)	1	9.507	.001–.01
City (C)			
1948–1953	1	0.002	.95–.98
1954–1955	1	0.008	.90–.95
Sum	2	0.010	>.99
Mother (M)			
1948–1953	2	0.082	.95–.98
1954–1955	2	2.810	.20–.30
Sum	4	2.892	.50–.70
Father (F)			
1948–1953	2	0.126	.90–.95
1954–1955	2	0.449	.70–.80
Sum	4	0.575	.95–.98

We are indebted to Dr. Robert Holmes, current director of the ABCC, and Dr. Lowell Woodbury, chief of the Statistics Unit of the ABCC, for their cooperation in connection with these later aspects of the sex-ratio program. We also acknowledge with pleasure Mrs. Jean Okumoto's efficient supervision of the collection of these data.

Chapter VIII

ANALYSIS OF THE MALFORMATION DATA

DATA on major malformations were collected at the examination shortly following birth, and again at the 9-months examination. For the reasons advanced in Section 2.6, it was deemed advisable to consider the results of these examinations separately. Moreover, to permit pooling the information from the two examinations, if such pooling was warranted, all infants found on the "at-birth" examination to have one or more major malformations were excluded from consideration in the analysis of the 9-months data.

8.1 *The trait.* — In any consideration of the relation of parental irradiation history to the occurrence of major congenital malformations in the offspring, it is of course first necessary to define major congenital malformation. Since there is no absolute dividing line between "major" and "minor" malformation, an attempt to formulate an *a priori* definition encounters certain difficulties. In actual practice, the examining physicians were required to record all abnormalities, down to and including such variations as minor haemangiomas (excluding the very common occipital-nuchal type) and auricular pits. When the final report on each infant was coded, a decision was made as to the presence of major malformation. Any one type of malformation was consistently coded as either major or minor.

Table 8.1 is a listing of the malformations which in the course of this investigation were coded as major. A discussion of the relative frequencies of the various malformations, and the various combinations in which they occurred, will be presented elsewhere. For our present purposes, then, we shall define major malformation simply as any condition found on that list. Obviously there are many other major congenital malformations which do not appear on that list, simply because they either did not occur in this series, or, if they did occur, were not recognized at birth. This listing deals solely with conditions recognizable at birth by the usual physical examination. A later chapter will deal with autopsy findings.

8.2 *Reliability of diagnosis.* — A question which arose repeatedly during the course of this study — and which will undoubtedly occur to the reader — concerned the completeness with which the significant malformations occurring in the populations under study were recognized and reported. This question has already been commented on briefly (Sec. 2.6). It should at the outset be made quite clear that no study organized as this one was can be expected to yield an accounting of *all* significant congenital defects present in the newborn population. Thus, the diagnosis of mental defect, or severe visual or hearing loss, is notoriously unreliable during the first several months of life. From the standpoint of the present study, while for obvious reasons one should strive for as high a level of diagnostic accuracy as possible, the primary danger to be guarded against is that different diagnostic standards obtain for the various segments of the population under study. There is no reason to believe that diagnostic standards were not uniform with respect to infants in one exposure subclass as compared with infants in another. We have already commented briefly on some of the evidence suggesting that no bias has been introduced by differences in the degree to which various groups cooperated in the study (Sec. 5.7).

It is important to reiterate that each infant recorded in this study, with the exception of a certain number of stillbirths or neonatal deaths, as recorded in Section 2.5, was examined by a Japanese physician. It was recognized from the outset that both the loyalties and training of the midwife were such that she might at times fail to record the obvious. This suspicion was well justified by instances in which a child with

harelip has been reported as normal by the attending midwife. Beginning in early 1950, the mothers of all living infants found to have either a major or a minor malformation in the course of the home visit were invited to bring marily occasioned by the disposal of the body of an infant alleged to have a major defect before examination by an ABCC physician, every malformation recognized at birth was verified before the defect was accepted. Prior to early

TABLE 8.1 AN ALPHABETICAL LISTING OF THOSE MALFORMATIONS OBSERVED IN THIS STUDY WHICH OCCURRING ALONE OR IN COMBINATION WITH ONE ANOTHER WERE GRADED AS MAJOR CONGENITAL DEFECT

Achondroplasia
Albinism
Amputation, congenital
Anencephaly
Anonychia
Anophthalmos
Arthrogryposis
Atresia ani or rectum
Atresia of the external nares

Blepharophimosis
Brachydactyly

Cataracts
Central nervous system defect, severe
Claw hand
Cleft palate
Club foot
Coloboma iridis
Corneal defect
Cryptophthalmia
Cystic hygroma

Diaphragm, defect of
Diastasis recti, severe
Dislocations, multiple, congenital
Dysplasia of acetabulum, congenital

Ectodermal defect, congenital
Ehlers-Danlos syndrome
Encephalocele
Exostosis, severe, of bone
External ear, major malformation of

Facial cleft, oblique
Flexion deformity of fingers and toes
Foetus amorphous
Foetus papyraceous
Funnel chest

Gastroschisis

Harelip
Heart disease, congenital, cyanotic
Hemimelus
Hydrocephalus

Hypertelorism
Hypospadias, marked

Ichthyosis congenita
Inguinal hernia (females)
Intestinal obstruction, ? congenital defect

Lymphangioma

Macroglossia
Major bone, absence of
Malformation, complex and ill-defined
Microcephaly
Micropenis
Microphthalmus
Microtia
Mongolism

Neurofibromatosis (von Recklinghausen's disease)
Nystagmus, congenital

Oligodactyly
Omphalocele

Polydactyly
Polyostotic fibrous dysplasia
Polyoty
Ptosis of eyelids
Pupillary membrane, persistent

Radio-ulnar synostosis
Recto-vaginal fistula
Rhinencephaly

Situs inversus viscerum
Spina bifida with or without meningocele or myelomeningocele
Status Bonnevie-Ullrich
Sympodia
Syndactyly
Synorchism

Teratoma
Thyroglossal duct, persistent
Tumor, type undetermined

their children to an ABCC "Verification Clinic" at some convenient time shortly after the diagnosis. Transportation to and from the clinic was supplied by the ABCC. There the child was seen by another Japanese physician and an American physician. Thus, with exceptions pri- 1951, major malformations were usually verified by a second home visit on the part of either one of the more experienced Japanese physicians or an American pediatrician. An analysis of the results of these home visits and Verification Clinic indicate a high degree of accuracy of

diagnosis for those defects recognized as present.

The crucial question is, how many defects which should have been diagnosed at birth went

our control material with roughly comparable material collected by other investigators. The most extensive and reliable Japanese series to come to our attention is that compiled by Dr.

TABLE 8.2 THE TYPES AND FREQUENCY OF VARIOUS MAJOR CONGENITAL MALFORMATIONS OBSERVED AT THE TOKYO RED CROSS MATERNITY HOSPITAL DURING THE YEARS 1922 THROUGH 1940

(The over-all frequency is 456 malformations among 49,645 births [0.92%]. The table is based on data made available by Drs. S. Mitani and V. Kuji.)

Description of abnormality	No. of cases
Acrania or hemicrania (anencephaly)	32
Hydrocephalus	14
Congenital ascites	2
Spina bifida with club foot	3
Sacral teratoma	3
Umbilical cord hernia (omphalocele)	9
Spina bifida	8
Club foot (T. varus, T. equino-varus)	63
Harelip with cleft palate	53
Polydactyly (fingers)	38
Cleft palate	30
Club foot (T. valgus)	18
Hemangioma or lymphangioma	16
Congenital teeth	13
Malformation of ear lobe	13
Micromelus	11
Atresia ani et vaginalis (including imperforate anus)	10
Syndactyly of toes	10
Microphthalmia (and others)	7
Luxation of knee joint	6
Polydactyly (toes)	6
Microcephaly	4
Pseudohermaphroditism	4
Hydrocephalus, cleft palate, gastroschisis	1
Atresia ani, umbilical hernia	1
Atresia ani, malformation of external auditory canal	1
Atresia ani, cleft palate (mild degree?)	1
Atresia ani, sympodia	1
Atresia ani, cleft palate	1
Cleft palate, oblique facial cleft	1
Cleft palate, malformation of fingers	1
Cleft palate, malformation of ear	1
Harelip, pseudohermaphroditism	1
Harelip, oblique facial cleft, absent nose, syndactyly (fingers)	1
Harelip, polydactyly (fingers), syndactyly (toes)	1
Fissured palate, club foot	1
Fissured palate, microphthalmia	1
Club hand, club foot, small thumb, missing right labium	1
Elbow contraction, club foot, luxation of knee joint	1

Description of abnormality	No. of cases
Inguinal hernia, micromelus, hydrocephalus	1
Gigantism of feet and arms	1
Polydactyly (toes), harelip	1
Polydactyly (fingers), syndactyly (fingers)	1
Syndactyly (fingers), harelip	1
Torticollis, abdominal hernia	1
Torticollis, club foot, facial asymmetry	1
Monophthalmia, polydactyly (toes)	1
Monophthalmia, polydactyly (fingers)	1
Hydrocele	4
Cleft hand (lobster claw)	4
Polydactyly (toes) with syndactyly (toes)	4
Ankylo-glossum (tongue-tie)	3
Club foot (T. calcaneus)	3
Abnormal elbow joint	3
Malformation of external auditory canal	3
Amelus (arm or arms missing)	3
Inguinal and scrotal hernia	2
Meningocele (cranium)	2
Club hand	2
Brachydactyly (2 fingers)	2
Contraction of the knee joint	2
Polydactyly (toes) with syndactyly (fingers)	2
Malformation of fingers	2
Absence of nose	1
Club foot (T. equinus)	1
Parasitic teratoma of jaw	1
Tumor of "waist"	1
Bifid uvula	1
Hypospadias or epispadias	1
Cleft sternum (lower portion)	1
Phimosis (extreme)	1
White hair (partial albinism)	1
Tumor (small) of vagina	1
Anencephaly, polydactyly (fingers)	1
Club foot, syndactyly (toes)	1
Club foot, malformation of fingers	1
Malformation of face	1
Monster (amorphous)	1
Polydactyly (fingers and toes)	1
Missing fingers, syndactyly (toes)	1

unrecognized? There are three general approaches to the evaluation of this question which we shall now consider.

(A) The first approach involves a comparison of the frequency of major malformation in

V. Kuji (personal communication) at the Tokyo Red Cross Maternity Hospital. The findings are shown in Table 8.2. The over-all frequency of malformed individuals in this series (0.92 per cent) and in the children of the non-irradiated

parents included in this study (0.85 per cent) is quite comparable. The agreement is all the more striking in view of two facts: (1) certain of the "malformations" on the Kuji list, such as congenital teeth, hydrocoele, and partial albinism, would be considered minor in our terminology, and (2) a hospital series in Japan cannot be considered random, but probably includes a disproportionate representation of the complications of pregnancy. In view of the tendency of gross malformations to be associated with such complications as hydramnios (Prindle, Ingalls, and Kirkwood, 1955), this should bias the Red Cross Hospital series in an upward direction. There is no evidence from this comparison that any considerable number of congenital defects diagnosable at birth are being overlooked in Hiroshima and Nagasaki.

(B) A second approach to the question of missed diagnoses of significant defect consists of checking the ABCC findings against those of another medical group in Hiroshima or Nagasaki. Such a spot check was carried out for the last three months of 1951 in Nagasaki. Out of twelve major "malformations" seen at the two largest hospitals in the city (the Nagasaki Medical School Hospital and the Mitsubishi Hospital), five were already known to the ABCC. The remaining seven were the following: congenital dislocation of the hip — 3 cases; congenital heart disease — 2 cases; idiocy with funnel chest — 1 case; hydrocephalus — 1 case. In view of the frequent delayed appearance of congenital dislocation of the hip and hydrocephalus, these cases are not thought to constitute "missed" diagnoses. As noted earlier (Sec. 2.6) congenital heart disease was not included in the "at-birth" series of the ABCC. Only the funnel chest perhaps should have been detected in the ABCC series. Again, then, there is no evidence that any substantial proportion of major defect diagnosable at birth by the techniques of physical examination was missed.

(C) A third approach to the question of accuracy of diagnosis comes from a comparison of the "at-birth" with the "9-months" series. In Section 2.6 we have recorded the results of a "spot check" in Hiroshima in 1952 concerning the clerical and medical problems which can arise in a program concerned with congenital malformations such as this, and the steps taken to meet some of these problems. Every effort was made to insure that these errors did not recur in the remainder of the study. It might be noted, however, that, as can be seen in Section 2.6, most of the errors were ones which would *overestimate* the number of those diagnoses which were made at birth and not subsequently verified at nine months. We may now compare the final figures for major malformations present in those infants examined at age 9 months with the findings recorded for these same infants at birth. Such a comparison, reproduced in Table 8.3, provides the best means available for reaching an opinion concerning the confidence to be placed in the "at-birth" data on congenital malformation. Perusal of the table brings out a number of points of interest. Out of a total of 563 infants with one or more major defects diagnosed either at the examination shortly after birth or at the examination at an age of approximately 9 months, or on both occasions, only 112 were clearly verifications at age 9 months of an "at-birth" diagnosis. However, there were altogether 130 infants with one or more major defects in whom the diagnosis could not be verified at 9 months because of the death of the child in question; it seems likely that the great majority of these children did indeed suffer from a major malformation. The 2 cases of defect at birth which were not verified at age 9 months were both cases of polydactyly which we assume to have undergone surgery. Of the total number of defective infants detected by these two approaches, then, the figures taken at face value indicate that only 43 per cent of them were observed at birth. However, it is clear that in many of the 319 infants with one or more defects not diagnosed shortly following birth, the defects are of such a nature that their certain detection at that time would have been difficult if not impossible. This is especially true of the two diagnoses which contribute most to this list (dysplasia of the acetabulum — 133 cases; and inguinal hernia in females — 56 cases; total — 189 cases, 33.6 per cent of the grand total of 563 infants or 59.3 per cent of the additional infants diagnosed at 9 months). The relatively high frequency of acetabular dysplasia may reflect a particular interest in this condition on the part of several of the American pediatricians associated with the project. We have attempted to indicate in the table by a single asterisk those diagnoses which should in all probability have been made at the time of the

Analysis of the Malformation Data

TABLE 8.3 A COMPARISON, FOR HIROSHIMA AND NAGASAKI, OF THE FINDINGS AS REGARDS MAJOR CONGENITAL MALFORMATION IN INFANTS EXAMINED AT APPROXIMATELY AGE 9 MONTHS, AND IN THOSE SAME INFANTS WHEN EXAMINED SHORTLY AFTER BIRTH

(The figures are obtained from 14,768 infants in Hiroshima and 12,324 infants in Nagasaki. Infants born of consanguineous marriages have not been excluded from this comparison. Since no diagnoses of congenital heart disease were accepted on the "at-birth" examination, this diagnosis does not appear in the comparison. There were 50 such diagnoses made in Hiroshima in the "9-months" examination and 55 in Nagasaki. Explanation of asterisks in text.)

Comparison	Hiroshima	Nagasaki	Total
1. Total number of infants in whom a diagnosis of one or more major malformations was made on the "at-birth" and the "9-months" examination.	57	55	112
2. Total number of infants in whom a diagnosis of one or more major malformations was made on the "at-birth" examination, but who could not be re-examined at age 9 months because of the death of the child.	74	56	130
3. Total number of infants in whom a diagnosis of one or more major malformations was made in the "at-birth" examination, but in whom no major defect was detected at age 9 months.	2	0	2
4. Total number of infants in whom *no* diagnosis of major malformation was made at birth, but in whom one or more major defects were detected in the "9-months" examination [a]	165	154	319
Achondroplasia *	1	2	3
Anterior synechiae	0	1	1
Branchial cyst	0	1	1
Cataracts **	1	0	1
Club foot **	2	2	4
Coloboma of eyelids *	1	1	2
Coloboma of iris *	2	0	2
Cyst, thyroglossal	1	0	1
Complex and ill-defined condition *	0	1	1
Cryptorchism	0	2	2
Deformity of bone	0	1	1
Dextrocardia *	0	1	1
Dysplasia of acetabulum	76	57	133
Ectodermal defect, congenital	0	1	1
Funnel chest **	8	5	13
Hammer toe	0	1	1
Harelip and cleft palate *	1	2	3
Hernia, inguinal, in females	35	21	56
Hydrocephalus	4	1	5
Hypertelorism **	0	1	1
Hypoplasia of bone	0	1	1
Ichthyosis (congenital type) **	3	0	3
Laryngeal stridor, congenital	0	1	1
Lymphangioma	0	1	1
Maldevelopment of central nervous system	0	1	1
Mental defect, severe	2	18	20
Microcephaly	4	2	6
Micrognathia *	0	1	1
Micropenis *	1	0	1
Microphthalmus *	1	0	1
Microtia *	0	2	2
Mongolism **	8	7	15
Nystagmus, congenital	0	6	6
Polydactyly *	1	0	1
Ptosis	1	2	3

THE CHILDREN OF ATOMIC BOMB SURVIVORS

TABLE 8.3—Continued

Comparison	Hiroshima	Nagasaki	Total
Situs inversus viscerum	1	3	4
Spastic infantile paralysis	4	0	4
Spina bifida *	1	0	1
Strabismus	5	3	8
Syndactyly *	1	1	2
Teeth, malformation of	0	1	1
Tumor, cutaneous	0	1	1
Wilm's tumor	0	1	1
Xeroderma pigmentosum	0	1	1
	165	154	319
			563

ᵃ A defective infant, regardless of number of major defects, is entered under Item 4 in the table only once. In case two or more major defects were detected for the first time at the 9-months examination, only the most severe of the defects is listed under Item 4.

"at-birth" examination, and by a double asterisk those diagnoses which as a rule are more difficult to make in newborn infants but which *might* well have been made at that time. (However, each infant was entered into Table 8.3 but once, and the numbers entered opposite specific diagnoses under item 4 refer to the number of infants in whom that diagnosis was considered the most severe one). There are 35 diagnoses in the first category, and 22 in the second. All of the remaining defects may either not have been diagnosable at birth or are of such a nature that uncertainty might surround them.

In Section 2.6 the possibility was raised that such relatively common diagnoses as acetabular dysplasia and pilonidal sinus "unbalance" the 9-months data. These are both findings which border on being normal variations, in their less extreme forms. Ideally, in an analysis of this type one would introduce a weighting constant to take into account the genetic component in the etiology of the various malformations under study, but in view of the obvious impossibility of this, there seemed no alternative to giving each diagnosis, including the common ones, equal weight.

Despite the many observations which have been published on congenital malformation in man, it is a surprising fact that to our knowledge there exists only one series comparable to this one, in terms of re-examination at some later date of a series of infants examined shortly after birth. McIntosh et al. (1954), in a study of 5,739 products of conception weighing over 500 grams who were followed, unless death intervened, until 12 months of age, and some even longer, found that only 43.2 per cent of all malformations detected in the course of the study were observable at birth. Although for a number of reasons, to be dealt with elsewhere, these figures are not directly comparable to our own, the general agreement between the two series is striking.

Table 8.3 also serves another purpose, in that it permits a partial comparison of the diagnoses being made in Hiroshima and Nagasaki. With one possible exception, the findings in the two cities appear as uniform as might be hoped for under these circumstances. It will be noted that severe mental defect is listed 18 times for Nagasaki, but only twice for Hiroshima. It seems quite likely that this reflects a difference in approach in the two cities, rather than a real difference.

8.3 *The genetic argument for radiation-induced changes.* — Within the spectrum of mutations induced by the irradiation of experimental animals, the most dramatic, but least common, group of mutations are the so-called "visibles." These mutations produce, as the name implies, visible alterations in the phenotype. Although direct evidence is lacking, it seems probable that in man a certain fraction of these mutations would lead to gross physical departures from the norm readily detectable at birth. Accordingly, one measure of genetic damage consequent to atomic bombing would entail determining whether differences in the frequency of congenital malformation arose

consistent with parental exposure. On an *a priori* basis, we might expect that the frequency of malformation would increase with increasing parental exposure.

The extent to which the frequency of malformation would be altered by parental exposure would, naturally, be a function of the component in the etiology of such malformations attributable to genetic factors. While this component cannot be exactly specified, that it is not inappreciable is indicated by (1) the number of specific malformations known to be inherited, and (2) the finding that the total malformation rate within the Japanese data increases with increasing parental relatedness (Schull, unpublished).

8.4 *Concomitant variation influencing the indicator.* — Among the numerous agents for which there exists some evidence regarding a role in the etiology of congenital malformations in man, and which may prove important in the analysis of these data, are such diverse factors as maternal age, parity, nutrition, and viral infections in the first trimester of pregnancy. No really complete review of the literature on the etiology of human malformations has been published. Certain aspects of the etiology have been treated by Warkany (1947), Gruenwald (1947), and Landtman (1948).

Within the data here presented, there appear to exist only two sources of concomitant variation which require comment. These are: (a) known differences between exposure categories in maternal age and parity, and (b) possible nutritional differences between the exposure categories.

Of these concomitant variables maternal age and parity are undoubtedly the most important. An extensive literature exists describing the effects of these variables on pregnancy outcome. In the main, however, this literature is concerned with the effect of maternal age and/or parity on specific malformations rather than on the total malformation rate. Two exceptions are the work of Carter and McCarthy (1951) and Coffey and Jessop (1955), who find a positive association between maternal age and total malformation rate.

The relationship of maternal age and parity to malformation rate in the present data is summarized in Tables 8.4 and 8.5 and Figures 8.1 and 8.2. Table 8.4 presents the effect of age for specific parities. The data have been analyzed according to the method of Krooth (1955). In the column headed "A.D." have been entered by parity class, the absolute differences between the proportion of all normal infants born at a specific age interval and the proportion of all malformed infants born at the same age interval. A χ^2 test has been applied to the significance of the findings for each parity class, with the usual convention of indicating significance at the 5 per cent level with one asterisk, and at the 1 per cent level with two asterisks.

For each parity, an index of absolute difference (I_{AD}) has been calculated as suggested by Krooth (1955), namely,

$$I_{AD} = \tfrac{1}{2}\sum_i |d_i|$$

where d_i corresponds to the individual entries in the column headed "A.D." For each parity the proportion of the total number of births which occurred at this parity (p_T) has been calculated. The I_{AD} for maternal age, all parities considered, is simply the weighted mean of the I_{AD}'s for the individual parities. The same material is presented graphically in Figure 8.1. In the entire data, there is a highly significant age effect, although only two of the individual parity classes show effects at the level of significance. It is noteworthy that by inspection, age effects seem to be limited to the four highest parity classes, suggesting an age-parity interaction. A much more detailed analysis of these data will be presented elsewhere.

Table 8.5 presents a similar analysis with respect to parity effects. There is a highly significant relationship between parity and the frequency of malformation in the data as a whole, consistent in all six age classes, although reaching the level of significance in just one of these six classes. The findings are shown graphically in Figure 8.2. These results, like the age effects, will be discussed in detail elsewhere. Suffice it to say for the present that the effects of age and parity appear to follow different patterns, the age effect, as noted previously, emerging strongly at the higher ages, the parity effects being more uniformly distributed.

Carter and McCarthy (1951) were inclined to attribute the positive association between maternal age and total malformation rate almost entirely to the known association of mongolism with increasing maternal age. This interpretation would not be valid for the Japanese data, since there are included only three children with

TABLE 8.4 THE EFFECT OF MATERNAL AGE AT FIXED PARITY ON THE FREQUENCY OF MALFORMED INFANTS
(Detailed explanation in text.)

	Total births	Normal births	Malformed infants	Per cent malformed	AD
Parity 1					
< 21	1,929	1,913	16	0.83	.02982
21–25	8,697	8,636	61	0.70	.02119
26–30	3,168	3,151	17	0.54	.04538
31–35	621	617	4	0.64	.00211
36–40	180 ⎱	198	1 ⎱	0.50	.00354
41 +	19 ⎰		— ⎰		
Total	14,614	14,515	99	—	.10204
$\chi^2_4 = 1.771$		$I_{AD} = 0.0510$		$p_T = 0.2664$	
Parity 2					
< 21	424	421	3	0.71	.00541
21–25	6,392	6,334	58	0.91	.01817
26–30	6,156	6,101	55	0.89	.01079
31–35	1,431	1,420	11	0.77	.01145
36–40	357 ⎱	405	2 ⎱	0.49	.01209
41 +	50 ⎰		— ⎰		
Total	14,810	14,681	129	—	.05791
$\chi^2_4 = 1.117$		$I_{AD} = 0.0290$		$p_T = 0.2700$	
Parity 3					
< 21	37	36	1	2.70	.00577
21–25	2,016	1,999	17	0.84	.02999
26–30	5,869	5,810	59	1.01	.00001
31–35	2,422	2,396	26	1.07	.01517
36–40	531 ⎱	525	6 ⎱	1.17	.00609
41 +	67 ⎰	66	1 ⎰		.00300
Total	10,942	10,832	110	—	.06003
$\chi^2_3 = 0.618$		$I_{AD} = 0.0300$		$p_T = 0.1995$	
Parity 4					
< 21	5 ⎱	484	— ⎱	1.63	.08157
21–25	487 ⎰		8 ⎰		
26–30	2,872	2,853	19	0.66	.05963
31–35	2,537	2,523	14	0.55	.10750
36–40	761	752	9	1.18	.06080
41 +	87	85	2	2.30	.02577
Total	6,749	6,697	52	—	.33527
$\chi^2_4 = 11.091*$		$I_{AD} = 0.1676$		$p_T = 0.1230$	
Parity 5					
21–25	77 ⎱	—	— ⎱	1.20	.02400
26–30	1,005 ⎰	1,069	13 ⎰		
31–35	1,783	1,758	25	1.40	.03295
36–40	847	839	8	0.94	.06290
41 +	102	98	4	3.92	.05396
Total	3,814	3,764	50	—	.17381
$\chi^2_3 = 6.466$		$I_{AD} = 0.0869$		$p_T = 0.0695$	

Analysis of the Malformation Data

TABLE 8.4—Continued

	Total births	Normal births	Malformed infants	Per cent malformed	AD
Parity 6					
21–25	11 ⎱	—	— ⎱	1.59	.00587
26–30	241 ⎰	248	4 ⎰		
31–35	989	979	10	1.01	.15231
36–40	794	780	14	1.76	.06145
41 +	147	142	5	3.40	.08548
Total	2,182	2,149	33	—	.30511

$\chi^2_3 = 5.535$ $\qquad I_{AD} = 0.1526 \qquad$ $p_T = 0.0398$

Parities 7 +					
21–25	4 ⎱	—	— ⎱	3.57	.04849
26–30	52 ⎰	54	2 ⎰		
31–35	455	449	6	1.32	.02196
36–40	935	928	7	0.75	.26142
41 +	293	283	10	3.41	.23489
Total	1,739	1,714	25	—	.56676

$\chi^2_3 = 13.046^{**}$ $\qquad I_{AD} = 0.2834 \qquad$ $p_T = 0.0317$

All parities					
< 21	2,395	2,375	20	0.84	.00354
21–25	17,684	17,540	144	0.81	.03355
26–30	19,363	19,194	169	0.87	.01378
31–35	10,238	10,142	96	0.94	.00617
36–40	4,405	4,358	47	1.07	.01420
41 +	765	743	22	2.88	.03051
Total	54,850	54,352	498	—	—

$\chi^2_5 = 36.398^{**}$ $\qquad I'_{AD}$ (for mother's age) $= \Sigma p_T I_{AD} = 0.0691$

FIGURE 8.1 — The distribution of the frequency of infants with major malformation by maternal age for specified parities.

THE CHILDREN OF ATOMIC BOMB SURVIVORS

FIGURE 8.2 — The distribution of the frequency of grossly malformed infants by parity for specified maternal ages.

TABLE 8.5 THE EFFECT OF MATERNAL PARITY AT FIXED AGE ON THE FREQUENCY OF MALFORMED INFANTS
(Detailed explanation in text.)

	Total births	Normal births	Malformed infants	Per cent malformed	AD
Mother's age: < 21					
1	1,929	1,913	16	0.82	.00547
2	424	421	3	0.71	.02726
3	37 }	41	1 }	2.38	.03274
4	5 }		— }		
Total	2,395	2,375	20	—	.06547
$\chi^2_2 = 1.296$		$I_{AD} = 0.0327$		$P_T = 0.0437$	
Mother's age: 21–25					
1	8,697	8,636	61	0.70	.06875
2	6,392	6,334	58	0.91	.04166
3	2,016	1,999	17	0.84	.00409
4	487 }		8 }		
5	77 }	571	— }	1.38	.02301
6	11 }		— }		
7+	4 }		— }		
Total	17,684	17,540	144	—	.13751
$\chi^2_3 = 4.387$		$I_{AD} = 0.0688$		$P_T = 0.3224$	

TABLE 8.5—Continued

	Total births	Normal births	Malformed infants	Per cent malformed	AD
Mother's age: 26–30					
1	3,168	3,151	17	0.54	.06358
2	6,156	6,101	55	0.89	.00758
3	5,869	5,810	59	1.01	.04641
4	2,872	2,853	19	0.66	.03621
5	1,005	992	13	1.29	.02524
6	241	237	4	1.66	.01132
7+	52	50	2	3.85	.00923
Total	19,363	19,194	169	—	.19957

$\chi^2_6 = 15.935^*$ $I_{AD} = 0.0998$ $p_T = 0.3530$

Mother's age: 31–35					
1	621	617	4	0.64	.01917
2	1,431	1,420	11	0.77	.02543
3	2,422	2,396	26	1.07	.03458
4	2,537	2,523	14	0.55	.10294
5	1,783	1,758	25	1.40	.08707
6	989	979	10	1.01	.00763
7+	455	449	6	1.32	.01823
Total	10,238	10,142	96	—	.29505

$\chi^2_6 = 10.472$ $I_{AD} = 0.1475$ $p_T = 0.1867$

Mother's age: 36–40					
1	180	179	1	0.56	.01979
2	357	355	2	0.56	.03891
3	531	525	6	1.13	.00719
4	761	752	9	1.18	.01893
5	847	839	8	0.94	.02231
6	794	780	14	1.76	.11889
7+	935	928	7	0.75	.06400
Total	4,405	4,358	47	—	.29002

$\chi^2_6 = 6.095$ $I_{AD} = 0.1450$ $p_T = 0.0803$

Mother's age: 41+					
1	19 ⎫		— ⎫		
2	50 ⎬	135	— ⎬	0.74	.13625
3	67 ⎭		1 ⎭		
4	87	85	2	2.30	.02349
5	102	98	4	3.92	.04992
6	147	142	5	3.40	.03615
7+	293	283	10	3.41	.07366
Total	765	743	22	—	.31947

$\chi^2_4 = 3.184$ $I_{AD} = 0.1597$ $p_T = 0.0139$

All ages:					
1	14,614	14,515	99	0.68	.06826
2	14,810	14,681	129	0.87	.01107
3	10,942	10,832	110	1.01	.02159
4	6,749	6,697	52	0.77	.01880
5	3,814	3,764	50	1.31	.03115
6	2,182	2,149	33	1.51	.02673
7+	1,739	1,714	25	1.44	.01867
Total	54,850	54,352	498	—	

$\chi^2_6 = 32.594^{**}$ $I''_{AD}\ (\text{parity}) = 0.1002$

THE CHILDREN OF ATOMIC BOMB SURVIVORS

diagnoses of mongolism, and of these three, one was born at a maternal age of 25 and the other two at 34 and 40 years of age. The diagnosis of mongolism, while not particularly uncommon at 9 months of age (1 out of 1,580 children), was singularly infrequent in the at-birth data. This is not surprising in view of the not-uncommon failure to diagnose mongolism at birth in Caucasian infants, and the added diagnostic difficulties among Japanese.

In view of the greater average age and parity of the mother as parental exposure increases, a spurious irradiation effect could arise if the data were unadjusted for this fact. It might also be pointed out, however, that if no differences are demonstrable without age-parity adjustments, it is unlikely that differences will be detected following adjustment because of the direction of the bias.

It was felt that while some adjustment for age-parity effects was indicated, any attempt to subdivide the present data in such a manner as to take into consideration *both* age and parity effects would jeopardize the validity of the statistical tests employed. Somewhat arbitrarily the decision was reached to take into consideration *age but not parity* effects. Accordingly, for this analysis the terminations have been classified not only with respect to sex, city, and parental exposure, but also maternal age. Five age categories were recognized, namely, 15–20, 21–35, 36–38, 39–41, and 42 and over. Since the selection of the age intervals is purely arbitrary, the intervals were so selected as to be of the shortest span consistent with significant numbers in those areas where maternal age appears to be most important. It must be borne in mind, however, that the "age effects" which emerge from this analysis are also in part "parity effects." Some bias may be introduced as a consequence of ignoring parity but this bias would again appear to favor a higher rate of malformation among the exposed population.

The necessity for accounting for possible differences between exposure cells in maternal nutrition is more difficult to appraise. Warkany's (1947) extensive studies on the effect of severe dietary stresses on pregnancy outcome in gravid rats and mice would lead to the conclusion that diet is of considerable importance. In these animals, the degree of dietary stress needed to induce malformation is, however, far more severe than could have obtained even in the war torn cities of Japan. Moreover, neither Smith (1947) nor Antonov (1947), in their studies of the populations of Rotterdam and Leningrad following severe dietary stress during World War II, could demonstrate an increase in the frequency of congenital malformations. Whatever dietary stresses may have existed in the populations of Hiroshima and Nagasaki during the earlier stages of this study were very probably not so severe as those which obtained in Rotterdam and Leningrad. With respect to possible differences in nutritional status between exposure categories, detailed dietary histories on random samples of the population are not available. However, in a marginal economy such as existed in post-war Japan, there is undoubtedly a high correlation between economic status and nutritional level; we have already seen (Sec. 5.3) that in neither city is there a demonstrable relationship between economic level and exposure history. It will be assumed in this analysis that there are no important differences in maternal nutrition with respect to the various exposure subclasses.

Lastly, by way of completing a discussion of the concomitants, it might be mentioned that during the course of this study no exceptional epidemics of German measles or other viral infections were known to have occurred in Hiroshima or Nagasaki.

8.5 The "at-birth" data. — In Tables 8.6 and 8.7 are presented the "at-birth" frequencies of major malformation by parental exposure, sex of infant, and city, with and without classification of the termination by maternal age. Inspection of these tables reveals no striking, general differences in the frequency of infants with at least one major malformation by sex, city or parental exposure. The results of the analysis of these two tables are given in Tables 8.8 and 8.9. From Table 8.8 (a and b) we note no demonstrable differences when age is ignored, among the cities, sexes, or parental exposures. It is worth noting at this point that the frequency of malformation in the unexposed groups in Hiroshima and Nagasaki is very similar to the frequency of malformation in Kure where during the years control studies were in effect there 57 malformed infants were reported in 7,544 registered terminations (0.76%).

In Table 8.9, with the data further partitioned to take into account age differences among the parents, the only "main effect"

TABLE 8.6 FREQUENCY OF MALFORMED INFANTS BY PARENTAL EXPOSURE, SEX OF INFANT, AND CITY
(Unrelated parents)

Male infants
Hiroshima

			Mothers				
Fathers			1	2	3	4–5	Total
	1	n	9,005	2,832	1,114	571	13,522
		m	73	30	4	6	113
		p	.0081	.0106	.0036	.0105	.0084
	2	n	772	941	220	109	2,042
		m	11	14	—	1	26
		p	.0142	.0149	—	.0092	.0127
	3	n	313	197	268	54	832
		m	4	2	1	2	9
		p	.0128	.0102	.0037	.0370	.0108
	4–5	n	209	122	80	56	467
		m	1	1	—	—	2
		p	.0048	.0082	—	—	.0043
	Total	n	10,299	4,092	1,682	790	16,863
		m	89	47	5	9	150
		p	.0086	.0115	.0030	.0114	.0089

Male infants
Nagasaki

			Mothers				
Fathers			1	2	3	4–5	Total
	1	n	7,608	4,849	360	279	13,096
		m	68	39	4	3	114
		p	.0089	.0080	.0111	.0108	.0087
	2	n	1,120	2,112	134	56	3,422
		m	7	23	2	—	32
		p	.0063	.0109	.0149	—	.0094
	3	n	129	140	51	18	338
		m	4	—	1	—	5
		p	.0310	—	.0196	—	.0148
	4–5	n	75	103	14	15	207
		m	2	—	—	—	2
		p	.0267	—	—	—	.0097
	Total	n	8,932	7,204	559	368	17,063
		m	81	62	7	3	153
		p	.0091	.0086	.0125	.0082	.0090

TABLE 8.6—Continued

Female infants

Hiroshima

			Mothers				
			1	2	3	4–5	Total
Fathers	1	n	8,289	2,536	1,071	546	12,442
		m	87	21	10	7	125
		p	.0105	.0083	.0093	.0128	.0100
	2	n	728	909	204	93	1,934
		m	6	12	2	2	22
		p	.0082	.0132	.0098	.0215	.0114
	3	n	283	188	253	56	780
		m	3	2	4	1	10
		p	.0106	.0106	.0158	.0179	.0128
	4–5	n	186	122	77	61	446
		m	3	1	1	1	6
		p	.0161	.0082	.0130	.0164	.0135
	Total	n	9,486	3,755	1,605	756	15,602
		m	99	36	17	11	163
		p	.0104	.0096	.0106	.0146	.0104

Female infants

Nagasaki

			Mothers				
			1	2	3	4–5	Total
Fathers	1	n	7,002	4,467	387	280	12,136
		m	66	31	5	3	105
		p	.0094	.0069	.0129	.0107	.0087
	2	n	1,050	2,032	145	60	3,287
		m	4	13	3	—	20
		p	.0038	.0064	.0207	—	.0061
	3	n	114	133	43	17	307
		m	1	—	—	—	1
		p	.0088	—	—	—	.0033
	4–5	n	64	75	21	13	173
		m	—	2	—	—	2
		p	—	.0267	—	—	.0116
	Total	n	8,230	6,707	596	370	15,903
		m	71	46	8	3	128
		p	.0086	.0069	.0134	.0081	.0080

TABLE 8.7 THE FREQUENCY OF MALFORMED INFANTS BY PARENTAL EXPOSURE, CITY, AND MATERNAL AGE
(Unrelated parents)

Hiroshima

Age	Exposure[a]		11	12	13	14 15	Total F₁	21	22	23	24 25	Total F₂	31	32	33	34 35	Total F₃	41 51	42 52	43 53	44 45 54 55	Total F₄,₅	M₁	M₂	M₃	M₄,₅	Total age
15–20		n m p	801 5 .0062	265 5 .0189	121 3 .0248	55 1 .0182	1,242 14 .0113	83 — —	44 1 .0227	12 — —	8 — —	147 1 .0068	39 — —	21 1 .0476	17 — —	5 — —	82 1 .0122	14 — —	8 — —	3 — —	3 — —	28 — —	937 5 .0053	338 7 .0207	153 3 .0196	71 1 .0141	1,499 16 .0107
21–35		n m p	15,476 140 .0091	4,766 41 .0086	1,921 10 .0052	985 11 .0111	23,148 202 .0087	1,309 16 .0122	1,462 19 .0130	347 2 .0058	165 3 .0182	3,283 40 .0122	497 7 .0141	302 3 .0099	382 3 .0079	85 3 .0353	1,266 16 .0126	347 3 .0086	206 2 .0097	128 1 .0078	87 1 .0115	768 7 .0091	17,629 166 .0094	6,736 65 .0096	2,778 16 .0058	1,322 18 .0136	28,465 265 .0093
36–38		n m p	681 8 .0117	233 2 .0086	93 — —	42 1 .0238	1,049 11 .0105	64 — —	212 3 .0142	44 — —	16 — —	336 3 .0089	36 — —	31 — —	68 1 .0147	11 — —	146 1 .0068	24 1 .0417	15 — —	18 — —	9 — —	66 1 .0152	805 9 .0112	491 5 .0102	223 1 .0045	78 1 .0128	1,597 16 .0100
39–41		n m p	254 3 .0118	85 2 .0235	38 1 .0263	26 — —	403 6 .0149	34 1 .0294	105 2 .0190	15 — —	11 — —	165 3 .0182	17 — —	19 — —	37 — —	5 — —	78 — —	5 — —	14 — —	5 — —	14 — —	38 — —	310 4 .0129	223 4 .0179	95 1 .0105	56 — —	684 9 .0132
42+		n m p	82 4 .0487	19 1 .0526	12 — —	9 — —	122 5 .0410	10 — —	27 1 .0370	6 — —	2 — —	45 1 .0222	7 — —	12 1 .0588	17 1 .0588	4 — —	40 1 .0250	5 — —	1 — —	3 — —	4 — —	13 — —	104 4 .0385	59 2 .0339	38 1 .0263	19 — —	220 7 .0318
Total		n m p	17,294 160 .0093	5,368 51 .0095	2,185 14 .0064	1,117 13 .0116	25,964 238 .0092	1,500 17 .0113	1,850 26 .0141	424 2 .0047	202 3 .0149	3,976 48 .0121	596 8 .0117	385 4 .0104	521 5 .0096	110 4 .0273	1,612 19 .0118	395 4 .0101	244 2 .0082	157 1 .0064	117 1 .0085	913 8 .0088	19,785 188 .0095	7,847 83 .0106	3,287 22 .0067	1,546 20 .0129	32,465 313 .0096

Nagasaki

Age	Exposure[a]		11	12	13	14 15	Total F₁	21	22	23	24 25	Total F₂	31	32	33	34 35	Total F₃	41 51	42 52	43 53	44 45 54 55	Total F₄,₅	M₁	M₂	M₃	M₄,₅	Total age
15–20		n m p	432 — —	488 6 .0123	42 — —	29 2 .0690	991 8 .0081	109 — —	173 3 .0173	24 — —	5 — —	311 3 .0096	11 — —	9 — —	2 — —	— — —	22 — —	3 — —	5 — —	— — —	1 — —	9 — —	555 6 —	675 9 .0133	68 — —	35 2 .0571	1,333 17 .0083
21–35		n m p	12,616 111 .0088	7,900 53 .0067	625 9 .0144	475 4 .0084	21,616 177 .0082	1,711 10 .0058	2,932 19 .0065	186 5 .0269	89 — —	4,918 34 .0069	189 5 .0265	206 — —	67 — —	25 — —	487 5 .0103	111 2 .0180	143 1 .0070	29 — —	22 — —	305 3 .0098	14,627 128 .0088	11,181 73 .0065	907 14 .0154	611 4 .0065	27,326 219 .0080
36–38		n m p	979 14 .0143	549 4 .0073	45 — —	33 — —	1,606 18 .0112	193 — —	494 4 .0081	39 — —	12 — —	738 4 .0054	20 — —	25 — —	14 — —	5 — —	64 — —	10 — —	10 1 .1000	3 — —	1 — —	24 1 .0417	1,202 14 .0116	1,078 9 .0083	101 — —	51 — —	2,432 23 .0095
39–41		n m p	451 6 .0133	303 6 .0198	28 — —	18 — —	800 12 .0150	117 — —	366 7 .0191	19 — —	5 — —	507 7 .0138	15 — —	16 1 —	8 1 .1250	5 — —	44 1 .0227	8 — —	12 — —	2 — —	3 — —	25 — —	591 6 .0102	697 13 .0187	57 1 .0175	31 — —	1,376 20 .0145
42+		n m p	132 3 .0227	76 1 .0132	7 — —	4 — —	219 4 .0183	40 1 .0250	179 3 .0168	11 — —	5 — —	235 4 .0170	8 — —	17 — —	3 — —	— — —	28 — —	7 1 —	8 — —	1 — —	1 — —	17 — —	187 4 .0214	280 4 .0143	22 — —	10 — —	499 8 .0160
Total		n m p	14,610 134 .0092	9,316 70 .0075	747 9 .0120	559 6 .0107	25,232 219 .0087	2,170 11 .0051	4,144 36 .0087	279 5 .0179	116 — —	6,709 52 .0078	243 5 .0206	273 1 .0106	94 1 .0106	35 — —	645 6 .0093	139 2 .0144	178 2 .0112	35 — —	28 — —	380 4 .0105	17,162 152 .0089	13,911 108 .0078	1,155 15 .0130	738 6 .0081	32,966 281 .0085

[a] Father's exposure is indicated first.

emerging at the level of significance is the already discussed effect of maternal age on the frequency of congenital malformation. Among the first order interactions none is significant save mother's exposure-mother's age. However, since the effect of mother's exposure is not independent of age no generalizations can be made with respect to the effect of maternal irradiation based on the main effects test alone. In order to interpret the effect of maternal exposure, we must consider separately the maternal age classifications. Table 8.10 presents the distribution of the frequency of malformation by mother's age and exposure. For purposes of comparison and its bearing on a possible somatic maternal effect, a comparable table (Table 8.13) is given for father's exposure. In Table 8.11 are given the results of analyzing the effect of mother's exposure at each of the five different maternal age levels. For these five age levels mother's exposure seems to exert an effect only in the 15–20 year age interval. From inspection of Table 8.10 we note that the frequency of malformation tends to increase with increasing maternal exposure with an eight-fold difference between mother's "1" and mother's "4–5," or a two-fold difference between mother's "2" and mother's "4–5." This difference ceases to be significant, however, when consideration is

TABLE 8.8 CHI-SQUARE ANALYSIS OF THE FREQUENCY OF CONGENITALLY MALFORMED INFANTS BY SEX, CITY, AND PARENTAL EXPOSURE
(Unrelated parents)

a. All exposure cells (4 × 4)

Source	DF	χ^2	P
Total	63	63.077	.50–.60
Interactions, first order			
CS	1	1.039	.30–.50
CM	3	9.427	.02–.05
CF	3	2.054	.50–.70
SM	3	5.210	.10–.20
SF	3	3.057	.30–.50
MF	9	6.107	.70–.80
Main effects			
City (C)	1	2.269	.10–.20
Sex (S)	1	0.169	.50–.70
Mothers (M)	3	1.907	.50–.70
Fathers (F)	3	1.232	.70–.80

b. Excluding parents in Category 1 (3 × 3)

Source	DF	χ^2	P
Total	35	28.806	.80–.90
Interactions, first order			
CS	1	1.027	.30–.50
CM	2	6.411	.02–.05
CF	2	3.282	.10–.20
SM	2	3.052	.20–.30
SF	2	2.364	.30–.50
MF	4	4.080	.30–.50
Main effects			
City (C)	1	2.406	.10–.20
Sex (S)	1	0.015	.90–.95
Mothers (M)	2	0.222	.80–.90
Fathers (F)	2	0.480	.80–.90

TABLE 8.9 CHI-SQUARE ANALYSIS OF THE FREQUENCY OF CONGENITALLY MALFORMED INFANTS BY CITY, MATERNAL AGE, AND PARENTAL EXPOSURE
(Unrelated parents)

All exposure cells

Source	DF	χ^2	P
Total	156	164.205	.50–.60
Interactions, first order			
CM	3	9.657	.02–.05
CF	3	2.053	.50–.70
CA	4	1.914	.70–.80
AM	12	23.786	.02–.05
AF	12	7.907	.70–.80
MF	9	6.107	.70–.80
Main effects			
Age (A)	4	18.057	.001–.01
City (C)[a]			
Age intervals			
15–20	1	0.438	.50–.70
21–35	1	2.720	.05–.10
36–38	1	0.032	.80–.90
39–41	1	0.062	.80–.90
42 +	1	1.863	.10–.20
Sum	5	5.115	.30–.50
Mothers (M)[a, b]			
Age intervals			
15–20	3	14.581	.001–.01
21–35	3	4.463	.20–.30
36–38	3	2.277	.50–.70
39–41	3	3.111	.30–.50
42 +	3	1.462	.50–.70
Sum	15	25.894	.02–.05
Fathers (F)[a]			
Age intervals			
15–20	3	0.411	.90–.95
21–35	3	2.619	.30–.50
36–38	3	3.555	.30–.50
39–41	3	1.303	.70–.80
42 +	3	1.400	.70–.80
Sum	15	9.288	.80–.90

[a] Adjusted for age (see Sec. 6.5).
[b] See Table 8.11.

limited to only those cells in which both parents are exposed. Moreover, it is not significant if only mothers who were unexposed are excluded from consideration ($\chi^2 = 1.062$; DF $= 2$). The significance of the difference seems to be largely a function of the inordinately low frequency of malformation observed among the unexposed mothers in the 15–20-year age interval. In

amination necessitated two departures from the usual analytical approach. Firstly, parental exposure categories 3, 4, and 5 were pooled. Secondly, all extraneous variation was ignored.

TABLE 8.10 THE DISTRIBUTION OF FREQUENCY OF MALFORMED INFANTS CLASSIFIED BY MOTHER'S AGE AND EXPOSURE ONLY

Age		1	2	3	4–5	Total
15–20	n	1,492	1,013	221	106	2,832
	m	5	16	3	3	27
	p	.0034	.0158	.0136	.0283	.0095
21–35	n	32,256	17,917	3,685	1,933	55,791
	m	294	138	30	22	484
	p	.0091	.0077	.0081	.0114	.0087
36–38	n	2,007	1,569	324	129	4,029
	m	23	14	1	1	39
	p	.0115	.0089	.0031	.0078	.0097
39–41	n	901	920	152	87	2,060
	m	10	17	2	—	29
	p	.0111	.0185	.0132	—	.0141
42 +	n	291	339	60	29	719
	m	8	6	1	—	15
	p	.0275	.0177	.0167	—	.0209
Total	n	36,947	21,758	4,442	2,284	65,431
	m	340	191	37	26	594
	p	.0092	.0088	.0083	.0114	.0091

TABLE 8.11 CHI-SQUARE ANALYSIS OF THE EFFECT OF MOTHER'S EXPOSURE ON THE FREQUENCY OF MALFORMED INFANTS AT EACH OF THE FIVE DIFFERENT AGE LEVELS

Age group	DF	χ^2	P
15–20	3	14.581	.001–.01
21–35	3	4.463	.20–.30
36–38	3	2.276	.50–.70
39–41	2[a]	2.421	.20–.30
42 +	2[a]	1.196	.50–.70

[a] Groups 3, 4, and 5 have been pooled.

Table 8.12 is presented the distribution of malformed infants among mothers in the 15–20-year age interval by parity and maternal exposure. We note that unexposed mothers (in the 15–20 year age interval) having their first or second pregnancies experienced an extremely low malformation rate. An explanation for this extremely low rate is not immediately apparent.

8.6 *The "9-months" data.* — The sparcity of the material available from the 9-months ex-

TABLE 8.12 THE DISTRIBUTION BY MATERNAL EXPOSURE AND PARITY OF MALFORMED INFANTS BORN TO MOTHERS OF AGES 15–20

Parity		1	2	3	4–5	Total
1	n	1,204	805	168	77	2,254
	m	3	13	1	3	20
	p	.0025	.0161	.0060	.0390	.0089
2	n	266	185	50	24	525
	m	1	3	2	—	6
	p	.0038	.0162	.0400	—	.0114
3–4	n	22	23	3	5	53
	m	1	—	—	—	1
	p	.0455	—	—	—	.0189
Total	n	1,492	1,013	221	106	2,832
	m	5	16	3	3	27
	p	.0034	.0158	.0136	.0283	.0095

TABLE 8.13 THE DISTRIBUTION OF FREQUENCY OF MALFORMED INFANTS CLASSIFIED AT BIRTH BY MOTHER'S AGE AND FATHER'S EXPOSURE ONLY

Age		1	2	3	4–5	Total
15–20	n	2,233	458	104	37	2,832
	m	22	4	1	—	27
	p	.0099	.0087	.0096	—	.0095
21–35	n	44,764	8,201	1,753	1,073	55,791
	m	379	74	21	10	484
	p	.0085	.0090	.0120	.0093	.0087
36–38	n	2,655	1,074	210	90	4,029
	m	29	7	1	2	39
	p	.0109	.0065	.0048	.0222	.0097
39–41	n	1,203	672	122	63	2,060
	m	18	10	1	—	29
	p	.0150	.0149	.0082	—	.0141
42 +	n	341	280	68	30	719
	m	9	5	1	—	15
	p	.0264	.0179	.0147	—	.0209
Total	n	51,196	10,685	2,257	1,293	65,431
	m	457	100	25	12	594
	p	.0089	.0094	.0111	.0093	.0091

These considerations must be borne in mind in appraising Tables 8.14 and 8.15.[1] In Table 8.14

[1] The discrepancy in terms of total number of infants recorded in Tables 8.3 and 8.13 is due to the fact that in the former any child seen in the "9-months follow-up" is scored (these infants will vary in age from 7½ to 17 months), whereas in the latter only those infants whose ages were 8–10 months in-

TABLE 8.14 THE FREQUENCY OF MALFORMED INFANTS AMONG ALL INFANTS RE-EXAMINED AT 9-MONTHS OF AGE, BY CITY AND PARENTAL EXPOSURE

(Malformed infants are defined here as being those infants found to have one or more major defects at 9 months of age not diagnosed at birth.)

Hiroshima

			Mothers		
Fathers		1	2	3–5	Total
1	n	5,424	1,718	1,016	8,158
	m	116	25	13	154
	p	.0214	.0146	.0128	.0189
2	n	458	501	180	1,139
	m	5	12	2	19
	p	.0109	.0239	.0111	.0167
3–5	n	303	178	250	731
	m	5	2	3	10
	p	.0165	.0112	.0120	.0137
Total	n	6,185	2,397	1,446	10,028
	m	126	39	18	183
	p	.0204	.0163	.0124	.0182

Nagasaki

			Mothers		
Fathers		1	2	3–5	Total
1	n	4,016	2,474	371	6,861
	m	97	45	5	147
	p	.0242	.0182	.0135	.0214
2	n	563	990	115	1,668
	m	16	22	4	42
	p	.0284	.0222	.0348	.0252
3–5	n	116	142	61	319
	m	1	2	3	6
	p	.0086	.0141	.0492	.0188
Total	n	4,695	3,606	547	8,848
	m	114	69	12	195
	p	.0243	.0191	.0219	.0220

are presented the frequencies of malformed infants, diagnosed at 9 months but not so diagnosed at birth, distributed by city and parental exposure. The analysis of these data is presented in Table 8.15. Inspection of the analysis reveals no significant differences between cities or father's exposure categories, but an effect of mother's exposure significant at the 5 per cent level. This effect, however, is contrary to what one might expect from radiation, since there are clusive are scored. The purpose of the more rigid definition in the latter table was to minimize another source of extraneous variation, namely, age of infant.

TABLE 8.15 CHI-SQUARE ANALYSIS OF THE FREQUENCY OF MALFORMED INFANTS AT 9 MONTHS OF AGE BY CITY AND PARENTAL EXPOSURE
(Unrelated parents)

Source	DF	χ^2	P
Total	17	22.543	.10–.20
Interactions, first order			
CM	2	1.062	.50–.70
CF	2	1.011	.50–.70
MF	4	3.942	.30–.50
Main effects			
City (C)	1	3.441	.05–.10
Mothers (M)	2	6.070	.02–.05
Fathers (F)	2	1.643	.30–.50

TABLE 8.16 THE DISTRIBUTION BY MATERNAL EXPOSURE OF THE SEVEN MOST COMMON MAJOR CONGENITAL MALFORMATIONS IN THE JAPANESE, EXCLUSIVE OF CONGENITAL HEART DISEASE

(The incidences of the malformations are given in terms of number of malformations per 1,000 births.)

Type of malformation	1 No.	1 No./1,000 births	2 No.	2 No./1,000 births	3 No.	3 No./1,000 births	4–5 No.	4–5 No./1,000 births	Total
Anencephaly	26	0.70	17	0.78	4	0.90	2	0.88	49
Cleft palate, single	32	0.87	7	0.32	3	0.68	—	—	42
Club foot	52	1.41	28	1.29	5	1.13	3	1.31	88
Harelip with cleft palate	55	1.49	34	1.56	3	0.68	6	2.63	98
Harelip without cleft palate	37	1.00	18	0.83	5	1.13	2	0.88	62
Polydactyly	37	1.00	28	1.29	3	0.68	2	0.88	70
Syndactyly	23	0.62	10	0.46	3	0.68	—	—	36
Sum of malformations	262	7.09	142	6.53	26	5.85	15	6.57	445
Total Births	36,947		21,758		4,442		2,284		65,431

THE CHILDREN OF ATOMIC BOMB SURVIVORS

relatively more malformations among the children of category 1 mothers, and this despite the fact that age-parity effects, which should introduce a bias in the direction of hypothesis, have not been allowed for.

8.7 *Analysis by specific malformation type.* — The genetic basis of most of the common types of malformations is poorly understood. Accordingly, the remote possibility had to be considered that the frequency of children with certain kinds of malformations was affected in a rather specific way by parental radiation. Despite the improbability of the existence of such an effect in the absence of an effect of radiation exposure in the material as a whole, an analysis of this possibility has been undertaken. The results, with respect to the seven most common malformation types, are given in Table 8.16. When the data are partitioned in this manner there are relatively few malformations of any specific type. For this reason, the classification according to radiation history is by mother's exposure only. Inspection of Table 8.16 reveals no noteworthy trends.

8.8 *Summary.* — Analysis of the frequency of malformed infants by city and parental exposure reveals no significant, consistent effect of parental exposure. The small departures from expectation which are observed cannot be said to be in the direction of genetic hypothesis.

Chapter IX

ANALYSIS OF THE STILLBIRTH DATA

9.1 *The trait.* — The definition of a stillbirth employed in this study is an infant who shows no signs of life at birth, following a period of gestation of at least 20 weeks. Any infant that moves, makes respiratory attempts, or is born with a beating heart is not in the strict sense "stillborn." This definition, which is that employed in United States vital statistics, runs counter to current Japanese usage, under which infants who showed signs of life at birth but never succeeded in establishing a pattern of regular respiration are frequently reported as stillborn. This usage was in part a matter of convenience, since a stillbirth required completing only one official form, whereas a livebirth dying an hour following delivery required completing two. The subject of the ABCC's usage of the term "stillbirth" was repeatedly taken up at conferences with the midwives.

The frequency of stillbirths encountered in our control material (both parents category 1) was in Hiroshima 1.27 per cent of all births (17,189 births) and in Nagasaki 1.31 per cent (14,450 births). In comparison, in 1950 the "official" stillbirth rate (stillbirths per 1,000 livebirths) was in Hiroshima prefecture 79.6 and in Nagasaki prefecture 87.7, or, expressed as percentage of all births, 7.37 per cent and 8.07 per cent respectively (Public Health and Welfare in Japan, 1950). Official statistics for other years are similar. There is apparent a marked discrepancy between official statistics and our own. There are at least five factors to be taken into account in an evaluation of these differences. Firstly, there is the matter of registration practices referred to above. Secondly, the official stillbirth data are for stillbirths after the third month of uterogestation, whereas by and large our figures refer to events following the fifth month of gestation. Thirdly, since the official figures are for an entire prefecture, there is the possibility of urban-rural differentials. This, however, works in the opposite direction than the first two factors, since in 1950 the stillbirth rate for all *shi* (cities) was 134.0 per 1,000 livebirths, whereas for all *gun* (villages and rural areas) the rate was 70.4. Fourthly, the official figures may include some induced terminations, whereas these were excluded from our own data. Lastly, those stillborn infants with gross malformations were excluded from our figures. The reasons for this exclusion were given in Section 6.2.

When approximate allowance is made for the exclusion from our figures of stillborn, malformed infants, the stillbirth rates observed in this study are quite similar to those currently obtaining in many parts of the United States (e.g., Michigan, 1950, 1.9 per cent). The possibility that the ABCC program was failing to obtain data on any substantial number of stillbirths seems precluded by the efficiency of pregnancy registration (Sec. 2.1) and the system of careful follow-up for all registered terminations.

9.2 *The genetic argument for radiation-induced changes.* — Animal experimentation on the genetic effects of irradiation indicates that the largest group of induced mutations having clear-cut effects detectable in the first generation of offspring following exposure are the autosomal lethals. As detected in the first generation of offspring these mutations would consist largely of the dominant lethals, although a few recessive lethals would be recovered due to the fortuitous combination in some individuals of an induced lethal mutant with an allelic lethal mutant of spontaneous origin. Presumably one of the manifestations of the presence of these mutant genes would be fetal death. Accordingly, we might expect, as one of the changes symptomatic of irradiation-induced genetic damage, an increase in the frequency with which infants are stillborn as parental exposure increases.

9.3 *Concomitant variation known to affect*

the stillbirth rate. — No one of the indicators with which we have dealt or shall deal is more complex with respect to concomitant variation than the one presently under consideration. No less than seven sources of concomitant variation known to affect the stillbirth rate could, conceivably, be influencing the data which we shall analyze. These seven sources of extraneous variation are congenital syphilis, birth injury, maternal age, parity, nutrition, rate of reproduction, and paternal age.

Congenital syphilis has long been one of the major causes of fetal mortality. In Hiroshima and Nagasaki the frequency of positive serologies is from 5 to 7 per cent among post-parturient women. To assess the amount of infantile mortality attributable to congenital syphilis among the births described here would have required an expenditure of time and energy which, because of the limitations of personnel by syphilitic stillborn infants (and we have no evidence that it is not) then, because of the known differential in maternal age among the parental exposure cells, herein lies a possible source of bias. The age distribution among the exposure cells is such that this bias would tend to inflate the stillbirth rate among those parents less heavily exposed or not exposed at all. However, these potential effects would appear to be quite small, of a magnitude which could safely be ignored in this analysis.

A second major factor affecting infant survival (which it was not possible to control) is the occurrence of an injury during parturition. In the United States, it has been estimated that birth injury accounts for no less than 13.6 per cent of infantile mortality (Stander, 1941). Of the various types of injuries which may be sustained during the passage of the infant through the birth canal, by far the most common

TABLE 9.1 CONGENITAL SYPHILIS AMONG LIVING INFANTS BORN IN NAGASAKI IN 1951
(Incidence and relation to maternal age [after Wright, S.W. et al., 1952])

	Totals	15–19	20–24	25–29	30–34	35–39	40–44	45–50
Number of living infants	6,673	181	1,691	2,435	1,456	735	171	4
Number of cases of congenital syphilis	47	4	17	16	5	4	1	0
Incidence × 1,000	7	22	10	6.5	3.4	5.4	5.8	0

$\chi^2 = 11.160$ DF = 5 P < 0.05

and facilities, was not possible. However, a small study was made in Nagasaki on infants born in the year 1951 in an attempt to determine the incidence of congenital syphilis among the infants being examined in connection with the genetics program (Wright, S. W. et al., 1952). Initially it was planned that this study would estimate the incidence of congenital syphilis among living infants, stillbirths, and neonatal deaths. Because of the formidable difficulties posed in estimating the latter incidences in the absence of an exhaustive post-mortem program in Nagasaki, attention had to be focused on living infants alone. Two items of interest here emerged from this study. Firstly, some seven out of every thousand liveborn infants were shown to have congenital syphilis either by clinical, serologic, or roentgenologic examination, or any combination thereof. Secondly, a gradual decline in the rate of transmission with increasing maternal age was noted. These latter data are reproduced in Table 9.1. If this same effect of maternal age is exhibited

is intra-cranial injury with hemorrhage. It has been reported that 40–80 per cent of infants coming to autopsy will show evidence of this form of birth injury (Stander, 1941). In our own data among 50 autopsied infants, selected at random, 26 showed intra-cranial hemorrhages ranging from moderate to severe in degree. While the importance of birth injury in infantile death cannot be gainsaid, it is difficult to visualize circumstances which would lead to a non-random distribution of birth injuries among the parental exposure categories, and this possibility will be ignored in the analysis.

A third major factor affecting infant survival is maternal nutrition. The importance of this variable is amply attested to by an extensive literature. In the main, this literature is in agreement in indicating that rather drastic changes in maternal nutrition must occur before clearly demonstrable changes in the stillbirth rate can be shown (see, for example, Antonov, 1947, and Smith, 1947). There is no evidence that during the course of this study the limitations

imposed on maternal nutrition by economic conditions prevailing in Japan reached the levels necessary to produce measurable changes in the stillbirth rate, or that (arguing again from the rough classification of economic status presented in Sec. 5.3) such differences in maternal nutrition as existed were non-randomly distributed.

Among the variables affecting the stillbirth rate previously enumerated, the effects of maternal age and parity are by far the most complex and general in scope. These effects as manifested in the present material are summarized in Tables 9.2 and 9.3 and Figures 9.1 and 9.2. The data are presented and analyzed in the same manner as age and parity effects with respect to major congenital malformations (cf. Sec. 8.4). Again, a detailed discussion of the findings will be deferred for a subsequent publication. There is a significant effect on stillbirth frequency of both age and parity, with, in both instances, an apparent increase in stillbirth frequency at the extremes of the scale. This in-

TABLE 9.2 THE EFFECT OF MATERNAL AGE AT FIXED PARITY ON THE FREQUENCY OF STILLBORN INFANTS
(The meaning of the symbols employed is explained in Sec. 8.4.)

	Total births	Normal births	Stillborn infants	Per cent stillborn	AD
Parity 1					
<21	1,912	1,882	30	1.57	.01995
21–25	8,623	8,479	144	1.67	.05687
26–30	3,192	3,125	67	2.10	.03101
31–35	614	595	19	3.09	.02920
36–40	178 }	189	7 }	4.06	.01661
41+	19 }		1 }		
Total	14,538	14,270	268	—	.15364
$\chi^2_5 = 14.046*$		$I_{AD} = 0.0768$		$P_T = 0.2612$	
Parity 2					
<21	421	415	6	1.43	.01095
21–25	6,639	6,572	67	1.01	.00615
26–30	6,354	6,294	60	0.94	.03259
31–35	1,415	1,399	16	1.13	.00983
36–40	366	360	6	1.64	.01460
41+	47	46	1	2.13	.00336
Total	15,242	15,086	156	—	.07748
$\chi^2_5 = 3.174$		$I_{AD} = 0.0387$		$P_T = 0.2738$	
Parity 3					
<21	36 }	2,224	— }	1.02	.03634
21–25	2,211 }		23 }		
26–30	5,799	5,729	70	1.21	.01513
31–35	2,390	2,357	33	1.38	.02444
36–40	525 }	580	11 }	1.86	.02703
41+	66 }		— }		
Total	11,027	10,890	137	—	.10294
$\chi^2_3 = 3.153$		$I_{AD} = 0.0515$		$P_T = 0.1981$	
Parity 4					
<21	5 }	476	1 }	1.45	.02010
21–25	478 }		6 }		
26–30	2,848	2,828	20	0.70	.16468
31–35	2,521	2,487	34	1.35	.07112
36–40	750	736	14	1.87	.07286
41+	84	83	1	1.19	.00060
Total	6,686	6,610	76	—	.32936
$\chi^2_4 = 9.770*$		$I_{AD} = 0.1647$		$P_T = 0.1201$	

Analysis of the Stillbirth Data

TABLE 9.2—Continued

	Total births	Normal births	Stillborn infants	Per cent stillborn	AD
Parity 5					
21–25	78	76	2	2.56	.01723
26–30	991	975	16	1.61	.03879
31–35	1,756	1,737	19	1.08	.11020
36–40	837	823	14	1.67	.04208
41+	97	95	2	2.06	.01211
Total	3,759	3,706	53	—	.22041

$\chi^2_4 = 3.116$ $\quad I_{AD} = 0.1102$ $\quad P_T = 0.0675$

	Total births	Normal births	Stillborn infants	Per cent stillborn	AD
Parity 6					
21–25	12 ⎫	270	1 ⎫	1.46	.00158
26–30	262 ⎭		3 ⎭		
31–35	956	943	13	1.36	.03585
36–40	794	785	9	1.13	.08678
41+	141	135	6	4.26	.12421
Total	2,165	2,133	32	—	.24842

$\chi^2_3 = 8.208*$ $\quad I_{AD} = 0.1242$ $\quad P_T = 0.0389$

	Total births	Normal births	Stillborn infants	Per cent stillborn	AD
Parity 7+					
21–25	4 ⎫	58	1 ⎫	4.92	.02803
26–30	57 ⎭		2 ⎭		
31–35	488	479	9	1.84	.05538
36–40	1,195	1,167	28	2.34	.02451
41+	498	483	15	3.01	.05188
Total	2,242	2,187	55	—	.15980

$\chi^2_3 = 3.015$ $\quad I_{AD} = 0.0799$ $\quad P_T = 0.0403$

	Total births	Normal births	Stillborn infants	Per cent stillborn	AD
All parities					
<21	2,374	2,337	37	1.56	.00507
21–25	18,045	17,801	244	1.35	.01032
26–30	19,503	19,265	238	1.22	.04472
31–35	10,140	9,997	143	1.41	.00189
36–40	4,645	4,556	89	1.92	.03153
41+	952	926	26	2.73	.01659
Total	55,659	54,882	777	—	—

$\chi^2_5 = 26.547**$ $\quad I_{AD}'$ (for age) $= 0.0761$

crease at the lower end of the scale is more definite with respect to parity than with respect to age.

One of the more exhaustive previous studies of this relationship of maternal age and parity to infant survival is that of Yerushalmy (1945), who found that (1) "the lowest rates do not occur in one particular age group irrespective of parity. There is a consistent shifting of the minimum rate to the older ages with increasing parity." (2) "When the stillbirth rates by age of mother are compared for the different parity groups the increase in the rate is not proportionate for the various age groups. The increase is very much higher for the younger age of mother groups than for the older." The present data appear to agree with the first of these two conclusions, but of themselves would scarcely permit drawing the second conclusion — this may be in part a matter of the numbers involved. From the standpoint of controlling extraneous variation, these findings, taken in conjunction with our own, imply that the only really adequate control is one which takes into account not only maternal age but also parity. Among the Japanese data, to effect an adequate control would require recognizing approximately 25 parity-specific age intervals (five intervals for parity, and five for age, or 25 in all). The data, because of the already existing numerous ways

TABLE 9.3 THE EFFECT OF MATERNAL PARITY AT FIXED AGE ON THE FREQUENCY OF STILLBORN INFANTS
(The meaning of the symbols employed is explained in Sec. 8.4.)

	Total births	Normal births	Stillborn infants	Per cent stillborn	AD
Mother's age: < 21					
1	1,912	1,882	30	1.57	.00551
2	421	415	6	1.43	.01542
3+	41	40	1	2.44	.00991
Total	2,374	2,337	37	—	.03084
$\chi^2_2 = 0.257$		$I_{AD} = 0.0154$		$p_T = 0.0427$	
Mother's age: 21–25					
1	8,623	8,479	144	1.67	.11384
2	6,639	6,572	67	1.01	.09460
3	2,211	2,188	23	1.04	.02865
4	478	472	6	1.26	.00193
5	78 ⎫	90	4	4.26	.01133
6+	16 ⎭				
Total	18,045	17,801	244	—	.25035
$\chi^2_4 = 19.969^{**}$		$I_{AD} = 0.1252$		$p_T = 0.3242$	
Mother's age: 26–30					
1	3,192	3,125	67	2.10	.11930
2	6,354	6,294	60	.94	.07461
3	5,799	5,729	70	1.21	.00326
4	2,848	2,828	20	.70	.06276
5	991	975	16	1.61	.01662
6	262	259	3	1.15	.00083
7+	57	55	2	3.51	.00555
Total	19,503	19,265	238	—	.28293
$\chi^2_6 = 34.577^{**}$		$I_{AD} = 0.1415$		$p_T = 0.3504$	
Mother's age: 31–35					
1	614	595	19	3.09	.07335
2	1,415	1,399	16	1.13	.02805
3	2,390	2,357	33	1.38	.00500
4	2,521	2,487	34	1.35	.01101
5	1,756	1,737	19	1.08	.04088
6	956	943	13	1.36	.00342
7+	488	479	9	1.84	.01503
Total	10,140	9,997	143	—	.17674
$\chi^2_6 = 15.444^{*}$		$I_{AD} = 0.0884$		$p_T = 0.1822$	
Mother's age: 36–40					
1	178	171	7	3.93	.04112
2	366	360	6	1.64	.01160
3	525	514	11	2.10	.01077
4	750	736	14	1.87	.00435
5	837	823	14	1.67	.02334
6	794	785	9	1.13	.07118
7+	1,195	1,167	28	2.34	.05846
Total	4,645	4,556	89	—	.22082
$\chi^2_6 = 8.111$		$I_{AD} = 0.1104$		$p_T = 0.0835$	

Analysis of the Stillbirth Data

TABLE 9.3—*Continued*

	Total births	Normal births	Stillborn infants	Per cent stillborn	AD
Mother's age: 41 +					
1	19	18	1	5.26 ⎫	
2	47	46	1	2.13 ⎬	.06346
3	66	66	—	— ⎭	
4	84	83	1	1.19	.05117
5	97	95	2	2.06	.02567
6	141	135	6	4.26	.08499
7 +	498	483	15	3.01	.05533
Total	952	926	26	—	.28062

$\chi^2_4 = 3.030$ $\qquad I_{AD} = 0.1403$ $\qquad p_T = 0.0171$

All ages					
1	14,538	14,270	268	1.84	.08491
2	15,242	15,086	156	1.02	.07411
3	11,027	10,890	137	1.24	.02211
4	6,686	6,610	76	1.14	.02263
5	3,759	3,706	53	1.41	.00068
6	2,165	2,133	32	1.48	.00231
7 +	2,242	2,187	55	2.45	.03094
Total	55,659	54,882	777	—	

$\chi^2_6 = 59.978^{**}$ $\qquad I_{AD}''$ (for parity) $= 0.1186$

FIGURE 9.1 — The distribution of the frequency of stillborn infants by age of mother for specified parities.

FIGURE 9.2 — The distribution of the frequency of stillborn infants by parity for specified maternal ages.

of classification, would not permit so elegant an approach to the extraneous variation occasioned by mother's age and parity. Accordingly, it became a matter of judgment as to which of these two factors was to be controlled. A decision was reached to disregard age but to recognize, for analytical purposes, five parity classes, namely, parity 1, parities 2–3, 4–5, 6–7, and 8 and higher.

9.4 The data. — Tables 9.4 and 9.5 present the distribution and analysis of stillborn infants by sex, city, and parental exposure without further subdivision by parity. Inspection of Table 9.4 suggests a difference in the stillbirth rate when the mother was not exposed as opposed to when she was. From the analytical table (9.5a) we note a significant effect of mother's exposure but not of sex or father's exposure. Attention is called to the fact that the sex-mother's exposure interaction is not significant, in contrast to expectation in view of the possible effect of exposure on the sex ratio described in Chapter VII. With regard to the reality of the effect of mother's exposure on the stillbirth rate, it should be borne in mind that these data are uncorrected for parity and the more heavily exposed cells are biased in a way which would increase the stillbirth rate in these cells. If in the analysis of Table 9.4 we limit our attention to those exposure cells in which *both* parents were exposed, which would reduce the differences between cells with regard to parity, the apparent effect of maternal exposure disappears (see Table 9.5b). This observation is of considerable importance in the interpretation of these data since (a) maternal exposure confounds the effects of genetic and somatic damage and hence a significant effect of maternal exposure not reflected in paternal exposure might well be a somatic rather than a genetic effect, and (b) there is evidence which has been presented in Chapter V and elsewhere in this report that the comparison of pregnancy terminations occurring to exposure category 1 parents with terminations occurring to parents in any other exposure category is biased.

Let us now turn to a consideration of what these data reveal when the sex of the infant is ignored but terminations are classified with respect to parity. Tables 9.6 and 9.7 present the

TABLE 9.4 FREQUENCY OF STILLBIRTHS BY SEX, PARENTAL EXPOSURE AND CITY
(Unrelated parents)

Male infants
Hiroshima

			Fathers				
			1	2	3	4–5	Total
Mothers	1	n	8,919	761	309	208	10,197
		s	114	10	5	2	131
		p	.0128	.0131	.0162	.0096	.0128
	2	n	2,796	924	195	121	4,036
		s	36	15	5	5	61
		p	.0129	.0162	.0256	.0413	.0151
	3	n	1,108	220	266	80	1,674
		s	27	1	8	1	37
		p	.0244	.0045	.0301	.0125	.0221
	4–5	n	565	108	52	56	781
		s	14	—	1	—	15
		p	.0248	—	.0192	—	.0192
	Total	n	13,388	2,013	822	465	16,688
		s	191	26	19	8	244
		p	.0143	.0129	.0231	.0172	.0146

Male infants
Nagasaki

			Fathers				
			1	2	3	4–5	Total
Mothers	1	n	7,527	1,108	125	73	8,833
		s	110	21	1	1	133
		p	.0146	.0190	.0080	.0137	.0151
	2	n	4,809	2,087	140	103	7,139
		s	77	32	3	—	112
		p	.0160	.0153	.0214	—	.0157
	3	n	355	132	50	13	550
		s	6	1	—	—	7
		p	.0169	.0076	—	—	.0127
	4–5	n	276	56	18	15	365
		s	2	1	—	—	3
		p	.0072	.0179	—	—	.0082
	Total	n	12,967	3,383	333	204	16,887
		s	195	55	4	1	255
		p	.0150	.0163	.0120	.0049	.0151

TABLE 9.4—*Continued*

Female infants
Hiroshima

		Fathers				
		1	2	3	4-5	Total
Mothers 1	n	8,190	720	280	183	9,373
	s	104	9	9	6	128
	p	.0127	.0125	.0321	.0328	.0137
2	n	2,512	897	186	121	3,716
	s	46	11	4	3	64
	p	.0183	.0123	.0215	.0248	.0172
3	n	1,060	202	248	76	1,586
	s	21	3	2	1	27
	p	.0198	.0149	.0081	.0132	.0170
4-5	n	539	91	55	60	745
	s	6	1	—	2	9
	p	.0111	.0110	—	.0333	.0121
Total	n	12,301	1,910	769	440	15,420
	s	177	24	15	12	228
	p	.0144	.0126	.0195	.0273	.0148

Female infants
Nagasaki

		Fathers				
		1	2	3	4-5	Total
Mothers 1	n	6,923	1,040	112	64	8,139
	s	80	17	—	—	97
	p	.0116	.0163	—	—	.0119
2	n	4,430	2,016	133	73	6,652
	s	57	49	2	3	111
	p	.0129	.0243	.0150	.0411	.0167
3	n	382	142	43	21	588
	s	9	3	—	—	12
	p	.0236	.0211	—	—	.0204
4-5	n	276	60	17	13	366
	s	4	2	1	—	7
	p	.0145	.0333	.0588	—	.0191
Total	n	12,011	3,258	305	171	15,745
	s	150	71	3	3	227
	p	.0125	.0218	.0098	.0175	.0144

TABLE 9.5 CHI-SQUARE ANALYSIS OF THE FREQUENCY OF STILLBIRTHS BY SEX, PARENTAL EXPOSURE, AND CITY
(Unrelated parents)

a. All exposure cells (4 × 4)

Source	DF	χ^2	P
Total	63	92.343	.01–.02
Interactions, first order			
CM	3	.620	.80–.90
CF	3	10.626	.01–.02
CS	1	0.200	.50–.70
SM	3	1.565	.50–.70
SF	3	5.499	.10–.20
MF	9	5.548	.70–.80
Main effects			
Cities (C)[a]			
Mother's exposure — 1	1	0.069	.70–.80
Mother's exposure — 2	1	0.001	.95–.98
Mother's exposure — 3	1	0.393	.50–.70
Mother's exposure — 4, 5	1	0.140	.70–.80
Sum	4	0.603	.95–.98
Sex (S)[a]			
Mother's exposure — 1	1	0.725	.30–.50
Mother's exposure — 2	1	0.661	.30–.50
Mother's exposure — 3	1	0.202	.50–.70
Mother's exposure — 4, 5	1	0.065	.70–.80
Sum	4	1.653	.70–.80
Mothers (M)	3	12.797	.001–.01
Fathers (F)[a]			
Mother's exposure — 1	3	3.943	.20–.30
Mother's exposure — 2	3	6.769	.05–.10
Mother's exposure — 3	3	4.192	.20–.30
Mother's exposure — 4, 5	3	0.187	.95–.98
Sum	12	15.091	.20–.30

b. Excluding parents in category 1 (3 × 3)

Source	DF	χ^2	P
Total	35	14.186	.50–.70
Interactions, first order			
CM	2	2.001	.30–.50
CF	2	4.121	.10–.20
CS	1	5.737	.01–.02
SM	2	2.083	.30–.50
SF	2	4.825	.05–.10
MF	4	0.781	.90–.95
Main effects			
Sex (S)	1	1.884	.10–.20
Cities (C)	1	1.150	.20–.30
Mothers (M)	2	2.816	.20–.30
Fathers (F)	2	0.388	.80–.90

[a] Adjusted for mothers.

TABLE 9.6 THE FREQUENCY OF STILLBIRTHS BY PARENTAL EXPOSURE, CITY, AND PARITY
(Unrelated parents)



[a] Father's exposure is given first.

distribution and analysis of stillborn infants by city and parental exposure with further subdivision by parity. It should be noted that, as expected, there are significant differences in the stillbirth rate among different parities. From Table 9.7 we further note that when all exposure cells are considered with reference to parity, there is no effect of city or of maternal exposure. However, an effect of paternal exposure significant at the 5 per cent level now appears. Inexplicably this latter effect seems to be largely, if not entirely, attributable to those fathers whose infants were firstborn. The frequency of stillbirths among firstborn infants by city and paternal exposure is presented in Table 9.8. The data suggest a regular increase in stillbirth frequencies with increasing paternal exposure. To the end of determining whether this apparent effect was explicable in terms of the residual variation in maternal age not removed by the parity classification, Table 9.9 was prepared. An examination of this table reveals a number of points of interest. Of the five age groups investigated, the stillbirth rate increases with paternal exposure in three and decreases in two. In one of the three age groups (21–25) the increase with paternal exposure primarily reflects a "1" vs. "non-1" difference. In substance, then, when maternal age as well as parity is taken into account, there emerges no clearly consistent relationship between paternal exposure and the stillbirth rate.

9.5 Summary. — Analysis of the stillbirth data, taking into account differences between exposure cells in the parity distributions, fails to reveal significant differences between cities or consistent significant effects of parental exposure. The deviations from expectation observed are so small and inconsistent that they can scarcely be labeled as favoring or not favoring genetic hypothesis.

TABLE 9.7 CHI-SQUARE ANALYSIS OF THE FREQUENCY OF STILLBIRTHS BY PARENTAL EXPOSURE, CITY, AND PARITY
(Unrelated parents)

Source	DF	χ^2	P
Total	159	221.950	.001–.01
Interactions, first order			
CM	3	0.620	.80–.90
CF	3	10.626	.01–.02
CP	4	8.925	.05–.10
MF	9	5.548	.70–.80
PM	12	9.060	.50–.70
PF	12	16.085	.10–.20
Main effects			
Parity (P)	4	53.599	< .001
City (C)[a]	5	8.965	.10–.20
Mothers (M)[a]			
Parity Class 1	3	4.879	.10–.20
Class 2	3	7.537	.05–.10
Class 3	3	4.008	.20–.30
Class 4	3	4.269	.20–.30
Class 5	3	2.003	.50–.70
All classes	15	22.696	.05–.10
Fathers (F)[a]			
Parity Class 1	3	21.749	< .001
Class 2	3	1.865	.50–.70
Class 3	3	0.518	.90–.95
Class 4	3	3.521	.30–.50
Class 5	3	1.117	.70–.80
All classes	15	28.770	.01–.02

[a] Adjusted for parity.

TABLE 9.8 THE FREQUENCY OF STILLBIRTHS AMONG FIRSTBORN INFANTS BY CITY AND PATERNAL EXPOSURE
(Unrelated parents)

Fathers		Hiroshima	Nagasaki	Total
1	n	8,162	5,937	14,099
	s	159	89	248
	p	.0195	.0150	.0176
2	n	803	1,125	1,928
	s	18	34	52
	p	.0224	.0302	.0270
3	n	344	113	457
	s	13	2	15
	p	.0378	.0177	.0328
4–5	n	213	52	265
	s	11	1	12
	p	.0516	.0192	.0453
Total	n	9,522	7,227	16,749
	s	201	126	327
	p	.0211	.0174	.0195

TABLE 9.9 THE FREQUENCY OF STILLBIRTHS AMONG FIRSTBORN INFANTS BY PATERNAL EXPOSURE AND MATERNAL AGE
(Unrelated parents)

Maternal age		Fathers 1	2	3	4–5	Total	χ^2_2 ‡
≤ 20	n	1,772	356	78	29	2,235	1.055
	s	29	8	1	—	38	
	p	.0164	.0225	.0128	—	.0170	
21–25	n	8,505	1,031	247	141	9,924	5.908
	s	138	26	7	3	174	
	p	.0162	.0252	.0283	.0213	.0175	
26–30	n	3,100	372	84	56	3,612	11.962**
	s	63	8	5	4	80	
	p	.0203	.0215	.0595	.0714	.0221	
31–35	n	564	109	29	26	728	5.768
	s	13	7	—	1	21	
	p	.0230	.0642	—	.0385	.0288	
36–40	n	144	51	16	11	222	11.047**
	s	4	3	1	4	12	
	p	.0278	.0588	.0625	.3636	.0541	
41 +	n	14	9	3	2	28	—
	s	1	—	1	—	2	
	p	.0714	—	.3333	—	.0714	
Total	n	14,099	1,928	457	265	16,749	—
	s	248	52	15	12	327	
	p	.0176	.0270	.0328	.0453	.0195	

‡ Classes 3 and 4–5 are pooled.

Chapter X

THE ANALYSIS OF THE BIRTHWEIGHT DATA

10.1 *The trait.* — Birthweights were obtained for all infants. However, for the reasons given in Chapter VI we shall limit our attention to birthweights obtained on liveborn infants only. Shortly after the initiation of the study it became apparent that there was a remarkable diversity among the practicing midwives in the apparatus used to weigh newborn infants. Accordingly, early in 1949 each practicing midwife in the two cities was presented with a simple, spring type scale; prior to distribution, all scales were checked for accuracy. Weight was recorded by the midwives in *momme* (a Japanese unit of weight; 1 momme = 3.756 grams) and later converted to grams.

10.2 *The genetic argument for irradiation effects.* — Thus far in our consideration of irradiation effects we have concerned ourselves with only those induced mutations which would, presumably, occasion fairly large departures from the norm among individuals in whom the gene(s) finds manifestation. There exists in addition to these classes a much larger group of induced mutant genes, the so-called "invisibles" or "detrimentals," which give rise to much more subtle changes. These mutants, on the average, reduce the fitness of the organism. This loss of fitness may take the form of a decrease in the life span, a reduction in physical vigor, or similar changes. It seems logical to assume that in most instances a marked reduction in physical vigor would be reflected in certain body measurements, among these measurements being the birthweight of an infant. Such a reduction might be manifested as a change in the mean value, an increase in the variance of the measurement, or both. Accordingly, changes in mean birthweight or in the variance of the weights of newborn infants associated with parental exposure could be interpreted as evidence for irradiation-induced genetic damage. It should be pointed out that radiation-induced genetic changes in birthweight would, of course, be a function of the extent to which variation in birthweight is under genetic control. The available evidence (Robson, 1955; Morton, 1955) suggests that the component of variation ascribable to heredity is small. However, this does not vitiate attempts to determine the effect of irradiation-induced mutations on birthweight.

10.3 *Concomitant variables known to affect birthweight.* — A host of variables are known to, or thought to, influence birthweight. The most important ones, seemingly, are maternal age and parity (*inter alia*, Karn and Penrose, 1951), and maternal nutrition and economic status (*inter alia*, Millis, 1952). A word or two regarding these variables seems appropriate.

Maternal nutrition and economic status may be considered simultaneously for our purposes since it seems quite possible that the effect of the latter is due largely to the former. While it seems axiomatic that maternal nutrition is important in birthweight, the amount of variation in birthweights attributable to maternal nutrition under various circumstances is not accurately known. However, there exists evidence (Antonov, 1947, and Smith, 1947) that, barring a totally inadequate diet, maternal nutrition has relatively little effect on body size. It is in the area of marked decrease (near starvation levels) that maternal nutrition seems most important. Unfortunately, maternal nutrition is a variable extremely difficult to measure even under the best of circumstances, and nigh impossible under the conditions in which the present study was conducted. That conditions in postwar Japan may have been such as to produce maternal nutritional deficiencies cannot be gainsaid. However, there is no evidence to suggest that during the course of this study nutrition was at near starvation levels for an appreciable number of mothers. At no time during the period 1947–1953 was the daily caloric

intake assured by the rationing system less than 1,775 calories per day. The average caloric intake for adults during the period 1947–1953, as recorded by the Public Health Section of the Hiroshima municipal government, was

	Calories
1947	1,775
1948	1,900
1949	1,912
1950	1,995
1951	2,027
1952	2,050
1953	2,073

This "official" intake was undoubtedly augmented, but to an extent impossible to determine, by unofficial ("black-market") foodstuffs. Moreover, the minimal caloric intake available for women five months or more pregnant was augmented by approximately 240 calories (70 gms.) of rice. Caloric intake is, admittedly, but one facet of the appraisal of the adequacy of maternal nutrition, but in the absence of more specific information it is the only yardstick by which we can appraise the diets of the mothers involved in this study. Conceivably, some measure of the possible effect of maternal nutrition could be gained by an inspection of the data available on economic status of the parents involved in this study, or, in view of the gradual betterment of the diet during the interval of this study, an analysis of year of birth effects might be informative. From the standpoint of the present analysis, it is important to recognize that while the diet of the pregnant Japanese woman may by Western standards have been sub-optimal during a portion of this study, there is no reason to believe that there were dietary differences with respect to radiation categories.

Maternal age and parity are variables more readily measured than maternal nutrition and economic status. Their importance is amply attested to by, for example, Karn and Penrose's finding (1951) that as much as a full pound difference may exist between first- and seventh-born children. Moreover, they have shown that maternal age exerts an effect not explicable in terms of the correlation between maternal age and parity. More recently, Millis and Seng (1954), studying infants born in Singapore during the period 1950 to 1951, have extended Karn and Penrose's findings. These workers limited their attention to normal, non-premature, liveborn infants. In two of these respects Millis and Seng's data are not unlike the data here presented. Millis and Seng find that (a) parity has a positive effect, and (b) maternal age a slight negative effect on birthweight. These are findings essentially similar to Karn and Penrose's. Millis and Seng have, however, proceeded one step further, and find that the relationship of parity to birthweight, for fixed maternal age, is curvilinear and fairly consistent over differing age groups. On the other hand, the relationship of maternal age to birthweight, for fixed parity, is neither consistent nor simply stated.

In our data, we find that taken conjointly maternal age and parity can be related, fairly consistently, to birthweight by a regression of the form:

where

$$\hat{y} = m + b_1 x + b_2 w + b_3 z$$

\hat{y} = birthweight
m = mean birthweight
x = parity
w = parity squared
z = maternal age

and b_1, b_2, and b_3 are, respectively, the regression coefficients associated with parity, parity-squared, and maternal age.

Since our interests here are limited to removing the effects of concomitant variation in maternal age and parity, a detailed presentation of the relationship of these variables to birthweight in Japanese children will be presented elsewhere. To afford the reader some measure of this relationship Figures 10.1–10.2 are presented.

10.4 *The data and their analysis.* — The extent of the data on the birthweights, and the complexity of the problem posed by their analysis, precludes presentation of the raw data in detail. It is hoped, however, that the presentation of these data will be sufficiently complete to illustrate the argument underlying the analysis.

In Table 10.1 are presented the mean birthweights in decagrams unadjusted for maternal age and parity and the number of observations on which these means are based for each of the 64 sex-city-mother-father cells. In Table 10.2 are given the results of two analyses of variance seeking to determine whether there exist significant differences among these means prior to adjustment for maternal age and parity differences. The two analyses differ only with respect to the inclusion or exclusion of the category 1 parents. A perusal of the analysis reveals little

in the way of significant differences in mean birthweight between parental exposure categories. The large number of significant interactions when the category 1 parents are included attests to the heterogeneity of the data, and vitiates attempts at generalizing with respect to parental exposure. In this connection, it is interesting to note that when these parents, that is, the category 1 parents, are excluded, the bulk of the heterogeneity disappears (see Table 10.2b). This preliminary analysis of the data does not suggest the existence of large differences in mean birthweight induced by paren-

Hiroshima, and then indicating where the four city-sex cells agree or disagree.

In an analysis of these data, we are interested in obtaining answers to the following questions:

1. Is the regression of birthweight on maternal age and parity significant?
2. Are the regressions based on the individual mother-father exposure cells within a given city-sex cell homogeneous?
3. Assuming a significant regression, do the adjusted means differ significantly among parental exposures?
4. Does there exist significant heterogeneity among the variances in the mother-father cells?

FIGURE 10.1 — The distribution of mean birthweight in decagrams by parity for all maternal ages, and for maternal age 30 only.

FIGURE 10.2 — The distribution of mean birthweight in decagrams by maternal age for fixed parities (parities 1 and 4).

tal exposure. It seemed pointless, however, to pursue the meaning of the interactions and main effects until adequate compensation had been made for the concomitant variables known to differ among the city-exposure cells. Before we consider the effects of concomitant variables, the reader's attention is called to Table 10.3 which presents the residual mean squares (estimated variances) by parental exposure weighted for sex and city differences. The heterogeneity is apparent on inspection.

In the analysis in which concomitant variation was to be accounted for, the data were partitioned into four parts corresponding to the four possible sex-city cells, namely, males-Hiroshima, females-Hiroshima, males-Nagasaki, and females-Nagasaki. The results of the analysis of these four city-sex cells are presented in Tables 10.4 to 10.24. To present the analytical argument, we shall content ourselves with examining in detail a single city-sex cell, namely, males-

5. How heterogeneous are the observations within a mother-father cell?
6. Assuming the answer to question (5) is "appreciable," what are the variables which may account for the within-cell heterogeneity?

These questions have been posed in the order in which we shall treat them in the following paragraphs.

In Table 10.4 are given the results of the analysis of covariance leading to the determination of the regression coefficients b_1, b_2, and b_3 associated with parity (x), parity squared (w), and maternal age (z) for each of the individual mother-father cells within the males-Hiroshima cell as well as the estimates of b_1, b_2, and b_3 based on the pool of the exposure cells.[1] From

[1] By "pool" we refer to the results obtained when we sum the sums of squares and cross products of deviations over all 16 exposure cells, and estimate the regression coefficients therefrom. By "sum" we refer to the results obtained when a regression is fitted to each of the individual exposure cells.

TABLE 10.1 MEAN BIRTHWEIGHT BY PARENTAL EXPOSURE, SEX, AND CITY
(Unrelated parents. Means expressed in decagrams.)

Hiroshima—Males

		Fathers				
Mothers		1	2	3	4-5	Total
1	n	8,673	744	295	199	9,911
	x̄	311.36	311.43	311.85	310.60	311.36
2	n	2,715	898	190	114	3,917
	x̄	312.19	314.57	308.49	308.36	312.44
3	n	1,065	219	256	77	1,617
	x̄	310.97	312.57	316.33	315.84	312.27
4-5	n	541	108	48	55	752
	x̄	312.65	315.91	301.46	310.18	312.22
Total	n	12,994	1,969	789	445	16,197
	x̄	311.55	313.23	311.86	310.88	311.75

Hiroshima—Females

		Fathers				
Mothers		1	2	3	4-5	Total
1	n	7,999	700	266	175	9,140
	x̄	303.87	301.02	302.27	300.26	303.54
2	n	2,434	874	181	118	3,607
	x̄	302.94	308.02	303.04	309.03	304.38
3	n	1,031	198	246	75	1,550
	x̄	302.21	303.39	304.91	305.20	302.93
4-5	n	526	89	55	58	728
	x̄	300.42	299.21	299.82	312.41	301.18
Total	n	11,990	1,861	748	426	15,025
	x̄	303.39	304.47	303.14	305.21	303.56

Nagasaki—Males

		Fathers				
Mothers		1	2	3	4-5	Total
1	n	7,287	1,068	121	72	8,548
	x̄	312.15	313.15	313.40	315.28	312.32
2	n	4,652	2,024	135	103	6,914
	x̄	310.91	311.89	313.04	317.60	311.34
3	n	339	130	50	13	532
	x̄	311.73	321.18	326.80	328.46	315.86
4-5	n	272	53	17	15	357
	x̄	310.27	310.57	327.71	318.67	311.50
Total	n	12,550	3,275	323	203	16,351
	x̄	311.64	312.65	316.08	317.55	312.00

Nagasaki—Females

		Fathers				
Mothers		1	2	3	4-5	Total
1	n	6,761	1,011	110	63	7,945
	x̄	303.63	300.64	311.85	306.03	303.38
2	n	4,313	1,941	129	70	6,453
	x̄	302.70	306.08	302.05	300.14	303.68
3	n	367	138	42	21	568
	x̄	308.01	306.03	310.71	312.90	307.91
4-5	n	267	57	16	13	353
	x̄	302.07	306.49	318.38	315.54	304.02
Total	n	11,708	3,147	297	167	15,319
	x̄	303.39	304.34	307.78	305.17	303.69

TABLE 10.2 ANALYSIS OF VARIANCE OF BIRTHWEIGHT BY PARENTAL EXPOSURE AND CITY
(Unrelated parents)

a. All exposure categories (4 × 4)

Source	Sum of squares of deviations	DF	Mean square	F
Main effects				
Cities (C)	675.79	1	675.792	2.640
Sexes (S)	1,070,650.05	1	1,070,650.049	600.077**
Fathers (F)	14,154.42	3	4,718.141	2.644*
Mothers (M)	4,346.49	3	1,448.831	1.231
Interactions, first order				
CS	51.98	1	51.981	34.324
CF	12,253.02	3	4,084.340	2.289
CM	20,235.97	3	6,745.323	3.781**
SF	289.84	3	96.613	18.467*
SM	54,757.96	3	18,252.652	10.230**
MF	39,647.35	9	4,405.261	2.469**
Higher orders	19,162.58	33	580.684	3.073**
Between	1,235,883.20	63	19,617.194	10.995**
Within	112,097,004.72	62,828	1,784.189	—
Total	113,332,887.92	62,891	—	—

b. Excluding parents with exposure 1 (3 × 3)

Source	Sum of squares of deviations	DF	Mean square	F
Main effects				
Cities (C)	9.64	1	9.638	207.725
Sexes (S)	114,255.51	1	114,255.511	57.072**
Fathers (F)	2,940.19	2	1,470.096	1.362
Mothers (M)	5,698.70	2	2,849.350	1.423
Interactions, first order				
CS	.37	1	.369	5,425.377*
CF	8,149.72	2	4,074.861	2.035
CM	15,069.70	2	7,534.852	3.764*
SF	2,018.93	2	1,009.464	1.983
SM	8,825.61	2	4,412.807	2.204
MF	6,995.23	4	1,748.807	1.145
Higher orders	—	16	—	—
Between	190,786.49	35	5,451.043	2.723**
Within	17,597,261.25	8,790	2,001.964	—
Total	17,788,047.74	8,825		

TABLE 10.3 THE DISTRIBUTION BY PARENTAL EXPOSURE OF THE WEIGHTED MEAN SQUARES OF DEVIATIONS

(Weighting is for city and sex. The degrees of freedom associated with the mean squares are indicated in parentheses.)

	Fathers			
Mothers	1	2	3	4-5
1	1,755.07 (30,716)	1,721.13 (3,519)	1,955.18 (788)	2,033.93 (505)
2	1,738.81 (14,104)	1,831.01 (5,733)	1,806.98 (631)	1,771.12 (401)
3	1,723.19 (2,799)	1,971.81 (681)	2,082.23 (590)	1,780.03 (182)
4-5	1,615.35 (1,602)	1,817.70 (303)	1,963.66 (132)	1,455.26 (137)

$\chi^2 = 38.584^{**}$ DF = 15

TABLE 10.4 ANALYSIS OF COVARIANCE OF BIRTHWEIGHTS: MALES, HIROSHIMA

Exposures	n	Σw^2	Σwx	Σwz	Σwy	Σx^2	Σz^2	Σxy	Σzy	Σxz	Σy^2	b_1	b_2	b_3
11	8,673	1,124,219.56	138,880.91	261,471.83	447,243.92	19,549.80	191,779.09	77,696.17	124,180.73	38,536.17	15,553,497.35	10.2497	−.7907	−.3340
12	2,715	345,543.81	43,319.37	88,747.85	149,610.99	6,085.20	61,367.84	25,338.63	47,117.52	12,778.05	4,604,035.72	10.6399	−.8417	−.2304
13	1,065	188,243.46	19,771.95	36,960.44	31,975.17	2,512.58	25,765.41	6,117.30	36,987.24	5,021.97	1,934,287.95	2.3726	−.3764	1.5131
14-15	541	69,686.06	8,652.62	16,287.33	44,559.87	1,208.96	14,100.53	6,092.07	14,047.08	2,356.91	888,360.96	3.5659	.1413	.2370
21	744	77,869.87	10,612.89	20,010.06	57,822.36	1,601.97	20,310.95	9,666.10	21,150.79	3,130.80	1,378,712.06	11.1748	−.8106	.1175
22	898	292,923.55	28,155.65	48,061.41	66,693.34	3,084.14	25,980.87	10,083.13	21,587.29	5,785.63	1,644,962.40	9.8669	−.7129	−.0475
23	219	69,836.15	7,028.43	12,419.27	968.94	1,138.21	6,451.01	1,138.21	2,528.63	1,433.06	387,381.69	13.0767	−.1.3028	.0036
24-25	108	40,882.92	4,106.75	6,629.42	655.06	446.23	3,048.91	485.16	378.04	712.25	200,465.10	12.6548	−1.2294	−.1592
31	295	36,121.99	4,777.23	9,674.97	16,948.50	693.72	8,717.94	3,534.36	6,943.98	1,441.47	573,295.52	21.3298	−2.3058	−.1713
32	190	60,367.94	6,492.79	14,017.62	45,665.14	764.36	6,374.61	5,438.74	10,785.17	1,703.75	403,533.48	9.6044	−.1499	−.5455
33	256	110,826.15	10,765.79	15,741.55	14,443.04	1,137.28	8,317.60	2,418.05	5,727.78	1,802.32	466,334.83	10.7297	−.9294	.1225
34-35	48	26,757.98	2,361.52	4,674.04	7,429.79	234.98	2,223.92	1,225.20	1,280.42	502.96	95,397.93	27.5060	−1.8389	−1.7800
45-1	199	18,816.59	2,652.97	5,051.68	11,997.62	409.20	5,054.75	1,979.79	5,986.81	746.87	476,325.83	7.5328	−.6063	.6773
45-2	114	11,891.62	1,700.29	2,851.02	2,832.43	258.29	3,191.86	447.78	−393.45	417.69	217,788.26	3.1101	−.1010	−.4400
45-3	77	31,939.95	3,092.57	4,464.17	1,538.31	325.71	1,900.70	78.56	−347.00	510.14	117,070.12	−1.6768	.2577	−.3379
45-45	55	33,122.18	2,980.73	5,064.45	3,651.82	287.71	1,903.63	329.28	−615.46	450.82	83,698.19	−.2310	.2901	−1.0405
Total	DF = 16,133	2,539,049.78	295,352.47	552,127.11	904,036.28	39,387.02	386,489.64	152,068.53	297,345.57	77,330.85	29,025,347.38	9.5652	−.7365	−.0923

y = birthweight w = parity squared
x = parity z = maternal age

Table 10.5a we note that the estimates of b_1, b_2, and b_3 based on the pool remove a significant amount of variation from the birthweights. In the light of the work of other investigators this is not an unexpected finding. However, if we return to Table 10.4, we note on inspection considerable heterogeneity among the b's associated with the various exposure cells. That this variation is significant is borne out by Table 10.5b.

At this point, it is of interest to inquire into the amount of variation in the birthweights accounted for by variation in the individual regressions after removing the portion associated with the common regression (see Sec. 6.3). From the data in Table 10.5, we form the mean

anticipated from irradiation, variation, accountable for on other grounds, of as small as 3 per cent could be sufficient to obfuscate irradiation differences.

For the analysis of variance on the adjusted data set out in Table 10.6, adjustment was to the common regression because adjustment to the individual exposure cell regressions removed a seemingly negligible additional amount of variation (for a discussion of the computational procedure see Wishart, 1950). For the reader not familiar with covariance analysis it might be pointed out that the computations in Table 10.6 effectively transform through the pooled regression the observed array of exposure cell means into the array of exposure cell means

TABLE 10.5 TESTS OF THE SIGNIFICANCE AND HOMOGENEITY OF THE REGRESSIONS OF BIRTHWEIGHT ON MATERNAL AGE AND PARITY: MALES, HIROSHIMA

(a) Test of the significance of the regression based on within-cells (pooled) regression coefficients

Source	SS	DF	MS	F
Variation removed by regression	761,262.72	3	253,754.240	145.246**
Residual within cells (pool)	28,264,084.66	16,178	1,747.069	—

(b) Test of the homogeneity of the regressions (all exposure cells considered)

Source	SS	DF	MS	F
Differences in regressions	116,630.11	45	2,591.780	1.485*
Residual within cells (sum)	28,147,454.55	16,133	1,744.713	—

(c) Test of the homogeneity of the regressions (only those cells in which both parents were exposed are considered)

Source	SS	DF	MS	F
Difference in regressions	38,429.76	24	1,601.240	1.124
Residual within cells (sum)	3,473,234.85	1,929	1,800.536	—

square ratio of the "mean square residual (pool)" to the "mean square residual (sum)." The value obtained is 1.001350. Accordingly, we may assert that the variation in regressions accounts for 0.14 per cent of the variation in birthweight not accounted for by the common regression. This value obviously attains perspective only if we know the amount of variation in birthweight accounted for by the common regression. The latter we obtain from the mean square ratio

$$\frac{28264084.66 + 761262.72}{16181(1747.069)}$$

which is 1.02674. The common regression, then, accounts for 2.7 per cent of the variation in y. It is natural to inquire here whether this amount, 2.7 per cent, is of importance. To this we can only answer that in view of the small effects

which one would obtain if each exposure cell had the same maternal age and parity distribution. The mean squares obtained following this transformation are wholly analogous to those obtained in a simple analysis of variance, and may be interpreted in the same sense. Inspection of this table fails to reveal significant differences among maternal or paternal exposure categories or evidence from the interaction mean square of nonadditive effects of parental exposure. The adjusted birthweight means are given in Table 10.7.

The next question to be asked of these data was, "Does there exist significant heterogeneity among the variances in the mother-father cells?" The residual mean squares for the sixteen exposure cells within males-Hiroshima are given in Table 10.8. Bartlett's test (cf. Rao, 1952) of the heterogeneity of the variances is not

TABLE 10.6 ANALYSIS OF VARIANCE ON THE ADJUSTED BIRTHWEIGHT MEANS: MALES, HIROSHIMA

	DF	Σw^2	Σwx	Σwz	Σwy	Σx^2	Σz^2	Σxy	Σzy	Σxz	Σy^2	b^2	d^2	Diff.	DF	MS	F
Mothers	3	27,465	3,647	7,642	9,682	486	2,252	1,317	2,576	1,006	3,978	—	—	782.86	3	260.95	6.695
Fathers	3	63,769	8,583	24,940	12,098	1,159	9,820	1,696	7,074	3,370	5,187	—	—	8,944.73	3	2,981.58	1.707
Interaction	9	34,028	4,465	11,194	9,886	592	4,206	1,319	878	1,512	18,494	—	—	16,541.55	9	1,837.95	1.052
Between	15	125,262	16,695	43,775	31,666	2,238	16,279	4,332	10,527	5,888	27,658	—	—	30,428.97	15	2,028.60	1.161
Within	16,181	2,539,050	295,352	552,127	904,036	39,387	386,490	152,069	297,346	77,331	29,025,347	761,353	28,264,085	—	16,178	1,747.07	—
W+B	16,196	2,664,311	312,047	595,903	935,702	41,625	402,768	156,400	307,873	83,219	29,053,005	758,492	28,294,514	—	—	—	—
W+M	16,184	2,566,514	299,000	559,769	913,718	39,873	388,742	153,385	299,922	78,337	29,029,325	764,457	28,264,868	—	—	—	—
W+F	16,184	2,602,819	303,935	577,067	916,134	40,546	396,310	153,765	304,419	80,701	29,030,534	757,505	28,273,029	—	—	—	—
W+I	16,190	2,573,078	299,818	563,321	913,923	39,979	390,696	153,387	298,223	78,842	29,043,841	763,215	28,280,626	—	—	—	—

y = birthweight
x = parity
w = parity squared
z = maternal age

b^2 = sum of squares removed by regression
d^2 = residual sum of squares

TABLE 10.7 THE ADJUSTED BIRTHWEIGHT MEANS: MALES, HIROSHIMA

	w	x	z	Δw	Δx	Δz	\bar{y} unadjusted	\bar{y} adjusted [a]
Father's exposure								
1	8.184	2.434	27.020	−0.970	−0.132	−0.385	311.551	312.063
2	12.711	3.072	28.838	3.557	0.506	1.433	313.233	311.146
3	14.418	3.217	29.279	5.264	0.651	1.874	311.864	309.689
4 + 5	12.422	3.029	28.964	3.268	0.463	1.559	310.881	309.004
Mean	9.154	2.566	27.405	—	—	—	311.753	311.753
Mother's exposure								
1	8.157	2.431	27.133	−0.997	−0.135	−0.272	311.361	311.892
2	10.307	2.735	27.648	1.153	0.169	0.243	312.442	311.698
3	11.650	2.879	28.079	2.496	0.313	0.674	312.266	311.174
4 + 5	10.918	2.785	28.263	1.764	0.219	0.858	312.223	311.507
Mean	9.154	2.566	27.405	—	—	—	311.753	311.753

$b_1 = 9.565$
$b_2 = -0.737$
$b_3 = -0.092$

[a] \bar{y} adjusted = \bar{y} unadjusted $- b_1 \Delta x - b_2 \Delta w - b_3 \Delta z$.

$\Delta x = (x_1 - \bar{x})$
$\Delta w = (w_1 - \bar{w})$
$\Delta z = (z_1 - \bar{z})$

x = parity
w = parity squared
z = maternal age

significant. Thus when all exposure cells within the males-Hiroshima cell are considered, the only evidence for an association of birthweight with parental exposure is the heterogeneity of the individual cell regressions. But, this association is neither consistent nor does it persist to a significant extent when the category 1 parents are excluded (see Table 10.5c).

Before we summarize and consider the possible interpretations of the findings on the males-Hiroshima data let us turn to the remaining three sex-city cells. Tables corresponding to those for males-Hiroshima are given for females-Hiroshima (Tables 10.9–10.13), males-

TABLE 10.8 THE RESIDUAL MEAN SQUARES FROM THE INDIVIDUAL CELL REGRESSIONS: MALES, HIROSHIMA

(The degrees of freedom are given in parentheses.)

	Mothers			
Fathers	1	2	3	4–5
1	1,747.87 (8,669)	1,649.29 (2,711)	1,768.00 (1,061)	1,595.93 (537)
2	1,777.14 (740)	1,783.05 (894)	1,738.38 (215)	1,876.84 (104)
3	1,849.41 (291)	1,957.12 (186)	1,798.85 (252)	1,764.54 (44)
4–5	2,382.73 (195)	1,968.26 (110)	1,598.47 (73)	1,609.30 (51)

Bartlett's test for between-cell heterogeneity of mean squares:

$\chi^2 = 20.43$ DF = 15

Bartlett's test for between-cell heterogeneity of mean squares when category 1 parents are excluded:

$\chi^2 = 2.12$ DF = 8

Nagasaki (Tables 10.14–10.18) and females-Nagasaki (Tables 10.19–10.23). In none of these three sex-city cells does the analysis of variance on the adjusted data reveal significant differences as regards mean birthweight between maternal or paternal exposure categories or evidence of heterogeneity as judged by the interactions. In Table 10.24 are set out the principal findings with regard to the four sex-city cells. From this table when all exposure cells are considered we note (a) significant heterogeneity in the individual father-mother regressions in two of the four sex-city cells, (b) a reasonable measure of constancy in the amount of variation removed by the common regression in each sex-city cell, and (c) significant heterogeneity between the residual mean squares within a sex-

TABLE 10.9 ANALYSIS OF COVARIANCE OF BIRTHWEIGHTS: FEMALES, HIROSHIMA

Exposures	n	Σw^2	Σwx	Σwz	Σwy	Σx^2	Σz^2	Σxy	Σzy	Σzx	Σy^2	b_1	b_2	b_3
11	7,999	835,919.59	111,570.50	220,103.48	336,837.57	16,603.80	171,582.75	60,986.73	95,799.68	33,656.40	13,226,625.00	10.1601	−.8688	−.3201
12	2,428	298,923.97	37,959.21	75,897.32	94,143.87	5,417.28	54,188.52	15,886.19	31,052.98	11,168.64	3,853,461.85	6.8276	−.5281	−.0946
13	1,032	127,817.90	15,701.76	31,374.94	52,687.91	2,200.10	24,105.97	8,200.33	7,330.74	4,494.45	1,764,388.79	7.9051	−.3993	−.6500
14–15	526	55,536.37	7,679.18	14,971.78	37,395.22	1,158.01	12,718.44	6,097.82	15,517.32	2,335.11	881,768.31	8.6399	−.6190	.3625
21	700	91,135.17	11,405.43	22,876.16	32,390.31	1,615.68	16,919.39	4,978.63	7,804.03	3,328.35	1,162,122.76	5.5564	−.2746	.2606
22	874	322,582.41	31,454.72	55,211.36	11,792.31	3,352.26	25,623.02	2,788.74	−771.56	6,181.33	1,564,204.59	7.2271	−.5776	−.5289
23	198	48,637.32	5,369.65	10,366.50	27,827.17	644.87	5,797.50	3,599.94	9,365.83	1,270.17	451,477.06	8.5504	−.5120	.6578
24–25	89	33,335.19	3,503.84	5,861.43	8,748.24	398.25	3,230.72	1,025.63	2,682.29	713.47	150,319.06	2.1625	−.0395	.4243
31	266	18,875.05	2,773.11	5,796.59	943.53	448.15	7,756.72	657.20	−2,716.80	927.52	472,924.51	14.0005	−1.7979	.6807
32	181	51,561.91	5,592.86	11,573.41	30,745.18	658.60	7,293.36	3,733.28	5,373.18	1,406.51	253,315.65	9.5062	−.2951	−.6314
33	246	90,168.13	8,722.98	13,686.21	24,111.55	927.32	7,655.09	2,607.15	−2,049.10	1,558.59	537,616.85	6.9939	−.2092	−1.3176
34–35	55	15,703.93	1,743.02	1,994.00	−682.96	209.75	1,466.00	−3.51	−879.00	202.00	80,838.99	3.7354	−.3827	−.5938
45–1	175	18,257.35	2,548.46	5,287.00	1,443.77	385.71	4,056.00	747.14	4,276.00	813.00	347,909.43	15.5270	−2.3971	1.0666
45–2	118	18,738.11	2,348.06	4,696.59	−5,738.68	319.26	3,992.27	−359.29	196.12	672.63	152,657.86	13.8117	−2.0791	.1680
45–3	75	29,933.15	2,932.29	5,614.83	9,985.25	314.69	2,711.39	1,050.51	1,698.97	645.65	156,448.35	3.7749	.0243	−.3226
45–45	58	14,956.16	1,628.16	3,614.10	1,320.90	196.16	2,094.07	320.90	−404.07	434.10	79,370.07	11.4599	−.9238	−.9742
Total	DF = 14,956	2,072,081.76	252,933.23	488,925.70	663,951.14	34,849.79	351,191.21	112,317.39	174,276.61	69,807.92	25,135,449.13	8.6196	−.6620	−.2954

w = parity squared y = birthweight
z = maternal age x = parity

TABLE 10.10 TESTS OF THE SIGNIFICANCE AND HOMOGENEITY OF THE REGRESSIONS OF BIRTHWEIGHT ON MATERNAL AGE AND PARITY: FEMALES, HIROSHIMA

(a) Test of the significance of the regression based on within-cells (pooled) regression coefficients

Source	SS	DF	MS	F
Variation removed by regression	477,085.92	3	159,028.640	96.746**
Residual within cells (pool)	24,658,363.21	15,001	1,643.781	—

(b) Test of the homogeneity of the regressions (all exposure cells considered)

Source	SS	DF	MS	F
Differences in regressions	80,722.24	45	1,793.827	1.092
Residual within cells (sum)	24,577,640.97	14,956	1,643.329	—

(c) Test of the homogeneity of the regressions (only those cells in which both parents were exposed are considered)

Source	SS	DF	MS	F
Difference in regressions	44,117.13	24	1,838.214	1.025
Residual within cells (sum)	3,333,536.37	1,858	1,794.153	—

TABLE 10.11 ANALYSIS OF VARIANCE ON THE ADJUSTED BIRTHWEIGHT MEANS: FEMALES, HIROSHIMA

	DF	Σw^2	Σwz	Σwy	Σx^2	Σz^2	Σxy	Σzy	Σzx	Σy^2	b^2	d^2	Diff.	DF	MS	F
Mothers	3	26,168	7,821	2,578	442	2,338	32	810	1,016	3,798	—	—	10,705.92	3	3,568.64	2.171
Fathers	3	64,137	24,392	13,521	1,149	9,448	1,807	5,252	3,272	5,788	—	—	4,831.08	3	1,610.36	1.020
Interaction	9	26,145	7,043	29,487	500	2,085	4,372	9,107	989	39,481	—	—	13,689.02	9	1,521.00	1.081
Between	15	116,450	39,256	45,586	2,091	13,871	6,211	15,169	5,277	49,067	477,085.92	24,658,363.21	—	15	1,955.99	1.190
Within	15,004	2,072,082	488,926	663,951	34,850	351,191	112,317	174,277	69,808	25,135,449	—	—	29,339.85	15,001	1,643.78	—
W + B	15,019	2,188,532	528,181	709,546	36,940	365,062	118,536	189,452	75,083	25,184,523	496,819.94	24,687,703.06	—	—	—	—
W + M	15,007	2,098,250	496,747	666,529	35,292	353,529	112,349	175,087	70,824	25,139,247	470,177.87	24,669,069.13	—	—	—	—
W + F	15,007	2,136,219	513,318	677,472	35,999	360,639	114,124	179,529	73,080	25,141,237	478,042.71	24,663,194.29	—	—	—	—
W + I	15,013	2,098,227	495,969	693,438	35,350	353,276	116,689	183,384	70,797	25,174,930	502,877.77	24,672,052.23	—	—	—	—

y = birthweight w = parity squared b^2 = sum of squares removed by regression
x = parity z = maternal age d^2 = residual sum of squares

TABLE 10.12 THE ADJUSTED BIRTHWEIGHT MEANS: FEMALES, HIROSHIMA

	x	w	z	Δx	Δw	Δz	\bar{y} unadjusted	\bar{y} adjusted [a]
Father's exposure								
1	2.413	7.942	26.991	−.138	−1.023	−.397	303.374	303.769
2	3.157	13.548	28.979	+.606	+4.583	+1.591	304.619	302.899
3	3.068	12.710	29.124	+.517	+3.745	+1.736	303.424	301.959
4 + 5	2.852	11.138	28.573	+.301	+2.173	+1.185	306.350	305.543
Mean	2.551	8.965	27.388	—	—	—	303.615	303.615
Mother's exposure								
1	2.414	7.910	27.074	−.137	−1.055	−.314	303.539	303.929
2	2.771	10.694	27.913	+.220	+1.729	+.525	304.324	303.727
3	2.779	10.623	27.870	+.228	+1.658	+.482	303.088	302.362
4 + 5	2.694	10.117	27.724	+.143	+1.152	+.336	302.176	301.805
Mean	2.551	8.965	27.388	—	—	—	303.615	303.615

[a] \bar{y} adjusted = \bar{y} unadjusted − $b_1 \Delta x$ − $b_2 \Delta w$ − $b_3 \Delta z$.

$b_1 = 8.620$ $\Delta x = (x_i - \bar{x})$ x = parity
$b_2 = -0.662$ $\Delta w = (w_i - \bar{w})$ w = parity squared
$b_3 = -0.295$ $\Delta z = (z_i - \bar{z})$ z = maternal age

TABLE 10.13 THE RESIDUAL MEAN SQUARES FROM THE INDIVIDUAL CELL REGRESSIONS: FEMALES, HIROSHIMA

(The degrees of freedom are given in parentheses.)

	Mothers			
Fathers	1	2	3	4–5
1	1,617.30 (7,995)	1,566.69 (2,424)	1,678.38 (1,028)	1,621.86 (522)
2	1,645.67 (696)	1,782.13 (870)	2,210.23 (194)	1,733.04 (85)
3	1,769.35 (262)	1,300.74 (177)	2,155.90 (242)	1,569.98 (51)
4–5	1,960.28 (171)	1,277.68 (114)	2,151.95 (71)	1,417.02 (54)

Bartlett's test for between-cell heterogeneity of mean squares:

$\chi^2 = 51.216**$ DF = 15

Bartlett's test for between-cell heterogeneity of mean squares when category 1 parents are excluded:

$\chi^2 = 26.156**$ DF = 8

TABLE 10.14 Analysis of Covariance of Birthweights: Males, Nagasaki

Exposures	n	Σw^2	Σwx	Σwz	Σwy	Σx^2	Σz^2	Σxy	Σzy	Σxz	Σy^2	b_1	b_2	b_3
11	7,287	1,389,119.45	161,723.58	314,315.43	480,228.41	20,886.70	184,893.18	75,852.32	116,296.05	42,009.31	13,556,842.46	10.8732	−.8182	−.4505
12	4,652	955,543.11	109,914.48	236,427.77	332,533.80	14,054.71	131,744.45	51,284.94	92,702.11	31,021.35	8,576,413.51	10.3081	−.7397	−.3961
13	339	63,038.71	7,774.28	18,577.04	17,944.67	1,033.85	10,229.93	3,518.13	4,165.76	2,494.48	618,708.66	20.3862	−1.9029	−1.1082
14–15	272	64,306.69	7,190.94	16,008.91	30,479.16	882.94	7,576.12	4,478.40	19,375.60	2,004.41	475,875.84	5.5533	−.8818	2.9915
21	1,068	330,263.67	37,431.43	71,299.88	102,253.75	4,386.20	38,639.35	14,061.88	44,118.17	8,730.28	1,775,226.77	6.1557	−.5048	.6825
22	2,024	996,175.37	95,300.21	190,797.48	224,408.22	9,980.00	76,618.41	27,122.65	53,836.41	20,940.05	3,840,498.91	7.5305	−.4532	−.2269
23	130	53,682.28	5,564.86	12,544.57	11,703.51	625.57	6,328.89	1,533.23	5,137.86	1,461.21	283,522.93	5.0203	−.4122	.4697
24–25	53	10,153.02	1,204.08	2,855.06	2,153.02	154.30	422.07	1,471.17	−20.94	378.23	93,883.04	22.2678	−1.7938	−2.2579
31	121	54,988.74	5,312.18	13,530.60	6,840.36	566.00	4,856.61	1,199.91	2,768.75	1,400.64	287,166.98	10.0363	−.8689	.0963
32	135	58,189.08	5,728.33	13,139.44	29,150.98	623.30	5,827.53	3,091.93	6,855.26	1,531.75	306,843.77	5.0448	.0777	−.3248
33	50	20,619.78	2,008.62	4,291.96	−59.20	210.98	2,098.72	265.71	385.60	466.84	111,288.00	−5.0305	.3761	.5336
34–35	17	4,557.53	558.71	1,259.35	1,796.53	70.94	520.24	347.50	85.35	155.47	33,927.53	19.6774	−1.3243	−2.5107
45–1	72	21,783.99	2,113.38	3,616.85	495.28	231.88	2,764.32	685.87	−221.94	426.13	101,994.43	12.7953	−1.1214	−.5855
45–2	103	36,359.51	3,731.65	6,394.21	4,666.63	406.39	3,202.46	−173.08	2,590.91	746.87	210,474.66	6.6694	−.6548	.5610
45–3	13	7,734.92	627.85	640.62	−2,491.54	53.69	221.08	47.31	45.23	16,069.24	16.0262	2.1553		
45–45	15	11,931.73	1,014.53	1,542.60	−868.33	90.93	402.40	−44.33	−451.00	135.80	13,537.33	8.1597	−.5267	−1.8553
Total	DF = 16,287	4,098,447.59	447,399.10	907,241.76	1,241,235.24	54,258.38	477,394.86	183,590.33	347,671.25	113,948.05	30,302,274.06	9.5693	−.6859	−.2523

y = birthweight
x = parity
w = parity squared
z = maternal age

TABLE 10.15 Tests of the Significance and Homogeneity of the Regressions of Birthweight on Maternal Age and Parity: Males, Nagasaki

(a) Test of the significance of the regression based on within-cells (pooled) regression coefficients

Source	SS	DF	MS	F
Variation removed by regression	817,737.83	3	272,579.277	150.986**
Residual within cells (pool)	29,484,536.23	16,332	1,805.323	—

(b) Test of the homogeneity of the regressions (all exposure cells considered)

Source	SS	DF	MS	F
Differences in regressions	96,880.39	45	2,152.897	1.193
Residual within cells (sum)	29,387,655.84	16,287	1,804.363	—

(c) Test of the homogeneity of the regressions (only those cells in which both parents were exposed are considered)

Source	SS	DF	MS	F
Differences in regressions	19,231.75	24	801.323	2.384
Residual within cells (sum)	4,784,382.86	2,504	1,910.696	—

The Analysis of the Birthweight Data

TABLE 10.16 ANALYSIS OF VARIANCE ON THE ADJUSTED BIRTHWEIGHT MEANS: MALES, NAGASAKI

	DF	Σw^2	Σwx	Σwz	Σwy	Σx^2	Σz^2	Σxy	Σzy	Σxz	Σy^2	b^2	d^2	Diff.	DF	MS	F
Mothers	3	18,001	1,767	623	−2,491	174	221	−194	121	70	9,454	—	—	10,912.36	3	3,637.45	2.015
Fathers	3	146,809	16,942	42,205	41,650	1,955	12,314	4,850	13,056	4,878	18,559	—	—	7,015.68	3	2,338.56	1.295
Interaction	9	16,562	2,578	8,830	9,309	387	3,989	1,328	3,562	1,243	16,246	—	—	12,091.69	9	1,343.52	1.344
Between	15	181,373	21,288	51,658	48,467	2,517	16,525	5,983	16,740	6,192	44,260	—	—	29,985.30	15	1,999.02	1.107
Within	16,335	4,098,447	447,399	907,241	1,241,235	54,258	477,394	183,590	347,671	113,948	30,302,274	817,737.83	29,484,536.23	—	16,332	1,805.32	—
W + B	16,350	4,279,820	468,687	958,900	1,289,703	56,775	493,919	189,574	364,411	120,140	30,346,534	832,013.35	29,514,521.53	—	—	—	—
W + M	16,338	4,116,448	449,166	907,864	1,238,743	54,432	477,616	183,395	347,792	114,018	30,311,728	816,280.39	29,495,448.59	—	—	—	—
W + F	16,338	4,245,257	464,341	949,447	1,282,885	56,213	489,709	188,440	360,727	118,826	30,320,833	829,281.43	29,491,551.91	—	—	—	—
W + I	16,344	4,115,010	449,977	916,072	1,250,544	54,645	481,384	184,918	351,233	115,191	30,318,520	821,892.77	29,496,627.91	—	—	—	—

y = birthweight w = parity squared b^2 = sum of squares removed by regression
x = parity z = maternal age d^2 = residual sum of squares

TABLE 10.17 THE ADJUSTED BIRTHWEIGHT MEANS: MALES, NAGASAKI

	x	w	z	Δx	Δw	Δz	\bar{y} unadjusted	\bar{y} adjusted[a]
Father's exposure								
1	2.834	10.975	28.156	−.190	−1.643	−.478	311.639	312.210
2	3.669	18.222	30.183	.645	5.604	1.549	313.307	311.370
3	3.594	17.551	30.344	.570	4.933	1.710	316.077	314.438
4 + 5	3.448	15.931	30.419	.424	3.313	1.785	317.552	316.217
Mean	3.024	12.618	28.634	—	—	—	312.134	312.134
Mother's exposure								
1	2.929	11.641	28.623	−.095	−.977	−.011	312.319	312.555
2	3.134	13.744	28.681	.110	1.126	.047	311.651	311.383
3	3.177	13.985	28.695	.153	1.367	.061	315.865	315.354
4 + 5	2.944	12.154	27.880	−.080	−.464	−.754	311.499	311.756
Mean	3.024	12.618	28.634	—	—	—	312.134	312.134

[a] \bar{y} adjusted = \bar{y} unadjusted $- b_1 \Delta x - b_2 \Delta w - b_3 \Delta z$.

$b_1 = 9.569$ $\Delta x = (x_1 - \bar{x})$ x = parity
$b_2 = -0.686$ $\Delta w = (w_1 - \bar{w})$ w = parity squared
$b_3 = -0.252$ $\Delta z = (z_1 - \bar{z})$ z = maternal age

144 Genetic Effects of Atomic Bombs — Chapter X

TABLE 10.18 THE RESIDUAL MEAN SQUARES FROM THE INDIVIDUAL CELL REGRESSIONS: MALES, NAGASAKI

(The degrees of freedom are given in parentheses.)

Fathers \ Mothers	1	2	3	4–5
1	1,809.34 (7,283)	1,792.27 (4,648)	1,748.51 (335)	1,569.76 (268)
2	1,607.31 (1,064)	1,856.52 (2,020)	2,208.22 (126)	1,802.02 (49)
3	2,400.01 (117)	2,222.96 (131)	2,409.10 (46)	2,407.12 (13)
4–5	1,440.79 (68)	2,095.98 (99)	1,583.62 (9)	1,145.90 (11)

Bartlett's test for between-cell heterogeneity of mean squares:

$\chi^2 = 28.41^*$ DF = 15

Bartlett's test for between-cell heterogeneity of mean squares when category 1 parents are excluded:

$\chi^2 = 7.47$ DF = 8

TABLE 10.19 ANALYSIS OF COVARIANCE OF BIRTHWEIGHTS: FEMALES, NAGASAKI

Exposures	n	Σw^2	Σwx	Σwz	Σwy	Σx^2	Σz^2	Σxy	Σzy	Σxz	Σy^2	b_1	b_2	b_3
11	6,761	1,305,166.15	151,614.33	301,721.08	371,343.39	19,597.27	174,556.06	54,375.77	77,599.88	40,438.68	11,571,699.77	6.8287	−.4094	−.4298
12	4,313	802,990.82	93,239.70	193,745.88	266,789.61	12,025.72	113,640.70	44,114.00	87,162.46	25,627.39	7,509,228.35	11.4655	−.9518	−.1959
13	367	104,545.60	11,255.56	24,562.54	20,644.54	1,326.39	10,754.53	2,886.17	3,498.16	2,977.75	505,846.00	8.1685	−.4899	−.8176
14–15	267	66,981.47	7,010.55	13,604.74	18,809.93	832.25	7,446.54	2,874.21	7,334.50	1,693.98	341,792.56	8.4752	−.6594	.2616
21	1,011	316,989.78	34,407.41	67,007.55	99,391.65	4,065.32	33,574.29	13,453.36	24,573.92	8,188.38	1,740,610.23	8.6865	−.5816	−.2259
22	1,941	1,108,303.96	100,741.96	189,478.39	183,038.29	10,055.71	74,280.06	21,104.17	44,011.29	20,047.08	3,447,535.57	5.0964	−.2913	−.0399
23	138	54,991.54	5,726.67	11,942.43	−2,003.23	636.67	5,429.22	−291.68	210.22	1,410.00	220,421.84	−3.9583	.2759	.4598
24–25	57	23,652.32	2,352.00	5,823.21	−4,793.17	260.67	2,666.14	−483.33	−1,835.44	681.67	106,098.25	2.6242	−.2789	−.7502
31	110	23,494.69	2,798.33	5,471.89	12,691.05	359.42	3,541.49	1,573.25	−4,708.25	729.53	207,304.34	.0657	.3528	−.7708
32	129	45,913.60	4,829.47	11,508.78	9,780.87	548.39	5,424.20	1,491.28	582.89	1,355.78	176,515.69	17.7422	−1.2144	−1.7505
33	42	10,940.57	1,174.00	2,553.00	14,138.57	137.33	1,533.33	1,810.00	4,500.00	312.33	113,078.59	24.9062	−1.4413	.2612
34–35	16	5,494.94	595.44	1,405.75	−1,411.38	67.94	585.00	15.63	82.50	157.75	49,039.75	48.0674	−5.6732	.8119
45–1	63	9,215.43	1,087.24	2,674.67	−1,316.19	144.32	2,012.22	−221.59	−1,427.78	369.56	100,907.94	−1.5065	.2614	−.7803
45–2	70	23,730.99	2,464.67	5,860.71	5,889.85	275.44	2,786.29	846.72	−62.85	674.43	129,298.58	19.0624	−1.2205	−2.0694
45–3	21	18,746.57	1,405.43	2,430.57	−6,645.71	117.24	634.57	−255.62	38.29	193.43	34,377.81	18.3827	−2.0143	2.1722
45–45	13	4,599.69	526.46	1,357.08	−252.92	64.31	663.23	−68.61	755.23	173.38	22,765.23	−34.4031	2.2538	5.5210
Total	DF = 15,255	3,925,758.10	421,229.21	841,148.26	986,095.14	50,514.37	439,527.89	143,223.72	251,732.22	105,031.12	26,276,520.50	7.8464	−.5283	−.2912

y = birthweight
x = parity
w = parity squared
z = maternal age

The Analysis of the Birthweight Data

TABLE 10.20 Tests of the Significance and Homogeneity of the Regressions of Birthweight on Maternal Age and Parity: Females, Nagasaki

(a) Test of the significance of the regression based on within-cells (pooled) regression coefficients

Source	SS	DF	MS	F
Variation removed by regression	529,505.87	3	176,501.957	104.885**
Residual within cells (pool)	25,747,014.63	15,300	1,682.811	—

(b) Test of the homogeneity of the regressions (all exposure cells considered)

Source	SS	DF	MS	F
Differences in regressions	114,038.63	45	2,534.192	1.508*
Residual within cells (sum)	25,632,976.01	15,255	1,680.300	—

(c) Test of the homogeneity of the regressions (only those cells in which both parents were exposed are considered)

Source	SS	DF	MS	F
Differences in regressions	72,541.81	24	3,022.575	1.729*
Residual within cells (sum)	4,180,899.46	2,391	1,748.599	—

TABLE 10.21 Analysis of Variance on the Adjusted Birthweight Means: Females, Nagasaki

	DF	Σw²	Σwx	Σwz	Σwy	Σx²	Σz²	Σxy	Σzy	Σxz	Σy²	b²	d²	Diff.	DF	MS	F
Mothers	3	16,873	1,652	87	6,977	162	179	743	−173	12	10,900	—	—	7,479.31	3	2,493.10	1.482
Fathers	3	145,136	16,357	40,962	22,597	1,845	11,591	2,636	6,726	4,624	7,706	—	—	5,039.30	3	1,679.77	1.002
Interaction	9	34,115	4,584	13,186	27,291	609	5,020	3,460	8,826	1,724	32,985	—	—	18,239.02	9	2,026.56	1.204
Between	15	196,126	22,594	54,236	56,866	2,617	16,792	6,840	15,380	6,362	51,592	529,505.87	25,747,014.63	30,882.15	15	2,058.81	1.223
Within	15,303	3,925,758	421,229	841,148	986,095	50,514	439,527	143,223	251,732	105,031	26,276,320	—	—	—	15,300	1,682.81	—
W + B	15,318	4,121,884	443,823	895,384	1,042,961	53,131	456,320	150,064	267,112	111,393	26,328,112	550,215.99	25,777,896.78	—	—	—	—
W + M	15,306	3,942,601	422,881	841,235	993,072	50,676	439,707	143,967	251,558	105,043	26,287,420	532,926.60	25,754,493.95	—	—	—	—
W + F	15,306	4,070,894	437,586	882,110	1,008,692	52,360	451,119	145,859	258,458	109,655	26,284,227	532,173.21	25,752,053.93	—	—	—	—
W + I	15,312	3,959,873	425,814	854,334	1,013,386	51,123	444,548	146,684	260,558	106,755	26,309,506	544,252.45	25,765,253.65	—	—	—	—

y = birthweight
x = parity
w = parity squared
z = maternal age

b² = sum of squares removed by regression
d² = residual sum of squares

TABLE 10.22 THE ADJUSTED BIRTHWEIGHT MEANS: FEMALES, NAGASAKI

	x	w	z	Δx	Δw	Δz	\bar{y} unadjusted	\bar{y} adjusted [a]
Father's exposure								
1	2.820	10.848	28.093	−.191	−1.689	−.480	303.391	303.858
2	3.664	18.379	30.200	+.653	+5.842	+1.627	304.339	302.774
3	3.475	15.960	29.842	+.464	+3.423	+1.269	307.781	306.317
4 + 5	3.305	14.790	29.311	+.294	+2.253	+.738	305.168	304.266
Mean	3.011	12.537	28.573	—	—	—	303.690	303.690
Mother's exposure								
1	2.916	11.547	28.579	−.095	−.990	+.006	303.384	303.608
2	3.110	13.574	28.606	+.099	+1.037	+.033	303.677	303.457
3	3.229	14.472	28.542	+.218	+1.935	−.031	307.910	307.212
4 + 5	3.011	12.756	27.875	0	+.219	−.698	304.023	303.936
Mean	3.011	12.537	28.573	—	—	—	303.690	303.690

[a] \bar{y} adjusted = \bar{y} unadjusted − $b_1 \Delta x$ − $b_2 \Delta w$ − $b_3 \Delta z$.

$b_1 = 7.846$ $\Delta x = (x_1 - \bar{x})$ x = parity
$b_2 = -0.528$ $\Delta w = (w_1 - \bar{w})$ w = parity squared
$b_3 = -0.291$ $\Delta z = (z_1 - \bar{z})$ z = maternal age

TABLE 10.23 THE RESIDUAL MEAN SQUARES FROM THE INDIVIDUAL CELL REGRESSIONS: FEMALES, NAGASAKI

(The degrees of freedom are given in parentheses.)

	Mothers			
Fathers	1	2	3	4-5
1	1,685.03 (6,757)	1,688.20 (4,309)	1,364.31 (363)	1,246.83 (263)
2	1,675.37 (1,007)	1,757.11 (1,937)	1,639.73 (134)	1,974.58 (53)
3	1,878.25 (106)	1,303.64 (125)	2,294.75 (38)	2,872.48 (14)
4-5	1,691.59 (59)	1,821.47 (66)	1,506.31 (17)	1,867.23 (9)

Bartlett's test for between-cell heterogeneity of mean squares:

$\chi^2 = 29.99*$ DF = 15

Bartlett's test for between-cell heterogeneity of mean squares when category 1 parents are excluded:

$\chi^2 = 9.40$ DF = 8

The Analysis of the Birthweight Data

TABLE 10.24 A SUMMARY OF THE SALIENT FINDINGS OF THE COVARIANCE ANALYSIS

(a) All exposure cells

Cell	Heterogeneity of regressions [a]	Variation in regressions after removing the portion associated with the common regression %	Total variation accounted for by the common regression %	Ratio of smallest to largest mean square (as per cent) after adjustment for individual cell regressions %	Bartlett's test for between-cell heterogeneity of mean squares
Male—Hiroshima	Exceeds 5% level	0.14	2.7	67	Not significant
Female—Hiroshima	Not significant	0.03	1.9	58	Exceeds 1% level
Male—Nagasaki	Not significant	0.05	2.8	48	Exceeds 5% level
Female—Nagasaki	Exceeds 5% level	0.15	2.0	43	Exceeds 5% level

(b) Only those cells in which both parents were exposed

Cell	Heterogeneity of regressions [a]	Variation in regressions after removing the portion associated with the common regression %	Total variation accounted for by the common regression %	Ratio of smallest to largest mean square (as per cent) after adjustment for individual cell regressions %	Bartlett's test for between-cell heterogeneity of mean squares
Male—Hiroshima	Not significant	0	2.8	81	Not significant
Female—Hiroshima	Not significant	0.03	1.3	58	Exceeds 1% level
Male—Nagasaki	Not significant	0	2.1	48	Not significant
Female—Nagasaki	Exceeds 5% level	0.72	0.9	45	Not significant

[a] All regressions of the form $y = m + b_1x_1 + b_2x_1^2 + b_3x_3$.

city cell for three of the four sex-city cells. But, from this same table (10.24b), we note that when exposure category 1 parents are eliminated both the heterogeneity in regressions and in residual mean squares largely disappears. Before we consider the implications of this, let us first consider what might account for the differences observed when all exposure cells are utilized.

Table 10.24a, then, provides evidence, albeit inconclusive at this stage, that birthweight may be associated with parental exposure. Two points are involved, namely:

1. In two of the four sex-city cells, the relation of birthweight to parity and maternal age varies in a minor but significant fashion among father-mother cells.
2. In three of the four sex-city cells, the residual mean squares vary substantially (and significantly) among the father-mother cells.

Let us consider these points. In point (2), under the assumption that within a sex-city cell the observations within a father-mother cell have homogeneous variance, common mean, and normal distributions, we should conclude that the variance differs between the exposure cells, and that this obtains for most sex-city cells. Normality may be safely assumed, but there exists the possibility that the other assumptions are invalid here. Accordingly, two alternative or supplementary explanations of the findings in point (2) must be entertained, namely:

(a) The variances are different between exposure cells, and presumably due to irradiation.
(b) The variances are different between exposure cells due to unaccounted-for concomitant variation. For example, if there were year-of-birth effects, or a socio-economic status effect, then differences in the residual mean squares might reflect unequal representations with respect to year of birth, or social background.

Similarly, for point (1) we must entertain more than one explanation, namely:

(a) The relation of birthweight to parity and age of mother is different among exposure cells and presumably due to irradiation.
(b) Factors partially correlated with age and parity may differ between exposure cells and thus give rise to an apparent variation in the relation of birthweight to parity and mother's age.

What evidence is available which might permit a choice among these alternatives?

In addition to data covering maternal age and parity, information is available on two other concomitant variables which are relevant to the questions raised in the preceding paragraph. These variables are year of birth (for all infants) and the parental socio-economic status (for a random 10 per cent of infants). Let us consider first the effect of year of birth on the residual mean squares.

Two alternatives which might be entertained here are the following: Firstly, we could assume that the relationship of mother's age and parity to birthweight is constant over the years within a given exposure cell, but that the intercept (mean) of the regression may vary from year to year. This would lead us to fit a model asserting that

$$E(y_{ijrk}) = m_{ijr} + b_1(x_{ijrk} - \bar{x}_{ij}) + b_2(w_{ijrk} - \bar{w}_{ij}) + b_3(z_{ijrk} - \bar{z}_{ij}).$$

That is to say, the expected value of the k^{th} observation on birthweight in the r^{th} year of the ij^{th} exposure cell (sex and city are fixed) is a function of the mean of the r^{th} year in the ij^{th} cell, of parity, of parity-squared, and of maternal age. Secondly, we might prefer not to assume that the relationship of mother's age and parity to birthweight is constant over years within a given exposure cell, and hence to fit a model of the form

$$E(y_{ijrk}) = m_{ijr} + b_{1r}(x_{ijrk} - \bar{x}_{ijr}) + b_{2r}(w_{ijrk} - \bar{w}_{ijr}) + b_{3r}(z_{ijrk} - \bar{z}_{ijr}).$$

The latter approach is, of course, equivalent to fitting a regression in each of the 112 year-exposure combinations within a given sex-city cell. It is questionable whether the increase in precision of the latter approach over the former is sufficient to justify the not insignificant additional labor in computing some 448 regressions. As a first approximation, then, we have elected to proceed using the simpler (the first) model.

To determine whether year of birth contributes significantly to the variation in birthweights, within exposure cells, following the removal of maternal age and parity effects, we may employ the mean square ratio test. The latter test consists of forming the likelihood ratio, say L, of the "mean square following removal of age and parity" to the "mean square following removal of age, parity, and year."

We assert that year of birth contributes significantly if $L > L_o$ where the critical value, L_o, is given by

$$L_o = \frac{f + (f' - f)F_o}{f'}$$

and where f and f' are, respectively, the degrees of freedom associated with the lesser and the greater mean square, and F_o is the critical F

In Tables 10.25 to 10.28 are given the distributions by parental exposure of the residual mean squares following removal of the "year-of-birth effect" for the four sex-city cells. In three of the four sex-city cells the residual mean squares do not, now, differ significantly one from another (two of these three previously revealed significant heterogeneity among the mean squares).

TABLE 10.25 THE DISTRIBUTION BY PARENTAL EXPOSURE OF THE RESIDUAL MEAN SQUARES AFTER REMOVAL OF VARIATION DUE TO YEAR OF BIRTH OF THE INFANT: MALES, HIROSHIMA

(Degrees of freedom are given in parentheses.)

Mothers	Fathers 1	2	3	4–5
1	1,709.60 (8,662)	1,793.66 (735)	1,835.06 (286)	2,347.31 (190)
2	1,639.79 (2,707)	1,786.50 (889)	1,982.83 (182)	1,982.08 (105)
3	1,790.77 (1,056)	1,747.22 (210)	1,764.08 (247)	1,647.70 (68)
4–5	1,597.92 (532)	1,914.44 (99)	1,811.33 (39)	1,745.37 (46)

$\chi^2 = 21.147$ DF = 15

TABLE 10.26 THE DISTRIBUTION BY PARENTAL EXPOSURE OF THE RESIDUAL MEAN SQUARES AFTER REMOVAL OF VARIATION DUE TO YEAR OF BIRTH OF THE INFANT: FEMALES, HIROSHIMA

(Degrees of freedom are given in parentheses.)

Mothers	Fathers 1	2	3	4–5
1	1,611.26 (7,990)	1,646.48 (691)	1,773.60 (257)	1,941.68 (166)
2	1,559.88 (2,419)	1,786.51 (865)	1,309.50 (172)	1,163.41 (109)
3	1,682.01 (1,023)	2,224.61 (189)	2,181.14 (237)	2,219.93 (66)
4–5	1,603.27 (517)	1,802.62 (80)	1,445.23 (46)	1,321.28 (49)

$\chi^2 = 45.110**$ DF = 15

TABLE 10.27 THE DISTRIBUTION BY PARENTAL EXPOSURE OF THE RESIDUAL MEAN SQUARES AFTER REMOVAL OF VARIATION DUE TO YEAR OF BIRTH OF THE INFANT: MALES, NAGASAKI

(Degrees of freedom are given in parentheses.)

Mothers	Fathers 1	2	3	4–5
1	1,805.73 (7,278)	1,712.89 (1,059)	2,359.97 (112)	1,465.77 (63)
2	1,782.86 (4,644)	1,854.71 (2,015)	2,133.38 (126)	2,051.48 (94)
3	1,771.84 (330)	2,166.44 (121)	2,544.20 (41)	1,542.99 (5)
4–5	1,643.25 (263)	1,824.87 (45)	3,360.16 (9)	1,002.24 (6)

$\chi^2 = 20.442$ DF = 15

TABLE 10.28 THE DISTRIBUTION BY PARENTAL EXPOSURE OF THE RESIDUAL MEAN SQUARES AFTER REMOVAL OF VARIATION DUE TO YEAR OF BIRTH OF THE INFANT: FEMALES, NAGASAKI

(Degrees of freedom are given in parentheses.)

Mothers	Fathers 1	2	3	4–5
1	1,680.50 (6,752)	1,753.30 (1,003)	1,892.83 (101)	1,591.88 (54)
2	1,688.58 (4,305)	1,753.37 (1,932)	1,344.54 (120)	1,865.09 (61)
3	1,675.05 (358)	1,662.10 (129)	2,280.34 (33)	1,752.83 (13)
4–5	1,505.33 (259)	2,050.00 (49)	4,305.18 (7)	2,363.29 (5)

$\chi^2 = 22.003$ DF = 15

(here $F_{.05}$) at $(f' - f)$ and f degrees of freedom. In all sex-city cells, year of birth can be shown to exert an effect; this effect is, however, significant in only two instances, namely, Hiroshima-females and Nagasaki-males. Under the circumstances, then, to prevent ambiguity it seems appropriate to remove the year-of-birth effects in all cells.

In view of the fact that the variances were heterogeneous in only one sex-city cell (Hiroshima-females) when year of birth was taken into account, no attempt was made to exploit the data with regard to socio-economic status. The latter decision stemmed primarily from the fact that analysis of the socio-economic data would, since such data are available on only 10

per cent of the observations, be applicable only inferentially to the problem of whether residual concomitant variation could account for the heterogeneity of the variance estimates within the Hiroshima-females cell.

What, now, may we conclude with regard to the effect of parental exposure on birthweights? Three points are involved, namely,

1. Does parental exposure affect birthweight means?
2. Does parental exposure affect the relationship between birthweight and concomitant variables, notably maternal age and parity?
3. Does parental exposure affect birthweight variances?

With regard to birthweight means, we cannot demonstrate a significant effect of parental exposure in any one of the four sex-city cells. With regard to the regressions of birthweight on parity and maternal age, significant differences obtain among exposure cells in two sex-city cells. In one of these instances, significance does not obtain if attention is limited to only those infants born to parents both of whom were exposed. In view of the known somatic effects of irradiation, it would seem most likely that the regression differences, if real, reflect either a "disaster" effect or a direct (non-genetic) effect of irradiation. Lastly, with regard to the effect of parental exposure on birthweight variances, we note that, when maternal age and parity are removed, the variance estimates are heterogeneous among exposure cells (all exposure cells considered) in three of the four sex-city cells. If, however, year of birth is also removed, only one sex-city cell continues to exhibit significant heterogeneity in the variance estimates among exposure cells. Moreover, we note that if attention is limited to those terminations to parents both of whom were exposed, the variance estimates among the exposure cells are significantly heterogeneous in only one sex-city cell even if year of birth is not removed. These facts raise some doubt as to the reality of the differences in mean squares as an irradiation effect, especially if one assumes that a consistent pattern is the *sine qua non* of a radiation-induced change. Whether consistency is a valid assumption is difficult to appraise. One can postulate certain parental interactions or relationships between fetal resorption and degree of radiation damage, both admissible hypotheses, from which it would not necessarily follow that the variance would be similarly affected in all sex-city cells. The conservative interpretation of the data with regard to the birthweight variances would be one which asserts no clearly demonstrable effect of parental irradiation on the spread of birthweights.

10.5 *Summary.* — There exist no consistent findings suggestive of an effect of parental exposure on (a) birthweight means, (b) the relationship of maternal age and parity to birthweight, or (c) birthweight variances.

Chapter XI

ANALYSIS OF THE DATA CONCERNING DEATH DURING THE NINE-MONTH PERIOD FOLLOWING DELIVERY

11.1 *The trait.* — Although from its very first usage the Genetics Short Form included a question concerning death following delivery, the collection of data with reference to death after the first week following delivery was not originally systematic. However, the instruction to the midwives regarding promptly notifying the ABCC in the event of an abnormal pregnancy termination included a statement regarding death during the first days of life. As the autopsy program developed, the midwives often voluntarily informed the ABCC of death at any time during and after the first week postpartum. The routine follow-up visit was usually made between the seventh and fourteenth postpartum days — sometimes later, usually in consequence of recurrent transportation crises in the "motor pool," seldom earlier, due largely to the system of collecting Genetics Short Forms from the midwives and also a natural and proper reluctance on the part of Japanese mothers, especially during the winter months, to permit the examination of their children at an earlier date. At the time of this follow-up visit, any deaths which had occurred were noted. This system led to quite complete information regarding death during the first week of life. In the actual analysis, we have confined our attention to the occurrence of death during the first six days postpartum. In the strict sense, this is not the neonatal period, which is usually thought of as including the first month of life, but we shall apply the term "neonatal" to it as a matter of convenience.

In addition to this information regarding neonatal death, there is available supplementary information from the "9-months program." As noted earlier, infants were selected for follow-up studies at age 9 months on the basis of the terminal digit in their registration numbers. Thus, depending on the personnel and facilities available, one month an attempt might be made to re-examine all infants whose registration number ended in 0 or 5, the next month 0, 5, or 9, etc. At the time of the contacts necessary to arrange for this examination, it could be determined if the infant had died in the interval between our initial examination and the time of the second contact. There is thus available supplementary information for 21,788 infants concerning the occurrence of death between birth and the ninth month following delivery. The latter figure is somewhat approximate because of variations in the "contact day," but no bias exists with reference to parental radiation history.

11.2 *The genetic argument for radiation-induced changes in the neonatal death rate.* — The argument from which one might expect radiation-induced genetic changes in the neonatal death rate is wholly analogous to the argument for differences in the stillbirth rate with increasing parental exposure (see Sec. 9.2). Mutations detected in the first generation of offspring by a study of the neonatal death rate would consist largely of the dominant lethals (or possibly in a more strict sense the sublethals) although some recessive lethals might also be recovered because of the fortuitous combination in some individuals of an induced lethal mutation with an allelic lethal mutation of spontaneous origin. Accordingly, we might expect as one manifestation of radiation-induced genetic damage an increase in the frequency of neonatal deaths with increasing parental exposure.

151

11.3 *Concomitant variables known to influence the occurrence of a neonatal death.* — The sources of extraneous variation affecting the frequency of neonatal deaths are not unlike those variables influencing the frequency of stillbirths (see Sec. 9.3). Among the concomitant variables common to both of the aforementioned indicators are maternal disease (principally syphilis), nutrition, birth injury, maternal age, order of birth, and rate of reproduction. In the case of neonatal death this list must be augmented by the addition of at least one more extraneous variable, namely, infantile infectious diseases. In the first six days of life, the principal infectious diseases, which could and frequently do lead to the death of an infant, are respiratory infections and infantile diarrhea.

The observations and comments with regard to maternal disease, nutrition, and birth injury made in Chapter IX with regard to stillbirths are no less pertinent here. To recapitulate briefly, we noted that:

1. There existed a gradual but significant decline in the rate of transmission of syphilis from mother to infant with increasing maternal age. The age distribution among the exposure cells is such that this bias would tend to inflate the neonatal death rate among the non-exposed or less heavily exposed parents. It was felt that these effects could be ignored.

2. There is ample evidence attesting to the importance of nutrition and birth injury in the neonatal death rate, but these two variables will be ignored in our presentation because there are no demonstrable or probable differences among exposure cells.

FIGURE 11.1 — The distribution of the frequency of neonatal deaths by age of mother for specified parities.

With regard to the effect of maternal age and parity on the neonatal death rate the reader is referred to Figures 11.1 and 11.2 and Tables 11.1 and 11.2. The latter figure gives the distribution of the frequency of neonatal death by parity for specified maternal ages. The former figure gives the distribution of the frequency of neonatal death by maternal age for specified parities. From these two figures and tables we note that the effect of maternal age and parity on the frequency of neonatal death is strikingly similar to the effect of age and parity on the

stillbirth rate.[1] We find that the age-specific parity distributions or the parity-specific age distributions of the frequency of neonatal death tend to be U-shaped. However, unlike the stillbirth data the minima of the age-specific parity distributions appear quite similar, whereas the minima of the parity-specific age distributions tend to occur at a higher level with increasing exposure cells would require also taking into account variation in maternal age. However, as in the case of stillbirths, a partitioning of the data as elegant as that required to effect the "best" control would seriously jeopardize the validity of the tests because of (a) the large number of empty cells, and (b) the very small expected values in the bulk of the occupied cells.

FIGURE 11.2 — The distribution of the frequency of neonatal deaths by parity for specified maternal ages.

parity. These findings would suggest a significant effect of parity which varies by mother's age on the frequency of neonatal death (a rate of reproduction effect), but no significant independent effect of mother's age. This is consistent with Yerushalmy's (1938) observations in the United States. Now since the effect of parity is not independent of maternal age, adequate adjustment to control variation in parity between

[1] A full description of these data is to be published elsewhere.

Accordingly, the data were further partitioned only with regard to parity. We recognized five parity classes, namely, parity 1, parities 2–3, 4–5, 6–7, and 8 and higher. The bias which might remain after this partitioning would be one difficult to specify adequately but probably would lead to an increase in neonatal mortality among the more heavily exposed cells because of increased maternal age within these latter cells.

The possibility of a spurious irradiation effect

TABLE 11.1 THE EFFECT OF MATERNAL AGE AT FIXED PARITY ON THE FREQUENCY OF NEONATAL DEATHS

	Total births	Normal births	Neonatal deaths	% neonatal	AD
Parity 1					
<21	1,882	1,849	33	1.75	.00808
21–25	8,479	8,356	123	1.45	.07422
26–30	3,125	3,062	63	2.02	.04877
31–35	595	587	8	1.34	.00793
36–40	171	164	7	4.09	.01797
41+	18	16	2	11.11	.00733
Total	14,270	14,034	236	—	.16430

$\chi^2_5 = 21.293^{**}$ $I_{AD} = .08215$ $P_T = .26001$

Parity 2					
<21	415	413	2	0.48	.01684
21–25	6,572	6,482	90	1.37	.05416
26–30	6,294	6,225	69	1.10	.04273
31–35	1,399	1,383	16	1.14	.00585
36–40	360	353	7 }	1.72	.01127
41+	46	46	— }		
Total	15,086	14,902	184	—	.13086

$\chi^2_4 = 4.818$ $I_{AD} = .06543$ $P_T = .27488$

Parity 3					
<21	36	35	1	2.78	.00374
21–25	2,188	2,151	37	1.69	.05859
26–30	5,729	5,658	71	1.24	.02997
31–35	2,357	2,330	27	1.15	.02799
36–40	514	508	6	1.17	.00531
41+	66	65	1	1.52	.00094
Total	10,890	10,747	143	—	.12654

$\chi^2_5 = 3.864$ $I_{AD} = .06327$ $P_T = .19843$

Parity 4					
<21	4	4	— }	2.52	.05783
21–25	472	460	12 }		
26–30	2,828	2,788	40	1.41	.00230
31–35	2,487	2,455	32	1.29	.03262
36–40	736	728	8	1.09	.02569
41+	83	82	1	1.20	.00183
Total	6,610	6,517	93	—	.12028

$\chi^2_4 = 5.087$ $I_{AD} = .06014$ $P_T = .12044$

TABLE 11.1—*Continued*

	Total births	Normal births	Neonatal deaths	% neonatal	AD
Parity 5					
< 21	—	—	—	—	—
21–25	76	70	6	7.89	.08992
26–30	975	960	15	1.54	.00979
31–35	1,737	1,712	25	1.44	.01437
36–40	823	814	9 ⎫	0.98	.08534
41 +	95	95	— ⎭		
Total	3,706	3,651	55	—	.19942

$\chi^2_3 = 22.999^{**}$ $I_{AD} = .09971$ $p_T = .06753$

Parity 6					
< 21	—	—	—	—	—
21–25	11	10	1	9.09	.03229
26–30	259	254	5	1.93	.06458
31–35	943	936	7	0.74	.18519
36–40	785	775	10	1.27	.00237
41 +	135	131	4	2.96	.08594
Total	2,133	2,106	27	—	.37038

$\chi^2_4 = 11.484^{*}$ $I_{AD} = .18519$ $p_T = .03887$

Parity 7 +					
< 21	—	—	—	—	—
21–25	3	3	— ⎫		.00052
26–30	55	54	1 ⎭	1.72	
31–35	479	471	8	1.67	.00285
36–40	1,167	1,146	21	1.80	.03454
41 +	483	476	7	1.45	.03221
Total	2,187	2,150	37	—	.07012

$\chi^2_3 = .254$ $I_{AD} = .03506$ $p_T = .03985$

All parities					
< 21	2,337	2,301	36	1.54	.00392
21–25	17,801	17,532	269	1.51	.02307
26–30	19,265	19,001	264	1.37	.01053
31–35	9,997	9,874	123	1.23	.02378
36–40	4,556	4,488	68	1.49	.00480
41 +	926	911	15	1.62	.00252
Total	54,882	54,107	775	—	—

$\chi^2_5 = 4.641$ I'_{AD} (for age) $= .07447$

Heterogeneity $\chi^2 = 65.158^{**}$ DF = 23

TABLE 11.2 THE EFFECT OF PARITY AT FIXED MATERNAL AGE ON THE FREQUENCY OF NEONATAL DEATHS

Mother's age: < 21	Total births	Normal births	Neonatal deaths	% neonatal	AD
1	1,882	1,849	33	1.75	.11310
2	415	413	2	0.48	.12393
3	36	35	1 ⎫	2.50	.01083
4+	4	4	— ⎭		
Total	2,337	2,301	36	—	.24786

$\chi^2_2 = 3.872$ $\quad I_{AD} = .12393 \quad$ $p_T = .04258$

Mother's age: 21–25					
1	8,479	8,356	123	1.45	.01937
2	6,572	6,482	90	1.37	.03515
3	2,188	2,151	37	1.69	.01486
4	472	460	12	2.54	.01837
5	76	70	6	7.89	.01831
6	11	10	1 ⎫	7.14	.00298
7+	3	3	— ⎭		
Total	17,801	17,532	269	—	.10904

$\chi^2_5 = 28.736^{**}$ $\quad I_{AD} = .05452 \quad$ $p_T = ..32435$

Mother's age: 26–30					
1	3,125	3,062	63	2.02	.07749
2	6,294	6,225	69	1.10	.06625
3	5,729	5,658	71	1.24	.02883
4	2,828	2,788	40	1.41	.00479
5	975	960	15	1.54	.00629
6	259	254	5	1.93	.00557
7+	55	54	1	1.81	.00095
Total	19,265	19,001	264	—	.19018

$\chi^2_6 = 14.791^*$ $\quad I_{AD} = .09509 \quad$ $p_T = .35103$

Mother's age: 31–35					
1	595	587	8	1.34	.00559
2	1,399	1,383	16	1.14	.00998
3	2,357	2,330	27	1.15	.01646
4	2,487	2,455	32	1.29	.01153
5	1,737	1,712	25	1.44	.02987
6	943	936	7	0.74	.03788
7+	479	471	8	1.67	.01734
Total	9,997	9,874	123	—	.12866

$\chi^2_6 = 3.589$ $\quad I_{AD} = .06433 \quad$ $p_T = .18215$

Mother's age: 36–40					
1	171	164	7	4.09	.06640
2	360	353	7	1.94	.02429
3	514	508	6	1.17	.02496
4	736	728	8	1.09	.04456
5	823	814	9	1.09	.04902
6	785	775	10	1.27	.02562
7+	1,167	1,146	21	1.80	.05348
Total	4,556	4,488	68	—	.28834

$\chi^2_6 = 11.456$ $\quad I_{AD} = .14417 \quad$ $p_T = .08301$

Data Concerning Death During Period Following Delivery

TABLE 11.2—Continued

	Total births	Normal births	Neonatal deaths	% neonatal	AD
Mother's age: 41 +					
1	18	16	2⎫	3.13	.06528
2	46	46	—⎭		
3	66	65	1	1.52	.00468
4	83	82	1⎫	0.56	.12763
5	95	95	—⎭		
6	135	131	4	2.96	.12287
7 +	483	476	7	1.45	.05584
Total	926	911	15	—	.37630

$\chi^2_4 = 3.781$ $I_{AD} = .18815$ $P_T = .01687$

All ages:					
1	14,270	14,034	236	1.65	.04514
2	15,086	14,902	184	1.22	.03800
3	10,890	10,747	143	1.31	.01411
4	6,610	6,517	93	1.41	.00045
5	3,706	3,651	55	1.48	.00349
6	2,133	2,106	27	1.27	.00408
7 +	2,187	2,150	37	1.69	.00801
Total	54,882	54,107	775	—	—

$\chi^2_6 = 12.462$ I''_{AD} (for parity) $= .08320$

Heterogeneity $\chi^2 = 53.763^{**}$ $DF = 23$

on the neonatal death rate due to a non-random distribution (by parental exposure) of contacts with the agents responsible for the more serious infectious infantile diseases such as the respiratory infections and infantile diarrhea is difficult to evaluate. We are not aware of any widespread outbreaks of respiratory infections or diarrhea among the newborn infants of Hiroshima and Nagasaki during the interval 1948–1953.

11.4 *The data.* — In Tables 11.3 and 11.4 are given the distribution of neonatal deaths by sex of infant, parental exposure and city without and with allowance for parity. The corresponding analytical results are given in Tables 11.5 and 11.6. Let us consider first the evidence for an irradiation effect from the data prior to partitioning by parity. From inspection of Table 11.3 and from the results given in Table 11.5 we note a significant mother-father exposure interaction. The occurrence of this interaction considerably complicates further analysis of these data since, as in the non-orthogonal case of the analysis of variance, the disproportion among the numbers of observations in the various ways of classification leads to distortion of main effects tests based upon the marginal totals, due to the unequal contributions to the marginal class totals. Under the circumstances it would seem that the most appropriate tests for those ways of classification other than those involved in the interactions, here mother's exposure and father's exposure, would be the sum of the chi-squares obtained by testing a given way of classification at each level of the ways of classification involved in the interaction. Thus, in this instance, the test of the city effect would be obtained by summing the sixteen chi-squares and their degrees of freedom obtained by testing the effect of city on the variate at each of the sixteen mother-father levels. The disadvantage of this procedure is that all sixteen tests contribute equally to the sum, whereas the power of the individual tests will vary considerably. Ideally, perhaps, one would weight the individual contributions of these tests to the sum by their power functions. This has not been done, and hence the sum tests will probably be somewhat biased.

When the data are analyzed as just outlined, for only one of the eight exposure tests when the exposure of one parent is held constant does a significant effect of mother's or father's exposure emerge. The weight this one positive finding should be given in the face of seven non-significant, similar tests is debatable. There are significant differences between the sexes but

TABLE 11.3 THE FREQUENCY OF NEONATAL DEATHS BY PARENTAL EXPOSURE, CITY, AND SEX OF INFANT
(Unrelated parents)

Male infants
Hiroshima

		Fathers				
Mothers		1	2	3	4–5	Total
1	n	8,805	751	304	206	10,066
	d	132	7	9	7	155
	p	.0150	.0093	.0296	.0340	.0154
2	n	2,760	909	190	116	3,975
	d	45	11	—	2	58
	p	.0163	.0121	—	.0172	.0146
3	n	1,081	219	258	79	1,637
	d	16	—	2	2	20
	p	.0148	—	.0078	.0253	.0122
4–5	n	551	108	51	56	766
	d	10	—	3	1	14
	p	.0181	—	.0588	.0179	.0183
Total	n	13,197	1,987	803	457	16,444
	d	203	18	14	12	247
	p	.0154	.0091	.0174	.0263	.0150

Male infants
Nagasaki

		Fathers				
Mothers		1	2	3	4–5	Total
1	n	7,417	1,087	124	72	8,700
	d	130	19	3	—	152
	p	.0175	.0175	.0242	—	.0175
2	n	4,732	2,055	137	103	7,027
	d	80	31	2	—	113
	p	.0169	.0151	.0146	—	.0161
3	n	349	131	50	13	543
	d	10	1	—	—	11
	p	.0287	.0076	—	—	.0203
4–5	n	274	55	18	15	362
	d	2	2	1	—	5
	p	.0073	.0364	.0556	—	.0138
Total	n	12,772	3,328	329	203	16,632
	d	222	53	6	—	281
	p	.0174	.0159	.0182	—	.0169

TABLE 11.3—*Continued*

Female infants
Hiroshima

		Fathers				
		1	2	3	4–5	Total
Mothers	1 { n	8,086	711	271	177	9,245
	d	87	11	5	2	105
	p	.0108	.0155	.0185	.0113	.0114
	2 { n	2,466	886	182	118	3,652
	d	32	12	1	—	45
	p	.0130	.0135	.0055	—	.0123
	3 { n	1,039	199	246	75	1,559
	d	8	1	—	—	9
	p	.0077	.0050	—	—	.0058
	4–5 { n	533	90	55	58	736
	d	7	1	—	—	8
	p	.0131	.0111	—	—	.0109
Total	n	12,124	1,886	754	428	15,192
	d	134	25	6	2	167
	p	.0111	.0133	.0080	.0047	.0110

Female infants
Nagasaki

		Fathers				
		1	2	3	4–5	Total
Mothers	1 { n	6,843	1,023	112	64	8,042
	d	82	12	2	1	97
	p	.0120	.0117	.0179	.0156	.0121
	2 { n	4,373	1,967	131	70	6,541
	d	60	26	2	—	88
	p	.0137	.0132	.0153	—	.0135
	3 { n	373	139	43	21	576
	d	6	1	1	—	8
	p	.0161	.0072	.0233	—	.0139
	4–5 { n	272	58	16	13	359
	d	5	1	—	—	6
	p	.0184	.0172	—	—	.0167
Total	n	11,861	3,187	302	168	15,518
	d	153	40	5	1	199
	p	.0129	.0126	.0166	.0060	.0128

TABLE 11.4 THE FREQUENCY OF NEONATAL DEATHS BY PARENTAL EXPOSURE, CITY, AND PARITY
(Unrelated parents)

TABLE 11.5 CHI-SQUARE ANALYSIS OF THE FREQUENCY OF NEONATAL DEATHS BY PARENTAL EXPOSURE, CITY, AND SEX
(Unrelated parents)

(a) 4 × 4

Source	DF	χ^2	P
Total	63	83.974	.05–.10
Interactions, first order			
CS	1	0.070	.70–.80
CM	3	3.192	.30–.50
CF	3	5.336	.10–.20
SM	3	2.202	.50–.70
SF	3	5.224	.10–.20
MF	9	22.986	.001–.01
Main effects			
Mothers: Fathers 1	3	1.223	.70–.80
Fathers 2	3	4.365	.20–.30
Fathers 3	3	12.480	.001–.01
Fathers 4,5	3	4.378	.20–.30
Fathers: Mothers 1	3	6.258	.10–.20
Mothers 2	3	5.280	.10–.20
Mothers 3	3	7.132	.05–.10
Mothers 4,5	3	2.479	.30–.50
Sex:[a] M_1F_1	1	13.296	< .001
M_1F_2	1	0.051	.80–.90
$M_1F_{3,4,5}$	1	1.840	.10–.20
M_2F_1	1	2.504	.10–.20
M_2F_2	1	0.077	.70–.80
$M_2F_{3,4,5}$	1	0.070	.80–.90
$M_{3,4,5}F_1$	1	2.080	.10–.20
$M_{3,4,5}F_2$	1	0.204	.50–.70
$M_{3,4,5}F_{3,4,5}$	1	6.266	.01–.02
Sum	9	26.388	.001–.01
City:[a] M_1F_1	1	2.048	.10–.20
M_2F_1	1	0.091	.70–.80
$M_{3,4,5}F_1$	1	1.837	.10–.20
M_1F_2	1	0.361	.50–.70
M_2F_2	1	0.168	.50–.70
$M_{3,4,5}F_2$	1	3.265	.05–.10
$M_1F_{3,4,5}$	1	0.780	.30–.50
$M_2F_{3,4,5}$	1	1.093	.20–.30
$M_{3,4,5}F_{3,4,5}$	1	0.036	.80–.90
Sum	9	9.679	.30–.50

[a] To increase the numbers of observations, categories 3, 4, and 5 have been pooled in these tests.

TABLE 11.5—Continued

(b) 3 × 3

Source	DF	χ^2	P
Total	35	39.587	.50–.60
Interactions, first order			
CS	1	0.161	.50–.70
CM	2	0.425	.80–.90
CF	2	4.354	.10–.20
SM	2	2.258	.30–.50
SF	2	5.007	.05–.10
MF	4	5.441	.20–.30
Main effects			
Sex (S)[a]			
Mother's exposure — 2	1	0.106	.70–.80
Mother's exposure — 3	1	0.432	.50–.70
Mother's exposure — 4,5	1	2.604	.10–.20
Sum	3	3.142	.30–.50
City (C)[a]			
Mother's exposure — 2	1	1.005	.30–.50
Mother's exposure — 3	1	0.454	.50–.70
Mother's exposure — 4,5	1	0.980	.30–.50
Sum	3	2.439	.30–.50
Mother (M)	2	6.216	.02–.05
Father (F)[a]			
Mother's exposure — 2	2	3.708	.10–.20
Mother's exposure — 3	2	1.108	.50–.70
Mother's exposure — 4,5	2	2.421	.20–.30
Sum	6	7.237	.20–.30

[a] Adjusted for mothers.

TABLE 11.6 CHI-SQUARE ANALYSIS OF THE FREQUENCY OF NEONATAL DEATHS BY PARENTAL EXPOSURE, CITY, AND PARITY
(Unrelated parents)

Source	DF	χ^2	P
Total	159	189.601	.05–.10
Interactions, first order			
CM	3	3.192	.30–.50
CF	3	4.855	.10–.20
CP	4	14.401	.001–.01
MF	9	22.986	.001–.01
MP	12	21.436	.02–.05
FP	12	13.334	.30–.50
Main effects (only CP interaction taken into account)			
Parity (P)			
Hiroshima (H)	4	5.872	.20–.30
Nagasaki (N)	4	32.045	< .001
City (C)			
Parity			
Class — 1	1	18.585	< .001
Class — 2	1	0.458	.30–.50
Class — 3	1	0.420	.50–.70
Class — 4	1	1.720	.10–.20
Class — 5	1	0.200	.50–.70
Mother (M)			
H: Parity			
Class — 1	3	11.753	.001–.01
Class — 2	3	3.424	.30–.50
Class — 3	3	0.809	.80–.90
Class — 4	3	2.882	.30–.50
Class — 5	3	0.775	.80–.90
N: Parity			
Class — 1	3	5.235	.10–.20
Class — 2	3	1.977	.50–.70
Class — 3	3	2.552	.30–.50
Class — 4	3	2.409	.30–.50
Class — 5	3	4.662	.10–.20
Sum	30	36.478	.10–.20
Father (F)			
H: Parity			
Class — 1	3	4.260	.20–.30
Class — 2	3	2.387	.30–.50
Class — 3	3	10.236	.01–.02
Class — 4	3	2.559	.30–.50
Class — 5	3	1.976	.50–.70
N: Parity			
Class — 1	3	1.085	.70–.80
Class — 2	3	4.322	.20–.30
Class — 3	3	3.156	.30–.50
Class — 4	3	2.322	.50–.70
Class — 5	3	0.846	.80–.90
Sum	30	33.149	.30–.50

not the cities. Omission of the category 1 parents from the analysis results in the disappearance of the M-F interaction (Table 11.5b). As before, there is no significant effect of father's exposure but now a barely significant effect of mother's exposure appears.

From Tables 11.4 and 11.6, we note that when neonatal deaths are further partitioned by parity rather extensive heterogeneity in the data is revealed. Both the city-parity and mother-father interactions are significant. To adjust the main effects tests as indicated in Chapter VI leads, in this instance, to such extensive partitioning as to seriously jeopardize the validity of the tests. However, one could adjust for one of the interactions at a time. Accordingly, two sets of "main effects" tests could be generated. If one ignores the mother-father interaction, the main effects tests of city, parity, and exposure would be as given in Table 11.6. If the city-parity interaction were ignored, the tests of exposure and city would be as given in Table 11.5a, and there would, in addition, be a test of parity. One must obviously view the results of these two sets of tests as approximate. However, in neither instance does evidence emerge indicating a significant, consistent effect of parental exposure.

The results from the "at-birth" examinations are corroborated by the findings at 9 months of age, presented in Table 11.7. The reader should note the following departures from the usual classifications of the 9-months data: (1) sex of infant is ignored, (2) exposure categories 3, 4, and 5 have been pooled, and (3) deaths from birth to 9 months are recorded. The purpose of these latter procedures was to bring to bear on the problem of infant mortality following dissimilar exposure experiences on the part of the parents, the largest possible number of observations. No attempt has been made to control concomitant variation which would presumably be no less of a problem here than elsewhere in these data, and would, in the main, tend to inflate the death rate among infants born to the more heavily exposed parents. The analysis of these data is presented in Table 11.8. We note no effect of city, maternal or paternal exposure, or evidence of heterogeneity between the cities or exposure cells.

11.5 Summary. — No consistent, significant effect of parental exposure on neonatal mortality emerges from the data obtained in Hiroshima and Nagasaki on deaths occurring in the first six days postpartum. Analysis of neonatal and infantile mortality obtained at the time of the 9-months examination fails to disclose an effect of parental exposure on infant survival. No consistent exposure trend is exhibited by these data.

Data Concerning Death During Period Following Delivery

TABLE 11.7 THE FREQUENCY OF DEATHS BETWEEN BIRTH AND NINE MONTHS OF AGE BY PARENTAL EXPOSURE AND CITY
(Unrelated parents)

			Hiroshima						Nagasaki			
			Mothers						Mothers			
Fathers			1	2	3-4-5	Total	Fathers		1	2	3-4-5	Total
	1	n	5,678	1,803	1,064	8,545		1 { n	4,225	2,599	396	7,220
		d	254	85	48	387		d	209	125	25	359
		p	.0447	.0471	.0451	.0453		p	.0495	.0481	.0631	.0497
	2	n	482	528	188	1,198		2 { n	583	1,051	117	1,751
		d	24	27	8	59		d	20	61	2	83
		p	.0498	.0511	.0426	.0492		p	.0343	.0580	.0171	.0474
	3-4-5	n	319	186	264	769		3-4-5 { n	120	149	66	335
		d	16	8	14	38		d	4	7	5	16
		p	.0502	.0430	.0530	.0494		p	.0333	.0470	.0758	.0478
	Total	n	6,479	2,517	1,516	10,512		Total { n	4,928	3,799	579	9,306
		d	294	120	70	484		d	233	193	32	458
		p	.0454	.0477	.0462	.0460		p	.0473	.0508	.0553	.0492

TABLE 11.8 ANALYSIS OF THE FREQUENCY OF DEATHS BETWEEN BIRTH AND NINE MONTHS OF AGE, BY PARENTAL EXPOSURE AND CITY
(Unrelated parents)

Source	DF	χ^2	P
Total	17	13.146	.70–.80
Interactions, first order			
CM	2	0.382	.80–.90
CF	2	0.658	.70–.80
MF	4	4.647	.30–.50
Main effects			
City	1	1.097	.20–.30
Mothers	2	1.081	.50–.70
Fathers	2	0.087	.95–.98

Chapter XII

THE ANALYSIS OF THE ANTHROPOMETRIC DATA

12.1 *The measurements obtained at nine months.* — As was pointed out in Section 2.6, one feature of the "9-months" program involved obtaining certain physical measurements on the infants seen. These measurements were felt to supply the best index to general physical vigor obtainable under the circumstances. The present chapter is concerned with an analysis of these data. It should be pointed out that because of scheduling problems, it was not possible to examine each infant at precisely nine months of age; the data here presented are based on babies varying in age from eight to ten months inclusively. Since the system of selection for study by terminal registration digit precludes an age bias with respect to irradiation history, this age variation becomes a problem only insofar as it may enhance "within-cell" variation to the point of obscuring tests on interactions and main effects. We shall consider this possibility in detail in Section 12.4.3. Caution is also indicated in utilizing certain aspects of these data for standards of normality.

The measurements to be analyzed are the interdependent variables, weight, body length, head circumference, and chest circumference. In each city, all measurements were taken by one of two nurses, both of whom devoted much of their working time to this procedure. Originally the weight of the baby was obtained by weighing mother and (stripped) child together on a beam-type scale, then weighing the mother separately, and calculating the weight of the baby from the difference. During the latter part of 1952 the practice of weighing the baby directly in a basket on the same scale was instituted. Body length was measured by a stadiometer which was wider than the width of the baby. The readings were taken directly off a scale on the instrument. Head circumference was taken with a cloth tape at the maximum girth of the head in a plane passing through the nasion. Chest circumference was measured by means of the same tape, the measurement representing the chest girth in the plane of the nipples, taken midway between inspiration and expiration. This midpoint was estimated after a brief period of observation.

12.2 *The genetic argument for irradiation effects.* — The genetic argument for irradiation effects is essentially the same as the argument advanced for changes in the birthweights of infants born to exposed parents. The existence of irradiation-induced changes in body measurements is predicated on the assumptions that (a) the largest class of gene mutations are the recessive "detrimentals," and (b) the presence of "detrimentals" would be reflected in changes in body measurements consequent to abnormalities in growth pattern and rate in the first nine months of life.

The differences to be expected would be (a) changes in the multivariate means with changing parental exposure, (b) a change, quite possibly an increase, in the generalized variance with increasing parental exposure, or (c) both. The likelihood of the occurrence of these changes would obviously be a function not only of parental exposure but of the component of variation in measurements at nine months of age ascribable to genetic factors. Little information is available from which to estimate this component.

12.3 *Concomitant variables known to affect growth and development during the first year of life.* — The concomitant variables affecting growth and development during the first year of life are, in some respects, less well known than the variables which influence birthweight.

We do not know, for example, whether the maternal age and parity effects demonstrable at birth persist throughout the first year of life or not. That the effects of these two variables on measurements at nine months would be negligible would seem to follow from (a) the small amount of the total variation (approximately 3%) which they account for in birthweights, (see Table 10.24) and (b) the ever decreasing correlation between weight at birth and at age x (by nine months of age this correlation is 0.319 as estimated from these data). For the purposes of this analysis we shall assume that the various exposure sub-populations do not differ significantly as regards any factor influencing growth and development.

12.4 *The data.* — Presented in Tables 12.1 to 12.4 are the means and the number of observations on which they are based for the 64 sex-city-mother-father cells for the measurements of weight, height, head girth, and chest girth. Inspection of these data reveals no striking trends associated with parental exposure. It should be noted, however, that, save for uncommon exceptions, infants born to parents one or both of whom are in exposure category 1 are larger in all measurements than infants born to parents both of whom were exposed.

Let us turn now to the questions to be asked of these data, namely:

1. Are there significant differences between the multivariate means associated with parental exposure?

2. Are there significant differences between ex-

TABLE 12.1 DISTRIBUTION OF MEAN WEIGHT IN DECAGRAMS AT 9 MONTHS OF AGE BY CITY, SEX, AND PARENTAL EXPOSURE
(Unrelated parents. The numbers of observations on which the means are based are given in parentheses.)

Hiroshima

		Males Mothers						Females Mothers				
Fathers		1	2	3	4–5	Total		1	2	3	4–5	Total
	1	814.35 (2,765)	819.51 (871)	812.03 (350)	813.51 (169)	815.20 (4,155)	1	767.39 (2,543)	766.33 (822)	766.65 (331)	762.46 (153)	766.90 (3,849)
	2	808.64 (230)	789.89 (236)	804.36 (69)	802.34 (29)	799.95 (564)	2	769.30 (223)	756.16 (253)	758.78 (54)	751.00 (26)	761.44 (556)
	3	820.98 (91)	796.66 (58)	790.38 (73)	776.00 (18)	802.42 (240)	3	771.45 (100)	757.48 (52)	755.07 (67)	801.71 (17)	765.90 (236)
	4–5	797.27 (59)	797.70 (33)	782.95 (22)	811.42 (12)	796.23 (126)	4–5	766.85 (48)	757.27 (33)	796.55 (22)	734.88 (16)	765.39 (119)
	Total	813.80 (3,145)	811.97 (1,198)	806.68 (514)	809.02 (228)	812.43 (5,085)	Total	767.66 (2,914)	763.46 (1,160)	765.51 (474)	762.12 (212)	766.18 (4,760)

Nagasaki

		Males Mothers						Females Mothers				
Fathers		1	2	3	4–5	Total		1	2	3	4–5	Total
	1	797.52 (2,074)	794.62 (1,261)	774.13 (99)	789.22 (81)	795.63 (3,515)	1	753.03 (1,845)	749.06 (1,168)	763.16 (106)	744.29 (80)	751.70 (3,199)
	2	788.81 (283)	788.79 (486)	794.94 (35)	762.23 (13)	788.64 (817)	2	750.17 (264)	738.09 (482)	736.21 (47)	708.13 (16)	741.33 (809)
	3	782.78 (36)	795.63 (48)	791.21 (14)	774.67 (6)	789.38 (104)	3	749.14 (29)	720.85 (40)	734.35 (17)	694.50 (4)	731.34 (90)
	4–5	785.22 (23)	771.64 (36)	768.25 (4)	784.60 (5)	776.99 (68)	4–5	735.89 (27)	708.69 (16)	791.00 (4)	774.00 (4)	734.67 (51)
	Total	796.15 (2,416)	792.65 (1,831)	780.34 (152)	784.83 (105)	793.94 (4,504)	Total	752.41 (2,165)	744.92 (1,706)	753.71 (174)	737.95 (104)	749.03 (4,149)

TABLE 12.2 DISTRIBUTION OF MEAN HEIGHT IN MILLIMETERS AT 9 MONTHS OF AGE BY CITY, SEX, AND PARENTAL EXPOSURE
(Unrelated parents. The numbers of observations on which the means are based are given in parentheses.)

Hiroshima

		Males						Females				
		Mothers						Mothers				
Fathers		1	2	3	4-5	Total		1	2	3	4-5	Total
	1	698.20 (2,765)	698.85 (871)	696.14 (350)	699.16 (169)	698.20 (4,155)	1	683.83 (2,543)	682.84 (822)	681.92 (331)	684.02 (153)	683.46 (3,849)
	2	697.21 (230)	693.31 (236)	693.94 (69)	697.86 (29)	695.21 (564)	2	682.65 (223)	681.11 (253)	681.67 (54)	676.81 (26)	681.58 (556)
	3	699.07 (91)	691.59 (58)	691.93 (73)	691.00 (18)	694.49 (240)	3	683.88 (100)	681.31 (52)	684.42 (67)	685.35 (17)	683.57 (236)
	4-5	691.34 (59)	693.21 (33)	697.50 (22)	705.58 (12)	694.26 (126)	4-5	681.77 (48)	687.36 (33)	696.32 (22)	667.00 (16)	684.02 (119)
	Total	698.02 (3,145)	697.25 (1,198)	695.30 (514)	698.69 (228)	697.59 (5,085)	Total	683.71 (2,914)	682.52 (1,160)	682.77 (474)	681.96 (212)	683.25 (4,760)

Nagasaki

		Males						Females				
		Mothers						Mothers				
Fathers		1	2	3	4-5	Total		1	2	3	4-5	Total
	1	696.86 (2,074)	695.36 (1,261)	694.31 (99)	692.12 (81)	696.14 (3,515)	1	682.05 (1,845)	679.94 (1,168)	683.01 (106)	684.03 (80)	681.36 (3,199)
	2	695.92 (283)	691.85 (486)	695.54 (35)	694.92 (13)	693.47 (817)	2	680.56 (264)	677.76 (482)	677.28 (47)	676.06 (16)	678.61 (809)
	3	696.44 (36)	693.71 (48)	691.93 (14)	693.83 (6)	694.42 (104)	3	675.24 (29)	677.65 (40)	680.82 (17)	670.75 (4)	677.17 (90)
	4-5	696.13 (23)	694.25 (36)	690.00 (4)	686.00 (5)	694.03 (68)	4-5	678.41 (27)	676.81 (16)	697.75 (4)	673.75 (4)	679.06 (51)
	Total	696.74 (2,416)	694.36 (1,831)	694.26 (152)	692.27 (105)	695.58 (4,504)	Total	681.73 (2,165)	679.24 (1,706)	681.59 (174)	681.90 (104)	680.70 (4,149)

TABLE 12.3 DISTRIBUTION OF MEAN HEAD GIRTH IN MILLIMETERS AT 9 MONTHS OF AGE BY CITY, SEX, AND PARENTAL EXPOSURE
(Unrelated parents. The numbers of observations on which the means are based are given in parentheses.)

Hiroshima

		Males						Females				
		Mothers						Mothers				
		1	2	3	4–5	Total		1	2	3	4–5	Total
Fathers	1	444.63 (2,765)	444.28 (871)	443.88 (350)	443.79 (169)	444.46 (4,155)	Fathers 1	433.82 (2,543)	433.38 (822)	434.11 (331)	433.10 (153)	433.72 (3,849)
	2	444.05 (230)	442.04 (236)	444.22 (69)	442.38 (29)	443.14 (564)	2	434.65 (223)	433.38 (253)	435.19 (54)	431.27 (26)	433.97 (556)
	3	447.10 (91)	443.69 (58)	441.67 (73)	445.72 (18)	444.52 (240)	3	435.53 (100)	431.63 (52)	430.85 (67)	438.94 (17)	433.59 (236)
	4–5	441.47 (59)	442.18 (33)	444.91 (22)	448.33 (12)	442.91 (126)	4–5	433.04 (48)	433.58 (33)	439.91 (22)	426.00 (16)	433.51 (119)
	Total	444.60 (3,145)	443.75 (1,198)	443.66 (514)	444.00 (228)	444.28 (5,085)	Total	433.93 (2,914)	433.31 (1,160)	434.04 (474)	432.81 (212)	433.74 (4,760)

Nagasaki

		Males						Females				
		Mothers						Mothers				
		1	2	3	4–5	Total		1	2	3	4–5	Total
Fathers	1	452.48 (2,074)	452.50 (1,261)	450.23 (99)	450.69 (81)	452.38 (3,515)	Fathers 1	441.22 (1,845)	441.27 (1,168)	441.97 (106)	438.03 (80)	441.18 (3,199)
	2	452.40 (283)	452.82 (486)	455.46 (35)	454.23 (13)	452.81 (817)	2	440.17 (264)	439.82 (482)	441.74 (47)	438.94 (16)	440.03 (809)
	3	451.42 (36)	453.73 (48)	454.86 (14)	460.67 (6)	453.48 (104)	3	440.28 (29)	439.73 (40)	443.47 (17)	439.75 (4)	440.61 (90)
	4–5	451.91 (23)	449.44 (36)	449.25 (4)	447.20 (5)	450.10 (68)	4–5	438.48 (27)	435.50 (16)	449.75 (4)	440.50 (4)	438.59 (51)
	Total	452.45 (2,416)	452.56 (1,831)	451.83 (152)	451.53 (105)	452.45 (4,504)	Total	441.05 (2,165)	440.77 (1,706)	442.23 (174)	438.33 (104)	440.92 (4,149)

posure cells in the generalized variances, that is, the determinants of the variance-covariance matrices?

3. Does there exist demonstrable within-cell (exposure cell, that is) heterogeneity which would limit or augment the conclusions which can be drawn from these data?

the cell numbers are unequal and disproportionate. The analysis of dispersion is, of course, merely the multivariate equivalent of the analysis of variance.

As in the case of birthweights, we shall discuss in detail only one of these analyses and then summarize the information from the two

TABLE 12.4 DISTRIBUTION OF MEAN CHEST GIRTH IN MILLIMETERS AT 9 MONTHS OF AGE BY CITY, SEX, AND PARENTAL EXPOSURE
(Unrelated parents. The numbers of observations on which the means are based are given in parentheses.)

Hiroshima

Males

Fathers	Mothers 1	2	3	4–5	Total
1	427.32 (2,765)	428.92 (871)	427.51 (350)	427.18 (169)	427.67 (4,155)
2	425.41 (230)	423.38 (236)	427.12 (69)	424.34 (29)	424.71 (564)
3	430.70 (91)	426.09 (58)	423.29 (73)	420.83 (18)	426.59 (240)
4–5	422.59 (59)	429.85 (33)	421.59 (22)	432.17 (12)	425.23 (126)
Total	427.19 (3,145)	427.72 (1,198)	426.60 (514)	426.58 (228)	427.23 (5,085)

Females

Fathers	Mothers 1	2	3	4–5	Total
1	416.96 (2,543)	417.39 (822)	416.38 (331)	415.96 (153)	416.96 (3,849)
2	417.74 (223)	415.34 (253)	418.56 (54)	415.46 (26)	416.62 (556)
3	418.82 (100)	419.21 (52)	413.64 (67)	422.94 (17)	417.73 (236)
4–5	415.42 (48)	415.45 (33)	426.77 (22)	415.13 (16)	417.40 (119)
Total	417.06 (2,914)	416.96 (1,160)	416.72 (474)	416.40 (212)	416.97 (4,760)

Nagasaki

Males

Fathers	Mothers 1	2	3	4–5	Total
1	439.23 (2,074)	438.44 (1,261)	434.36 (99)	437.21 (81)	438.76 (3,515)
2	438.14 (283)	438.75 (486)	441.97 (35)	432.38 (13)	438.58 (817)
3	433.19 (36)	438.31 (48)	442.21 (14)	439.67 (6)	437.14 (104)
4–5	436.00 (23)	431.75 (36)	446.00 (4)	432.20 (5)	434.06 (68)
Total	438.98 (2,416)	438.39 (1,831)	437.14 (152)	436.51 (105)	438.62 (4,504)

Females

Fathers	Mothers 1	2	3	4–5	Total
1	429.07 (1,845)	428.75 (1,168)	430.50 (106)	424.20 (80)	428.88 (3,199)
2	428.91 (264)	427.69 (482)	430.13 (47)	417.38 (16)	428.03 (809)
3	428.69 (29)	421.95 (40)	428.76 (17)	425.75 (4)	425.58 (90)
4–5	424.07 (27)	421.44 (16)	431.50 (4)	431.50 (4)	424.41 (51)
Total	428.98 (2,165)	428.22 (1,706)	430.25 (174)	423.49 (104)	428.58 (4,149)

12.4.1 The multivariate means. — The analyses of the multivariate means are given in Tables 12.5 to 12.11. Tables 12.5 to 12.8 cover the analysis when all categories of parental exposure are considered but variation in age at examination is ignored. Tables 12.9 to 12.11 present a comparable analysis save for the exclusion of all parents of exposure category 1. In both instances, analysis has been by Rao's (1955) analysis of dispersion for the case when (i.e., with and without category 1 parents) sources regarding differences in the multivariate means. In Table 12.5 are given the data appropriate to the analysis of the first order (two-factor) interactions. Inspection of the mean squares for the individual analyses of variance reveals that in most instances the interaction mean squares are smaller than the within-cell mean squares. Because of these low values Wilks' test (cf. Rao, 1952), which tests the

The Analysis of the Anthropometric Data

TABLE 12.5 ANALYSIS OF DISPERSION
(All exposure cells)

(a) Sums of squares and cross products of deviations for the two-factor interactions

Source (DF)	Total (18,497)	Within (18,434)	C×S (1)	F×S (3)	M×S (3)	F×C (3)	M×C (3)	F(M₁−M₂) (3)	F×Mᵃ F(M₃−M₄,₅) (3)	F(M₁−M₄,₅) (3)
y^2	3,804,932	2,955,778	960	235	663	121	538	403	1,887	369
x^2	11,272,708	10,208,953	108	1,046	1,171	122	2,375	3,069	4,135	929
z^2	8,212,037	7,088,914	83	596	686	2,760	2,051	980	1,551	1,180
w^2	151,099,462	139,259,440	5,915	15,418	33,424	10,786	12,298	58,323	13,202	17,085
yx	2,808,860	2,131,955	322	−180	788	−31	−419	856	1,185	−332
yz	2,920,623	1,979,555	−283	−103	560	316	−141	552	1,174	17
yw	11,774,995	10,021,620	−2,383	−200	3,584	550	−534	4,403	4,141	−574
xz	4,416,701	3,829,089	−95	637	753	384	1,984	1,693	711	108
xw	28,370,723	24,957,896	−800	3,610	5,851	825	4,389	10,345	4,465	3,052
zw	22,649,636	21,316,400	703	2,872	4,390	5,439	4,876	6,714	4,154	3,102

(b) Mean squares for individual analyses of variance

Source (DF)	Total (18,497)	Within (18,434)	C×S (1)	F×S (3)	M×S (3)	F×C (3)	M×C (3)	F(M₁−M₂) (3)	F×Mᵃ F(M₃−M₄,₅) (3)	F(M₁−M₄,₅) (3)
y^2	—	160.34	960*	78.33	221.00	40.33	179.33	134.33	629.00*	123.00
x^2	—	553.81	108	348.67	390.33	40.67	791.67	1,023.00	1,378.33	309.67
z^2	—	384.56	83	198.67	228.67	920.00	683.67	326.67	517.00	393.33
w^2	—	7,554.49	5,915	5,139.33	11,141.33	3,595.33	4,099.33	19,441.00	4,400.67	5,695.00

y = head circumference
w = weight
x = height
M = mother
C = city
F = father
S = sex
z = chest girth

ᵃ The father-mother interaction has been partitioned into three orthogonal components each of which compares the fathers' exposure as a whole with the two designated mothers' groups.

significance of the ratio of two determinants, say, $|W|$ and $|W+Q|$ where $|W|$ is the error matrix (within cells) and $|Q|$ is the matrix due to any other category, was not performed (this ratio is, of course, the multivariate analogue of the L statistic used in Chapter X). It may be that the variation in age at examination has inflated the within-cell mean squares to the point where small interactions would be obscured. For the moment, suffice it to say that at this stage there is no evidence for the non-additivity of the main effects. The significance of this evidence is, of course, limited if there exists within-cell heterogeneity.

Turning now to Table 12.6 we note evidence from 12.6c that the multivariate means differ significantly between (1) sex-city-mother-father cells, (2) categories of paternal exposure, (3) categories of maternal exposure, (4) cities, and (5) sexes, but the pooled interactions ("Residual" in Table 12.6) are not significant. The precise nature of these differences are more clearly given in Tables 12.7 and 12.8, to which we now turn.

For each class of father's exposure, mother's exposure, sex, and city, there can be assigned a constant estimated in a fashion such that a comparison of two constants from differing classes of a single classification reflects differences between these classes in any one of the four measurements. These estimates are derived from a linear model assuming additivity of the main effects, and specifying, without loss of generality, that the constant associated with the highest class within a given classification is zero; for example, the constant for father's exposure category 4–5 ($f_{4, 5}$) is assigned the value zero.[1] Comparison of the difference between two con-

TABLE 12.6 ANALYSIS OF DISPERSION
(All exposure cells)

(a) Sums of squares and cross products of deviations for main effects and additivity

Source (DF)	Between cells (63)	Fathers (3)	Mothers (3)	Sex (1)	City (1)	Residual [a] (55)
y^2	849,154	854	949	558,451	261,037	12,392
x^2	1,063,755	12,890	8,069	978,425	18,163	31,591
z^2	1,123,123	3,496	1,577	475,161	577,066	28,252
w^2	11,840,022	227,141	55,354	9,567,378	1,327,332	451,750
yx	676,905	2,119	1,850	739,190	−68,857	9,435
yz	941,068	1,518	1,045	515,125	388,118	11,066
yw	1,753,375	10,554	6,643	2,311,473	−588,628	43,488
xz	587,612	6,064	984	681,843	−102,379	124,722
xw	3,412,827	52,788	18,096	3,059,570	155,270	69,818
zw	1,333,236	27,420	6,821	2,132,146	−875,191	92,923

(b) Mean squares for individual analyses of variance (on main effects and additivity)

Source (DF)	Between cells (63)	Fathers (3)	Mothers (3)	Sex (1)	City (1)	Residual [a] (55)
y^2	13,478.63**	284.67	316.33	558,451**	261,037**	225.31
x^2	16,885.00**	4,296.67**	2,689.67**	978,425**	18,163**	574.38
z^2	17,827.35**	1,165.33**	525.67	475,161**	577,066**	513.67
w^2	187,936.86**	75,713.67**	18,451.33	9,567,378**	1,327,332**	8,213.64

(c) Analysis of dispersion, Wilks' test (using Bartlett's approximation)

	Significant (by inspection)	40.5**	24.0*	Significant for sex and city (by inspection)	Not significant (by inspection)
χ^2					
(DF)	—	12	12	4	—

[a] Sum of interaction terms.

[1] It is convenient when only two classes exist for a classification to indicate their difference rather than the constants associated with each of the two classes. For example, in Table 12.7, the constants designated (N − H) and (♂ − ♀) are in fact the difference between N and H, and males and females.

The Analysis of the Anthropometric Data

stants from a given classification with the variance of the difference affords a test of the significance of the difference. For example, if we were interested in testing whether there existed evidence for a significant difference in the mean of the w measurement (weight) between father's 2 and father's 3, the test would be

$$t = \frac{w_{f_2} - w_{f_3}}{\sqrt{\sigma_w^2 (\sigma_{f_2}^2 + \sigma_{f_3}^2 - 2\sigma_{f_2 f_3})}}$$

where w_{f_2} and w_{f_3} are the w constants for father's exposure categories 2 and 3, σ_w^2 is the variance of w (obtained from "within-cell"), and $\sigma_{f_2}^2$, $\sigma_{f_3}^2$, and $\sigma_{f_2 f_3}$ are respectively the variances of the f_2 and f_3 constants and their covariance. The results of all possible contrasts for the data here presented are given in Table 12.8. It should be noted that in all instances where a significant difference exists it consists of a contrast involving the comparison of infants from category 1 parents with some other class of parental exposure. We shall return to the implication of this finding, but first let us consider some explanations for the significant findings in Table 12.6. The findings requiring some statement are the differences between the cities and the parental exposure classes.

We note first that the cities differ with respect to all four measurements, with Nagasaki infants being shorter and lighter but greater in circumferential measurements than their Hiroshima counterparts. This difference while not precisely unexpected was not predicted. The findings are, however, compatible with Matsumura's and Hasebe's (see Hulse, 1943) classification of the Japanese. These authors recognize four physical types of Japanese, namely, the Ishikawa, Chikuzen, Okayama, and Satsuma. Only the Okayama and Satsuma types, whose centers of distribution are respectively the shores of the Inland Sea (and central Japan), and southern Kyushu (presumably including Nagasaki prefecture), concern us. The Okayama type is described as an individual taller than the average Japanese, and possessing a head shorter than usual but of average breadth. The Satsuma type is described as a short individual with a broad head of average length. Granted the reality of this distinction and the distribution ascribed to these types, the city differences are readily explicable. As an historical aside, it might be pointed out that Kaempfer (1728; 1906 edi-

TABLE 12.7 ESTIMATES OF CONSTANTS AND THEIR VARIANCES FOR TESTS OF EQUALITY
(Linear model assuming only main effects)

Internal variances [a]	y 160.34	x 553.84	z 384.58	w 7,554.90	Variance—covariance matrix of estimates for any character without the internal variance. Variances are italicized.
	Constants				
N — H	7.729818**	−2.038997**	11.492928**	−17.430408**	*0.0002288947* — — —
♂ — ♀	10.998079**	14.557551**	10.144835**	45.521958**	*0.0002165952* — — —
					0 0 0 0
M_1	1.078390	0.852651	1.594916	6.566452	0.0016468914 0.0015516228 0.0015516228 0.0015359959
M_2	0.757550	−0.557520	1.504906	3.730330	0.0015516228 0.0017367076 0.0015516228 0.0015365419
M_3	0.910047	−0.516337	1.323011	2.850223	0.0015359959 0.0015365419 0.0015359959 0.0023100881
$M_{4,5}$	0	0	0	0	0 0 0 0
F_1	1.284078	1.654174	2.073144	12.074681	0.0028351553 0.0027620818 0.0027620818 0.0027425144
F_2	0.925659	−0.584108	1.112317	3.448576	0.0027620818 0.0031414861 0.0027620818 0.0027436738
F_3	1.435112	−0.145078	1.367756	4.406972	0.0027425144 0.0027436738 0.0027436738 0.0042519096
$F_{4,5}$	0	0	0	0	0 0 0 0

[a] Mean square (within cells). N = Nagasaki H = Hiroshima

tion, Vol. II, p. 372) commented on the relatively small stature of the inhabitants of Hizen (an area which then included Nagasaki).

The findings with respect to parental exposure are less simply resolved. As has been repeatedly stated, to assert, without qualification, that a given change is the consequence of irradiation would entail at least the demonstration that (1) a difference exists between the pregnancies occurring to "1" and "non-1" parents, and (2) differences exist among "non-1" parents consistent with increasing exposure. Let us consider to what extent these requirements are satisfied within the Japanese anthropometric data. From Tables 12.7 and 12.8 we have evidence that "1" terminations differ from "non-1" terminations. However, from Table 12.10 we note no demonstrable differences between the "non-1" terminations. The absence of changes between exposure categories 2, 3, and 4–5 could be interpreted as evidence that the observed differences, when all categories of exposure are considered, are not due to irradiation. That this is the most likely interpretation stems from the following considerations:

1. In the main, exposure category 1 parents are repatriates from Korea, Manchuria, and elsewhere, and migrants from the rural areas surrounding Hiroshima and Nagasaki. Shapiro (1939) has shown that the Japanese who migrated to Hawaii were appreciably larger than their neighbors who remained in Japan. It is reasonable to assume that similar selection would have occurred among the migrants to Korea, Manchuria, and elsewhere in eastern and southeastern Asia in the years preceding World War II. The repatriation of the Japanese migrants to Korea, etc., following the war, was both forcible and almost exhaustive since the Japanese had become unwelcome throughout much of Asia. As a consequence of this repatriation, there may well have been settled in Hiroshima and Nagasaki in the years immediately following the bombings a group of non-exposed persons appreciably larger than the natives. Whenever possible, repatriates were settled in the areas from which they or their families originally stemmed.

2. Another factor of possible importance is nutrition. To what extent a differential in nutrition may have contributed to the observed differences is admittedly a matter of speculation. However, it seems likely that migrants from the rural areas adjacent to these cities may have had an opportunity, not available to other urban dwellers, to supplement their diet with foodstuffs from the family farm.

However, it may be argued that the requirement that a demonstrable difference obtain among the remaining exposure classes when the

TABLE 12.8 A Summary of the Significance of Tests Comparing All Possible Pairs of Exposure for Each Parent with Respect to the Variables w, x, y, and z [a]

		F₁	F₂	F₃	F₄,₅			F₁	F₂	F₃	F₄,₅
			Variable y						Variable z		
F₁	Variable x		n.s.	n.s.	n.s.	F₁	Variable w		*	n.s.	*
F₂		**		n.s.	n.s.	F₂		**		n.s.	*
F₃		**	n.s.		n.s.	F₃		*	n.s.		n.s.
F₄,₅		n.s.	n.s.	n.s.		F₄,₅		**	n.s.	n.s.	

		M₁	M₂	M₃	M₄,₅			M₁	M₂	M₃	M₄,₅
			Variable y						Variable z		
M₁	Variable x		n.s.	n.s.	*	M₁	Variable w		n.s.	n.s.	*
M₂		**		n.s.	n.s.	M₂		n.s.		n.s.	n.s.
M₃		n.s.	n.s.		n.s.	M₃		n.s.	n.s.		n.s.
M₄,₅		n.s.	n.s.	n.s.		M₄,₅		n.s.	n.s.	n.s.	

n.s. = not significant
* = significant at 5%
** = significant at 1%

[a] These tables are designed so that the entries *above* the diagonal refer to tests on the variable above the diagonal, whereas entries *below* the diagonal refer to the variable below the diagonal.

The Analysis of the Anthropometric Data

TABLE 12.9 ANALYSIS OF DISPERSION
(Only those cells where both parents were exposed)

(a) Sums of squares and cross products of deviations for the two-factor interactions

Source (DF)	Within (2,331)	F×S (2)	M×S (2)	M×C (2)	F×C (2)	S×C (1)	F×(M₂ −M₃) (2)	F×(M₂ −M₄,₅) (2)
w²	17,996,341	2,073	195	11,371	4,608	42,666	9,140	3,259
x²	937,693	612	2,915	62	197	1,006	971	229
y²	380,665	162	318	132	608	1,812	1,749	664
z²	835,406	88	39	1,187	2,154	1,648	7,213	328
wx	3,287,294	−71	62	736	302	6,550	2,111	−788
wy	1,283,995	413	76	−863	708	8,794	3,989	592
wz	2,767,807	216	73	−3,533	3,060	8,384	8,107	1,021
xy	282,414	−235	938	−25	344	1,350	981	−289
xz	501,296	−207	−164	−265	347	1,287	1,769	−229
yz	254,348	115	−31	191	717	1,728	3,526	118

(b) Mean squares for individual analyses of variance (on two-factor interactions)

Source (DF)	Within (2,331)	F×S (2)	M×S (2)	M×C (2)	F×C (2)	S×C (1)	F×(M₂ −M₃) (2)	F×(M₂ −M₄,₅) (2)
w²	7,720.44	1,037	98	5,686	2,304	42,666*	4,570	1,630
x²	402.27	306	1,458*	31	99	1,006	486	115
y²	163.31	81	159	66	304	1,812**	875*	332
z²	358.39	44	20	594	1,077*	1,648*	3,607**	164

TABLE 12.10 ANALYSIS OF DISPERSION
(Only those cells where both parents were exposed)

(a) Sums of squares and products of deviations for main effects and additivity

Source (DF)	Total (2,366)	Within (2,331)	Between (35)	Sex (1)	City (1)	Mother (2)	Father (2)	Residual[a] (29)
w^2	19,505,898	17,996,341	1,509,557	1,157,230	102,046	6,403	1,331	228,733
x^2	1,062,802	937,693	125,109	101,698	2,826	1,184	1,687	15,390
y^2	508,234	380,665	127,569	77,087	38,987	611	58	7,670
z^2	997,555	835,406	162,149	57,031	79,680	1,048	243	15,215
wx	3,692,611	3,287,294	405,317	343,057	16,983	2,753	−114	35,867
wy	1,542,974	1,283,995	258,979	298,675	−63,075	1,411	229	31,325
wz	2,973,035	2,767,807	205,228	256,900	−90,173	2,571	496	47,545
xy	365,478	282,414	83,064	88,541	−10,497	609	−196	7,470
xz	564,292	501,296	62,996	76,157	−15,007	1,106	272	5,154
yz	389,232	254,348	134,884	66,305	55,736	635	52	6,764

(b) Mean squares for individual analyses of variance for main effects and additivity

Source (DF)	Total (2,366)	Within (2,331)	Between (35)	Sex (1)	City (1)	Mother (2)	Father (2)	Residual[a] (29)
w^2	—	7,720.44	—	1,157,230**	102,046**	3,201.50	665.50	7,887.34
x^2	—	402.27	—	101,698**	2,826**	592.00	843.50	530.69
y^2	—	163.31	—	77,087**	38,987**	305.50	29.00	264.48*
z^2	—	358.39	—	57,031**	79,680**	524.00	121.50	524.66
Bartlett's test				significant by inspection		8.74	14.43	234.99**
(DF)						(8)	(8)	(116)

[a] Includes all interaction terms.

"1" terminations are removed becomes too restrictive in practice, and that because of the great preponderance of parents one or both of whom are "1," exclusion of the "1" parents may so reduce the sample size that we could not detect differences of the magnitude expected even if such differences did, in fact, exist. This consideration would favor drawing inferences regarding irradiation effects from the "1" vs. "non-1" contrast, however, only if there existed no *a priori* basis for viewing this contrast as biased. We have advanced arguments in the preceding pages which suggest that this contrast may well be biased. Accordingly, we are unwilling to draw inferences regarding the effects of irradiation in the absence of demonstrable differences among the "non-1" terminations. Before pursuing this matter further, let us consider the data with regard to the generalized variances, and the effects of within-cell heterogeneity among the observations.

12.4.2 *The equality of the generalized variances.* — The term "generalized variance" was coined by Wilks (1932) and is defined as the determinant of the variance-covariance matrix. The generalized variance plays the same role relative to the generalized mean (multivariate mean) as that played by the variance relative to the univariate mean. The heterogeneity of a series of generalized variances may be (1) demonstrated directly by the multivariate analogue of Bartlett's test, or (2) inferred from tests on elements of the variance-covariance matrix.

Let us consider first the latter of these two approaches. It may be argued that if randomly chosen, corresponding elements in the variance-covariance matrices associated with the exposure cells are significantly different, then the generalized variances are most probably significantly different (or heterogeneous). To infer significance, of course, assumes that the elements to be tested are chosen on an *a priori* basis. It may be pointed out also that if the elements chosen for testing are not significantly different it does not follow that the generalized variances are not. The elements of the exposure cell variance-covariance matrices with which we shall concern ourselves are the variances of weight (w), height (x), head girth (y), and chest girth (z). Moreover, in view of the problem of the comparability of "1" and "non-1" terminations, we shall routinely perform four tests, namely, a test on the homogeneity of the variances when (1) all exposure cells are considered, (2) only those exposure cells in which both parents were exposed are considered, (3) only those exposure cells in which one or both parents are in exposure category 1, and (4) only those exposure cells in which only one parent is in exposure category 1. In Tables 12.12 to 12.15 are given the mean squares and mean products of deviations (the elements of the mean product [MP] matrices) for each exposure cell for the four city-sex cells. Inspection of the first four columns of these tables (corresponding to the estimates of the variances of w, x, y, and z) reveals no striking evidence of heterogeneity. In Tables 12.16 and 12.17 are set out the results of testing these estimates for heterogeneity by Bartlett's method (cf. Rao, 1952). From the latter tables emerges little that can be construed as evidence for the heterogeneity of the generalized variances. A comment on the two "signifi-

TABLE 12.11 ESTIMATES OF CONSTANTS AND THEIR VARIANCES FOR TEST OF EQUALITY
(Linear model assuming only main effects)

Internal variances[a]	w 7,720.44	x 402.27	y 163.31	z 358.39	Variance — covariance matrix of estimates for any character without σ^2	
		Constants				
H — N	14.047595	2.337845	−8.682824	−12.413078	0.00193378	—
♂ — ♀	44.24880	13.117449	11.420425	9.823059	0.00169194	—
F_2	0.855447	−3.012068	0.456508	−0.231801	0.00569837	0.00481708
F_3	−1.261074	−3.104190	0.121017	−1.071249	0.00481708	0.00734644
F_{4-5}	0	0	0	0	0	0
M_2	2.488047	1.058013	−0.797087	0.691081	0.00698914	0.00604162
M_3	6.372319	2.733161	0.607450	2.376684	0.00604162	0.00845267
M_{4-5}	0	0	0	0	0	0

N = Nagasaki H = Hiroshima

[a] Mean square (within cells).

TABLE 12.12 MEAN PRODUCT MATRICES: MALES, HIROSHIMA

Exposure	DF	w²	x²	y²	z²	wx	wy	wz	xy	xz	yz	Δ
1, 1	2,765	7,620.455	552.993	148.211	325.089	1,386.441	571.191	1,154.828	120.031	213.599	89.573	35,887,067,819
1, 2	871	8,296.716	588.975	168.083	355.139	1,484.541	635.730	1,249.869	146.260	224.932	94.347	51,960,551,297
1, 3	350	8,160.352	631.971	133.905	371.235	1,519.269	555.215	1,306.991	150.221	240.722	80.029	43,067,905,321
1, 4–5	169	6,489.417	421.851	173.214	270.554	1,049.208	650.190	879.345	115.690	141.500	104.440	26,231,272,155
2, 1	230	7,948.563	587.651	135.751	327.362	1,377.362	501.424	1,257.044	105.825	214.424	78.087	36,812,134,059
2, 2	236	7,207.711	507.451	145.489	300.626	1,362.464	528.694	1,064.885	98.340	207.770	67.013	29,131,761,372
2, 3	69	7,441.118	464.765	167.059	367.191	1,285.588	471.000	1,268.971	123.544	256.956	82.279	33,958,664,592
2, 4–5	29	8,922.964	771.750	220.821	253.107	2,120.966	787.286	1,137.071	201.857	317.357	104.500	35,915,464,679
3, 1	91	8,095.289	588.422	171.844	341.078	1,370.644	685.644	1,203.922	159.056	195.156	113.344	50,247,117,865
3, 2	58	11,331.316	683.860	150.175	407.561	2,060.526	791.000	1,845.228	168.228	323.579	128.719	34,745,901,957
3, 3	73	7,069.958	520.764	109.361	348.125	1,007.111	241.764	1,121.681	92.389	132.361	42.917	41,152,637,949
3, 4–5	18	6,635.059	400.000	152.118	291.765	982.471	593.529	1,223.471	163.176	208.647	122.294	8,153,661,713
4–5, 1	59	8,089.379	704.466	162.121	288.483	1,626.310	749.276	1,164.103	134.448	249.448	103.569	33,764,536,704
4–5, 2	33	5,870.344	537.875	143.844	302.875	972.094	370.000	1,007.344	146.156	194.531	56.000	28,295,829,672
4–5, 3	22	7,648.714	624.143	211.333	346.714	1,740.048	128.762	1,319.952	39.524	346.810	45.048	39,133,623,191
4–5, 4–5	12	6,429.727	491.727	119.000	348.545	1,170.273	581.727	1,078.364	78.455	211.182	138.000	15,896,068,251

TABLE 12.13 MEAN PRODUCT MATRICES: FEMALES, HIROSHIMA

Exposure	DF	w²	x²	y²	z²	wx	wy	wz	xy	xz	yz	Δ
1, 1	2,542	6,909.437	501.892	131.432	311.724	1,217.874	455.588	1,050.528	97.090	195.205	74.754	30,171,743,998
1, 2	821	6,517.340	485.317	138.680	307.536	1,162.258	464.248	1,014.384	105.876	180.347	72.183	28,120,583,348
1, 3	330	6,401.730	471.964	124.897	288.242	1,092.021	413.473	977.348	89.321	169.815	71.742	24,518,913,254
1, 4–5	152	7,088.237	533.862	136.224	381.487	1,370.257	538.671	1,139.816	122.559	245.230	97.099	34,970,342,093
2, 1	222	6,655.982	475.248	145.266	285.707	1,205.333	489.919	971.752	97.050	191.856	84.428	26,354,505,356
2, 2	252	7,101.714	587.317	154.651	335.714	1,440.147	575.397	1,170.258	143.163	212.877	87.802	31,157,206,688
2, 3	53	6,857.340	511.094	107.774	355.189	1,286.585	558.660	1,163.415	104.434	221.962	100.491	18,104,432,779
2, 4–5	25	10,564.320	669.360	267.240	331.360	2,219.520	1,225.560	1,507.680	257.680	313.960	180.320	31,097,527,853
3, 1	99	7,919.091	534.859	173.283	339.707	1,536.727	616.747	1,246.495	130.980	277.303	97.475	32,297,311,094
3, 2	51	5,861.039	466.451	154.784	330.647	946.471	489.706	925.196	139.765	120.961	80.961	34,013,378,637
3, 3	66	9,430.985	603.091	158.955	324.742	1,763.212	890.136	1,409.879	200.818	277.667	135.424	43,090,442,696
3, 4–5	16	6,072.125	545.250	101.938	285.063	1,203.813	318.563	1,036.688	143.438	209.250	53.500	12,905,135,192
4–5, 1	47	7,292.979	538.298	201.702	264.511	1,364.617	698.043	1,122.468	178.957	224.426	125.830	39,618,486,089
4–5, 2	32	4,754.781	346.688	81.125	225.625	500.281	190.094	625.875	46.688	63.031	13.188	14,049,782,530
4–5, 3	21	6,661.762	470.714	191.333	317.429	1,105.810	551.524	963.857	65.048	186.381	32.619	44,268,653,408
4–5, 4–5	15	6,008.400	893.467	105.867	228.133	1,687.533	382.000	811.333	107.400	252.200	61.133	23,867,794,617

The Analysis of the Anthropometric Data 177

TABLE 12.14 MEAN PRODUCT MATRICES: MALES, NAGASAKI

Exposure	DF	w²	x²	y²	z²	wx	wy	wz	xy	xz	yz	Δ
1, 1	2,073	8,095.760	585.185	194.300	388.619	1,434.229	607.827	1,221.924	120.012	205.034	144.661	73,332,447,415
1, 2	1,260	8,067.952	609.648	182.477	389.864	1,454.123	576.646	1,217.760	124.183	191.383	132.031	73,951,720,354
1, 3	98	7,590.929	480.439	215.876	432.724	1,241.480	620.184	1,409.459	131.663	229.520	173.867	51,018,491,679
1, 4–5	80	8,494.100	539.562	224.188	482.962	1,403.075	797.838	1,473.438	168.562	205.788	185.300	77,464,577,095
2, 1	282	8,061.277	631.528	161.582	371.582	1,575.266	587.287	1,254.394	125.851	247.993	130.376	50,550,017,916
2, 2	485	9,020.023	627.437	189.726	442.814	1,718.153	656.796	1,394.984	130.307	264.800	155.511	79,302,548,120
2, 3	34	8,408.294	571.618	160.206	391.853	785.294	638.118	1,333.412	41.176	161.206	156.118	69,440,782,051
2, 4–5	12	8,596.167	538.417	94.000	387.917	1,548.500	126.750	1,000.833	−0.333	110.000	125.833	26,226,652,145
3, 1	35	11,121.943	929.057	147.857	463.314	2,490.829	564.543	1,848.971	161.800	346.914	141.971	56,486,730,809
3, 2	47	9,479.128	525.064	193.213	397.830	1,545.574	476.809	1,402.915	106.787	186.915	112.404	78,948,705,745
3, 3	13	8,980.615	816.231	223.385	260.769	2,059.923	372.615	1,281.692	74.462	314.538	38.615	48,605,779,892
3, 4–5	5	12,353.800	526.200	167.000	425.800	2,249.800	891.000	2,149.200	215.000	426.000	164.000	4,486,865,579
4–5, 1	22	8,942.000	394.773	161.818	580.364	1,426.727	487.455	1,877.091	64.227	263.500	79.682	34,825,009,727
4–5, 2	35	7,975.143	559.571	198.771	349.971	1,386.114	738.286	1,090.800	150.514	192.886	156.543	57,967,354,402
4–5, 3	3	551.667	66.667	32.333	19.333	−76.667	−117.000	−38.000	33.333	−10.000	8.667	40,779
4–5, 4–5	4	4,436.250	330.500	228.250	251.250	1,126.000	558.250	847.500	83.500	192.000	214.000	78,359,578

TABLE 12.15 MEAN PRODUCT MATRICES: FEMALES, NAGASAKI

Exposure	DF	w²	x²	y²	z²	wx	wy	wz	xy	xz	yz	Δ
1, 1	1,844	7,582.454	539.471	161.323	382.431	1,324.596	475.523	1,165.667	94.134	213.564	121.948	56,650,054,372
1, 2	1,167	6,969.566	573.116	177.429	376.459	1,239.410	539.459	1,089.111	112.345	202.589	135.873	61,898,108,594
1, 3	105	6,968.324	490.429	163.133	296.390	1,199.562	642.876	1,048.038	133.943	169.505	127.790	25,873,071,979
1, 4–5	79	7,496.405	714.684	162.684	424.063	1,471.063	579.658	1,227.443	107.608	243.241	137.025	78,227,211,737
2, 1	263	8,180.243	634.468	193.700	381.578	1,577.532	662.597	1,248.232	164.114	254.122	152.259	63,325,567,668
2, 2	481	6,867.426	490.775	163.541	333.008	1,159.262	462.407	1,029.865	103.181	169.119	108.083	44,013,362,272
2, 3	46	7,076.652	618.239	131.152	363.804	1,209.696	462.196	1,050.587	128.783	217.217	111.109	52,312,861,309
2, 4–5	15	9,057.733	521.667	93.000	329.867	1,655.400	147.200	622.000	64.400	201.067	44.067	41,442,800,443
3, 1	28	8,214.107	562.750	122.500	248.357	897.929	452.429	964.107	80.107	101.393	56.607	49,236,001,942
3, 2	39	7,524.282	569.615	103.385	377.333	1,453.359	414.359	1,091.564	91.872	178.385	135.897	24,556,887,976
3, 3	16	7,271.500	405.750	287.875	493.063	1,010.188	804.625	1,056.313	179.063	72.188	163.313	103,562,595,360
3, 4–5	3	6,013.667	177.667	70.333	315.000	889.667	509.333	−175.333	47.667	−116.000	2.333	132,897,583
4–5, 1	26	10,203.808	410.885	141.962	486.308	1,214.538	331.615	1,722.692	26.692	180.269	113.423	59,941,806,872
4–5, 2	15	7,157.400	914.667	220.400	435.333	2,002.733	787.533	1,394.333	221.533	337.533	189.067	49,612,844,266
4–5, 3	3	11,324.000	705.000	217.000	979.667	2,509.667	−273.667	3,282.000	−2.000	778.000	−15.667	326,002,045
4–5, 4–5	3	7,266.667	356.333	567.000	537.667	916.667	2,018.000	1,972.000	292.667	232.667	546.000	6,965,953

TABLE 12.16 TESTS OF THE HOMOGENEITY OF THE VARIANCES OF THE ANTHROPOMETRIC MEASUREMENTS OVER ALL EXPOSURE CELLS, AND SUBDIVISIONS THEREOF FOR SPECIFIED SEX AND CITY (HIROSHIMA)
(The tabular entries are chi-squares.)

(a) Males — Hiroshima

Variable	All exposure cells (DF = 15)	Cells where *neither* parent is exposure category 1 (DF = 8)	Cells where *one* or *both* parents are exposure category 1 (DF = 6)	Cells where *only one* parent is exposure category 1 (DF = 5)
w	15.201	7.390	7.712	4.096
x	18.675	6.073	11.627	10.223
y	23.041	7.593	14.965*	11.773*
z	13.472	3.971	9.272	7.411

(b) Females — Hiroshima

Variable	All exposure cells (DF = 15)	Cells where *neither* parent is exposure category 1 (DF = 8)	Cells where *one* or *both* parents are exposure category 1 (DF = 6)	Cells where *only one* parent is exposure category 1 (DF = 5)
w	12.716	8.751	3.182	2.529
x	12.581	7.305	1.395	1.621
y	23.908	15.472	4.678	1.462
z	9.580	3.284	5.875	5.951

TABLE 12.17 TESTS OF THE HOMOGENEITY OF THE VARIANCES OF THE ANTHROPOMETRIC MEASUREMENTS OVER ALL EXPOSURE CELLS, AND SUBDIVISIONS THEREOF FOR SPECIFIED SEX AND CITY (NAGASAKI)
(The tabular entries are chi-squares.)

(a) Males — Nagasaki

Variable	All exposure cells (DF = 15)	Cells where *neither* parent is exposure category 1 (DF = 8)	Cells where *one* or *both* parents are exposure category 1 (DF = 6)	Cells where *only one* parent is exposure category 1 (DF = 5)
w	11.978	7.012	2.516	2.330
x	16.186	6.279	9.328	8.907
y	14.373	5.947	8.086	5.525
z	15.940	9.476	4.917	4.854

(b) Females — Nagasaki

Variable	All exposure cells (DF = 15)	Cells where *neither* parent is exposure category 1 (DF = 8)	Cells where *one* or *both* parents are exposure category 1 (DF = 6)	Cells where *only one* parent is exposure category 1 (DF = 5)
w	15.743	1.193	13.736*	12.859*
x	14.577	6.345	8.413	5.910
y	23.280	15.003	7.664	4.124
z	13.595	4.821	7.102	6.050

cant" findings in Table 12.17b is necessary. Box (1949) has shown that Bartlett's χ^2 test overestimates significance, particularly when the degrees of freedom are large. An alternative F test which Box has shown to be relatively unbiased fails to confirm the significance of the two starred entries.

Let us, therefore, turn now to the other of the two approaches listed above, a direct test of the homogeneity of the generalized variances. The basic theory for the generalized test of homogeneity was advanced by Wilks (1932). The test employed on the data presented here proceeds from Wilks' original considerations of the problem. Briefly the test is as follows: The test criterion, M, is the generalized form of Bartlett's criterion for the univariate case, and

$$M = N \log_e |s_{ij}| - \sum_l (v_l \log_e |s_{ijl}|)$$

where s_{ijl} is the unbiased estimate of the variance or covariance, σ_{ijl}, between the i^{th} and j^{th} variable in the l^{th} sample based on v_l degrees of freedom, and suppose $l = 1, \ldots, k$, and s_{ij} is the average variance or covariance, that is,

$$s_{ij} = (\sum_l v_l s_{ijl})/N$$

and $\quad N = \sum_l v_l.$

Now Box (1949) has shown that M/b is distributed as F with n_1 and n_2 degrees of freedom where

$$b = \frac{n_1}{1 - A_1 - (n_1/n_2)}$$
$$n_1 = [(k-1)p(p+1)]/2$$
$$n_2 = \frac{n_1 + 2}{A_2 - A_1^2}$$
$$A_1 = \frac{2p^2 + 3p - 1}{6(k-1)(p+1)} \left[\sum_l \frac{1}{v_l} - \frac{1}{N} \right]$$
$$A_2 = \frac{(p-1)(p+2)}{6(k-1)} \left[\sum_l \frac{1}{v_l^2} - \frac{1}{N^2} \right]$$

and where p is the number of variates.

The results of testing the generalized variances in this fashion are presented in Table 12.18. In no one of the sex-city cells can the generalized variances be shown to be significantly different. There is then no evidence that parental irradiation has significantly altered the variances associated with these measures of physical vigor.

12.4.3 Within-cell heterogeneity. — We have previously indicated that there may exist extraneous sources of variation which could lead to the observations within an exposure cell having dissimilar expectations with respect to the multivariate mean and the generalized variance. In other words, the observations within an ex-

TABLE 12.18 TEST OF THE GENERALIZED VARIANCES OF THE ANTHROPOMETRIC MEASUREMENTS BY SPECIFIED SEX AND CITY

(The tabular entries are F-values. All tests are against the alternative that the variance increases with increasing parental exposure.)

City — sex	All exposure cells		Only those exposure cells where both parents were exposed	
Hiroshima — males	F = 1.128	DF = 150, ∞	F = 1.003	DF = 80, ∞
Hiroshima — females	F < 1.000	DF = 150, ∞	F < 1.000	DF = 80, ∞
Nagasaki — males	F = 1.109	DF = 150, ∞	F < 1.000	DF = 80, ∞
Nagasaki — females	F = 1.222*	DF = 150, ∞	F = 1.283	DF = 80, ∞

posure cell may be heterogeneous in the sense that they do not represent observations drawn from the same parent population. The most apparent possible extraneous source of variation which should be considered is difference in age at examination. To "control" the age variation we elected to recognize another way of classification, namely, age at examination. Three age groups were defined; these age groups were 7.5–8.5, 8.5–9.5, and 9.5–10.5 months at examination. The results of analyzing the data following this further partitioning are given in Tables 12.19 to 12.22. As might be surmised, the within-cell variation is reduced. However, aside from the demonstration that weight, height, head girth, and chest girth vary with the age of the infant, the results of the analysis are essentially unchanged from those results obtained when the age variation was ignored. That is to say, we still find significant differences between sexes and cities, no evidence for interaction, and that the exposure differences are explicable in terms of the differences between the "1" and "non-1" parental exposure categories.

12.5 Summary. — Analysis of the multivariate means and the generalized variances asso-

TABLE 12.19 ANALYSIS OF DISPERSION, WITH AGE (A) AT EXAMINATION INCLUDED AS A WAY OF CLASSIFICATION
(All exposure cells)

(a) Sums of squares and cross products of deviations for the two-factor interactions

Source (DF)	Total (18,501)	Within (18,323)	S×A (2)	A×M (6)	S×C (1)	A×F (6)	S×M (3)	S×F (3)	M×F (9)
w^2	151,043,104	137,204,748	1,090	4,782	7,107	11,359	60,454	13,321	120,207
x^2	11,256,624	9,814,982	715	4,226	535	19,349	1,373	965	5,240
y^2	3,805,126	2,881,395	233	66	643	855	794	332	3,418
z^2	7,597,178	6,405,833	61	5,264	14,707	665	965	956	8,590
wx	28,057,399	23,287,603	717	1,645	1,599	7,590	8,304	3,005	21,369
wy	14,587,130	12,525,804	−30	160	867	−1,540	5,883	−778	13,691
wz	22,520,300	20,913,544	4	−672	1,326	1,202	7,507	3,532	22,742
xy	2,806,433	1,983,136	218	224	502	−1,535	1,030	−461	1,577
xz	4,413,292	3,728,298	−120	−2,887	−1,294	−815	953	725	3,050
yz	2,918,752	1,933,991	120	110	−2,625	223	666	−137	5,080

(b) Mean squares for individual analyses of variance

Source (DF)	Total (18,501)	Within (18,323)	S×A (2)	A×M (6)	S×C (1)	A×F (6)	S×M (3)	S×F (3)	M×F (9)
w^2	—	7,488.12	545.0	797.0	7,107	1,893.1	20,151.3*	4,440.3	—
x^2	—	535.66	357.5	704.3	535	3,224.8**	457.6	321.7	—
y^2	—	157.26	116.5	11.0	643*	142.5	264.7	110.7	—
z^2	—	349.61	30.5	877.3*	14,707*	110.8	321.7	318.7	—

The Analysis of the Anthropometric Data

TABLE 12.20 ANALYSIS OF DISPERSION WITH AGE AT EXAMINATION INCLUDED AS A WAY OF CLASSIFICATION

(a) Sums of squares and products of deviations for main effects and additivity

Source (DF)	Total (18,501)	Within (18,323)	Between (179)	Sex (1)	City (1)	Mother (3)	Father (3)	Age (2)	Residual (169)
w^2	151,043,104	137,204,748	13,838,356	9,451,588	1,071,950	55,725	222,652	1,194,302	1,278,871
x^2	11,256,624	9,814,982	1,441,642	962,798	5,781	7,101	12,348	318,396	91,848
y^2	3,805,126	2,881,395	923,731	553,377	279,875	1,074	848	54,314	32,898
z^2	7,597,178	6,405,833	1,191,345	472,027	586,157	1,681	3,456	13,600	84,037
wx	28,057,399	23,287,603	4,769,796	3,016,615	78,721	17,186	50,678	615,837	815,882
wy	14,587,130	12,525,804	2,061,326	2,286,982	−547,733	6,995	10,025	254,511	99,758
wz	22,520,300	20,913,544	1,606,756	2,112,203	−792,673	7,518	26,825	126,822	243,073
xy	2,806,433	1,983,136	823,297	729,925	−40,224	1,809	1,977	130,984	25,220
xz	4,413,292	3,728,298	684,994	674,141	−58,212	1,240	5,724	65,730	44,883
yz	2,918,752	1,933,991	984,761	511,086	405,031	1,203	1,464	26,926	28,499

(b) Mean squares for individual analyses of variance for main effects and additivity

Source (DF)	Total (18,501)	Within (18,323)	Between (179)	Sex (1)	City (1)	Mother (3)	Father (3)	Age (2)	Residual (169)
w^2	—	7,488.12	—	9,451,588**	1,071,950**	18,575.00	74,217.33**	597,151**	7,567
x^2	—	535.66	—	962,798**	5,781**	2,367.00**	4,116.00**	159,198**	543
y^2	—	157.26	—	553,377**	279,875**	358.00	282.67	27,157**	195*
z^2	—	349.61	—	472,027**	586,157**	560.33	1,152.00*	6,800**	497**
Bartlett's (DF)	—	—	—	significant by inspection	significant by inspection	24.047* (12)	46.245** (12)	significant by inspection	175.68 [a] (676)

[a] Not significant on one-tailed test.

TABLE 12.21 ESTIMATES OF CONSTANTS AND THEIR VARIANCES FOR TESTS OF EQUALITY

(Linear model assuming only main effects)

Internal variances[a]	w 7,488.12	x 421.35	y 157.26	z 349.61	Variance-covariance matrix of estimates for any character without σ^2
	Constants				
N — H	−15.78924	−1.15952	8.06781	11.67564	0.0002325668 — — —
♂ — ♀	45.24432	14.44040	10.94768	10.11102	0.0002165825 — — —
F_1	12.22589	1.50499	1.22862	2.15255	0.0028202429 0.0027471830 0.0027275426
F_2	3.66296	−0.73132	0.87364	1.22502	0.0027471830 0.0031261248 0.0027286450
F_3	5.07801	−0.03436	1.49375	1.49768	0.0027275426 0.0027286450 0.0042372410
$F_{4,5}$	0	0	0	0	0 0 0
M_1	7.09807	1.09708	1.20416	1.65223	0.0016477453 0.0015524865 0.0015371326
M_2	4.37406	−0.22320	0.90545	1.55274	0.0015524865 0.0017375721 0.0015376733
M_3	3.67542	−0.11254	1.09558	1.39793	0.0015371326 0.0015376733 0.0023114987
$M_{4,5}$	0	0	0	0	0 0 0
A_1	−28.31300	−14.23984	−6.14185	−2.86362	0.0008248126 0.0001840021 0
A_2	−13.49855	−7.27893	−2.78079	−1.55995	0.0001840021 0.0002729890 0
A_3	0	0	0	0	0 0 0

[a] Mean square within cells.

TABLE 12.22 A SUMMARY OF THE SIGNIFICANCE TESTS COMPARING ALL POSSIBLE PAIRS OF EXPOSURE FOR EACH PARENT WITH RESPECT TO THE VARIABLES x, y, w, AND z AFTER ALLOWANCE IS MADE FOR AGE AT EXAMINATION.[a]

	F_1	F_2	F_3	$F_{4,5}$			F_1	F_2	F_3	$F_{4,5}$
	\multicolumn{4}{c}{Variable y}			\multicolumn{4}{c}{Variable z}						
F_1 (Variable x)		n.s.	n.s.	n.s.	F_1 (Variable w)		*	n.s.	*	
F_2	**		n.s.	n.s.	F_2	**		n.s.	n.s.	
F_3	n.s.	n.s.		n.s.	F_3	*	n.s.		n.s.	
$F_{4,5}$	n.s.	n.s.	n.s.		$F_{4,5}$	**	n.s.	n.s.		

	M_1	M_2	M_3	$M_{4,5}$			M_1	M_2	M_3	$M_{4,5}$
	\multicolumn{4}{c}{Variable y}			\multicolumn{4}{c}{Variable z}						
M_1 (Variable x)		n.s.	n.s.	*	M_1 (Variable w)		n.s.	n.s.	*	
M_2	**		n.s.	n.s.	M_2	n.s.		n.s.	*	
M_3	n.s.	n.s.		n.s.	M_3	n.s.	n.s.		n.s.	
$M_{4,5}$	n.s.	n.s.	n.s.		$M_{4,5}$	*	n.s.	n.s.		

n.s. = not significant
* = significant at 5%
** = significant at 1%

[a] These tables are designed so that the entries *above* the diagonal refer to tests on the variable above the diagonal whereas entries *below* the diagonal refer to the variable below the diagonal.

ciated with the anthropometric measurements fails to reveal differences between exposure cells which are unequivocally due to parental irradiation. The only significant differences to emerge involve those contrasts utilizing infants born to parents whose exposure category is 1. These differences are in the direction of genetic expectation. In view of the numerous differences known to exist between the "1" and "non-1" parents, these differences are not thought to constitute unbiased evaluations of parental exposure.

Chapter XIII

THE AUTOPSY FINDINGS

ONE facet of the attempt to characterize the kinds and frequencies of abnormal terminations occurring to exposed and non-exposed persons in Hiroshima and Nagasaki was as ambitious an infant necropsy program as space and personnel would permit. The performance of infant autopsies began in Hiroshima in 1948 although it was not until May, 1949 that a concentrated effort was made to obtain the bodies of as many stillborn registered infants or registered infants dying during the first six days of life as was possible. Autopsies were not performed in Nagasaki until somewhat later (October, 1949), due to lack of personnel. The program in Nagasaki never attained the standard of exhaustiveness which we felt necessary to permit valid inferences. The reasons for this are somewhat complex but revolve largely around the fact that both the ABCC and Dr. I. Hayashi of the Nagasaki University Medical School were attempting to conduct an infant autopsy program. The efforts of Dr. Hayashi were motivated both by the requirements of an active teaching program in anatomy and an interest in a possible relationship between irradiation and malformation. Under the circumstances obtaining in Nagasaki, we could not assure ourselves that the infants whom we obtained for necropsy represented a random sample. Accordingly, the data obtained in Nagasaki will be presented later only with a view towards supplementing the data of Dr. Hayashi (see Hayashi, 1955; Sevitt, 1955).

The genetic argument which would lead us to expect differences between exposure cells in the frequency with which major congenital anomalies would be detected at autopsy is the same, in all major particulars, as the argument advanced in Section 8.3.

13.1 *The randomness of the Hiroshima autopsies.* — In the interval 1948–1953 approximately 750 infant cadavers were collected for autopsy in Hiroshima. Of this number, 717 were actually autopsied and the remainder were rejected because of advanced autolysis. Among the 717 infants autopsied only 431 represented registered pregnancies. The 286 unregistered infants who were autopsied have been excluded from the data to be presented in this chapter because they are known to be a non-random sample of all unregistered pregnancies. During the interval, March, 1948–May, 1949, only a few autopsies were performed on registered infants, the majority of these involving "special interest" material, and certainly non-random. In May of 1949, the autopsy program was greatly expanded. We have felt it wise to exclude from consideration all autopsy results obtained prior to this expansion of the program, as well as autopsy material not meeting the restrictions placed on all data and described in Section 6.4. The actual analysis is thus confined to 406 autopsies conducted on registered infants between May, 1949, and December, 1953. Tables 13.1 through 13.4 present an analysis of the randomness of this autopsy sample.

From Table 13.1 we note that: (1) the infants coming to autopsy were distributed by exposure in the same fashion as those registered infants not coming to autopsy; (2) a significantly higher proportion of males than females was obtained; and (3) in all, some 62 per cent of the infants who were stillborn or who died in the first six days of life were examined at necropsy. In view of the absence of an interaction between sex and exposure and also the absence of a significant difference in the exposure distribution of autopsied and non-autopsied infants, it would appear that valid inferences could be drawn from these data. However, this body of data is no less affected by concomitant variation than those portions of the total data presented in preceding chapters. Accordingly, it is of some interest to determine

The Autopsy Findings

TABLE 13.1

A. The randomness of the distribution of autopsied infants by sex of infant and parental exposure

Sex		Neither parent exposed	Mother exposed; father not	Father exposed; mother not	Both parents exposed	Total
Male	n[a]	178	103	23	39	343
	x[b]	120	67	18	27	232
Female	n	156	98	24	26	304
	x	91	60	13	10	174
Total	n	334	201	47	65	647
	x	211	127	31	37	406

[a] Total number of infants stillborn or dying during the first 6 days of life.
[b] Number of infants stillborn or dying during the first 6 days of life who were autopsied.

B. Chi-square analysis of the frequency of autopsy by sex of infant and parental exposure

Source	χ^2	DF	P
Total	13.6767	7	.05–.10
Sex-exposure interaction	6.2765	3	.05–.10
Sex (unadjusted)	7.4599	1	.01–.001
Exposure (adjusted for sex)	6.1008	6	.30–.50

whether the autopsied infants differ from the non-autopsied infants in such sources of extraneous variation as the age of the mother at the birth of the infant, the economic status of the parents, and the presence of a positive serology on the part of the mother at the time of delivery.

Let us consider first the evidence with regard to maternal age. In Figure 13.1, the frequency of malformation among autopsied infants is distributed by maternal age. Mother's age is, we see, a rather important factor in the frequency of malformation among infants coming to autopsy. Krooth's (1955) index of absolute difference which may be used to evaluate the significance of these data is, in this instance, 10.61 per cent (variance 1.17%). Maternal age differences could, then, be an important source of bias if maternal ages are dissimilarly distributed among exposure cells. From Table 13.2, we note (a) no evidence of interaction between the sex, autopsy, or exposure categories, (b) no effect of sex of infant nor the occurrence of an autopsy on the age of the mother, but

FIGURE 13.1 — The distribution of the frequency of autopsied infants with major malformation, relative to all autopsied infants, by age of mother with parity ignored.

(c) a significant difference in mothers' ages between exposure cells. This latter effect is principally characterized by a pronounced increase in mean maternal age when both parents were exposed as opposed to those cases where only one or neither parent was exposed. The is not significant. The apparent trend is not unexpected in view of the fact that the ABCC assumed all expenses in the cremation, etc. of the infant cadaver. It seems quite probable that this would be a consideration of some moment to parents in the lower economic strata. The

TABLE 13.2 Mean Maternal Age of Infants Stillborn or Dying During the First 6 Days of Life by Sex of Infant, Parental Exposure, and the Occurrence of Autopsy

Class of infant	Neither parent exposed	Mother exposed; father not	Father exposed; mother not	Both parents exposed	Total
Autopsied					
Males	27.16 (120)	26.91 (67)	26.39 (18)	28.96 (27)	27.24 (232)
Females	26.70 (91)	26.90 (60)	27.54 (13)	31.70 (10)	27.12 (174)
Not autopsied					
Males	26.02 (58)	26.08 (36)	25.60 (5)	28.42 (12)	26.28 (111)
Females	27.26 (65)	27.84 (38)	28.18 (11)	28.44 (16)	27.65 (130)
Total					
Males	26.79 (178)	26.62 (103)	26.22 (23)	28.79 (39)	26.93 (343)
Females	26.94 (156)	27.27 (98)	27.83 (24)	29.69 (26)	27.35 (304)

Analysis of variance

Source	SS	DF	MS	F
Sex	41.872	1	41.872	1.68
Autopsy	11.715	1	11.715	2.12
Exposures	312.946	3	104.315	4.20**
Interactions	188.875	10	18.888	1.32
Between	537.415	15	35.828	1.44
Within	15,707.445	631	24.893	—
Total	16,244.860	646	—	—

TABLE 13.3 The Distribution of Economic Status among Autopsied and Non-Autopsied Infants

Infant	Very poor	Poor	Average	Well-to-do	Rich	Unclassifiable	Total
Autopsied	5	56	325	20	—	—	406
Not autopsied	2	18	210	11	—	—	241
Total	7	74	535	31	—	—	647

$\chi^2 = 6.474$ DF = 3 P = .05–.10

difference in maternal age distributions will lead chiefly to a bias which results in an increase in the frequency of congenital malformations when both parents are exposed relative to the case when neither or only one parent is exposed.

With respect to the distribution of economic status we note that the autopsied infants appear more frequently to represent the lower economic groups (see Table 13.3) although this trend rise in frequency among the well-to-do parents, if real, is quite possibly correlated with higher levels of education and an appreciation of the information to be gained from a postmortem examination. In view of the absence of a significant difference in economic classes it seems unlikely that the autopsy data with regard to malformation are grossly biased by parental economic status.

In Table 13.4 is given the distribution of positive serologies by parental exposure and the occurrence of an autopsy. It is apparent from the analysis of these data that the autopsied infants differ somewhat from those not autopsied in the frequency with which the mother of the infant has a positive postpartum serology. The reason for this is not apparent. However, there is no evidence of an exposure effect with respect to serology, nor is the autopsy-exposure interaction significant. Accordingly, we conclude that there is no evidence of significant exposure than in Nagasaki, and as a consequence, an irradiation effect should be more pronounced in Hiroshima than in Nagasaki (other factors being equal, of course).

2. The sampling scheme underlying those infants coming to autopsy in Hiroshima is reasonably well known.

3. The data from Hiroshima are confined to infants where the age range is known; the infants were all of no less than 21 weeks' gestation and were either stillborn or died in the first six days following birth.

TABLE 13.4 THE DISTRIBUTION OF POSITIVE SEROLOGIES BY PARENTAL EXPOSURE AMONG AUTOPSIED AND NON-AUTOPSIED INFANTS

(Serology is scored as the number of infants whose mother's serology was positive or negative.)

Infant	Neither parent exposed	Mother exposed; father not	Father exposed; mother not	Both parents exposed	Total
Autopsied					
Positive	11	6	0	2	19
Negative	200	121	31	35	387
Total	211	127	31	37	406
Not autopsied					
Positive	15	4	2	1	22
Negative	108	70	14	27	219
Total	123	74	16	28	241
Total					
Positive	26	10	2	3	41
Negative	308	191	45	62	606
Total	334	201	47	65	647

Chi-square analysis [a]

Source	χ^2	DF	P
Total	9.5635	5	.05–.10
Autopsy-exposure interaction	3.8209	2	.10–.20
Autopsy	5.0427	1	.02–.05
Exposure (unadjusted)	2.4682	2	.20–.30

[a] Father exposed and both parents exposed pooled.

traneous variation with respect to the frequency of possibly syphilitic mothers among the mothers of autopsied infants.

13.2 The data. — Two bodies of data exist pertinent to the occurrence of major malformations among infants born to exposed and nonexposed parents and coming to autopsy in Hiroshima and Nagasaki. The first body of data to be presented here is the data obtained in Hiroshima under the auspices of the Atomic Bomb Casualty Commission. We present these data first for the following reasons:

1. Proportionately more "exposed" individuals in Hiroshima received significant amounts of

In Table 13.5 are presented the data with regard to the frequency of congenital malformations among infants autopsied in Hiroshima, distributed by parental exposure without adjustment for differences in maternal age. Four comparisons have been made, namely, the equality of the frequency of congenital malformations (a) among all four exposure cells, (b) between infants born to non-exposed parents and those born to parents one or both of whom were exposed, (c) between infants born to non-exposed parents and to exposed parents one or both of whom were exposed and in exposure classes 4 or 5, and (d) between infants

born to exposed parents with exposures less than 4 with infants born to parents in exposure classes 4 or 5. In no one of these instances does there emerge a significant difference. In view of the effect of maternal age, adjustment for the age differences among exposure groups would lead to differences between exposure categories deviating in a greater fashion from what is to be expected under an hypothesis of radiation-induced genetic changes. Obviously the sensitivity, as it were, of the comparisons given in Table 13.5 is a function of the number of parents who received significant amounts of irradiation. The distribution of the exposed parents by exposure category is given in Table 13.6, from which we note that about 53.6 per cent of the parents received inappreciable amounts of irradiation (exposure class 2). It seems doubtful, therefore, that the autopsy data are sufficiently extensive to detect small irradiation effects such as *a priori* considerations would lead us to expect.

TABLE 13.5 THE DISTRIBUTION BY PARENTAL EXPOSURE OF INFANTS BORN IN HIROSHIMA AND FOUND TO BE GROSSLY ABNORMAL AT AUTOPSY

(a) Distribution of abnormal infants by parental exposure when exposure is defined as "present in city at the time of the bombing"

Class of infant	Neither parent exposed	Mother only exposed	Father only exposed	Both parents exposed	Total
Normal	174	107	24	28	333
Abnormal	37 (17.5%)	20 (15.8%)	7 (22.6%)	9 (24.3%)	73
Total	211	127	31	37	406

(b) Distribution of abnormal infants by parental exposure when exposure is defined in terms of the categories given in Chapter IV

Class of infant	One or both parents in exposure categories 4 or 5	Both parents exposed but neither in exposure categories 4 or 5
Normal	33	25
Abnormal	5 (13.2%)	8 (24.2%)
Total	38	33

(c) Analysis of data in Tables 13.5a and 13.5b

Contrast	χ^2	DF	P
All exposure cells in 13.5a	1.912	3	.50–.70
Neither parent vs. one or both parents exposed in 13.5a	0.059	1	.80–.90
Neither parent (13.5a) vs. one or both parents exposed as 4 or 5 (13.5b)	0.440	1	.50–.70
All exposure cells in 13.5b	1.451	1	.10–.20

TABLE 13.6 THE DISTRIBUTION BY EXPOSURE CLASS OF THE EXPOSED PARENTS GIVEN IN TABLE 13.5

Exposure class	Father not exposed; mother in class shown in first column	Mother not exposed; father in class shown in first column	Both parents exposed; more heavily exposed parent in class shown in first column	Total
2	72 (56.7%)[a]	14 (45.2%)	18 (48.6%)	104 (53.3%)
3	38	13	7	58
4	4	1	4	9
5	13	3	8	24
Total	127	31	37	195

[a] Percentage of total.

A second body of autopsy data bearing on the problem of radiation-induced genetic changes in the malformation rate is the data collected by Professor Ichiro Hayashi of the Nagasaki University Medical School. We feel compelled to present and discuss these data in some detail because of the recent grossly over-simplified, biased presentation of these data by Sevitt (1955). Professor Hayashi has kindly furnished us with a copy of a preliminary report on these

data given as an address at a meeting of the Japanese Atomic Bomb Investigation Group and members of the Japan Science Council, held at the Nagasaki Medical School on 3 October 1955. At the outset of a discussion of Hayashi's data, it should be pointed out that his data differ in two important respects from the data we have presented from Hiroshima, namely, (1) the fact that the age range among the infants coming to autopsy is not clearly defined (they are described as fetuses or newborn infants), and (2) the definition of exposure. Hayashi's exposed cases are defined as those infants who were born to parents one or both of whom *lived within* 10 kilometers of ground zero at the time of the bombing (August 9, 1945). It should be noted that the definition specifies *lived within*, which does not necessarily imply *present within*. Hayashi asserts that the location of these parents with regard to the hypocenter is still under investigation. If we assume, however, that all of the individuals who lived within were in fact present within 10 kilometers of the hypocenter at the time of the bombing, Hayashi's definition of exposure is still, in our estimation, a most unfortunate one for a variety of reasons. The most important of these reasons is that 10 km. extends some 7 km. beyond the area of significant irradiation (5r or higher) except possibly in the "fallout" sector (principally the Nishiyama reservoir area). A definition as all encompassing as this can only lead to a dilution of radiation effects, if such exist, because of the inclusion of large numbers of essentially unexposed persons. An additional disadvantage of Hayashi's definition which is worth mentioning is that a radius of 10 km. extends far beyond the limits of the city of Nagasaki (particularly to the north, east, and west) and hence creates difficulty in any attempt to determine the completeness and/or randomness of sampling. While we believe there is ample justification for a conservative definition of exposure, Hayashi's definition would appear overly conservative.

It is extremely difficult to appraise the real significance of Hayashi's data. Among the factors which make it difficult is, as Professor Hayashi has pointed out, the lack of information regarding (a) the randomness of the sample, (b) the degree of exposure of the parents, and (c) other etiological agents, such as maternal age, of importance in congenital malformations.

Some measure of the importance and effect of these factors can be gleaned from the corpus of data collected by the Commission. With regard to maternal age, we have commented repeatedly on both its importance and the fact that maternal ages are so distributed among the terminations occurring in Hiroshima and Nagasaki as to bias upwards an estimate of the effect of irradiation.

With respect to the degree of exposure most probably experienced by the parents in Hayashi's series, the following observations are pertinent:

1. All dosage estimates, both Japanese and American, of the bombs detonated over Hiro-

TABLE 13.7 HAYASHI'S DATA ON CONGENITAL ABNORMALITIES IN RELATION TO EXPOSURE OF PARENTS

(After Sevitt, 1955)

Category	Number of infants	Number malformed	% malformed
One or both parents exposed to A-bomb	497[a]	92	18.9
Both parents exposed	149	27	18.1
Father exposed	80	18	22.5
Mother exposed	259	47	18.1
Neither parent exposed	363	40	11.0
Both parents unknown exposure; information not available	27	10	—

$\chi^2 (2 \times 4) = 10.718$ DF = 3 $P < 0.02$
$\chi^2 (2 \times 2) = 9.062$ DF = 1 $P < 0.01$

[a] Includes 9 children "mother exposed, father unknown."

shima and Nagasaki indicate that the instantaneous irradiation at 3,000 meters was less than 5r and consisted of gamma irradiation alone.

2. The total irradiation received by an individual is the sum of the instantaneous irradiation and the residual irradiation to which he or she may have been exposed. The major component of residual irradiation arises from the so-called "fallout" following the bombing. The area of fallout in Nagasaki is reasonably well defined, and is, as has been indicated, east of ground zero in the region of the Nishiyama reservoir.[1] Save for a very small area immediately adjacent to the reservoir, the estimated residual irradiation from one hour to infinity

[1] The fallout area could be described as a band of about 3,000 meters' width (on its north-south axis) extending eastward from a point 1,500 meters from the hypocenter.

following the bombing is a dose probably less than 6r, and possibly less than 2r. The area encompassed by the fallout is in the main relatively sparsely populated, and consists principally of hills and rice paddies. It is a safe conclusion that relatively few persons residing outside 3 km. from ground zero received exposures of the order of magnitude of 5r from the fallout.

3. In the interval 1949–1953, some 363 registered infants came to autopsy in Nagasaki under the auspices of the ABCC. The distribution by exposure of these infants is compared with those autopsied by Hayashi in Table 13.8. The two distributions are not significantly different and by inspection appear to be very nearly identical one with the other. In Table 13.9, we have given the distribution of exposed parents

Hiroshima data, derived from a group whose average exposure was greater, is also 18.5 per cent (36 infants out of 195). Be that as it may, the more important consideration is that Hayashi's control group has a frequency of congenital malformation of only 11 per cent (40 out of 363), whereas the Hiroshima control group has a frequency of 17.5 per cent (37 out of 211). This is a significant difference, yet the evidence from physical examinations presented in Tables 8.6 and 8.7 suggests no difference in the malformation rates between these two cities. (It need hardly be pointed out that if Hayashi's exposed group is contrasted with the control group from Hiroshima no significant difference emerges.) Further evidence which suggests that Hayashi's control is low stems from the following considerations: (1) Among registered

TABLE 13.8 A COMPARISON OF THE EXPOSURE DISTRIBUTION OF HAYASHI'S DATA AND THAT OF THE ATOMIC BOMB CASUALTY COMMISSION COLLECTED IN NAGASAKI

Source	Neither parent exposed	Mother exposed; father not	Father exposed; mother not	Both parents exposed	Total [a]
Hayashi	363 (42.66%)	259 (30.43%)	80 (9.40%)	149 (17.51%)	851
ABCC	155 (42.70%)	115 (31.68%)	34 (9.37%)	59 (16.25%)	363
Total	518	374	114	208	1,214

$\chi^2 = 0.364 \quad P = .9-.95$

[a] See text for differences between Hayashi's and ABCC's exposure definitions.

in our series in terms of the categories of exposure defined in Chapter IV. We note that the large majority of exposed parents (87.5 per cent) are in exposure category 2 wherein the average exposure did not exceed 5r. Since Hayashi's definition of exposure must encompass proportionately more individuals outside the 3 km. zone than are presently residing in Nagasaki (and hence would enter our study) we can only conclude that an even greater number of Hayashi's exposed cases are infants born to parents whose exposure was insignificant. In summary, then, it would appear that the great bulk of Hayashi's exposed parents must have received exposures of less than 5r.

The representativeness of Hayashi's series is, as has been indicated, a matter of conjecture; however, certain observations which we shall now discuss suggest that the series may be biased. We note that Hayashi's data (see Table 13.7) reveal a frequency of congenital malformation of 18.5 per cent among the exposed group (92 out of 497 infants) whereas the

TABLE 13.9 DISTRIBUTION BY EXPOSURE CLASS OF EXPOSED PARENTS WHOSE INFANTS CAME TO AUTOPSY AT ABCC IN NAGASAKI

Exposure class	Father not exposed; mother in class shown in first column	Mother not exposed; father in class shown in first column	Both parents exposed; more heavily exposed parent in class shown in first column
2	101 (87.8%) [a]	33 (97.1%)	48 (81.4%)
3	12	1	9
4	—	—	—
5	2	—	2
Total	115	34	59

[a] Percentage of total.

infants who were stillborn or who died in the first six days of life and were born to non-exposed parents in Nagasaki (both parents exposure category 1), no less than 9.4 per cent (42 out of 444) had clearly visible major malformations. (2) Experience in Hiroshima suggests that for every eight infants coming to

autopsy on whom a prior clinical diagnosis of major malformation has been made (verifiable at autopsy, of course), six cases will come to autopsy and a major malformation will be found on which no prior clinical diagnosis exists. In short, among infants coming to autopsy, clinical examination will reveal only 58 per cent of the infants with major malformations. These two observations taken conjointly suggest that a more reasonable figure for Hayashi's control would be no less than 16 per cent.

From the preceding considerations we are led to the conclusion that the more obvious explanation for Hayashi's findings lies either in extraneous variation or in non-representativeness of the control, rather than in irradiation (for example, a maternal age effect similar to the one noted in Figure 13.1 could easily produce a spurious radiation effect in a larger series of cases).

13.3 *Summary.* — The data with regard to the frequency of congenital malformations among the pregnancy terminations to exposed and non-exposed persons in Hiroshima afford no evidence of a significant irradiation effect, although the small differences observed are in the direction of genetic expectation. Reasons are advanced for believing that the "irradiation effect" reported by Sevitt (1955) from Hayashi's data in Nagasaki is more likely due to extraneous variation (or sampling biases) between the exposure cells than to irradiation.

Chapter XIV

RECAPITULATION

The present chapter will be devoted to a brief recapitulation of the findings of this study, and a comparison of these results with those of other studies utilizing mammalian material.

14.1 *The findings.* — Information has been collected concerning the relationship between irradiation and the outcome of 76,626 pregnancies terminating in Hiroshima and Nagasaki between 1948 and 1953. The following aspects of pregnancy outcome were considered: sex of infant, occurrence of major malformation, viability at birth, birthweight, and occurrence of death during the first six postpartum days. In addition, on a random sample of 19,818 pregnancy terminations followed up at age nine months, data were obtained on infant survival up to nine months postpartum, malformations which had become apparent since the previous examination, and growth of the child as reflected in certain anthropometric measurements. The variation in these potential indicators of radiation-induced genetic change has been analyzed in relation to the experience of the parents at the time of the atomic bombings. For analytical purposes, five categories of radiation, defined in Table 4.7, have been recognized. Each infant studied may thus fall into any one of 25 categories when classified according to the radiation history of both father and mother.

In evaluating the significance of the findings, it is important to bear in mind that there are, in effect, two sets of control parents, namely, those parents (group 1) who were not in either city at the time of the bombings, and those parents (group 2) who, although in one or the other of the two cities, were at distances of 3,000 or more meters from the hypocenter or, if nearer, were significantly shielded, and presumably received negligible amounts of irradiation.

In Chapters III and V an effort has been made to explore dissimilarities between Hiroshima and Nagasaki parents which might be responsible for differences in indicator findings in the two cities. It is noteworthy that despite the several demonstrated or possible dissimilarities, the two cities did not differ significantly with respect to any of the indicators save birthweight and the anthropometric measurements at age nine months. This may be taken as some measure of "relative stability," within this range of environmental variation, on the part of the indicators of potential genetic damage utilized in this study. This has some bearing on one's attitude towards the appropriateness of radiation category 1 parents as the source of control material. As mentioned in Section 5.10, there is a relatively high proportion of repatriates or persons of rural background among category 1 parents. It seems unlikely, however, that category 1 parents, as a group, differ biologically from the "exposed" group any more than the inhabitants of Hiroshima differ from those of Nagasaki. If this is correct, then theoretically the differences, if any, between the offspring of category 1 parents and of categories 2, 3, 4, and 5 parents may be used as a measure of combined radiation and "disaster" effect, with the radiation effect then best measured by progressive differences between the offspring of parents falling into categories 2, 3, 4, and 5.

There are significant sex differences as regards the indicators birthweight, neonatal death rate, and anthropometric measurements at age nine months, but not with respect to major malformations or stillbirths. Such sex differences are well recognized in the literature. It is perhaps worth pointing out that the level of the significance with which these previously recognized differences emerge in this analysis increases one's confidence that valid differences of appreciable magnitude between the various analytical subclasses would be detected.

The reader's attention is directed to Figures 14.1 and 14.2 which present graphically the

Recapitulation 193

FIGURE 14.1 — A graphical representation of the effect of parental exposure on: (1) the sex ratio; (2) the frequencies of malformed infants, stillborn infants, and infants dying in the neonatal period; (3) birthweight means; and (4) birthweight variances.

distributions of the indicators by parental exposure, and to Table 14.1 for a summarization of the tests of significance with reference to exposure. With regard to the effect of parental exposure on the indicators, we find the following:

(1) *Sex ratio.* — The one indicator wherein differences of opinion with regard to the effect of parental exposure may arise is the sex ratio.

terminations where the father was in class 1, provided the mother was not also in this class, there emerges a significant effect, on a one-tailed test, of mother's exposure, but not of father's exposure. Two questions immediately arise, namely, (a) is this a legitimate comparison, and (b) if so, does there exist ancillary evidence which supports the reality of this difference?

FIGURE 14.2 — A graphical representation of the effect of parental exposure on: (1) the frequency of malformed infants alive at age nine months; (2) the frequency of death in the first nine months of life; and (3) the anthropometric measurements of weight, height, head circumference and chest circumference.

This stems from the fact that significance or non-significance of a maternal exposure effect is, in part, a function of that portion of the data which one elects to analyze. If all exposure classes are used, there is no demonstrable effect of mother's exposure or father's exposure on a two-tailed, or for that matter, a one-tailed test of significance. Similarly, if all terminations occurring to parents one or both of whom are in exposure class 1 are rejected, there is no effect of mother's exposure or father's exposure. If, however, one rejects those terminations where the mother was in class 1 but retains the

It is difficult to appraise the legitimacy of this comparison; however, two observations seem pertinent to any appraisal. Firstly, there is an element of arbitrariness in a procedure which rejects as unsuitable mothers in one exposure class but not fathers on such vague grounds as dissimilarity in place of origin, in the absence of demonstrable differences with respect to concomitant variables which are known or can be shown to appreciably affect the indicator. Secondly, rejection of mothers in class 1 leads to the use of a "control" approximately 25 per cent of which is contributed by

a single exposure cell (mothers 2, fathers 1, Hiroshima) exhibiting a sex ratio which is higher than any observed in these cities (or all of Japan) for any one of the fourteen years in the interval 1935–1952 for which data exist.

Figure 14.1 fails to reveal consistent changes in the "exposure surface."

Data continue to be collected which bear on the relationship of maternal exposure to the sex ratio.[1]

TABLE 14.1 A SUMMARIZATION OF THE COMPARISONS OF THE VARIOUS INDICATORS WITH PARENTAL EXPOSURE WHEN (a) ALL EXPOSURE CELLS ARE CONSIDERED (THE 4 × 4 CASE), AND (b) ONLY THOSE CELLS WHERE BOTH PARENTS WERE EXPOSED ARE CONSIDERED (THE 3 × 3 CASE)

(The general direction of change is indicated by an arrow pointing upwards if the observed frequency of departure from "normality" increases with increasing parental exposure, and by an arrow pointing downwards if the frequency of the event decreases with increasing exposure. The tabular entries are probabilities.)

	Fathers 4 × 4 case	Fathers 3 × 3 case	Mothers 4 × 4 case	Mothers 3 × 3 case
Sex ratio	.30–.50 ↑	.90–.95 ↑	.10–.20 ↓	.95–.98 ↓
Malformation				
At birth	.70–.80 ↑	.80–.90 ↓	.50–.70 ↑	.80–.90 ↑
At 9 months	—	.30–.50 ↓	—	.02–.05 ↓
Stillbirth	.20–.30 ↑	.80–.90 ↑	.001–.01 ↑	.20–.30 ↓
"Neonatal" death	a	.20–.30 ↓	a	.02–.05 ↑
Death in 9 months	—	.95–.98 ↑	—	.50–.70 ↑
Birthweight means				
Males–Hiroshima	.10–.25 ↓	—	> .25 ↓	—
Females–Hiroshima	> .25 ↓	—	.05–.10 ↓	—
Males–Nagasaki	> .25 ↑	—	.10–.25 ↓	—
Females–Nagasaki	> .25 ↑	—	.10–.25 ↑	—
Anthropometrics generalized means	< .001	.25–.50	.02–.05	.05–.10

a No general test (see Chapter XI).

	Combined parental exposure 4 × 4 case	Combined parental exposure 3 × 3 case
Indicator		
Birthweight variances		
Males–Hiroshima	.10–.25	—
Females–Hiroshima	< .001	—
Males–Nagasaki	.10–.25	—
Females–Nagasaki	.10–.25	—
Anthropometrics generalized variances		
Males–Hiroshima	.10–.25	> .25
Females–Hiroshima	.10–.25	> .25
Males–Nagasaki	.10–.25	> .25
Females–Nagasaki	.02–.05	.05–.10

With regard to whether there exists ancillary evidence suggesting an effect of mother's exposure, it may be stated that neither the limited observations available on early pregnancy terminations, nor the stillbirth data, nor the neonatal death data, indicate an interaction of sex with mother's exposure such as one might expect if the death of male infants is a function of mother's exposure. Moreover, inspection of

(2) *Malformation.* — There is no significant effect of mother's or father's exposure on the frequency of congenitally malformed infants irrespective of whether exposure class 1 parents are or are not included in the analysis. Further classification and analysis of these data to take

[1] The reader may wish at this point to refer to the discussion of the findings in this supplementary material, at the end of Chap. VII.

into account known differences between exposure classes in the age of the mother at the birth of the infant fails to reveal consistent, significant differences in the frequency of congenital malformation attributable to parental exposure. There is evidence, however, of an effect of mother's exposure on the frequency of malformation among the children of very young mothers (<21 years of age). This effect, however, seems largely one of an inordinately low frequency of malformation among young, unexposed mothers rather than an elevation of the frequency among exposed mothers since there is no demonstrable difference among mothers in exposure classes 2, 3, and 4–5. Finally, analysis of additional data on the frequency of congenital malformation obtained at nine months of age discloses no significant effect of parental exposure.

(3) *Stillbirths.* — Analysis of the data unadjusted for differences in parity and maternal age between exposure cells fails to reveal an effect of paternal exposure, regardless of whether category 1 fathers are or are not included in the analysis. However, there does exist an effect of maternal exposure (significant at the 5 per cent level) when category 1 mothers are included in the analysis. This effect does not persist when mothers in exposure category 1 are excluded. Further classification of the data by parity and subsequent analysis reveals no effect of mother's exposure when the category 1 mothers are included. It would seem that the apparent mother's exposure effect is explicable in terms of differences between exposure classes in birth ranks. The latter analysis suggests an effect of paternal exposure limited to first-born infants. An explanation of this finding is not immediately apparent.

(4) *Neonatal death.* — Both with and without allowance for differences in parity among the mothers falling into the various radiation categories, there is no significant effect of mother's or father's exposure on the frequency of neonatal death. Moreover, the analysis of deaths during the first nine months of life fails to reveal a significant effect of parental exposure.

(5) *Birthweight.* — When allowance is made for differences between exposure cells in maternal age and parity, there exists no significant difference between classes of father's exposure or classes of mother's exposure in mean birthweight. Furthermore, it was not possible to demonstrate a consistent effect of parental exposure on (a) the relationship between birthweight and concomitant variables, notably maternal age and parity, or (b) the residual birthweight variances, that is, on the birthweight variances following removal of maternal age and parity effects. Figure 14.1 illustrates the lack of a consistent exposure effect on the birthweight means or birthweight variances.

(6) *Anthropometrics.* — Neither the generalized means nor the generalized variances can be shown to differ among classes of paternal or classes of maternal exposure. This remains true following removal of differences between exposure cells in the age of the infant at the time the measurements were obtained.

14.2 *The question of evaluating the over-all direction of the indicators.* — As noted in Section 6.2, the plan of analysis adopted ensured non-overlapping indicators of genetic effects. By suitable transformations it is readily possible to combine the results of such independent tests, given a theory which permits specifying the direction of the differences between control and irradiated material. In this particular study, however, the findings are sufficiently small and inconsistent, that we have not felt an attempt at a combined treatment to be justifiable.

There is in the analysis of these data another problem, closely related to the foregoing. There are segments of the analysis which suggest the possibility of a radiation effect (e.g., certain selected comparisons within the sex ratio data; frequency of malformations in relation to maternal radiation history when only very young mothers are considered; frequency of stillbirths in relation to paternal radiation exposure history, when only the first-born children of these fathers are considered). Such findings tend to stand out in one's mind as he thinks back over the analysis as a whole. It is important in this connection not to lose sight of the fact that even in the absence of any effects whatsoever, by definition 1 in 20 independent tests or subtests may be expected to exceed the 5 per cent level of significance. Thus, while these "hints" cannot be entirely disregarded, they must be viewed against the background provided by the totality of the significance tests to which the data were subjected.

14.3 *The confidence limits determined by these observations.* — Thus far in our analysis we have been concerned with attempts to demonstrate a positive effect of exposure to the

atomic bombs on the indicators selected for study. There is, however, another aspect of these data. They permit us to place upper limits on the effects which may have been induced but not demonstrated by these studies. In other words, we can place confidence limits on our observations. One may proceed to establish these limits by several different approaches. To avoid misunderstanding, it should be clearly pointed out at this juncture that we are not primarily concerned with placing limits on a series of estimates, since estimates of abnormalities in pregnancy terminations derived from these data would probably have little meaning to other populations, in view of the racial differences and the variety of "experimental" conditions which could arise in man. Our primary concern here is to determine the adequacy of our test procedures. The ideal solution, perhaps, would be the computation of the power curves associated with each of the individual tests of significance. In multi-way analyses, however, the computation of the power function is, to say the least, tedious, and for a number of the tests performed here the power function is not known. An approximate solution to the problem of the adequacy of our data, sufficient for our purposes, can, however, be obtained. We shall begin by restricting our attention to the adequacy of the data for detecting differences of a specified size between a control population and a single, moderate-to-heavily irradiated population. Thus we shall seek to determine the probability of rejecting the null hypothesis (no difference between groups), given a fixed sample size, when the true situation is one wherein the exposed group differs from the control group by some amount, say d, for varying values of d. The groups with which we shall be concerned will be those terminations occurring to parents (a) neither of whom had exposures in excess of class 2 — the control group, and (b) one or both of whom had exposures equal to or greater than class 3 — the moderate-to-heavily irradiated group. We shall make the following simplifying assumptions:

1. The proportion of "successes" in the control population is known without error.
2. The normal approximation to the binomial distribution has the requisite accuracy over the critical range, which, for this study, includes those values for the exposed group which deviate from the control by a factor less than two.

The latter assumption requires no special comment, and seems warranted in this situation. The former assumption is obviously not strictly satisfied by these data; however, the sample number available on the control group is sufficiently large that the error associated with the estimate of the true proportion of successes for this group is small under all circumstances, and can for our purposes be disregarded when contrasted with the error associated with an estimate based on the sample sizes available for the exposed groups.

Now to determine the probability of rejecting the null hypothesis under a variety of alternative situations we proceed as follows: We shall assume that the level of confidence with regard to the null hypothesis is 0.05. There exist, then, two limiting values corresponding to this level of confidence such that the number of successes, say x, among N trials will lie within these limits 95 times in 100 if the null hypothesis is, in fact, true, that is, if there is no difference between the proportion of successes in the control, say p_c, and the proportion of successes in the heavily exposed group, say p_e. These limiting values, L_1 and L_2, are chosen so that

$$P(L_1 \leq x \leq L_2 | H_0) = \sum_{x=L_1}^{x=L_2} \binom{N}{x} p_c^x (1-p_c)^{N-x} = 0.95.$$

If now N and p_c (or p_e) are sufficiently large such that $Np_c > 5$, this probability can be reasonably approximated by the normal distribution. Under the latter circumstance, we obtain L_1 and L_2 from the relations

$$L_1 = -1.96\sqrt{Np_c(1-p_c)} + Np_c$$
$$L_2 = 1.96\sqrt{Np_c(1-p_c)} + Np_c.$$

If $p_e \neq p_c$, but the number of successes, x, among N trials from a population characterized by p_e should lie within the limits L_1 and L_2, we would erroneously accept the null hypothesis as true (the type II error). The probability of the latter can be computed for any value of p_e, and is merely

$$P(L_1 \leq x \leq L_2 | H_1) = \sum_{x=L_1}^{x=L_2} \binom{N}{x} p_e^x (1-p_e)^{N-x}$$

and, again, this probability can be reasonably approximated by the normal distribution. If this procedure were repeated for every possible value of p_e, we would generate a continuous curve resembling either the normal cumulative probability curve (if we were interested solely in one-tailed tests of significance) or the normal frequency curve (if we were interested in two-tailed tests of significance). These curves are termed the power function of the test or the tests of significance. For the remaining operating characteristic curves the sample size was the number of parents both of whom experienced exposures equalling or exceeding class 3. Before we turn to a consideration of these five curves, it is worth noting that the amount of information actually available with regard to these four indicators is in excess of that used to compute the OC curves. Accordingly, we may view these curves as underestimating the adequacy of the

FIGURE 14.3 — The adequacy of the data with regard to sex ratio as indicated by the operating characteristic (OC) curves for analyses based on sample sizes of 5,629 mothers and 2,453 fathers, where the true proportion of successes is assumed to be 0.5198, the value observed in the control population.

operating characteristic of the test depending upon whether we are interested in the probability of rejecting the null hypothesis when it is, in fact, false, or the probability of erroneously accepting the null hypothesis when some alternative hypothesis is true.

Figures 14.3 and 14.4 present five operating characteristic curves for sex ratio, malformation, stillbirth, and neonatal death. The two curves for sex ratio are based on sample sizes corresponding to the number of mothers in classes 3, 4, and 5 whose spouses were in classes 1 and 2, and the number of fathers in classes 3, 4, and 5 whose spouses were in classes 1 and 2. In both instances, we are concerned with one-tailed data to detect departures of specified size from the null hypothesis.

Before turning to Figures 14.3 and 14.4, we must define what is meant by an adequate test. We shall say that a test is adequate with regard to the alternatives, $p_e - p_c \geqq |s|$, if the prior probability of detecting departures from the null hypothesis of size s, or greater, equals, or exceeds, some specified probability. Otherwise stated, a test is adequate with reference to all those alternatives for which the power (the probability of correctly rejecting the null hypothesis when it is, in fact, false) is not less than some specified amount. The reader may, of course, impose any power restriction he

Recapitulation

wishes, but we shall define adequate power, here, as being 0.90. That is to say, if the prior probability of detecting a difference of size *s* is at least nine chances in ten we shall say that the test is adequate with regard to departures of size *s* or greater. This value may well be too high, since, as we have indicated, our procedure takes into account appreciably less than the total number of observations which are germaine. Within this frame of reference, we can say that our data are adequate to give reasonable assurance that we would be able to detect the following:

1. A decrease in the sex ratio, following maternal exposure, in excess of an absolute change of 1.6 per cent;
2. An increase in the sex ratio, following paternal exposure, in excess of an absolute change of 4 per cent;
3. An alteration of the malformation rate in excess of two times the control value;
4. An alteration of the stillbirth rate (and neonatal death rate) in excess of approximately 1.8 times the control value.

Alternatively stated, we should be unable to detect:

1. Changes of less than 1.6 per cent (absolute) in the sex ratio when the mother only was heavily exposed, or 4 per cent if the father only was exposed;
2. Changes of less than 100 per cent (absolute) in the frequency of malformed infants;
3. Changes of less than 80 per cent (absolute) in the frequency of stillborn infants or infants dying during the neonatal period.

14.4 *A résumé of work on mammalian material pertinent to the interpretation of these findings.* — Before proceeding to a final interpretation of these findings, we would do well to consider at some length the results of more or less comparable studies on other animal species. Although the many important studies which have been carried out on the genetic effect of irradiation on Drosophila clearly indicate the types of results to be expected in mammals, the differences between mammalian and insect physiology are such that any semiquantitative extrapolation is at best hazardous. Accordingly, we will for the purposes of this discussion confine our attention to studies on mammalian material. The current status of our knowledge of mammalian radiation genetics has recently been comprehensively summarized by Russell (1954). There is actually a rather striking dearth of information available for com-

FIGURE 14.4 — The adequacy of the data with regard to malformations, stillbirths, and neonatal deaths as indicated by the operating characteristic (OC) curves for analyses based on samples of size 1,097 parents, where the true proportion of successes are assumed to be 0.0090 for malformations and 0.0142 for stillbirths and neonatal deaths, the control values.

THE CHILDREN OF ATOMIC BOMB SURVIVORS

parison with the findings of the present study. Such information as there is will be presented in the following six subsections.

In most of the work to be cited, the house mouse has been the experimental animal. Some two to four weeks after male mice are irradiated with a dosage in excess of 400r, they become sterile, the time of onset of sterility depending on dosage. In those animals which survive, fertility usually returns in four to twelve weeks, again depending on dosage. There are important quantitative and qualitative differences between the observable genetic effects associated with sperm released in the pre-sterile period — corresponding to the result of irradiation of spermatocytes, spermatids, and spermatozoa — and the effects associated with sperm released during the post-sterile period — corresponding to the irradiation of spermatogonia. We will consider primarily data collected during the post-sterile period, since this corresponds to the period during which observations have been made in Japan.

14.4.1 *Effects on sex ratio.* — It will be recalled that genetic theory suggests that the irradiation of males should result in an increase in the proportion of males in the next generation. Parkes (1925), studying the offspring of male mice exposed to doses of X-ray below those which produce temporary sterility, reported (a) a moderate but non-significant increase in the proportion of males born to mice mated 0–4 days after irradiation (59.4 per cent of 133 offspring of irradiated males vs. 51.6 per cent of 735 controls), (b) a significant decrease in the proportion of males born from matings made 5–18 days after irradiation, and (c) no change in the sex ratio among the offspring of matings made 19–57 days post irradiation, this group apparently reflecting the results of the irradiation of spermatogonia. The observations of Hertwig (1938) on the results of irradiation of spermatogonial stages in the house mouse are reproduced in Table 14.2. Although the difference is in the postulated direction, the results of a comparison of "all irradiated" vs. "controls" are not at the level of significance nor, for that matter, were the results obtained during the pre-sterile period significant. However, when only the two groups receiving the most irradiation are considered, there is a significant difference. Russell (1954) reports that in his experiments with the exposure of male mice to 600r, there were 50.35 per cent males in 72,472 post-sterile period offspring, as against 51.00 per cent males in 55,828 controls. Finally, Kalmus, Metrakos, and Silverberg (1952) reported a significant decrease in the frequency of females among the offspring of male mice mated immediately following treatment with 150 or 300r, but were unable to confirm this observation in a later study (Trasler and Metrakos, 1953).

Even if clear-cut effects had been demonstrated for the house mouse, the advisability of extrapolating to man would be rendered questionable by the much higher apparent proportion of inherited defects which are sex-linked in

TABLE 14.2 THE EFFECT OF IRRADIATION OF MALE MICE ON THE SEX RATIO OF OFFSPRING CONCEIVED DURING THE POST-STERILE PERIOD
(After Hertwig, 1938)

Dose in r	Number of offspring	Number of males	% males
400	415	197	47.47
500	167	86	51.50
600	167	86	51.50
800	683	347	50.80
1,000	574	291	50.69
1,200–1,400	171	102	59.65
1,500–1,600	63	37	58.72
All doses	2,240	1,146	51.16
Controls	2,595	1,290	49.71

χ^2 (all doses vs. controls) = 1.011
DF = 1 $0.50 > P > 0.30$

man than in the mouse, suggesting the possibility of a genetically more active differential segment of the X-chromosome in man. The corollary of this is, of course, a probable higher expectation of induced sex-linked lethals per unit irradiation in man.

Recently Macht and Lawrence (1955) have published a detailed comparison of the sex ratio, the frequency of twinning, the frequency of fetal death, and the occurrence of congenital defect among the children of radiologists as contrasted with the children of physicians in other specialties (pathologists, psychiatrists, anesthesiologists, plastic surgeons, and ophthalmologists). The information was obtained by questionnaires mailed to the subjects. Answers were obtained from 74.1 per cent of the radiologists queried, with information on 5,461 children, but from only 53.8 per cent of controls, with data on 4,484 children. Because of the

publicity given the matter of the genetic risks of radiation, one immediately wonders if radiologists as a group were especially motivated to reply, perhaps particularly if the reproductive history was in some way unusual. The authors are aware of this source of bias, and quite properly point out that inasmuch as the distribution of single persons, married couples without children, and married couples with children is similar in the two groups, this suggests that radiologists with (abnormal) children were not especially motivated to reply. On the other hand, they present data suggesting that recall for abnormal terminations is better for later than earlier pregnancies. Despite this, they combine the (early) reproductive history of radiologists prior to entering this field with the control data. A study of their Table 13 suggests that this may be an important source of bias.

A second source of bias in this study stems from the fact that "persons who did not respond (to the questionnaire) within approximately two months were sent a second questionnaire together with a further appeal that it be returned." Of the total completed questionnaires available for analysis, 78.5 per cent of those completed by radiologists were obtained on first writing, as contrasted to 63.8 per cent of those from non-radiologists. The possible significance of this to the study lies in the fact that there was, in general, a higher percentage of abnormal outcomes reported on "second-appeal" questionnaires than on "first-appeal" questionnaires. Specification of the direction of the bias introduced is difficult, since it depends on one's interpretation of the reasons for the different results obtained on the two appeals. But that this may be an important source of bias is indicated by the fact that whereas after correction for parity differences, there were 3.10 per cent more "abnormal offspring (fetal death and congenital defects)" in the children of radiologists than non-radiologists on the basis of first questionnaires, this difference dropped to 0.76 per cent on second questionnaires. This is a striking and disturbing discrepancy.

Inasmuch as their study will be widely quoted, a detailed comparison between their findings and our own seems indicated. With regard to the indicator under consideration, sex ratio, the findings are as shown in Table 14.3. With respect to livebirths, the proportion of

TABLE 14.3 THE SEX RATIO AMONG LIVEBIRTHS AND FETAL DEATHS IN THE STUDY OF MACHT AND LAWRENCE (1955)

Offspring of persons	Livebirths				Fetal deaths			
	Males	Females	Unknown	Males per 100 females	Males	Females	Unknown	Males per 100 females
Exposed throughout marriage	1,104	1,115	293	99.01	53	23	316	230.43
Exposed part of marriage	986	922	299	106.94	55	19	300	289.47
Unexposed part of marriage	526	461	152	114.10	22	14	102	157.14
Unexposed throughout marriage	1,240	1,163	404	106.62	42	23	345	182.61
All exposed	2,090	2,037	592	102.60	108	42	616	257.14
All unexposed	1,766	1,624	556	108.74	64	37	447	172.97
Total	3,856	3,661	1,148	105.33	172	79	1,063	217.72

males is *decreased*, although not significantly so, a finding opposite in direction to expectation when fathers are irradiated unless — as they suggest — the male is more sensitive than the female to the effects of induced mutations in the autosomes. Among fetal deaths (their term; includes miscarriages and stillbirths), the data are in the direction of expectation but so very limited in number that even the large difference observed is not significant; it is worth noting that sex was not recorded for 81 per cent of this material.

14.4.2 *Effects on malformation frequency.* — Dominantly inherited mutant forms, some of which in our terminology would be classified as congenital malformations, have been detected in small numbers by Charles (1950) among the offspring of male mice exposed to 60r (7 in 3,072 offspring; 0 in 2,755 controls), and by Russell (1954) among the post-sterile period

offspring of male mice exposed to 600r (5 in approximately 30,000 offspring; number of controls not stated). The examination given the offspring by Russell was much less searching than that by Charles since Russell's work was primarily oriented in other directions. From the quantitative standpoint, these findings have debatable carry-over value for man if only because of the tendency observed in most laboratories for female mice to devour dead and defective embryos immediately following parturition.

The older medical literature contains a number of studies, based on the questionnaire approach, of the relationship between therapeutic irradiation and the "health" of children conceived subsequently (review in Murphy, 1928). For a variety of reasons, and particularly with reference to the problems of diagnostic standards and control observations, these data do not lend themselves to critical inferences. More recently, however, Macht and Lawrence (1955), in the study described earlier, after correcting for parity differences between the two sets of data, record 5.99 per cent congenital defect among 5,461 children of radiologists and 4.84 per cent among 4,484 control children. The figures cannot be compared with those of the present study because (a) minor defects are included, (b) autopsy findings are included with no analysis of comparability of autopsy frequency in the two groups, (c) congenital conditions not usually regarded as malformations, such as erythroblastosis fetalis and atelectasis of the lung, are included in the list, (d) multiple defects in one individual are scored separately, and (e) disease states are listed which very probably developed and were detected sometime postpartum, such as muscular dystrophy, Oppenheim's disease, spondylolisthesis, celiac syndrome, melanoma, lymphosarcoma, retinoblastoma, Tay-Sach's disease, missing teeth, and defective night vision. There are differences between the children comprising the two groups which are difficult to understand: There are 14 cases of "erythroblastosis" and one "died 4 days Rh positive problem" among the children of radiologists but only three cases of "erythroblastosis" among the controls. The interpretation of this discrepancy is complicated by the fact that the children born to radiologists before they entered this specialty are included among the controls, and, as is well-known, erythroblastotic infants are found among later pregnancies. This fact would tend to bias the results in the observed direction, although whether to the extent observed is debatable. There are nine cases of atelectasis of the lungs among the exposed group but only three among the controls. These two differences alone account for approximately one-third of the observed 1.15 per cent increase in malformation among the children of radiologists. Although Macht and Lawrence believe the difference between the two groups is significant, we do not feel they have established this fact.

14.4.3 *Effects on stillbirth frequency.* — Mutations with a dominant lethal effect can,

TABLE 14.4 THE EFFECT OF IRRADIATION OF MALE MICE ON THE FREQUENCY OF STILLBIRTHS AMONG OFFSPRING CONCEIVED DURING THE POST-STERILE PERIOD
(After Hertwig, 1938)

Dose in r	Number of offspring	Stillborn	% stillborn
400	423	21	4.97
500	170	11	6.47
600	175	12	6.85
800	693	12	1.73
1,000	590	36	6.10
1,200–1,400	171	0	0.0
1,500–1,600	65	11	16.92
All doses	2,287	103	4.50
Controls	2,616	90	3.45

χ^2 (all doses vs. controls) = 3.649
DF = 1 $0.10 > P > 0.05$

by definition, produce death any time from shortly following fertilization to just prior to reproduction (cf. Hadorn, 1955). If the dominant lethal exerts an effect during the last trimester of pregnancy, the result would be a stillbirth. Evidence concerning an increase in stillbirths among the pre- and post-sterile period progeny of irradiated male mice is unsatisfactory because of the tendency, referred to in the preceding section, for females following parturition to devour stillborn young. Hertwig (1938) has published data, reproduced in Table 14.4, which show a small and insignificant increase in stillbirths in the offspring of post-sterile period irradiated male mice, and Strandskov (1932), in a pioneer experiment in which the dose varied from 173 to 2,592r, has recorded similar data for the guinea pig, although the latter's results must be interpreted with caution

because the effect was present in post-sterile period but not pre-sterile period offspring. However, the number of the latter observed was quite small (39).

Macht and Lawrence (1955) report 13.77 per cent "fetal death" in the children of the exposed, as contrasted to 12.50 per cent in the controls. The term fetal death "includes all reported cases where the product of conception was not born alive. It therefore includes miscarriages and stillbirths irrespective of the length of uterogestation and regardless of whether or not the embryo or fetus was otherwise normal at this stage" (p. 447). From the frequency of the event in the controls, as well as the wording of the questionnaire, this term must include abortions as well. The difference between the two series was not statistically significant. Crow (1955), in a similar questionnaire-type study, also failed to detect any statistically significant differences between frequencies of stillbirths and miscarriages among the children of radiologists and in the children of a control group of pathologists.

14.4.4 *Effects on birthweight.* — Strandskov (1932) observed that when correction was made for the smaller mean litter size of females mated to irradiated male guinea pigs, mean birthweight was decreased in the post-sterile period offspring of his experiment from 90.93 gms. in the controls to 86.35 gms. in the post-sterile period offspring. The number of animals involved is not stated but can be estimated from the data in the paper to be approximately 260 for the controls and 75 for the irradiated. The data are not presented in such a way that the significance of the findings can be analyzed. The interpretation is further complicated by the fact that only the data for animals surviving at least 30 days are presented. Variances are not given for the birthweights of guinea pigs conceived during the post-sterile period, a point of especial interest in view of the findings of our study. There do not seem to be comparable data for the house mouse.

14.4.5 *Effects on neonatal death rates.* — Hertwig (1938) observed a slightly but not significantly increased death rate during the first 75 days following birth in the offspring of male mice receiving 800–1,600r, the survival figures being 79.14 ± 1.80 per cent in 508 progeny of irradiated mice as contrasted to 82.43 ± 1.38 per cent in 757 control offspring. Strandskov's data, on the other hand, involving approximately the numbers referred to earlier, reveal no difference in post-natal death rate in the offspring of irradiated guinea pigs up to 30 days postpartum. Crow (1955), in the questionnaire study referred to earlier, found that "the infant mortality rates also were not significantly different in the children of the two groups, but the numbers were very small."

14.4.6 *Effects on growth and development.* — The only pertinent data appear to be those of Strandskov (1932). At 30 days postpartum the mean weight of the post-sterile period offspring of irradiated male guinea pigs, corrected for litter size, was 265.73 gms., against a control value of 281.66 gms. The number of animals involved is presumably the same as for the birthweight figures given above.

14.5 *Interpretation of the findings.* — The interpretation of the findings of the present study can now be very simply stated. The foregoing section has brought out the fact that when one considers data of a type which it is feasible to collect on human populations, the available information on other animals is scattered, fragmentary, and often contradictory. However, this information, taken in conjunction with the much more extensive data on Drosophila, certainly does not suggest a high likelihood of demonstrating clear-cut effects in Hiroshima and Nagasaki. The divergence of opinion with regard to what one might expect in human populations is sufficiently large as to make it unprofitable to explore all presently held opinions with regard to all of the indicators. However, it does seem worth-while to explore a single class of mutations, say the sex-linked lethals. Evidence from Drosophila suggests that the yield in sex-linked lethals is approximately 3 per cent per 1,000 roentgens (Timoféeff-Ressovsky, 1937; Spencer and Stern, 1948; Muller, 1954). The yield in the mouse, on the basis of the data on induced autosomal mutations, might, if the mouse X-chromosome had the same number of loci capable of giving rise to lethal mutations as the Drosophila X, be greater by a factor of 15–20. Now, if the dosage estimates advanced for the various exposure classes are reasonably accurate, then the average dose in *rep*'s received by the 5,629 mothers with exposure 3 or greater (whose spouses were in

classes 1 or 2) is approximately 100 *rep*'s. If we impose the restrictions indicated in the previous paragraphs with regard to an adequate test, then differences in the sex ratio between the control (average exposure 0 *rep*'s) and the exposed group (100 *rep*'s) as large as 1.6 per cent (absolute change) would quite probably not be detected. Accordingly, we may estimate that the yield, in man, might be as high as roughly 2 per cent per 100 *rep*'s and we would not detect it. This is a value six times that in Drosophila, but approximately one-half to one-third the value to be expected if human genes were as sensitive to irradiation as the small series of tested mouse genes (extrapolating from autosomal visibles to sex-linked lethals) and the X-chromosome of man had the same genetic length as that of Drosophila. This is, of course, the upper limit; the yield could be, and quite probably is, much lower. Similar conjectures could be made for a number of the other indicators.

Accordingly, we can say of the present study that *under circumstances where, on the basis of what is known concerning the radiation genetics of mammals, it appeared unlikely that conspicuous genetic effects of the atomic bombs could be demonstrated, such effects have in fact not been demonstrated*. The present study can in no way be interpreted to mean that there were no mutations induced in the survivors of the atomic blasts. Neither, on the other hand, is the reverse interpretation — that of mutation production — permissible from this series of observations, although, on the basis of all that is known of radiation genetics, there is no real reason to doubt that mutations were produced in Hiroshima and Nagasaki. We are left with inconclusive findings, albeit findings which permit us to set confidence limits.

Chapter XV

PERMISSIBLE INFERENCES

15.1 *Basic data necessary to reaching permissible inferences.* — There would at this juncture be ample justification, judging from the genetic literature of recent years, to utilize the findings of this study, taken in conjunction with other available information, as the basis for a semi-quantitative treatment of the problem of the genetic risks of increased radiation of the human species. Such calculations proceed in general along the following lines: On the basis of existing data, one estimates the average spontaneous mutation rate of human genes and the number of genes in man. From this one obtains an estimate of the average number of mutations per human gamete per generation (a "total" mutation rate). Then, again on the basis of existing data, one estimates the probability of mutation/locus/r in man, and from this the amount of radiation required to double, triple, or otherwise increase the mutation rate over the spontaneous baseline. Next, then, by one approach or the other, one estimates the ratio of "recessive" mutations which have been accumulated in the population to "recessive" mutations arising anew each generation, the so-called "accumulation factor." Given this, one can estimate by how much, on the basis of the preceding calculations, this frequency will be altered by any arbitrary increase in radiation. Finally, on the basis of certain assumptions concerning gene physiology and the operation of natural selection in man, one attempts to evaluate the phenotypic impact of this increase in radiation.

There are, then, five estimates which are basic to these semi-quantitative treatments:

1. The spontaneous mutation rate in man.
2. The induced mutation rate/locus/r in man.
3. The total number of genes in man.[1]
4. The "accumulation factor."
5. The manner in which selection operates on the total gene complex.

It is our contention, which we will now proceed to document, that the available data on which these five estimates are based are so inadequate that semi-quantitative treatments are ill-advised, since except to the relatively few who have made a detailed study of the problem, they impart an air of mathematical exactitude and scientific accuracy to an area where the errors are sometimes large and often indeterminate.

15.2 *The spontaneous mutation rate in man.* — Current thinking concerning the rate of mutation of mammalian genes is for obvious reasons strongly influenced by what is known concerning Drosophila rates. We will accordingly first consider briefly what seem to us to be some of the more pertinent data concerning this species. For methodological reasons it is customary to distinguish on the basis of their physiological effects between three categories of mutations, namely, those associated with visible effects, those associated with lethal effects, and those which express themselves through a reduction of viability in the absence of detectable somatic effects, the so-called semi-lethal mutations (1–10 per cent viability) and deleterious mutations (over 10 but less than 100 per cent viability). Terminology in this field leaves something to be desired. Thus, the "deleterious" mutations must have an organic basis, so that many of them would be found on careful study to be also "visibles." By the same token, most "visibles" are also "deleterious." Finally, the dividing line between "lethals" and "semi-lethals" may be altered by culture conditions. Be that as it may, the division into

[1] The product of (1)×(3), or (2)×(3), is the rate of mutation per gamete, spontaneous or induced, as the case may be. It is possible in suitably designed experiments to estimate this directly (cf. Muller, 1955), and so decrease the number of variables involved in the calculation.

these three categories has an operational usefulness, as we shall now see.

Beginning with the pioneer attempts of Muller (1934; see also Kerkis, 1935, and Timoféef-Ressovsky, 1935), a number of efforts have been made to establish the relative frequencies with which these types of mutations are represented among all mutations. These attempts have involved radiation-induced rather than spontaneous mutations because of the much more laborious nature of the problem if attacked through the study of spontaneously occurring mutations. In view of the possibility that the relative frequency of lethals is higher among the radiation-induced mutations because of the increased proportion of minute deletions, the estimate of the ratio, (semi-lethals + deleterious)/lethals, may be a minimum estimate. Muller (1954; see also Falk, 1955) places this ratio at 3–5 to 1. This same author goes on to state that ". . . the ratio may indeed be considerably higher than this, since the technique was hardly refined enough for the detection of detrimentals with a viability greater than some 85 per cent of normal. Other studies have shown that 'invisible' mutants causing sterility or lowered fertility of some degree also form a very large group. This group, however, overlaps, to an extent not yet well investigated, that of the detrimental mutations" (p. 396).

The significance of information concerning the relative frequency of mutants with viability in the 85–99 per cent range in attempts to quantitate the genetic risks of radiation is of course enormous. A related problem concerns the frequency of mutations for which the organism *at the time* is able to compensate completely, the undetectable mutations. Lately considerable attention has been directed towards the genetic basis and evolutionary implications of physiological homeostasis (refs. in Lerner, 1955). The possibility cannot be ruled out that the principle of homeostasis enables some organisms to compensate entirely, under particular sets of circumstances, for the effects of certain mutations.

It may be argued that there is no reason to be concerned about the relative frequency of mutants with undetectable effects in a consideration of the deleterious effects of radiation. However, these mutations are undetectable only under the conditions set by the observer. Under other conditions, set by nature and not by man, they might have decided effects. It is not at all difficult to argue that the mutants-with-over-85% viability which cannot now be studied in Drosophila may in evolutionary importance far outweigh the visibles.

Muller (1950) in a discussion of the question of the numerical relationship between lethals, on the one hand, and semi-lethals and deleterious mutations, on the other hand, has made the following statement: "However, studies carried on in *Drosophila* during the past year by Meyer, Edmondson, and the writer indicate that in this organism the assumption of an equal distribution of detrimental mutations throughout all i_{ho} values [2] (when represented on an arithmetic scale) does not hold. Instead, it appears that, following the high but descending peak formed by complete lethals ($i_{ho} = 100\%$) and nearly complete lethals (i_{ho} = between 98% and 100%), there is a marked drop in the frequency of mutations. The mutations studied were induced in an autosome (the second chromosome) by ultraviolet light acting on an interphase stage (in the polar cap). Along with 208 complete lethals there were 20 mutants found in the range of i_{ho} between 98% and 100%, and again only 20 in the range of i_{ho} between 90% and 98%, although this range is four times as wide as the preceding one. If the rest of the distribution, as far as $i_{ho} = 10\%$, had only the same frequency of mutations as in the range between 90% and 98% there would have been only 240 detrimentals in the entire interval between 100% and 10%, to set against the 208 complete lethals found. But since we know from other work, previously cited, that the detrimentals in this interval are in reality several (about 5) times as numerous as the complete lethals, it is evident that their frequency must, at lower degrees of detriment (lower i_{ho}), rise very much above that existing in the 90% to 98% range. The distribution of frequencies of i_{ho} therefore forms a bimodal curve with one peak at the left origin, lethality ($i_{ho} = 100\%$), and another peak somewhere to the right.

"Little more than this is yet known definitely about the shape of the curve in question, important though this genetic question is. However, there are grounds, both theoretical and observational, for regarding it as very unlikely that the second peak is near the first or that the rise towards it is sharp. Hence it is probable that

[2] i_{ho} — the amount of impairment produced by a gene when homozygous.

detrimental mutations, instead of having an even distribution with respect to values of i_{ho}, form a curve which, except for its peak of near-lethals at the left end, is massively skewed towards the right, with its mean at a value of i_{ho} significantly beyond the middle (0.5)" (pp. 140–141).

If we consider these remarks of Muller in conjunction with the possibility of "invisible" mutants discussed earlier, then the problem of estimating the relative frequency of lethal mutants vs. those viable to some degree assumes new complexity. Figure 15.1 attempts to present some of this complexity graphically. The abscissa of this figure represents viability of the homozygous genotype in some arbitrary environment. In this connection, it is apparent that the term "lethal" is relative, some lethal mutations having effects under no known circumstances compatible with life, other lethal mutations having far lesser effects. Likewise, the term "normal" as applied to viability is relative, some normals being more normal than others, with the differences brought out only under unusual circumstances. Thus far, observations have been limited to the range of lethality and 1–85 per cent viability. As Muller has pointed out in the statement quoted above, there is great doubt concerning the shape of the curve of numerical relationships within this range. We have indicated two of the principal alternatives. Curve A assumes a mode at 60–70 per cent viability, from which it would seem likely that the proportion of mutations in the 85–100 per cent and normal viability range is small. Curve B assumes that the mode is farther to the right with the corollary that there is a considerable group of mutations not now being detected. How large that group is depends of course on the shape of the curve.

The question of the relative frequency of lethal mutations as contrasted to visibles, is on somewhat more secure footing than the question of the ratio of lethal mutations to mutations reducing viability to a lesser degree. In tabulating the results of radiation experiments by five different workers, Schultz (1936) found this ratio to be 7.4:1. In view of the well recognized differences in the ability of individuals to recognize mutant phenotypes, the true ratio is probably somewhat lower. We prefer, for instance, the ratio of 5.2:1 which obtained in the extensive and meticulous experiments of Spencer and Stern (1948). For all three groups, the ratio of visibles: lethals: semi-lethals and deleterious has been set at approximately 1:5:20, although, as noted above, the third figure is certainly an underestimate.

The important question of the mutation spectrum at individual loci remains in its early stages

FIGURE 15.1 — A schematic representation of two different "mutation spectra" with reference to degree of viability, both compatible with the present data. Further explanation in text.

because of the amount of labor involved in securing reliable data. There is evidence that in Drosophila some few genetic loci which are associated with visible mutants are not absolutely essential to life, although in the absence of these loci viability is usually markedly reduced. On the other hand, there are numerous illustrations of lethal and visible mutations arising at what seems to be the same locus. It should also be pointed out that the question of the total relative frequency of mutation at different loci is in a very unsettled state. Although there seems no doubt that the rate of recovery of mutations differs from locus to locus, care must be exercised in reasoning to the magnitude of the true differences (cf. Neel and Schull, 1954). In the following discussion of mutation rates at specific loci, the fact that these are *selected* loci must constantly be borne in mind.

mutations per generation, a rate some fourfold that ordinarily observed. Since the numbers involved are not given, the reliability of this apparent fourfold increase cannot be evaluated. Muller et al. argue, from this fourfold increase in recovered lethals, that "the frequency of gene-mutations at the nine loci would *ordinarily* average between 10^{-5} and 7×10^{-6} per locus in females" (p. 125).[3] For our present purposes, it will be sufficient to average all three of these series together, with no attempt to introduce the correction suggested by Muller et al. (which would in any event then appear to create a possible discrepancy between their findings and those of Glass and Ritterhoff). In a total of 1,665,345 locus tests, at least 32 mutations were recovered, a rate of 1.9×10^{-5}. However, *this applies only to "visible mutations."* If lethals, semi-lethals, and deleterious

TABLE 15.1 FREQUENCY OF OCCURRENCE OF SPONTANEOUS "VISIBLE MUTATIONS" IN VARIOUS SPECIES

Author	Chromosome	No. of organisms	No. of loci	Total locus tests	Mutations	μ
Drosophila						
Muller, Valencia, and Valencia, 1950	X	±60,000	9	540,000	15	2.8×10^{-5}
Alexander, 1954	III	45,504	8	364,032	0	0
Glass and Ritterhoff, unpublished	Multiple (males)	102,759	4	359,657	16–21	$4.5–5.8 \times 10^{-5}$
	Multiple (females)	100,414	4	401,656	1	3.1×10^{-6}
House mouse						
Russell, 1954	Several	37,868	7	265,076	2	$.8 \times 10^{-5}$

Three recent studies on the rate of occurrence of spontaneous "visible mutations" in Drosophila appear outstanding. The results, *in terms of frequency of appearance of visible mutations*, are summarized in Table 15.1. There appear to be significant differences between the studies with respect to the observed frequencies, although it must be kept in mind that different loci are involved in the three studies. Furthermore, the work (unpublished) of Glass and Ritterhoff suggests that the mutation rates of males and females differ, although different loci were being tested in the two sexes. This observation may provide a clue to the apparent difference between the findings of Muller, Valencia, and Valencia (1950) and Alexander (1954), since the former were studying mutation rates in females and the latter in males. It is also noteworthy that in other experiments with the same strain used for the "visible mutation" studies, Muller, Valencia, and Valencia (1950) recovered 0.7 per cent sex-linked lethal

mutations are arising at these same loci, the mutation rate must be higher. If, for instance, the ratio of visibles: (undetected lethal + semi-lethal + deleterious mutations) at these loci was a conservative 1:4, the mutation rate per locus becomes 1×10^{-4}!

Two other studies involving individual loci should be quoted: Lefevre (1955), in a paper which contains an excellent discussion of the problem of estimating spontaneous mutation rates, reports that the rate of appearance of mutants with visible or lethal effects at the *y* locus is "about 1 per 75,000" (p. 379). On the other hand, Bonnier and Lüning (1949), in a paper criticized by Muller (1954) because the rate of recovery of spontaneous mutations ap-

[3] The question of the frequency of these "high mutation rate" lines has been discussed by Ives (1950; see also Neel, 1942), who suggests that the mutator genes which are responsible for these high rates "are the major cause of both gene mutations and inversions in natural populations" (p. 251).

peared to be too low in comparison with certain other findings, observed only one mutation at the white and forked loci among 153,579 flies tested for visible mutations, a rate of $.4 \times 10^{-5}$. Both of these estimates do not take into account the semi-lethal and deleterious mutations. The same conservative 1:5 ratio applied in the preceding paragraph would bring these estimates well within the range of the others quoted earlier.

Utilizing a somewhat different approach, Dobzhansky, Spassky, and Spassky (1952) have estimated the average rate of mutation to lethals, semi-lethals (1–20 per cent viability), and visibles *per lethal producing locus* in different species. These estimates, which they feel are more likely to be overestimates than underestimates, are reproduced in Table 15.2. Again, the estimate does not include the deleterious mutants.

TABLE 15.2 ESTIMATED AVERAGE MUTATION RATES PER LETHAL-PRODUCING LOCUS IN SEVERAL DROSOPHILA SPECIES
(After Dobzhansky, Spassky, and Spassky [1952]. These estimates are felt by the investigators to be more likely overestimates than underestimates.)

Species	Second chromosome	Third chromosome
D. melanogaster	1.1×10^{-5}	—
D. pseudo-obsura	—	1.1×10^{-5}
D. willistoni	2.2×10^{-5}	3.0×10^{-5}
D. prosaltans	1.1×10^{-5}	2.1×10^{-5}

In summary, then, it would appear that depending on one's view of the representativeness of the loci studied, and the problem of the relative frequency of mutations not detected by current techniques, there is room for a wide divergence of opinion concerning the average rate of mutation of Drosophila genes, with the range of possibilities perhaps extending from 0.5×10^{-5} to 5×10^{-5}. It is in our opinion not permissible to apply even this wide range of estimates directly to human problems, since even within species of the same genus mutation rates appear to differ significantly (cf. Table 15.2).

Turning now to mammals, we find that significant studies are available for only two species, the house mouse and man himself. The figures for the house mouse were derived in much the same fashion as the figures quoted for Drosophila, namely, through a search for mutant individuals among animals simultaneously heterozygous at multiple loci. Russell (1954), in connection with the important observations on radiation-induced mutations in the house mouse which we will refer to shortly, has now amassed the control data shown in Table 15.1. The rate of appearance of *visible* mutations in 265,076 locus tests was 0.8×10^{-5}. The observational error is of course large. Again it must be recognized that these tests detect only a fraction of the mutations occurring at these loci.

The available estimates for the frequency of occurrence in man of mutations with certain visible effects are shown in Table 15.3, prepared by Dr. T. E. Reed in collaboration with one of us (J.V.N.). Many of the problems involved in estimating human mutation rates have been discussed elsewhere (Haldane, 1948, 1949; Neel, 1952; Neel and Schull, 1954; Nachtsheim, 1954). Because of the particular problems associated with the study of incompletely recessive genes, no estimates based on such genes are included. The average of the estimates in Table 15.3 is approximately 3×10^{-5}. This estimate, now, is entirely limited to dominant mutations with visible effects. Even what would seem a very conservative allowance for recessive visibles and for the lethal, semi-lethal, and deleterious mutations would bring the average estimate up to 1×10^{-4} for these loci.

The representativeness of these estimates has been repeatedly challenged. There can be no doubt that there is definite selection in the loci studied. How this influences our estimates is not at all clear. As we have pointed out elsewhere (Neel and Schull, 1954), mutation at any particular locus may be thought of in terms of these aspects: (1) the frequency of mutation at that locus; (2) the number of alternative forms of the gene which may occur at any locus, i.e., the number of multiple alleles; and (3) the ease with which the effect associated with each of these multiple alleles can be detected. We assume that some loci are more mutable than others because we detect the results of mutation more frequently at these loci. However, making allowance for "unstable loci," the hypothesis has not been disproven that the inherent instability of all the genes is, by virtue of their biochemical complexity, very similar, but that the results of mutation are more readily detected at some loci than at others because of the role of that particular locus in the animal's physiology. It is entirely conceivable that the

loci[4] thus far selected for study in man are those at which a high proportion of all possible alleles at that locus results in readily detectable effects, but at which the per locus mutation rate is fairly representative of the human species. This problem is discussed further in Neel and Schull (1954).

For purposes of calculation, estimates of the rate of mutation of human genes have included 10^{-5} (Evans, 1949), 10^{-7} (Wright, 1950), and 2×10^{-5} (Muller, 1950; Slatis, 1955). In the current state of our knowledge students of the problem can select and justify estimates differing from one another by a factor of 100.

15.3 *The radiation-induced mutation rate in man.*

[4] In point of accuracy, we do not know but what any particular mutation rate study in man is detecting mutation at several loci.

— If we turn now to the estimation of the sensitivity of human genes to irradiation, we must first of all take cognizance of the same technical problems that exist in the estimation of spontaneous rates, as well as additional problems discussed by Muller (1954). For reasons brought out earlier, we are interested in effects on spermatogonia rather than mature sperm. For obvious reasons, no precise estimates are available for man. Until relatively recently, there was no mammalian material whatsoever. Now, however, Russell's (1954) figures, based on specific locus tests, indicate a visible mutation rate in spermatogonial cells of $(25.0 \pm 3.7) \times 10^{-8}$/gene/r. Russell has emphasized the danger of generalizing from these limited results, a point with which we heartily agree. But as long as these are the only figures available, they can-

TABLE 15.3 FREQUENCY OF OCCURRENCE OF NINE DIFFERENT DOMINANT OR SEX-LINKED MUTATIONS IN MAN

Dominant genes

Character	Method of estimation [a]	Mutations per gene per generation	Source
Epiloia	Direct	0.4–0.8×10^{-5}	M. Gunther and L. S. Penrose (1935, *J. Genet.* 31: 413), L. S. Penrose (1936, *Ann Eugen.* 7: 1).
Chondrodystrophy [b]	Direct	4.2×10^{-5} [c]	E. T. Mørch (1941, *Opera ex Domo Biologiae Hereditariae Humanae Universitatis Hafniensis*, Vol. 3).
	Indirect	4.3×10^{-5}	J. B. S. Haldane (1949, Proc. 8th Int. Cong. Genet., *Hereditas*, suppl. : 267) from data of Mørch (1941, vide supra).
	Direct	4.9×10^{-5} [c]	J. B. S. Haldane (1949, vide supra from data of Mørch, 1941, vide supra).
	Direct	7×10^{-5}	J. A. Böök (1952, *J. Genet. Humaine* 1: 24).
Pelger's nuclear anomaly	Direct	2.7×10^{-5}	H. Nachtsheim (1954, *Naturwiss.* 17: 385).
Aniridia	Direct	0.5×10^{-5}	C. J. Møllenbach (1947, *Opera ex Domo Biologiae Hereditariae Humanae Universitatis Hafniensis*, Vol. 15).
Retinoblastoma	Direct	1.4×10^{-5}	U. Philip and A. Sorsby (unpublished, quoted by Haldane, 1949, vide supra).
	Direct	2.3×10^{-5}	J. V. Neel and H. F. Falls (1951, *Science* 114: 419).
	Direct	4.3×10^{-6} [d]	F. Vogel (1954, *Ztschr. menschl. Vererbungs-u. Konstitutionslehre* 32: 308).
Waardenburg's syndrome	Direct	3.7×10^{-6}	P. J. Waardenburg (1951, *Amer. J. Hum. Genet.* 3: 195).
Neurofibromatosis	Direct	1.3–2.5×10^{-4}	F. W. Crowe, W. J. Schull, and J. V. Neel (1956, Multiple Neurofibromatosis, *American Lectures in Dermatology* series, C. C. Thomas).
	Indirect	0.8–1.0×10^{-4}	F. W. Crowe, W. J. Schull, and J. V. Neel, loc. cit.

TABLE 15.3—Continued
Sex-linked recessive genes

Character	Method of estimation [a]	Mutations per gene per generation	Source
Hemophilia	Indirect	3.2×10^{-5} [e]	J. B. S. Haldane (1947, *Ann. Eugen.* 13: 262) from data of M. Andreassen (1943, *Opera ex Domo Biologiae Hereditariae Humanae Universitatis Hafniensis*, Vol. 6).
Childhood progressive muscular dystrophy	Direct [f]	1×10^{-4}	F. E. Stephens and F. H. Tyler (1951, *Amer. J. Hum. Genet.* 3: 111).
	Indirect	1×10^{-4}	F. E. Stephens and F. H. Tyler (1951, loc. cit.).
	Indirect	4.5–6.5×10^{-5}	A. C. Stevenson (1953, *Ann. Eugen.* 18: 50).

[a] Estimation is considered to be direct when based on observed mutations, indirect when not so based. All indirect estimation makes use of estimates of the relative fitness and frequency at birth of the trait, and assumes that the population is in equilibrium.

[b] Slatis (1955, *Amer. J. Hum. Genet.* 7: 76) has suggested that these estimates may be spuriously high in this case because of some evidence for the occurrence of phenocopies.

[c] Independent estimates from different data contained in the paper of Mørch (1951).

[d] This estimate is based on the proposition that approximately 75 per cent of all sporadic cases of retinoblastoma are phenocopies. There are reasons to believe that this is an overestimate of the proportion of phenocopies. On the other hand, the estimates of Philip and Sorsby, and Neel and Falls, which treated all sporadic cases as due to mutation, are undoubtedly too high.

[e] This study may have included three distinct types of hemophilia: (a) classical sex-linked hemophilia resulting from deficiency of anti-hemophilic globulin (AHG), (b) a sex-linked clotting defect from lack of "plasma thromboplastin component" (PTC), and (c) an autosomally-inherited clotting defect from lack of "plasma thromboplastin antecedent" (PTA). Frick (1954, *J. Lab. & Clin. Med.* 43: 860) found that of 55 patients with coagulation defects, 45 had AHG deficiency, 6 had PTC deficiency, and 4 had PTA deficiency. Probably, therefore, Andreassen's study was mainly concerned with AHG deficiency; the mutation rate estimate for this deficiency may be about 15 per cent too high.

[f] Strictly, not a true direct estimate but an approximation which overestimates the mutation rate.

not help but strongly influence current thought. For what it is worth this average rate is about 15–20 times the figure of 1.5×10^{-8}/gene/r obtained for *Drosophila melanogaster* spermatogonia under comparable conditions (Alexander, 1954). Again, we emphasize that these estimates are undoubtedly based on the recovery of only a portion of the mutations occurring at these loci, with no assurance that the relative recovery rates are the same in the two species. Even so, taken at face value, the data would seem to indicate that mouse genes are decidedly more sensitive to irradiation than Drosophila genes. However, Ives (1954) has recently drawn attention to a number of factors which make it difficult to say precisely how much (if any) more sensitive mouse genes are than Drosophila genes, concluding that "for the present the radiation-induced mutation rate per r per locus appears to be similar in flies and mice" (p. 364). Although Ives' paper was written prior to the appearance of the paper by Alexander, quoted above, some of his objections would doubtless still stand, particularly the point concerning the disproportionate contribution of mutation at one particular locus to Russell's conclusions. Many of Ives' objections seem to have been met by a recent paper by Russell (1956). Early discussions of the genetic effects of radiation on human populations used an estimate of the frequency of induced mutation (3×10^{-8}/gene/r) based on work on mature Drosophila sperm (Evans, 1949; Wright, 1950); more recent treatments (Muller, 1955; Slatis, 1955) have used a figure (2×10^{-7}/gene/r) based on Russell's work on mouse spermatogonia, although Sturtevant (1955) prefers to retain the figures based on Drosophila.

In a discussion of radiation hazards, it is sometimes convenient to think in terms of the amount of radiation necessary to double the spontaneous mutation rate. This is obviously a function of the average spontaneous mutation rate per locus and the probability of mutation per locus per r. For purposes of discussion, Muller (1954; 1955), on the basis of Russell's (1954) work, has utilized figures ranging between 37 and 80r as the doubling dose. Haldane (1955), on the other hand, appears to feel that the doubling dose is closer to 3r, which would seem to imply a belief that the average spontaneous mutation rate per locus is in the neigh-

borhood of 1×10^{-6}. Westergaard (1955) also stresses the possibility that the doubling dose may be low, in the neighborhood of 3–6r.

15.4 *Estimates of the number of genes in man.* — The haploid gene number in Drosophila is commonly placed at 5,000 to 10,000 (cf. Muller, 1935). The existence of the relatively enormous salivary gland cell chromosomes in Drosophila, and the usual correspondence between a genetic locus and a visible "band" in these chromosomes, provides an unusual opportunity for gene number estimates in this species. No corresponding situation has been established in man or any other mammal. It is commonly argued that man, because of his greater physiological complexity, must have more genes than Drosophila. There exists only one piece of objective evidence in support of this argument. Spuhler (1948) has pointed out that the mean total length of the chromosomes in late prophase nuclei in man is some 8.5 times that of Drosophila. If one assumes that the spacing of the genes in the chromosomes is similar in the two species, this permits an estimate of the haploid gene number in man of between 42,000 and 85,000. Current treatments of the problem use estimates of the haploid gene number of from 100,000 (Evans, 1949) to 5,000 (Muller, 1950; Slatis, 1955). The data, then, permit competent students of the problem to make assumptions varying by a factor of 20.

15.5 *The "accumulation factor."* — The term "accumulation factor" is applied to the ratio of "recessive" genes already present in the population to those arising spontaneously each generation through mutation. Evans (1949), the first to make use of this factor in calculations, assumed, "somewhat arbitrarily," a value of 50. No discussion of the basis for this estimate is given. With his assumptions concerning average mutation rate (1×10^{-5}) and haploid gene number (1×10^5), the total mutation rate per zygote would be 2.0, and the average number of accumulated "recessive" genes per individual would be 100.

Muller (1950), on the other hand, after an extensive discussion of the evidence regarding the average degree of dominance of Drosophila genes, arrives at an "accumulation factor" — his p value — of 40, writing at the same time that there is "such a lamentable paucity of numerical values for human beings, or for any vertebrates, of a type which would throw light on the actual values of these factors, that we cannot feel too secure in regarding even 40 as a lower limit for p" (p. 141). The total mutation rate per gamete (μ_t), i.e., the product of gene number × average mutation rate, is set for purposes of calculation at 0.1. Introducing a factor of 2 to make allowance for the fact that selection probably operates predominantly on heterozygotes, the average number of "recessive" deleterious genes carried by a human being is placed at $2 \times 0.1 \times 40 = 8$. In a later calculation, Muller (1954) adopts a figure of 0.3 as the total mutation rate, and suggests a figure of 100 for p. This leads to an estimate of 60 for the average number of recessive deleterious genes for which each individual is heterozygous.

Slatis (1954) has attempted a direct calculation of the frequency of abnormal autosomal recessive genes in man, from a consideration of the outcome of first cousin marriages. Data are presented on 17 sibships, in nine of which defects occur that are attributed to recessive inheritance. This leads to the estimate that "the average person is heterozygous for eight abnormal genes" (p. 418). Although the method is a valuable contribution, the data are unfortunately so biased as to be practically worthless for a calculation of this type. As Slatis points out, five of these nine families were ascertained *because* of the abnormality. Furthermore, two of the four remaining traits (polydactyly, distal webbing of the digits) are frequently due to dominant genes of irregular penetrance. If one discards those sibships, the data remaining are insufficient for a calculation of this type. The extensive data on the children born to consanguineous parents which were collected in the course of the present study, although still incompletely analyzed, are at marked variance with those of Slatis as regards frequency of appearance of genetic defect in the offspring.

In his later discussion of the results of the induction of mutations by irradiation, Slatis (1955) uses this figure of 8 for the average number of recessive deleterious genes present in man. Muller's above-quoted estimate of 60 is dismissed as "obviously too high," on the basis of a calculation which assumes that each of these recessive genes has a clear-cut and readily demonstrable effect. While we hold no particular brief for either Evans' estimate of 50 nor Muller's of 60, Slatis' rejection of estimates of this magnitude appears to be based on the mis-

conception that all deleterious genes of any importance to the problem should result in clearcut, readily diagnosable defect.

15.6 *The nature of natural selection.* — We come now to the last of the factors which enter into attempts to quantitate the genetic effects of the irradiation of human populations. All treatments of this subject, implicitly or explicitly, make certain assumptions concerning the mode of operation of natural selection. This problem has been treated most extensively by Muller (1950, 1954). In his earlier treatment, Muller (1950) writes: "Moreover since, firstly, the extinction rate due to an individual locus is exceedingly small and since, secondly, in the great majority of cases the extinctions caused by any locus are probably only in small proportion (relatively to the whole) determinately connected with those caused by particular genes at other loci, we may tentatively and as a first approximation consider the genetic deaths associated with different loci as occurring independently" (p. 121). Later (p. 151) we read: "We have seen (p. 138) that if μ_t, the total mutation rate per gamete, is 0.1, and if the dominance is 'effective,' then n_t, the frequency of newly manifested cases per individual, is the same as the frequency of individuals eliminated at equilibrium under natural selection, and has a value only slightly less than 0.2, namely, 0.18, provided we assume (1) the independence of distribution and (2) the independence of detrimental action of the mutant genes. Although the second assumption is certainly inaccurate, as pointed out previously, nevertheless provisional consideration of the matter indicates that it is unlikely for the frequency and strength of synergistic action of mutant genes to be so great as to reduce the frequency of elimination from a value of 0.18 to below, say, 0.15." We interpret this passage to mean that, given Muller's assumptions, at genetic equilibrium, at a minimum, the reproductive performance of the population will be, as a result of *natural selection,* some 15 per cent below what it would have been in the absence of these mutations. Somewhat later, however, it is pointed out that, given the assumption of an average of 8 deleterious genes per individual, artificial selection which prohibited that 3 per cent of the population with the highest concentration of deleterious genes from breeding could balance the total mutation rate of 0.1 per individual which is assumed. Thus, the operation of natural selection is postulated to involve the elimination at equilibrium of approximately five times as many individuals as a "completely efficient" system of artificial selection.

As Neel and Falls (1951) have emphasized in considering this problem, the survival of an individual under competition is as a rule not determined by single genes but by constellations of genes. These constellations of genes may be considered as having either additive or non-additive effects — a variety of observations suggest that non-additive effects are rather common where combinations of mutant genes are concerned (e.g., in Drosophila, Neel, 1941, 1943; House, 1953a, b; in the guinea pig, Wright, 1949). Non-additive effects suggest that an individual with 12 deleterious "recessives" does not necessarily carry double the handicap of one with 6. The concept of synergistic gene action was given the term "dependent overlapping" by Muller (1954). It is to be distinguished from "independent overlapping," the latter illustrated by the case of an individual who meets genetic extinction through the effect of a given gene who would have met extinction anyway through the effect of one or more other independently acting genes which he also happened to carry. Muller (1954) in his most recent treatment of dependent overlapping modifies the earlier position stated in the preceding paragraph, and now considers it very unlikely that dependent overlapping "would reduce the elimination rate of individuals by a factor of more than 2 or 3," but after a consideration of the probable distribution of mutant genes in the individuals of a population, does not introduce this factor into his most recent calculation of the average number of deleterious genes per individual (60), although introduction of the factor would of course reduce the estimate to a half or third. It is to us not clear how our present knowledge permits assigning a definite numerical value to the role of dependent overlapping. This question must certainly be considered one of the most important in the field of natural selection.

The problem of defining the balance between selection and mutation is extremely complex. Thus, Muller (1950) on pages 149–50 argues as follows: "Whatever the values finally found, it is evident that the natural rate of mutation of man is so high, and his natural rate of repro-

duction so low, that not a great deal of margin is left for selection. Thus if μ_t has the minimal value of 0.1 ($n_t = 0.18$) an average reproductive rate of 2.4 children per individual would be necessary to compensate for individuals genetically eliminated, without taking any account whatever of all the deaths and failures to reproduce due to non-genetic causes. But when these are taken into account as well (even though we allow only that reduced number of them that occur under modern conditions) it becomes perfectly evident that the present number of children per couple cannot be great enough to allow selection to keep pace with a mutation rate of 0.1. If, to make matters worse, μ_t should be anything like as high as 0.5, a possibility that cannot yet be ignored, our present reproductive practices would be utterly out of line with human requirements."

Now, if at equilibrium, and invariably, over a long period of time, each member of one generation contributes genetically to, on the average, two members of the next, then there is no question whatever of a decrease in the size of the population. This condition is both necessary and sufficient to define the restriction on a population of unaltering size, and the mortality experience of the population — intrinsic or extrinsic in source — will effect no change in number provided the condition is met. Suppose, for example, there are N individuals in one generation. These N individuals will, under the condition, contribute $2N$ gametes to the next generation. Since the number of individuals equals $\frac{1}{2}$ the number of gametes, there will be N individuals in the next generation.[5]

Therefore, Muller's statement that there must be on the average 2.4 children per individual to compensate for those lost through selection is somewhat ambiguous, and his assertion that still further compensation is necessary for additional sources of mortality is even more difficult to interpret. Without meaning to belabor the point, suppose, to take a simple and extreme case, there are N individuals born into one generation, and $\frac{2}{3}$'s of them are completely sterile. If the condition for unaltering size, given in the above paragraph, is met, there will *still* be $2N$ gametes contributed to the next generation, and therefore N individuals.

[5] We are indebted to Dr. Robert Krooth for calling our attention to the line of reasoning developed in this and the following several paragraphs.

Let us accept the assumption that selection acts primarily through the heterozygote. If there are p_i heterozygotes at locus i (and let us consider but one mutant allele), and if W_i is the mean number of children to whom persons heterozygous at locus i contribute genetically ("fitness"), \overline{W} is the mean fitness of the whole population, μ_i is the mutation rate per chromosome per generation from the normal allele to the mutant gene, and if, finally, there are but N individuals in the population, then there will be Np_i heterozygotes in one generation who will contribute Np_iW_i gametes so that there will be $\frac{1}{2}Np_iW_i$ "affected" genes from "affected" parents. There will be $N\overline{W}$ loci (of the type in question) in the whole of the next generation, and of these (near enough) $N\overline{W}\mu_i$ will be represented by "affected" genes newly mutated from the wild type.

Since the population is in equilibrium, there will be no change in gene frequency from the first generation to the second, hence

$$\frac{N\overline{W}\mu_i + \frac{1}{2}Np_iW_i}{N\overline{W}} - \frac{Np_i}{2N} = 0, \text{ or}$$

$$N\overline{W}2\mu_i = N\overline{W}p_i - NW_ip_i.$$

Therefore, $2\mu_i$ is the *ratio* of gametes lost through selection acting on the parents of the second generation to the number of gametes actually present in the form of zygotes in the second generation.

Now in the passage quoted, Muller assumes that μ_t — a quantity which he considers to be roughly, though not exactly, equivalent to $2\sum_i \mu_i$ (the sum being over all loci) — is equal to 0.18. Thus, ignoring overlapping effects, as Muller does,[6] the ratio of gametes lost through selection to gametes actually realized in zygotes is 0.18. If W_o is the mean fitness which would prevail in the absence of this gametic loss, and if \overline{W} is the fitness which in fact does prevail, then

$$\mu_t = \frac{W_o - \overline{W}}{\overline{W}}.$$

Substituting 0.18 for μ_t and letting \overline{W} equal 2, the minimum fitness which will enable the population to maintain its numbers, we find $W_o = 2.36$. This then is presumably the origin of

[6] And we shall not in this section consider whether this is permissible.

Muller's 2.4 — it is a virtual fertility, not an actual one. It is the fertility which *would* prevail if there were no gametic loss due to mutation and if the population were just maintaining itself; it is not the fertility which must prevail to maintain the size of the population. If the *actual* fertility of the population were 2.4, the population of the United States would increase by 20 per cent within a single generation.

If now we substitute 2.4 for \overline{W}, the actual fertility, and 0.18 for μ_t, we find the virtual fertility of the population to be 2.832. Substituting $W_o = 2.832$ and $\overline{W} = 2$ into the formula, we can determine the maximum total mutation rate which would permit the population to maintain its numbers. This quantity equals 0.416, which is slightly more than twice as large as the figure given above for μ_t $(= 0.18)$. In other words, if the "average reproductive rate of 2.4 children per individual" actually did prevail, a human population could experience a *permanent doubling* of Muller's estimate of the total natural mutation rate without loss in number.

Finally, and in summary, if one really did take all the genetic *and* non-genetic factors into account in the computations, a minimum *virtual* fertility far higher than those considered above would result, but the population of the country would still maintain itself (and it is in fact increasing exponentially) provided \overline{W} averages 2, *even if* $\dfrac{\overline{W}}{W_o}$ averaged, say, but 10 per cent!

In addition to the question of the extent to which selection may be considered to involve single genes as contrasted to constellations of genes, and the relationship at equilibrium between fertility and mutation rate, there is perhaps an even more basic assumption as regards selection which enters into treatments of this problem. It is customary to assume that the mutations produced by irradiation have with very rare exceptions harmful effects; one accordingly calculates the increase in "undesirable recessive traits," "congenital malformations," etc. In point of fact, however, irradiation, insofar as it increases the appearance of mutations which would occur anyway, accelerates a process which modern biology accepts as the cornerstone of evolution. Although the results of mutation are from the standpoint of the organism usually unfortunate, they may sometimes be beneficial. Natural selection, it is assumed, sorts out the beneficial from the harmful. Treatments of the problem of the genetic effects of irradiation tend to assume that modern civilization has so blunted the effectiveness of selection that the ill-effects of radiation-induced mutation far outweigh the potential good. In view of our abysmal ignorance of the actual selective forces which have shaped mankind, there is no way to reach a valid opinion concerning the extent to which Western culture does in fact negate the selective mechanisms which are responsible for *Homo sapiens* as we know him today. There is also no real information on the extent to which certain characteristics which are of great importance to the species and which have a strong genetic component, such as intelligence, are maintained by "balanced polymorphic" systems (cf. Penrose, 1950), which systems would be relatively resistant to increased mutation pressure. Of all the gaps in the background information necessary to an accurate quantitative approach to the genetic effects of increased radiation, the deficiency of data on this point is perhaps the most striking.

Muller (1954, pp. 436–37) has spelled out the conditions under which radiation might speed the evolutionary advance of man (or any other animal) as follows: ". . . the first is that the spontaneous mutation rate should not be already so high that when irradiation is applied mutations occur too frequently to allow an equilibrium elimination rate and/or a genetic load low enough to be tolerated by the population. A second condition is that the advantageous mutants should multiply fast enough to replace the original type at a rate commensurate with their increased rate of origination. A third requirement is that the organism should not be at the limit of an evolutionary blind end, i.e., that pathways of advantageous change still remain open to it. Such opportunities will be present in greater abundance, allowing more of the mutations that occur to be helpful in the given situation, if the population has been placed in an environment, and subjected to conditions of living, somewhat different from those previously natural to it; for it must already have become so highly adapted to its natural conditions as to make further progress difficult. Advance is also achieved more readily if the population is one which has to some extent lost, through genetic changes or recombinations, its original nicety of adaptation. This may have

come about through the prior establishment of some more or less harmful mutations, the effects of which can now be overcome by reverse or counteracting mutations. Such prior retrogression is likely to have occurred if the given population has recently been derived from one or from a mixture of a few more or less inbred lines, or from relatively few progenitors; in that case, moreover, the population will also start out with unusually restricted genetic variability, which the application of radiation will tend to remedy."

With respect to the first two of these conditions, Muller argues that very probably "human beings are already near if not at or beyond the mutation rate which, in relation to their conditions of living and breeding, is the 'critical' one." With respect to the second condition, Muller contends that the reproductive differentials between members of the human species are insufficient to allow for the necessary rate of incorporation of advantageous mutations into the genotype. For both of these arguments, it is important to recognize that although they may be in essence correct, we are in no position to supply the precise figures on which quantitative treatments of the problem of the genetic risks of irradiation must be based. Finally, with respect to the third condition, one can argue that in view of the striking changes in the circumstances under which man lives which have occurred during the past century, man is entering a period of biological readjustment, during which he will need an increased store of genetic variability on which to draw. Wallace (1951, et seq.) has recently irradiated Drosophila populations under circumstances which, as Muller (1954) points out, fulfill these very conditions. At the end of 100 generations of exposure to 2,000r each generation, viability and fertility appeared unimpaired. While extrapolation to human populations can under no circumstances be justified, this does remain an illustration of the amount of irradiation which *under special circumstances* a species can tolerate.

In a recent paper, Keosian (1955) has taken the extreme point of view that on the basis of the existing evidence, it can actually be argued that increased radiation will lead to genetic betterment. The gist of the argument is that with respect to human populations much room exists for biological progress, so that "increased genetic variability can lead to an accelerated evolution along beneficial lines." This approach challenges the validity of the concept that man is near a "critical mutation load," and also assumes the normal operation of natural selection, or at least sufficient natural selection, plus improved medical care, that the potential beneficial effects of radiation-induced mutations will outweigh the ill effects.

In order to avoid possible misunderstanding, in closing this section we would emphasize that we do not mean, directly or by inference, to suggest that radiation will lead to the genetic betterment of human populations. Thus, quoting Keosian's speculations does not constitute endorsement of his position. Our purpose is solely to emphasize the present inadequacies in our knowledge of the operation of natural selection on human populations, inadequacies which permit widely different viewpoints.

15.7 *Concluding remarks.*— It seems wise at this point to phrase as succinctly as possible the present position of the authors regarding the genetic risks of radiation. On the basis of extensive plant and animal work, it seems reasonable to conclude that all levels of irradiation of human populations will result in mutation production. There is a high probability (but not certainty) that under the conditions of Western culture, such mutations will act to the detriment of the populations concerned. However, the present data do not permit, in the authors' opinion, a satisfactory quantitative approach to this problem.

Radiation is but one of a number of dysgenic influences at work in human populations. Two others commonly mentioned are war and differential birth rates. The quantification of these influences is just as unsatisfactory as that of radiation.

As coal and oil resources dwindle, increasing recourse will be had to atomic fuel as a source of energy. Our concern as to the undesirable consequences of increased radiation due to industrial installations, as well as other sources, *must* find a frame of reference. If it were possible to assess the sum total of the influences which would be considered dysgenic, does increased radiation of all types at the present and at foreseeable levels constitute 1%, 10%, or 50% of that total — and what is the size of the total?

It will seem to many that to attempt to phrase the problem in those terms is hopelessly im-

practical. At the moment, yes. But it is the belief of the authors that the stage is now set for very substantial advances in our knowledge of the genetics of man. Given the facilities and the investigators, great progress can be made towards a better understanding of the problem of the genetic effects of the irradiation of human populations, and, on the time scale of human evolution, within a relatively short period of time.

We have seen that great uncertainty exists concerning the value of each of the five factors which must be known to quantitate the genetic effects of the irradiation of human populations. The data presented in this monograph have *potential* value in the clarification of these factors. Given independent approaches which lead to the elucidation of, for example, gene number or gene sensitivity to irradiation, these data then enable the next step to be taken. At the moment, however, we are reluctant to infer more than that the data remove the remote possibility of a conspicuous sensitivity of human genes to irradiation (i.e., marked mutability). Even this conclusion is open to challenge, depending on the definition given "conspicuous." As the term is here employed, we mean only that under circumstances where a group of geneticists felt that a clear-cut demonstration of genetic radiation effects was unlikely (Genetics Conference, 1947), no effects have been conclusively demonstrated. Conversely, had clearly significant effects on sex-ratio and malformation and stillbirth frequencies emerged, these would, on the basis of current knowledge, have qualified as "conspicuous" effects. More specifically, we feel that the present data render it improbable that, as has been suggested by some (Haldane, 1955), human genes are so sensitive that as little as $3r$, or even $10r$, will double the present mutation rate, although it must be admitted that a rigorous demonstration of this belief would be difficult.

But, it may be argued, the urgency of the problem of setting "permissible" individual radiation dosages is such that we must be guided by the data at hand. There is no doubt concerning the urgency of the problem. There *is* doubt concerning the advisability of calculations which have the appearance of mathematical exactitude to persons not thoroughly indoctrinated in genetics and unfamiliar with the shaky basis of the primary assumptions. Exposure to irradiation of all types should undoubtedly be minimized until we have a clearer idea of just how harmful these effects are. Until the day of this better understanding, it is as unfortunate, on the one hand, to deny the possibility that low doses are dysgenic at all as it is, on the other hand, to assert that a serious threat to the genetic integrity of mankind is involved. The practical difficulty in this position is the apparent necessity of setting tolerance limits for, e.g., workers in situations involving the production of atomic energy, or military personnel involved in the use of atomic weapons. However, it is well worth bearing in mind that even if the present "limits" are ultimately found to be too high, there are few who would argue that in the period it takes to establish that fact with certainty, man will have suffered serious genetic harm. There is, on the other hand, the possibility that by refusing to be drawn into premature speculative calculations which in the nature of things will be "used" as soon as they have been set to paper, and by insisting on all possible occasions that the work that should be done actually be carried forward, the geneticist in the long run will arrive more quickly at the goal of a lasting, valid appraisal of this problem.

But although we entertain reservations concerning the inferences which can be drawn from the data presented in this report, this should not be construed to indicate doubts concerning the wisdom of collecting the material. A problem as complex as the evaluation of the genetic risks of human irradiation will only finally be solved through the combined efforts of many investigators. This study will have justified itself if in that final synthesis it proves of value.

APPENDIX

PROCEDURE FOR CODING GENETICS SHORT-FORM QUESTIONNAIRES

		Column
1.	*Genetics Questionnaire number*...	1–6

(First digit indicates location; subsequent five, the case number.)

 Hiroshima H 00000
 Kure K 00000
 Nagasaki N 00000

2. *Husband's Master File number*.... 7–12

(First digit indicates location.)

 Hiroshima 2 00000
 Kure 5 00000
 Nagasaki 0 00000

3. *Wife's Master File number*....... 13–18

(First digit indicates location.)

4. *Age* 19–22

(Coded as 00 to 99, two columns for husband, two for wife.)

5. *In Hiroshima or Nagasaki at time of bombing* 23

(YES or NO for husband and wife.)

 1 Neither
 2 Husband only in Hiroshima
 3 Wife only in Hiroshima
 4 Both in Hiroshima
 5 Husband only in Nagasaki
 6 Wife only in Nagasaki
 7 Both in Nagasaki
 8 Husband in Hiroshima; wife in Nagasaki
 9 Wife in Hiroshima; husband in Nagasaki

6. *Distance from hypocenter*........ 24–27

(In hundreds of meters, with maximum coded 98 and final entry being 99 or greater; two columns for husband, two for wife.)

7. *Indoors or outside*............. 28

(Coded as both if only one exposed.)[1]

Neither sheltered by hill
 2 Both outside
 3 Husband outside, wife indoors
 4 Wife outside, husband indoors
 5 Both indoors

Husband sheltered by hill
 B Both outside
 C Husband outside, wife indoors
 D Wife outside, husband indoors
 E Both indoors

Wife sheltered by hill
 K Both outside
 L Husband outside, wife indoors
 M Wife outside, husband indoors
 N Both indoors

Both sheltered by hill
 S Both outside
 T Husband outside, wife indoors
 U Wife outside, husband indoors
 V Both indoors

8. *Type of building*.............. 29

(Answered as Japanese or other, the latter including concrete buildings of all types, brick, air raid shelters, etc. Coded as both if only one exposed.)

 1 Husband in Japanese type bldg., wife outdoors

[1] It should be noted that in this and a number of items to follow, special provision is not made for the case where only one parent is exposed. When this obtained, the unexposed parent was treated as having given the same response as the exposed parent. Thus, if the husband were exposed and the wife not, and if the husband asserted that he was outdoors at the time of the bombing, item 7 would be coded as 2, namely, both outside. No confusion should exist between this situation and the one where both parents were in fact exposed and outside because of items 5 and 6 in the code.

		Column
2	Husband in other type bldg., wife outdoors	
3	Wife in Japanese type bldg., husband outdoors	
4	Wife in other type bldg., husband outdoors	
5	Both in Japanese type bldg.	
6	Both in other type bldg.	
7	Husband in Japanese type bldg., wife in other type	
8	Husband in other type, wife in Japanese type bldg.	

9. *Petechiae* 30
 (Coded as both if only one exposed. Coded as:)
Husband	Wife
1 yes	yes
2 yes	no
3 no	yes
4 no	no

10. *Gingivitis* 31
 (Coded as both if only one exposed. Coded as:)
Husband	Wife
1 yes	yes
2 yes	no
3 no	yes
4 no	no

11. *Bloody diarrhea* 32
 (Coded as both if only one exposed. Coded as:)
Husband	Wife
1 yes	yes
2 yes	no
3 no	yes
4 no	no

12. *Epilation* 33
 (Coded as both if only one exposed. Coded as:)
Husband	Wife
1 yes	yes
2 yes	no
3 no	yes
4 no	no

13. *Fever* 34
 (Coded as both if only one exposed. Coded as:)
Husband	Wife
1 yes	yes
2 yes	no
3 no	yes
4 no	no

14. *Burns* 35
 (Coded as both if only one exposed. Coded as:)
Husband	Wife
1 yes	yes
2 yes	no
3 no	yes
4 no	no

15. *Trauma* 36
 (Coded as both if only one exposed. Coded as:)
Husband	Wife
1 yes	yes
2 yes	no
3 no	yes
4 no	no

16. *Total duration of cohabitation*.... 37–39
 (In months coded from 1 to 999.)

17. *Months cohabitation prior to August 1945* 40–42
 (Coded from 1 to 999.)

18. *Conceptions prior to August 1945*. 43–44
 (Coded from 0 to 99.)

19. *Stillbirths prior to August 1945*... 45
 (Coded from 0 to 9.)

20. *Therapeutic abortions prior to August 1945* 46
 (Coded as 0 to 9.)

21. *Months cohabitation after August 1945* 47–49
 (Coded from 1 to 999.)

22. *Conceptions after August 1945*... 50–51
 (Coded as 1 to 99.)

23. *Stillbirths after August 1945*..... 52
 (Coded as 0 to 9.)

24. *Therapeutic abortions after August 1945* 53
 (Coded as 0 to 9.)

25. *Total conceptions* 54–55
 (Coded as 0 to 99.)

26. *Total stillbirths* 56
 (Coded as 0 to 9.)

27. *Total therapeutic abortions*....... 57
 (Coded as 0 to 9.)

28. *Consanguinity* 58
 (Coded as:)
 1 None
 2 First cousins
 3 One and one-half cousins
 4 Second cousins
 5 Other
 9 Unknown

29. *Date of termination*............ Column 59–60

Month		Year	
January	1	1948	1
February	2	1949	2
March	3	1950	3
April	4	1951	4
May	5	1952	5
June	6	1953	6
July	7	1954	7
August	8		
September	9		
October	0		
November	X		
December	V		

30. *Duration of pregnancy in weeks*... 61–62
(Coded as 00 to 99, unknown is 99.)

31. *Onset of labor*............... 63
(Coded as:)
1 Spontaneous less than 24 hours
2 Spontaneous more than 24 hours
3 Induced less than 24 hours
4 Induced more than 24 hours
5 Operative

32. *Course of labor*............... 64
(Coded as:)
1 Normal
2 Operative
3 Instrumental
4 Miscellaneous complications

33. *Termination* 65
(Coded as:)
1 Livebirth after 38th week
2 Premature birth under or including 38th week
3 Stillbirth 20 weeks or under
4 Stillbirth 21–29 weeks
5 Stillbirth 30–38 weeks
6 Stillbirth after 38th week
7 Unknown livebirth
8 Unknown stillbirth

34. *Multiple birth* 66
(Coded as:)
1 Single birth
2 First twin
3 Second twin
4 First triplet
5 Second triplet
6 Third triplet

35. *Sex of newborn*................ 67
(Coded as:)
1 Male
2 Female
3 Undetermined

36. *Weight of newborn*............ 68–70
(In tens of grams. 999 for unknown.)

37. *Abnormality* 71
(Coded as:)
1 Normal
2 Congenital disease
3 Birth injury
4 Major malformation
5 Minor malformation
6 Congenital disease, birth injury
7 Congenital disease, major malformation
8 Congenital disease, minor malformation
9 Birth injury, major malformation
0 Birth injury, minor malformation
X Congenital disease, birth injury, major malformation
V Congenital disease, birth injury, minor malformation

38. *Type of abnormality*........... 72–77
(Classified by code. See Part II.)

39. *Neonatal death* 78–79
(Coded as age at death.)
00 Under 1 day
01 1 day
02 2 days
03 3–6 days
04 7–13 days (1 week)
05 14–20 days (2 weeks)
06 21–27 days (3 weeks)
07 1 month
08 2 months etc. to 11 months

40. *Other pregnancies since 1948*..... 80
(Number for which questionnaire was completed.)

Appendix

CODING INSTRUCTIONS FOR GENETICS FOLLOW-UP QUESTIONNAIRES

	Column
1. *Genetics number*	1–6

(As item 1, columns 1 to 6, of Genetics Short-Form Code, i.e., first digit indicates location, subsequent five, the case number.)

2. *Parental radiation history* 7–14

(As items 5, 6, 9, 10 and 12, columns 23, 24–27, 30, 31, and 33 of the Genetics Short-Form Code.)

(a) *Location:*
(In Hiroshima or Nagasaki at time of bombing. Yes or no for husband and wife.)
1. Neither
2. Husband only in Hiroshima
3. Wife only in Hiroshima
4. Both in Hiroshima
5. Husband only in Nagasaki
6. Wife only in Nagasaki
7. Both in Nagasaki
8. Husband in Hiroshima; wife in Nagasaki
9. Wife in Hiroshima; husband in Nagasaki

(b) *Distance:*
(Distance from hypocenter in hundreds of meters, with maximum coded 98 and final entry being 99 or greater, two columns for husband, two for wife.)

(c) *Petechiae:*
(Coded as both if only one exposed.)

	Husband	Wife
1	yes	yes
2	yes	no
3	no	yes
4	no	no

(d) *Gingivitis:*
(Coded as both if only one exposed.)

	Husband	Wife
1	yes	yes
2	yes	no
3	no	yes
4	no	no

(e) *Epilation:*
(Coded as both if only one exposed.)

	Husband	Wife
1	yes	yes
2	yes	no
3	no	yes
4	no	no

3. *Date of child's birth* 15–16

(As item 29, columns 59–60 of Genetics Short-Form Code. Coded as:)

Month		Year	
January	1	1948	1
February	2	1949	2
March	3	1950	3
April	4	1951	4
May	5	1952	5
June	6	1953	6
July	7	1954	7
August	8		
September	9		
October	0		
November	X		
December	V		

4. *Duration of pregnancy in weeks* ... 17–18

(As item 30, columns 61–62 of Genetics Short-Form Code. Coded as 00 to 99; 99 is unknown.)

5. *Onset of labor* 19

(As item 31, column 63 of Genetics Short-Form Code. Coded as:)
1. Spontaneous less than 24 hours
2. Spontaneous more than 24 hours
3. Induced less than 24 hours
4. Induced more than 24 hours
5. Operative

6. *Termination* 20

(As item 33, column 65 of Genetics Short-Form Code. Coded as:)
1. Livebirth after 38th week
2. Premature birth under or including 38th week
3. Stillbirth 20 weeks or under
4. Stillbirth 21–29 weeks
5. Stillbirth 30–38 weeks
6. Stillbirth after 38th week
7. Unknown livebirth
8. Unknown stillbirth

		Column
7.	*Sex*	21

(As item 35, column 67 of Genetics Short-Form Code. Coded as:)
1. Male
2. Female
3. Unknown

		Column
8.	*Weight*	22–24

(As item 36, column 68 to 70, of Genetics Short-Form Code. Weight of newborn coded in tens of grams. 999 for unknown.)

(NOTE: Items 1 to 8 were punched on the Long Form IBM card by machine transfer from the Short-Form IBM card.)

		Column
9.	*Multiple birth*	25

(Coded as:)
1. Single birth
2. First twin
3. Second twin
4. First triplet
5. Second triplet
6. Third triplet

		Column
10.	*Abnormality*	26

(Coded as:)
- A Normal to clinical examination and autopsy
- 1 Normal to clinical examination
- 2 Congenital disease (including autopsy data)
- 3 Birth injury (clinical diagnosis *only*)
- 4 Major malformation — clinical or autopsy diagnosis
- 5 Minor malformation — clinical or autopsy diagnosis
- 6 Congenital disease, birth injury
- 7 Congenital disease, major malformation
- 8 Congenital disease, minor malformation
- 9 Birth injury, major malformation
- 0 Birth injury, minor malformation
- X Congenital disease, birth injury, major malformation
- V Congenital disease, birth injury, minor malformation

		Column
11.	*Clinical and autopsy diagnosis*	27–44

(Maximum of three diagnoses of 6 digits; first digit of each diagnosis coded as:)
1. Clinical diagnosis of defect limited to one region or system
2. Clinical diagnosis of multiple defects, with most important defect coded first, and successive diagnoses introduced by a 1...
3. Clinical diagnosis of a complex and ill-defined defect
4. Clinical diagnosis of a double monster
- A Autopsy confirmation of 1.., coded as A followed by five diagnostic digits
- B Autopsy confirmation of 2.., coded as B followed by five diagnostic digits; successive diagnoses as A...
- C Autopsy confirmation of 3.., coded as C followed by five diagnostic digits
- D Autopsy confirmation of 4.., coded as D followed by five diagnostic digits
- J Autopsy diagnosis alone of a 1.., coded as J followed by five diagnostic digits
- K Autopsy diagnosis alone of a 2.., coded as K followed by five diagnostic digits; successive diagnoses as J...
- L Autopsy diagnosis alone of a 3.., coded as L followed by five diagnostic digits
- M Autopsy diagnosis alone of a 4.., coded as M followed by five diagnostic digits
- S Autopsy fails to confirm a clinical diagnosis of a 1.., coded as S followed by five diagnostic digits
- T Autopsy fails to confirm a clinical diagnosis of a 2.., coded as T followed by five diagnostic digits; successive diagnoses as S...

U Autopsy fails to confirm a clinical diagnosis of a 3.., coded as U followed by five diagnostic digits

V Autopsy fails to confirm a clinical diagnosis of a 4.., coded as V followed by five diagnostic digits

12. *Neonatal and infant death* 45–46
 (Coded as age at death; 99 is unknown.)
 - 00 Under 1 day
 - 01 1 day
 - 02 2 days
 - 03 3–6 days
 - 04 7–13 days
 - 05 14–20 days
 - 06 21–27 days
 - 07 1 month
 - 08 2 months
 - 09 3 months
 - etc. to 11 months

13. *Age of father at termination of pregnancy* 47–48
 (Coded as 00 to 99. Unknown coded as 99.)

14. *Family history of father* 49
 (Coded as:)
 1 Negative family history
 2 Father alone with same defect
 3 Father alone with other defect
 4 Other paternal relatives with same defect
 5 Other paternal relatives with other defect
 6 Father and other paternal relatives with same defect
 7 Father same defect, paternal relatives with other defect
 8 Father with other defect, paternal relatives with same defect
 9 Father and paternal relatives with other defect
 0 Unknown

 (NOTE: Same or other defect refers to the same or a different defect from that present in the infant.)

15. *Consanguinity of parents* 50
 (Coded as:)
 1 None
 2 First cousins
 3 One and one-half cousins
 4 Second cousins
 5 Other
 6 Unknown

16. *Consanguinity of grandparents* 51
 (Coded as:)
 1 Both negative
 2 Paternal positive, maternal negative
 3 Paternal negative, maternal positive
 4 Paternal and maternal positive
 5 Paternal unknown, maternal positive
 6 Paternal unknown, maternal negative
 7 Paternal positive, maternal unknown
 8 Paternal negative, maternal unknown
 9 Paternal unknown, maternal unknown

17. *Economic status* 52
 (Coded as:)
 1 Very poor
 2 Poor
 3 Average
 4 Well-to-do
 5 Rich
 6 Not indicated

18. *Age of mother at termination of pregnancy* 53–54
 (Coded as 00 to 99. Unknown coded as 99.)

19. *Family history of mother* 55
 (Coded as:)
 1 Negative family history
 2 Mother alone with same defect
 3 Mother alone with other defect
 4 Other maternal relatives with same defect
 5 Other maternal relatives with other defect
 6 Mother and other maternal relatives with same defect
 7 Mother with same defect, maternal relatives with other defect
 8 Mother with other defect, maternal relatives with same defect

9 Mother and maternal relatives with other defect
0 Unknown
(NOTE: Same or other defect refers to the same or a different defect from that present in the infant.)

20. *Menstrual history*
 1 Age of onset in years to nearest year, coded as 00 to 99; 99 is unknown.......... 56–57
 2 Duration in days, coded as 0 to 9 (no entry is unknown; greater than 9 coded as 9).. 58
 3 Interval in days between periods coded as 00 to 39 (coded as 00 if entry is "irregular," 99 if unknown).......... 59–60

21. *Irregularity of menstruation*...... 61
 (Coded as:)
 1 None
 2 Periods of amenorrhea exceeding 90 days
 3 Periods of amenorrhea of less than 90 days
 4 Cycles of 3 weeks or less
 5 Ill-defined or nature not stated

22. *Pelvic operations, treatment, or diseases* 62
 (Coded as:)
 1 None
 2 Pelvic operations, abdominal approach
 3 Pelvic operations, D and C
 4 Radium or X-ray therapy
 5 Other
 6 Pelvic operations, D and C, and abdominal
 7 Unknown

23. *Illness during pregnancy*......... 63
 (Coded as:)
 1 None
 2 Febrile disease, clearly defined
 3 Other illness, may be febrile but no clear indication
 4 Unknown

24. *Uterine bleeding during first six months of pregnancy*........... 64
 (Coded as:)
 1 None
 2 Present
 3 Unknown

25. *Presentation* 65
 (Coded as:)
 1 Occiput
 2 Face
 3 Breech
 4 Footling
 5 Arm
 6 Cesarean section
 7 Unknown

26. *Maternal and infant serology*..... 66
 (NOTE: For Nagasaki, the prepartum Hokensho serological tests are coded as the alphabetical equivalents of 1 to 8. The prepartum serology is to be coded if postpartum testing by ABCC did not occur.)
 1 Mother negative, child untested
 2 Mother negative, child negative
 3 Mother negative, child positive
 4 Mother positive, child untested
 5 Mother positive, child negative
 6 Mother positive, child positive
 7 Mother and child untested
 8 Mother doubtful, child untested

27. *Total number of pregnancies this marriage* 67–68
 (Coded as 00 to 99. Multiples scored as one pregnancy; 99 is unknown.)

28. *Number of liveborn infants*...... 69–70
 (Coded as 00 to 99. Multiples scored as 2, 3, etc., or in combination.)

29. *Number of premature infants*..... 71
 (Liveborn or stillborn exclusive of abortions, coded as 0 to 9. Multiples scored as 2, 3, etc., or in combination.)

30. *Number of stillborn infants*...... 72
 (Coded as 0 to 9. Multiples scored as 2, 3, etc., or in combination; artificial interruption after 20 weeks ending in stillborn fetus is scored as stillborn infant.)

Appendix

	Column		Column
31. *Number of pregnancies terminating as spontaneous abortions, i.e., gestation of less than or equal to 20 weeks* . (Coded as 0 to 9. Multiples scored as 2, 3, etc., or in combination.)	73	36. *Relationship of informant to infant* . (Coded as:) 1 Mother 2 Father 3 Maternal grandmother 4 Maternal grandfather 5 Paternal grandmother 6 Paternal grandfather 7 Not related 8 Other 9 Unknown A Father and mother B Other combination	80
32. *Number of pregnancies terminating as induced abortions, i.e., gestation of less than or equal to 20 weeks*. . (Coded as 0 to 9.)	74		
33. *Number of pregnancies terminating as multiple births*. (Coded as 0 to 9.)	75		
34. *Number of infants dying before one year of age*. (Coded as 0 to 9.)	76		
35. *Defective terminations in sibs*.	77–79		
(Coded as:)			

	Code Column 77	Code Column 78	Code Column 79
Child abnormal clinically*; negative sibship history	1		
Child abnormal; negative sibship history with respect to this defect but other defect	2	No. of sibs with other *major* defect, 1...9	
(NOTE: A history of another minor defect is scored as 2, column 77, with no entry in columns 78–79.)			
Child abnormal; positive sibship history with respect to this defect	3	No. of sibs with same defect, including this sibling, 1...9	
Child abnormal; positive sibship history with respect to this defect and other defect present in sibship	4	No. of sibs with same defect, including this sibling, 1...9	No. of sibs with other *major* defect, 1...9
Child normal; sibship history negative	5		
Child normal; sibship history positive for defect	6	No. of sibs with *major* defect, 1...9	

(NOTE: In the event a previous Long Form exists on this family, it will be indicated by an X in column 79 if some number other than 4 is entered in column 77. If 4 is entered in column 77, then M is to be punched in column 79.)

* The term "abnormal" as used here does not apply to stillbirths or abortions, but only to visible defects.

CODING INSTRUCTIONS FOR PEDIATRICS FOLLOW-UP QUESTIONNAIRE

	Column		Column
1. *Genetics Questionnaire number*. . . (As item 1, columns 1–6 of Genetics Short-Form Code. Coded as: first digit indicates location; subse-	1–6	quent five, the case number.) Hiroshima H 00000 Nagasaki N 00000 Kure K 00000	

THE CHILDREN OF ATOMIC BOMB SURVIVORS

		Column
2.	*Mother's Master File number*	7–12

(Coded as: first digit indicates location.)

 Hiroshima 2 00000
 Nagasaki 0 00000
 Kure 5 00000

3. *Parental age at time of termination.* 13–16
 (As calculated from Genetics Short-Form Code. Coded as 00 to 99; 99 is unknown; two columns for husband, two for wife.)

4. *Parental radiation history* 17–30
 (As items 5–15, columns 23–36, of Genetics Short-Form Code. Coded as:)

 (a) *Location:*
 (In Hiroshima or Nagasaki at time of bombing; yes or no for husband and wife. Coded as:)
 1 Neither
 2 Husband only in Hiroshima
 3 Wife only in Hiroshima
 4 Both in Hiroshima
 5 Husband only in Nagasaki
 6 Wife only in Nagasaki
 7 Both in Nagasaki
 8 Husband in Hiroshima; wife in Nagasaki
 9 Wife in Hiroshima; husband in Nagasaki

 (b) *Distance:*
 (From hypocenter in hundreds of meters, with maximum coded as 98 and final entry being 99 or greater, two columns for husband, two for wife.)

 (c) *Indoors or outside:*
 (Coded as both if only one exposed.)
 Neither sheltered by hill
 2 Both outside
 3 Husband outside, wife indoors
 4 Wife outside, husband indoors
 5 Both indoors
 Husband sheltered by hill
 B Both outside
 C Husband outside, wife indoors
 D Wife outside, husband indoors
 E Both indoors
 Wife sheltered by hill
 K Both outside
 L Husband outside, wife indoors
 M Wife outside, husband indoors
 N Both indoors
 Both sheltered by hill
 S Both outside
 T Husband outside, wife indoors
 U Wife outside, husband indoors
 V Both indoors

 (d) *Type of building:*
 (Answered as Japanese or other, the latter including concrete buildings of all types, brick, air raid shelters, etc. Coded as both if only one exposed. Coded as:)
 1 Husband in Japanese type bldg., wife outdoors
 2 Husband in other type bldg., wife outdoors
 3 Wife in Japanese type bldg., husband outdoors
 4 Wife in other type bldg., husband outdoors
 5 Both in Japanese type bldg.
 6 Both in other type bldg.
 7 Husband in Japanese type bldg., wife in other type
 8 Husband in other type bldg., wife in Japanese type

 (e) *Petechiae:*
 (Coded as both if only one exposed. Coded as:)

	Husband	Wife
1	yes	yes
2	yes	no
3	no	yes
4	no	no

Appendix

(f) *Gingivitis:*
(Coded as both if only one exposed. Coded as:)

	Husband	Wife
1	yes	yes
2	yes	no
3	no	yes
4	no	no

(g) *Bloody diarrhea:*
(Coded as both if only one exposed. Coded as:)

	Husband	Wife
1	yes	yes
2	yes	no
3	no	yes
4	no	no

(h) *Epilation:*
(Coded as both if only one exposed. Coded as:)

	Husband	Wife
1	yes	yes
2	yes	no
3	no	yes
4	no	no

(i) *Fever:*
(Coded as both if only one exposed. Coded as:)

	Husband	Wife
1	yes	yes
2	yes	no
3	no	yes
4	no	no

(j) *Burns:*
(Coded as both if only one exposed. Coded as:)

	Husband	Wife
1	yes	yes
2	yes	no
3	no	yes
4	no	no

(k) *Trauma:*
(Coded as both if only one exposed. Coded as:)

	Husband	Wife
1	yes	yes
2	yes	no
3	no	yes
4	no	no

5. *Date of birth* Column 31–32
 (As item 29, columns 59–60 of Genetics Short-Form Code. Coded as:)

Month		Year	
January	1	1948	1
February	2	1949	2
March	3	1950	3
April	4	1951	4
May	5	1952	5
June	6	1953	6
July	7	1954	7
August	8		
September	9		
October	0		
November	X		
December	V		

6. *Duration of pregnancy in weeks* ... 33–34
 (As item 30, columns 61–62 of Genetics Short-Form Code. Coded as 00 to 99; 99 is unknown.)

7. *Course of labor* 35
 (As item 32, column 64 of Genetics Short-Form Code. Coded as:)
 1. Normal
 2. Operative
 3. Instrumental
 4. Miscellaneous complications

8. *Termination* 36
 (As item 33, column 65 of Genetics Short-Form Code. Coded as:)
 1. Livebirth after 38th week
 2. Premature birth under or including 38th week
 3. Stillbirth 20 weeks or under
 4. Stillbirth 21–29 weeks
 5. Stillbirth 30–38 weeks
 6. Stillbirth after 38th week
 7. Unknown livebirth
 8. Unknown stillbirth

9. *Sex* 37
 (As item 35, column 67 of Genetics Short-Form Code. Coded as:)
 1. Male
 2. Female
 3. Undetermined

10. *Weight* 38–40
 (As item 36, columns 68–70, of Genetics Short-Form Code. Weight of newborn coded in tens of grams; 000 to 999; 999 is unknown.)
 (NOTE: Items 1 to 10 are to be a machine transfer from the Short-Form IBM card to the Pediatrics Follow-up IBM card.)

11. *Master File number* 41–46
 (Of infant as given on Pediatrics Follow-up form.)

	Column
12. *Age*	47–48

(At examination in months given on Pediatrics Follow-up form. Coded as 00 to 99; 99 is unknown.)

13. *Multiple birth* 49

(Coded as:)
1. Single birth
2. First twin
3. Second twin
4. First triplet
5. Second triplet
6. Third triplet

14. *Abnormality* 50

(As diagnosed either on home visit [initial], or when child is brought to clinic for confirmation of a diagnosis of malformation, or on 9-months follow-up but not autopsy diagnosis.)
1. Normal
2. Congenital disease
3. Birth injury
4. Major malformation
5. Minor malformation
6. Congenital disease, birth injury
7. Congenital disease, major malformation
8. Congenital disease, minor malformation
9. Birth injury, major malformation
0. Birth injury, minor malformation
X. Congenital disease, birth injury, major malformation
V. Congenital disease, birth injury, minor malformation

15. *Diagnosis on home visit and follow-up* 51–74

(Maximum of four diagnoses each of six digits, first digit coded as:)
1. Home visit and/or Pediatrics Clinic diagnosis at early age of defects limited to one system or region, confirmed on 9-months follow-up.
2. Home visit and/or Pediatrics Clinic diagnosis at early age of multiple defects, with most important defect coded first, and successive defects introduced by a 1..., confirmed on 9-months follow-up.
3. Home visit and/or Pediatrics Clinic diagnosis at early age of a complex and ill-defined defect, confirmed on 9-months follow-up.
4. Home visit and/or Pediatrics Clinic diagnosis at early age of a double monster or teratoma, confirmed on 9-months follow-up.
A. Home visit and/or Pediatrics Clinic diagnosis at early age of a 1..., not confirmed on 9-months follow-up (child alive).
B. Home visit and/or Pediatrics Clinic diagnosis at early age of a 2..., not confirmed on 9-months follow-up (child alive); successive diagnoses introduced by A....
C. Home visit and/or Pediatrics Clinic diagnosis at early age of a 3..., not confirmed on 9-months follow-up (child alive).
D. Home visit and/or Pediatrics Clinic diagnosis at early age of a 4..., not confirmed on 9-months follow-up (child alive).
J. Home visit and/or Pediatrics Clinic diagnosis at early age of a 1..., cannot now be confirmed because child dead.
K. Home visit and/or Pediatrics Clinic diagnosis at early age of a 2..., cannot now be confirmed because child dead; successive diagnoses introduced by J....
L. Home visit and/or Pediatrics Clinic diagnosis at early age of a 3..., cannot now be confirmed because child dead.
M. Home visit and/or Pediatrics Clinic diagnosis at early age of a 4..., cannot now be

Appendix

Column

 confirmed because child dead.
- S Nine-months follow-up diagnosis of a defect limited to one region or system not previously diagnosed.
- T Nine-months follow-up diagnosis of multiple defects not previously diagnosed, most important coded first with diagnoses after the first introduced by S....
- U Nine-months follow-up diagnosis of a complex and ill-defined defect not previously diagnosed.
- V Nine-months follow-up diagnosis of a double monster or teratoma not previously diagnosed.

16. *Infant and neonatal mortality*..... 75–76
 (Coded as:)
 - 00 Under 1 day
 - 01 1 day
 - 02 2 days
 - 03 3–6 days
 - 04 7–13 days
 - 05 14–20 days
 - 06 21–27 days
 - 07 28–55 days
 - 08 56–83 days
 - 09 84–111 days
 - 10 112–139 days
 - 11 140–167 days
 - 12 168–195 days
 - 13 196–223 days
 - 14 224–251 days

Column

 - 15 252–279 days
 - 16 280–307 days
 - 17 308–325 days
 - 18 326–353 days
 - 19 Over 1 year
 - 99 Unknown — (see L. F. code)

17. *No punch — seen in clinic*....... 77
 (Coded as:)[1]
 1. Dead on L. F.
 2. Found dead by P. C.
 3. Moved or unable to locate
 4. Infant ill 3 successive weeks
 5. Mother ill 3 successive weeks
 6. Both or other P or F
 7. Other reasons

18. *Consanguinity* 78
 (Coded as:)
 1. Not related
 2. First cousins
 3. One and one-half cousins
 4. Second cousins
 5. Other
 6. Unknown

19. *Exposure category* 79–80
 Father's Column 79
 Mother's Column 80

[1] The purpose of this item in the code is severalfold; but primarily it is designed to detect biases which might arise due to unequal attrition ratio among the exposure subclasses. It also affords some measure of the accuracy of the reports of death. Selection was by terminal registration digit and hence some of the infants selected were known to be stillborn or to have died during the neonatal period. For the latter cases, the statement at 9 months could be collated with the information available on the at-birth records.

REFERENCES

ALEXANDER, M. L. 1954. Mutation rates at specific autosomal loci in the mature and immature germ cells of *Drosophila melanogaster*. *Genetics* **39**:409–428.

ANTONOV, A. N. 1947. Children born during the siege of Leningrad in 1942. *J. Pediat.* **30**:250–259.

ATOMIC BOMB CASUALTY COMMISSION SEMI-ANNUAL REPORT. January 1, 1952–June 30, 1952. Appendix 18.

ATOMIC BOMB CASUALTY COMMISSION SEMI-ANNUAL REPORT. January–June, 1954.

BALFOUR, M. I. 1944. Supplementary feeding in pregnancy. *Lancet* **1**:208–210.

BARTLETT, M. S. 1935. Contingency table interactions. *J. Roy. Stat. Soc.* **2**:248–252.

BEARDSLEY, R. K. 1955. Japan before history: a survey of the archaeological record. *Far Eastern Quarterly* **14**:317–346.

BERNSTEIN, M. E. 1948. Changes in sex ratio, upper social strata. *Human Biol.* **20**:182–194.

BONNIER, G., and LÜNING, K. G. 1949. Studies on X-ray mutations in the white and forked loci of *Drosophila melanogaster*. I. A statistical analysis of mutation frequencies. *Hereditas* **35**:163–189.

BOX, G. E. P. 1949. A general distribution theory for a class of likelihood criteria. *Biometrika* **36**:317–346.

BOX, G. E. P. 1953. Non-normality and tests on variances. *Biometrika* **40**:318–335.

BOX, G. E. P. 1954a. Some theorems on quadratic forms applied in the study of analysis of variance problems. I. Effect of inequality of variance in the one-way classification. *Ann. Math. Stat.* **25**:290–302.

BOX, G. E. P. 1954b. Some theorems on quadratic forms applied in the study of analysis of variance. II. Effects of inequality of variance and of correlation between errors in the two-way classification. *Ann. Math. Stat.* **25**:484–498.

BOXER, C. R. 1950. *Jan Compagnie in Japan, 1600–1850*. The Hague: Martinus Nijhoff. Pp. xii and 198.

BOYD, W. C. 1939. Blood groups. *Tabulae Biol.* **17**:113–240.

BRINKLEY, F. 1915. *A History of the Japanese People*. New York: Encyclopedia Britannica Co. Pp. xi and 784.

BRITISH MISSION TO JAPAN. 1946. *The Effects of Atomic Bombs at Hiroshima and Nagasaki*. London: His Majesty's Stationery Office. 39 pp.

BURKE, B. S., STEVENSON, S. S., WORCESTER, J., and STUART, H. C. 1949. Nutrition studies during pregnancy: V. Relation of maternal nutrition to condition of infant at birth: study of siblings. *J. Nutrition* **38**:453–467.

CARTER, C. 1950. Maternal states in relation to congenital malformations. *J. Obst. & Gynec., Brit. Emp.* **57**:897–911.

CARTER, C., and MCCARTHY, D. 1951. Incidence of mongolism and its diagnosis in the newborn. *Brit. J. Soc. Med.* **5**:83–90.

CHARLES, D. R. 1950. Radiation-induced mutations in animals. *Radiology* **55**:579–581.

CIOCCO, A. 1938. Variations in the sex ratio at birth in the United States. *Hum. Biol.* **10**:36–64.

COCHRAN, W. G. 1952. The chi-square test of goodness of fit. *Ann. Math. Stat.* **23**:315–345.

COFFEY, V. P., and JESSOP, W. J. E. 1955. Congenital abnormalities. *Irish J. Med. Sci., Sixth Series*, No. 349. Pp. 30–48.

COGAN, D. G., MARTIN, S. F., and KIMURA, S. J. 1949. Atom bomb cataracts. *Science* **110**:654–655.

COLLINS, V. P., and LOEFFLER, R. K. 1956. The therapeutic use of single doses of total body radiation. *Am. J. Roent.* **75**:542–547.

CRONKITE, E. P., BOND, V. P., and DUNHAM, C. L. (eds.) The effects of ionizing radiation on human beings: a report on the Marshallese and Americans accidentally exposed to radiation from fallout and a discussion of radiation injury in the human being. (Unpublished manuscript.)

CROW, J. F. 1955. A comparison of fetal and infant death rates in the progeny of radiologists and pathologists. *Am. J. Roent.* **73**:467–471.

DEAN, R. F. A. 1950. Effect of undernutrition on size of baby at birth and on ability of mother to lactate. *Proc. Roy. Soc. Med.* **43**:273–274.

DOBZHANSKY, TH., SPASSKY, B., and SPASSKY, N. 1952. A comparative study of mutation rates in two ecologically diverse species of Drosophila. *Genetics* **37**:650–664.

EBBS, J. H., BROWN, A., TISDALL, F. F., MOYLE, W. J., and BELL, M. 1942b. The influence of improved prenatal nutrition upon the infant. *Canad. Med. Assoc. J.* **46**:6–8.

EBBS, J. H., SCOTT, W. A., TISDALL, F. F., MOYLE, W. J., and BELL, M. 1942a. Nutrition in pregnancy. *Canad. Med. Assoc. J.* **46**:1–6.

EISENHART, C. 1947. In: Eisenhart, C., Hastay, M. W., and Wallis, W. A. *Techniques of Statistical Analysis*. New York: McGraw-Hill Co. Pp. xiv and 473.

EISENHART, C. 1947. The assumptions underlying the analysis of variance. *Biometrics* **3**:1–21.

EVANS, R. D. 1949. Quantitative inferences concerning the genetic effects of radiation on human beings. *Science* **109**:299–304.

FALK, R. 1955. Studies on X-ray induced viability mutations in the third chromosome of *Drosophila melanogaster*. *Hereditas* **41**:259–278.

FILLMORE, P. G. 1952. The medical examination of Hiroshima patients with radiation cataract. *Science* **116**:322–323.

FISHER, R. A. 1939. The comparison of samples with possibly unequal variances. *Ann. Eugen.* **9**:174–180.

FISHER, R. A. 1949. *The Design of Experiments*. 5th edition. Edinburgh: Oliver & Boyd. Pp. xi and 242.

FOLLEY, J. H., BORGES, W., and YAMAWAKI, T. 1952. Incidence of leukemia in survivors of the atomic bomb in Hiroshima and Nagasaki, Japan. *Am. J. Med.* **13**:311–321.

FURUHATA, T. 1929. The distribution of blood groups of the Japanese. *Ukrainisches Zentralblatt für Blutgruppenforschung.* Vol. 3.

GENETICS CONFERENCE. 1947. Genetic effects of the atomic bombs in Hiroshima and Nagasaki. *Science* **106**:331–333.

GRUENWALD, P. 1947. Mechanisms of abnormal development. I. Causes of abnormal development in the embryo. *Arch. Path.* **44**:398–436.

HADORN, E. 1955. *Letalfaktoren.* Stuttgart: G. Thieme. 338 pp.

HALDANE, J. B. S. 1948. The formal genetics of man. *Proc. Roy. Soc. Lond.,* Series B. **135**:147–170.

HALDANE, J. B. S. 1949. The rate of mutation of human genes. Proc. VIII Int. Cong. Genetics. *Hereditas Suppl.* Pp. 267–273.

HALDANE, J. B. S. 1955. Genetical effects of radiation from products of nuclear explosions. *Nature* **176**:115.

HARRIS, H., and SMITH, C. A. B. 1947. The sib-sib age of onset correlation among individuals suffering from a hereditary syndrome caused by more than one gene. *Ann. Eugen. Lond.* **14**:309–318.

HASEBE, K. 1917. Geographical distribution of stature of the Japanese. *Tohoku Journal of Medicine* **2**:152 (quoted from Matsumura, 1925).

HAYASHI, I. 1955. Pathological research on the influence of atomic bomb exposure upon fetal development. An address given at a joint meeting of the Japanese Atomic Bomb Investigation Group and members of the Japan Science Council at Nagasaki on October 3, 1955.

HEGNAUER, H. 1951. Incidence of congenital defects and age of mother. *Geburtshilfe und Frauenheilkunde* **11**:771–777.

HEMPELMANN, L. H., LISCO, H., and HOFFMAN, J. G. 1952. The acute radiation syndrome: a study of nine cases and a review of the problem. *Ann. Int. Med.* **36**:279–510.

HERTWIG, P. 1938. Unterschiede in der Entwicklungsfähigkeit von F_1 Mäusen nach Röntgen-bestrahlung von Spermatogonien, fertigen und unfertigen Spermatozoen. *Biol. Zentr.* **58**:273–301.

HOUSE, V. L. 1953a. The interaction of mutants affecting venation in *Drosophila melanogaster*. I. Interaction of Hairless, engrailed, and cubitus interruptus. *Genetics* **38**:199–215.

HOUSE, V. L. 1953b. The interaction of mutants affecting venation in *Drosophila melanogaster*. II. Additive and pattern effects of combinations of Hairless, engrailed, and cubitus interruptus. *Genetics* **38**:309–327.

HULSE, F. S. 1943. Physical types among the Japanese. In: Coon, C. S. and Andrews, J. M. IV., editors. Studies in the Anthropology of Oceania and Asia. *Papers of the Peabody Museum* **20**:122–133.

IVES, P. T. 1950. The importance of mutation rate genes in evolution. *Evolution* **4**:236–252.

IVES, P. T. 1954. Radiation-induced mutation rates in Drosophila and mice. *Amer. Nat.* **88**:361–364.

IZUMI, TETSU 1921. *Kokusai-hō gairon (Outline of the International Law)*. Tokyo: Yūhi-kaku. Pp. xxiii and 591.

JAPAN SCIENCE COUNCIL. 1951. *Summary Report on Atomic Bomb Casualties and Damages*. Translated and reproduced by the Atomic Bomb Casualty Commission. 140 pp.

KAEMPFER, E. 1728. *The History of Japan*, translated by J. G. Scheuchzer. New York: The Macmillan Co. 1906. 3 vols., pp. xc and 337; pp. x and 397; and pp. ix and 386.

KALMUS, H., METRAKOS, J. D., and SILVERBERG, M. 1952. Sex ratio of offspring from irradiated male mice. *Science* **116**:274–275.

KARN, M. N., and PENROSE, L. S. 1951. Birthweight and gestation time in relation to maternal age, parity, and infant survival. *Ann. Eugen.* **16**:147–164.

KENDALL, M. G. 1946. *The Advanced Theory of Statistics*. Vols. I and II. London: Charles Griffin and Co., Ltd. Pp. xii and 457; pp. vii and 521.

KEOSIAN, J. 1955. Speculations on hazards of exposure to radiations. *Science* **122**:586–587.

KERKIS, J. J. 1935. The preponderance of "physiological mutations." *Summ. Commun. XV Intern. Physiol. Congr.* (Leningrad-Moscow) (quoted from Muller, 1954). Pp. 198–200.

KITANO, KOJI. 1911. *Nagasaki kyōdo shi (History of Nagasaki)*. Nagasaki: Nagasaki Shōgakkō Shokuin-kai. 354 pp. (see p. 162).

KIYONO, KENJI. 1949. The origin of the Japanese race. *Kagaku Asahi*, August (in Japanese).

KOYA, Y. 1953. The program for family planning in Japan. *Eug. News* **38**:1–3.

KOYA, Y. 1954. A study of induced abortion in Japan and its significance. *Milbank Mem. Fund Quart.* **32**:282–293.

KROOTH, R. S. 1955. Some problems in maternal age. *Am. J. Hum. Genet.* **7**:147–162.

LANDTMAN, B. 1948. Relationship between maternal conditions during pregnancy and congenital malformations. *Arch. Dis. Child.* **23**:237–246.

LANGE, R. D., MOLONEY, W. C., and YAMAWAKI, T. 1954. Leukemia in atomic bomb survivors. I. General observations. *Blood* **9**:574–585.

LANGE, R. D., WRIGHT, S. W., TOMONAGA, M., KURASAKI, H., MATSUOKA, S., and MATSUNAGA, H. 1955. Refractory anemia occurring in survivors of the atomic bombing in Nagasaki, Japan. *Blood* **10**:312–324.

LASKER, G. W. 1946. Migration and physical differentiation. *Am. J. Phys. Anthropol.* **4**:273–300.

LEFEVRE, G., JR. 1955. Mutation isoalleles at the yellow locus in *Drosophila melanogaster*. *Genetics* **40**:374–387.

LERNER, I. M. 1955. Buffered genotypes and improvement in egg production. *Amer. Nat.* **89**:29–34.

LEROY, G. V. 1950. Hematology of atomic bomb casualties. *Arch. Int. Med.* **86**:691–710.

LORENZ, E., and collaborators. 1954. Effects of long-continued total-body gamma irradiation on mice, guinea pigs, and rabbits. Chaps. I, II, III, IV, V, and VI. In: *Biological Effects of External X and Gamma Radiation* (R. E. Zirkle, ed.). New York: McGraw-Hill Book Co. Pp. xxvi and 530.

LOS ALAMOS SCIENTIFIC LABORATORY. 1950. *The Effects of Atomic Weapons*. Washington, D.C.: U.S. Govt. Printing Office. Pp. x and 456.

LOWE, C. R., and MCKEOWN, T. 1950. The sex ratio of human births related to maternal age. *Brit. J. Soc. Med.* **4**:75–85.

MACHT, S. H., and LAWRENCE, P. S. 1955. National survey of congenital malformations resulting from exposure to roentgen radiation. *Amer. J. Roent.* **73**:442–466.

MacMahon, B., and Gordon, J. E. 1953. Epidemiologic inferences derived from maternal age. *Am. J. Med. Sci.* **226**:326–349.

Matsumoto, Y. S. 1954. Patient rapport in Hiroshima. *Amer. J. Nursing* **54**:69–72.

Matsumura, A. 1925. Cephalic index and stature of Japanese and their local differences. *Journal of Faculty of Science* (Tokyo University). Section V; Vol. I; Part I; pp. 1–312.

McIntosh, R., Merritt, K. K., Richards, M. R., Samuels, M. H., and Bellows, M. T. 1954. The incidence of congenital malformations: a study of 5,694 pregnancies. *Pediatrics* **14**:505–522.

Millis, J. 1952. A study of the effect of nutrition on fertility and the outcome of pregnancy in Singapore in 1947 and 1950. *Med. J. Malaya* **6**:157–172.

Millis, J. and Seng, Y. P. 1954. The effect of age and parity of the mother on birthweight of the offspring. *Ann. Hum. Genet.* **19**:58–73.

Moloney, W. C., and Lange, R. D. 1954. Leukemia in atomic bomb survivors. II. Observations on early phases of leukemia. *Blood* **9**:663–685.

Moloney, W. C., and Kastenbaum, M. A. 1955. Leukemogenic effects of ionizing radiation on atomic bomb survivors in Hiroshima City. *Science* **121**:308–309.

Morton, N. E. 1953. Present Status of the Genetics Analysis (mimeographed publication of the Atomic Bomb Casualty Commission).

Morton, N. E. 1955. The inheritance of human birthweight. *Ann. Hum. Genet.* **20**:125–134.

Morton, N. E. Empirical risks in consanguineous marriage: II. Birthweight, gestation time, and measurements of infants (in manuscript).

Mourant, A. E. 1954. *The Distribution of the Human Blood Groups.* Springfield: C. C. Thomas. Pp. xxi and 438.

Muller, H. J. 1934. Radiation genetics. *Proc. IV Intern. Radiologen Kongr.* **2**:100–102.

Muller, H. J. 1935. On the dimensions of chromosomes and genes in Dipteran salivary glands. *Amer. Nat.* **69**:405–411.

Muller, H. J. 1950. Our load of mutations. *Am. J. Hum. Genet.* **2**:111–176.

Muller, H. J. 1954. The nature of the genetic effects produced by radiation. In: *Radiation Biology* (A. Hollaender, ed.). New York: McGraw-Hill Book Co. Vol. I, pp. 351–453.

Muller, H. J. 1955. How radiation changes the genetic constitution. *Bull. Atomic Scientists* **11**:329–338.

Muller, H. J., Valencia, J. I., and Valencia, R. M. 1950. The frequency of spontaneous mutations at individual loci in Drosophila. *Genetics* **35**:125–126.

Munro, N. G. 1908. *Prehistoric Japan.* Yokohama. Pp. xvii and 705.

Murdoch, J. 1926. *A History of Japan.* New York: Greenberg. 3 vols.

Murphy, D. P. 1928. Ovarian irradiation, its effect on the health of subsequent children. *Surg., Gyn., and Obst.* **47**:201–215.

Murphy, D. P. 1947. *Congenital Malformations.* Second edition. Philadelphia: J. B. Lippincott Co. 127 pp.

Myers, R. J. 1954. The effect of age of mother and birth order on sex ratio at birth. *Milbank Mem. Fund Quart.* **32**:275–281.

Nachtsheim, H. 1954. Die Mutationsrate menschlicher Gene. *Naturwissen* **17**:385–392.

Nagasaki-Shi Shōgakkō Shokuin-Kai. 1925. *Meiji Ishinigo no Nagasaki (Nagasaki since the Restoration).* 692 pp. (see pp. 86–89).

Naikaku Tōkei-Kyoku (Cabinet Statistical Bureau) Census Report. 1930. Vol. 4, p. 40.

Naikaku Tōkei-Kyoku. Nihon Teikoku Tōkei Nenkan (Japanese Empire Statistical Annual). 1899–1938. Nos. 19–58.

Neel, J. V. 1941. Studies on the interaction of mutations affecting the chaetae of *Drosophila melanogaster*. I. The interaction of hairy, polychaetoid, and Hairy wing. *Genetics* **26**:52–68.

Neel, J. V. 1942. A study of a case of high mutation rate in *Drosophila melanogaster*. *Genetics* **27**:519–536.

Neel, J. V. 1943. Studies on the interaction of mutations affecting the chaetae of *Drosophila melanogaster*. II. The relation of character expression to size in flies homozygous for polychaetoid, hairy, Hairy wing, and the combinations of these factors. *Genetics* **28**:49–68.

Neel, J. V. 1952. The study of human mutation rates. *Amer. Nat.* **86**:129–144.

NEEL, J. V., and FALLS, H. F. 1951. The rate of mutation of the gene responsible for retinoblastoma in man. *Science* **114**:419–422.

NEEL, J. V., KODANI, M., BREWER, R., and ANDERSON, R. C. 1949. The incidence of consanguineous matings in Japan. *Am. J. Hum. Genet.* **1**:156–178.

NEEL, J. V., SCHULL, W. J., McDONALD, D. J., MORTON, N. E., KODANI, M., TAKESHIMA, K., ANDERSON, R. C., WOOD, J., BREWER, R., WRIGHT, S., YAMAZAKI, J., SUZUKI, M., and KITAMURA, S. 1953. The effect of exposure to the atomic bombs on pregnancy termination in Hiroshima and Nagasaki: Preliminary report. *Science* **118**:537–541.

NEEL, J. V., and SCHULL, W. J. 1954. *Human Heredity.* Chicago: University of Chicago Press. 361 pp.

NICKSON, J. J. 1951. Blood changes in human beings following total-body irradiation. *National Nuclear Energy Series* **IV-20**:308–337.

NIXON, W. C. W. 1950. Social factors in obstetrics. In: *Modern Trends in Obstetrics and Gynecology.* (K. Bowes, ed.). New York: Paul B. Hoeber. Pp. 201–218.

NOVITSKI, E. 1953. The dependence of the secondary sex ratio in humans on the age of the father. *Science* **117**:531–533.

OUGHTERSON, A. W., LEROY, G. V., LIEBOW, A. A., HAMMOND, E. C., BARNETT, H. L., ROSENBAUM, J. D., and SCHNEIDER, B. A. 1951. Medical effects of atomic bombs: The report of the Joint Commission for the Investigation of the Effects of the Atomic Bomb in Japan. In 5 vols. Issued by Technical Information Service, U.S. Atomic Energy Comm., Oak Ridge, Tenn.

PARKES, A. S. 1925. The effects on fertility and sex-ratio of sub-sterility exposures to X-rays. *Proc. Roy Soc. Lond.,* Series B, **98**:415–436.

PEARSON, E. S. 1936. The efficiency of statistical tools and a criterion for the rejection of outlying observations. *Biometrika* **28**:308–320.

PENROSE, L. S. 1939. Maternal age, order of birth and developmental abnormalities. *J. Ment. Sci.* **85**:1141–1150.

PENROSE, L. S. 1950. Propagation of the unfit. *Lancet* **2**:425–427.

POPULATION REFERENCE BUREAU, INC. 1953. Falling birth rates: what do they mean? *Population Bull.* **9**:13–23.

PRINDLE, R. A., INGALLS, T. H., and KIRKWOOD, J. B. 1955. Maternal hydramnios and congenital anomalies of the central nervous system. *N. E. J. Med.* **252**:555–561.

PUBLIC HEALTH AND WELFARE SECTION, GHQ, SCAP. 1948, 1949, 1950. *Public Health and Welfare in Japan.*

RAO, C. R. 1952. *Advanced Statistical Methods in Biometric Research.* New York: Wiley and Sons, Inc. Pp. xvii and 390.

RAO, C. R. 1955. Analysis of dispersion for multiply classified data with unequal numbers in cells. *Sankya* **15**:253–280.

RECORD, R. G., and McKEOWN, T. 1949–1950. Congenital malformations of the central nervous system. Parts I–III. *Brit. J. Soc. Med.* **3**:183–219; **4**:26–50; **4**:217–220.

ROBSON, E. B. 1955. Birthweight in cousins. *Ann. Hum. Genet.* **19**:262–268.

ROY, S. N., and KASTENBAUM, M. A. 1956. A generalization of analysis of variance and multivariate analysis to data based on frequencies in qualitative categories or class intervals. *Ann. Math. Stat.* **27**:749–757.

RUSSELL, W. L. 1954. Genetic effects of radiation in mammals. In: *Radiation Biology* (A. Hollaender, ed.). New York: McGraw-Hill Book Co. Vol. I, pp. 825–859.

RUSSELL, W. L. 1956. Comparison of X-ray-induced mutation rates in Drosophila and mice. *Amer. Nat.* **90**:67–80.

SALBER, E. J., and BRADSHAW, E. S. 1953. Birthweights of South African babies. IV. Association with maternal age. *Brit. J. Prev. Soc. Med.* **7**:20–23.

SANSOM, G. B. 1943. *Japan: A Short Cultural History.* New York: D. Appleton-Century Co. Pp. xviii and 554.

SASANO, AKIKUNI. 1921. *Nihon Kokusai shihō (Japan's International Private Laws).* Tokyo: Nichiei-do. 247 pp. (see p. 162).

SCHULL, W. J. 1953. The effect of Christianity on consanguinity in Nagasaki. *Amer. Anthrop.* **55**:74–88.

SCHULL, W. J. Empirical risks in consanguineous marriage: I. Sex ratio, malformation, stillbirth, and neonatal and infantile death (in manuscript).

SCHULTZ, J. 1936. Radiation and the study of mutations in animals. In: *Biological Effects*

of Radiation (B. M. Duggar, ed.). New York: McGraw-Hill Book Co. Vol. II, pp. 1209–1261.

SEVITT, S. 1955. The bombs. *Lancet* **269**:199–201.

SHAPIRO, H. L. 1939. *Migration and Environment*. New York: Oxford University Press. Pp. xi and 594.

SINSKEY, R. M. 1955. The status of lenticular opacities caused by atomic radiation. *Am. J. Ophth.* **39**:285–293.

SLATIS, H. M. 1954. A method of estimating the frequency of abnormal autosomal recessive genes in man. *Am. J. Hum. Genet.* **6**:412–418.

SLATIS, H. M. 1955. Current status of information on the induction of mutations by irradiation. *Science* **121**:817–821.

SMITH, C. A. 1947. Effects of maternal undernutrition upon the newborn infant in Holland 1944–1945. *J. Pediat.* **30**:229–243.

SNEDECOR, G. W. 1946. *Statistical Methods*. Ames: Iowa State College Press. Pp. xvi and 485.

SPENCER, W. P., and STERN, C. 1948. Experiments to test the validity of the linear r-dose/mutation frequency relation in Drosophila at low dosage. *Genetics* **33**:43–74.

SPUHLER, J. N. 1948. On the number of genes in man. *Science* **108**:279–280.

STANDER, H. J. 1941. *Williams' Obstetrics*. 8th Edition. New York: D. Appleton-Century Co. Pp. xix and 1401.

STRANDSKOV, H. H. 1932. Effect of X-rays in an inbred strain of guinea-pigs. *J. Exper. Zool.* **63**:175–202.

STRANDSKOV, H. H. 1945. Birth sex ratios in the total, the "white" and the "colored" U.S. populations. *Am. J. Phys. Anthrop. N. S.* **3**:165–175.

STURTEVANT, A. H. 1955. The genetic effects of high energy irradiation of human populations. *Engrg. & Sci. Mo.* January, 1955.

SUTHERLAND, I. 1949. *Stillbirths: Their Epidemiology and Social Significance*. New York: Oxford Univ. Press. 95 pp.

SYMPOSIUM: Some biological effects of radiation from nuclear detonations. *Amer. Nat.* **88**: 209–314. 1954.

THUNBERG, K. P. 1795. *Travels in Europe, Africa, and Asia, Made between the Years 1770 and 1779*. 4 vols. 3rd edition. London: F. and C. Rivington.

THUNBERG, K. P. 1796. *Voyage de C. P. Thunberg au Japon*. 2 vols. Paris: B. Dandré.

TIETZE, C. 1948. A note on the sex ratio of abortions. *Hum. Biol.* **20**:156–160.

TIMOFÉEFF-RESSOVSKY, N. W. 1935. Auslosung von Vitalitätsmutationen durch Röntgen-bestrahlung bei *Drosophila melanogaster*. *Nachr. Ges. Wis. Göttingen* (Math.-physik. Klasse, Fachgruppen VI, Biol.) **1**:163–180.

TIMOFÉEFF-RESSOVSKY, N. W. 1937. *Mutationsforschung in der Vererbungslehre*. Dresden: Th. Steinkopff. Pp. x and 181.

TRASLER, D. G., and METRAKOS, J. 1953. Sex ratio of offspring from irradiated male mice. *Genetics* **38**:697.

UNITED STATES STRATEGIC BOMBING SURVEY. 1946. *The Effects of Atomic Bombs on Hiroshima and Nagasaki*. Washington: U.S. Govt. Printing Office. Pp. ix and 91.

VOR DER BRUEGGE, C. F. 1952. Radiation injury following A-bomb explosion. *Ann. Int. Med.* **36**:1444–1458.

WALLACE, B. 1951. Genetic changes within populations after X-irradiation. *Genetics* **36**:612–628.

WALLIS, W. A. 1951. Personal communication.

WARKANY, J. 1947. Etiology of congenital malformations. In: *Advances in Pediatrics*. Vol. II, pp. 1–63. New York: Academic Press. Pp. x and 409.

WARREN, S. 1946. Memorandum to Chief of Naval Operations. Subject: Target report — atomic bombs, Hiroshima and Nagasaki, Art. 2 — Medical Effects, Supplementary Studies. Dated May 29, 1946.

WARREN, S., and BOWERS, J. Z. 1950. The acute radiation syndrome in man. *Ann. Int. Med.* **32**:207–216.

WARREN, S. L. 1945. Memorandum to Maj. Gen. L. R. Groves. Subject: Preliminary report — atomic bomb investigation. Dated November 27, 1945.

WESTERGAARD, M. 1955. Man's responsibility to his genetic heritage. *Impact of Science on Society* **6**:63–87.

WILKS, S. S. 1932. Certain generalizations in the analysis of variance. *Biometrika* **24**:471–491.

WILKS, S. S. 1938. The analysis of variance and covariance in non-orthogonal data. *Metron* **13**:141–154.

WILSON, R. R. 1951. Report on nuclear radiation at Hiroshima and Nagasaki, submitted to the Committee on Atomic Casualties of the National Research Council.

WISHART, J. 1950. *Field Trials II. The Analysis of Covariance.* Technical Communication No. 15. Commonwealth Bureau of Plant Breeding and Genetics. Cambridge (see p. 35).

WOODBURY, L. A. 1953. Analysis of the relationship of shielding and distance factors to radiation cataracts. *ABCC Semi-Annual Report,* January 1, 1953–June 30, 1953, Part 1, Appendix 15.

WOOLLEY, W. A. 1881. Historical notes on Nagasaki. *Asiatic Soc. of Japan, Transactions.* 9:125–151.

WORCESTER, J., STEVENSON, S. S., and RICE, R. G. 1950. 677 congenitally malformed infants and associated gestational characteristics. II. Parental factors. *Pediatrics* 6:208–222.

WRIGHT, S. 1949. Estimates of the amounts of melanin in the hair of diverse genotypes of the guinea pig, from transformation of empirical grades. *Genetics* 34:245–271.

WRIGHT, S. 1950. Discussion on population genetics and radiation. *J. Cell. and Comp. Physiol.,* 35, Suppl. 1, pp. 187–204.

WRIGHT, S. W., et al. 1952. A study of the incidence of congenital syphilis among living Japanese infants. *ABCC Semi-Annual Report,* January 1, 1952–June 30, 1952.

YATES, F. 1934. The analysis of multiple classifications with unequal numbers in the different classes. *J. Am. Stat. Assoc.* 29:51–66.

YERUSHALMY, J. 1938. Neonatal mortality by order of birth and age of parents. *Am. J. Hyg.* 28:244–270.

YERUSHALMY, J. 1945. On the interval between successive births and its effect on survival of infant. I. An indirect method of study. *Hum. Biol.* 17:65–106.

SUBJECT INDEX

Abortions, reporting of, 9
　spontaneous, 18
　therapeutic (induced), 19, 61, 76
Acetabulum, dysplasia of, 15, 102, 104
"Accumulation factor," in man, 212
Age, maternal
　allowance for, in analysis, 76
　differences in irradiation subclasses, 55
Age effects, maternal
　on birthweight, 132, 148
　on death during first 9 months, 152
　on growth and development, 165
　on malformation frequency, 105, 189
　on sex ratio, 89
　on stillbirth frequency, 119
Ainu, 21
Americans, in Nagasaki, 23
Analytic methods, general
　attribute data
　　"interactions," 78, 162
　　"main effects," 78, 162
　measurement data, 83, 84
Analytic methods, specific
　analysis of dispersion, 168
　Bartlett's test, 137, 175, 179
　concomitant variation, approach to, 73
　confidence limits, 197
　covariance analysis, 75, 133, 137
　determinants of matrices, 168
　F test, 179
　heterogeneity, within-cell, 85, 133, 137, 168, 175, 179
　index of absolute difference, 105, 185
　L statistic, 148, 170
　means, multivariate, 164, 165, 168, 170, 175, 179
　"omnibus" tests, 84
　"portmanteau" tests, 84
　power curves, 197, 198
　pyramidal handling of data, 73
　regression, 82, 132, 133, 148
　significance test, one-tailed, 89, 198
　variance, generalized, 164, 168, 175, 179
　variance analysis, 82, 86, 133, 137, 150, 157, 175
　Wilks' test, 168
Anamnestic data, reliability of, 14
Anthropometric studies
　genetic component in, 164
　measurements obtained, 164
　variation, effect of maternal age on, 165
　　parity, 165
　　radiation, 165, 196
Armed Forces Institute of Pathology, 47
Atomic Bomb Casualty Commission, 2
Atomic bombs, mortality from, 28
　radiation from, 33
Austrians, in Nagasaki, 23
Autopsy program, 9
　description of, 184
　of Hayashi, 189
　randomness of autopsies, 184
　type of data, 187
Biases, of sampling, 191
　see also "comparability of irradiation subclasses"
Birth injury
　death during first 9 months, 152
　effect on stillbirth rate, 119
Birth rates, Japanese, 19
Birthweight
　accuracy of, 131
　effect on, of maternal age, 132, 148
　　economic status, 131, 148
　　nutrition, 131
　　parental radiation, 132, 196
　　parity, 132, 148
　　year of birth, 148
　genetic component in, 131
"Black market," 4, 132
Blood group (A-B-O) frequencies, 25
Cataracts, radiation, 33, 69
Chest circumference of child, in relation to parental radiation, 164
Chikuzen type, of Japanese, 171
China, 21
Chinese, in Nagasaki, 22, 28
Chi-square, 82
　factorial, 82
　in analysis of age and parity effects on malformations, 105
Christianity, ban against, 22, 24
Codes for data, 18
　Genetics Long Form, 221
　Genetics 9-Months Follow-up, 225
　Genetics Short Form, 218
Committee for the Investigation of the Effects of the Atomic Bombs (Japanese), 2, 20
Committee on Atomic Casualties, 1, 19, 87
Comparability of irradiation subclasses
　age, 55, 73
　background, 71, 192
　consanguinity, 53, 73
　dilatation and curettage of uterus, 61, 73
　economic status, 59, 73
　induced abortions, 61, 73
　parental cooperation, 63
　parity, 55, 73
　positive serological test for syphilis, 61, 73
　repeat registrations, 63, 73
　sequelae of exposure, 69
　year to year changes in proportions, 69
Confidence limits, of attribute data, 196
Congenital defect—see "malformation, congenital"
Consanguinity
　allowance for, in analysis, 76
　differences, in exposure subgroups, 53, 73
　effect of Christianity on, 55
Controls, use of exposed persons as, 86
　see also "comparability of irradiation subclasses"
Cooperation, of parents, 63

237

Subject Index

Covariance analysis, 75, 133, 137
Cytogenetic studies, 18
Danes, in Nagasaki, 23
Death, from birth to 9 months of age
 frequency of, as affected by birth injury, 152
 congenital syphilis, 152
 maternal age, 152
 maternal nutrition, 152
 parental radiation, 157, 196
 parity, 152, 162
 genetic component of, 151
 neonatal, definition of, 151
 operating characteristic curve of, 198
Degrees of freedom
 in analysis of attribute data, 81
Deshima, 22, 24
"Disaster effect," vs. "radiation effect," 33, 150, 192
Distance, in relation to symptoms, 41
Distance-dosage relationship, 36, 46
Drosophila, 199, 203, 204, 208, 209, 212, 213, 216
 mutation rate, induced, 211
 mutation rate, spontaneous, 206
 number of genes in, 212
Dutch, in Nagasaki, 22, 23, 24
Economic status, allowance for, in analysis, 76
 effect on birthweight, 131
 effect on frequency of autopsy, 186
 of control and irradiated, 59
English, in Nagasaki, 22, 23
Errors, clerical, 15
 diagnostic, 15
Exposure to radiation, sequelae of
 cataracts, 33, 69
 leukemia, 33, 69
Finland, 21
Formosa, 21
Franciscans, in Nagasaki, 22
French, in Nagasaki, 23
Funnel chest, 102
Gamma rays, importance in dosage, 52
Genes—see also "accumulation factor"
 additive effects of, 213
 number in Drosophila, 212
 number in man, 212, 217
Genetic damage, indicators of, 3, 5, 72
Genetics Conferences
 first, 2, 217, 231
 second, 19
Genetics Long Form, 9, 14, 61, 93, 221
Genetics 9-Months Follow-up, 15, 225
Genetics Short Form, 5, 9, 44, 53, 61, 218
Guinea pig, 213
 radiation of, effect on birthweight, 203
 on growth and development, 203
 on neonatal death rate, 203
Goa, 21
Government, Japanese, 3
Hawaii, 28, 173
Head circumference of child, in relation to parental radiation, 164
Heart disease, congenital, 15, 102
Hernia, inguinal, 15, 102
Heterogeneity, within exposure cells, 85, 133, 137, 168, 175, 179
Honshu, 21

Indonesia, 21
Ishikawa type, of Japanese, 171
Japan Science Council, 34, 189
Japanese, origin of, 21
 physical types of, 171
Jesuits, in Nagasaki, 21
Korea, 21, 173
"Korean colony," in Hiroshima, 28
Kyushu, 21
LD_{50}, man, 45
Length of child, in relation to parental radiation, 164
Leucopenia, following irradiation, 46, 50
Leukemia, 33, 69
Macao, 21, 24
Malaysia, 21
Malformation, congenital—see also under specific type
 accuracy of diagnosis, 99
 autopsy studies of, 184
 definition of
 major, 99
 minor, 99
 frequency, in relation to parental radiation, 195
 at age 9 months, 115
 at birth, 110
 specific types, 117
 frequency of, as affected by maternal age, 105
 nutrition, 110
 parity, 105
 viral infections, 110
 operating characteristic curve of, 198
 opportunities for concealment, 63
 study of Macht and Lawrence, 202
 types encountered, in ABCC study, 100
 comparison, in Hiroshima and Nagasaki, 102
 in other studies on Japanese, 101
Manchuria, 21, 172
Marshall Islands, 49
Meiji era, 22
Mental defect, severe, 102, 104
Midwives, 4, 5, 9
Midwives Association, 5
Ministry of Welfare, Japanese, 2
Miscarriages, reporting of, 9
Mongolism, 110
Mouse
 radiation of, effect on malformation frequency, 201
 on neonatal death rate, 203
 on sex ratio, 200
 on stillbirth frequency, 202
 sterility following irradiation, 200
Mutation rates, induced by radiation, 3
 in Drosophila, 211
 in man, 211, 217
 in the mouse, 210
Mutation rates, spontaneous, 3, 205
 in Drosophila, 208
 in man, 209
 in the mouse, 209
Mutations, types of
 "detrimentals," 164
 "invisibles," 131
 lethal, 88, 118, 204, 205
 ratio of types, 206, 207
 sex-linked, 88, 203
 "visibles," 104, 204, 205

Subject Index

Nagasaki Medical School, 29, 102, 184, 188, 189
National Academy of Sciences-National Research Council (U.S.), 1, 3, 19, 20
National Institute of Health, Japanese, 2, 5, 20
National Research Council, Japanese, 2
Neonatal death—see "death, neonatal"
Neutrons, importance in dosage, 51, 52
"Nine-months program," description of, 14
Non-orthogonality, 86, 157
Nutrition, maternal, effect on
 birthweight, 132
 death during first 9 months, 152
 malformation frequency, 105
 size of adult, 172
 stillbirth frequency, 119
Okayama type, of Japanese, 171
Operating characteristic curves, 198
Parity
 allowance for, in analysis, 76
 differences, in irradiation subclasses, 55
Parity effects
 on birthweight, 132, 148
 on death during first 9 months, 152
 on growth and development, 165
 on malformation frequency, 105
 on sex ratio, 89
 on stillbirth frequency, 120
Plutonium-239, 28
Polynesia, 21
Portuguese, in Japan, 21, 24
Prefecture, definition of, 23
Pregnancy registration, Japanese, 5
 completeness of, 7
 repeat, 63, 77
Radiation, indicators of genetic effects of, 3, 5, 72
Radiation, residual, 33, 189
Radiation, symptoms of
 diarrhea, 36
 epilation, 29, 36, 47
 oropharyngeal lesions, 29, 36, 47
 petechiae, 29, 36, 47
Radiation categories
 definition of
 by ABCC, 44
 by Hayashi, 189
 dosage in, 45, 50
Radiation Census, 44
Radiation history, 5, 9, 38
 validity of, 44
Radiation sickness, syndrome of, 34
Ration system, 4
Registration, of pregnancy—see "pregnancy registration"
Rejected observations, justification, 77
Rockefeller Foundation, 20
Roentgen, 45
Roentgen equivalent physical, 45, 46, 51, 203
Russia, 21
Russians, in Nagasaki, 23, 25
Ryukyu Islands, 21

Sampling, balanced, 75
Satsuma type, of Japanese, 171
Selection, natural, in man, 213
Sequelae, late, of exposure to bombs, 69
Sex differences, in anthropometrics, 170
 in birthweight, 135
 in indicator values, 192
 in malformation frequency (absent), 110
 in neonatal death rate, 157
 in stillbirth frequency (absent), 124
Sex ratio
 effect of concomitant variation on, 89
 effect of irradiation of parents on, 89, 194
 genetic control of, 88
 operating characteristic curves of, 198
 study of Macht and Lawrence, 201
Shielding, 36, 50
 in relation to symptoms, 41, 42
Siberia, 21
Sinus, pilonidal, 15, 104
Spanish, 22
Sterility, in mouse following radiation, 200
Stillbirth
 definition of, 118
 frequency of, as affected by
 birth injury, 119
 congenital syphilis, 119
 maternal age, 120
 nutrition, 119
 parental radiation, 124, 196
 parity, 120
 paternal age, 120
 genetic component in, 118
 operating characteristic curve of, 198
 study of Macht and Lawrence, 203
Sweden, 21
Syphilis, maternal, 9
 allowance for, in analysis, 76
 different rates, in irradiation subclasses, 61
 effect on death during first 9 months, 152
 effect on stillbirth frequency, 119
 relation to frequency of autopsy, 187
"Ten-per cent sample," 9
Tokugawa Shogunate, 22, 24
U.S. Army of Occupation, 3
U.S. Army-Navy Joint Commission, 2, 47, 49
U.S. Atomic Energy Commission, 1, 3, 19, 87
"Unregistered Series," 7
Uranium-235, 28
Uterus, dilatation and curettage of, 61
 allowance for, in analysis, 76
Variance analysis, 82, 86, 133, 137, 150
Variances, test for equality of, 84
Viral diseases
 effect on malformation frequency, 105
Weight of child, in relation to parental radiation, 164
X-chromosome, 88, 204
Y-chromosome, 88

NAME INDEX

Alexander, M. L., 208, 211, 230
Anderson, R. C., 53, 234
Antonov, A. N., 59, 110, 131, 230
Armitage, P., 20
Balfour, M. I., 59, 230
Barnett, H. L., 234
Bartlett, M. S., 175, 179, 230
Beadle, G. W., 1, 2, 19
Beardsley, R., 21, 230
Bell, M., 231
Bellows, M. T., 233
Bernstein, M., 89, 230
Block, M., 1
Bond, V. P., 50, 230
Bonnier, G., 208, 230
Borges, W., 33, 69, 231
Bowers, J. Z., 36, 235
Box, G. E. P., 84, 179, 230
Boxer, C. R., 24, 230
Boyd, W. C., 28, 230
Bradshaw, E. S., 55, 234
Brandt, A. E., 20, 82
Brewer, R., 53, 234
Brinkley, F., 21, 230
Bronk, D. W., 1
Brown, A., 231
Brues, A., 1
Burke, B. S., 59, 230
Butterfly, Madame, 25
Carter, C., 55, 105, 230
Charles, D. R., 2, 19, 201, 202, 230
Ciocco, A., 55, 89, 230
Cochran, W. G., 82, 230
Coffey, V. P., 105, 230
Cogan, D. G., 33, 69, 230
Collins, V. P., 49, 230
Connell, F. H., 19
Cox, G., 86
Craig, C. C., 19, 20
Cronkite, E. P., 50, 230
Crow, J. F., 203, 231
Danforth, C. H., 2
Davidson, F., 20
Dean, R. F. A., 59, 231
Dobzhansky, T., 209, 231
Dunham, C. L., 50, 230
Dwyer, P. S., 20
Ebbs, J. H., 59, 231
Eisenhart, C., 82, 231
Evans, R. D., 210, 211, 212, 231
Falk, R., 206, 231
Falls, H. F., 213, 234
Fillmore, P. G., 69, 231
Fisher, R. A., 84, 231
Folley, J. H., 33, 69, 231
Forrestal, J. T., 1
Glass, H. B., 208
Gordon, J. E., 55, 233
Green, E., 20

Gruenwald, P., 105, 231
Hadorn, E., 202, 231
Haldane, J. B. S., 209, 211, 217, 231
Hammond, E. C., 234
Harris, H., 85, 231
Harris, T., 22
Hasebe, K., 171, 231
Hayashi, I., 184, 188, 189, 190, 191, 231
Hechter, H., 50
Hegnauer, H., 55, 231
Hempelmann, L. H., 36, 45, 49, 231
Henshaw, P. S., 1
Hertwig, P., 200, 202, 203, 231
Hill, B., 20
Hoffman, J. G., 36, 45, 49, 231
Holmes, R., 98
House, V. L., 213, 231
Hsiao, B., 20
Hulse, F. S., 171, 232
Ingalls, T. H., 102, 234
Ives, P. T., 208, 211, 232
Izumi, T., 23, 232
Jessop, W. J. E., 105, 230
Johnson, H., 20
Kaempfer, E., 22, 24, 171, 232
Kalmus, H., 200, 232
Karn, M. N., 55, 131, 232
Kastenbaum, M., 20, 69, 233, 234
Kelley, H. C., 20
Kendall, M. G., 86, 232
Keosian, J., 216, 232
Kerkis, J. J., 206, 232
Kimura, S., 33, 69, 230
Kirk, N. T., 1
Kirkwood, J. B., 102, 234
Kitamura, S., 234
Kiyono, K., 21, 232
Kobayashi, R., 20
Kodani, M., 18, 53, 234
Komai, T., 20
Koya, Y., 19, 232
Krooth, R., 20, 75, 105, 185, 214, 232
Kuji, V., 101
Kurasaki, H., 69, 232
Lamphiear, D. E., 20
Landtman, B., 55, 105, 232
Lange, R. D., 33, 69, 232, 233
Lawrence, P. S., 200, 202, 203, 232
Lefevre, G., 208, 232
Lerner, I. M., 206, 232
LeRoy, G. V., 47, 232, 234
Liebow, A. A., 234
Lisco, H., 36, 45, 49, 231
Loeffler, R. K., 49, 230
Lorenz, E., 69, 232
Lowe, C. R., 55, 89, 232
Lucas, H. C., 20, 74
Lüning, K. G., 208, 230
Lyon, G. M., 1

Name Index

Macht, S. H., 200, 202, 203, 232
MacMahon, B., 55, 233
Maki, H., 20
Martin, S. F., 33, 69, 230
Matsubayashi, I., 3
Matsumoto, S., 4, 233
Matsumura, A., 171, 233
Matsunaga, H., 69, 232
Matsuoka, S., 69, 232
McCarthy, D., 105, 230
McDonald, D. J., 234
McIntosh, R., 104, 233
McKeown, T., 55, 89, 232, 234
Meerdervoort, P., 22
Merritt, K. K., 233
Metrakos, J. D., 200, 232, 235
Millis, J., 131, 233
Mitani, S., 101
Moloney, W. C., 33, 69, 232, 233
Morton, J., 19
Morton, N., 131, 233, 234
Mourant, A. E., 28, 233
Moyle, W. J., 231
Muller, H. J., 2, 203, 206, 207, 208, 210, 211, 212, 213, 215, 216, 233
Munro, N. G., 21, 233
Murdoch, J., 21, 24, 233
Murphy, D. P., 202, 233
Myers, R. J., 55, 233
Nachtsheim, H., 209, 233
Nagai, I., 20
Neel, J. V., 1, 2, 19, 45, 53, 55, 96, 208, 209, 210, 213, 233, 234
Nickson, J. J., 49, 234
Nixon, W. C. W., 59, 234
Novitski, E., 55, 89, 234
Olkin, I., 20
Oughterson, A. W., 1, 33, 233
Parkes, A. S., 200, 234
Penrose, L. S., 20, 55, 131, 215, 232, 234
Perry, Commodore, 22, 25
Phelps, L. V., 7, 20
Plough, H., 19
Prindle, R. A., 102, 234
Rao, C. R., 20, 82, 86, 137, 168, 175, 234
Record, R. G., 55, 234
Reed, T. E., 209
Rhoads, C. P., 1
Rice, R. G., 55, 236
Richards, M. R., 233
Ritterhoff, R. K., 208
Rivers, T. M., 1
Robson, E., 131, 234
Rosenbaum, J. D., 234
Roy, S. N., 78, 234
Russell, W. L., 199, 200, 201, 202, 210, 211, 234
Salber, E. J., 55, 234
Sams, C. F., 2, 20
Samuels, M. H., 233
Sansom, G. B., 21, 24, 234
Sasano, A., 23, 234
Schneider, B. A., 234
Schneidewind, J., 20
Schull, W. J., 19, 54, 105, 208, 209, 210, 234

Schultz, J., 207, 234
Scott, W. A., 231
Seijas, B., 20
Seng, Y. P., 132, 233
Sevitt, S., 184, 188, 191, 235
Shapiro, H. L., 172, 235
Silverberg, M., 200, 232
Sinskey, R. M., 33, 69, 235
Slatis, H. M., 210, 211, 212, 235
Smith, C. A., 59, 110, 131, 235
Smith, C. A. B., 85, 231
Smith, H. F., 20
Snedecor, G. W., 86, 235
Snyder, L. H., 2, 19
Spassky, B., 209, 231
Spassky, N., 209, 231
Spencer, W. P., 203, 207, 235
Spuhler, J. N., 212, 235
Stern, C., 19, 20, 203, 207, 235
Stevenson, S. S., 55, 230, 236
Strandskov, H. H., 89, 202, 203, 235
Stuart, H. C., 230
Sturtevant, A. H., 211, 235
Sutherland, I., 55, 235
Suzuki, M., 234
Takeshima, K., 234
Taylor, G., 19
Tessmer, C. F., 19
Thunberg, K. P., 25, 235
Tietze, C., 88, 235
Timoféeff-Ressovsky, N. W., 203, 206, 235
Tisdall, F. F., 231
Tomonaga, M., 69, 232
Trasler, D. G., 200, 235
Tsuzuki, M., 20
Ullman, J., 20
Ullrich, F., 1
Valencia, J. I., 208
Valencia, R. M., 208
Vor der Bruegge, C. F., 36, 235
Wallace, B., 216, 235
Wallis, W. A., 73, 235
Warkany, J., 105, 110, 235
Warren, S. L., 1, 34, 235
Warren, Sh., 1, 34, 36
Weed, L. H., 1
Westergaard, M., 212, 235
Whipple, G. H., 1
Wilks, S. S., 86, 179, 235
Wilson, R. R., 34, 46, 236
Wishart, J., 137, 236
Wood, J., 234
Woodbury, L., 43, 98, 236
Woolley, W., 24, 236
Worcester, J., 55, 230, 236
Wright, S., 210, 211, 213, 236
Wright, S. W., 69, 232, 234, 236
Yamawaki, T., 33, 69, 231, 232
Yamazaki, J., 234
Yates, F., 86, 232
Yerushalmy, J., 55, 153, 232
Yesley, G., 20
Zelle, M., 19
Zirkle, R. E., 1

In a study of births to related parents who had been exposed to radiation in the bombings of Hiroshima and Nagasaki no consistent effect on the frequency of malformed infants or perinatal deaths was demonstrable. However, this "negative" study helps to make more precise the area where the search for genetic differences must be carried on.

ATOMIC BOMB EXPOSURE AND THE PREGNANCIES OF BIOLOGICALLY RELATED PARENTS

A PROSPECTIVE STUDY OF THE GENETIC EFFECTS OF IONIZING RADIATION IN MAN

William J. Schull, Ph.D., and James V. Neel, M.D., Ph.D.

THE CONCEPTS and particularly the technics of modern epidemiology are surprisingly similar to those which the geneticist associates with what he terms population genetics. It is one of the purposes of this presentation to indicate this parallelism in approaches through a consideration of a study of the genetic effects of ionizing radiation in man. While the data to be reported have not previously appeared in the literature, we shall emphasize methodology here rather than data.

The prospects for genetic damage following exposure to ionizing radiations have been admirably summarized in the reports of the U. S. National Academy of Sciences-National Research Council (1956), the British Medical Research Council (1956), the World Health Organization (1957), and more recently, the United Nations Scientific Committee on the Effects of Radiation (1958). We shall assume that the substance of these reports is now a matter of common knowledge and, we shall, therefore, not review the experimental observations from which one argues that a genetic risk will accrue from the exposure of human beings to ionizing radiations. Assessment of this risk in man has been hampered by a lack of basic information regarding such variables as spontaneous rates of mutation, selection pressure, average overdominance of nominally recessive mutant genes, and others. Furthermore, and fortunately so, the number of human populations exposed to high or moderately high amounts of ionizing radiations are few. One such population, however, comprises the survivors of the atomic bombings of Hiroshima and Nagasaki. The bulk of the results of a comprehensive study of the effect of exposure to the atomic bombs on this population has been reported elsewhere.[1,2] In the major report, attention was confined to observations on pregnancies occurring to unrelated parents. During the same interval of time, namely, 1948-1953, observations were also obtained on some 5,163 registered pregnancies occurring

Reproduced, with permission, from the *American Journal of Public Health*, vol. 49, © 1959 by the American Public Health Association, Inc.

in Hiroshima and Nagasaki to parents who were related as first cousins, first cousins once removed, or second cousins. There are a number of reasons why the geneticist is inclined to view the latter pregnancies of related parents as potentially a more sensitive index of radiation-induced genetic damage. In its simplest terms, the main line of the argument is that the addition of radiation-induced mutants to the more homozygous, and presumably less elastic, genetic background of inbred children may produce a relatively greater effect than would be apparent if the same mutants were superimposed on the more heterozygous genetic background of noninbred children, that is, children born to unrelated parents. It seemed unwise, therefore, in the analysis of the data from Hiroshima and Nagasaki, to simply pool the present observations with those on noninbred children, particularly since the inbred children were not uniformly distributed over the various exposure classes to be described shortly. The present report serves then to complete the picture of the effects of parental exposure to the atomic bombing of these two cities.

Brief Description of the Program—In the years immediately following World War II, a ration system existed in Japan which permitted pregnant women to acquire certain rationed items by registering the fact of their pregnancy some time after the fifth month of gestation. The economic milieu of Japan in this period was such that registration was virtually complete. There existed, then, a set of circumstances which afforded an opportunity to launch a comprehensive prospectively oriented study of pregnancy terminations following parental irradiation. With the cooperation of the municipal authorities in Hiroshima and Nagasaki, a system was instituted in 1948 whereby at the time of her registration for rations, each pregnant woman or her representative also registered with the Atomic Bomb Casualty Commission (ABCC), an agency of the National Academy of Sciences-National Research Council working in cooperation with the National Institute of Health of Japan. At the time of registration, the first two-thirds of a questionnaire was completed which included such items as identifying information, a brief radiation history of the husband and wife, a short summary of the past reproductive performances, and pertinent details concerning the present pregnancy. Upon the termination of the pregnancy, the attending midwife or physician notified the ABCC, and completed the previously mentioned questionnaire. Irrespective of the type of termination, a Japanese physician in the employ of the ABCC or the Japanese National Institute of Health called to examine the child as promptly after birth as possible. The completeness of this system of reporting and follow-up was checked periodically; these checks revealed that approximately 93 per cent of the births occurring in Hiroshima, and a somewhat higher figure in Nagasaki, were known to the commission. A large proportion of the 7 per cent not ascertained through this scheme subsequently came to our attention through other channels. The latter, unregistered births, while not included in the results to be reported, permit an appraisal of the representativeness of the registered births.

In the event that a pregnancy terminated abnormally as in a stillbirth or a child with a congenital malformation, a supplementary questionnaire was completed in the patient's home by the examining physician. This questionnaire was designed to obtain more detailed information on the gynecologic history, maternal illness during pregnancy, past reproductive performance, and economic status. In addition to this questionnaire, blood was drawn from the mother for a serological test for syphilis (on the average, some 5 per cent of these tests were positive). The same supple-

mentary questionnaire was routinely completed on every registration where the terminal digit in the registration number was zero.

A more complete description of the program will be found in Neel and Schull.[1] Information bearing on the reliability of the consanguinity data, and the importance of certain socioeconomic concomitants on the data to be presented has been given elsewhere.[3]

Parental Radiation Exposure—Clearly to extract the maximum information from the situation we have described, an experimental design more sophisticated than the simple dichotomy, exposed-not-exposed, was needed. Available to estimate the exposure of a given individual was information on (1) his or her distance from ground zero at the time of the bombing, (2) the occurrence of epilation, gingivitis and petechiae, symptoms highly correlated with radiation sickness, and (3) the shielding he or she may have experienced. Available also was some knowledge regarding the neutron and gamma dose, in air, at specified distances from ground zero. The latter information was considerably less precise than could be desired, or than is frequently imagined. The distance-dosage relationships have been repeatedly revised in the intervening 13 years, and will most probably undergo further revision. This, then, was the background on which one had to decide how best to recognize varying degrees of radiation exposure.

In general, two solutions to the problem of estimating degree of exposure were available. On the one hand, one could attempt to assign a score to each exposed individual, this score being a function of distance, shielding, and symptomatology. If applicable, this approach would permit the use of somewhat more elegant procedures for estimating the dose-genetic effect relationships than could be used with the second alternative. It has the disadvantage, in practice, of requiring a value judgment in the assignment of almost every score, and as a consequence, the apparent precision in estimating the genetic effect may well be spurious. On the other hand, one could adopt a classificatory scheme. The principal advantage of the latter, perhaps, is that it is the more conservative of the two procedures, and less subject to dating by revisions in the basic distance-dosage relationships. The classificatory scheme evolved to take into account distance, shielding, and symptomatology was as follows:

1. Not present in Hiroshima or Nagasaki at the time of the bombings.
2. Present in one or the other of the two cities, but asymptomatic, and at a distance from ground zero (a) greater than 3,000 meters, or (b) 0-3,000 meters and heavily shielded, or (c) 1,500-3,000 meters and moderately shielded, or (d) 2,000-3,000 meters and lightly shielded.
3. Present but asymptomatic, and at a distance of (a) 2,000-3,000 meters and unshielded, or (b) 1,000-2,000 meters and lightly shielded, or (c) 0-1,000 meters and moderately shielded.
4. Present but asymptomatic, and at a distance of (a) less than 2,000 meters and unshielded, or (b) less than 1,000 meters and lightly shielded.
5. Present but less than 3,000 meters from ground zero, and reporting epilation a/o gingivitis, a/o petechiae.

The structures which were defined as affording heavy, moderate, or light shielding are given in Neel and Schull.[1, p. 44] From a variety of sources of evidence, it has been estimated that these five categories of exposure correspond to doses of approximately 0, 5-10, 40-80, 100-150, and 200-300 roentgens equivalent physical respectively.

The Data—Of the 5,163 registered pregnancies of related parents which were observed, 382 were rejected prior to analysis for a variety of reasons, among these reasons being (1) incomplete information on birth weight, birth rank, maternal age, or parental exposure, (2) relationship of uncertain degree, more remote than second cousins, or closer than first cous-

Table 1—The Distribution of Births to Parents Related as First Cousins, First Cousins Once Removed, or Second Cousins by City and Parental Exposure

a. Hiroshima

		Mother's Exposure				
		1	2	3	4-5	Total
Father's Exposure	1	1,141	293	89	71	1,594
	2	91	108	22	9	230
	3	40	23	16	6	85
	4-5	22	16	3	5	46
Total		1,294	440	130	91	1,955

b. Nagasaki

		Mother's Exposure				
		1	2	3	4-5	Total
Father's Exposure	1	1,481	670	65	33	2,249
	2	204	286	13	10	513
	3	13	22	7	–	42
	4-5	8	8	4	2	22
Total		1,706	986	89	45	2,826

ins, (3) induced pregnancy terminations, and (4) multiple births. Table 1 gives the distribution of the remaining 4,781 births to related parents by parental exposure and city. Because of the paucity of individuals falling into exposure categories 4 and 5, these individuals are combined with group 3 in all subsequent tables (the estimated mean exposure for the combined groups (3-5) is about 100 reps).

Ideally in the analysis of these data, the precise relationship of the parents would be taken into account by a procedure, say, such as regressing the variable in question on the coefficient of inbreeding within each of the various exposure cells, and then testing the homogeneity of the intercepts and regression coefficients so obtained. Unfortunately, the data are insufficient, particularly at the higher combined parental exposures, to make this approach feasible. A less sensitive procedure is to pool the various types of parental relationship within each of the exposure cells, and then contrast the pooled observations. The validity of this latter approach rests on several assumptions, the most important of which is that the average coefficient of inbreeding, as well as the variances of these coefficients, is the same in each of the exposure cells. Table 2 presents the weighted average coefficient of inbreeding for the nine exposure cells. It will be noted that these values range from 0.0391 to 0.0501, but in general, cluster fairly tightly around the mean over-all exposure cells, namely, 0.0459. The within exposure cell variances exhibit a comparable range of variation to that observed among the means. Thus, it would seem that the related marriages encountered in the exposure cells are sufficiently similar in type and frequency that to pool the observations on the various degrees of relationship within exposure cells would not lead to significant confounding of inbreeding effect with the effects due to parental exposure.

Information is available on the following variables presumably indicative of

Table 2—Mean Coefficient of Inbreeding Among the Offspring of Related Parents by Parental Exposure. The Numbers of Observations on Which These Means Rest Are Given in Parentheses

		Mother's Exposure			
		1	2	3-5	Total
Father's Exposure	1	0.0457 (2,622)	0.0466 (963)	0.0451 (258)	0.0459 (3,843)
	2	0.0453 (295)	0.0468 (394)	0.0391 (54)	0.0457 (743)
	3-5	0.0444 (83)	0.0501 (69)	0.0421 (43)	0.0459 (195)
Total		0.0457 (3,000)	0.0468 (1,426)	0.0438 (355)	0.0459 (4,781)

genetic radiation damage, the sex ratio, the frequency of major congenital malformations, the frequency of stillbirths, the frequency of death in the first week of life among liveborn children, birth weight, and certain anthropometric data obtained approximately nine months post partum on a sample of those surviving to this age. The observations on the sex ratio have been presented elsewhere[2] as has the information on birth weight and the anthropometric measurements.[4] No significant differences among exposure cells in the birth weights (means or variances) or among the anthropometric measurements were demonstrable. The sex ratio data on the offspring of the related parents alone did not reveal a significant exposure effect. However, when taken in conjunction with the data presented on the unrelated parents, there emerged small, but consistent differences in the sex ratio compatible with the effects to be expected if sex-linked lethal genes were induced by the radiation. We present here the observations on the frequencies of major congenital anomalies and perinatal mortality (stillbirths plus deaths in the first week of life).

The specific malformations defined as major are tabulated in Neel and Schull[1]; in substance, any malformation which was incompatible with life, or, if compatible, seriously limited the function of the individual, was defined as a major malformation. The distribution by parental exposure and city of birth of children with major congenital malformations born to related parents is given in Table 3. In the analysis of these data, observations from Hiroshima and Nagasaki have been pooled to increase the numbers of entries in the exposure cells. This assumes that there is no heterogeneity between cities, an assumption which may not be strictly true. However, in an analysis of the effects of inbreeding on pregnancy termination in Japan based upon children born to parents who were either unexposed or lightly exposed (category 2), Schull[3] found no evidence of heterogeneity between Hiroshima and Nagasaki in the regression coefficients when the frequency of major congenital malformations was regressed on the coefficient of inbreeding F (including F = 0, that is, the offspring of unrelated parents). There was, however, evidence of heterogeneity in the regression of "early deaths" on the coefficient of inbreeding. The term "early death" does not cover a period of time equal to that encompassed by "perinatal death." The former includes death from the seventh day of life to the end of the first month. Since most deaths in the first month occur, in fact, in the first week, it is reasonable to assume that heterogeneity would also exist between cities with regard to perinatal mortality, and this can be shown to be true.

Analysis of the observations on major congenital anomalies by either the method of Roy and Kastenbaum[5] or regression technics failed to reveal a significant effect of maternal or paternal exposure on the frequency of major congenital malformations. The regression model used was of the following general form:

$$E(p_{ij}) = \bar{p} + b_1 (F_i - \bar{F}) + b_2 (M_j - \bar{M})$$

where $E(p_{ij})$ is the expected proportion of malformed infants in the ij^{th} exposure cell, \bar{p} is the average proportion of malformed infants, b_1 and b_2 are regression coefficients, F_i and \bar{F} (M_j and \bar{M}) are respectively the average exposure of fathers (mothers) in the i^{th} (j^{th}) exposure category and the mean paternal (maternal) exposure. In actual computation, a weighted regression was fitted using the arc sin transformation rather than the observed proportions. It must, of course, be borne in mind that the application of regression methods to these data involve an element of approximation. As might be surmised from inspection of Table 3, the regression co-

Table 3—The Distribution by City of Birth and Parental Exposure of Infants with Major Congenital Anomalies Born to Parents Who Were Biologically Related

a. Hiroshima

			Mother's Exposure 1	2	3–5	Total
Father's Exposure	1	N	1,141	293	160	1,594
		m	18	4	1	23
		p	0.0158	0.0137	0.0063	0.0144
	2	N	91	108	31	230
		m	0	1	0	1
		p	–	0.0093	–	0.0043
	3–5	N	62	39	30	131
		m	1	0	0	1
		p	0.0161	–	–	0.0076
Total		N	1,294	440	221	1,955
		m	19	5	1	25
		p	0.0147	0.0014	0.0045	0.0128

b. Nagasaki

			Mother's Exposure 1	2	3–5	Total
Father's Exposure	1	N	1,481	670	98	2,249
		m	25	7	1	33
		p	0.0169	0.0104	0.0102	0.0147
	2	N	204	286	23	513
		m	3	4	0	7
		p	0.0147	0.0140	–	0.0136
	3–5	N	21	30	13	64
		m	0	0	1	1
		p	–	–	0.0769	0.0156
Total		N	1,706	986	134	2,826
		m	28	11	2	41
		p	0.0164	0.0112	0.0149	0.0145

efficients are both negative in sign although neither is significantly different from zero. The surprise with which one greets this failure to demonstrate significant differences among the exposure groups will be in part a function of whether one views natural selection as primarily favoring homozygotes or heterozygotes. At the present time such is the paucity of data on man that one can defend either of these alternatives with equal vigor. In this connection,

Neel[6] has recently presented evidence which can be interpreted as suggesting that some congenital malformations in man may represent the phenodeviants from balanced homeostatic systems. There may, then, be no contradiction between the apparent effect of parental exposure on the sex ratio, and the failure to observe an effect of parental exposure on the frequency of congenital defects.

Table 4 presents the distribution, by city of birth and parental exposure, of

Table 4—The Distribution by City of Birth and Parental Exposure of Infants Dying in the Perinatal Period Born to Parents Who Were Biologically Related

a. Hiroshima

Father's Exposure			Mother's Exposure 1	2	3–5	Total
	1	N	1,123	289	159	1,571
		d	48	15	4	67
		p	0.0427	0.0519	0.0252	0.0426
	2	N	91	107	31	229
		d	4	3	4	11
		p	0.0440	0.0280	0.1290	0.0480
	3–5	N	61	39	30	130
		d	2	8	2	12
		p	0.0328	0.2051	0.0667	0.0923
Total		N	1,275	435	220	1,930
		d	54	26	10	90
		p	0.0424	0.0598	0.0455	0.0466

b. Nagasaki

Father's Exposure			Mother's Exposure 1	2	3–5	Total
	1	N	1,456	663	97	2,216
		d	41	22	1	64
		p	0.0282	0.0332	0.0103	0.0289
	2	N	201	282	23	506
		d	8	3	0	11
		p	0.0398	0.0106	–	0.0217
	3–5	N	21	30	12	63
		d	0	1	0	1
		p	–	0.0333	–	0.0159
Total		N	1,678	975	132	2,785
		d	49	26	1	76
		p	0.0292	0.0267	0.0076	0.0273

infants dying in the perinatal period. It is worth emphasizing that perinatal deaths include those pregnancies terminating after 21 weeks of gestation in a stillborn infant or a child failing to survive the first week of life, and where the child in question did not have a visible, gross abnormality. The elimination of stillborn children or those dying in the first seven days of life with known abnormality precludes the possibility of measuring any genetic damage twice, a situation which might otherwise prevail in view of the high probability that a grossly deformed child will be either stillborn or succumb in the first week of life.

On inspection, the most striking aspect of Table 4 is the apparent heterogeneity between cities. One not only observes a fairly marked difference in perinatal death rates in the two cities, but the increase in mortality with parental exposure which seems to occur in Hiroshima is countered by a decreasing mortality with increasing exposure in Nagasaki. Analysis of these data readily confirms that the heterogeneity is significant. When these data are explored further, one finds that the differences between the nine exposure groups in Nagasaki are not significant whereas in Hiroshima the groups are significantly different. However, when the model previously described is fitted to the Hiroshima data, one finds that though the regression removes a significant amount of variation the variation not removed by the regression is also significantly large. In short, the linear model would appear inadequate to account for more than a small fraction of the observed variation. Interesting too, is the finding that on the linear model the regression coefficients associated with mother's exposure and father's exposure differ in sign. In most circles, the latter would not be considered consistent with our present knowledge of genetic damage following parental radiation. It must be borne in mind that these seemingly anomalous findings may be due to confounded concomitant variation. Unfortunately, such is the paucity of data that it seems unlikely that uncontrolled concomitant variation can be satisfactorily removed. To briefly summarize the findings, there is no demonstrable, consistent effect of parental exposure on the frequency of malformed infants or perinatal deaths among the children born to related parents.

Before we examine the conclusions which seem justified from these data, permit us to restate, for comparative purposes, the findings with reference to the frequency of congenital malformations and of perinatal mortality obtained from the study of offspring of unrelated parents. These were as follows[1]: "Analysis of the frequency of malformed infants by city and parental exposure reveals no significant, consistent effect of parental exposure." "Analysis of the stillbirth data fails to reveal significant differences between cities or consistent significant effects of parental exposure." Finally, "No consistent, significant effect of parental exposure on neonatal mortality emerges from the data obtained in Hiroshima and Nagasaki on deaths occurring in the first six days post partum." The present findings are not, then, at variance with those previously reported for pregnancy terminations to unrelated parents.

What now are the conclusions which we can draw from these data? Can we assert that no genetic damage has accrued from exposure of the populations of Hiroshima and Nagasaki to ionizing radiations? The answer to this question is "no" since it can easily be shown that, with the present numbers, the differences which would have to obtain among exposure groups for significance to result are considerably larger than those expected on a genetic basis. This body of data does not, then, afford an adequate test of the genetic hypothesis. In much the same fashion, one is led to the con-

clusion that these data do not afford a very sharp test of the basic premise that inbred children because of their increased homozygosity are a more sensitive indicator of radiation induced genetic damage than noninbred children since a several-fold difference in response between these groups could go undetected. Must we conclude, then, that these data contribute little, if anything to our understanding of the genetic hazards of ionizing radiation in man? Fortunately no; in point of fact, each "negative" study serves to further isolate the "critical range," that is, that area wherein we must search for genetic differences. A corollary of this would be that each "negative" study further refines the experimental design from which we may ultimately answer our basic problem. Advances in this area of human biology, important and as necessary as they are, will not come easily nor inexpensively, but they will come. Unhappily, however, we are called upon even now to make decisions regarding "permissible doses." These decisions can affect not only the practice of medicine in this country but our national safety as well. Our thinking on these matters is not aided either by those who with head in sand deny the existence of a danger nor those who see in each new increment of fallout x new monsters.

Wisdom would certainly dictate that in so far as is consistent with national defense and the sagacious practice of medicine we should keep at a minimum the exposures to which we are subject.

While it may seem that in these closing comments we have strayed somewhat afield from our objective of indicating the similarity of approaches in epidemiology and in population genetics, this is really not so. In both areas, we are frequently confronted with problems of such moment that objectivity is difficult to attain. Yet only in an atmosphere of impartiality can we hope to sort fact from fancy, be this with reference to smoking and lung cancer, or ionizing radiation and our "load of mutations."

REFERENCES

1. Neel, James V., and Schull, William J. The Effect of Exposure to the Atomic Bombs on Pregnancy Termination in Hiroshima and Nagasaki. National Academy of Sciences-National Research Council Publ. 461, 1956.
2. Schull, William J., and Neel, James V. Radiation and the Sex Ratio in Man. Science 128:343 (Aug.), 1958.
3. Schull, William J. Empirical Risks in Consanguineous Marriages: Sex Ratio, Malformation, and Viability. Am. J. Human Genet. 10:294 (Sept.), 1958.
4. Morton, Newton E. Empirical Risks in Consanguineous Marriages: Birth Weight, Gestation Time, and Measurements of Infants. Ibid. 10:344 (Sept.), 1958.
5. Roy, S. N., and Kastenbaum, Marvin A. A Generalization of Analysis of Variance and Multivariate Analysis to Data Based on Frequencies in Qualitative Categories or Clan Intervals. Ann. Math. Statist. 27:749 (Sept.), 1956.
6. Neel, James V. A Study of Major Congenital Defects in Japanese Infants. Am. J. Human Genet. 10:398 (Dec.), 1958.

The authors are associated with the Department of Human Genetics, University of Michigan Medical School, Ann Arbor, Mich.

This paper was presented before a Joint Session of the Epidemiology and Maternal and Child Health Sections of the American Public Health Association at the Eighty-Sixth Annual Meeting in St. Louis, Mo., October 29, 1958.

This work was sponsored by the Atomic Bomb Casualty Commission, Field Agency of the National Academy of Sciences-National Research Council of the United States and the National Institute of Health of Japan. Analysis of these data was sponsored by the Atomic Energy Commission under a grant to the University of Michigan (Contract AT (11-1) –405).

THE CHILDREN OF ATOMIC BOMB SURVIVORS

Some Further Observations on the Sex Ratio Among Infants Born to Survivors of the Atomic Bombings of Hiroshima and Nagasaki

WILLIAM J. SCHULL,[1] JAMES V. NEEL,[1] AND A. HASHIZUME[2]

[1]Department of Human Genetics,
University of Michigan Medical School,
Ann Arbor 48104.
[2]Atomic Bomb Casualty Commission,[*]
Hijiyama-koen, Hiroshima, Japan.

SOME YEARS AGO, we presented the evidence then existing pertinent to the association in man of the sex ratio with parental exposure to ionizing radiation (Schull and Neel, 1958). At that time, we tentatively concluded that "the sex of children born to the survivors of the atomic bombings of Hiroshima and Nagasaki reveals significant changes in the sex ratio of these children, changes in the direction to be expected if exposure had resulted in the induction of sex-linked lethal mutations." This conclusion was reached through a somewhat tortuous line of reasoning, for no single study, in Japan or elsewhere, offered unequivocal evidence of a radiation-induced change. In the years which have intervened, there have been two developments which warrant a reinspection of this association and, of course, the earlier conclusion. First, further data are available. Second, we now recognize additional complexities in the problem, in particular that nondisjunction of the sex chromosomes can occur in man and, if increased by ionizing radiation, could complicate the interpretation of any changes in the sex ratio. Specifically, a radiation-induced increase in the frequency of nondisjunction conceivably could either obscure an increase in sex-linked lethal mutants or lead to a spuriously high estimate of their frequency.

The purpose of this paper is to record some 47,624 observations made in Hiroshima and Nagasaki in the years 1956 through 1962 relevant to the functional relationship between the sex ratio and parental exposure to atomic radiation and, in the light of these further observations, to review the evidence presently available on this issue. In terms of the total radiation dose represented, these new data increase the observations previously available by about 70%.

STUDIES OF THE SEX RATIO IN HIROSHIMA AND NAGASAKI

The data from Hiroshima and Nagasaki concerning the effects of radiation on the sex ratio fall into four categories, as follows, three of which have been

Received January 26, 1966.

[*]The Commission is a co-operative research agency of the U. S. National Academy of Sciences-National Research Council and the Japanese National Institute of Health of the Ministry of Health and Welfare, with funds provided by the U. S. Atomic Energy Commission, the Japanese Institute of Health, and the U. S. Public Health Service.

328

Reproduced, with permission, from the *American Journal of Human Genetics*, vol. 18, © 1966 by the American Society of Human Genetics.

presented previously: (1) the sex ratio in the infants born to *unrelated* parents in the years 1948 through 1953 (Neel and Schull, 1956), (2) the sex ratio in infants born to *related* parents in the years 1948 through 1953 (Schull and Neel, 1958, 1959), (3) the sex ratio in infants born to unrelated parents in the years 1954 through 1955 (Neel and Schull, 1956), and (4) the sex ratio in infants born to parents (related and unrelated) in the years 1956 through 1962. The rationale for analyzing pregnancy terminations of biologically related parents separately from unrelated parents has been presented elsewhere (Schull and Neel, 1959). The principal distinctions between these groups, aside from consanguinity in the parents, are as follows: Categories (1) and (2) differ from (3) and (4) in that pregnancies included in (1) and (2) were ascertained through a system of registration carried out in conjunction with the special ration registration for pregnant women which existed in the postwar period in Japan; whereas (3) and (4) were ascertained at the time that the city office of vital statistics was notified of the birth of a child. The details of these two methods of ascertainment have been given elsewhere (Neel and Schull, 1956; Schull and Neel, 1958). All children comprising categories (1) and (2) were examined by physicians either in the employ of the Atomic Bomb Casualty Commission (ABCC) or the Japanese National Institute of Health; among the functions of these physicians was the verification of the sex of the child as reported by the individual attending the delivery of the child, generally a midwife. In groups (3) and (4), the sex ascribed to a given infant is that reported at the registration of the birth and was not verified by a physician employed by either of the agencies previously mentioned. Categories (1) and (2) differ from one another only insofar as the children are the offspring of unrelated spouses in one instance, (1), and not in the other, (2). Groups (3) and (4) differ in two respects. First, group (3) represents *all* of the births reported to the municipal authorities in the years 1954 through 1955 except those where the parents were related; whereas group (4) represents only those births reported to the municipal authorities in the years 1956 through 1962 where either both parents were included in the so-called Master Sample which forms the basis of the Life Span Study of the Atomic Bomb Casualty Commission (Beebe, Ishida, and Jablon, 1962) or, if the exposure status of one parent was unknown, the other was stated to have been within 2,000 meters of ground zero. In the former instance, the exposure status of both parents was known to ABCC; in the latter, further effort was necessary to complete the exposure histories of the spouses.

In the years 1954 through 1955, whenever the exposure status of an individual was unknown, an effort was made routinely, either by mail or by interview, to ascertain whether the person had been exposed to the atomic bombing of Hiroshima and/or Nagasaki. If the answer was affirmative, a trained field investigator obtained an exposure history which included such items as the distance from ground zero at the time of the bombing (ATB), the occurrence of symptoms indicative of radiation sickness, etc. Since the files of ABCC included all individuals, irrespective of place of residence, who had reported themselves to be exposed either at the time of the national census of

Japan in 1950 or at an earlier *ad hoc* radiation census of these two cities, rarely did an individual whose exposure status was not known prove to have been exposed. Thus, the vast majority of inquiries, undertaken at considerable cost in time and effort, failed to augment the group of children of greatest interest, namely, those born to exposed parents. Accordingly, when the burden of field interviews exceeded the manpower available, as was inevitable since the relative proportion of births to nonexposed parents—those least likely to be included in the Master File—increased with each year, the decision was reached retroactive to 1956 to accept for study only those births where either the exposure status of both parents could be ascertained by a file check or where the exposure of only one parent had to be determined by interview, the other parent having been in the so-called "proximally exposed" group, i.e., within 2,000 meters of the hypocenter ATB. Clearly, this decision, from which the present body of data stems, resulted in the loss to the series of some children one or both of whose parents were exposed, but this number is presumably small, for a loss would occur only if the parents had not been included in one of the aforementioned censuses or if, at the time of these censuses, they had elected to deny their exposure.

As previously stated, the data comprising categories (1), (2), and (3) have been analyzed and the results presented elsewhere (summary in Schull and Neel, 1958). Briefly, the analysis proceeded as follows: Within each of the three categories, the pregnancies were viewed as divisible into three sets, namely, those where the mother was exposed and the father not, where the father was exposed and the mother not, and where both parents were exposed. Within each of these sets, there existed three or more dosage levels. Thus, for each of the categories (1), (2), and (3), it was possible to fit three linear regressions of the frequency of male births on the dose of radiation received by the parent(s). Two of these, corresponding to those cases where only one parent was exposed, were of the form

$$E(p_i) = p_0 + bd_i,$$

where $E(p_i)$ is the expected proportion of males in the i^{th} exposure class; p_0 is the proportion of males expected at zero dose, i.e., in the nonexposed group; d_i is the dose in the i^{th} exposure class; and b is the regression coefficient. The third regression was of the form

$$E(p_{ij}) = p_0 + b_1 F_i + b_2 M_j$$

where b_1 and b_2 are now partial regression coefficients, F_i and M_j are, respectively, the doses in the i^{th} paternal and j^{th} maternal exposure groups, and p_0 is, again, the proportion of males expected in the nonexposed class. The regressions which were, in fact, fitted were weighted to allow for the differences in the numbers of observations at the various exposure levels. The weights which were used were the reciprocals of the variances of the proportion of males at the different dosage levels; this variation was assumed to be solely binomial in origin, an assumption which is not wholly justified in view of the occurrence of siblings in the data and the effects of the extraneous variables which have been

TABLE 1. MEANS AND REGRESSION COEFFICIENTS OBTAINED BY FITTING A WEIGHTED LINEAR REGRESSION OF THE PROPORTION OF MALE BIRTHS TO AVERAGE GROUP EXPOSURE IN r IN THE HIROSHIMA AND NAGASAKI DATA

The values in parentheses are those obtained when unexposed parents are rejected.
(Extended from Schull and Neel, 1958)

	Father only exposed		Mother only exposed		Both parents exposed		
	p_0	b*	p_0	b	p_0	Father b	Mother b
1948–1953							
Unrelated parents	0.5202	0.0058 (0.0094)	0.5213	−0.0101† (−0.0111)	0.5102	0.0039	−0.0037
Related parents	0.5307	0.0188 (0.0423)	0.5204	−0.0116 (0.0386)	0.5310	0.0024	−0.0179
1954–1955							
Unrelated parents	0.5211	0.0039 (0.0047)	0.5186	0.0090 (0.0141)	0.5484	0.0137	−0.0269
1956–1962	0.5150	−0.0016 (−0.0036)	0.5141	0.0072 (0.0077)	0.5137	−0.0129	0.0044
All data combined	0.5191	−0.0001	0.5185	0.0027	0.5166	−0.0020	−0.0035

*Increase or decrease per 100 rep.
†Significant at the 5% level.

mentioned. The problems inherent in treating nonbinomial variation and the types of bias it may introduce have been discussed by Cochran (1940). The final weights were obtained by iteration, starting with the observed proportions as trial values. The intercepts and regression coefficients which were obtained for the first three categories of data are given in Table 1.

We present in Tables 2 and 3, along with the earlier data, the 47,624 observations in Hiroshima and Nagasaki which have become available since the analysis summarized in the previous paragraphs. These observations are distributed by maternal and paternal exposure; for a justification of the average exposures assigned to the various groups, the reader is referred to an earlier publication (Neel and Schull, 1956, Chapter 4).* The years and cities have been pooled, but only after a search was made for significant heterogeneity within cities between years and within years between cities. No consistent, significant differences emerged either between cities or years. The results of an analysis of these more recent data, in the manner previously

*Since this paper was submitted, there has appeared a re-estimation of the gamma and neutron distance-dose relationships associated with the nuclear devices exploded over Hiroshima and Nagasaki (see Auxier et al., 1966. Free-field radiation-dose distributions from the Hiroshima and Nagasaki bombings. *Health Physics* 12: 425–429). These new estimates suggest that the values previously assumed to hold for Hiroshima may be in error (upwards) by a factor of two, on the average; there are no substantial changes in the doses in Nagasaki. Clearly, if true, the regression coefficients given in Table 1 are too high; however, there remain many unresolved problems associated with individual exposures in these cities which counsel caution in a too hasty extrapolation from these new curves to the biological situation. Thus, we are still uncertain of the appropriate attenuation factors for various forms of shielding and the integrated fall-out dose, and, moreover, biological dosimeters such as frequency of epilation, etc., are at variance in the middle distances with the purely physical estimates.

TABLE 2. SUMMARY OF THE FINDINGS IN HIROSHIMA AND NAGASAKI WITH REGARD TO THE ASSOCIATION OF THE FREQUENCY OF MALE BIRTHS AND PARENTAL EXPOSURE
Only one parent exposed.

Father only exposed			Estimated mean exposure (rep)	Mother only exposed			Estimated mean exposure (rep)
Total births	Male births	p		Total births	Male births	p	
1948–1953, PARENTS UNRELATED							
31,904	16,613	0.5207	0	31,904	16,613	0.5207	0
3,670	1,892	0.5155	8	14,684	7,681	0.5231	8
839	442	0.5268	75	2,932	1,474	0.5027	75
534	284	0.5318	200	1,676	850	0.5072	200
1954–1955, PARENTS UNRELATED							
11,640	6,067	0.5212	0	11,640	6,067	0.5212	0
1,498	774	0.5167	8	4,926	2,512	0.5099	8
387	211	0.5452	75	1,026	562	0.5478	75
219	113	0.5160	200	592	311	0.5253	200
1948–1953, PARENTS RELATED							
2,622	1,396	0.5324	0	2,622	1,396	0.5324	0
295	152	0.5153	8	963	466	0.4839	8
83	46	0.5542	100	258	134	0.5194	100
1956–1962, ALL PARENTS							
20,382	10,483	0.5143	0	20,382	10,483	0.5143	0
4,841	2,505	0.5175	8	9,284	4,788	0.5157	8
1,730	892	0.5156	75	2,087	1,060	0.5079	75
1,199	612	0.5104	200	1,191	638	0.5357	200

outlined, are also given in Table 1. With the addition of these data, there are now sixteen regression coefficients associated with the four categories of observations. Only one, namely, mothers only exposed, 1948 through 1953, unrelated parents, can be shown to be significantly different from zero. Furthermore, inspection of the array of regression coefficients with respect to the direction and magnitude of the changes with parental exposure fails to suggest an unequivocal pattern; we note six of eight regression coefficients associated with paternal exposure are positive whereas three of eight associated with maternal exposure are positive. Neither of these arrays of signs is particularly unlikely under the hypothesis that plus and minus signs are equally probable. It will be noted that the regression coefficients observed in the new data tend to be opposite in sign from those obtained previously, thereby essentially destroying the consistency in the data upon which so much of the earlier argument for an effect of radiation on the sex ratio was based. Finally, if the differences between the various sets of observations are ignored and the aggregate analyzed, the data fail to disclose a significant association of either maternal or paternal exposure with the sex ratio.

It will be noted from Table 2 that certain observations, those where both parents were unexposed, enter into the estimation of the maternal effect when the father was not exposed as well as the paternal effect when the mother was

TABLE 3. SUMMARY OF THE FINDINGS IN HIROSHIMA AND NAGASAKI WITH REGARD TO THE ASSOCIATION OF THE FREQUENCY OF MALE BIRTHS AND PARENTAL EXPOSURE

Total births	Male births	p	Estimated mean exposure (rep) Mother	Estimated mean exposure (rep) Father
		1948–1953, PARENTS UNRELATED		
5994	3053	0.5093	8	8
658	337	0.5122	8	75
422	225	0.5332	8	200
703	354	0.5036	75	8
615	319	0.5187	75	75
192	94	0.4896	75	200
318	165	0.5189	200	8
145	72	0.4966	200	75
145	71	0.4896	200	200
		1954–1955, PARENTS UNRELATED		
1474	806	0.5468	8	8
220	129	0.5864	8	75
174	101	0.5805	8	200
212	111	0.5236	75	8
107	53	0.4953	75	75
66	35	0.5303	75	200
89	48	0.5393	200	8
43	20	0.4651	200	75
33	18	0.5455	200	200
		1948–1953, PARENTS RELATED		
394	208	0.5279	8	8
69	38	0.5507	8	100
54	29	0.5370	100	8
43	21	0.4884	100	100
		1956–1962, ALL PARENTS		
3899	1996	0.5119	8	8
756	374	0.4947	8	75
572	283	0.4948	8	200
608	334	0.5493	75	8
314	160	0.5096	75	75
144	68	0.4722	75	200
374	183	0.4893	200	8
124	70	0.5645	200	75
119	59	0.4958	200	200

unexposed. These regression coefficients are, therefore, not wholly independent of one another, and one might argue that we are not dealing here with 16 independent regression coefficients but some lesser number. To obtain estimates of the effect of maternal and paternal exposure, when only one parent was exposed, which are fully independent, we need merely omit the observations on both parents not exposed. Although this achieves statistical independence, it must be clearly borne in mind that the loss of data which is involved leads also to a loss in the precision of the estimates. The values in

parentheses in Table 1 are those obtained when unexposed parents are rejected.

Thus far we have emphasized our efforts to demonstrate a significant association of the sex ratio with parental exposure; however, there is another aspect to the evaluation of radiation induced effects to which these data can contribute, and this is the specification of the upper limits of an effect which may have been induced but was not demonstrated. An approximate solution of this problem adequate for our purposes proceeds somewhat as follows: For convenience, we restrict our attention to the adequacy of these data for demonstrating differences between a single moderate-to-heavily exposed population and a suitable comparison population. More specifically, we seek to determine the probability of rejecting the hypothesis of no difference, given a fixed sample size, when the true situation is one wherein the exposed group differs from the standard by some amount, say d, for varying values of d. The groups of concern to us here are those terminations occurring to parents (1) neither of whom had exposures exceeding 8 rep, on the average, *the comparison group*, and (2) one or both of whom had exposures equaling or exceeding 75 rep, on the average, *the exposed group*. To simplify the calculations further, we shall assume that

1. The proportion of male births in the comparison population is known without error.

2. The normal approximation to the binomial distribution has the requisite accuracy over the critical range; here this includes those values for the exposed group which deviate from the comparison by not more than 5%.

3. The level of confidence with regard to the null hypothesis is 0.05.

We shall, in addition, limit ourselves to the cases of one parent exposed and assume that in these instances the direction of the deviation is specifiable in advance. The one-sided power functions or operating characteristics (see Paulson and Wallis, 1947) which one generates under these restrictions and for sample sizes of 9,762 (mothers) and 4,991 (fathers) are set out in Fig. 1. Before considering this figure, a comment or two on the assumptions enumerated above seems in order. Obviously the first assumption, namely, that the true proportion of male births in the comparison population is known without error, is not strictly satisfied by these data; however, the sample available on which to base the comparison value is sufficiently large, in excess of 75,000 births, that the error associated with the estimate of the true proportion of male births in this group is small under all circumstances and can be disregarded when contrasted with the error associated with an estimate based on the sample sizes available for the irradiated groups. The normal approximation requires no special justification since the sample size is not small and neither the proportion of male births in the comparison or the irradiated group is close to 0 or 1. Finally, the level of confidence is the conventional one.

To utilize Fig. 1, we must first define what constitutes an "adequate test." We shall say that a test is adequate with regard to the alternatives, $p_e - p_c > d$, where p_e and p_c are, respectively, the frequencies of male births in the exposed and comparison populations if the prior probability of detecting de-

FIG. 1. The probability of rejecting the null hypothesis if the true proportion of male births, p_e, among the offspring of exposed fathers or mothers is as specified and the proportion of male births, p_c, in the comparison population is taken to be 0.5189. The curves are based upon sample sizes of 9,762 infants born to exposed mothers and 4,991 to exposed fathers.

partures from the null hypothesis of size d or greater equals or exceeds some specified probability. The choice of this latter probability depends in large measure upon the importance one attaches to a type II error, that is, to accepting the null hypothesis as true when in fact it is false. In the present instance, it would seem to be the more cautious course to minimize such errors. Accordingly, we shall set the frequency of such errors to be 0.10, and this is tantamount to asserting that adequate power is 0.90. This value may well be too high in view of the simplifications we have introduced and the fact that not all of the observations which are germane enter into the specification of the curves in Fig. 1. However, the error is a conservative one insofar as the adequacy of these data is, in fact, greater than that which is here claimed for them. It is apparent from Fig. 1 that these data are adequate to detect a departure of approximately 1.5% (absolute) in the case of exposed mothers, and 2.2% in the case of exposed fathers.

DISCUSSION

Other Studies of the Effect of Radiation on the Sex Ratio in Man

We have elsewhere summarized the results of the various studies through 1962 on the relationship between exposure to radiation and the sex ratio in man (Neel, 1963; Schull, 1963). In four of the six studies involving paternal radiation, the sex ratio was increased; in all three instances involving maternal radiation, the sex ratio was decreased. A study missed in those summaries involved a small series of women exposed in the Joachimstal mines, in whose offspring the sex ratio was lower than in controls (Muller, Kubat, and Marsalek, 1962). A further study by Cox (1964) reports a decrease in the sex ratio in the children of Canadian patients receiving repeated diagnostic radiation because of congenital dislocation of the hip. Thus, in four of six other studies involving paternal radiation, the sex ratio is increased, while in five of five studies involving maternal radiation, the sex ratio is decreased. Although the apparent effect is statistically significant in only one of these studies, the manner in which the findings parallel expectation on the basis of simple sex-linked inheritance has made it tempting to accept their validity.

The circumstances and "information content" of the various studies to date are extremely uneven, rendering precise comparisons difficult. However, although an exact calculation is impossible, it would appear that the radiation dose represented in the children of exposed parents described in this and previous reports on Hiroshima and Nagasaki is greater than the sum total of the radiation doses represented by the children included in all other published studies to date. In addition to the numbers and doses involved, there is the added value in the present study of a system of ascertainment which ensures uniform treatment of all offspring regardless of the exposure status of their parents. On the basis of the operational curves described earlier, the present data are incompatible with effects of the magnitude reported in the papers of Musil (1962), Muller, Kubat, and Marsalek (1962), and Cox (1964). Contrariwise, they are consistent with the results of the remaining investigations.

On an Interpretation of the Present Studies

Ionizing radiation can and undoubtedly does produce lethal mutations associated with the X chromosome, and under the simplest of circumstances these mutations may lead to an altered sex ratio among progeny born subsequent to exposure. There is, we believe, more or less universal acceptance of the truth of this assertion, but this is not the issue as we see it. Rather it is a question of the best possible estimate of the rate of induction of sex-linked lethal mutations and the realization that other forces may be at work which make the simplistic point of view just set forth of dubious value. In particular, two questions concern us. First, what explanation, if any, can be advanced for the apparent reversal of an effect pereeived, albeit dimly, in the earlier data from Hiroshima and Nagasaki? Second, what interpretation is to be placed upon the totality of data presently available?

With respect to the first of these questions, one possible explanation, of course, is that there was in fact no effect and that the apparent effect in the

earlier data was fortuitous. But, if one assumes that the apparent effect was real, then obviously one is obliged to conclude that the later data, that is, the observations from 1956 to 1962, are not *in pari materia* with the earlier. This could be so because of a confounding of extraneous variation, but we have no evidence to support such a supposition. There is, however, another possibility. Some experimental evidence suggests that in female mice the yield in mutations, following an exposure experience, may diminish with time either as a consequence of the repair of certain mutations, of cell selection, or both (see Russell, 1965). It is conceivable, therefore, that the number of sex-linked lethal mutations potentially recoverable from the populations of Hiroshima and Nagasaki has diminished with time. However, a comparison of the regression coefficients for the three time intervals in which the data have been analyzed fails to reveal clear evidence for a time trend. It is possible, of course, that if a tendency of the kind implied exists, it is obscured by the accumulation of spontaneous sex-linked lethal mutations to be anticipated with time, especially in males (cf. Novitski and Sandler, 1956; Novitski and Kimball, 1958). However, if the difference in age between exposed and nonexposed parents is increasing with the passing years, then one might expect the increased frequency of spontaneous mutations to dampen the reduction in induced mutations. The net effect would be to minimize a trend. If, on the other hand, the difference in parental ages between the comparison groups is diminishing with time, possibly even to the point where the exposed parents are younger on the average, then the trend might be maximized. Although improbable perhaps, the sign of the regression coefficients might even be altered if the exposed parents were actually younger than the nonexposed.

One possible complication in the Japanese data as well as that from elsewhere in the world concerns the occurrence of chromosomal abnormalities involving the X or Y chromosome. While it has not been rigorously demonstrated that such abnormalities in man increase with exposure to ionizing radiation, there is evidence from other organisms which makes a presumption to this effect reasonable. Be this as it may, any effort to assay the impact of chromosomal abnormalities upon the sex ratio ultimately becomes an attempt to ascertain the relative frequencies of gametes lacking a sex chromosome and those having an accessory one among the gametes produced by an exposed father on the one hand and by an exposed mother on the other. Suffice it to state that current information precludes an exact analysis of this problem, but, to the extent that an evaluation is possible, it would seem that chromosomal abnormalities are not likely to alter the direction of change in the sex ratio anticipated on the basis of the segregation of sex-linked lethals but will almost certainly alter the magnitude of the latter change (see Neel, 1963).

We have repeatedly stressed the unsatisfactory nature of the sex ratio as a variable. It is apparently influenced by any number of factors, e.g. maternal age, paternal age, parity, etc. While the effects of these variables are generally small, adequate explanation of their origin has not been advanced despite the existence of bodies of data far larger than those pertinent to the radiation problem. Unfortunately, these unexplained perturbations are often lost sight of, and only the elegance of the genetic argument is seen. As matters stand—

and we sincerely hope this is our last word on the subject—the Hiroshima-Nagasaki data fail to provide unequivocal evidence for an effect of radiation on the sex ratio, although they are consistent with a small effect in the early post-bomb years which has since disappeared. The manner in which these data can be combined with the other findings from Hiroshima and Nagasaki to shed light on the "doubling dose" of radiation for man will be dealt with elsewhere (Kato, Schull, and Neel, 1966).

SUMMARY

Data are presented on the sex ratio of 47,624 children born in Hiroshima and Nagasaki in the years 1956 through 1962. The total number of births in these two cities for which information is available is now 140,542, and of this number in 73,994 instances one or both parents were exposed to the atomic bombings. The suggestion of an effect of exposure on sex ratio in the earlier data is not borne out by the present findings. One can argue either that a small early effect has disappeared or that the original observation had no biological significance.

REFERENCES

BEEBE, G. W., ISHIDA, M., AND JABLON, S. 1962. Studies of the mortality of A-bomb survivors. I. Plan of study and mortality in the medical subsample (selection 1), 1950–1958. *Radiat. Res.* 16: 253–280.

COCHRAN, W. G. 1940. The analysis of variance when experimental errors follow the Poisson or binomial laws. *Ann. Math. Stat.* 11: 335–347.

COX, D. W. 1964. An investigation of possible genetic damage in the offspring of women receiving multiple diagnostic pelvic X-rays. *Amer. J. Hum. Genet.* 16: 214–230.

KATO, H., SCHULL, W. J., AND NEEL, J. V. 1966. A cohort-type study of survival in the children of parents exposed to atomic bombings. *Amer. J. Hum. Genet.* 18: 339–373.

MULLER, C., KUBAT, M., AND MARSALEK, J. 1962. Der Einfluss des Arbeitsrisiko auf die Generationsfunktionen der beim Fördern und Aufbereiten von radioaktiven Rohstoffen beschäftigten Frauen. *Zbl. Gynäk.* 84: 561–568.

MUSIL, J. 1962. Vliv opakovaných plicních rtg vyšentřeni zen na vyvoj příští generace. *Rozhledy Tuberk.* 22: 164–169.

NEEL, J. V. 1963. *Changing Perspectives on the Genetic Effects of Radiation*. Springfield, Ill.: Charles C Thomas, 97 pp.

NEEL, J. V., AND SCHULL, W. J. 1956. *The Effect of Exposure to the Atomic Bombs on Pregnancy Termination in Hiroshima and Nagasaki*. Washington D. C.: National Academy of Sciences-National Research Council Publ. 461, 241 pp.

NOVITSKI, E., AND SANDLER, L. 1956. The relationship between parental age, birth order, and the secondary sex ratio in humans. *Ann. Hum. Genet.* (Lond.) 21: 123–131.

NOVITSKI, E., AND KIMBALL, A. W. 1958. Birth order, parental age, and sex of offspring. *Amer. J. Hum. Genet.* 10: 268–275.

PAULSON, E., AND WALLIS, W. A. 1947. Planning and analyzing experiments for comparing two percentages. In *Techniques of Statistical Analysis*, C. Eisenhart, M. W. Hastay, and W. A. Wallis, eds. New York: McGraw-Hill, pp. 249–265.

RUSSELL, W. L. 1965. Effect of the interval between irradiation and conception on mutation frequency in female mice. *Proc. Nat. Acad. Sci.* (U. S.) 54: 1552–1557.

SCHULL, W. J. 1963. Hereditary effects. *Nucleonics* 21: 54–57.

SCHULL, W. J., AND NEEL, J. V. 1958. Radiation and the sex ratio in man. *Science* 128: 343–348.

SCHULL, W. J., AND NEEL, J. V. 1959. Atomic bomb exposure and the pregnancies of biologically related parents. *Amer. J. Public Health* 49: 1621–1629.

A Cohort-Type Study of Survival in the Children of Parents Exposed to Atomic Bombings

HIROO KATO,[1] WILLIAM J. SCHULL,[2] AND JAMES V. NEEL[2]

[1]Department of Statistics,
Atomic Bomb Casualty Commission,* Hiroshima, and
Hiroshima Branch,
Japanese National Institute of Health,
Hiroshima, Japan.
[2]Department of Human Genetics,
University of Michigan Medical School,
Ann Arbor 48104.

AMONG THE GENETIC EFFECTS to be expected in the first generation progeny of mammals exposed to radiation is a shortening of the life span due to the action of deleterious mutations. Although in recent years a considerable body of data concerned with such experimental species as mice, rats, and swine has become available, pertinent observations in man remain, for obvious reasons, rare. The purpose of this paper is to describe a continuing study in Hiroshima and Nagasaki, the objective of which is to determine through a cohort approach whether the life span of liveborn children one or both of whose parents were exposed to the mixed radiation spectrum of the atomic bombings differs from that of suitable control children. The data will then be employed to approximate a "minimum doubling dose" for man.

The observations may be regarded as both a follow-up and an extension of the earlier study of Neel and Schull (1956), in which there was no evidence of an increased death rate among the children of survivors during the first six postpartum days, or, in a subsample, during the first nine months of life expectancy. Malformed children were excluded from these calculations, but the frequency of malformations was not elevated. Inasmuch as the interval since birth of the surviving subjects of the present study ranged at the time of investigation between three and 15 years, with a mean of nine, the results to be reported cover the age interval in which maximum prereproductive mortality occurs but still must be regarded as preliminary. A consideration of the pertinent literature on other mammals, including man, will be delayed until the Discussion.

Factors Determining the Design of the Study

Two different albeit related approaches to the question of the effect of parental exposure to ionizing radiation upon the survival of their progeny can

Received February 17. 1966.

*The Commission is a co-operative research agency of the U. S. National Academy of Sciences-National Research Council and the Japanese National Institute of Health of the Ministry of Health and Welfare, with funds provided by the U. S. Atomic Energy Commission, the Japanese National Institute of Health, and the U. S. Public Health Service.

Reproduced, with permission, from the *American Journal of Human Genetics*, vol. 18, © 1966 by the American Society of Human Genetics.

be envisaged. On the one hand, an effort can be made to determine whether the frequencies of death within some specified period of time differ in the offspring of exposed and control groups. Virtually all of the studies thus far reported are of this nature, and most involve rather short time periods, often only the first year of life. On the other hand, an effort can be made to determine the cumulative mortality experience in cohorts, some from exposed parents and some not, with a view toward defining survivorship curves. Patently, in the latter approach, death rates can be compared and hence this type of study includes the first. We are aware of no studies of mortality among the children of exposed parents of the life-expectancy kind except for the one to be reported here, but the merits of this type of investigation appear to us overwhelming. In the present instance, once it had been determined that a cohort analysis was feasible and the decision had been reached to adopt such an approach, the ultimate shape which the study assumed stemmed largely from certain administrative and scientific considerations to which we briefly turn.

The potential size of the study groups. The largest study group which can be envisaged is that population consisting of all children born alive subsequent to May 1, 1946, one or both of whose parents were present either in Hiroshima or Nagasaki or both cities at the time of the atomic bombings. This population is not now fully definable and will not be so until every person exposed to either or both of the bombs has passed beyond the reproductive age span. This will not occur for possibly another 30 years. Clearly, then, it is not yet possible to attempt to study the maximal population. The population to be dealt with in this presentation consisted of all children born between May 1, 1946, and December 31, 1958, to parents whose residence at the birth of their child was Hiroshima or Nagasaki and whose exposure status was either known to the Atomic Bomb Casualty Commission (ABCC) or could readily be obtained. The former date corresponds to the earliest time at which pregnancies conceived subsequent to the bombing would be expected to terminate normally; the latter date defines the maximal sample size possible when this cohort was established in 1959. Since it can be shown that at least 82% of the children ever to be born to exposed parents residing in Hiroshima and Nagasaki were born prior to January 1, 1959, relatively little appeared to be gained by further delay in initiating the study.

The probable duration of the study. Ideally in a study of life expectancy in a cohort, survivorship is followed until such time as all or virtually all members of the cohort have succumbed. In the present instance, this represents some 80 or 90 years, and it is obviously unlikely that the study will, in fact, continue for that length of time. Accordingly, it seemed important to emphasize the collection of data adequate to define with the requisite precision the earlier portions of the life curve, on the assumption that any change in life expectancy would be reflected in the early as well as the late portions of the survival curve.

The existence of the koseki system. There has existed in Japan since the last quarter of the nineteenth century a system of compulsory family registration. Vital events affecting the composition of a given family or the status of members of that family must be reported to the office having custody of the family's

record. The latter is termed the *koseki*. An indispensable part of the system is the legal or permanent address of the family, known as the *honseki*. Changes in the latter must also be reported to the *koseki* office. Because of the care with which Japanese family records are maintained, knowledge of the last, or any recent, *honseki* of an individual is a virtual guarantee that the survival status of that individual can be determined indefinitely. More complete accounts of the system are to be found in Naruge (1956), Ohkura (1960), and Yanase (1962); see also Schull and Neel (1965).

The availability through previous or current studies of ABCC of ancillary data of great relevance to the interpretation of an apparent effect of radiation. In the years 1948–1953, one facet of the activities of ABCC involved a comprehensive attempt to determine whether the pregnancy terminations occurring to parents, one or both of whom were exposed to the atomic bombs, differed from those occurring to a suitable control. This study, the Genetics Program, has been described in detail elsewhere (Neel and Schull, 1956). In these years the outcome of some 93% of all pregnancies occurring in Hiroshima and Nagasaki and lasting for 21 weeks or more of gestation was investigated. Available on these children, in addition to the results of one and frequently two clinical examinations, are data on a variety of variables of potential interest in the interpretation of an apparent radiation effect or the meaningfulness of the lack of such an effect. These observations range from parental ages at the birth of the children in question and the occurrence of consanguineous marriage to an appraisal of the socioeconomic status of the family at the time of the delivery. For a variety of reasons (see Neel and Schull, 1956), this study terminated early in 1954, but at that time it was decided that further information on the relationship of parental exposure to the sex ratio among children should be collected. This decision gave rise to a second study of interest here, the Sex Ratio Program. The latter involves an attempt to link the sex of children as reported on the birth records filed with the city with the exposure of their parents as known to ABCC. This study began in 1954 and continues. Through these two studies, exposure status can be determined for the parents of the majority of children born in these cities since 1948 without further field investigations, an important administrative consideration.

DESCRIPTION OF STUDY PROCEDURES

Three samples were utilized. The first consists of livebirths occurring between May 1, 1946, and January 31, 1948, i.e., prior to the previously mentioned Genetics Program, and was derived from municipal birth records. The second stems from the livebirths recorded in these cities in the interval from February 1, 1948, through December 31, 1953, under the Genetics Program. The third was drawn from livebirths recorded in the Sex Ratio Program in the years from January 1, 1954, through December 31, 1958. Within each of the three samples, in order to facilitate follow-up and to reduce extraneous sources of variation, the following selection criteria were imposed:

1. Parental residence *at the birth of the child* must have been within Hiroshima or Nagasaki. This restriction stemmed from the knowledge that (a)

infant and childhood mortality varies appreciably from urban to urban-rural areas in Japan and (b) establishing proper controls for exposed persons who had left the two cities prior to childbirth would be extremely difficult.

2. *Honseki* was not restricted to Hiroshima or Nagasaki. Although restriction of *honseki* to these two cities would have materially simplified the follow-up, the loss of data which this step would have entailed seemed prohibitive.

3. Multiple births were excluded on the grounds that the mortality rates are appreciably different from those of single births, and they are too infrequent to warrant the added analytic complication they introduce.

4. The samples were to include *all* infants born to parents one or both of whom were known to the ABCC to have been within 2000 meters of the hypocenter at the time of the bombing. In addition, from among births occurring in the two cities to parents whose exposure status was on file with ABCC, equal numbers of children were to be randomly drawn from among births to parents (a) one or both of whom were exposed in the interval 2500–9999 meters (neither parent was, of course, to be nearer) and (b) neither of whom were in the cities at the time of the atomic bombings. These latter two samples were to be matched with the former group by sex and year of birth. This procedure yields three major or, if joint parental exposure is considered, nine minor exposure groups, as shown in Fig. 1. In view of some findings to be brought out later, it should be emphasized that the matching by sex and year of birth was by major but not by minor exposure group.

Fact of death was established by routine, periodic inspection of the appropriate *koseki*. A home visit by a trained "contactor" was utilized to establish *honseki* in those cases where the *koseki* could not be located by routine methods. The first cycle of *koseki* checking was begun in January, 1961, and completed in January, 1964, and it is the results of this cycle which are here reported. Since this cycle embraced some three years of record checking, we shall restrict our attention to deaths prior to December 31, 1961, and the initiation of the first cycle, to obviate the possibility, remote but nonetheless real, that exposure class may have been correlated with the ease or difficulty of follow-up and hence the number of years at risk of death. We present in Table 1 a summary of the results of this cycle of record reviews. It will be noted that among some 54,243 cases, the predominant reason for failure to determine survival status was non-Japanese parentage (i.e., no *koseki* record); in all except 53 instances where the child was Japanese, the survival status of the child in 1961 could be determined. Failure of follow-up because of lack of *koseki* can scarcely be regarded as a source of bias; the 53 children of Japanese parents whose survival status is unknown are too few to influence the data significantly, even if all the cases involved children no longer alive. In subsequent tables, the 810, largely non-Japanese, children in Hiroshima and Nagasaki for whom koseki records were not available have been excluded, as have 14 births in the Sex Ratio Program for whom the information was incomplete. The resulting distribution of children on the basis of the radiation histories of their parents is shown in Table 2. Where death had occurred, a record of the

FIG. 1. A diagrammatic representation of the parental exposures, as implied by distance from hypocenter, associated with the three major and nine minor exposure groups.

entry in the "vital statistics" portion of the death certificate was obtained through the co-operation of the Japanese National Institute of Health.

SOURCES OF VARIATION IN MORTALITY OTHER THAN RADIATION

It is common knowledge that fetal, infantile, and childhood mortality are influenced by numerous variables other than radiation. As a consequence, the contribution of this or any other study to the ultimate evaluation of the radiation hazard is directly proportional to the degree to which extraneous sources of variation can be delineated and controlled. Accordingly, we turn now to an account of the possible role of extraneous variables in the mortality data to be presented shortly.

A priori, one recognizes that the survival of a liveborn infant is a function of certain attributes of the infant and, in addition, certain characteristics of the parents and pregnancy. Ample evidence exists attesting to the role in early mortality of parental ages, birth rank, birth weight, duration of pregnancy, and, in a changing society, year of birth. To this list can be added a number of variables commonly termed socioeconomic, such as parental education, parental occupation, the occurrence of consanguineous marriage, social class, and nutrition. Other variables of importance exist, e.g., presence of congenital defect, but of greater importance than an exhaustive enumeration is the recognition that few if any of these concomitant variates are entirely independent of

TABLE 1. RESULTS OF THE FIRST *Koseki* CHECK

	Hiroshima	Nagasaki
Total number in sample	35,289	18,954
Total for whom no *koseki* record is available	744	66
Non-Japanese	710	47
Japanese, *koseki* cannot be located	34	19

TABLE 2. TABULATION OF THE FINAL SAMPLE BY MAJOR AND MINOR EXPOSURE GROUP, PROGRAM COMPONENT, SEX, AND CITY

Sex	Source	Total	Exposure Major Group I Subtotal	a	b	c	d	e	Exposure Subgroup Subtotal	f	g	h	III i
								Hiroshima					
Male	Pre-GE-3	3679	1231	288	180	140	185	438	1235	528	123	584	1213
	GE-3	9007	2976	359	258	417	320	1622	3027	665	509	1853	3004
	Sex Ratio	5201	1720	107	194	381	185	853	1732	280	512	940	1749
Female	Pre-GE-3	3522	1169	280	180	93	167	449	1194	508	141	545	1159
	GE-3	8422	2784	360	248	393	276	1507	2812	630	496	1686	2826
	Sex Ratio	4701	1547	108	155	353	163	768	1566	211	455	900	1588
								Nagasaki					
Male	Pre-GE-3	1521	507	58	100	56	87	206	507	235	56	216	507
	GE-3	4544	1513	98	329	229	163	694	1520	439	240	841	1511
	Sex Ratio	3425	1139	55	201	212	129	542	1142	297	211	634	1144
Female	Pre-GE-3	1502	501	72	137	56	55	181	501	230	56	215	500
	GE-3	4514	1508	119	272	236	150	731	1506	406	245	855	1500
	Sex Ratio	3381	1126	53	190	228	137	518	1128	283	226	619	1127
Total		53419	17721	1957	2444	2794	2017	8509	17870	4712	3270	9888	17828

every other. Thus, for example, we recognize that parental ages and birth rank are significantly correlated, and, since both apparently affect mortality rates, it is not immediately obvious whether their effects are attributable to the same basic phenomenon or to a different one. In this particular instance, the latter appears to be true, for there exists an effect of parity upon the frequency of death in the postnatal period not accountable for in terms of variation in parental ages (see Neel and Schull, 1956, Chapter 11). Relationships of this nature considerably complicate efforts to control the effects of extraneous variables on mortality through selection of the comparison groups, since samples matched on multiple variables are often not feasible. We shall return to this issue after a consideration of the relationship to exposure groups of the concomitant variables previously enumerated.

It will be recalled that three samples were selected for study, namely, a sample (1) for the years May 1, 1946, through January 31, 1948, derived from the birth information on file at the city offices of Hiroshima and Nagasaki, (2) for the years February 1, 1948, through December 31, 1953, derived from the so-called Genetics Program of the ABCC, and finally (3) for the years January 1, 1954, through December 31, 1958, derived from the Sex Ratio Program. Through city birth records and the records associated with the two programs previously mentioned, it has been possible to obtain information on virtually all births with respect to maternal age, paternal age, parity, birth weight, duration of pregnancy, and year of birth. The most conspicuous exception to this statement involves birth weight, which was not routinely recorded on the Japanese birth certificate in the years immediately following World War Two (routine recording began in the latter half of 1947). Let us consider, now, the relationship of each of these five concomitants to the *major* exposure groups.

Year of birth. Substantial changes have been effected in infant mortality in Japan in the years since 1945. When one considers that the major portion of the mortality in the first two decades of life expectancy occurs in the first year, patently variation in year of birth among the exposure classes could obscure a radiation effect or produce a spurious one. While the method of sample selection ostensibly precludes significant differences within each of the three samples, the possibility exists that an equal degree of completeness of follow-up is not achieved in all radiation groups, and some variation, perhaps nonnegligible, is thereby introduced. In Table 3, we present a summary of the results of comparing the distributions of the six concomitant variables previously enumerated within sex-city-major exposure groups. Reference to this table reveals that year of birth does not differ significantly among exposure groups within either city or within samples within cities.

Maternal age. The significant effect of maternal age upon early mortality in this sample has been well documented by Neel and Schull (1956). The nature of this effect is complex but, in general, one finds that early mortality in age-specific parity distributions or parity-specific age distributions tends to be J or U-shaped (see also Yerushalmy, 1938; Tsunoda, 1956; Newcombe, 1964; Newcombe and Tavendale, 1965). This of course implies an optimal age associated

346 RADIATION AND OFFSPRING SURVIVAL

TABLE 3. SUMMARY OF CONTINGENCY χ^2 COMPARISONS OF THE DISTRIBUTION OF CONCOMITANT VARIABLES IN EXPOSURE MAIN GROUPS

		Male				Female			
Variables	City	Pre-GE-3	GE-3	Sex Ratio	Total	Pre-GE-3	GE-3	Sex Ratio	Total
Year of birth	H	NS	NS	NS	NS	NS	NS	NS	NS
	N	NS	NS	NS	NS	NS	NS	NS	NS
Maternal age	H	**	**	*	**	**	**	NS	**
	N	**	**	**	**	**	**	*	**
Parity	H	**	**	NS	**	**	**	**	**
	N	**	**	**	**	**	NS	**	**
Duration of pregnancy	H	NS	NS	NS	NS	NS	NS	NS	NS
	N	NS	NS	NS	NS	NS	NS	NS	NS
Birth weight	H	NS	NS	NS	NS	NS	NS	NS	NS
	N	NS	NS	NS	NS	NS	*	NS	NS
Paternal age	H	**	**	**	**	**	**	**	**
	N	**	**	**	**	**	**	**	**

*Significant at 5% level.
**Significant at 1% level.
NS = Not significant.

with a given parity, or an optimal parity for a given age. Be this as it may, the effect is sufficiently pronounced that failure to take into account this concomitant could have serious consequences. We note from Table 3 that the exposure groups differ significantly in maternal age within virtually all samples in both cities. In general, the more "heavily" exposed mothers are older than the more "lightly" exposed, and the latter are, in turn, older than the nonexposed mothers.

Paternal age. Parental ages are highly correlated, and in view of the known effect of maternal age upon early mortality, one would anticipate an effect of paternal age when maternal age is ignored. It is not clear, however, whether paternal age exerts an effect not explicable in terms of maternal age. Accordingly, it seemed appropriate to determine whether the exposure groups within the various samples differed in paternal age as was to be expected in view of the differences in maternal age. They do (Table 3). Mean paternal age is greater for "heavily" exposed than for "lightly" exposed or nonexposed fathers.

Parity. Parity and maternal age are closely correlated; one would, therefore, expect the exposure groups to differ in mean parities. This is, in fact, the case; on the average, parity is directly related to exposure, the more "heavily" exposed having had more pregnancies. As noted above, there exists an effect of parity not explicable in terms of the correlation between parity and maternal age (see Neel and Schull, 1956, for further details).

Duration of pregnancy. Mortality rates are strikingly and significantly elevated among pregnancies which terminate prematurely; a much lesser effect is associated with postmaturity. It follows, then, that if duration of pregnancy

was significantly related to exposure groups, differences in mortality among the exposure classes might arise. As Table 3 reveals, exposure groups do not differ significantly with respect to duration of pregnancy.

Birth weight. There is no significant association of birth weight with exposure status, in confirmation of the earlier finding of Neel and Schull (1956) on a partially overlapping sample.

To summarize briefly, among the six sources of extraneous variation for which data exist on virtually all births and which are known to be significantly associated with mortality, three, namely, maternal age, paternal age, and parity, differ significantly among the *major* exposure groups within most samples. These differences are such as to lead to augmented mortality rates within the more heavily exposed groups. Thus, an apparent effect of radiation on mortality could not be accepted before these differences were controlled, but on the other hand, if there is no apparent effect even when these differences are ignored, it seems most unlikely one would appear *after* they were controlled.

There are, of course, other important sources of extraneous variation for which data were not available on the birth certificate or through one of the previously mentioned programs. Thus, data were not available on socioeconomic class, yet we know that infant mortality, for example, is higher in the lower socioeconomic classes of Japan. Moreover, the Child Health Survey (Schull and Neel, 1965) revealed a significant effect of socioeconomic status upon physical and mental growth and development of a magnitude which might lead to serious biases in the analysis of mortality data if the socioeconomic concomitants were dissimilarly distributed between the comparison groups. To ascertain whether the latter might or might not be true, a pilot mail survey was undertaken on a small random sample in 1962. The primary purposes of this preliminary survey were to ascertain probable response rates and the quality of responses in a projected larger study. As a consequence of this preliminary study and the experience of the Child Health Survey of Schull and Neel (1965), a larger study involving a random 10% of the pregnancy terminations included in the total sample was initiated in 1964.

A mail questionnaire was sent to the last address known to the ABCC of the parents of 5,294 children, regardless, of course, of present status. If the initial questionnaire was not returned, a second and if need be a third was sent at intervals of two weeks. When a letter was returned as undeliverable, the *koseki* was checked and the questionnaire sent to the most recent address indicated on the record of residence filed with the *koseki*. For those cases where no response was obtained to any one of the three possible mailings, a field follow-up was conducted if the family lived in Hiroshima or Nagasaki or areas immediately adjacent thereto (see Table 4). It will be noted that data were collected on all save some 4% of the sample; it seems unlikely that the failure to obtain information on so small a group could markedly bias the results to be presented. Information was sought with respect to parental occupation, parental schooling, survival status of the parents, the size of the home (in mats) and the number of persons normally residing therein, food expenditures per

TABLE 4. ACCUMULATED RESPONSE RATE IN THE MAIL SURVEY ON SOCIOECONOMIC STATUS BY CITY

	First mailing		Second mailing		Third mailing		Field investigation	
	H	N	H	N	H	N	H	N
Total number	3416 (100%)	1878 (100%)	3416 (100%)	1878 (100%)	3416 (100%)	1878 (100%)	3416 (100%)	1878 (100%)
Responses	1130 (33.1%)	802 (42.7%)	2129 (62.3%)	1120 (59.6%)	2397 (70.2%)	1247 (66.4%)	3306 (96.8%)	1774 (94.5%)

person per month, and, finally, a brief indication of the child's nutrition in terms of the consumption of certain dietary staples, namely, fish, meat, eggs, and milk.

Table 5 summarizes the tests of significance with regard to the distributions of the aforementioned variables among the exposure groups within samples and cities. We note that number of mats per person, food expenditures, and consumption of dietary staples do not differ among the comparison groups. However, parental industry (and possibly occupation within industry), schooling, and survival status are significantly different. Thus, we observe relatively more children one or both of whose parents are dead in the proximally exposed group than in either of the other two samples. Exposed parents are less well educated than nonexposed, and relatively more farmers and fishermen are to be found in the distally exposed groups than in either of the other two comparison groups. These findings are, in general, in accord with those for the other concomitant variables to which attention has been directed. Thus, for example, we know that the exposed parents are older on the average than nonexposed parents (see Table 3) and that the education received by the average individual in Japan has steadily increased in the post-war era. We would expect, therefore, the exposed to be less well educated than the younger, nonexposed population. *In general, these differences in distribution of extraneous socioeconomic variables, like those associated with the variables discussed earlier, are such as to lead to an inflation of mortality rates in the children of the "heavily" exposed group as contrasted with the other two comparison groups.*

THE RELATIONSHIP OF MORTALITY TO EXPOSURE

The various tabulations, analyses, and comparisons of the data have been quite extensive; only a very condensed account can be presented. Table 6 sets out the data subclassified by city, sex, main exposure group, and year of risk. A profusion of comparisons is possible. Age-specific rates are given only for those years to which all the individuals in a defined cohort have been exposed. However, the figure for total deaths includes not only those occurring during the risk years experienced by the entire group but also those occurring during risk years experienced by only a proportion of the cohort. Accordingly, while the figure may be used in comparisons between groups of the three segments of the study, it has no normative value in the strict sense. Inspection of the

TABLE 5. SUMMARY OF CONTINGENCY χ^2 COMPARISONS OF THE DISTRIBUTION OF SOCIOECONOMIC VARIABLES IN EXPOSURE MAIN GROUPS

Socioeconomic variables	City	Male Pre-GE-3	Male GE-3	Male Sex Ratio	Male Total	Female Pre-GE-3	Female GE-3	Female Sex Ratio	Female Total
Father's education	H	NS	NS	NS	°	NS	NS	NS	°°
	N	NS	NS	NS	NS	NS	NS	NS	NS
Mother's education	H	°	NS	°	°°	NS	NS	NS	°°
	N	NS	NS	NS	°	NS	NS	NS	NS
Father's occupation	H	NS	NS	NS	°	NS	NS	NS	NS
	N	NS	NS	NS	NS	NS	NS	NS	NS
Father's industry	H	NS	NS	NS	°°	NS	NS	NS	NS
	N	NS	NS	°°	°°	NS	NS	NS	NS
Mats per person	H	NS	NS	NS	NS	NS	NS	NS	NS
	N	NS	NS	NS	NS	NS	NS	NS	NS
Food cost per person	H	NS	NS	NS	NS	NS	NS	NS	NS
	N	NS	NS	NS	NS	°	NS	NS	NS
Survival status of parents	H	NS	°°	NS	°°	°	NS	NS	NS
	N	NS	NS	NS	NS	NS	°	NS	NS
Eating habit: Fish	H	NS	NS	NS	NS	NS	NS	NS	NS
	N	°	NS	NS	NS	NS	NS	NS	NS
Meat	H	NS	NS	NS	NS	NS	NS	NS	NS
	N	NS	NS	NS	NS	°°	NS	NS	NS
Eggs	H	NS	NS	NS	NS	NS	NS	NS	NS
	N	NS	NS	NS	NS	NS	NS	NS	NS
Milk	H	°	NS	NS	NS	NS	NS	NS	NS
	N	NS	NS	NS	NS	NS	NS	NS	NS

° Significant at 5% level.
°° Significant at 1% level.
NS = Not significant.

table fails to reveal evidence of a systematic change in mortality experience related to parental major exposure group within either sex in either city. None of the χ^2 values derived from 12 3 × 2 contrasts based on the totals for each city-sex-program component approaches significance. This is in spite of the fact, it will be recalled, that extraneous sources of variation are so distributed as to lead to an inflation of mortality in major exposure group 1. The data on the first year of life expectancy support the earlier report of Neel and Schull (1956) that no significant association was observable between mortality in the first year of life and parental exposure. The data are, of course, not wholly independent of those presented in the aforementioned publication.

In view of these data it seemed unlikely that analysis of cause of death would be particularly revealing. However, such analyses were performed,

TABLE 6. TOTAL AND AGE-SPECIFIC DEATH RATES (ALL CAUSES OF DEATH) BY EXPOSURE MAIN GROUP (PER 1,000)

Source of data	Age at death	Total N*	Total D*	Total P*	Exposure 1 N	Exposure 1 D	Exposure 1 P	Exposure 2 N	Exposure 2 D	Exposure 2 P	Exposure 3 N	Exposure 3 D	Exposure 3 P	χ^2 (df = 2)
					Hiroshima male									
	Total	3679	290	78.8	1231	85	69.0	1235	102	82.6	1213	103	84.9	2.49
	0	3679	160	43.5	1231	50	40.6	1235	51	41.3	1213	59	48.6	
	1	3519	48	13.6	1181	12	10.2	1184	18	15.2	1154	18	15.6	
	2	3471	21	6.1	1169	6	5.1	1166	10	8.6	1136	5	4.4	
	3	3450	20	5.8	1163	5	4.3	1156	7	6.1	1131	8	7.1	
	4	3430	14	4.1	1158	4	3.5	1149	6	5.2	1123	4	3.6	
	5	3416	6	1.8	1154	3	2.6	1143	2	1.8	1119	1	0.9	
Pre-GE-3	6	3410	3	0.9	1151	—	—	1141	1	0.9	1118	2	1.8	
	7	3407	3	0.9	1151	1	0.9	1140	2	1.8	1116	—	—	
	8	3404	2	0.6	1150	1	0.9	1138	1	0.9	1116	—	—	
	9	3402	1	0.3	1149	1	0.9	1137	—	—	1116	—	—	
	10	3401	5	1.5	1148	—	—	1137	—	—	1116	5	4.5	
	11	3396	2	0.6	1148	—	—	1137	2	1.8	1111	—	—	
	12	3394	—	—	1148	—	—	1135	—	—	1111	—	—	
	13	3394	5	1.5	1148	2	1.7	1135	2	1.8	1111	1	0.9	
	14–15		0			0			0			0		
	Total	9007	610	67.7	2976	185	62.2	3027	232	76.6	3004	193	64.2	5.86
	0	9007	388	43.1	2976	126	42.3	3027	148	48.9	3004	114	38.0	
	1	8619	62	7.2	2850	18	6.3	2879	24	8.3	2890	20	6.9	
	2	8557	43	5.0	2832	11	3.9	2855	15	5.3	2870	17	5.9	
	3	8514	28	3.3	2821	5	1.8	2840	14	4.9	2853	9	3.2	
GE-3	4	8486	22	2.6	2816	5	1.8	2826	7	2.5	2844	10	3.5	
	5	8464	22	2.6	2811	8	2.9	2819	7	2.5	2834	7	2.5	
	6	8442	12	1.4	2803	4	1.4	2812	4	1.4	2827	4	1.4	
	7	8430	9	1.1	2799	1	0.4	2808	4	1.4	2823	4	1.4	
	8	8421	8	1.0	2798	2	0.7	2804	3	1.1	2819	3	1.1	
	9–12		16			5			6			5		
	Total	5201	197	37.9	1720	70	40.7	1732	67	38.7	1749	60	34.3	1.01
	0	5201	134	25.8	1720	48	27.9	1732	44	25.4	1749	42	24.0	
Sex Ratio	1	5067	23	4.5	1672	6	3.6	1688	9	5.3	1707	8	4.7	
	2	5044	13	2.6	1666	4	2.4	1679	6	3.6	1699	3	1.8	
	3	5031	10	2.0	1662	5	3.0	1673	3	1.8	1696	2	1.2	
	4–7		17			7			5			5		

*N = number of subjects, D = number of deaths, P = death rate per 1,000.

TABLE 6. *(Continued)*

Source of data	Age at death	Total N	Total D	Total P	Exp 1 N	Exp 1 D	Exp 1 P	Exp 2 N	Exp 2 D	Exp 2 P	Exp 3 N	Exp 3 D	Exp 3 P	χ^2 (df=2)
							Hiroshima female							
Pre-GE-3	Total	3522	262	74.4	1169	85	72.7	1194	98	82.1	1159	79	68.2	1.73
	0	3522	141	40.0	1169	43	36.8	1194	61	51.1	1159	37	31.9	
	1	3381	43	12.7	1126	19	16.9	1133	12	10.6	1122	12	10.7	
	2	3338	29	8.7	1107	7	6.3	1121	9	8.0	1110	13	11.7	
	3	3309	11	3.3	1100	2	1.8	1112	3	2.7	1097	6	5.5	
	4	3298	15	4.6	1098	3	2.7	1109	5	4.5	1091	7	6.4	
	5	3283	11	3.4	1095	5	4.6	1104	4	3.6	1084	2	1.9	
	6	3272	2	0.6	1090	1	0.9	1100	1	0.9	1082	—	—	
	7	3270	—	—	1089	—	—	1099	—	—	1082	—	—	
	8	3270	2	0.6	1089	1	0.9	1099	1	0.9	1082	—	—	
	9	3268	2	0.6	1088	2	1.8	1098	—	—	1082	—	—	
	10	3266	2	0.6	1086	—	—	1098	1	0.9	1082	1	0.9	
	11	3264	1	0.3	1086	—	—	1097	1	0.9	1081	—	—	
	12	3263	1	0.3	1086	1	0.9	1096	—	—	1081	—	—	
	13	3262	2	0.6	1085	1	0.9	1096	—	—	1081	1	0.9	
	14-15		0			0			0			79		
GE-3	Total	8422	483	57.3	2784	168	60.3	2812	167	59.4	2826	148	52.4	1.97
	0	8422	293	34.8	2784	90	32.3	2812	108	38.4	2826	95	33.6	
	1	8129	64	7.9	2694	20	7.4	2704	23	8.5	2731	21	7.7	
	2	8065	49	6.1	2674	23	8.6	2681	15	5.6	2710	11	4.1	
	3	8016	21	2.6	2651	9	3.4	2666	6	2.3	2699	6	2.2	
	4	7995	24	3.0	2642	12	4.5	2660	8	3.0	2693	4	1.5	
	5	7971	12	1.5	2630	7	2.7	2652	1	0.4	2689	4	1.5	
	6	7959	3	0.4	2623	2	0.8	2651	1	0.4	2685	—	—	
	7	7956	5	0.6	2621	1	0.4	2650	2	0.8	2685	2	0.7	
	8	7951	2	0.3	2620	1	0.4	2648	—	—	2683	1	0.4	
	9-12		10			3			3			4		
Sex Ratio	Total	4701	139	29.6	1547	50	32.3	1566	38	24.3	1588	51	32.1	2.31
	0	4701	106	22.6	1547	39	25.2	1566	28	17.9	1588	39	24.6	
	1	4595	12	2.6	1508	6	4.0	1538	3	2.0	1549	3	1.9	
	2	4583	8	1.8	1502	2	1.3	1535	3	2.0	1546	3	1.9	
	3	4575	6	1.3	1500	1	0.7	1532	2	1.3	1543	3	1.9	
	4-7		7			2			2			3		

°N = number of subjects, D = number of deaths, P = death rate per 1,000.

TABLE 6. (Continued)

Source of data	Age at death	Total N*	Total D*	Total P*	Exposure 1 N	Exposure 1 D	Exposure 1 P	Exposure 2 N	Exposure 2 D	Exposure 2 P	Exposure 3 N	Exposure 3 D	Exposure 3 P	χ^2 (df = 2)
							Nagasaki male							
Pre-GE-3	Total	1521	136	89.4	507	42	82.8	507	46	90.7	507	48	94.7	0.45
	0	1521	73	48.0	507	23	45.4	507	26	51.3	507	24	47.3	
	1	1448	26	18.0	484	7	14.5	481	7	14.6	483	12	24.8	
	2	1422	10	7.0	477	2	4.2	474	5	10.6	471	3	6.4	
	3	1412	11	7.8	475	3	6.3	469	4	8.5	468	4	8.6	
	4	1401	5	3.6	472	4	8.5	465	1	2.2	464	—	—	
	5	1396	1	0.7	468	1	2.1	464	—	—	464	—	—	
	6	1395	3	2.2	467	1	2.1	464	2	4.3	464	—	—	
	7	1392	2	1.4	466	—	—	462	1	2.2	464	1	2.2	
	8	1390	1	0.7	466	—	—	461	—	—	463	1	2.2	
	9	1389	1	0.7	466	—	—	461	—	—	462	1	2.2	
	10	1388	—	—	466	—	—	461	—	—	461	—	—	
	11	1388	2	1.4	466	1	2.1	461	—	—	461	1	2.2	
	12	1386	1	0.7	465	—	—	461	—	—	460	1	2.2	
	13	1385	—	—	465	—	—	461	—	—	459	—	—	
	14–15		0			0			0			0		
GE-3	Total	4544	308	67.8	1513	104	68.7	1520	110	72.4	1511	94	62.2	1.27
	0	4544	201	44.2	1513	71	46.9	1520	72	47.4	1511	58	38.4	
	1	4343	29	6.7	1442	8	5.6	1448	12	8.3	1453	9	6.2	
	2	4314	28	6.5	1434	9	6.3	1436	9	6.3	1444	10	6.9	
	3	4286	12	2.8	1425	5	3.5	1427	5	3.5	1434	2	1.4	
	4	4274	16	3.7	1420	3	2.1	1422	7	4.9	1432	6	4.2	
	5	4258	4	0.9	1417	1	0.7	1415	1	0.7	1426	2	1.4	
	6	4254	4	0.9	1416	3	2.1	1414	1	0.7	1424	—	—	
	7	4250	—	—	1413	—	—	1413	—	—	1424	—	—	
	8	4250	11	2.6	1413	3	2.1	1413	2	1.4	1424	6	4.2	
	9–12		3			1			1			1		
Sex Ratio	Total	3425	149	43.5	1139	51	44.8	1142	56	49.0	1144	42	36.7	2.15
	0	3425	118	34.5	1139	43	37.8	1142	42	36.8	1144	33	28.9	
	1	3307	10	3.0	1096	4	3.7	1100	4	3.6	1111	2	1.8	
	2	3297	6	1.8	1092	—	—	1096	2	1.8	1109	4	3.6	
	3	3291	8	2.4	1092	2	1.8	1094	4	3.7	1105	2	1.8	
	4–7		7			2			4			1		

°N = number of subjects, D = number of deaths, P = death rate per 1,000.

TABLE 6. (Continued)

Source of data	Age at death	Total N*	Total D*	Total P*	Exposure 1 N	Exposure 1 D	Exposure 1 P	Exposure 2 N	Exposure 2 D	Exposure 2 P	Exposure 3 N	Exposure 3 D	Exposure 3 P	χ^2 (df = 2)
							Nagasaki female							
Pre-GE-3	Total	1502	133	88.5	501	47	93.8	501	48	95.8	500	38	76.0	1.48
	0	1502	65	43.3	501	22	43.9	501	25	50.0	500	18	36.0	
	1	1437	30	20.9	479	12	25.1	476	10	21.0	482	8	16.6	
	2	1407	13	9.2	467	3	6.4	466	6	12.9	474	4	8.4	
	3	1394	7	5.0	464	3	6.5	460	2	4.4	470	2	4.3	
	4	1387	9	6.5	461	3	6.5	458	2	4.4	468	4	8.6	
	5	1378	3	2.2	458	1	2.2	456	2	4.4	464	—	—	
	6	1375	1	0.7	457	1	2.2	454	—	—	464	—	—	
	7	1374	—	—	456	—	—	454	—	—	464	—	—	
	8	1374	1	0.7	456	—	—	454	1	2.2	464	1	2.2	
	9	1373	1	0.7	456	—	—	454	—	—	463	—	—	
	10	1372	—	—	456	—	—	453	—	—	463	—	—	
	11	1372	—	—	456	—	—	453	—	—	463	—	—	
	12	1372	1	0.7	456	1	2.2	453	—	—	463	—	—	
	13	1371	—	—	455	—	—	453	—	—	463	—	—	
	14–15		2						0			1		
GE-3	Total	4514	286	63.4	1508	93	61.7	1506	85	56.4	1500	108	72.0	3.17
	0	4514	174	38.6	1508	59	39.1	1506	49	32.5	1500	66	44.0	
	1	4340	32	7.4	1449	9	6.2	1457	11	7.6	1434	12	8.4	
	2	4308	25	5.8	1440	9	6.3	1446	9	6.2	1422	7	4.9	
	3	4283	17	4.0	1431	4	2.8	1437	5	3.5	1415	8	5.7	
	4	4266	15	3.5	1427	5	3.5	1432	4	2.8	1407	6	4.3	
	5	4251	9	2.1	1422	3	2.1	1428	4	2.8	1401	2	1.4	
	6	4242	4	0.9	1419	1	0.7	1424	1	0.7	1399	2	1.4	
	7	4238	4	0.9	1418	—	—	1423	2	1.4	1397	2	1.4	
	8	4234	2	0.5	1418	1	0.7	1421	—	—	1395	1	0.7	
	9–12		4						0			2		
Sex Ratio	Total	3381	104	30.8	1126	34	30.2	1128	35	31.0	1127	35	31.1	0.01
	0	3381	79	23.4	1126	26	23.1	1128	26	23.1	1127	27	24.0	
	1	3302	9	2.7	1100	1	0.9	1102	3	2.7	1100	5	4.6	
	2	3293	6	1.8	1099	3	2.7	1099	3	2.7	1095	—	—	
	3	3287	3	0.9	1096	1	0.9	1096	2	1.8	1095	—	—	
	4–7		7			3			1			3		

*N = number of subjects, D = number of deaths, P = death rate per 1,000.

TABLE 7. CAUSE OF DEATH, EXPRESSED AS SPECIFIC DEATH RATES (PER 10,000) BY MAJOR EXPOSURE GROUP

The numbers on which these rates are based are also given; they correspond to the appropriate totals in Table 2.

Cause of death	Male Total	Male Exp. 1	Male Exp. 2	Male Exp. 3	Female Total	Female Exp. 1	Female Exp. 2	Female Exp. 3
Hiroshima								
All causes	613.3	573.6	669.0	596.7	531.1	550.9	545.8	498.8
Infectious disease	66.0	47.2	73.4	77.1	91.3	96.4	89.7	87.9
Neoplasm	*4.5	3.4	10.0	0	6.6	5.5	5.4	9.0
Allergy or nutrition	15.1	15.2	15.0	15.1	10.2	10.9	9.0	10.8
Blood disease	0.6	0	1.7	0	0.6	1.8	0	0
Nervous, sense organ	24.0	25.3	23.4	23.5	18.6	20.0	19.7	16.1
Circulatory system	4.5	6.7	5.0	1.7	7.2	3.6	5.4	12.6
Respiratory system	91.1	86.0	91.8	95.5	65.5	65.5	75.4	55.6
Digestive system	78.8	81.0	85.1	70.4	66.1	85.5	57.4	55.6
Genitourinary system	3.9	1.7	1.7	8.4	3.0	1.8	3.6	3.6
Skin	2.2	3.4	3.3	0	1.2	3.6	0	0
Bone	1.1	1.7	0	1.7	0.6	1.8	0	0
Congenital malformation	24.0	16.9	31.7	23.5	*15.0	12.7	25.1	7.2
Newborn disease	114.0	118.1	135.1	88.8	123.2	120.0	125.6	123.8
Ill-defined	110.1	114.7	105.1	110.6	85.9	81.8	91.5	84.3
Accident	67.6	52.3	80.1	70.4	31.2	36.4	34.1	23.3
Total number of subjects	17887	5927	5994	5966	16645	5500	5572	5573
Nagasaki								
All causes	624.9	623.6	669.0	581.9	556.6	555.0	535.9	578.8
Infectious disease	99.1	79.1	132.5	85.4	92.6	89.3	79.7	108.7
Neoplasm	12.6	12.7	6.3	19.0	6.4	6.4	3.2	9.6
Allergy or nutrition	15.8	19.0	12.6	15.8	13.8	15.9	12.8	12.8
Blood disease	3.2	0	9.5	0	1.1	0	3.2	0
Nervous, sense organ	12.6	22.2	3.2	12.7	17.0	19.1	15.9	16.0
Circulatory system	3.2	3.2	0	6.3	7.4	3.2	9.6	9.6
Respiratory system	104.3	110.8	104.1	98.0	94.7	98.9	92.5	92.7
Digestive system	69.5	72.8	72.6	63.3	61.7	73.4	60.6	51.2
Genitourinary system	8.4	12.7	9.5	3.2	4.3	3.2	9.6	0
Skin	6.3	9.5	6.3	3.2	3.2	0	3.2	6.4
Bone	0	0	0	0	0	0	0	0
Congenital malformation	23.2	22.2	31.6	15.8	18.1	3.2	25.5	25.6
Newborn disease	154.9	148.8	179.9	136.0	118.1	118.0	108.5	127.9
Ill-defined	71.7	76.0	75.7	63.3	88.3	95.7	82.9	86.3
Accident	32.7	31.7	25.2	41.1	22.3	22.3	25.5	19.2
Total number of subjects	9490	3159	3169	3162	9397	3135	3135	3127

*Difference of death rates among the three exposure groups are statistically significant at 5% level.

more with a view toward the completeness befitting a scientific problem of this gravity than because of prior expectations. Table 7 sets forth deaths per 10,000 for the more common categories of causes of death within sexes, cities, and the three major exposure groups. All three components of the data have been combined in this analysis. For each row, the significance of the difference between the observed number of deaths and the number expected on the basis of random distribution by exposure has been tested by χ^2. We observe only two significant associations, both at the five per cent level and both in Hiroshima. Firstly, deaths from congenital defects differ in females, but as a consequence of relatively more malformations in the lightly exposed than in the heavily or nonexposed groups. Secondly, the frequency with which death results from a neoplasm varies in males, but again the difference is one of

relatively more deaths in the lightly exposed group than in either of the other two. This latter effect does not appear in Nagasaki, although with regard to deaths due to congenital defects a trend similar in some respects to the one observed in Hiroshima does obtain.

It will be recalled that the major exposure groups differ in maternal age, paternal age, and parity, in a manner which might be expected to result in "false" radiation effects. In the face of failure to demonstrate any apparent effects despite these known biases, it is improbable that significance would emerge after analyses which took these variables into account. Nevertheless, because we are interested not only in significance but in a best estimate, which might at some time in the future be combined with the results of others, we have also analyzed the data on a maternal age specific basis, a paternal age specific basis, a parity specific basis, and various combinations of paternal and maternal age and parity. Publication of all the voluminous tables is out of the question. However, Tables 8, 9, and 10 present deaths per thousand by sex, city, and major exposure group for specified maternal ages, specified paternal ages, and specified parities. Perusal of these three tables again fails to suggest any consistent, significant effect of parental exposure. Thus, for example, we note from Table 8 that when mothers are 25–29 years of age (the largest maternal age group), death rates in the lightly exposed exceed those in the heavily exposed group in two of four possible instances, the nonexposed rates exceed those of the heavily exposed group in three of four, and the nonexposed exceed the lightly exposed three out of four times. In the two comparisons where there is significance at the 5% level, the death rates in males born to exposure group 1 parents are lower than those observed for one or both of the control groups. In Tables 9 and 10, there are three additional comparisons at the 5% level of significance. Again exposure group 1 males have *lower* mortality rates. These three tables involve 92 comparisons; the fact that 5 comparisons are significant at the 5% level can be explained by chance. We are intrigued by the *better* survival of group 1 male children in all 5 cases but can scarcely interpret this as a radiation effect! Since, had the deviation been in the opposite direction, it would have been tempting to read greater significance into it, this is a reminder of the need for caution in interpreting data of this type.

There exists the remote possibility, because of the correlations which exist between parental ages and parity, that simultaneous control of two or more of these variates might be informative. Control of all three variables through classification of the data so diminishes the numbers of observations as to make any conclusion hazardous. We do present in Tables 11 and 12, however, for both sexes, cities, and exposure groups, parity specific death rates and paternal age specific death rates for a fixed maternal age, namely, 25–29, chosen for presentation because this encompasses the largest body of data for any such comparison. No trends apparently related to radiation exposure group are to be observed; none of the χ^2's for the row comparisons is significant.

As Fig. 1 illustrates, when the exposure status of both parents is considered, children may fall into any one of nine minor exposure categories. The genetic

TABLE 8. THE RELATIONSHIP BETWEEN MATERNAL AGE SPECIFIC DEATH RATES (PER 1000) AND MAJOR EXPOSURE GROUP. (TOTAL SAMPLE)

The numbers in parentheses are the numbers of children in that category. The number of children whose mother's age at birth was unknown is also indicated.

Maternal age	Male Total	Male Exp. 1	Male Exp. 2	Male Exp. 3	Female Total	Female Exp. 1	Female Exp. 2	Female Exp. 3
Hiroshima								
15–19	94.1 (255)	90.0 (100)	70.6 (85)	128.6 (70)	81.2 (234)	107.5 (93)	54.8 (73)	73.5 (68)
20–24	58.0* (4928)	57.6 (1494)	68.4 (1651)	48.3 (1783)	53.1 (4554)	52.7 (1442)	54.3 (1455)	52.5 (1657)
25–29	55.0 (7085)	54.8 (2334)	52.5 (2267)	57.6 (2484)	44.3 (6563)	44.2 (2102)	47.6 (2145)	41.5 (2316)
30–34	71.3* (3662)	56.7 (1253)	80.0 (1263)	77.7 (1146)	59.7 (3419)	58.0 (1155)	63.3 (1185)	57.5 (1079)
35–39	68.2 (1554)	66.4 (572)	79.9 (576)	54.2 (406)	60.7 (1482)	74.5 (537)	51.7 (561)	54.7 (384)
40–44	72.9 (384)	37.0 (162)	108.8 (147)	80.0 (75)	92.1 (369)	92.6 (162)	90.9 (143)	93.8 (64)
45+	105.3 (19)	166.7 (12)	0.0 (5)	0.0 (2)	166.7 (24)	222.2 (9)	100.0 (10)	200.0 (5)
Total	61.3 (17887)	57.4 (5927)	66.9 (5994)	59.7 (5966)	53.1 (16645)	55.1 (5500)	54.4 (5572)	49.9 (5573)
Unknown	0	0	0	0	0	0	0	0
Nagasaki								
15–19	87.0 (138)	63.8 (47)	142.9 (56)	28.6 (35)	62.5 (144)	35.1 (57)	62.5 (48)	102.6 (39)
20–24	78.1 (2088)	86.9 (725)	72.8 (701)	74.0 (662)	59.8 (2125)	57.5 (765)	67.0 (687)	55.0 (673)
25–29	55.2 (3572)	49.2 (1241)	57.6 (1128)	59.0 (1203)	49.6 (3409)	53.2 (1128)	40.8 (1127)	54.6 (1154)
30–34	57.6 (2151)	54.3 (645)	64.7 (696)	54.3 (810)	58.2 (2216)	60.4 (695)	54.0 (722)	60.1 (799)
35–39	60.9 (1149)	77.4 (349)	67.6 (429)	37.7 (371)	57.8 (1125)	60.0 (350)	56.4 (408)	57.2 (367)
40–44	68.1 (367)	56.3 (142)	81.1 (148)	64.9 (77)	67.2 (357)	38.8 (129)	80.9 (136)	87.0 (92)
45+	80.0 (25)	.0 (10)	181.8 (11)	.0 (4)	.0 (19)	.0 (9)	.0 (7)	.0 (3)
Total	62.5 (9490)	62.4 (3159)	66.9 (3169)	58.2 (3162)	55.7 (9395)	55.5 (3133)	53.6 (3135)	57.9 (3127)
Unknown	0	0	0	0	0	0	0	0

*Difference of death rates among the three exposure groups are statistically significant at 5% level.

complement of children both of whose parents were within 2000 meters of the hypocenter at the time of the blast (minor group *a*) received substantially more radiation than that of children only one of whose parents was within 2000 meters but the other unexposed or lightly exposed (minor groups *b, c, d,* and *e*). (See Discussion for actual estimates.) Although highly unlikely in view of the foregoing, the possibility had to be considered that a radiation effect

TABLE 9. THE RELATIONSHIP BETWEEN PATERNAL AGE SPECIFIC DEATH RATES (PER 1000) AND MAJOR EXPOSURE GROUP. (TOTAL SAMPLE)

The numbers in parentheses are the number of subjects. The number of children whose father's age at birth of child was unknown is also indicated.

Paternal age	Male Total	Male Exp. 1	Male Exp. 2	Male Exp. 3	Female Total	Female Exp. 1	Female Exp. 2	Female Exp. 3
Hiroshima								
15–19	125.0	200.0	.0	.0	142.9	166.7	83.3	250.0
	(16)	(10)	(5)	(1)	(28)	(12)	(12)	(4)
20–24	62.3	51.3	71.4	64.0	53.4	56.3	50.4	53.5
	(1140)	(390)	(406)	(344)	(1011)	(355)	(357)	(299)
25–29	55.2	54.6	59.5	51.5	47.2	49.8	41.4	50.5
	(5018)	(1575)	(1715)	(1728)	(4685)	(1446)	(1594)	(1645)
30–34	58.7	58.0	63.2	55.3	48.5	47.0	58.4	41.4
	(5743)	(1776)	(1851)	(2116)	(5315)	(1680)	(1678)	(1957)
35–39	64.4	60.2	67.3	65.5	53.9	55.9	58.7	47.4
	(3403)	(1129)	(1144)	(1130)	(3115)	(1001)	(1039)	(1075)
40–44	72.0*	51.8	82.9	86.6	72.3	64.7	73.9	80.8
	(1666)	(637)	(567)	(462)	(1646)	(618)	(595)	(433)
45–49	71.3	54.4	91.7	73.5	69.8	92.6	37.7	75.0
	(659)	(294)	(229)	(136)	(602)	(270)	(212)	(120)
50–54	107.7	105.3	133.3	75.0	61.1	60.2	58.8	69.0
	(195)	(95)	(60)	(40)	(180)	(83)	(68)	(29)
55+	65.2	95.2	.0	111.1	114.8	121.2	176.5	.0
	(46)	(21)	(16)	(9)	(61)	(33)	(17)	(11)
Total	61.3	57.4	66.9	59.7	53.1	55.1	54.4	49.9
	(17886)	(5927)	(5993)	(5966)	(16643)	(5498)	(5572)	(5573)
Unknown	1	0	1	0	2	2	0	0
Nagasaki								
15–19	71.4	.0	142.9	.0	.0	.0	.0	.0
	(14)	(6)	(7)	(1)	(5)	(1)	(3)	(1)
20–24	96.1	98.7	82.1	113.6	50.8	40.9	52.6	62.9
	(562)	(223)	(207)	(132)	(591)	(220)	(228)	(143)
25–29	67.3	72.7	66.1	62.6	54.8	50.6	56.2	58.0
	(2407)	(825)	(847)	(735)	(2392)	(849)	(801)	(742)
30–34	54.4	46.5	59.4	56.9	49.7	53.4	44.7	50.8
	(2940)	(924)	(909)	(1107)	(2795)	(880)	(851)	(1064)
35–39	60.1	59.0	67.9	54.4	60.8	66.9	56.8	59.2
	(1865)	(593)	(574)	(698)	(1893)	(583)	(634)	(676)
40–44	50.0	54.8	59.1	35.1	60.6	60.4	53.6	68.0
	(1079)	(365)	(372)	(342)	(1090)	(364)	(373)	(353)
45–49	64.0	80.0	58.5	49.5	66.4	67.1	68.6	61.9
	(422)	(150)	(171)	(101)	(452)	(164)	(175)	(113)
50–54	110.3	83.3	133.3	108.1	66.7	18.9	94.3	103.4
	(145)	(48)	(60)	(37)	(135)	(53)	(53)	(29)
55+	115.4	41.7	210.5	111.1	71.4	100.0	.0	166.7
	(52)	(24)	(19)	(9)	(42)	(20)	(16)	(6)
Total	62.4	62.4	66.6	58.2	55.7	55.5	53.6	57.9
	(9486)	(3158)	(3166)	(3162)	(9395)	(3134)	(3134)	(3127)
Unknown	4	1	3	0	2	1	1	0

*Difference of death rates among the three exposure groups are statistically significant at 5% level.

TABLE 10. THE RELATIONSHIP BETWEEN PARITY SPECIFIC DEATH RATES (PER 1000) AND MAJOR EXPOSURE GROUP. (TOTAL SAMPLE)

The numbers in parentheses are the number of subjects. The number of children whose mother's parity at birth of child was unknown is also indicated.

Parity	Male Total	Exp. 1	Exp. 2	Exp. 3	Female Total	Exp. 1	Exp. 2	Exp. 3
				Hiroshima				
1	51.5 (5675)	50.0 (1839)	58.4 (1763)	46.8 (2073)	45.1 (5345)	44.1 (1747)	44.5 (1706)	46.5 (1892)
2	56.7 (5151)	59.5 (1698)	58.1 (1688)	52.7 (1765)	48.4 (4713)	44.1 (1518)	52.9 (1512)	48.1 (1683)
3	64.9 (3114)	59.3 (1046)	72.9 (1043)	62.4 (1025)	46.0 (2979)	56.4 (975)	45.8 (1005)	36.0 (999)
4	73.1* (1751)	56.6 (565)	69.9 (658)	94.7 (528)	74.9 (1563)	90.2 (543)	56.1 (535)	78.4 (485)
5	72.9 (850)	79.5 (302)	66.0 (318)	73.9 (230)	66.6 (826)	68.3 (278)	84.1 (321)	39.6 (227)
6	84.8 (460)	73.6 (163)	100.0 (170)	78.7 (127)	78.8 (406)	62.5 (144)	96.4 (166)	72.9 (96)
7	99.8* (481)	48.9 (184)	134.6 (208)	123.6 (89)	83.7 (418)	87.5 (160)	65.9 (182)	118.4 (76)
Total	60.8 (17482)	57.3 (5797)	66.5 (5848)	58.6 (5837)	52.0 (16250)	54.1 (5365)	52.9 (5427)	49.1 (5458)
Unknown	405	130	146	129	395	135	145	115
				Nagasaki				
1	70.6 (2352)	72.7 (825)	78.3 (792)	59.9 (735)	52.5 (2360)	45.7 (809)	48.0 (792)	64.6 (759)
2	52.4 (2269)	46.7 (728)	58.8 (714)	52.0 (827)	53.2 (2256)	53.8 (743)	59.9 (718)	46.5 (795)
3	62.7 (1835)	68.9 (595)	48.4 (579)	69.6 (661)	49.5 (1839)	55.8 (591)	43.5 (598)	49.2 (650)
4	54.3 (1235)	57.1 (403)	60.7 (412)	45.2 (420)	58.8 (1157)	47.4 (401)	60.2 (349)	68.8 (407)
5	57.8 (744)	39.4 (254)	69.1 (246)	65.6 (244)	53.5 (785)	57.7 (260)	54.4 (294)	47.6 (231)
6	61.8 (421)	87.0 (138)	65.9 (167)	25.9 (116)	66.2 (393)	102.4 (127)	42.9 (140)	55.6 (126)
7	83.7 (502)	73.0 (178)	103.6 (222)	58.8 (102)	76.4 (471)	57.3 (157)	75.4 (199)	104.3 (115)
Total	61.8 (9358)	61.8 (3121)	66.4 (3132)	57.0 (3105)	54.7 (9261)	53.8 (3088)	53.4 (3090)	57.1 (3083)
Unknown	132	38	37	57	136	47	45	44

*Difference of death rates among the three exposure groups are statistically significant at 5% level.

might emerge if the analysis was extended to minor category, and this has been done.

However, when the nine minor exposure groups are compared with respect to the six concomitants for which the three major exposure groups were contrasted (Table 3), not only is there again significant heterogeneity with respect to paternal and maternal ages and parity, but now there is also very significant

TABLE 11. DEATH RATES FOR A FIXED MATERNAL AGE (25–29) BY PARITY PER 1000). (TOTAL SAMPLE)

The numbers in parentheses are the number of subjects. The number of children whose mother's age at birth was known but the parity was not is indicated.

Parity	Male Total	Exp. 1	Exp. 2	Exp. 3	Female Total	Exp. 1	Exp. 2	Exp. 3
Hiroshima								
1	46.0 (1828)	42.5 (636)	51.8 (541)	44.5 (651)	33.3 (1830)	25.3 (592)	36.1 (609)	38.2 (629)
2	52.5 (2706)	52.0 (885)	52.5 (857)	52.9 (964)	40.6 (2485)	42.4 (778)	46.8 (790)	33.8 (917)
3	54.1 (1571)	61.0 (508)	40.9 (514)	60.1 (549)	38.7 (1473)	49.7 (463)	34.3 (496)	33.1 (514)
4+	84.4 (794)	78.7 (254)	75.9 (290)	100.0 (250)	84.8 (672)	79.5 (239)	88.8 (214)	86.8 (219)
Total	54.8 (6899)	54.3 (2283)	52.7 (2202)	57.2 (2414)	42.7 (6460)	43.4 (2072)	45.0 (2109)	39.9 (2279)
Unknown	186	51	65	70	103	30	36	37
Nagasaki								
1	56.8 (345)	51.5 (291)	66.0 (288)	52.6 (266)	43.9 (843)	32.1 (280)	34.7 (288)	65.5 (275)
2	49.5 (1172)	30.5 (393)	52.3 (363)	64.9 (416)	50.4 (1130)	52.2 (364)	42.1 (380)	57.0 (386)
3	54.8 (949)	61.2 (343)	45.1 (288)	56.6 (318)	39.0 (872)	50.2 (279)	42.3 (284)	25.9 (309)
4+	54.1 (555)	50.0 (200)	72.6 (179)	39.8 (176)	71.4 (532)	82.5 (194)	47.3 (169)	82.8 (169)
Total	53.4 (3521)	47.3 (1227)	57.2 (1118)	56.1 (1176)	49.2 (3377)	51.9 (1117)	41.0 (1121)	54.4 (1139)
Unknown	51	14	10	27	32	11	6	15

heterogeneity in most of the comparisons with respect to year of birth and birth weight, although to a lesser degree than for the first three concomitants. Fig. 2 illustrates for male births the year-of-birth effect during the 13 years of the program in Hiroshima. It will be noted that the line dividing groups *a, b, c, d,* and *e* collectively (major exposure group 1) from groups *f, g,* and *h* (major exposure group 2) and the line dividing *f, g,* and *h* (group 2) from *i* (group 3) is relatively straight, thus attesting to the effectiveness of the original matching procedure. The significant year-of-birth effects emerge from the changes with time of the proportions of the minor exposure categories within the respective major exposure groups, for which the data were of course unmatched. At least two reasons for the significant year-of-birth effect are apparent from inspection of Fig. 2. On the one hand, the representation of certain subgroups appears to increase (group *c*) or diminish (groups *a* and *f*) with time. On the other hand, for some exposure groups there appear to be increases followed by decreases (groups *e* and *h*).

The several significant differences (none exceeding the 0.01 level) between minor exposure classes with respect to birth weight are in general due to

TABLE 12. DEATH RATES FOR A FIXED MATERNAL AGE (25–29) BY PATERNAL AGE (PER 1000). (TOTAL SAMPLE)

The numbers in parentheses are the number of subjects. The number of children whose mother's age at birth was known but the parity was not is indicated.

Paternal age	Male Total	Male Exp. 1	Male Exp. 2	Male Exp. 3	Female Total	Female Exp. 1	Female Exp. 2	Female Exp. 3
Hiroshima								
–24	59.9 (167)	45.5 (66)	72.7 (55)	65.2 (46)	59.3* (118)	20.0 (50)	29.4 (34)	147.1 (34)
25–29	50.2 (1971)	51.2 (645)	48.1 (665)	51.4 (661)	34.9 (1891)	36.1 (581)	30.9 (680)	38.1 (630)
30–34	55.1 (3485)	53.6 (1101)	53.9 (1095)	57.4 (1289)	43.5 (3193)	43.3 (992)	53.8 (1004)	35.1 (1197)
35–39	55.6 (1206)	56.0 (411)	46.8 (385)	63.4 (410)	56.8 (1056)	57.5 (348)	64.3 (342)	49.2 (366)
40+	85.9 (256)	90.1 (111)	89.6 (67)	76.9 (78)	62.7 (303)	62.0 (129)	47.1 (85)	78.7 (89)
Total	55.0 (7085)	54.8 (2334)	52.5 (2267)	57.6 (2484)	44.4 (6561)	44.3 (2100)	47.6 (2145)	41.5 (2316)
Unknown	0	0	0	0	2	2	0	0
Nagasaki								
–24	87.5 (80)	0.0 (32)	148.1 (27)	142.9 (21)	62.5 (80)	74.1 (27)	31.3 (32)	95.2 (21)
25–29	57.4 (1081)	55.8 (394)	51.8 (386)	66.4 (301)	44.4 (1082)	35.8 (363)	49.5 (384)	47.8 (335)
30–34	49.9 (1762)	44.1 (589)	46.9 (533)	57.8 (640)	51.0 (1568)	60.4 (497)	36.6 (492)	55.3 (579)
35–39	54.7 (530)	59.1 (186)	68.5 (146)	40.4 (198)	49.6 (564)	50.3 (199)	39.1 (179)	59.1 (186)
40+	93.2 (118)	51.3 (39)	166.7 (36)	69.8 (43)	69.6 (115)	119.0 (42)	25.0 (40)	60.6 (33)
Total	55.2 (3571)	49.2 (1240)	57.6 (1128)	59.0 (1203)	49.6 (3409)	53.2 (1128)	40.8 (1127)	54.6 (1154)
Unknown	1	1	0	0	0	0	0	0

*Difference of death rates among the three exposure groups are statistically significant at 5% level.

heavier birth weights among the groups where exposure was greatest and are thought to be secondary to the age and parity effects which have already been discussed. A "duration of pregnancy" heterogeneity is observed in a single comparison among twelve and may be disregarded. Because of the declining infant and childhood mortality in Japan during the years of this study, it was felt the data should be adjusted for the year-of-birth effect. Accordingly, we present in Table 13 an analysis wherein by city, sex, and program component the observed number of deaths in each minor exposure category is compared with expectation adjusted for differences in years-at-risk per individual for the exposure subgroups. Table 13 reveals two differences at the 5% level of significance, both (again) resulting from a deficiency of deaths among the children of the more heavily irradiated parents.

FIG. 2. The contributions of the nine minor exposure groups to the samples available for each year in the interval 1946–1958. The different minor exposure groups are indicated by the same letters as those given in Fig. 1.

Before we enlarge further on these findings, it seems appropriate to inquire into the magnitude of the differences between cohorts which should be detectable with samples of this size. It is conceivable that our failure to find significant differences reflects samples too small to afford a reasonable test of whether the groups under discussion are or are not different. What constitutes a reasonable test involves an intuitive appraisal of the risks to be associated with the two types of judgmental errors which can be made, namely, to assert that there is an effect of radiation when, in fact, there is none, the so-called Type I error, or to assert that there is no effect when, in fact, there is one, the Type II error. The frequency of the former is specified by the level of con-

TABLE 13. A COMPARISON OF OBSERVED AND EXPECTED DEATHS BY MINOR EXPOSURE GROUPS, EXPECTATION BEING AGE-ADJUSTED BECAUSE OF HAVING MEAN PERIODS OF RISK OF DEATH IN THE EXPOSURE GROUPS

		a	b	c	d	e	f	g	h	i	χ^2 (df = 8)
Male						*Hiroshima*					
Pre-GE-3	Observed death	18	13	19	13	22	38	10	54	103	18.11*
	Person-year	3945.0	2423.0	1744.0	2548.5	6034.5	7169.0	1647.5	7765.0	16208.5	
	Expected death	22.9	14.1	10.1	14.8	35.0	41.6	9.6	45.0	94.0	
	Death rate	4.6	5.4	10.9	5.1	6.1	3.6	5.3	7.0	6.4	
GE-3	Observed death	27	20	31	16	91	55	43	134	193	13.76
	Person-year	3867.5	2719.0	4307.5	3439.0	17021.0	7200.5	5126.5	19184.5	31695.0	
	Expected death	24.8	17.4	27.6	22.0	108.9	46.1	32.8	122.8	202.8	
	Death rate	7.0	7.4	7.2	4.7	5.3	7.6	8.4	7.0	6.1	
Sex Ratio	Observed death	7	8	12	10	33	16	20	31	60	8.98
	Person-year	603.5	1048.0	2099.5	1014.5	4862.5	1540.0	2794.0	5397.0	9837.5	
	Expected death	4.0	7.0	14.1	6.8	32.6	10.3	18.7	36.2	65.9	
	Death rate	11.6	7.6	5.7	9.9	6.8	10.4	7.2	5.7	6.1	
Female											
Pre-GE-3	Observed death	25	16	8	11	25	41	13	44	79	7.05
	Person-year	3735.0	2403.0	1230.5	2308.5	6170.5	6837.0	1861.5	7298.5	15737.5	
	Expected death	20.5	13.2	6.8	12.7	33.9	37.6	10.2	40.1	86.6	
	Death rate	6.7	6.7	6.5	4.8	4.1	6.0	7.0	6.0	5.0	
GE-3	Observed death	23	14	32	17	82	42	33	92	148	9.35
	Person-year	3901.0	2670.0	4033.5	2966.0	15954.5	6813.0	5127.0	17856.0	30197.0	
	Expected death	20.7	14.2	21.4	15.7	84.6	36.1	27.2	94.6	160.0	
	Death rate	5.9	5.2	7.9	5.7	5.1	6.2	6.4	5.2	4.9	
Sex Ratio	Observed death	4	4	7	3	32	1	16	21	51	13.39
	Person-year	630.0	847.5	1924.5	940.5	4362.0	1207.5	2449.5	5212.0	8929.0	
	Expected death	3.3	4.4	10.0	4.9	22.7	6.3	12.7	27.1	46.4	
	Death rate	6.3	4.7	3.6	3.2	7.3	0.8	6.5	4.0	5.7	

*Difference of death rates among the nine exposure groups are statistically significant at 5% level.

TABLE 13. (Continued)

Nagasaki

		a	b	c	d	e	f	g	h	i	χ^2 (df = 8)
Male											
Pre-GE-3	Observed death	3	10	—	7	22	16	10	20	48	17.69*
	Person-year	797.0	1308.0	788.0	1153.5	2646.0	3145.5	676.0	2812.0	6632.5	
	Expected death	5.4	8.9	5.4	7.8	18.0	21.4	4.6	19.1	45.1	
	Death rate	3.8	7.6	0.0	6.1	8.3	5.1	14.8	7.1	7.2	
GE-3	Observed death	8	22	17	6	51	31	12	67	94	8.20
	Person-year	1005.0	3372.5	2304.5	1684.5	6926.0	4542.5	2451.0	8293.5	15401.5	
	Expected death	6.6	22.3	15.2	11.1	45.7	30.0	16.2	54.7	101.6	
	Death rate	8.0	6.5	7.4	3.6	7.4	6.8	4.9	8.1	6.1	
Sex Ratio	Observed death	3	13	4	8	23	13	12	31	42	9.98
	Person-year	303.5	1088.5	1112.0	663.5	3106.0	1605.5	1146.5	3513.0	6358.0	
	Expected death	2.4	8.5	8.7	5.2	24.2	12.5	8.9	27.4	49.6	
	Death rate	9.9	11.9	3.6	12.1	7.4	8.1	10.5	8.8	6.6	
Female											
Pre-GE-3	Observed death	5	14	5	6	17	24	5	19	38	3.26
	Person-year	966.0	1783.5	738.0	711.5	2361.5	2990.0	739.0	2808.5	6660.0	
	Expected death	6.5	11.9	4.9	4.8	15.8	20.0	5.0	18.8	44.6	
	Death rate	5.2	7.8	6.8	8.4	7.2	8.0	6.8	6.8	5.7	
GE-3	Observed death	6	18	9	12	48	24	14	47	108	7.77
	Person-year	1270.5	2763.0	2452.0	1517.0	7354.5	4206.0	2457.5	8742.5	15154.0	
	Expected death	7.9	17.1	15.2	9.4	45.6	26.1	15.2	54.2	94.0	
	Death rate	4.7	6.5	3.7	7.9	6.5	5.7	5.7	5.4	7.1	
Sex Ratio	Observed death	3	3	11	4	13	9	4	22	35	7.82
	Person-year	281.5	1007.0	1213.0	754.5	2882.0	1493.5	1195.0	3452.5	6122.5	
	Expected death	1.6	5.6	6.8	4.2	16.1	8.4	6.7	19.3	34.3	
	Death rate	10.7	3.0	9.1	5.3	4.5	6.0	3.3	6.4	5.7	

*Difference of death rates among the nine exposure groups are statistically significant at 5% level.

fidence which characterizes the test of significance upon which one bases his judgment of whether there is or is not a difference. Conventionally, this is taken to be 0.05. In a fixed sample, when one compares two percentages, the frequency of the Type II error depends upon the size of the samples and the difference between the frequency of the event of interest in a "standard" population and a population to be compared with the standard. Presumably we should like this type of error to be as small as practicable, certainly not greater than one chance in ten, and possibly as small as one in twenty.

To determine whether the cohorts are adequate in the sense that the frequency of Type II errors is small for a variety of reasonable alternatives to the null hypothesis, we have made the following simplifying assumptions: We have assumed that the test of significance to be used to ascertain whether an irradiation effect has or has not occurred involves *either* a comparison of the frequency of death in Exposure Group 1 with that in Exposure Group 2, *or* a comparison of Exposure Group 1 with Exposure Group 3, *or*, finally, a comparison of Group 2 with Group 3. Clearly, the frequency of death in the comparison group, the standard, will depend upon the total number of years at risk of death for the cohort and this will vary with the passage of time. Rather, therefore, than arbitrarily specify some one mortality rate, we have concerned ourselves with the difference between the two groups which we believe it is important to detect whatever may be the frequency of the standard. To avoid the theoretical as well as the computational complications introduced by the fact that the variance of a proportion varies with the proportion, we have elected to replace the frequencies of death by their angular (arcsin) equivalents and thus to concern ourselves with the angular difference (as expressed in degrees) between two percentages which it would be desirable to detect.

Under the conditions set out above, it is possible to show (see Eisenhart, Hastay, and Wallis, 1947) that

$$\xi_D^2 = (1641.6/N)(K_a + K_b)^2$$

where ξ_D^2 is the angular difference expressed in degrees, N is the number of observations upon which each of the two estimates of the frequency of death is based, K_a is the normal deviate associated with a confidence level of 0.05 when one is interested in deviation in only one direction from the standard (hence $K_a = 1.645$), and K_b is the frequency of the second type of error previously discussed.

Figure 3 discloses the frequency of Type II errors, say β, expected for various angular differences when the two sample sizes are presumed to be 17,721, the size of the Exposure Group 1 cohort. We note from this figure that if we presume β to be 0.05, then the sample is large enough to detect an angular difference of 1.0. What this means in terms of percentages depends, of course, upon the frequency of deaths in the standard. If the frequency of death in the standard is 0.050, then from a table of the angular transformation (see, for example, Fisher and Yates, 1938, Table XII), we find that these samples are adequate to detect an increase to about 0.058; if the frequency is 0.100, then we could detect a shift wherein the frequency of deaths in the

FIG. 3. The operating characteristic of a test comparing two percentages derived from samples of equal size, where the samples consist of 17,721 observations. The ordinate is the power of the test, that is, $1 - \beta$, where β is the frequency of a type II error. The abscissa is the angular difference in degrees between the per cent mortality in the exposed group and in the comparisons group.

exposed cohort equaled or exceeded 0.110. To a close approximation, then, and within the range of interest to us, we are in a position to detect a 10% deviation from the standard, that is, the death rate in the comparison cohort.

DISCUSSION

There emerges from this study no indication that the exposure of parents to the atomic bombs has altered the life expectancy of their children. This failure to demonstrate significant changes in F_1 mortality as a function of parental exposure of course can not be construed as evidence that no genetic effects resulted from exposure to these nuclear devices. Unless one is willing to argue that man differs from all other forms of life thus far studied, lethal and semilethal mutations of a type which might manifest themselves as mortality in the first decade or so of life were induced. But the present study provides no evidence for their existence.

The exact importance to be attached to the present observations depends, of

course, on the estimated mean exposure to radiation experienced by the various exposure groups. The problem of estimating the amount of radiation received by persons exposed to the atomic bombs in Hiroshima and Nagasaki is still troublesome today, after 20 years of study of the question (see, for example, Neel and Schull, 1956; Ritchie and Hurst, 1959; Arakawa, 1960). At present the ABCC employs the so-called T57 Dose Scale, developed on the basis of the reports of Ritchie and Hurst (1959) and Arakawa (1960) and based on distance from the hypocenter and shielding. The estimates of individual dosage so obtained are considered to be accurate, for each city, only to a factor of 2, i.e. the true dose may be as low as 50% of the T57 estimate or it may be 100% greater than the T57 estimate (Jablon, Ishida, and Beebe, 1963). One of the uncertainties in the estimate stems from the fact that for neither of the atomic bombs detonated is the yield accurately known. Efforts to refine all of the factors entering into dose estimates are continuing, but for the present we must be content with this range of uncertainty. This T57 dose is for whole body radiation and fails to take all aspects of shielding into consideration; the gonad dose will presumably be less, although, because of the high energy of the radiation, the attenuation factor should be minor.

A dose estimate is not available for all parents of the subjects for this study. However, the majority of the parents exposed under 2000 m (exposure group 1) are also included (with additional persons of similar radiation background) in a life span study of the survivors being conducted by ABCC. In connection with this study, an average T57 dose has been derived for each 100 m zone, and, as a simple approximation, survivors in each 100 m zone have been assigned the average T57 dose for that zone. This procedure is less than ideal but it seemed likely that the inaccuracies would prove to be of minor importance compared to the 150-fold variations in dose as between 700 and 2000 m. The average dose received by both parents of children in major exposure group 1 thus estimated was 191 and 220 rad in Hiroshima and Nagasaki respectively. These averages are perhaps unduly influenced by a relatively few persons to whom the T57 scale assigns doses presumably incompatible with survival. The median doses, which might be more appropriate under these circumstances, are 65 rad for Hiroshima and 96 rad for Nagasaki. Since the distance-dose curve reveals that radiation is approximately 5 rad and diminishing steadily at 2500 meters, the average dose for exposure group 2 can be considered negligibly close to the background dose. It must be again emphasized, however, that in view of the various problems regarding dosimetry mentioned above, it would be unwise to regard the particular numbers used for dose as more than rough approximations. Using the T57 data, the mean joint exposures to be attributed to parents falling into the various minor categories of major group 1 may be estimated as: for Hiroshima, 270, 184, 187, 130, and 187 rad for groups *a*, *b*, *c*, *d*, and *e* respectively; for Nagasaki, 285, 212, 202, 210 and 223 rad for these same groups. These estimates, incidentally, do not differ to a significant degree from those derived by Neel and Schull in their earlier treatment of this problem.

Assuming the approximate correctness of these dose estimates, is the present

lack of findings consistent with the corpus of experimental data, as well as the other observations on human material? Much of the recent experimental work has been concerned with the effects of chronic radiation repeated over many generations. Both because of the higher genetic yield of x units of radiation given in a single rather than divided doses (Russell, Russell and Kelley, 1958) and the opportunities for selection under these conditions, direct comparisons with these investigations seem unwise. With respect to acute radiation whose effects were evaluated in the first generation following parental exposure, there is still no abundance of data for completely relevant comparisons with the findings of this investigation. In fact, because of differences in the length of life cycle and the "medical care" received by the "indisposed," the "litter size effect" (see below), and differing intervals between radiation and conception, there is room for equivocation regarding any comparison of data from other mammals with man. Perhaps the most germane comparison possible with the present material would be three-week or pre-weaning mortality or a reasonable approximation thereto in experimental mammals with deaths under one year of age in man. Since the present data are so predominantly concerned with offspring resulting from germ cells which in the male were in the spermatogonial stage at the time of radiation and in the female dictyate-stage oöcytes, and because of the relationship between stage of maturation at time of irradiation and genetic yield, comparisons should be restricted to the "post-sterile period" offspring of experimental mammals.

For mice, Russell (Russell, 1955; Russell and Russell, 1959), in extensive experiments, has reported a significant reduction in mean number of offspring at weaning (age 3 weeks) following acute paternal X irradiation. In the largest series, a paternal exposure of 300 r, with mating in the post-sterile period, resulted in a reduction in litter size at weaning of 3.8%. Mean number of liveborn offspring is not given. Kohn (1960) has reported a statistically non-significant decrease in liveborn litter size and pre-weaning survival in the offspring of male mice receiving 525 r in either a single or a divided dose and mated during the post-sterile period. Sugahara (1964) has reported significantly *larger* litter sizes and pre-weaning mortality among the offspring of male mice receiving an acute dose of 600 r and mated during both the pre- and post-sterile period. Finally Frölen (1965) found no significant differences in life span between 161 F_1 males whose fathers had received 500 r of X rays and a control group.

In rats, Chapman, Hansen, Havenstein, and Morton (1964) failed to observe a significant effect on survival up to 69 days in the offspring of males who received 450 r in three divided doses and for whom, because of the experimental design, it was assumed that the sperm responsible for fertilization were from irradiated spermatogonia. Litter size was affected differently by irradiation of the male and female parents; in the rat, unlike the mouse, probability of death appears to decrease with increasing litter size (McGregor, James, and Newcombe, 1960).

In studies utilizing swine, Cox and Willham (1962) reported that following an acute dose of 300 r of X rays to males mated during the post-sterile period,

mortality in their liveborn offspring during the first three weeks of life was significantly increased. However, a later report (D. F. Cox, 1964) found that this apparent effect was confounded by differences in litter size between irradiated and control animals; no statistically significant effect remained when this fact was taken into consideration. This report raises the question of a nonspecific litter size effect in other experimental work with mammals.

The studies on this problem which utilized human material have all been negative. Aside from our own previous reports, the only studies on the effect of acute radiation on survival of offspring involve very small series concerned with the progeny of seemingly infertile women whose infertility was treated with 60 to 65 r of ovarian X radiation (Kaplan, 1957) or of women who in the course of multiple diagnostic X rays received a dose between 7.5 and 20 rads (D. W. Cox, 1964). Not surprisingly, no effects were observed. Neither Crow (1955) nor Macht and Lawrence (1955) noted increased infantile mortality among the children of American radiologists. The gonad doses of chronic irradiation are quite uncertain. Finally, Tanaka and Ohkura (1958) report an insignificant increase in first year mortality among the children of Japanese X-ray technicians as contrasted with the children of pharmacy technicians, from 1.9% to 2.6%, the mean gonad dose in the former being estimated at between 230 and 575 r of X radiation delivered in small pulses over an average of 11.5 years.

In summary, for few of the observations on experimental mammals and none on man are the effects of irradiation on survival in the F_1 significant. However, for the experimental material the trend is in the direction suggested by genetic hypothesis, i.e. reduced survival. Unfortunately, until the confounding effect of litter size has been removed from the experimental data as well as certain other possible biases summarized by Roderick (1964), a direct comparison with the studies on man is not advisable. Although litter size is usually decreased following paternal irradiation, it is sometimes increased. Inspection of the published data suggests that the variance in litter size may also be increased after radiation, making simple corrections difficult. Litter-size specific mortality figures are an obvious step towards accurate comparisons. Whether, because of the possibility of a degree of prenatal selection due to intrauterine competition in litter bearing mammals, even these data can be directly compared with human data is a moot point. However, most investigators would probably agree that there is no conflict between the results of the present studies and those with experimental mammals, even with allowance for the inclusion in the radiation spectrum of the atomic bombs of a neutron component with a higher relative biological effectiveness than gamma radiation.

Russell (1965) has recently reported that the yield in mutations diminishes as the interval between irradiation and conception increases. Since this study reaches back to include the first post-exposure conceptions, it therefore includes the interval most appropriate to the demonstration of atomic-bomb induced mutations.

One wonders whether further study of this group of children is warranted. Certainly we recognize that some 85% of mortality prior to the age of maturity

(say, 20 years) has occurred before the age of 10, and that, therefore, the mortality to be experienced by these groups in the immediate future is minor compared to what has already been experienced. It seems doubtful indeed that further study is likely to be rewarding within the next decade. However, the issues are such that, in the face of any uncertainty, continued study is desirable and uncertainties do exist. In particular it may be argued that the level of medical care in man delays "genetic deaths" until relatively late in the life span. The present study with its orientation toward cohorts and *koseki* determination of fact of death does not require annual nor biennial ascertainment of survivorship status to maintain the necessary currency of address, etc., so that follow-up can continue. It is thus possible to permit a lapse of time such as to increase meaningfully the numbers of deaths before a further cycle of *koseki* checking is initiated.

EPILOGUE

The probability of the increasing exposure of the human species to ionizing radiation has led to much concern among geneticists and others concerning "permissible" levels of exposure. A guide figure in this discussion has been the doubling dose, i.e., the amount of radiation producing the same rate of mutation as that occurring spontaneously each generation. The estimates employed in the most authoritative pronouncements to date are 30–80 r (National Academy of Sciences, 1956) and 10–100 r (United Nations, 1958). The present data, when coupled with several other recent advances in our knowledge of mammalian radiation genetics, would appear to permit a refinement of that estimate. As noted earlier, this study excludes, at the 10% probability level, an increase in the death rate in the children of Group 1 parents in excess of 1%. In this material, some 7% of liveborn children succumb prior to age 8. In order to utilize these data to derive an estimate of the doubling dose, we need to know what proportion of these deaths is due to mutation in the preceding generation, i.e., due to newly arisen dominant mutations lethal prior to age 8. There is no figure presently available based on a detailed study of this mortality experience and, indeed, our present procedures are probably inadequate to the challenge. However, an indirect approach can help us set an approximate figure. Muller (1950) has suggested that a conservative estimate of the total mutation rate per gamete in man is of the order of 0.1. We have suggested that the true value may be several times this (Neel and Schull, 1954) but will work with the more conservative figure for the present. From this value, it follows that about 20% of the population carries a new mutation each generation. From the accumulated literature on the phenotypic effects of spontaneous mutation in *Drosophila*, mice, and man, it seems reasonable to assume that at present in man at least 1 in 40 of these "new" mutations results in an early death, i.e., that in civilized countries with high standards of living, such as Japan, 0.5% of liveborn children die prior to age 8 because of mutation (chromosomal and point) in the preceding generation. This is equivalent to saying that currently approximately 1 in 14 children in Japan dying before age 8 does so because of dominant mutation in the preceding generation. If this assump-

tion is accepted, and it is obviously a key assumption, and if we assign a total parental radiation dose of 100 rad units to Group 1 parents (again conservative),* then the minimal doubling dose for radiation of the Hiroshima-Nagasaki type, for mutations causing postnatal death, is simply

$$\frac{(\text{mean parental dose}) \times \begin{pmatrix}\text{mortality due to spontaneous}\\ \text{mutation in preceding generation}\end{pmatrix}}{(\text{increase in mortality excluded by present data})} = \frac{100 \times 0.5}{1.0} = 50\,\text{r}$$

The propriety of using the present data for this type of calculation is in our opinion reinforced by the fact that of the four other indicators of possible genetic damage studied in Hiroshima and Nagasaki (sex ratio, malformation rate, birth weight, frequency of stillbirth), only one (sex ratio) has shown even a suggestion of a change (Neel and Schull, 1956; Schull and Neel, 1959; Neel, 1963; Schull, Neel, and Hashizume, 1966). It should be pointed out that this calculation makes no assumptions concerning genetic equilibrium, requiring only that the proportion of newly arisen mutations manifesting themselves through death in the next generation is the same for spontaneous and induced mutations.

Two apparently well established facts from mammalian radiation genetics would seem to necessitate a revision of this estimate before it can be applied to a variety of human problems. Firstly, Russell, Russell, and Kelley (1958) and Russell, Russell, and Cupp (1959) have shown that for both male and female mice, the greater the dose rate the higher the yield of mutations for a given dose. The radiation received by the survivors of Hiroshima and Nagasaki was delivered at a relatively high dose rate. Secondly, Russell (1965) has demonstrated a marked decrease in the yield of mutations in the later litters as contrasted with the earlier litters of radiated female mice. Now, the chief sources of human exposure to ionizing radiation—medical and fall-out—involve substantially lower dose rates than in Hiroshima and Nagasaki and, in addition, the average interval between exposure and reproduction will be somewhat longer than the interval in this study. Although a direct and literal extrapolation from mouse to man seems unwise, the aforementioned results in our opinion would lead to a revision of the derived minimal estimate of 50 r to something of the order of 100 r.

*Since this manuscript was submitted, Auxier et al. (Free-field radiation-dose distributions from the Hiroshima and Nagasaki bombings. *Health Physics* 12: 425–429, 1966) have published revised distance-dose curves for Nagasaki and Hiroshima. That for the former city is essentially unchanged. For the latter, there is a significant revision downwards from former estimates, doses being approximately halved at most distances compatible with survival for subjects in the open. The precise effect of this revision on the estimates used in this paper cannot be determined until detailed calculations which include allowance for shielding have been completed. However, the purposely conservative figure of 100 rad for Group 1 parents used in this calculation is now almost certainly more appropriate for Hiroshima than the mean total parental dose of 191 rad given earlier.

SUMMARY

A cohort-type study of children born subsequent to the atomic bombings of Hiroshima and Nagasaki and divisible into three groups on the basis of parental exposure (a heavily exposed group, a lightly exposed one, and a nonexposed set) has failed to disclose significant variation in mortality ascribable to differences in radiation. The average number of years at risk is nine but varies from three to 15 years. Though the comparison groups differ significantly with respect to certain extraneous sources of variation in mortality, notably, parental ages and birth rank, the failure to observe significant mortality differences among the exposure groups is not attributable to the existence of this concomitant variation. In view of the number of years of death at risk on the average and the knowledge that the bulk of mortality prior to the twenty-first birthday is, in fact, prior to the tenth, it seems improbable that mortality prior to maturity can be shown to be a function of parental exposure in either Hiroshima or Nagasaki. Our failure to demonstrate significant differences among the exposure groups does not necessarily imply that there are no differences. It merely means that within these years and numbers of observations no effect is to be seen.

These data, considered in conjunction with the other data from Hiroshima and Nagasaki, may be utilized to approximate the minimal doubling dose for man, i.e., the amount of radiation necessary to produce the same number of mutations as occur spontaneously each generation. It is suggested that the present findings render it unlikely that the figure is less than 50 r for radiation of the type emitted by an atomic bomb and for mutations of the type resulting in early death in the first post-bomb generation. It is further suggested that under the more usual conditions of human exposure to radiation, the minimal doubling dose for mutations of this type is of the order of 100 r.

REFERENCES

ARAKAWA, E. T. 1960. Radiation dosimetry in Hiroshima and Nagasaki atomic bomb survivors. *New Engl. J. Med.* 263: 488–493.

CHAPMAN, A. B., HANSEN, J. L., HAVENSTEIN, G. B., AND MORTON, N. E. 1964. Genetic effects of cumulative irradiation on prenatal and early postnatal survival in the rat. *Genetics* 50: 1029–1042.

COX, D. F., AND WILLHAM, R. L. 1962. Genetic effects of irradiation on early mortality in swine. *Genetics* 47: 785–788.

COX, D. F. 1964. Effects of radiation on litter size in swine. *Genetics* 50: 1025–1028.

COX, D. W. 1964. An investigation of possible genetic damage in the offspring of women receiving multiple diagnostic pelvic X-rays. *Amer. J. Human Genet.* 16: 214–230.

CROW, J. F. 1955. A comparison of fetal and infant death rates in the progeny of radiologists and pathologists. *Amer. J. Roentgenol.* 73: 467–471.

EISENHART, C., HASTAY, M. W., AND WALLIS, W. A. 1947. *Techniques of Statistical Analysis*. New York: McGraw-Hill Co., xiv and 473 pp.

FISHER, R. A., AND YATES, F. 1938. *Statistical Tables for Biological, Agricultural, and Medical Research*. London: Oliver and Boyd, viii and 90 pp.

FRÖLEN, H. 1965. The effect on the length of life in the offspring of X-irradiated male mice. *Mutation Res.* 2/3: 287–292.

JABLON, S., ISHIDA, M., AND BEEBE, G. W. 1963. *JNIH-ABCC Life Span Study, Hiroshima and Nagasaki Report 2, Mortality in Selection, I and II*. October 1950–September 1959, ABCC Technical Report 01-63.

Kaplan, I. I. 1957. The treatment of female sterility with X-rays to the ovaries and the pituitary, with special reference to congenital anomalies of the offspring. *Can. Med. Assoc. J.* 76: 43–46.

Kohn, H. I. 1960. The effect of paternal X-ray exposure on the secondary sex ratio in mice (F_1 generation). *Genetics* 45: 771–778.

Macht, S. H., and Lawrence, P. S. 1955. National survey of congenital malformations resulting from exposure to roentgen radiation. *Amer. J. Roentgenol* 73: 442–466.

McGregor, J. F., James, A. P., and Newcombe, H. B. 1960. Mutation as a cause of death in offspring of irradiated rats. *Radiation Res.* 12: 61–66.

Muller, H. J. 1950. Our load of mutations. *Amer. J. Human Genet.* 2: 111–176.

Naruge, T. 1956. *Practice of koseki and its theory.* (In Japanese) Tokyo: Nihon Kajo-Shuppan, 988 pp.

National Academy of Sciences 1956. *The Biological Effects of Atomic Radiation: Summary Reports.* Washington: National Academy of Sciences–National Research Council, xiv and 108 pp.

Neel, J. V. 1963. *Changing Perspectives on the Genetic Effects of Radiation.* Springfield: C. C Thomas, 97 pp.

Neel, J. V., and Schull, W. J. 1954. *Human Heredity.* Chicago: University Press, 361 pp.

Neel, J. V., and Schull, W. J. 1956. *The Effect of Exposure to the Atomic Bombs on Pregnancy Termination in Hiroshima and Nagasaki.* Washington, D. C.: National Academy of Sciences–National Research Council Publ. 461.

Newcombe, H. B. 1964. Screening for effects of maternal age and birth order in a register of handicapped children. *Ann. Human Genet.* (Lond.) 27: 367–382.

Newcombe, H. B., and Tavendale, O. G. 1965. Effects of father's age on the risk of child handicap or death. *Amer. J. Human Genet.* 17: 163–178.

Ohkura, K. 1960. Use of the family registration in the study of human genetics in Japan. *Jap. J. Human Genet.* 5: 61–68.

Ritchie, R. H., and Hurst, G. S. 1959. Penetration of weapons radiation, application to the Hiroshma-Nagasaki studies. *Health Physics* 1: 390–404.

Roderick, T. H. 1964. Summary of the general discussion of the Symposium. *Genetics* 50: 1213–1217.

Russell, W. L. 1955. Genetic effects of radiation in mice and their bearing on the estimation of human hazards. In *Proc. Int. Conf. on the Peaceful Uses of Atomic Energy*, Vol. II. New York: United Nations, pp. 382–383.

Russell, W. L. 1965. Effect of the interval between irradiation and conception on mutation frequency in female mice. *Proc. Nat. Acad. Sci.* (U. S.) 54: 1552–1557.

Russell, W. L., and Russell, L. B. 1959. Radiation-induced genetic damage in mice. In *Progress in Nuclear Energy*, Vol. II, Series VI, Biological Sciences. London: Pergamon Press, pp. 179–188.

Russell, W. L., Russell, L. B., and Cupp, M. B. 1959. Dependence of mutation frequency on radiation dose rate in female mice. *Proc. Nat. Acad. Sci.* (U. S.) 45: 18–23.

Russell, W. L., Russell, L. B., and Kelley, E. M. 1958. Radiation dose rate and mutation frequency. *Science* 128: 1546–1550.

Schull, W. J., and Neel, J. V. 1959. Atomic bomb exposure and the pregnancies of biologically related parents. *Amer. J. Public Health* 49: 1621–1629.

Schull, W. J., and Neel, J. V. 1965. *The Effects of Inbreeding on Japanese Children.* New York: Harper and Row, 419 pp.

Schull, W. J., Neel, J. V., and Hashizume, A. 1966. Some further observations on the sex ratio among infants born to survivors of the atomic bombings of Hiroshima and Nagasaki. *Amer. J. Human Genet.* 18: 328–338.

Sugahara, T. 1964. Genetic effects of chronic irradiation given to mice through three successive generations. *Genetics* 50: 1143–1158.

Tanaka, K., and Ohkura, K. 1958. Evidence for genetic effects of radiation in offspring of radiological technicians. *Jap. J. Human Genet.* 3: 135–145.

TSUNODA, R. 1956. Follow-up from birth to infant death. (In Japanese) *Koseki no Shihyo–Index Health Welfare Statistics* 3: 2–33.

United Nations. 1958. Report of the United Nations Scientific Committee on the Effects of Atomic Radiation. *General Assembly Official Records*, Thirteenth Session. Supplement No. 17 (A/3838).

YANASE, T. 1962. The use of the Japanese family register for genetic studies. In *UN/WHO Seminar on the Use of Vital and Health Statistics for Genetic and Radiation Studies*. New York: United Nations, pp. 119–132.

YERUSHALMY, J. 1938. Neonatal mortality by order of birth and age of parents. *Amer. J. Hyg.* 28: 244–270.

MORTALITY IN THE CHILDREN OF ATOMIC BOMB SURVIVORS AND CONTROLS[1]

JAMES V. NEEL

Department of Human Genetics, University of Michigan Medical School, Ann Arbor, Michigan 48104

HIROO KATO

Department of Statistics, Atomic Bomb Casualty Commission, Hiroshima, and Hiroshima Branch, Japanese National Institute of Health, Hiroshima, Japan

WILLIAM J. SCHULL

Center for Demographic and Population Genetics, Graduate School of Biomedical Sciences at Houston, The University of Texas, Houston, Texas 77025

Manuscript received July 15, 1973
Revised copy received October 24, 1973

ABSTRACT

A continuing study of mortality rates among children born to survivors of the atomic bombings and a suitable group of controls has been updated; the average interval between birth and verification of death or survival is 17 years. The mortality experience is now based on 18,946 children liveborn to parents one or both of whom were proximally exposed, receiving jointly an estimated dose of 117 *rem*; 16,516 children born to distally exposed parents receiving essentially no radiation; and 17,263 children born to parents not in Hiroshima or Nagasaki at the time of the bombings. No clearly significant effect of parental exposure on child's survival can be demonstrated either by a contingency χ^2 type of analysis or regression analysis. On the basis of the regression data, the minimal gametic doubling dose of radiation of this type for mutations resulting in death during (on the average) the first 17 years of life among liveborn infants conceived 0–13 years after parental exposure is estimated at 46 *rem* for fathers and 125 *rem* for mothers. On the basis of experimental data, the gametic doubling dose for chronic, low-level radiation would be expected to be three to four times this value for males and as much as 1000 *rem* for females.

ONE of the possible manifestations of exposure to ionizing radiation is a decrease in the life expectancy of children born to radiated parents, because of the induction of deleterious mutations in the gonadal tissues of these parents. We have previously described in detail a study of the mortality through a mean follow-up period of nine years among a group of children liveborn between 1946 and 1958 to survivors of the atomic bombings of Hiroshima and Nagasaki, and a suitable group of controls, reviewing also the pertinent literature (KATO, SCHULL

[1] This study was undertaken as one aspect of the program of the Atomic Bomb Casualty Commission, a cooperative research agency of the U.S. National Academy of Sciences—National Research Council and the Japanese National Institute of Health of the Ministry of Health and Welfare, with funds provided by the U.S. Atomic Energy Commission, the Japanese National Institute of Health, and the U.S. Public Health Service.

Reproduced, with permission, from *Genetics*, vol. 76, © 1974 by the Genetics Society of America.

and NEEL 1966). No difference in mortality rates was demonstrated. The present report will extend the study for another eight years (through December 31, 1969), thus providing a near-complete accounting of what might be termed pre-reproductive mortality. Although the actual number of deaths since the prior review of the mortality experience is rather small, publication at this time is prompted by the additional fact that a significant revision of dose estimates has recently become available, thus facilitating a more meaningful estimate of the minimum dose of radiation necessary to increase by 100% the mutation rate resulting in this type of effect (i.e., the so-called doubling dose).

As originally defined, the study population consisted of three cohorts of children, drawn from both Hiroshima and Nagasaki, as follows:

THE STUDY POPULATION

Group I—either or both parents < 2000 meters from hypocenter at time of bombing (ATB),

Group II—neither parent < 2000 meters but either or both > 2500 meters from hypocenter ATB, and

Group III—both parents not in city ATB.

A detailed description of how these groups were assembled will be found in the earlier paper (KATO, SCHULL and NEEL 1966). The number of liveborn infants in each of the panels when they were established was reported as 17,721, 17,870, and 17,828, respectively. Although we are interested in total zygote loss, stillborn infants have been excluded from the present study because a definitive analysis of stillbirth frequency between 1948 and 1953, the period during which the majority of the study births occurred, has been presented elsewhere; it revealed no relation of the phenomenon to parental radiation (NEEL and SCHULL 1956). The results of that analysis will not be altered by the present findings. Parents of group III children of course received no radiation exposure ATB. Furthermore, it has been clear for some time that the exposure received by the parents of group II children was negligible, usually less than 1 *rad*. Thus both groups II and III are "control" groups. However, for a variety of reasons it has proven difficult to estimate the dosage received by the parents of group I children.

Early attempts to estimate for genetic purposes the amount of radiation received by survivors relied heavily on such medical parameters as epilation and petechiae, in addition to distance from hypocenter and shielding (NEEL and SCHULL 1956). The only unclassified distance-dose curve then available was that published in "The Effects of Atomic Weapons" (Los Alamos Scientific Laboratory, 1950), for a "nominal" atomic bomb. The studies of RITCHIE and HURST (1959) and ARAKAWA (1960), together with the declassification of distance-dose curves specifically estimated for Hiroshima and Nagasaki, then resulted in the adoption by the Atomic Bomb Casualty Commission (ABCC), the joint U.S.-Japan official follow-up agency, of the so-called T57 Dose Scale, which assigned individual doses solely on the criteria of distance from hypocenter and shielding. This was the basis for the estimates of radiation exposure in our first publication

on the mortality experience of these three groups of children (KATO, SCHULL and NEEL 1966); we simply adopted for group I and group II parents the estimate of the *mean* dose to individuals falling in these distance categories. However, AUXIER et al. (1966) then published a revision of the two distance-dose curves, even as much more detailed shielding data were assembled. The result was the adoption by the ABCC of the T65 Dose Scale, based again solely on distance and shielding (MILTON and SHOHOJI 1968), and each surviving individual within 2500 meters ATB has now been assigned a dose. In developing this dose scale, independent estimates were made of gamma and neutron doses, and total dose expressed in *rad*'s. The present analysis is based on these individual dose estimates.

However, not unexpectedly, some problems regarding the assignment of individual doses persist. Thus, a few individuals give apparently valid histories of radiation sickness when their T65 doses are well below those thought necessary to radiation sickness. Conversely, a few individuals who report proximity to the hypocenter with very little shielding ATB must on the T65 schedule be assigned doses clearly incompatible with survival. In both instances, this probably results from confusion about position or shielding or both at the time of the explosions. The distribution of estimated doses among the parents of group I children is such that the preponderance of estimated values fall in the range of 50 to 150 *rad*, although some few individuals have estimated values greater than 600 *rad*.

Among the parents of the subjects of this study there were, when the T65 dose schedule became available, instances where one or both parents were deceased or no longer resided in Hiroshima or Nagasaki, i.e., were not available for dosimetry studies. This was relatively unimportant in the case of parents of group II and III children, since it was clear that they were, for the purposes of this study, to be considered essentially unirradiated, but seriously impeded dose estimates for the parents of group I children. There were also a number of parents of group I children who were unable to supply adequate shielding or location data ATB. Unfortunately, for these various reasons, there were 2885 parents of group I or II children from whom no proper radiation history could be obtained. It is clear that the proximally exposed among these parents received considerable doses of radiation. The exclusion of their children from the study would result in a very real loss of data. Accordingly, such parents have been assigned a radiation dose on the basis of their own very approximate histories or of third-person histories. Where shielding data were available, they were incorporated into the dose estimate; where they were not, dose was assigned solely on the basis of distance from the hypocenter.

We turn now to the problem in dosimetry raised by the substantial neutron component in the Hiroshima spectrum but its lesser representation in Nagasaki (cf. JABLON et al. 1970). Three of the most reliable symptoms of radiation sickness are epilation, petechiae, and severe gingivitis. The Joint Commission (OUGHTERSON et al. 1951) reported that of 3506 survivors in Hiroshima living 20 days following the bombings, whose histories placed them within 2000 meters of the hypocenter (their exposure groups A, B, and C), 1223 (35%) reported

epilation and/or petechiae. Of 3471 similarly positioned survivors from Nagasaki, the corresponding figure was 1072 (31%). The Commission's data are not presented in such a manner that one can consider the frequency of individuals with one or more of the three cardinal symptoms just enumerated. Granted some confusion in the evaluation of radiation sickness, it seems the great majority of persons actually did develop these findings. The minimum dose thought to elicit these symptoms in man is the equivalent of 200 roentgens of gamma rays (Los Alamos Scientific Laboratory, 1950). These data suggest that the biologically effective dose (for these end points) was as great in Hiroshima as in Nagasaki.

Additional data on this point are available because the majority of the parents of group I children were drawn from the study of NEEL and SCHULL (1956). The mothers of these children registered their pregnancies well in advance of delivery, at which time a brief radiation history was obtained. Among 4245 mothers registering in Hiroshima who were within 2000 meters of the hypocenter at the time of the bombings, 660 persons (16%) reported epilation and/or petechiae and/or oral lesions, including necrotic gingivitis. The similar figure for Nagasaki for 1636 mothers was 329 persons (20%). We have discussed the inadequacy of these data in our study in some detail (NEEL and SCHULL 1956). Nevertheless, they provide a crude guide to biological dosimetry. Like the foregoing data collected just following the bombings, they suggest relatively little difference between the two cities.

If we calculate the *rad* dose in the two cities assuming an RBE for neutrons of 1, then the mean *rad* dose for individuals exposed within 2000 meters is estimated as 55.9 for Hiroshima and 136.7 for Nagasaki. This lower mean for Hiroshima is at variance with the facts, just presented, that within 2000 meters, the symptoms of radiation sickness were quite similar in the two cities. JABLON *et al.* (1970) have recently pointed out that there are striking discrepancies between the two cities in the relationship of symptomatology to T65 dose when the RBE for neutrons is taken as 1, but that these differences disappear when the RBE is placed at 4–5.

If we intend to recognize the neutron component in calculating the genetically effective dose, the RBE figure of greatest pertinence is specifically for genetic effects. The data for mammals are not so extensive as one might desire; we will accept RUSSELL's appraisal that "in general, for a given absorbed dose, neutrons prove to be far more mutagenic than X and gamma rays, namely, of the order of 5 or 6 times both for oocytes and for the rising part of the dose curve for spermatogonia. This is for acute radiation" (RUSSELL 1967; see also SEARLE 1967; SEARLE, EVANS and WEST 1969; and SEARLE and PHILLIPS 1971). The RBE undoubtedly varies with both the absolute levels and the proportions of neutrons and gamma rays in the total dose. The RBE therefore varies from survivor to survivor. In this complex situation we will for now set the average RBE at 5 for the types of genetic damage (point mutations, small deletions, unbalanced translocation, nondisjunction) which we may assume are being measured in this study. Incorporation of an RBE of 5 for neutrons into the distance-shielding data necessitates that we now measure dosage in *rem* units. All estimates were rounded down to the nearest integral value.

FATHER'S DOSE INTERVAL (REM)

		100 +	10-99	1-9	< 1	NE
MOTHER'S DOSE INTERVAL (REM)	100 +	a 437-670	f 308-426	k 262-338	p 272-335	u 275-376
	10-99	b 326-423	g 76-76	l 39-39	q 40-40	v 40-40
	1-9	c 288-483	h 43-43	m 10-10	r 4-4	w 4-4
	< 1	d 284-359	i 40-40	n 4-4	s 0-0	x 0-0
	NE	e 295-428	j 40-40	o 5-5	t 0-0	y 0-0

Mean dose for both parents combined, groups a-r, u-w, with 600 *rem* maximum, is estimated to be 117 *rem*.

FIGURE 1.—The total mean dose in *rem* received by *both* parents combined for the 25 subclasses of children defined in this study. With those exceptions described in the text, children falling into subdivisions a-r and u-w correspond to group I children of the previous report, those in subdivisions s, t, and x, to group II, and those in y to group III. The average parental dose for all parents of group I children is estimated to be 117 *rem*; of group II, 0 *rem*; and of group III, 0 *rem*. The lower figure in each box is mean dose after reduction of all dose estimates greater than 600 *rem* to 600 *rem*; the higher figure is the estimate before that adjustment.

In order to project some better idea of the distribution of doses, and to facilitate the statistical analysis (see below), the radiated parents of group I and II children have been subdivided into four classes, according to the *rem* equivalents of the T65 dose, namely, < 1 *rem*, 1-9 *rem*, 10-99 *rem*, and 100 + *rem*. There is also the category of unexposed parents. With this subdivision 25 subclassifications of the total material result, shown in Figure 1. In general, the parents of group I children fall into subclassifications a-r and u-w, the parents of group II children into subclassifications s, t, and x, and the parents of group III children into subclassification y. The fact that parents in subgroups s, t, and x are shown as receiving no radiation when in fact they often received a fraction of a *rem* is due to the practice of rounding down to the nearest integer. The first entry in each box in Figure 1 represents the estimated mean dose in *rems* received by the father and mother together. In computing this, whenever the estimated T65 *rem* dose for an individual exceeded 600, it has been reduced to that figure, on the grounds that this is the approximate maximum dose of whole or near-whole body radiation compatible with survival, and doses estimated as greater than that result from confusion over position ATB or shielding or both. However, it has not been possible to introduce a corresponding counter-correction, for parents who in their histories may inadvertently have exaggerated the amount of shielding or distance from hypocenter ATB. The magnitude of this correction is indicated by the second figure in each box, which is the estimate of the mean before estimated doses greater than 600 *rem* were reduced to that level.

Ascertainment of the survival status of the children in each panel utilized the system of compulsory family registration (*koseki*) which has existed in Japan since the last quarter of the nineteenth century. All deaths in a family are routinely entered in this record. The *koseki* record is kept in that civil administrative unit which the family regards as its ancestral home (*honseki*). At the time the study was undertaken, the *honseki* had been established for each father. Thus, the death of any child in the study groups could be ascertained by consulting personally or by mail the *koseki*. In case of death, cause of death was updated for mortality through 1969. Because of the manner in which the groups were assembled, the age of the living children at the end of 1969 ranged from 11 to 23. In only 51 instances where the child was of Japanese parentage was it impossible to obtain information on child's survival; these children have been omitted from the analysis.

Mortality, especially infant mortality, is influenced by many variables, such as parents' age, birthweight, and socioeconomic conditions. If the three study groups differ significantly in regard to these conditions, failure to take this into consideration might lead to spurious inferences concerning radiation effects. Through city birth records and the records of the earlier genetic study from which the present mortality study evolved (NEEL and SCHULL 1956) it was possible in 1963 to obtain data on virtually all births in the panel with respect to maternal age, paternal age, parity, birthweight, duration of pregnancy, and, of course, year of birth. In an effort to extend this information, in 1964 a mail survey was undertaken of a random 10% of the parents in the total sample, with respect to present parental occupation, parental schooling, survival status of the parents, the size of the home (as measured by the standard module of Japanese mats) and the number of persons normally residing therein, food expenditures per person per month, and child's nutrition, as measured by consumption of such dietary items as fish, meat, eggs, and milk. A detailed analysis of these findings has already been presented (KATO, SCHULL and NEEL 1966). It can be summarized with the simple statement that in general the unexposed control parents (who came to Hiroshima and Nagasaki following the bombings) were slightly younger and had a little more education and somewhat higher occupation ratings than the exposed. The differences as tested by contingency x^2 were of borderline statistical significance. In general, these differences in the distribution of extraneous socioeconomic variables, together with the fact that parents' ages are slightly older in the exposed groups than in the control groups, are such as might lead to an inflation of mortality rates in the children of the 'heavily' (proximally) exposed parents as contrasted to the other comparison groups. However, the data do not permit a rigorous treatment of this possible bias. We return to this subject later.

STATISTICAL CONSIDERATIONS

Two analyses of the data have been undertaken. The first is a simple contingency x^2 analysis of a 5 x 5 table defined by the radiation classes of father and mother as shown in Figure 1, the cell entries being the proportion of deceased among the children of parents with that radiation history. Expectations were corrected for age of child (i.e., year of birth), sex, and city.

The second has been a stepwise regression analysis (see, for example, DRAPER and SMITH 1966), based on 52,725 individuals, of the form

$$y = a + bx_1 + cx_2 + dx_3 + ex_4$$

where y = probability of death,
 x_1 = year of birth
 x_2 = sex of child
 x_3 = father's dose in *rem*, and
 x_4 = mother's dose in *rem*

The variables have been entered into the analysis in the order of the amount of variance each accounts for, with a recomputation of the contribution of each preceding variable as each successive variable is added to the regression. The analysis was performed on an IBM 1440 computer.

THE DATA

Table 1 displays the results of the contingency x^2 analysis. During the immediate post-war years, infant and childhood mortality declined significantly in Japan. In the composition of these study groups, the limiting factor on size of group was the number of children born to relatively heavily irradiated parents. There was an excess of potential controls among children falling into categories II and III. Accordingly in constituting the three groups, groups II and III were balanced with group I with respect to year of birth, and this factor does not confound this analysis. The expected number of deaths for each cell is calculated on the basis of the same probability of death for each age (and sex) balanced cell, based on the rate observed in the total sample. Traumatic (accidental) deaths (322 in number) have been included in the tabulations and analyses of Table 1, on the general thesis that such accidents often fall on prepared soil. However, an analysis which excludes them yields substantially the same result. The table also presents the total number of years at risk. The mean age of the subjects contributing to each cell can of course be derived by dividing the latter figure by the number of subjects. The contingency x^2 is well below the level of significance.

When the numbers in the data cells were assembled to correspond as closely as possible to the definition of the three study groups followed in our first treatment of this material (by distance rather than dosage), there were certain discrepancies from the numbers given earlier (KATO, SCHULL and NEEL 1966). In the course of preparing this analysis, it was discovered that in the previous analysis, the stillbirths occurring in one segment of the data (GE-3) had, contrary to tabulation instructions, not been excluded from the base (denominator) in the calculation of rates. Since there were 694 stillbirths, this resulted, with cumulative mortalities at the time of the study of 6–7%, in an underestimate of the true rate of approximately 0.1%. This error, which in no way alters the conclusions of the earlier paper, has been corrected. Beyond this, in the change from a classification based on location ATB as obtained at the time of pregnancy registration to one based on radiation as estimated from a special study conducted some ten years later, there have been transfers of parents from one group to another. Some parents who on the original questionnaire administered as long ago as 1948 indicated absence from either city ATB, have now on subsequent contacts altered their history to presence. In part this may stem from confusion regarding the

TABLE 1

The mortality experience of the 25 groups of children defined in this study
(Data are for both Hiroshima and Nagasaki, sexes combined. Doses given are in *rem* units)

Mother	Father 100+	10-99	1-9	<1	NE	Total	
100+	20	17	10	28	186	261	Observed
	22.60	11.08	7.06	32.85	180.40	253.99	Expected*
	0.88	1.53	1.42	0.85	1.03	1.03	O/E
	323	169	110	511	2920	4033	No. Subjects
	6409	3165	2053	9718	53866	75211	P.Y.†
10-99	18	61	19	55	239	392	Observed
	14.30	63.32	17.88	58.79	262.12	416.41	Expected*
	1.26	0.96	1.06	0.94	0.91	0.94	O/E
	220	855	266	880	4129	6350	No. Subjects
	4139	17569	5154	17007	77583	121452	P.Y.†
1-9	16	16	31	31	123	217	Observed
	11.77	17.42	29.59	36.50	127.94	223.22	Expected*
	1.36	0.92	1.05	0.85	0.96	0.97	O/E
	180	259	413	567	2024	3443	No. Subjects
	3406	5028	8276	10730	37837	65277	P.Y.†
<1	46	65	35	309	590	1045	Observed
	48.81	64.32	28.03	295.97	582.05	1019.18	Expected*
	0.94	1.01	1.25	1.04	1.01	1.03	O/E
	768	993	429	4268	9207	15665	No. Subjects
	14413	18806	8102	84069	172168	297558	P.Y.†
NE	73	67	37	195	1071	1443	Observed
	59.67	74.96	37.06	175.13	1098.37	1445.19	Expected*
	1.22	0.89	1.00	1.11	0.98	1.00	O/E
	1030	1281	619	3041	17263	23234	No. Subjects
	18240	23013	11234	53889	324390	430766	P.Y.†
Total	173	226	132	618	2209	3358	Observed
	157.15	231.10	119.62	599.24	2250.88	3357.99	Expected*
	1.10	0.98	1.10	1.03	0.98	1.00	O/E
	2521	3557	1837	9267	35543	52725	No. Subjects
	46607	67581	34819	175413	665844	990264	P.Y.†

$\chi^2 = 22.46$
d.f. = 24 (expectation set by grand total).
* Basis of survival of total sample.
† P.Y. = person years.

precise city limits (often somewhat vaguely defined in Japan) and, since the radiation dose at the periphery of the city was minimal, this is a shift from one type of control to another. However, there were also 78 changes from a report of

TABLE 2

The regression of death among liveborn infants on year of birth, sex, father's radiation dose, and mother's radiation dose, as determined by a stepwise regression procedure
(Draper and Smith 1966)

Factor	Partial correlation coefficient	Proportion of variation explained (R^2)	Coefficient of regression
Year of birth	-0.08274	0.00684	-0.00588 ± 0.000307**
Sex	0.02031	0.00725	0.00988 ± 0.002119**
Father's dose	0.008516	0.00733	0.000029 ± 0.000015
Mother's dose	0.000075	0.00733	0.000000 ± 0.000012

** $p < 0.01$.

"not in city" to a report of a position in city ATB at which more than 1 *rem* unit was received, including 14 in which the conjoint dose exceeded 100 *rem*. To some extent this may be due to the fact that our original histories of parental radiation were usually obtained from the mothers alone, or her surrogate, whereas the later histories have been obtained directly from the persons concerned. Anyone familiar with the problem of obtaining accurate histories concerning highly emotional subjects will not be too surprised at these minor changes.

The results of the stepwise regression analysis are shown in Table 2. A year-of-birth effect on mortality emerges as quite significant, as does the well-known sex effect (male death rates greater than female in the early years of life). The regression of proportion of death on mother's exposure is non-significant, as is the regression on father's exposure, but the latter term is just below the 5% level. One might in this situation elect to apply a one-tailed test of significance. In this case, the regression becomes significant at the 5% level.

As noted earlier, a ten percent sample survey in 1964, some ten years after the atomic bombings, revealed small socioeconomic differences between families where one or both parents were exposed and families for which neither parent was exposed. Socioeconomic status changed rapidly in postwar Japan and the sample was relatively small, but the possibility had to be considered that these differences were influencing the variable of interest. The most appropriate way to examine this possibility is through a regression analysis which excludes the cell (γ) in Figure 1 comprised of children both of whose parents were unexposed. The results of such an analysis are given in Table 3. Sample size is now reduced from 52,725 to 35,462. The differences between the results of the two analyses are negligible.

TABLE 3

The regression of death among liveborn infants on year of birth, sex, father's radiation dose, and mother's radiation dose after exclusion from the study material of those control children neither of whose parents were exposed ATB

Factor	Partial correlation coefficient	Proportion of variation explained (R^2)	Coefficient of regression
Year of birth	-0.08323	0.00693	-0.00598 ± 0.000378**
Sex	0.02008	0.00733	0.00983 ± 0.002600**
Father's dose	0.008578	0.00740	0.000031 ± 0.000019
Mother's dose	0.001916	0.00740	0.000006 ± 0.000016

** $p < 0.01$.

DISCUSSION

There is, then, no clear evidence thus far for an increased death rate in the children born to the survivors of the atomic bombings of Hiroshima and Nagasaki. These data are in keeping with most of the results on experimental mammals such as mice, rats, and pigs, where substantially larger amounts of radiation have been delivered, in some experiments over several generations, without the clear demonstration of net dominant deleterious effects (KOHN 1960; SUGAHARA 1964; FRÖLEN 1965; CHAPMAN et al. 1964; COX and WILLHAM 1962; D. F. COX 1964; RODERICK 1964; but see RUSSELL 1955, and RUSSELL and RUSSELL 1959). These data also continue to be in keeping with other, more limited data on man (MACHT and LAWRENCE 1955; CROW 1955; KAPLAN 1958; D. W. COX 1964; TANAKA and OHKURA 1958), although in most of these series the numbers are so small and the dosage so low that no real test of hypothesis is involved. As is our custom, we enter the usual caveat, that this failure to demonstrate an unequivocal effect of this type does not constitute evidence there were no effects.

In a previous paper (KATO, SCHULL and NEEL 1966) employing preliminary dose estimates, these "negative" findings were used to generate a minimum estimate of the doubling dose for mutations resulting in death in liveborn carriers in the first generation after their origin. Using an arcsin transformation, the angular difference (as expressed in degrees) which could be demonstrated between two samples of size 17,721 (the size of group I) with respect to death rate was derived. It was shown that if the Type II error (β) was set at 0.05, then on the hypothesis of an increased death rate among the children of heavily radiated parents (i.e., one-tailed significance test), the sample was large enough to detect an angular difference of 1.0. As the cumulative death rate in the two control groups among liveborn infants was close to 6.0% at the time of this evaluation, we could detect

a difference between the children in group I and either group II or III of approximately 0.8–1.0%.

In order to make a calculation of the doubling dose, we needed an estimate of the percent of liveborn children who each generation die before age 9 (the average age of that sample) because of mutation (point and chromosomal) in the preceding generation. This is an extremely difficult figure to generate in the present state of knowledge. In this study, pregnancies were generally registered during the sixth month of gestation, and if at the time of termination there was any sign of life whatsoever, the child was classified as liveborn. The data thus include terminations occurring during the last trimester. We have previously, on the basis of both theoretical and enumerative considerations, suggested that 0.005 of a cohort of liveborn children so defined will die before age 9 as a result of mutation in the preceding generation, and we will retain this estimate for this paper, although with improving standards of medical practice it may be on the high side. Half of this was assumed to be contributed by gross chromosomal mutation and half by so-called point mutation (which might include small deletions). Quite similar figures are found in the recent report of the Subcommittee on Genetic Effects of the Advisory Committee on the Biological Effects of Ionizing Radiations (1972). The calculation then proceeded as follows:

$$d = \frac{a \cdot b}{c}$$

where a = joint parental dose,
b = mortality due to spontaneous mutation in the preceding generation,
c = increase in mortality excluded by present data, (conservatively taken to be 0.010), and
d = minimal doubling dose.

In view of the possibility that per unit exposure, the recovery of mutations from the offspring of irradiated males and females differ, it should be pointed out that this calculation tacitly assumes equal exposures of fathers and mothers, which is only approximately correct (Table 1). In the previous situation, it was calculated by this means and with these assumptions and facts that the doubling dose of radiation of this type to either or both parents for dominant mutations resulting in the death of liveborn infants prior to reproduction was not less than 50 *rad*. Since the data on the neutron component of the radiation spectrum of the atomic bombs were not yet available, this calculation was in terms of *rad* rather than *rem* units, and so did not take into consideration the greater RBE of neutrons. In the present situation, with the new data on dosage, we can modify the previous estimate as follows:

$$d = \frac{117 \cdot 0.005}{0.010} = 58.5 \; rem.$$

The most vulnerable aspect of this calculation is the estimate that 0.005 of liveborn infants will die prior to maturity because of chromosomal and "point" mutation in the preceding generation. Given our present knowledge of dominant mutation and chromosomal abnormalities, this figure can scarcely be below 0.003. On the other hand, the radiation dose is probably conservative, because of

the correction downward to an estimate of 600 of all individual T65 *rem* dose estimates above 600, without a corresponding correction for spuriously low estimates. This correction avoids the embarrassment of assigning to individuals doses of acute whole-body radiation obviously incompatible with survival. However, without this correction, the estimate in *rem* for those cells where one or both parents received 100 + *rem* units would be 63 to 233 *rem* higher (Table 1). We have also employed the conservative figure of 5 for the genetic RBE of fast neutrons. Let us assume the average total dose was only 20 *rem* greater. Then, taken with the above mentioned 0.003, the minimal estimate of doubling dose becomes

$$d = \frac{137 \cdot 0.003}{0.010} = 41 \; rem.$$

Thus the two most probable sources of error in the estimate tend to be partially compensating.

It should perhaps be emphasized that this estimate is for *zygotic* doubling dose, i.e., the amount of radiation necessary to produce the number of mutations which as a result of spontaneous mutation are "normally" present in those zygotes which result in liveborn infants. We are indebted to Dr. James Crow for reminding us that the usual estimates of doubling dose, for such organisms as Drosophila and the mouse, are gametic doubling dose. The conversion is achieved by dividing the foregoing estimates by a factor of 2.

The fact that the present analysis is of the regression type permits an alternative and more powerful approach to the question of the doubling dose than the simple contrast of two groups. If the hypothesis is that parental radiation can only increase the death rate in children (one-tailed significance test), then from Table 2 we see that for fathers the data exclude at the 5% level a regression term greater than $0.000029 + 1.645 \times 0.000015 = 0.000054$, and on the assumption that the father contributes half of the mutations resulting in prereproductive death among liveborn infants, the minimum* gametic doubling dose is $\frac{0.005 \times .5}{0.000054} = 46 \; rem$. For mothers the corresponding regression value excluded at the 5% level is 0.000020, and the minimum gametic doubling dose is estimated to be 125 *rem*. In a situation where the parents are radiated equally, the minimum average conjoint zygotic doubling dose is 67 *rem*. Now the age span covered by this estimate is, on average, the first 17 years of life.

As described earlier, an analysis has also been performed excluding the children of unexposed parents. The estimates of minimum gametic doubling dose based on this analysis is essentially the same for the fathers (44 *rem*) but somewhat lower for mothers (77 *rem*). The differences in the magnitude of the coefficients of regression in the two analyses are well within random fluctuation, and because of the greater precision of estimation afforded by the larger numbers, we prefer the analysis of Table 2. However, a slightly more conservative estimate of minimal doubling dose (but with less precision) emerges from the analysis of Table 3.

* The term "minimum" as here employed is synonymous with "97.5% lower confidence limit".

As noted above, the effects of paternal radiation approach significance, and, in fact, with a one-tailed significance test, *are* significant at the 5% level. In a question of this importance, we believe a finding should be unequivocally significant before being trumpeted as such. It is worth noting, however, that if with the further accumulation of data this regression became significant, while not departing greatly from its present value, then the estimate of the gametic doubling dose for male radiation of this type, on the assumption that 0.0025 of liveborn infants die prior to the age of reproduction because of spontaneous mutation in their fathers in the preceding generation, becomes approximately $\frac{0.0025}{0.000029} = 86$ *rem*. For the present, it is more prudent to be guided by the doubling dose which the data exclude.

The approximate nature of these calculations is obvious. They apply to only one type of genetic effect, albeit the one which probably has maximum psychological effect on human populations, since early deaths of the type under study here are often accompanied by physical defect and chronic disease. We recognize that in the experimental investigations of the mouse, the doubling dose for genetic effects of the type here under consideration has been higher than for a variety of other types of genetic effects (summary in LÜNING and SEARLE 1971) and the same may yet be shown to be true for man. For the present we note only that these estimates of minimal doubling dose for males and females fall within the range of recent consensuses from the study of other mammals (UNITED NATIONS 1972; ADVISORY COMMITTEE 1972). The need for more readily quantifiable and more objective measures of genetic damage in man continues to be great, and may be met by recent technical developments which permit large-scale screening of exposed populations for mutations affecting protein structure (cf. NEEL 1971; NEEL, TIFFANY and ANDERSON 1973).

The subjects of this study were conceived within one month to thirteen years after parental exposure. RUSSELL (1965) has shown that although per unit exposure the recovery of mutations is as great or greater in the offspring of radiated females as of radiated males for offspring conceived during the first six weeks following exposure, there was no evidence of an increased mutation rate among later litters of these same females. It is unknown whether this same phenomenon applies to the human female or, if so, what the appropriate time scale is. The data of the present study are consistent with the male-female difference observed in mice, although far from the level of significance. It may be anticipated that in the case of man, there will on the average be a considerable interval between exposure and reproduction (assuming cumulative exposures during the prereproductive and reproductive periods). The present data therefore seem to be the type appropriate to the calculation of a doubling dose for deleterious genetic effects from high intensity, short burst radiation. However, it must also be borne in mind that in the mouse the mutational yield from spaced or low dose radiation appears to be less than that from single-dose, quickly administered radiation by a factor of 3 or 4 in the male and 20 in the female, the latter figure still based on relatively scanty data (RUSSELL, RUSSELL and KELLEY 1958;

Russell, Russell and Cupp 1959; Russell 1963; Searle and Phillips 1968; Bachelor, Phillips and Searle 1969; Lyon, Phillips and Bailey 1972). Thus, under the usual conditions of human exposure (small, widely spaced doses of radiation), if human spermatogonia and oocytes respond to radiation as do those of the mouse, with the present argument the gametic doubling dose of radiation for mutations resulting in pre-reproductive mortality in liveborn infants (sexes combined) is unlikely to be less than 138 *rem* for males and over 1000 *rem* for females.

LITERATURE CITED

Advisory Committee on the Biological Effects of Ionizing Radiations, National Academy of Sciences—National Research Council, 1972. The effects on populations of exposure to low levels of ionizing radiation. National Research Council, Washington.

Arakawa, E. T., 1960 Radiation dosimetry in Hiroshima and Nagasaki atomic bomb survivors. New Engl. J. Med. **263**: 488–493.

Auxier, J. A., J. S. Cheka, F. F. Haywood, T. O. Jones and J. H. Thorngate, 1966 Free-field radiation-dose distributions from the Hiroshima and Nagasaki bombings. Health Physics **12**: 425–429.

Bachelor, A. L., R. J. S. Phillips and A. G. Searle, 1969 The ineffectiveness of chronic irradiation with neutrons and gamma rays in inducing mutations in female mice. Brit. J. Radiol. **42**: 448–451.

Chapman, A. B., J. L. Hansen, G. B. Havenstein and N. E. Morton, 1964 Genetic effects of cumulative irradiation on prenatal and early postnatal survival in the rat. Genetics **50**: 1029–1042.

Cox, D. F., 1964 Effects of radiation on litter size in swine. Genetics **50**: 1025–1028.

Cox, D. F. and R. L. Willham, 1962 Genetic effects of irradiation on early mortality in swine. Genetics **47**: 785–788.

Cox, D. W., 1964 An investigation of possible genetic damage in the offspring of women receiving multiple diagnostic pelvic X-rays. Amer. J. Hum. Genet. **16**: 214–230.

Crow, J. F., 1955 A comparison of fetal and infant death rates in the progeny of radiologists and pathologists. Amer. J. Roentgenol. **73**: 467–471.

Draper, N. R. and H. Smith, 1966 *Applied Regression Analysis*. Wiley, New York.

Frölen, H., 1965 The effect on the length of life in the offspring of X-irradiated male mice. Mutation Res. **2/3**: 287–292.

Jablon, S., S. Fujita, K. Fukushima, T. Ishimaru and J. A. Auxier, 1970 RBE of neutrons in atomic bomb survivors: Hiroshima-Nagasaki. Atomic Bomb Casualty Commission Technical Report 12-70. p. 36. Also: Symposium on Neutrons in Radiobiology, Nov. 11-14, 1969. Oak Ridge, Tenn. USAEC Conf.—691106. pp. 547–577.

Kaplan, I. I., 1958 The treatment of female sterility with X-ray therapy directed to the pituitary and ovaries. Amer. J. Obstet. Gynecol. **76**: 447–453.

Kato, H., W. J. Schull and J. V. Neel, 1966 A cohort-type study of survival in the children of parents exposed to atomic bombings. Amer. J. Hum. Genet. **18**: 339–373.

Kohn, H. I., 1960 The effect of paternal X-ray exposure on the secondary sex ratio in mice (F_1 generation). Genetics **45**: 771–778.

Los Alamos Scientific Laboratory, 1950 *The Effects of Atomic Weapons*. G.P.O., Washington.

Lüning, K. G. and A. G. Searle, 1971 Estimates of the genetic risks from ionizing radiation. Mutation Res. **12**: 291–304.

Lyon, M. F., R. J. S. Phillips and H. J. Bailey, 1972 Mutagenic effects of repeated small radiation doses to mouse spermatogonia. Mutation Res. 15: 185–190.

Macht, S. H. and P. S. Lawrence, 1955 National survey of congenital malformations resulting from exposure to roentgen radiation. Amer. J. Roentgenol. 73: 442–466.

Milton, R. C. and T. Shohoji, 1968 Tentative 1965 radiation dose estimation for atomic bomb survivors. ABCC Technical Report 1–68.

Neel, J. V., 1971 The detection of increased mutation rates in human populations. Perspectives Biol. Med. 14: 522–537.

Neel, J. V. and W. J. Schull, 1956 *The Effect of Exposure to the Atomic Bombs on Pregnancy Termination in Hiroshima and Nagasaki*. National Academy of Sciences—National Research Council, Publ. 461: Washington, D.C.

Neel, J. V., T. O. Tiffany and N. G. Anderson, 1973 Approaches to monitoring human populations for mutation rates and genetic disease. pp. 105–150. In: *Chemical Mutagens: Principles and Methods for their Detection*, Vol. 3. Edited by A. Hollaender. Plenum Press, New York.

Oughterson, A. W., G. V. LeRoy, A. A. Lieben, E. C. Hammond, H. C. Barnett, J. D. Rosenbaum and B. A. Schneider, 1951 *Medical Effects of Atomic Bombs*. The Report of the Joint Commission for the investigation of the Effect of the Atomic Bomb in Japan. Army Institute of Pathology. 5 vols.

Ritchie, R. H. and G. S. Hurst, 1959 Penetration of weapons radiation, application to the Hiroshima-Nagasaki studies. Health Physics 1: 390–404.

Roderick, T. H., 1964 Summary of the general discussion of the symposium. Genetics 50: 1213–1217.

Russell, W. L., 1955 Genetic effects of radiation in mice and their bearing on the estimation of human hazards. pp. 382–383. In: *Proc. Int. Conf. on the Peaceful Uses of Atomic Energy*, Vol. II. United Nations, New York. ——, 1963 The effect of radiation dose rate and fractionation on mutation in mice. pp. 205–217. In: *Repair from Genetic Radiation*. Edited by F. Sobels. Pergamon Press, Oxford. ——, 1965 Effect of the interval between irradiation and conception on mutation frequency in female mice. Proc. Natl. Acad. Sci. U.S. 54: 1552–1557. ——, 1967 Factors that affect the radiation induction of mutations in the mouse. Ann. Acad. Brasil, Cienc 39(Suppl.): 66–75.

Russell, W. L. and L. B. Russell, 1959 Radiation-induced genetic damage in mice. pp. 179–188. In: *Progress in Nuclear Energy*, Vol. II, Series VI, Biological Sciences. Pergamon Press, London.

Russell, W. L., L. B. Russell and M. B. Cupp, 1959 Dependence of mutation frequency on radiation dose rate in female mice. Proc. Natl. Acad. Sci. U.S. 45: 18–23.

Russell, W. L., L. B. Russell and E. M. Kelley, 1958 Radiation dose rate and mutation frequency. Science 128: 1546–1550.

Searle, A. G., 1967 Progress in mammalian radiation genetics. pp. 469–481. In: *Intern. Congr. of Radiation Research 1966*. Edited by G. Silini. North Holland Publishing Co., Amsterdam.

Searle, A. G. and R. J. Phillips, 1968 Genetic insensitivity of the mouse dictyate oöcyte to chronic irradiation. pp. 17–25. In: *Effects of Radiation on Meiotic Systems*. International Atomic Energy Agency, Vienna. ——, 1971 The mutagenic effectiveness of fast neutrons in male and female mice. Mutation Res. 11: 97–105.

Searle, A. G., E. P. Evans and B. J. West, 1969 Studies on the induction of translocations in mouse spermatogonia. Mutation Res. 7: 235–240.

Subcommittee on Genetic Effects, Advisory Committee on the Biological Effects of Ionizing Radiations, 1972.

SUGAHARA, T., 1964 Genetic effects of chronic irradiation given to mice through three successive generations. Genetics **50**: 1143–1158.

TANAKA, K. and K. OHKURA, 1958 Evidence for genetic effects of radiation in offspring of radiological technicians. Jap. J. Hum. Genet. **3**: 135–145.

United Nations Scientific Committee on the Effects of Atomic Radiation. 1972 Ionizing radiation: Levels and Effects, Vol. 2. United Nations, New York.

Corresponding Editor: G. LEFEVRE

8 Cytogenetic Study of the Offspring of Atomic Bomb Survivors, Hiroshima and Nagasaki

A. A. Awa[1], T. Honda[2], S. Neriishi[3], T. Sufuni[1,4], H. Shimba[1], K. Ohtaki[1], M. Nakano[1], Y. Kodama[1], M. Itoh[2], and H. B. Hamilton[5]

8 Cytogenetic Study of the Offspring of Atomic Bomb Survivors, Hiroshima and Nagasaki

A. A. Awa[1], T. Honda[2], S. Neriishi[3], T. Sufuni[1,4], H. Shimba[1], K. Ohtaki[1], M. Nakano[1], Y. Kodama[1], M. Itoh[2], and H. B. Hamilton[5]

1 Introduction

A cytogenetic study of the children born to atomic bomb survivors in Hiroshima and Nagasaki and children born to unexposed parents was initiated in 1967 (Awa 1975; Awa et al. 1968). The study was expanded in 1976 as a part of the Genetic Platform Research Program at RERF (Radiation Effects Research Foundation), and has been continued to the present time in conjunction with the ongoing mortality study and biochemical genetics survey on the F_1 progeny (RERF Research Protocol 1975).

The main objective of the present study is to evaluate the radiation sensitivity of human germ-cell chromosomes by measuring the frequency of children with chromosome changes in structure or number induced by radiation in the germ cells of exposed parents. It is expected that stable chromosome aberrations, if induced in the germ cells, would be most likely transmitted to the offspring. Although there is no evidence of chromosome aneuploidy being induced by radiation exposure in humans, it is difficult to exclude the possibility that abnormalities, such as XYY and XXX, would be induced in the offspring.

The present chapter describes the results of somatic chromosome analysis of 8,322 children born to A-bomb survivors in Hiroshima and Nagasaki and 7,976 children born to parents who had received less than 1 rad (distally exposed) or were not in the cities (NIC) at the time of the bomb (ATB). Chromosome analyses were based mostly on nonbanded preparations throughout the study.

Because of the recent, extensive reassessment of A-bomb dosimetry by a US-Japan team of experts (1st and 2nd US-Japan Joint Workshop 1983, 1984), the present study samples have been divided into exposed and control groups based on the T65DR system that has been routinely used until recently at RERF (Milton and Shohoji 1968). The data base for the new DS86 dose system has been entered into the RERF

[1] Radiation Effects Research Foundation, 5-2 Hijiyama Park, Minami-ward Hiroshima 732, Japan
[2] Department of Radiobiology, Nagasaki, Japan
[3] Department of Clinical Studies, Nagasaki, Japan
[4] Biological Safety Research Center, National Institute of Hygienic Sciences, Tokyo, Japan
[5] RERF Consultant

Reproduced, with permission, from *Cytogenetics*, ed. by G. Obe and A. Basler, © Springer-Verlag, Berlin-Heidelberg, 1987.

Cytogenetic Study of the Offspring of Atomic Bomb Survivors

computer; however, calculations of the individual dose estimates for each survivor are now in progress, but are not available at this time. For this reason, no attempt has been made to analyze the present data in terms of parental radiation doses.

2 Materials and Methods

The sample subjects of the present survey were selected primarily from the RERF F_1 mortality study cohort (Kato et al. 1966). This cohort includes children born between 1 May 1946 and 31 December 1958 to parents, one or both of whom were the residents of Hiroshima and Nagasaki ATB. The samples were later expanded to include children who were born after 1959 through the end of 1972. The contribution of the extended samples was about 10% of the total.

In this analysis, the exposed group consists of children born to parents, one or both of whom were located within 2000 m from the hypocenter and who had T65DR dose estimates of more than 1 rad. The control group consisted of children born to parents (1) one or both of whom were exposed distally (2500 m or more from the hypocenter) with estimated doses of less than 1 rad, or (2) were not present in the city ATB.

About 40% of the total individuals in the original samples were not included in this study, because they had died (5%), or migrated outside the contactable areas of both cities (35%). Of the remaining individuals, approximately 74% agreed to participate in this study. Thus, the participation rate of this survey was 45% of the total original sample.

After obtaining consent from the F_1 participants and, if necessary, their parents, they were invited to visit the RERF clinic. At that time they were interviewed by nurses to obtain medically-related information and a blood sample was drawn. If requested, a physical examination was performed.

Heparinized blood specimens, 1 to 2 ml per person, collected from each participant, were cultured for 2 days, and then harvested for chromosome preparations using the conventional Giemsa staining methods, the details of which have been described elsewhere (Awa et al. 1978).

In each case, ten well-spread metaphases were examined directly under the microscope, and three of them were photographed for detailed karyotype analysis. Cases with less than ten scorable metaphases were regarded as culture failure. The rate of failure was about 0.1% of the total blood samples in the two cities. When an abnormality was suspected, 100 or more cells were examined.

In addition to the conventional stain, both Q- and C-band preparations (Caspersson et al. 1971; Sumner 1972) were used as a routine procedure. The G-banding method of Seabright (1971) was also applied to cases suspected of having abnormality. When family studies were performed on probands with structural rearrangements, high resolution banding techniques (Yunis et al. 1978; Pai and Thomas 1980) were employed for the precise identification of breakpoints of the chromosomes involved in the aberrations. A description of the type of chromosome abnormalities was made following the standardized nomenclature of ISCN (1978, 1981, 1985).

Table 1. Number of F_1 children examined

			Number of cases			Number of parental couples
			Males	Females	Total	
Hiroshima	Control		2,477	2,635	5,112	4,242
	Exposed:	Father	638	636	1,274	968
		Mother	1,348	1,490	2,838	2,018
		Both	274	330	604	467
		Total	2,260	2,456	4,716	3,453
Nagasaki	Control		1,205	1,659	2,864	2,231
	Exposed:	Father	543	623	1,166	802
		Mother	912	1,123	2,035	1,305
		Both	199	206	405	263
		Total	1,654	1,952	3,606	2,370
Total	Control		3,682	4,294	7,976	6,473
	Exposed:	Father	1,181	1,259	2,440	1,770
		Mother	2,260	2,613	4,873	3,323
		Both	473	536	1,009	730
		Total	3,914	4,408	8,322	5,823

3 Results

3.1 Characteristics of the Study Sample

As shown in Table 1, the total number of children examined was 16,298 in the two cities; 9,828 in Hiroshima (4,716 exposed and 5,112 controls), and 6,470 in Nagasaki (3,606 exposed and 2,864 controls). The number of females predominated over males in both cities as well as in both the exposed and control groups. The exposed group was further divided into three categories by parental exposure status; i.e., children born to parents in which only the father was exposed, only the mother was exposed, or both parents were exposed.

Since efforts were made to collect as many cases as possible in the exposed group, the number of children per parental couple is considerably higher in the exposed group than in the controls (Table 1); 1.4 children per couple (or 8,322 in 5,823) in the exposed, and 1.2 per couple (or 7,976 in 6,473) in the controls, when the two cities were combined.

The mean age of the participants at examination was 24 in Hiroshima, and 23 in Nagasaki with a range between 12 and 38.

Cytogenetic Study of the Offspring of Atomic Bomb Survivors

Table 2. Overall frequency of F_1 children with chromosome abnormalities (per 1,000)

		Sex chromosomes	Rearrangements Balanced	Rearrangements Unbalanced	Trisomy	Total	No. of cases examined
Hiroshima	Control	17 (3.33)	16 (3.13)	0	0	33 (6.46)	5,112
	Exposed: Father	4 (3.14)	5 (3.92)	2 (1.57)	1 (0.78)	12 (9.42)	1,274
	Mother	5 (1.76)	7 (2.47)	1 (0.35)	0	13 (4.58)	2,838
	Both	3 (4.97)	2 (3.31)	0	0	5 (8.28)	604
	Total	12 (2.54)	14 (2.97)	3 (0.64)	1 (0.21)	30 (6.36)	4,716
Nagasaki	Control	7 (2.44)	9 (3.14)	2 (0.70)	0	18 (6.28)	2,864
	Exposed: Father	3 (2.57)	3 (2.57)	1 (0.86)	0	7 (6.00)	1,166
	Mother	4 (1.97)	0	1 (0.49)	0	5 (2.46)	2,035
	Both	0	1 (2.47)	0	0	1 (2.47)	405
	Total	7 (1.94)	4 (1.11)	2 (0.55)	0	13 (3.61)	3,606
Total	Control	24 (3.01)	25 (3.13)	2 (0.25)	0	51 (6.39)	7,976
	Exposed: Father	7 (2.87)	8 (3.28)	3 (1.23)	1 (0.41)	19 (7.79)	2,440
	Mother	9 (1.85)	7 (1.44)	2 (0.41)	0	18 (3.69)	4,873
	Both	3 (2.97)	3 (2.97)	0	0	6 (5.95)	1,009
	Total	19 (2.28)	18 (2.16)	5 (0.60)	1 (0.12)	43 (5.17)	8,322
Newborn infants[a]		127 (2.23)	110 (1.93)	34 (0.60)	82 (1.44)	353 (6.20)	56,952

[a] Cited from Hook and Hamerton (1977).

3.2 Types and Frequencies of Chromosome Abnormalities

The results of the cytogenetic observations are shown in Table 2, and every abnormal case is fully described in the Appendix Table (see pp. 181–183).

Chromosome abnormalities are classified into the following three groups; (1) sex chromosome abnormalities, (2) autosomal structural rearrangements, and (3) autosomal trisomics.

3.2.1 Sex Chromosome Abnormalities

As shown in Table 3, there were 43 cases with sex chromosome anomalies; 19 of 8,322 in the exposed group (2.28 per 1,000) and 24 of 7,976 in the controls (3.01 per 1,000). No increased frequency of sex chromosome abnormalities, ascribable to parental radiation exposure, was observed.

The majority of the abnormalities was due to sex chromosome aneuploidy, mostly XYY and XXY in males and XXX in females, which constituted 75% of the total sex

Table 3. Frequency of F$_1$ children with sex chromosome abnormalities (per 1,000)[a]

			Males				Females				
		XYY	XXY	Mosaic	Other	Total	X	XXX	Mosaic	Other	Total
Hiroshima	Control	5 (2.02)	4 (1.61)	0	2 (0.81)	11 (4.44)	0	3 (1.14)	3 (1.14)	0	6 (2.28)
	Exposed: Father	0	1 (1.57)	1 (1.57)	0	2 (3.13)	0	2 (3.14)	0	0	2 (3.14)
	Mother	0	4 (2.97)	0	0	4 (2.97)	0	1 (0.67)	0	0	1 (0.67)
	Both	0	1 (3.65)	0	0	1 (3.65)	0	1 (3.03)	1 (3.03)	0	2 (6.06)
	Total	0	6 (2.65)	1 (0.44)	0	7 (3.10)	0	4 (1.63)	1 (0.41)	0	5 (2.04)
Nagasaki	Control	0	5 (4.15)	0	0	5 (4.15)	0	1 (0.60)	0	1 (0.60)	2 (1.21)
	Exposed: Father	2 (3.68)	0	0	0	2 (3.68)	0	1 (1.61)	0	0	1 (1.61)
	Mother	1 (1.10)	1 (1.10)	0	1 (1.10)	3 (3.29)	0	0	1 (0.89)	0	1 (0.89)
	Both	0	0	0	0	0	0	0	0	0	0
	Total	3 (1.81)	1 (0.60)	0	1 (0.60)	5 (3.02)	0	1 (0.51)	1 (0.51)	0	2 (1.02)
Total	Control	5 (1.36)	9 (2.44)	0	2 (0.54)	16 (4.35)	0	4 (0.93)	3 (0.70)	1 (0.23)	8 (1.86)
	Exposed: Father	2 (1.69)	1 (0.85)	1 (0.85)	0	4 (3.39)	0	3 (2.38)	0	0	3 (2.38)
	Mother	1 (0.44)	5 (2.21)	0	1 (0.44)	7 (3.10)	0	1 (0.38)	1 (0.38)	0	2 (0.77)
	Both	0	1 (2.11)	0	0	1 (2.11)	0	1 (1.87)	1 (1.87)	0	2 (3.73)
	Total	3 (0.77)	7 (1.79)	1 (0.26)	1 (0.26)	12 (3.07)	0	5 (1.13)	2 (0.45)	0	7 (1.59)
Newborn infants[b] No. of cases		35 (0.93)	35 (0.93)	14 (0.37)	14 (0.37)	98 (2.59) 37,779	2 (0.10)	20 (1.04)	7 (0.37)	0	29 (1.51) 19,173

[a] The rates for the sex chromosome abnormalities apply only to the affected sex.
[b] Cited from Hook and Hamerton (1977).

anomaly cases. Except for mosaic situations, no female with 45,X was detected in any of the groups studied.

When the data from the two cities were combined, there were no significant differences between the frequencies of sex chromosome abnormalities in the exposed group when compared to the controls.

The frequency of mosaic cases was higher in females (2 in 4,408 exposed and 3 in 4,294 controls) than in males (1 in 3,914 exposed). Among the mosaics, there was a woman showing a mosaic of 45,X/46,X,r(X) in the Hiroshima controls. A family study revealed that the abnormality was identified as a de novo mutant since both parents showed the normal karyotype.

There were three cases with structural rearrangements involving sex chromosomes, all of which belonged to the control group; two unrelated males, each with a pericentric inversion of the Y-chromosome in Hiroshima, and a female with a distally deleted long arm of one of the X-chromosomes in Nagasaki.

One unusual observation was made in the Nagasaki series. A male child, born to an exposed mother, was found to have 46 chromosomes with two X-chromosomes (46,XX).

3.2.2 Autosomal Structural Rearrangements

3.2.2.1 Balanced. The majority of structural rearrangements involving autosomes were translocations of the Robertsonian or reciprocal types and pericentric inversions. They would, therefore, result in no loss or gain of chromosome material, and would be genetically balanced without any phenotypic effect. In the present study sample, there were occasionally two or more sibs in a family showing the identical karyotypic abnormality.

Among these cases, there were ten children (seven D/D and three D/G) from eight families with Robertsonian translocations in the exposed group, and six children (all D/D) from three families with similar rearrangements in the controls. By the same token, there were seven children from five families in the exposed group, and 13 children from 12 families in the controls with reciprocal translocations. Only one child in the exposed, and six children from five families in the controls had inversions.

An examination of Table 4 shows that an unusually high incidence of pericentric inversions was observed in the Hiroshima controls, and in Nagasaki a strikingly high rate of reciprocal translocations was observed in the controls, while no such cases were found in the exposed group.

As shown in the Appendix Table, chromosomes 5 and 8 were found to be involved more frequently than the others in the formation of reciprocal translocations. Furthermore, there was a frequent involvement of a chromosome 2 in pericentric inversions in Hiroshima.

In Table 5, pooled data on the frequencies of balanced structural rearrangements are tabulated in terms of the number of families in which they are observed as a function of total families included in the study. Although there is no significant difference in the frequencies of families with stable autosomal rearrangements, a relatively low frequency of rearrangements was observed in the exposed group in Nagasaki. The data

Table 4. Frequency of F₁ children with structural rearrangements (per 1,000)

		Balanced[a]					Unbalanced[a]					
		rob(D/D)	rob(D/G)	rcp	inv	Total	rob	rcp	del	supern	other	Total
Hiroshima	Control	4 (0.78)	0	6 (1.17)	6 (1.17)	16 (3.13)	0	0	0	0	0	0
	Exposed:											
	Father	2 (1.57)	0	3 (2.35)	0	5 (3.92)	0	0	0	1 (0.78)	1 (0.78)	2 (1.57)
	Mother	3 (1.06)	1 (0.35)	2 (0.70)	1 (0.35)	7 (2.47)	0	0	0	0	1 (0.35)	1 (0.35)
	Both	0	0	2 (3.31)	0	2 (3.31)	0	0	0	0	0	0
	Total	5 (1.06)	1 (0.21)	7 (1.48)	1 (0.21)	14 (2.97)	0	0	0	1 (0.21)	2 (0.42)	3 (0.64)
Nagasaki	Control	2 (0.70)	0	7 (2.44)	0	9 (3.14)	0	0	0	0	2 (0.70)	2 (0.70)
	Exposed:											
	Father	2 (1.72)	1 (0.86)	0	0	3 (2.57)	0	0	0	1 (0.86)	0	1 (0.86)
	Mother	0	0	0	0	0	0	0	0	0	1 (0.49)	1 (0.49)
	Both	0	1 (2.47)	0	0	1 (2.47)	0	0	0	0	0	0
	Total	2 (0.55)	2 (0.55)	0	0	4 (1.11)	0	0	0	1 (0.28)	1 (0.28)	2 (0.55)
Total	Control	6 (0.75)	0	13 (1.63)	6 (0.75)	25 (3.13)	0	0	0	0	2 (0.25)	2 (0.25)
	Exposed:											
	Father	4 (1.64)	1 (0.41)	3 (1.23)	0	8 (3.28)	0	0	0	2 (0.82)	1 (0.41)	3 (1.23)
	Mother	3 (0.62)	1 (0.21)	2 (0.41)	1 (0.21)	7 (1.44)	0	0	0	0	2 (0.41)	2 (0.41)
	Both	0	1 (0.99)	2 (1.98)	0	3 (2.97)	0	0	0	0	0	0
	Total	7 (0.84)	3 (0.36)	7 (0.84)	1 (0.12)	18 (2.16)	0	0	0	2 (0.24)	3 (0.36)	5 (0.60)
Newborn infants[b] (56,952)		40 (0.70)	11 (0.19)	51 (0.90)	8 (0.14)	110 (1.93)	4 (0.07)	7 (0.12)	5 (0.09)	10 (0.18)	8 (0.14)	34 (0.60)

[a] Abbreviations: *rob* Robertsonian translocation; *rcp* reciprocal translocation; *inv* inversion; *del* deletion; *supern* supernumerary chromosome.
[b] Cited from Hook and Hamerton (1977).

Table 5. Frequency of autosomal structural rearrangements by parental couples

City	Group[a]	No. of parental couples	No. of rearrangements	Rate ($\times 10^{-3}$)
Hiroshima	E	3453	11	3.19
	C	4242	13	3.06
	T	7695	24	3.12
Nagasaki	E	2370	3	1.27
	C	2231	7	3.14
	T	4601	10	2.17
Combined	E	5823	14	2.40
	C	6473	20	3.09
	T	12296	34	2.77
Neonates[b]		59542	113	1.90

[a] Abbreviations: *E* exposed; *C* control; *T* total.
[b] Cited from Jacobs (1981).

were further compared with the frequencies of the same types of abnormalities derived from cytogenetic surveys on 59,542 consecutive newborn infants (Jacobs 1981). The results obtained in this study of both exposed and control groups were approximately 50% higher than that reported for neonatal surveys.

Family studies of abnormal cases were undertaken to determine whether the observed structural rearrangements arose de novo or were inherited from one or the other parent (Table 6). Family studies in all cases were not possible because of death of parents, or because parents did not wish to cooperate in this study. When two or more sibs in a family were found to carry the identical rearrangement, however, their abnormalities were judged to be inherited, even though a family study may not have been done.

The ample evidence indicates that the majority of cases with rearrangements are heritable. There were only two de novo mutants identified. Both mutants (one in the exposed and one in the control group) were seen in the Hiroshima group. None have been detected so far in Nagasaki. The gametic mutation rates on structural rearrangements derived from the combined data were estimated as 1.72×10^{-4} in the exposed group, and 1.40×10^{-4} in the controls, respectively. These values did not deviate significantly from the value of 1.88×10^{-4} per gamete per generation in the liveborn infant survey (Jacobs 1981) (see Addendum p. 178).

3.2.2.2 Unbalanced. There were seven abnormal cases which were placed in this category; five in the exposed and two in the controls. All of them were characterized by the presence of an extra-small metacentric element (or elements), termed as "mar" according to the standardized nomenclature system (ISCN 1978, 1981, 1985). The element (or elements) was present either as a "supernumerary" piece of chromosomal

Table 6. Parental origin of balanced structural rearrangements[a]

	Hiroshima[b]		Nagasaki[b]		H + N[b]		Neonates[c]
	E	C	E	C	E	C	
de novo	1	1	0	0	1	1	18
Inherited	5	5	1	5	6	10	73
Father	2	3	1	3	3	6	37
Mother	0	1	0	1	0	2	36
Undetermined	3	1	0	1	3	2	0
Total	6	6	1	5	7	11	91
Not studied	5	7	2	2	7	9	22
Mutation rate ($\times 10^{-4}$)	2.65	2.55	–	–	1.72	1.40	1.88

[a] Including reciprocal and Robertsonian translocations and pericentric inversions.
[b] Abbreviations: *E* exposed; *C* control.
[c] Cited from Jacobs (1981).

material, in addition to the normal chromosome complement in all cells, or was in the form of a mosaic (Table 4, Appendix Table). Five of the cases were mosaics (three exposed and two controls).

In one male observed in the Nagasaki controls, three extra minute elements, each of which differed in size and shape, existed as cell lines with different combinations of markers (Itoh et al. 1984). Family studies revealed that most of these abnormal cases of unbalanced type were found to arise as de novo mutants (see Appendix Table).

3.2.3 Autosomal Trisomics

There was only one male Downs syndrome case (47,XY,+21) born in 1966 to an exposed father in Hiroshima, whose age at cytogenetic examination was 15. No other trisomic cases, such as D- and E-trisomy, have been observed in the F_1 population.

A summary of the overall frequencies of cases with abnormalities by parental exposure and by city is shown in Table 7. It can be seen that for sex chromosome abnormalities and structural rearrangements, the frequencies are consistently higher in the controls than in the exposed, and also higher in Hiroshima than in Nagasaki. Yet none of these differences are significantly different.

The data from the two cities were combined, regardless of parental radiation exposure, and each of the frequencies was compared with that of corresponding abnormalities in the cytogenetic surveys on 56,952 consecutive liveborn infants (Hook and

Table 7. Frequency of cases with chromosome abnormalities in the F_1 population (per 1,000)

	Exposed	Control	Hiroshima	Nagasaki	Combined	Neonates[a]
No. of cases	8,322	7,976	9,828	6,470	16,298	56,952
Abnormalities:						
Sex chromosomes	2.28	3.01	2.95	2.16	2.64	2.23
Autosomal rearrangement						
Balanced	2.16	3.13	3.05	2.01	2.64	1.93
Unbalanced	0.50	0.25	0.31	0.62	0.43	0.60
Autosomal trisomy	0.12	0	0.10	0	0.06	1.44
Total	5.17	6.39	6.41	4.79	5.77	6.20
	(43)	(51)	(63)	(31)	(94)	(353)

[a] Cited from Hook and Hamerton (1977).

Hamerton 1977). Here again, the frequencies of both sex abnormalities and rearrangements were somewhat higher in the children included in this study as compared to the neonates. In contrast, the incidence of autosomal trisomics was strikingly decreased in our samples.

3.2.3.1 Heteromorphic Variants. In the course of the present screening, there was a variety of heterochromatic variants as detected with the conventional staining method. Some of these cases were reanalyzed by the application of C-, Q-, and other banding techniques, and the results have been already reported elsewhere (Sofuni et al. 1978, 1980; Sofuni and Awa 1982). Of the heteromorphic variants observed, cases with inv(9p+q-), 1qh+, 9qh+, 16qh+, (C-band variants at the constitutive heterochromatic regions), Dp+, Gp+, Dps+, Gps+, DP-, Gp- (Q-band variants either deleted or enlarged satellites and/or short arms of acrocentric chromosomes) were identified with high frequency.

4 Discussion

As mentioned previously, the main objective of this study was to demonstrate whether there is any measurable increase in the frequency of children with chromosome abnormalities that might be associated with A-bomb radiation exposure of parental germ-cell chromosomes. This cytogenetic evaluation was conducted by comparing the frequencies of children born to A-bomb survivors with chromosome abnormalities, especially structural rearrangements, with children born to nonexposed parents.

The present results show that there are no statistically significant increases in the frequency of chromosome abnormalities among the children of the exposed. In fact, the frequency of both sex chromosome aneuploidy and structural rearrangements was higher in the controls than in the exposed.

This does not imply that genetic effects ascribable to parental A-bomb exposure have not occurred. It simply indicates that such effects have not been detectable by the current cytogenetic methods used. The reason for the absence of cytogenetic effects of A-bomb radiation on the children of the survivors remains unresolved. One of the possible interpretations is that germ cells with chromosome damage could have been eliminated either in the course of gametogenesis or in the very early period of gestation as "unrecognized" spontaneous abortions.

Extensive cytogenetic surveys on consecutive liveborn infants, undertaken as international collaborative studies in several European and North American countries, have provided very useful and relevant information on the natural incidence of various types of constitutional chromosome abnormalities in the human population (Sergovich et al. 1969; Lubs and Ruddle 1970; Friedrich and Nielsen 1973; Jacobs et al. 1974; Hamerton et al. 1975; Nielsen and Sillesen 1975; Walzer and Gerald 1977; also refer to Hook and Hamerton 1977 for review).

Cytogenetic data based on these surveys were compared with our present findings (Tables 2 to 4, and 7). It is apparent that the frequencies of both sex chromosome abnormalities and balanced rearrangements are slightly higher in our study than in the neonatal surveys. The situation is reversed when one compares unbalanced rearrangements and autosomal trisomics. It is known that most, but not all, situations where autosomal trisomy as well as unbalanced rearrangements are associated with physical malformations, are incompatible with life. Since the mean age of our study subjects was 24 years at examination, it is conceivable that most conceptions with these abnormalities do not survive for 2 or 3 decades even if they survived to term.

Recently, attempts have been made to reevaluate cytogenetic surveys of neonates using banding techniques in order to obtain more precise information on the frequency of abnormalities associated with structural chromosomal changes in the general human population (Lin et al. 1976; Buckton et al. 1980; Hansteen et al. 1982; Nielsen et al. 1982). The results obtained in such studies are discordant. Some studies indicated an increase in the frequency of reciprocal translocations (Hansteen et al. 1982), and inversions (Hansteen et al. 1982; Nielsen et al. 1982), or an increase in the Q-band variants (Lin et al. 1976). In contrast, the frequency of all types of chromosome abnormalities detected when G-banding techniques were used have been found to be similar to that observed previously when conventional stain preparations were used (Buckton et al. 1980). In our experience, there seemed to be no difference in the detectability of structural rearrangements between conventional stain and G- and Q-banding methods, although the latter were found to be more efficient in detecting heteromorphic variants than the conventional method.

Maeda et al. (1978) reported the results of a cytogenetic survey on 2,626 consecutive liveborn infants in Japan. They found nine cases with sex chromosome abnormalities (0.30%) and ten autosomal abnormalities, including five with Robertsonian translocation of the balanced type (0.19%), one with 13-trisomy associated with Robertsonian translocation (0.04%), one with a supernumerary marker (0.04%), and three with 21-trisomy (0.11%). Although the sample size of this survey was rather small, the results agreed to a certain extent with our findings with respect to the frequency of sex chromosome abnormalities and types and patterns of autosomal abnormalities.

Table 8. Frequency of autosomal structural rearrangements

City	Group[a]	No. of children	No. of rearrangements	Rate ($\times 10^{-3}$)
Hiroshima	E	4716	14	2.97
	C	5112	16	3.13
	T	9828	30	3.05
Nagasaki	E	3606	4	1.11
	C	2864	9	3.14
	T	6470	13	2.01
Combined	E	8322	18	2.16
	C	7976	25	3.13
	T	16298	43	2.64
Neonates[b]		59542	113	1.90

[a] Abbreviations: *E* exposed; *C* control; *T* total.
[b] Cited from Jacobs (1981).

The origin and cytogenetic features of an additional marker chromosome, often designated as a supernumerary chromosome, have been studied extensively (refer to Buckton et al. 1985). This marker chromosome is chracterized by a small metacentric element. Often it has a satellite (or satellites) at one or both distal ends. With regard to the presence of cases in the F_1 carrying such a marker, it was interesting to observe that (1) the supernumerary was seen in all cells or existed in the form of a mosaic, and (2) the majority of the supernumeraries were do novo mutants without any phenotypic change. This finding led to the presumption that the origin of the markers in our cases might arise from a product of a Robertsonian translocation between the short arms of the D- and G-chromosomes.

There has been discrepancy concerning a possible association between maternal radiation and trisomy-21 (Uchida et al. 1961; Schull and Neel 1962; Uchida 1977). In the A-bomb exposed population, Schull and Neel (1962) reported that the frequency of children with Down syndrome born to exposed mothers (fathers not exposed) was 0.54 per 1,000 (3 in 5,579), and 1.27 per 1,000 (12 in 9,440) in children born to nonexposed mothers. These results suggest that no association exists between Down syndrome and maternal radiation. The present results do not indicate an increase in children with 21-trisomy or aneuploidy involving any other chromosome.

Finally, when the new A-bomb radiation dosimetry system (DS86) becomes available for estimating the dose for individual survivors, the present data will be reanalyzed to determine the relationship between the frequency of chromosome abnormalities observed in children as a function of parental gonadal dose.

Table 9. Parental origin of balanced structural rearrangements[a]

	Hiroshima E	Hiroshima C	Nagasaki E	Nagasaki C	H + N E	H + N C	Neonates[b]
de novo	1	1	0	0	1	1	18
Inherited	8	8	2	7	10	15	73
Father	2	4	2	4	4	8	37
Mother	0	1	0	1	0	2	36
Undetermined	6	3	0	2	6	5	0
Total	9	9	2	7	11	16	91
(Not studied)	(5)	(7)	(2)	(2)	(7)	(9)	(22)
Gametic mutation rate ($\times 10^{-4}$)	1.65	1.74	–	–	0.98	0.98	1.88

[a] Including reciprocal and Robertsonian translocations plus pericentric inversions.
[b] Cited from Jacobs (1981).
Abbreviations: *E* exposed; *C* control; *H* Hiroshima; *N* Nagasaki.

Addendum. In the test, we computed the gametic mutation rates of structural rearrangements of balanced type in our population in terms of the frequency of families with abnormality as a function of total families studied. However, it seems more appropriate to estimate the chromosomal mutation rate using the following formula:

$$\text{Mutation rate} = \frac{\text{Number of individuals with abnormality} \times \text{Proportion who are likely to have de novo mutations}}{\text{Total number examined}}$$

For the gametic mutation rate, the value derived from the above formula is further divided by 2 (Jacobs 1981). The gametic mutation rates on structural rearrangements derived from the combined data in Tables 8 and 9 were estimated as 0.98×10^{-4} in the exposed, and 0.98×10^{-4} in the controls, respectively. These values were much lower than the value of 1.88×10^{-4} per gamete per generation in the liveborn infant survey (Jacobs 1981).

Acknowledgments. We are grateful to the Hiroshima and Nagasaki citizens for their willingness to voluntarily participate in this survey. Without their cooperation this study would not have been possible. Our thanks are due also to the following RERF staff members for their continued support throughout this survey: Drs. M. Otake and S. Fujita, Department of Statistics, Dr. Y. Shimizu, Department of Epidemiology, for sample selection and statistical advices; Drs. T. Amano, M. Soda, R. Hazama, and their associaes in the Nursing Section, Department of Clinical Studies, for their clinical assistance; public health nurses and clinical contactors in the Department of Research Support who obtained consent of the participants and arrangements for their visit to RERF. We are very much indebted to Messrs. S. Iida, K. Tanabe, Mrs. Y. Urakawa, and their colleagues in the Cytogenetics Laboratories in Hiroshima and Nagasaki for their continued technical support including blood cultures, chromosome preparations, microscope and karyotype analysis, and photographic works. We wish to thank Dr. C.W. Edington, Vice-Chairman of RERF, for his encouragement and for going through the manuscript.

References

Awa AA (1975) Review of thirty years study of Hiroshima and Nagasaki atomic bomb survivors. II Biological effect. B Genetic effects. 2 Cytogenetic study. J Radiat Res 16 (Suppl):75–81

Awa AA, Bloom AD, Yoshida MC, Neriishi S, Archer PG (1968) Cytogenetic study of the offspring of atom bomb survivors. Nature (London) 218:367–368

Awa AA, Sofuni T, Honda T, Itoh M, Neriishi S, Otake M (1978) Relationship between the radiation dose and chromosome aberrations in atomic bomb survivors of Hiroshima and Nagasaki. J Radiat Res 19:126–140

Buckton KE, O'Riordan ML, Ratcliffe S, Slight J, Mitchell M, McBeath S, Keay AJ, Barr D, Short M (1980) A G-band study of chromosome in liveborn infants. Ann Human Genet 43:227–239

Buckton KE, Spowart G, Newton MS, Evans HJ (1985) Forty four probands with an additional "marker" chromosome. Human Genet 69:353–370

Caspersson T, Lomakka G, Zech L (1971) The 24 fluorescence pattern of the human metaphase chromosomes – distinguishing characters and variability. Hereditas 67:89–102

Friedrich U, Nielsen J (1973) Chromosome studies in 5,049 consecutive newborn children. Clin Genet 4:333–343

Hamerton JL, Canning N, Ray M, Smith S (1975) A cytogenetic survey of 14069 newborn infants. I Incidence of chromosome abnormalities. Clin Genet 8:223–243

Handsteen I-L, Varslot K, Steen-Johnsen, Langard S (1982) Cytogenetic screening of a newborn population. Clin Genet 21:309–314

Hook EB, Hamerton JL (1977) The frequency of chromosome abnormalities detected in consecutive newborn studies – differences between studies – results by sex and by severity of phenotypic involvement. In: Hook EB, Porter IH (eds) Population cytogenetics – Studies in humans. Academic Press, London New York, pp 63–79

ISCN (1978) An international system for human cytogenetic nomenclature (1978) In: Lindsten JE, Klinger HP, Hamerton JL (eds) Birth defects: original article series, vol 14 (8). Natl Found, New York; also in: Cytogenet Cell Genet 21:309–404

ISCN (1981) An international system for human cytogenetic nomenclature (1981) High resolution banding. In: Harnden DG, Lindsten JE, Buckton KE, Klinger HP (eds) Birth defects: original article series, vol 17 (5). March of Dimes, Birth Defects Found, New York; also in: Cytogenet Cell Genet 31:1–28

ISCN (1985) An international system for human cytogenetic nomenclature (1985) In: Harnden DG, Klinger HP (eds) puglished in collaboration with Cytogenet Cell Genet (Karger, Basel 1985); also in: Birth defects: original article series, vol 21 (1). March of Dimes Birth Defects Found, New York

Itoh M, Soda M, Honda T (1984) Unstable behavior and mosaicism of extra minute chromosome in man. Nagasaki Med J 59:441–445

Jacobs PA (1981) Mutation rates of structural chromosome rearrangements in man. Am J Human Genet 33:44–54

Jacobs PA, Melville M, Ratcliffe S, Keay AJ, Syme J (1974) A cytogenetic survey of 11680 newborn infants. Ann Human Genet 37:359–376

Kato H, Schull WJ, Neel JV (1966) A cohort-type study of survival in the children of parents exposed to atomic bombings. Am J Human Genet 18:339–373

Lin CC, Gedeon MM, Griffith P, Smink WK, Newton DR, Wilkie L, Sewell LM (1976) Chromosome analysis on 930 consecutive newborn children using quinacrine fluorescent banding technique. Human Genet 31:315–328

Lubs HA, Ruddle FH (1970) Chromosomal abnormalities in the human population: Estimation of rates based on New Haven newborn study. Science 169:495–497

Maeda T, Ono M, Takada M, Kato Y, Nishida M, Jobo T, Adachi H, Taguchi A (1978) A cytogenetic survey of consecutive liveborn infants – incidence and type of chromosome abnormalities. Jpn J Hum Genet 23:217–224

Milton RC, Shohoji T (1968) Tentative 1965 radiation dose (T65D) estimation for atomic bomb survivors; Hiroshima and Nagasaki. ABCC Teech Rep 1–68

Nielsen J, Sillesen I (1975) Incidence of chromosome aberrations among 11148 newborn children. Humangenetik 30:1–12

Nielsen J, Wohlert M, Faaborg-Andersen J, Hansen KB, Hvidman L, Klag-Olsen B, Moulvad I, Videbech P (1982) Incidence of chromosome abnormalities in newborn children. Comparison between incidences in 1969–1974 and 1980–1982 in the same area. Human Genet 61:98–101

Pai GS, Thomas GH (1980) A new R-banding technique in clinical cytogenetics. Human Genet 54: 41–45

RERF Research Protocol (1975) Research plan for RERF studies of the potential genetic effects of atomic radiation; Hiroshima and Nagasaki. RERF RP 4–75

Schull WJ, Neel JV (1962) Maternal radiation and mongolism. Lancet 1:537–538

Seabright M (1971) A rapid banding technique for human chromosomes. Lancet 2:971–972

Sergovich FR, Valentine GH, Chen ATL, Kinch RAH, Smout MS (1969) Chromosome aberrations in 2159 consecutive newborn babies. New Engl J Med 280:851–855

Sofuni T, Awa AA (1982) Chromosome heteromorphisms in the Japanese. III Frequency of C-band variants. RERF Tech Rep 4–82

Sofuni T, Tanabe K, Awa AA (1978) Chromosome heteromorphisms in the Japanese. I Banding patterns of Dp+ and Gp+ by Q- and C-staining methods. RERF Tech Rep 8–78

Sofuni T, Tanabe K, Awa AA (1980) Chromosome heteromorphisms in the Japanese. II Nucleolus organizer regions of variant chromosomes in D and G groups. Human Genet 55:265–270

Sumner AT (1972) A simple technique for demonstrating centromeric heterochromatin. Exp Cell Res 75:304–306

Uchida IA (1977) Maternal radiation and trisomy 21. In: Hook EB, Porter IH (eds) Population cytogenetics – Studies in humans. Academic Press, London New York, pp 285–299

1st US-Japan joint Worksh Reassessment of atomic bomb radiation dosimetry in Hiroshima and Nagasaki. Radiation Effects Research Foundation, Hiroshima, 1983

2nd US-Japan joint Worksh Reassessment of atomic bomb radiation dosimetry in Hiroshima and Nagasaki. Radiation Effects Research Foundation, Hiroshima, 1984

Walzer S, Gerald PS (1977) A chromosome survey of 13751 male newborns. In: Hook EB, Porter IH (eds) Population cytogenetics – studies in humans. Academic Press, London New York, pp 45–61

Yunis JJ, Sawyer JR, Ball DW (1978) The characterization of high resolution G banded chromosomes of man. Chromosoma 67:293–307

Appendix Table. Family studies of abnormal cases of the unbalanced type

Case No.	Sex	Age ATE	Year of birth	Age at birth Mo	Age at birth Fa	Exposure status	Type of abnormality	Remarks
Hiroshima								
FH3321	M	28	1948	32	38	C	A1. 47,XYY	
FH4305	M	22	1956	30	38	C	" "	
FH4998	M	29	1949	25	29	C	" "	
FH7308	M	17	1964	28	32	C	" "	
FH9913	M	15	1969	25	27	C	" "	
FH0757	M	17	1953	26	34	Mo	A2. 47,XXY	
FH0801	M	22	1948	23	27	Fa	" "	
FH0815	M	13	1957	31	48	B	" "	
FH4223	M	22	1956	34	31	Mo	" "	
FH4727	M	29	1949	33	38	Mo	" "	
FH5953	M	26	1953	21	24	C	" "	
FH6415	M	30	1950	32	40	C	" "	
FH6662	M	28	1952	27	31	C	" "	
FH8632	M	24	1958	24	28	Mo	" "	
FH8718	M	33	1949	35	41	C	" "	
FH8454	M	17	1965	25	30	Fa	A3. 46,XY/47,XYY	46:48 cells, 47:152 cells
FH6505	M	24	1956	32	38	C	A4. 46,X,inv(Y)(p11.2q11.2)	
FH7417	M	23	1958	33	31	C	" 46,X,inv(Y)(p11.2q11.23)pat	
FH0492	F	21	1948	25	29	Fa	A6. 47,XXX	
FH1870	F	15	1957	34	33	B	" "	
FH3886	F	30	1947	25	27	C	" "	F₂:one (liveborn)
FH5852	F	21	1958	30	36	Mo	" "	Sib (dizygotic co-twin)—normal
FH8291	F	24	1958	23	27	Fa	" "	
FH8912	F	16	1967	37	39	C	" "	
FH9339	F	15	1967	28	30	C	" "	
FH0092	F	18	1949	33	36	B	A7. 45,X/47,XXX	45:95 cells, 47:4 cells
FH3033	F	22	1954	33	38	C	" "	45:82 cells, 47:15 cells
FH8020	F	24	1957	24	28	C	" "	45:44 cells, 47:49 cells. F₂:one (liveborn), also pregnant at examination
FH8590	F	33	1948	25	38	C	" 45,X/46,X,r(x) §	§ *de novo* mutant. 45:141 cells, 46:59 cells

Appendix Table (continued)

Case No.	Sex	Age ATE	Year of birth	Age at birth Mo	Age at birth Fa	Exposure status	Type of abnormality	Remarks
FH0381	M	18	1951	21	30	Fa	B1. 45,XY,t(DqDq)	
FH1988	F	16	1956	27	32	Mo	45,XX,t(DqDq)	
FH2443	F	21	1952	23	28	Mo	" "	
FH4989	M	28	1951	29	39	Mo	45,XX, "	
FH6456	M	30	1949	24	27	C	45,XY,t(13q14q)	
FH6664	F	29	1951	26	29	C	45,XY,t(13q14q)	
FH6757	F	26	1954	29	32	C	45,XX, "	Sibs
FH6538	F	30	1950	20	23	C	45,XX, "	
FH8096	F	16	1966	30	35	C	45,XX,t(14q15q)	
FH3496	F	22	1954	26	34	Fa	45,XX,t(13q14q)	Sibs
FH0405	F	18	1951	29	43	Mo	B2. 45,XX,t(14q21q)	
FH8076	M	33	1949	27	41	Fa	B3. 46,XX,t(5;11)(q13;p15)	
FH2209	F	26	1947	34	36	Fa	46,XY, "	§de novo mutant
FH2210	F	26	1947	34	36	B	46,XX,t(5;17)(p13;q25)§	
FH2690	M	17	1957	23	29	B	46,XX,t(5;8)(q22;p11.2)pat	Sibs
FH4493	M	29	1949	23	27	Fa	46,XX,t(5;8)(q22;p11.2)pat	
FH4595	M	23	1955	32	34	C	46,XY,t(6;12)(q15;q22)	
FH6611	F	29	1951	27	30	C	46,XY,t(1;6)(q21;p21)	
FH6771	M	26	1954	30	32	C	46,XX,t(3;8)(q21;q24.1)	Sibs
FH7342	F	32	1949	27	35	Mo	46,XY, "	
FH9219	M	25	1957	25	36	Mo	46,XX,t(12;13)(q21.32;p12.3)§	§de novo mutant
FH9876	F	13	1970	29	31	C	B3. 46,XY,t(5;7)(q15;p21)pat	
FH9922	M	13	1971	25	30	C	46,XX,t(5;8)(q22;p11.2)pat	
FH1705	F	25	1946	34	30	C	46,XY,t(6;8)(q13;q22)	
FH5672	M	32	1947	27	35	C	B4. 46,XX,inv(2p+q−)	
FH6099	M	27	1953	27	32	C	46,XY,inv(18)(p11.32q11.2)mat	
FH7331	F	16	1964	27	38	Mo	46,XY,inv(2)(p11.2q13)	Sibs
FH7333	F	14	1966	29	33	C	46,XX,inv(2)(p11.2q13)pat	
FH9118	M	35	1948	39	45	C	46,XX, "	
FH9254	M	34	1949	27	34	C	46,XY,inv(2)(p11.2q14.2)	
FH7361	F	24	1957	24	25	Fa	46,XY,inv(5)(p15.3q13)	
FH6169	F	23	1956	28	28	Mo	B8. 47,XX,+mar*§ (*minute)	
FH8980	M	14	1969	25	28	Fa	B9. 46,XX/47,XX,+mar* (*minute)	de novo mutant. F₂:one (liveborn) 46:9 cells, 47:91 cells del(18p):181 cells, i(18q):7 cells
FH7594	M	15	1966	25	25	Fa	C1. 47,XY,+21§	§de novo mutant

ID	Sex	Age	City	Year	Age2	Exp	Karyotype	Notes
FN0217	M	13	Nagasaki	1956	37	33	Mo	A1. 47,XYY
FN0486	M	21		1948	30	35	Fa	" "
FN0970	M	19		1951	26	28	Fa	" "
FN1717	M	26		1948	31	39	C	A2. 47,XXY
FN2696	M	18		1957	27	27	C	" "
FN4025	M	30		1948	27	31	Mo	" "
FN4171	M	25		1953	22	19	C	" "
FN4317	M	28		1951	40	45	C	" "
FN6865	M	14		1970	25	29	C	" "
FN3338	M	21		1956	25	30	Mo	A4. 46,XX [XX male]
FN3252	F	22		1955	24	27	Fa	A6. 47,XXX
FN4237	F	31		1947	20	24	C	" "
FN2120	F	18		1956	26	24	Mo	A7. 46,XX/47,XXX 46:3 cells, 47:96 cells
FN3202	F	21		1956	33	37	C	A8. 46,XXq−
FN4935	F	32		1948	32	46	C	B1. 45,XX,t(13q14q)pat ⎫ Sibs
FN4936	F	28		1952	36	50	C	" 45,XX, " ⎭
FN5837	M	13		1969	32	34	Fa	" 45,XY,t(13q14q)pat ⎫ Sibs
FN5859	F	17		1965	29	31	Fa	" 45,XX, " ⎭
FN0870	M	18		1952	23	31	B	B2. 45,XY,t(DqGq)
FN6530	F	14		1969	29	36	Fa	" 45,XX,t(14q21q)
FN1638	M	24		1950	21	22	C	B3. 46,XY,t(Cq−;17q+)pat
FN3068	F	28		1948	40	44	C	" 46,XY,t(18q+;20q−)
FN3948	M	21		1957	33	41	C	" 46,XY,t(1p−;12q+)mat ⎫ Sibs
FN4419	F	32		1947	22	28	C	" 46,XX,t(2p−;8p+) ⎭
FN4891	F	31		1949	24	30	C	" 46,XX, "
FN4724	M	29		1951	24	29	C	" 46,XY,t(Bq−;Cq+)pat
FN5049	F	29		1951	28	35	C	" 46,XX,t(2q−;13q+)
FN2357	F	26		1949	23	46	Fa	B8. 47,XX,+mar* mat (*minute)
FN3990	M	22		1957	40	37	C	B9. 46,XY/47,XY,+mar* §(*minute) § de novo mutant. 46:18 cells, 47:31 cells
FN4121	M	30		1948	20	29	Mo	" 46,XY/47,XY,+mar* § (*minute) § de novo mutant. 46:21 cells, 47:9 cells
FN6023	M	31		1951	23	24	C	" 46,XY/47,XY,+mar/48,XY,+mar x2 § § de novo mutant. Complex mosaic consisting of cell lines with 2 or 3 different minute markers of varying combinations.

[Abbreviations] ATE: At the time of examination. Age at birth: Parental age. Mo: Mother. Fa: Father. Exposure status: B=Both parents exposed. Fa=Father exposed. Mo=Mother exposed. C=Control.

Search for Mutations Altering Protein Charge and/or Function in Children of Atomic Bomb Survivors: Final Report

James V. Neel,*,† Chiyoko Satoh,† Kazuaki Goriki,† Jun-ichi Asakawa,† Mikio Fujita,† Norio Takahashi,† Takeshi Kageoka,† and Ryuji Hazama†

*Department of Human Genetics, University of Michigan Medical School, Ann Arbor; and †Radiation Effects Research Foundation, Hiroshima and Nagasaki

Summary

A sample of (1) children whose parents had been proximally exposed (i.e., <2,000 m from the hypocenter) at the time of the atomic bombings of Hiroshima and Nagasaki and (2) a suitable comparison group have been examined for the occurrence of mutations altering the electrophoretic mobility or activity of a series of 30 proteins. The examination of the equivalent of 667,404 locus products in the children of proximally exposed persons yielded three mutations altering electrophoretic mobility; the corresponding figure for the comparison group was three mutations in 466,881 tests. The examination of a subset of 60,529 locus products for loss of enzyme activity in the children of proximally exposed persons yielded one mutation; no mutations were encountered in 61,741 determinations on the children of the comparison group. When these two series are compared, the mutation rate observed in the children of proximally exposed persons is thus 0.60×10^{-5}/locus/generation, with 95% confidence intervals between 0.2 and 1.5×10^{-5}, and that in the comparison children is 0.64×10^{-5}/locus/generation, with 95% intervals between 0.1 and 1.9×10^{-5}. The average conjoint gonad doses for the proximally exposed parents are estimated to be 0.437 Gy of gamma radiation and 0.002 Gy of neutron radiation. If a relative biological effectiveness of 20 is assigned to the neutron radiation, the combined total gonad dose for the parents becomes 0.477 Sv. (Organ absorbed doses are expressed in gray [1 Gy = 100 rad]; where dose is a mixture of gamma and neutron radiation, it is necessary because of the differing relative biological effectiveness of gamma and neutron radiation to express the combined gamma-neutron gonad exposures in sieverts [1 Sv = 100 rem]).

Introduction

Planning for a study on the genetic effects of the atomic bombs was initiated in 1946, and data collection began in 1948. The details of the program that evolved over the years have been presented on several occasions (see, e.g., Neel and Schull 1956; Neel et al. 1974; Schull et al. 1981, 1982; Satoh et al. 1982). In 1972 a pilot study was undertaken of the feasibility of incorporating into the program a search for mutations resulting in an altered electrophoretic mobility of a carefully chosen battery of proteins (Neel et al.

Received July 22, 1987; revision received December 2, 1987.

1978); on the basis of the results of the pilot study, a full-scale investigation of this phenomenon was added to the other ongoing studies in 1975. A preliminary report of this study was presented in 1980 (Neel et al. 1980). In 1979 the biochemical approach to the genetic effects of the bombs was expanded with the introduction of a search for mutations resulting in approximately half-normal values for a selected series of enzymes, i.e., a search for mutations resulting from any one of several routes in gene inactivation or loss. A preliminary report has also been issued on the results of this approach (Satoh et al. 1983). Both of these studies have now been terminated. In the present report the findings will be summarized and related to the revised dosage schedules which are just becoming available for the survivors of Hiroshima and Nagasaki, because of a reappraisal of the yields and transmission of radiation

from the atomic bombs (Radiation Effects Research Foundation [RERF] 1987).

The Protocol

The Indicators

The classical experimental studies on the genetic effects of radiation were conducted in the era before biochemical indicators were conveniently available. When the present study was conceived, in 1970, there were no published data on the production by radiation of germinal mutations resulting in transmitted electrophoretic variants in the mouse, the usual human surrogate in these matters. The present study was thus launched because, on theoretical grounds, it was difficult to believe that some fraction of the mutations resulting from radiation of germ-line cells would not be characterized by nucleotide substitutions. The probability of spontaneous loss of the purine moiety of a nucleotide in a mammalian cell has been estimated at 3×10^{-11}/s (Lindahl 1977). This low probability suggests a remarkable stability for any single nucleotide, but, because there are so many nucleotides in a mammalian nucleus, by this same calculation a mammalian cell should lose by spontaneous hydrolysis approximately 10,000 purines (and 500 pyrimidines) from its DNA during a 20-h generation period. Failure to repair each of these losses exactly would result in a mutation. Radiation might be expected not only to accelerate this spontaneous disintegration of nucleotides, with additional mutations as a result of misrepair, but to impair the repair process as well.

Furthermore, the policy of the genetic studies in Japan has been to examine as many valid indicators of the potential genetic effects of the bombs as possible. "Dogma" has had it that, in contrast to the results with Neurospora (Malling and de Serres 1973), the germinal mutations produced by the radiation of mammals are very predominantly deletions and other major chromosomal events. These should manifest themselves as children with congenital defects and/or decreased life spans. A search for such defects and for impaired survival was a major aspect of the early studies of the genetic effects of the bombs (Neel and Schull 1956). This dogma has been based primarily on the well-documented clastogenic effects of radiation and on the high frequency with which radiation-induced mutations are lethal when homozygous. A further, more recent argument has been that, in a variety of mammalian cell systems, ionizing radiation does not induce mutations to ouabain resistance, a type of mutation believed, on theoretical grounds, to arise from base-pair substitutions (see references in Liber et al. 1987). In our opinion, even if this viewpoint were essentially correct, it would have been a mistake, in any effort at a comprehensive study of the genetic effects of atomic bombs, to have concentrated only on phenotypes reflecting endpoints known to be especially sensitive to radiation.

There was a further point of principle for undertaking a study directed toward indicators reflecting mutations in single nucleotides. Studies of chemical mutagenesis—conducted some decades after the bulk of the studies on radiation mutagenesis—do demonstrate that the production of mutations resulting in electromorphs (as well as nulls) is an important component of the spectrum of mutations induced in mammalian systems by a variety of chemical mutagens. Lewis and Johnson (1986), in a summary of germinal mutations induced in the mouse by such agents as ethylene oxide, procarbazine, ethylnitrosourea, and triethylene-melamine, find that, in a system in which both loss of protein and/or enzyme activity and a change in electrophoretic mobility could be studied, 17 of the induced mutations were of the former type and 12 were of the latter type. Thus far, only a single spontaneous mutation has been observed in the controls of these experiments; it was an electromorph. In the ultimate, our ability to evaluate the relative genetic risks of the exposure of human populations to radiation and to chemical agents requires comparable data for the two types of mutagens.

Since the present study was initiated, two kinds of data directly relevant to the potential role of electrophoresis in evaluating X-ray-induced mutagenesis in humans have begun to become available. The first entails the molecular basis of the chromosomal lesions that occur spontaneously or are induced by X-rays; the second entails some empirical observations on the offspring of irradiated mice. With respect to the first development, we note that some 20%–40% of the spontaneous germinal and somatic cell mutations at the hypoxanthine phosphoribosyltransferase, adenine phosphoribosyltransferase, thymidine kinase, factor VIII, and Duchenne muscular dystrophy loci can be shown with the Southern blot technique to involve readily detectable genomic alterations, whereas the proportion of such readily detected lesions among X-ray-induced mutations at these loci in somatic cells is more like 50%–60%, the results dif-

fering by investigator and locus (see, e.g., Albertini et al. 1985; Stout and Caskey 1985; Turner et al. 1985; Grosovsky et al. 1986; Liber et al. 1986; van Ommen et al. 1986; Wilson et al. 1986; Yandell et al. 1986; Ashman and Davidson 1987; Gibbs and Caskey 1987; Monaco et al. 1987; Youssoufian et al. 1987). The relevance of these data in the context of electrophoretic variants must be interpreted with caution. On the one hand, the Southern blotting technique will not reliably detect small duplications or deletions (i.e., ≤100 bp) in restriction fragments of the usual 5–10-kb size such as were employed in the above-referenced studies; but such small lesions would result in the absence of an electrophoretically detectable gene product. Furthermore, as the analysis of the thalassemias shows so well (for review, see Antonarakis et al. 1985), even when the genome is intact, point mutations resulting in nucleotide substitutions may result in the absence of an electrophoretically detectable gene product. *On the other hand, the screening systems for mutations affecting these loci is highly selective, whether in the intact human or in cell lines,* depending on total or near-total loss of gene function. The mutations being detected must for the most part correspond to the thalassemias among the hemoglobin mutants. Thus, nucleotide substitutions resulting in mild or no loss of gene-product function, substitutions that might alter electrophoretic mobility, would not be detected in these systems. The recent extension of studies of this type to the sequencing of mutant loci failing to exhibit abnormalities on Southern blotting analysis should greatly clarify this question of the molecular basis for the mutations at these loci (see Grosovsky et al. 1988) but still will not compensate for the biased sampling of mutations resulting from the clinical or biochemical selective sieve.

Evidence of a different sort on the nature of X-ray-induced mutations comes from the studies of Liber et al. (1986). As noted, failure of X-rays to produce ouabain resistance in cultured cells has been seen as evidence that X-rays do not produce point mutations. However, Liber et al. (1986) find that at two other loci where the mutations selected should, for physiological reasons, be "point" (nucleotide-substitution) mutations, X-rays do produce mutations; these are a locus that codes for a mRNA synthesis factor (mutations selected through resistance to 5,6-dichlororibofuranosylbenzimidazole) and a locus that codes for tubulin (mutations selected as resistance to podophyllotoxin).

Recently, studies have also been initiated on the molecular basis of X-ray-induced mutations in mice (Rinchik et al. 1986; Russell 1986). These important studies are revealing predominantly deletion-type mutations, but to some extent this is inherent in the choice of some of the mutations studied (e.g., simultaneous involvement of the dilute and short-ear traits). Furthermore, it is by no means clear that the phenotypic screening involved in the specific locus system in mice (Russell 1951) would detect most nucleotide substitutions.

With respect to the second development mentioned above, since this investigation was launched the results of two studies searching for X-ray-induced electrophoretic variants in mice have been reported. Russell et al. (1976) examined the hemoglobin and albumin of 8621 F_1 of mice receiving 3–6 gray (Gy) of X-rays. The fact that the mice were heterozygous at both the hemoglobin loci, *Hba* and *Hbb,* (leading to a diffuse hemoglobin band) facilitated the recognition of quantitative variants. Three offspring exhibited hemoglobin bands of diminished intensity, attributed to simple loss-of-gene-activity mutants. One offspring had a hemoglobin band of increased activity, shown to be due to a duplication. Finally, one offspring exhibited a diminished hemoglobin band best explained on the basis of nondisjunctional events involving the chromosomes 7 of both P_1. No albumin variants were encountered. Malling and Valcovic (1977) irradiated the males of a cross between two lines heteromorphic with respect to the electrophoretic mobility of at least eight enzymes and the β-hemoglobin chain. The resultant F_1, as heterozygotes, should permit the recognition of both loss-of-activity and electromorphic mutations. Two doses of 5 Gy at a 24-h interval were employed. Four mutations in "somewhat over 2600 animals" were encountered. Two of the mutations involved loss of enzyme activity; two involved loss of hemoglobin production. Thus, none of nine X-ray-induced mutations encountered in the two studies was expressed as an electrophoretic variant. Although the data base is scanty, these findings are to be contrasted with those of the previously quoted studies, which had indicated that ~40% of the X-ray-induced mutations in somatic cell systems highly selective for reduced enzyme activity are not accompanied by alterations detectable on Southern blots. Note also the manner in which the data from the hemoglobin loci dominate these early findings.

Thus, the recent data on the molecular basis of

spontaneous and X-ray-induced mutations and the extent to which electrophoretically detectable variants are produced by X-rays are not as helpful in the present context as is desirable. The results (not reviewed in detail) appear to differ significantly both (1) from investigator to investigator working on a given locus and (2) from locus to locus. Given the prevailing thinking among radiobiologists, we may say that a surprising frequency of spontaneous mutations are being found to be deletions—and that an equally surprising frequency of X-ray-induced mutations, all in somatic cell systems, are not accompanied by evidences of DNA damage detectable with Southern blotting techniques. As already noted, this latter finding of course does not preclude small lesions incompatible with either mRNA synthesis or functional mRNA. There is thus a very limited basis for predicting what proportion of X-ray-induced mutations should manifest as electrophoretic variants; the arguments for proceeding with a study like this remain today much as they were 15 years ago.

As indicated in the Introduction, the second indicator of induced mutation employed in this study was the finding in a child of a half-normal level of any of a series of nine enzymes, when both parents exhibited normal levels for that enzyme. These enzymes had all been selected for study on the basis of exhibiting low coefficients of variation for their activity levels (Mohrenweiser 1981). Such a finding would result either from a gene-inactivation mutation of some type or from a chromosomal deletion. The preceding discussion should have justified the search for mutations of this type (see also Satoh et al. 1983). In principle, one-dimensional electrophoresis utilizing either protein or enzyme-activity stains should be useful for the detection of such variants, but in practice it has not appeared feasible, for any substantial subset of these indicators, to detect with the requisite accuracy samples characterized by levels 50% of normal. Unfortunately, the search, by other techniques, for variants characterized by half-normal enzyme levels is still relatively labor-intensive, so that this series of observations is limited.

We note, finally, that the previous studies of the genetic effects of the atomic bombs have been directed primarily toward what could be termed the "dominant" component of any genetic damage. There has in the past been much conjecture concerning a less apparent "recessive" component, expected to become manifest in later generations. The two indicators of the present study are an important aspect of any recessive component.

The Study Cohorts

The subjects of this study were drawn from two cohorts established in the course of the so-called F_1 Mortality Study of the RERF (see Neel et al. 1974; Schull et al. 1982). The composition of the original cohorts, consisting of children born between 1946 and 1959 (see Kato et al. 1966), has been updated on several occasions. Cohort 1 now consists of all the children born alive in Hiroshima and Nagasaki between May 1946 and December 1980 to parents in the various RERF study populations, one or both of whom had been "proximally exposed," i.e., <2,000 m from the hypocenter at the time of the Hiroshima or Nagasaki bombings (ATB). Cohort 2 consists of a sample of children born during the same period to "distally exposed" parents, i.e., (1) both parents >2,500 m from the hypocenter ATB, (2) one parent >2,500 m from the hypocenter and the other not in either city ATB, or, for a small fraction, (3) neither parent in either city ATB. It is assumed that the genes contributed by the fathers of these children were all in the spermatogonial stage at the time of the bombings.

Because the number of children born to distally exposed parents is much greater than the number born to proximally exposed parents, cohort 2 has been reduced to a manageable size for this study by selecting at random from the children of the distally exposed a subset matching cohort 1 as to sex and year of birth. At the time this study was designed, individuals >2,500 m from the hypocenter ATB were thought to have received no or negligible (<0.01 Gy) radiation ATB. With the current revision of the distance-dose curve (see below), it now appears that, if in the open, persons in the 2,500–3,000-m ring from the hypocenter ATB in Nagasaki may have received ≤0.02 Gy. Children born to parents 2,001–2,499 m from the hypocenter ATB have been excluded from the study because of the difficulty of evaluating the (small) doses received at this distance. In both of these cohorts, ~34% of the children are siblings to some members of the remaining 66%. Each child constitutes an independent test for mutation, however, and no correction of the data for the occurrence of siblings is indicated. The status of the cohorts is updated on a 4-year cycle.

The collection of the necessary blood samples extended from 1975 until 1985. The ongoing mortality study mentioned above supplied considerable information on those members of the cohorts who had died or left the city. Specially trained RERF personnel attempted to contact and explain the study to those

Table 1

Outcome of Effort to Conduct Family Studies (FS) on Individuals Exhibiting Rare Electrophoretic Variants

Exposure Category	Both P₁ Alive; FS Completed	Both P₁ Alive, Decline Study	One P₁ Unavailable,[a] the Other Cooperative	Both P₁ Deceased	Other[b]	Total
Parents proximally exposed	567 (81.0%)	45 (6.4%)	71 (10.1%)	4 (0.6%)	13 (1.9%)	700
Parents distally exposed	397 (74.5%)	44 (8.3%)	81 (15.2%)	0	11 (2.1%)	533
Total	964 (78.2%)	89 (7.2%)	152 (12.3%)	4 (0.3%)	24 (1.9%)	1233

[a] Includes those deceased, unwilling to participate in study, having moved from city, seriously ill, etc.
[b] Includes a miscellany of reasons for noncompletion of FS.

members of the cohorts who the records indicated were, at time of last contact, alive and residing in the city. No effort was made to obtain the necessary follow-up on, or blood samples from, children until they had attained the age of 13, so that children born subsequent to 1971 (~5% of the potential total sample) are not represented in the samples.

Whenever a "rare variant" (see below) was encountered in a child, examinations of both parents were necessary, to determine whether it represented an inherited variant or the result of a new mutation; and an effort was made to contact the parents and enlist their cooperation. As the study progressed, it became apparent that many parents of "children" exhibiting rare variants were either deceased or not in either city. (A "child" who entered the cohort in its early years could at the time of contact be as old as 37, and his/her parents, if alive, could be in their 70s.) Accordingly, during the latter two-thirds of the study, preference was given to obtaining specimens from children both of whose parents were known, on the basis of records available to the RERF, to be alive. The fact that both parents were alive of course did not ensure that they would wish to be examined if a family study seemed indicated.

Among a total of 16,702 children born to proximally exposed parents who were contacted for study, 13,052 (78.1%) agreed. For the children of the distally exposed, the corresponding numbers were 13,993 and 10,609 (75.8%). Table 1 presents an analysis of the availability of the parents in the event that a family study was necessary because a rare electrophoretic variant was detected in a child. When we consider below the results of the study (tables 2, 3), it will be apparent that there are substantially more locus tests on the children of proximally exposed parents than on the children of distally exposed parents. This is in part because the cohort of children of the distally exposed parents was somewhat smaller than the cohort having proximally exposed parents, but to a lesser extent it is because the children of the distally exposed parents (and the parents themselves) were aware from various sources of information that they were at low genetic risk and so were somewhat less inclined to participate in the study (table 1). Even so, overall the cooperation of the available F₁ in the studies was 77.1%, and that of their parents, when both were available, was 91.5%, levels to be considered exceptional in view of the fact that a venapuncture was involved. An additional reason for the difference in the number of determinations in the two series was a redefinition of the location of the hypocenter in Nagasaki during the course of the study (a redefinition that transferred some subjects from the distally to the proximally exposed category). Since, however, the biochemical indicators on which the study was based are not apparent to the subjects, and since this unequal participation was maintained throughout the study, we see no way in which this unequal participation could introduce bias into the study.

Biochemical Procedures

Although at the outset ammonium potassium oxalate was used as anticoagulant for the various blood samples, the majority of the samples were drawn into vacutainers in which formula A ACD solution was the anticoagulant. Most determinations were on samples that had been temporarily stored at −70 C or in liquid nitrogen. The processing of the samples for electrophoresis and for the enzyme activity measurements, as well as the techniques employed, have been

described by Ferrell et al. (1977), Ueda et al. (1977), Tanis et al. (1978), and Satoh et al. (1977, 1983). The 30 proteins examined for variants by starch-gel electrophoresis are listed in table 2; the nine proteins (enzymes) examined for activity levels are listed in table 3. The calculation of the number of gene products examined must take into consideration gene duplication (the α-hemoglobin locus) and multiple polypeptide subunits (i.e., lactate dehydrogenase, the hemoglobins). In earlier publications, we also treated the three esterase A1 isozymes as dimers of a shared polypeptide combined with three different independently coded gene products. Because all of the 34 variants of this isozyme encountered in this study appear to involve all three isozyme bands (K. Goriki, personal communication), we now feel it more appropriate to consider these isozymes as products of a single locus. The number of different gene products examined thus becomes 33. In employing electrophoretic and activity variants for studies of mutation under the present protocol, we make the implicit assumption that whatever negative selection (if any) may have operated on variants of the type under consideration since birth is equally distributed across the two classes of children.

A rare variant was defined as one that was represented by a gene frequency <.01. In nine of these systems, previously described genetic polymorphisms were encountered. Family studies were not performed on individuals apparently heterozygous or homozygous for one of these polymorphisms. It is, however, possible that mutation could result in a variant phenotypically indistinguishable from an established polymorphism; our protocol would not detect such "mimics." We have estimated that failure to perform family studies on all individuals exhibiting polymorphisms—an overwhelming task, given the personnel available—should result in an underestimate of the mutation rate of ~10% (Neel et al. 1983).

Special Studies of Putative Mutations

For the 1,280 determinations (tables 2, 3) in which a rare electrophoretic or activity variant was encountered, an effort was made to obtain blood samples from both parents to determine whether the variant was hereditary (nonmutational) or newly arisen. On the relatively few occasions when (1) neither parent exhibited the variant and (2) there was no information that the child was not the biological offspring of these parents, genetic typings appropriate for detecting discrepancies between legal and biological parentage were performed. These included A and B of the ABO system; M, N, S, and s of the MNSs system; C, D, E, c, and e of the RH system, Fy^a of the Duffy system; and the α_1-antitrypsin types. Among the 30 proteins studied for the occurrence of mutation, polymorphisms useful for the detection of parentage discrepancies occur in the following: HP, ACP1, 6PGD, PGM1, PGM3, ADA, ESD, GPT, and GOT1. In addition, major-histocompatibility-complex typings were obtained, involving 12 antigens of the A complex, 32 of the B complex, and four of the C complex. If the putative mutation involved one of the proteins included in the test battery, the findings with reference to that protein were excluded from the calculation of exclusion probabilities.

Dosimetry

Not until 1965 were the factors influencing the amount of radiation that was received by the survivors of the atomic bombings thought to be sufficiently well understood that individual dose estimates, termed Tentative 1965 Doses (T65), could be assigned to survivors (Milton and Shohoji 1968; Auxier 1975; Hashizume and Maruyama 1975). At this time the T65 dose estimates suggested that the proximally exposed included some individuals who, because of felicitous shielding, had apparently received no gonadal radiation. A subsequent Dose Revision resulted in T65DR estimates. Unfortunately, in the late 1970s serious questions concerning the accuracy of these estimates emerged (for review, see Kerr 1981). A 5-year reevaluation has now terminated, with agreement on a new (and presumably final) approach to the problem of assigning individual doses (Radiation Effects Research Foundation 1987). The most salient feature of this reevaluation has been with respect to the bomb detonated over Hiroshima, for which a sharp reduction in the estimated neutron yield is coupled with a less marked increase, in parts of the city, in the estimated gamma yield. In addition, however, the estimates of the shielding from radiation provided by the roof tiles of Japanese homes and by concrete walls have been revised, as well as the estimates of the attenuation, because of the shielding effect of intervening tissues, in the amount of radiation reaching the gonad. The previous estimates of individual doses were thought to be characterized by a possible error, in either direction, of ~30% (Jablon 1971); presumably the same error applies to the new dose estimates. Gilbert (1984) has developed reasons

why doses might be systematically overestimated. We will calculate the gonadal doses sustained by the parents of the children included in the present study by using the new formulas, which result in what are termed *Dosimetry System 1986* (DS86) doses. In the calculation of DS86 doses, any gonad radiation exposure to a survivor that is calculated to be >4 Gy is arbitrarily rounded down to 4; a midline dose >4 Gy is currently assumed to be lethal. This practice may underestimate the exposure of a few exceptionally hardy individuals. On the other hand, it does to some extent offset the source of bias identified by Gilbert (1984).

Unfortunately, at this writing DS86 doses are available only for those parents of the children under study who were in the open or in Japanese-style houses ATB. There remain a substantial number of parents who were shielded from the effects of the bombs in more complex ways, such as by their presence in concrete buildings or bomb shelters, and although estimates of the attenuating effect of this shielding are available in the T65DR schedule, they will only become available for DS86 over the course of the next several years. The estimation of radiation exposure for this group will be discussed below.

Because of the differing biological effectiveness of gamma and neutron radiation, the total dose must be expressed in sieverts (Sv). A critical factor in arriving at the Sv dose is the relative biological effectiveness (RBE) assigned to neutrons. The RBE of the latter increases as dose per unit time decreases. Primarily on the basis of the data of Grahn and his collaborators (Garriott and Grahn 1982; Grahn et al. 1983, 1984) on cytogenetic endpoints and the induction of dominant lethals and reciprocal translocations in mice (see also Sinclair 1985; International Commission on Radiation Units and Measurement 1986), we shall use an RBE of 20 for the relatively low neutron radiation doses to survivors' gonads. By presenting the estimated neutron exposures separately, we make it possible for others to analyze the data by using a different RBE.

Results

The findings with reference to electrophoresis are presented in table 2, and those with reference to enzyme activity are presented in table 3. Table 2 presents only the results of the 767,665 locus tests performed by electrophoresis on children of proximally exposed persons; the data on the children of distally exposed persons have been presented elsewhere (Neel et al. 1986). Table 3 presents the results of the 126,460 locus tests performed for quantitative variation on the children of both proximally and distally exposed persons. As noted earlier, it was not always possible to test both parents of a child with a rare variant. Tables 2 and 3 indicate, by system, the number of times that both parents of a child with a variant could be examined. Since it is only when both parents of a child with a variant have been tested that a rigorous treatment of mutation becomes possible, for each system we have estimated—by multiplying the fraction of variants for which family studies are complete (table 2, col. 7 ÷ col. 6) by the total number of determinations (table 2, col. 3)—the number of alleles that have been effectively screened for mutation. With this convention, all of the determinations of a polypeptide for which no variants have been encountered are credited as contributing to locus tests.

Among the children of proximally exposed persons, there were five with an electrophoretic variant not observed in either parent; but for two of these children, the test battery described earlier indicated a discrepancy between legal and biological parentage. We are thus left with three putative mutations among 667,404 equivalent locus tests, of the following description:

1. A slowly migrating variant of glutamate-pyruvate transaminase (GPT) was detected in a female child of proximally exposed Hiroshima parents. This enzyme is a dimer. The abnormal phenotype consisted of three bands, which are interpreted as corresponding to the normal homodimer; a heterodimer of one normal and one mutant polypeptide; and the homodimer of the mutant polypeptide. The phenotype is similar to that of a rare hereditary variant of this enzyme, encountered in Nagasaki, which we have termed 6NG1. The mobility of the mutant homodimer is identical with that of the homodimer of 6NG1, but the staining intensity of the former is weaker than that of the latter. The variant will be described in greater detail elsewhere (C. Satoh, unpublished data). Neither the parents nor a younger sister of the proposita showed the abnormal phenotype. There was no parentage exclusion with the complete battery of tests (the mother's gonad exposure was 0.03 Gy of γ; the father was not exposed).

2. A slowly migrating variant of phosphoglucomutase-2 (PGM2) was encountered in a male child of parents both of whom had been proximally exposed

Table 2

Results from Examinations for Occurrence of Mutations Altering Electrophoretic Mobility: Data from Children of Proximally Exposed Parents

Protein (E.C. No.), Locus Symbol	Total Loci Screened	Rare Variants No. of Types	Rare Variants Total	Variants (Both Parents Examined)	Equivalent Locus Tests	Exceptional Children
Haptoglobin, HP	25,734	9	29	25	22,184	1[a]
Transferrin, TF	26,096	12	132	108	21,351	
Ceruloplasmin, CP	26,068	5	17	13	19,934	
Albumin, ALB	26,102	4	34	27	20,728	
Hemoglobin A1, HBA1 + HBA2	52,168	2	7	7	52,168	
HBB	26,084	3	4	2	13,042	
Hemoglobin A2, HBD	26,088	0	0	0	26,088	
Adenosine deaminase (3.5.4.4), ADA	26,092	2	7	3	11,182	
6-Phosphogluconate dehydrogenase (1.1.1.44), PGD	26,054	7	11	10	23,685	
Adenylate kinase-1 (2.7.4.3), AK1	25,164	0	0	0	25,164	
PGM1 (2.7.5.1), PGM1	25,046	13	78	61	19,587	
PGM2 (2.7.5.1), PGM2	25,064	4	9	8	22,279	1
Phosphoglucomutase-3 (2.7.5.1), PGM3	5,962	0	0	0	5,962	
Acid phosphatase-1 (3.1.3.2), ACP1	25,014	1	1	1	25,014	
TPI (5.3.1.1), TPI	21,916	2	2	2	21,916	
NP (2.4.2.1), NP	24,160	4	24	20	20,133	1
Esterase A1 (3.1.1.1), ESA1	23,776	5	14	14	23,776	
Esterase B (3.1.1.1), ESB	22,878	0	0	0	22,878	
Esterase D (3.1.1.1), ESD	24,126	1	1	1	24,126	
Peptidase A (3.4.11.−), PEPA	26,002	1	18	15	21,668	
Peptidase B (3.4.11.−), PEPB	26,104	2	14	12	22,375	
Glucosephosphate isomerase (5.3.1.9), GPI	26,100	7	132	105	20,761	1[a]
Isocitrate dehydrogenase-1 (1.1.1.42), IDH1	26,050	4	25	20	20,840	
Lactate dehydrogenase (1.1.1.27), LDHA	26,068	1	1	1	26,068	
LDHB	26,068	3	4	3	19,551	
Malate dehydrogenase-1 (1.1.1.37), MDH1	26,098	2	3	3	26,098	
Carbonic anhydrase-1 (4.2.1.1), CA1	25,958	3	17	15	22,904	
Carbonic anhydrase-2 (4.2.1.1), CA2	26,060	0	0	0	26,060	
Glucose-6-phosphate dehydrogenase (1.1.1.49), G6PD:						
Male	2,488	3	5	3	1,493	
Female	5,774	3	9	6	3,849	
Glutamate-oxaloacetate transaminase-1 (2.6.1.1), GOT1	17,292	3	53	42	13,703	
GPT (2.6.1.2), GPT1	17,282	6	49	40	14,108	1
Phosphoglycerate kinase-1 (2.7.2.3), PGK1:						
Male	2,163	0	0	0	2,163	
Female	4,566	0	0	0	4,566	
Total	767,665		700	567	667,404	5

[a] Parentage exclusion.

in Nagasaki. The abnormal phenotype consisted of the three bands customarily associated with the PGM2 1 phenotype and a variant band that migrated identically or slightly cathodally to the d-band of PGM1. This mutant has been designated PGM2 9NG2 (Satoh et al. 1984). Only two other variants of PGM2 have been encountered. On starch-gel electrophoresis, the mutant variant migrates slightly anodal to the band of one hereditary variant, PGM2 9NG1, but to the same position as that of the other variant, PGM2 9HR1. However, on thin-layer polyacrylamide-gel electrofocusing, the phenotypes of PGM2 9NG1, PGM2 9NG2, and PGM2 9HR1 are clearly distinguishable. Neither the parents nor two siblings exhibited an abnormal phenotype. No parentage discrepancy was revealed by the complete battery of tests (the mother's gonad exposure was 0.74 Gy of γ; the father's gonad exposure was 0.05 Gy of γ).

3. A rapidly migrating variant of nucleoside phosphorylase (NP) was detected in a male child of proximally exposed Nagasaki parents. The abnormal phenotype consisted of both a set of bands associated with the NP 1 phenotype and a set of rapidly migrating bands that exhibited a mobility similar to that of the bands associated with the NP 2 phenotype, a he-

Table 3

Results of Examining Nine Erythrocyte Enzymes for Occurrence of Mutations Resulting in Loss of Activity in (*a*) Children Born to Parents Proximally Exposed to the Atomic Bombs and (*b*) a Suitable Comparison Group

| | CHILDREN OF PROXIMALLY EXPOSED PARENTS |||| CHILDREN OF DISTALLY EXPOSED PARENTS ||||
| | | Variants Confirmed || | | Variants Confirmed || |
ENZYME (E.C. No.), LOCUS SYMBOL	Total Loci Screened	Total	Both Parents Examined	Equivalent-Locus Tests	Total Loci Screened	Total	Both Parents Examined	Equivalent-Locus Tests
6-Phosphogluconate dehydrogenase (1.1.1.44), *PGD*	9,598	0	0	9,598	9,890	4	4	9,890
Adenylate kinase-1 (2.7.4.3), *AK1*	9,520	3	3	9,520	9,626	1	1	9,626
Triosephosphate isomerase (5.3.1.1), *TPI*	6,774	6	6	6,774	6,370	3	3	6,370
Glucosephosphate isomerase (5.3.1.9), *GPI*	9,978	11	9	8,164	10,052	3	3	10,052
Lactate dehydrogenase (1.1.1.27), *LDHB*	9,624	2	2	9,624	9,820	0	0	9,820
Glutamate-oxaloacetate transaminase-1 (2.6.1.1), *GOT1*	4,876	2	2	4,876	5,544	7	4	3,168
Phosphoglycerate kinase-1 (2.7.2.3), *PGK1*:								
Male	2,371	0	0	2,371	2,363	3	3	2,363
Female	4,810	0	0	4,810	4,996	0	0	4,996
Glyceralde-3-phosphate dehydrogenase (1.2.1.12), *GAPD*	560	1	1	560	0	722	0	722
Hexokinase (2.7.1.1), *HK*	4,232	1	1	4,232	4,734	0	0	4,734
Total	62,343	26	24	60,529	64,117	21	18	61,741

reditary variant. The variant will be described in detail elsewhere (C. Satoh, unpublished data). Neither of the parents exhibited the abnormal bands. There was no parentage exclusion with the full battery of tests (the mother's gonad exposure was 0 Gy; the father's gonad exposure was 0 Gy). (In this instance, although both parents had been <2,000 m from the hypocenter ATB, they had been so shielded that no radiation is thought to have reached their gonads.)

As shown in table 4, the estimated mutation rate on the basis of these findings is 0.45×10^{-5}/locus/generation, with the 95% confidence interval, calculated on the assumption that the number of mutations corresponds to a Poisson variant, being between 0.1×10^{-5} and 1.3×10^{-5}. The mutation rate in the children of the distally exposed (control) cohort, as based on three mutations in 539,170 effective locus tests, was previously reported to be 0.56×10^{-5}/locus/generation, with 95% confidence intervals between 0.1×10^{-5} and 1.6×10^{-5} (Neel et al. 1986). As a result of altering our treatment of the number of loci contributing to the ESA1 isozymes (see above) and, with continuing study, the reclassification of several previously ambiguous findings into true rare variants, we now estimate the number of locus tests in the control series as being 466,881 (see table 4) and the mutation rate as being 0.64×10^{-5}/locus/generation, with 95% confidence intervals between 0.1×10^{-5} and 1.9×10^{-5}.

The findings regarding enzyme activity will be presented in detail elsewhere (C. Satoh, unpublished data). As shown in table 4, there was one mutation in 60,529 tests conducted on the children of proximally exposed persons and no mutation in 61,741 tests on the children of distally exposed persons. The apparent mutant of triosephosphate isomerase (TPI) was encountered in a female child of Nagasaki parents. The TPI activity of the propositus was 65% of normal, 3.9 SDs below the mean. The mother, father, and a younger brother exhibited 92%, 100%, and 104% of normal activity, respectively. The electrophoretic patterns of all of them were normal. There was no parentage discrepancy with the complete battery of tests (the mother was not exposed; the father's gonad exposure was 0.03 Gy of γ.)

We wish to combine the results of these two approaches. In principle, this requires that the proportionate contribution of distally (or proximally) exposed persons be the same for the two data sets. For reasons already noted, in the electrophoretic data there is an excessive representation of determinations on the children of proximally exposed persons; these children are 58.7% of the total. On the other hand, since the data on activity variants are derived from a

Table 4

Summary of Data Concerning Mutation in the Two Series

Parameter	Proximally Exposed Parents	Distally Exposed Parents	Total
A. Electromorphs			
Children examined	13,052	10,609	23,661
Electrophoretic tests	347,040	288,984	636,024
Rare variants[a]	700	533	1,233
Variants (both parents examined)	567 (81%)	397 (74%)	964 (78%)
Equivalent-locus tests	667,404	466,881	
Exceptional children	5	6	11
Mutations	3	3	6
Mutation rate/locus/generation	0.45×10^{-5}	0.64×10^{-5}	0.53×10^{-5}
95% Confidence limits:			
Lower limit	0.09×10^{-5}	0.13×10^{-5}	0.19×10^{-5}
Upper limit	1.3×10^{-5}	1.88×10^{-5}	1.15×10^{-5}
B. Enzyme-Deficiency Variants			
Children examined	4,989	5,026	10,015
Tests	32,357	33,240	65,597
Deficiency variants (≤66% of normal)	26	21	47
Variants (both parents examined)	24	18	42
Equivalent-locus tests	60,529	61,741	122,270
Mutation	1	0	1
Mutation rate/locus/generation	1.65×10^{-5}	0	0.82×10^{-5}
95% Confidence limits:			
Lower limit	0.04×10^{-5}	0	0.02×10^{-5}
Upper limit	9.20×10^{-5}	4.85×10^{-5}	4.56×10^{-5}

[a] Allele frequency <.005.

subset of the persons contributing to the electrophoretic data, it was easier to regulate the numbers to the desired 1:1 ratio; now the children of proximally exposed persons contribute 49.5% of the data. This disparity would, under most circumstances, require some adjustment, but with the "positive events" so few in number, such an adjustment seems an overrefinement of the data, and we shall simply add the numerators of the enzyme activity series (1,0) to those of the electrophoretic series (3,3). This results in a mutation frequency of 4 in 667,404 locus tests conducted on the children of proximally exposed persons and of 3 in 466,881 locus tests among the children of distally exposed persons; expressed in conventional fashion, these are rates of 0.60 and 0.64×10^{-5}/locus/generation, respectively. The 95% confidence interval for the former rate is $0.2-1.5 \times 10^{-5}$, and that for the latter is $0.1-1.9 \times 10^{-5}$. It should be emphasized that these rates are not normative values in any respect. A normative value for nucleotide-substitutions-plus-deficiency-states would require (1) equal numbers of the two types of observations and (2) correction for the fact that only approximately one-third to one-half of all nucleotide substitutions result in variants detectable by electrophoresis.

Technically, those mutation rates must be corrected for the possibility that they are inflated by undetected parentage discrepancies, i.e., that the variant allele was contributed to the child by someone other than the father (nonmaternity is relatively much less likely). The probability of this event depends on three parameters. The first is the frequency of nonpaternity in this series. We are indebted to Dr. Howard Hamilton (personal communication) for data on ABO blood group typings in 3,295 mother-father-child trios drawn from this study, typings that enable us to calculate that the frequency of extramarital conceptions in this series is .0045. The second is the probability that the true father of the child transmitted a rare allele, of the type on which this study of mutation was based. In the total data set of the children of

both the proximally and distally exposed persons, this allele frequency is ~.001. The use of this frequency is a conservative procedure, in that all except one of the variants in this series that appear to be new mutations are unique in their electrophoretic mobility; that is, the pool of possible fathers who could contribute this allele is even smaller than the overall frequency of rare variants would suggest. The third parameter is the probability of *not* detecting that a child was conceived extramaritally. On the basis of the blood and enzyme typing described earlier, employing the probabilities supplied by Chakraborty et al. (1974), supplemented when necessary by our own calculations, we find this probability to be .141. On the basis of data in Ito et al. (1985), it can be calculated that with the battery of HLA typings that we have employed the frequency of failing to detect an extramarital conception would be .057. With the combination of serological, enzymatic, and HLA typing employed, the probability of not detecting an extramarital conception becomes .0085. The a priori probability that an apparent mutant in this series is the result of an extramarital relationship is .0045 × .001 × .0085, or ~4 × 10^{-8}. We note that the frequency with which our procedures detected discrepancies between legal and biological parentage was 2/663,494 locus tests, or 3.0×10^{-6}, with 95% confidence intervals between 0.3×10^{-6} and 10.0×10^{-6}. On the basis of the above parameters, the theoretical expectation is .0045 × .001 × .9915, or 4.5×10^{-6}. There is thus satisfactory agreement between observation and expectation.

We come now to the assignment of gonadal doses. Recall that for a child to be classified as born of proximally exposed parents, only one of the parents was required to be <2,000 m from the hypocenter ATB. In fact, of the parents of the 13,044 children in the proximally exposed–parent category, 37.0% were either >2,500 m from the hypocenter ATB or not in the city ATB—and so received no radiation. Of the remaining parents, 75.7% can be assigned doses under the new DS86 schedule, but for the remaining 24.3% the shielding data are still inadequate for the application of the DS86 procedures. This is due to the individual's presence ATB in a concrete building, a factory, or other structure more complex than a Japanese home. For this group we have adopted, for the time being, a hybrid procedure for assigning dose category, a procedure that calculates organ radiation dose on the basis of the DS86 Kerma in air and tissue transmission factors but that uses the T65 DR physical shielding factors. Combining the data from this procedure with the data from the DS86 procedure, we estimate doses in Gy for the proximally exposed parents as follows: Hiroshima fathers, .204 γ and .002 μ; Hiroshima mothers, .231 γ and .001 μ; Nagasaki fathers, .216 γ and .001 μ; and Nagasaki mothers, .223 γ and .0001 μ. Since, as noted, for a child to be classified as born to proximally exposed parents, only one parent need have been <2,000 m from the hypocenter ATB, in the calculation of this average some parents are included who were distally exposed—or even not in the city— ATB. Assigning, as discussed earlier, an RBE of 20 to the neutron component of these exposures, we find that the average radiation doses to gonads become the following: Hiroshima parents, 0.495 Sv; and Nagasaki parents, 0.459 Sv. For the cities combined, the estimated conjoint parental exposure is 0.477 Sv. The average gametic dose—the measure in which the genetic effects of radiation is usually expressed—would of course be half this, or 0.239.

Because of the complexity of the dosimetry problem, no attempt has been made to calculate standard statistical errors to attach to the doses. Especially critical is the evaluation of the neutron component of the exposures, as well as the evaluation of the radiation attenuation provided by complex shielding. Efforts to refine these estimates will undoubtedly continue over the next several years and will be taken into consideration when, later, we attempt to use these data and other endpoints to develop an estimate of the doubling dose of radiation for humans. It is important from the standpoint of radiation biology to emphasize that 48.4% of this dose is derived from paternal exposure and 51.6% from maternal exposure.

Discussion

As we have emphasized in the past, the studies in Hiroshima and Nagasaki are not designed to test the hypothesis that radiation produced mutations in the survivors of the atomic bombings. Radiation has produced mutations in every properly investigated plant or animal species to which it has been applied, and it can scarcely be doubted that this is true of the situation under study. Our challenge, rather, is to treat the results of each of the various studies that have sought to elucidate the genetic effects of the bombs as an approximation of the true effect, ultimately combin-

ing all these findings to derive a best estimate of the genetic doubling dose of radiation for humans.

With respect to the indicators employed in this study, the two rates—that is, the rate in the children of proximally exposed persons and that in the children of distally exposed persons—are essentially identical. However, the errors to be attached to the two estimates are such that it cannot be excluded with any degree of confidence that the rates in the two series actually differ by a factor of 2. Any attempt at estimating a genetic doubling dose of radiation will be delayed until there has been a final analysis of the data on those other six indicators of a possible genetic effect—i.e., untoward pregnancy outcomes, survival through childhood, sex-chromosome aneuploids and balanced structural rearrangements of chromosomes, sex ratio, and childhood malignancies in the F_1—that lend themselves to the doubling-dose approach. An estimate based on a combination of these seven sources of information will of course have greater stability than an estimate based on any one of them.

Given the fact that the mutation rates are so similar in the two series, we are entitled to suspect that the mutations in the children of proximally exposed persons are "spontaneous" rather than radiation related. One indication of this is that the conjoint parental exposure in the parents of these four mutants averages 0.213 Sv, below the mean for the proximally exposed group—rather than above the mean, as would be expected if the mutations were radiation induced. Given the assumption that the mutations encountered in the children of proximally exposed persons are really spontaneous in nature, we can pool these mutations with the mutations in the children of distally exposed (i.e., nonradiated) persons (Neel et al. 1986), to ask the question, Are mutations more common in the systems exhibiting the larger number of electrophoretic variants? Dividing in half the number of polypeptides being scored, into those showing the greater number of different variants per locus and those showing the lesser number, we find that four of the mutations occurred among the former group and two among the latter group. The single activity mutant also involved a more variable system. At this point we can only record the finding for future reference.

We note, finally, that during the 10-year duration of this study, a particular effort was made to hold techniques constant. Were one launching such a study today, the electrophoretic technology employed would undoubtedly differ significantly (see discussion in Neel et al., in press). Furthermore, one would have to ask whether, in biochemically oriented studies of mutation, this technology is about to be superseded by the use of two-dimensional PAGE coupled with computer algorithms for gel analysis (Neel et al. 1984; Skolnick and Neel 1986) or by the emerging DNA techniques (Delehanty et al. 1986; U.S. Congress, Office of Technology Assessment 1986).

Acknowledgments

This program was supported by the Radiation Effects Research Foundation, a United States–Japan binational foundation, and by U.S. Department of Energy contract AC-02-82-ER60089.

References

Albertini, R. J., J. P. O'Neill, J. A. Nicklass, N. H. Heintz, and P. C. Kelleher. 1985. Alterations of the *hprt* gene in human *in vivo*-derived 6-thioguanine-resistant T lymphocytes. Nature 316:369–371.

Antonarakis, S. E., H. H. Kazazian, and S. H. Orkin. 1985. DNA polymorphism and molecular pathology of the human globin gene clusters. Hum. Genet. 69:1–14.

Ashman, C. R., and R. L. Davidson. 1987. Sequence analysis of spontaneous mutations in a shuttle vector gene integrated into mammalian chromosomal DNA. Proc. Natl. Acad. Sci. USA 84:3354–3358.

Auxier, J. A. 1975. Physical dose estimates for a-bomb survivors—studies at Oak Ridge, U.S.A. J. Radiat. Res. (Tokyo) [Suppl.] 16:1–11.

Chakraborty, R., M. Shaw, and W. J. Schull. 1974. Exclusion of paternity: the current state of the art. Am. J. Hum. Genet. 26:477–488.

Delehanty, J., R. L. White, and M. L. Mendelsohn. 1986. Approaches to determining mutation rates in DNA. Mutat. Res. 167:215–232.

Ferrell, R. E., N. Ueda, C. Satoh, R. J. Tanis, J. V. Neel, H. B. Hamilton, T. Inamizu, and K. Baba. 1977. The frequency in Japanese of genetic variants of 22 proteins. I. Albumin, ceruloplasmin, haptoglobin, and transferrin. Ann. Hum. Genet. 40:407–418.

Garriott, M. L., and D. Grahn. 1982. Neutron and γ ray effects measured by the micronucleus test. Mutat. Res. Lett. 105:157–162.

Gibbs, R. A., and C. T. Caskey. 1987. Identification and localization of mutations at the Lesch-Nyhan locus by ribonuclease A cleavage. Science 236:303–305.

Gilbert, E. S. 1984. Some effects of random dose measurement errors on analyses of atomic bomb survivor data. Radiat. Res. 98:591–605.

Grahn, D., B. A. Carnes, B. H. Farrington, and C. H. Lee. 1984. Genetic injury in hybrid male mice exposed to low doses of ^{60}Co γ-rays and fission neutrons. I. Response to single doses. Mutat. Res. **129**:215–229.

Grahn, D., C. H. Lee, and B. F. Farrington. 1983. Interpretation of cytogenetic damage induced in the germ line of male mice exposed for over 1 year to ^{239}Pu alpha particles, fission neutrons, and ^{60}Co gamma rays. Radiat. Res. **95**:566–583.

Grosovsky, A. J., J. G. de Boer, P. J. de Jong, E. A. Drobetsky, and B. W. Glickman. 1988. Base substitutions, frameshifts, and small deletions comprise ionizing radiation induced point mutations in mammalian cells. Proc. Natl. Acad. Sci. USA **85**:185–188.

Grosovsky, A. J., E. A. Drobetsky, P. J. de Jong, and B. W. Glickman. 1986. Southern analysis of genomic alterations in gamma-ray-induced aprt⁻ hamster cell mutants. Genetics **113**:405–415.

Hashizume, T., and T. Maruyama. 1975. Physical dose estimates for a-bomb survivors—studies at Chiba, Japan. J. Radiat. Res. [Suppl.] **16**:12–23.

International Commission on Radiation Units and Measurements. 1986. The quality factor in radiation protection. Int. Commission Radiat. Units Rep. **40**:1–32.

Ito, H., N. Yasuda, and H. Matsumoto. 1985. The probability of parentage exclusion based on restriction fragment length polymorphisms. Jpn. J. Hum. Genet. **30**:261–269.

Jablon, S. 1971. Atomic bomb radiation dose estimation at ABCC. At. Bomb Casualty Commission Tech. Rep. 23-71:41.

Kato, H., W. J. Schull, and J. V. Neel. 1966. A cohort-type study of survival in the children of parents exposed to atomic bombings. Am. J. Hum. Genet. **16**:214–230.

Kerr, G. D. 1981. Review of dosimetry for the atomic bomb survivors. Pp. 501–513 in G. Burger and H. G. Ebert, eds. Proceedings of the Fourth Symposium on Neutron Dosimetry. EUR-7448-EN. Luxembourg: Commission on the European Communities Directorate, Munich.

Lewis, S. E., and F. M. Johnson. 1986. The nature of spontaneous and induced electrophoretically detected mutations in the mouse. Pp. 359–365 in K. Ramel, B. Lambert, and J. Magnussen, eds. Genetic toxicology of environmental chemicals: part B. Alan R. Liss, New York.

Liber, H. L., K. M. Call, and J. B Little. 1987. Molecular and biochemical analysis of spontaneous and X-ray induced mutants in human lymphoblastoid cells. Mutat. Res. **178**:143–153.

Liber, H. L., P. H. Leong, V. H. Terry, and J. B Little. 1986. X-rays mutate human lymphoblast cells at genetic loci that showed response only to point mutagens. Mutat. Res. **163**:91–97.

Lindahl, T. 1977. DNA repair enzymes acting on spontaneous lesions in DNA. Pp. 225–240 in W. W. Nichols and D. G. Murphy, eds. DNA repair processes. Symposia Specialists, Miami.

Malling, H. V., and F. J. de Serres. 1973. Genetic alterations at the molecular level in X-ray induced ad-3B mutants of *Neurospora crassa*. Radiat. Res. **53**:77–87.

Malling, H. V., and L. R. Valcovic. 1977. A biochemical specific locus mutation system in mice. Arch. Toxicol. **38**:45–51.

Milton, R. C., and T. Shohoji. 1968. Tentative 1965 radiation dose estimation for atomic bomb survivors. ABCC Tech. Rep. 1–68.

Mohrenweiser, H. W. 1981. Frequency of enzyme deficiency variants in erythrocytes of newborn infants. Proc. Natl. Acad. Sci. USA **78**:5046–5050.

Monaco, A. P., C. J. Bertelson, C. Colletti-Feener, and L. M. Kunkel. 1987. Localization and cloning of Xp21 deletion breakpoints in muscular dystrophy. Hum. Genet. **75**:221–227.

Neel, J. V., H. Kato, and W. J. Schull. 1974. Mortality in the children of atomic bomb survivors and controls. Genetics **76**:311–326.

Neel, J. V., H. W. Mohrenweiser, and H. Gershowitz. A pilot study of the use of placental cord blood samples in monitoring for mutational events. Mutat. Res. (in press).

Neel, J. V., H. Mohrenweiser, S. Hanash, B. Rosenblum, S. Sternberg, K. H. Wurzinger, E. Rothman, C. Satoh, K. Goriki, T. Krasteff, M. Long, M. Skolnick, and R. Krzesicki. 1983. Biochemical approaches to monitoring human populations for germinal mutation rates. I. Electrophoresis. Pp. 71–93 in F. de Serres and W. Sheridan, eds. Utilization of mammalian specific locus studies in hazard evaluation and estimates of genetic risk. Plenum, New York.

Neel, J. V., B. B. Rosenblum, C. F. Sing, M. M. Skolnick, S. M. Hanash, and S. Sternberg. 1984. Adapting two-dimensional gel electrophoresis to the study of human germ-line mutation rates. Pp. 259–306 in J. E. Celis and R. Bravo, eds. Two-dimensional gel electrophoresis of proteins. Academic Press, New York.

Neel, J. V., C. Satoh, K. Goriki, M. Fujita, N. Takahashi, J. Asakawa, and R. Hazama. 1986. The rate with which spontaneous mutation alters the electrophoretic mobility of proteins. Proc. Natl. Acad. Sci. USA **83**:389–393.

Neel, J. V., C. Satoh, H. B. Hamilton, M. Otake, K. Goriki, T. Kageoka, M. Fujita, S. Neriishi, and J. Asakawa. 1980. Search for mutations affecting protein structure in children of atomic bomb survivors: preliminary report. Proc. Natl. Acad. Sci. USA **77**:4221–4225.

Neel, J. V., and W. J. Schull. 1956. The effect of exposure to the atomic bombs on pregnancy termination in Hiroshima and Nagasaki. Publ. 461. National Academy of Sciences–National Research Council, Washington, DC.

Neel, J. V., N. Ueda, C. Satoh, R. E. Ferrell, R. J. Tanis, and H. B. Hamilton. 1978. The frequency in Japanese of ge-

netic variants of 22 proteins. V. Summary and comparison with data on Caucasians from the British Isles. Ann. Hum. Genet. 41:429–441.

Radiation Effects Research Foundation. 1987. US-Japan joint reassessment of atomic bomb radiation dosimetry in Hiroshima and Nagasaki: final report. Pp. 434 *in* W. C. Roesch, ed. Hiroshima: Radiation Effects Research Foundation. Radiation Effects Research Foundation, Hiroshima.

Rinchik, E. M., L. B. Russell, N. G. Copeland, and N. A. Jenkins. 1986. Molecular genetic analysis of the *dilute-short ear* (D-SE) region of the mouse. Genetics 112:321–342.

Russell, L. B. 1986. Information from specific-locus mutants on the nature of induced and spontaneous mutations in the mouse. Pp. 437–447 *in* C. Ramel, B. Lambert, and J. Magnusson, eds. Genetic toxicology of environmental chemicals. Part B. Genetic effects and applied mutagenesis. Alan R. Liss, New York.

Russell, L. B., W. L. Russell, R. A. Popp, C. Vaughan, and K. B. Jacobson. 1976. Radiation-induced mutations of mouse hemoglobin loci. Proc. Natl. Acad. Sci. USA 73:2843–2846.

Russell, W. L. 1951. X-ray-induced mutations in mice. Cold Spring Harbor Symp. Quant. Biol. 16:327–336.

Satoh, C., A. A. Awa, J. V. Neel, W. J. Schull, H. Kato, H. B. Hamilton, M. Otake, and K. Goriki. 1982. Genetic effects of atomic bombs. Pp. 267–276 *in* B. Bonne-Tamir, ed. Human genetics. Part A: The unfolding genome. Alan R. Liss, New York.

Satoh, C., R. E. Ferrell, R. J. Tanis, N. Ueda, S. Kishimoto, J. V. Neel, H. B. Hamilton, and K. Baba. 1977. The frequency in Japanese of genetic variants of 22 proteins. III. Phosphoglucomutase-1, phosphoglucomutase-2, 6-phosphogluconate dehydrogenase, adenylate kinase, and adenosine deaminase. Ann. Hum. Genet. 41:169–183.

Satoh, C., J. V. Neel, A. Yamashita, K. Goriki, M. Fujita, and H. B. Hamilton. 1983. The frequency among Japanese of heterozygotes for deficiency variants of 11 enzymes. Am. J. Hum. Genet. 35:656–674.

Satoh, C., N. Takahashi, J. Asakawa, N. Masunari, M. Fujita, and K. Iwamoto. 1984. Electrophoretic variants of blood proteins in Japanese. I. Phosphoglucomutase-2 (PGM2). Jpn. J. Hum. Genet. 29:89–104.

Schull, W. J., J. V. Neel, M. Otake, A. Awa, C. Satoh, and H. B. Hamilton. 1982. Hiroshima and Nagasaki: three and a half decades of genetic screening. Pp. 687–700 *in* T. Sugimura and S. Kondo, eds. Environmental mutagens and carcinogens. Alan R. Liss, New York.

Schull, W. J., M. Otake, and J. V. Neel. 1981. Genetic effects of the atomic bombs: a reappraisal. Science 213:1220–1227.

Sinclair, W. K. 1985. Experimental RBE values of high LET radiation at low doses and the implications for quality factor assignment. Radiat. Protection Dosimetry 13:319–326.

Skolnick, M. M., and J. V. Neel. 1986. An algorithm for comparing two-dimensional electrophoretic gels, with particular reference to the study of mutation. Adv. Hum. Genet. 15:55–160.

Stout, J. T., and C. T. Caskey. 1985. HPRT: gene structure, expression, and mutation. Ann. Rev. Genet. 19:127–148.

Tanis, R. J., N. Ueda, C. Satoh, R. E. Ferrell, S. Kishimoto, J. V. Neel, H. B. Hamilton, and N. Ohno. 1978. The frequency in Japanese of genetic variants of 22 proteins. IV. Acid phosphatase, NADP-isocitrate dehydrogenase, peptidase A, peptidase B and phosphohexose isomerase. Ann. Hum. Genet. 41:419–428.

Turner, D. R., A. A. Morley, M. Haliandros, R. Kutlaca, and B. J. San Jerson. 1985. *In vivo* somatic mutations in human lymphocytes frequently result from major gene alterations. Nature 315:343–345.

Ueda, N., C. Satoh, R. J. Tanis, R. E. Ferrell, S. Kishimoto, J. V. Neel, H. B. Hamilton, and K. Baba. 1977. The frequency in Japanese of genetic variants of 22 proteins. II. Carbonic anhydrase I and II, lactate dehydrogenase, malate dehydrogenase, nucleoside phosphorylase, triose phosphate isomerase, hemoglobin A and hemoglobin A2. Ann. Hum. Genet. 41:43–52.

U.S. Congress, Office of Technology Assessment. 1986. Technologies for detecting heritable mutations in human beings. OTA-H-298. U.S. Government Printing Office, Washington, DC.

van Ommen, G. J. B., J. M. H. Verkerk, M. H. Hofker, A. P. Monaco, L. M. Kunkel, P. Ray, R. Worton, B. Wieringa, E. Bakker, and P. L. Pearson. 1986. A physical map of 4 million bp around the Duchenne muscular dystrophy gene on the human X-chromosome. Cell 47:499–504.

Wilson, J. M., J. T. Stout, T. D. Palella, B. L. Davidson, W. N. Kelley, and C. T. Caskey. 1986. A molecular survey of hypoxanthine-guanine phosphoribosyltransferase deficiency in man. J. Clin. Invest. 77:188–195.

Yandell, D. W., T. P. Dryja, and J. B. Little. 1986. Somatic mutations at a heterozygous autosomal locus in human cells occur more frequently by allele loss than by intragenic structural alterations. Somatic Cell Mol. Genet. 12:255–263.

Youssoufian, H., S. E. Antonarakis, S. Aronis, G. Tsiftis, D. G. Phillips, and H. H. Kazazian, Jr. 1987. Characterization of five partial deletions of the factor VIII gene. Proc. Natl. Acad. Sci. USA 84:3772–3776.

Congenital Malformations, Stillbirths, and Early Mortality among the Children of Atomic Bomb Survivors: A Reanalysis

MASANORI OTAKE,* WILLIAM J. SCHULL,†,[1] AND JAMES V. NEEL‡

*Department of Statistics, Radiation Effects Research Foundation (RERF), Hiroshima, Japan; †RERF, and The Genetics Centers, Graduate School of Biomedical Sciences, University of Texas Health Science Center, Houston, Texas; and ‡Department of Human Genetics, University of Michigan, Ann Arbor, Michigan

OTAKE, M., SCHULL, W. J., AND NEEL, J. V. Congenital Malformations, Stillbirths, and Early Mortality among the Children of Atomic Bomb Survivors: A Reanalysis. *Radiat. Res.* 122, 1–11 (1990).

Of all the data sets pertinent to the estimation of the genetic risks to humans following exposure to ionizing radiation, potentially the most informative is that composed of the cohort of children born to atomic bomb survivors. We present here an analysis of the relationship between parental exposure history and untoward pregnancy outcomes within this cohort, using to the fullest extent possible the recently revised estimates of the doses received by their parents, the so-called DS86 doses. Available for study are 70,073 terminations, but DS86 doses have not been or presently cannot be computed on the parents of 14,770. The frequency of untoward pregnancy outcomes, defined as a pregnancy terminating in a child with a major congenital malformation, and/or stillborn, and/or dying in the first 14 days of life, increases with combined (summed) parental dose, albeit not significantly so. Under a standard linear model, when the sample of observations is restricted to those children whose parents have been assigned the newly established DS86 doses ($n = 55,303$), ignoring concomitant sources of variation and assuming a neutron RBE of 20, the estimated increase per sievert in the predicted frequency of untoward outcomes is 0.00354 (± 0.00343). After adjustment for concomitant sources of variation, the estimated increase per sievert in the proportion of such births is 0.00422 (± 0.00342) if the neutron RBE is assumed to be 20. A "one-hit" model with appropriate adjustments for extraneous sources of variation results in an almost identical value, namely, 0.00412 (± 0.00364). When the sample is extended to include parents lacking the full array of dose parameters necessary to calculate the DS86 dose, but sufficient for an empirical conversion of the previously employed T65DR dose system to its DS86 equivalent, we find under the linear model that the estimated increase per sievert in untoward pregnancy outcomes is some 31% higher than that published previously, 0.00264 (± 0.00277), assuming an RBE of 20, after adjustment for extraneous sources of variation. (Since a dose could not be calculated in 367 of the 70,073 outcomes, the $n = 69,706$.) The corresponding value with the one-hit model is 0.00262 (± 0.00294).

INTRODUCTION

Estimation of the genetic risks to humans following exposure to ionizing radiation has been and continues to be one of the most important and difficult tasks that confront radiobiologists. A variety of kinds of experiences and data are available on which to base estimates. These include (a) exposure to diagnostic and therapeutic doses of X rays and radioactive materials such as radium or cobalt-60; (b) occupationally incurred exposures, for example, in uranium mining or the maintenance of nuclear reactors; (c) geographic areas with "high" natural or manmade background radiation levels, and (d) the atomic bombings of Hiroshima and Nagasaki. Exposure from these sources varies substantially in both radiation quality and dose. The exposure may be acute or chronic, of a single or more qualities, wholebody or partial, and possibly confounded by the health of the exposed person. Indeed, so disparate are the exposures and the methods of case ascertainment that have been used that it is impossible to examine these data in any collective manner. Of all the possible data sets, the cohort of children born to atomic bomb survivors is potentially the most informative. Recognition of this fact led to the initiation of a joint United States–Japan program of investigation shortly after the war, conducted initially by the Atomic Bomb Casualty Commission (ABCC) and now by the Radiation Effects Research Foundation (RERF). Those studies have provided more data than all other similar studies combined. It needs to be noted, however, that although the sample sizes available are large by the standards of conventional epidemiological studies, the average gonadal doses are well below those employed experimentally. On the other hand, the study deals with a population of different socio-economic states and occupations, a variety of ages, and both sexes, and can be termed a "natural population." These data do not confound medical indications for the use of ionizing radiation with the effects of ionizing radiation.

Finally, the new system of dosimetry, the DS86, provides better estimates of individual doses than the earlier T65DR

[1] Permanent director, RERF.

dosimetry for the survivors. These dose estimates are unquestionably more reliable than those known for the majority of other populations.

Most of the methods of detection of mutations that are feasible in the study of human populations have been employed in the study of the offspring of A-bomb survivors. These include a search for changes in the frequency of (a) certain population characteristics, for example, the occurrence of major congenital defects or premature death (1–4); (b) sentinel phenotypes (5, 6); (c) chromosomal abnormalities (7, 8); and (d) biochemical variants of a structural or kinetic nature (9, 10). Although these changes are diverse, the aim of identifying them is the same—to estimate the probability of mutation per unit exposure to ionizing radiation and to ascertain the public health implications of an increase in the number of mutations.

The purpose of this paper is to reexamine one facet of these studies of the offspring of the survivors of the atomic bombing of Hiroshima and Nagasaki: the clinical data on adverse pregnancy outcomes obtained in the years 1948–53. This reanalysis has been prompted by the revised dose estimates. The results of this reanalysis, of interest in its own right, will in a subsequent publication be combined with the results of similar analyses of all of the other data now available on these children. The aim is to generate an estimate of the lowest amount of radiation (95% probability level) that will produce a mutational impact equal to that arising spontaneously each generation, i.e., a minimal doubling dose, and also an estimate of the actual doubling dose suggested by the analysis.[2]

MATERIALS AND METHODS

The Study Sample

We have described the nature of the birth registration process involved in the ascertainment between 1948 and 1953 of the 70,073 pregnancies lasting at least five lunar months which resulted in the potential study group, the completeness of this process, and the information that was collected (1–4). Briefly, a continuous genetic surveillance of the children born in Hiroshima and Nagasaki subsequent to the atomic bombing began in 1948. The initial study, the one of concern here, utilized the postwar Japanese rationing system as a case-finding mechanism. A provision of this system entitled pregnant women who registered their pregnancies after the fifth lunar month to supplementary foodstuffs. As a consequence of the stringency of the postwar economic circumstances, most women registered their pregnancies, and it was possible through this means to identify more than 95% of all pregnancies in these cities persisting for at least 20 weeks of gestation and arrange for the examination of the progeny by a physician. These observations were generally made in the home and were supported by an infant autopsy program and a second examination of some 30% of surviving infants at the clinical facilities of ABCC 8–10 months after their birth [for a fuller description of this study see (1)].

The examination of newly born infants, in a program such as that just described, can provide the following information: sex, birth weight, viability at birth, presence of major malformation, and occurrence of death during the first weeks of life, all of which are potentially confounded by a variety of extraneous factors. Of the extraneous factors for which data exist, the most important are year of birth, city, sex, birth rank, and the ages of the parents at the birth of the child.

In this analysis our interest is focused on three aberrant outcomes of pregnancy: the birth of a child with a major congenital anomaly, and/or who was stillborn, and/or who succumbed within 2 weeks following birth. [Note: A major congenital anomaly in this context is defined as one that is incompatible with survival, is life-threatening, or seriously compromises the individual's capacity to function normally in society. A detailed listing of the major congenital malformations encountered in this study has been presented previously (1, 11).] We designate a pregnancy which terminated in an infant exhibiting one or more of these potential indicators of radiation damage as an untoward pregnancy outcome. Elsewhere we will estimate what fraction of these untoward outcomes can be considered directly related to mutation in the preceding generation.[2]

Dosimetry

To avoid ambiguity, we define the four terms that we shall use to describe dose and exposure: the free-in-air (FIA) kerma (*k*inetic *e*nergy *r*eleased in *ma*terial), kerma in shielded areas (termed here "shielded kerma"), organ-absorbed dose, and organ dose equivalent. The first describes the kerma in tissue at a point in air over bare ground (i.e., not in or near a building); the second, the kerma of the individual with allowance for structural shielding; the third, the radiation (γ, neutron) absorbed by the organ or tissue under consideration; and the last, the sum of the products of the various absorbed doses times their quality factors, or relative biological effectiveness (RBE). The gonad neutron doses were low, and we assume an RBE of 20 for the genetic effect of neutrons (12) and an RBE of 1 for γ rays. Kerma and organ-absorbed doses are expressed in grays and the organ dose equivalent in sieverts.

Our previous analyses of radiation-related risks among the offspring of A-bomb survivors have been based on the estimated T65DR doses of their parents (13). This system of dosimetry estimated a kerma in air with allowance for shielding, where appropriate, for most individual survivors (98%) within the RERF Life Span Study (LSS) sample, which included most of the proximally exposed parents (14). Organ-absorbed doses were assigned using fixed coefficients to describe the attenuation of radiation through organs and tissues before it reached the specific organ or tissue of interest (15). Shielded kerma and organ-absorbed doses were estimated separately for neutrons and γ rays.

The work of a number of investigators (16–18) led to a reassessment of the A-bomb radiation dosimetry sponsored by the United States and Japan (19). This reappraisal has resulted in the Dosimetry System 1986 (DS86) (20).

Patently, evaluation of the genetic effects of the A-bombs requires precise knowledge regarding the radiation exposure of the pertinent individuals, in this instance, the parents of the children in the study sample. Accordingly we describe in some detail exactly how these doses were estimated. We begin by distinguishing between two sets of doses, one the DS86 system and the other an ad hoc dose system we have developed. DS86 in this context implies a dose calculated by one of the methods installed at the RERF by the Science Applications International Corp. The bases and manner of computing doses in these instances are described either in Roesch (20) or in Preston and Pierce (21). Recently, as an addition to the descriptions in (20) and (21), the system has been extended to include direct computation of doses for those survivors in Nagasaki who were terrain-shielded or exposed in factories. This involves approximately 1000 survivors in the LSS

[2] J. V. Neel, W. J. Schull, A. A. Awa, C. Satoh, H. Kato, M. Otake, and Y. Yoshimoto. The children of parents exposed to atomic bombs: Thoughts on the genetic doubling dose of radiation for humans. Manuscript in preparation.

TABLE I
A Comparison of the Shielding of Parents with Reference to Whether DS86 Doses Have Been Assigned

Shielding category	DS86 dose known Mothers (17,010)	DS86 dose known Fathers (17,010)	DS86 dose unknown Mothers (12,207)	DS86 dose unknown Fathers (12,207)
In open	3.4	5.9	4.1	6.1
Japanese building	86.4	81.7	71.2	65.2
Other	10.2	12.4	24.7	28.7

sample. Rules have also been formulated for the indirect estimation of doses for a larger number of survivors exposed beyond 2000 m. Specifically, this includes individuals exposed in the open or for whom no shielding information is available.

While this recent development has further reduced the number of survivors without DS86 doses, there are still groups of individuals for whom doses are not available. Many of these individuals were exposed within 2000 m and can therefore be presumed to have received significant doses. It is this latter group, which involves one or both of the parents for approximately 14,000 pregnancy terminations, that is the primary focus of the ad hoc doses. The exposure data on these parents were collected between 1948 and 1954 (cf. (1)), a decade before the present detailed exposure history was codified, and while a T65DR-type dose can usually be calculated, the data are simply not sufficiently detailed for the DS86 system.

The 14,000 children described above constitute a 26% addition to the sample for this study, and including them in the study will enable us to provide information applicable to other samples where a similar problem exists. We have therefore returned to the original sample tape for this study. Through matching parental master file numbers on the tape with the roster of individuals, we have divided the parents into two groups: those for whom a DS86 dose has been calculated, and those for whom it has not. Within each of the two groups, we have tabulated the categorical data on parental shielding recorded in the course of the study. The results are shown in Table I. The tabulated values are percentages of individuals reporting this particular type of shielding among 12,207 parents in the DS86 unknown group and 17,010 among the DS86 known. It should be noted that the most common shielding by far in each group is a Japanese wooden

TABLE II
A Comparison of the T65DR Doses (G4) Assigned when a DS86 Dose Can Be Assigned and when it Cannot, further Subdivided by Distance and by Presence or Absence of Symptoms

Distance	Symptoms		0.01–6.0	0.01–0.09	0.1–0.19	0.2–0.49	0.5–0.99	1.0+	0
			a. T65DR known, DS86 known						
<2.0 km	Symptoms	No.	2471	217	240	374	319	1321	15
		Dose	165.1	5.8	14.7	31.1	75.0	278.3	0.0
	No symptoms	No.	10235	2034	2555	2448	1612	1586	6276
		Dose	55.7	5.7	14.6	31.5	70.2	208.7	0.0
	Total	No.	12706	2251	2795	2822	1931	2907	6291
		Dose	77.0	5.7	14.6	31.5	71.0	240.3	0.0
≥2.0 km	Symptoms	No.	225	168	57				295
		Dose	5.3	2.1	14.7				0.1
	No symptoms	No.	4180	3632	547	1			19297
		Dose	3.3	1.6	14.3	19.6			0.0
	Total	No.	4405	3800	604	1			19592
		Dose	3.4	1.6	14.3	19.6			0.0
			b. T65DR known, DS86 unknown						
<2.0 km	Symptoms	No.	470	19	28	37	68	318	22
		Dose	204.2	4.5	15.7	34.7	74.7	280.2	0.0
	No symptoms	No.	1408	213	229	297	260	409	95
		Dose	90.2	5.4	15.5	32.1	70.8	230.6	0.0
	Total	No.	1878	232	257	334	328	727	117
		Dose	118.7	5.3	15.5	32.4	71.6	252.3	0.0
≥2.0 km	Symptoms	No.	15	9	6				2
		Dose	7.3	3.9	12.5				0.2
	No symptoms	No.	269	206	63				65
		Dose	5.4	2.7	13.9				0.2
	Total	No.	284	215	69				67
		Dose	5.5	2.8	13.8				0.2

Note. Further explanation in text.

TABLE III
The Distribution of Joint Parental DS86 Gonadal Dose Equivalents, Based on a Neutron RBE of 20, by City and Dose Category, for the Cohort of Children Examined for Untoward Pregnancy Outcome

City		Total	<0.01	0.01–0.09	0.10–0.49	0.50–0.99	1.00–2.49	2.50+	0.01+
Hiroshima	Subjects	31283	24449	3761	1958	586	416	113	6834
	N mean (Sv)[a]	0.01	0.00	0.00	0.02	0.08	0.23	0.87	0.04
	G mean (Gy)	0.06	0.00	0.04	0.22	0.61	1.29	2.84	0.26
	T mean (Sv)	0.07	0.00	0.04	0.24	0.69	1.52	3.71	0.30
Nagasaki	Subjects	24020	20760	1724	711	488	286	51	3260
	N mean (Sv)	0.00	0.00	0.00	0.01	0.03	0.09	0.32	0.02
	G mean (Gy)	0.05	0.00	0.03	0.25	0.69	1.36	3.34	0.34
	T mean (Sv)	0.05	0.00	0.03	0.25	0.73	1.46	3.66	0.36
Both cities	Subjects	55303	45209	5485	2669	1074	702	164	10094
	N mean (Sv)	0.01	0.00	0.00	0.01	0.06	0.18	0.70	0.03
	G mean (Gy)	0.05	0.00	0.03	0.23	0.65	1.32	3.00	0.29
	T mean (Sv)	0.06	0.00	0.03	0.24	0.70	1.49	3.70	0.32

[a] $N = 20 \times$ neutron dose, $G = \gamma$ dose, $T =$ total $(\gamma + 20 \times$ neutron$)$ dose.

building. Further, although more parents are in the "other" category when the DS86 is unknown, this category does not necessarily imply shielding in a concrete structure or air raid shelter, but includes parents stated to be behind fences or trees, under the eaves of buildings, and the like. Presumably a DS86 dose could have been computed on most of the DS86 unknown group in wooden structures had detailed shielding information been available, and possibly on some of the "others" as well.

To proceed further, we have divided the DS86 known and DS86 unknown groups into two mutually exclusive subgroups for those parents for whom γ and neutron exposures have been calculated by the T65DR dose system: survivors exposed at 2000 m or less and those exposed at more than 2000 m. These two groups were further sorted into the asymptomatic and the symptomatic, i.e., those survivors who reported one or more of three cardinal signs or symptoms of acute radiation sickness: epilation, subcutaneous bleeding, or oropharyngeal lesions. Individuals in the eight groups that resulted were then distributed over five dose intervals, and within each interval the mean T65DR dose was computed. The results are shown in Table II.

It will be noted that the mean T65DR is virtually the same for the DS86 unknown and the DS86 known groups within each dose interval, and this is true with respect to persons exposed at 2000 m or less, more than 2000 m, or classified as symptomatic. We also note that within a T65DR dose group the percentage of symptomatic individuals is approximately the same in the DS86 known and unknown groups. These facts suggest that for these individuals the distribution of T65DR doses is independent of factors that determine whether or not a DS86 dose can be computed. If this be so, then the distribution of DS86 doses within the subgroups where both doses are known can be used to provide an ad hoc

TABLE IV
The Distribution of Joint Parental Gonadal Absorbed Dose Equivalent by City and Dose Category in the Cohort of Births Seen in 1948–1953, Extended (DS86 + ad hoc Dose) Cohort[a]

City		Total	<0.01	0.01–0.09	0.10–0.49	0.50–0.99	1.00–2.49	2.50+	0.01+
Hiroshima	Subjects	34113	26300	4229	2230	667	507	180	7813
	N mean (Sv)[b]	0.01	0.00	0.00	0.02	0.08	0.24	0.93	0.05
	G mean (Gy)	0.07	0.00	0.04	0.22	0.61	1.29	2.89	0.28
	T mean (Sv)	0.08	0.00	0.04	0.23	0.69	1.53	3.82	0.33
Nagasaki	Subjects	35593	30997	2257	1088	737	409	105	4596
	N mean (Sv)	0.00	0.00	0.00	0.01	0.03	0.09	0.29	0.02
	G mean (Gy)	0.05	0.00	0.03	0.26	0.68	1.35	3.07	0.37
	T mean (Sv)	0.05	0.00	0.03	0.27	0.71	1.44	3.36	0.40
Both cities	Subjects	69706	57297	6486	3318	1404	916	285	12409
	N mean (Sv)	0.01	0.00	0.00	0.01	0.06	0.18	0.69	0.04
	G mean (Gy)	0.06	0.00	0.03	0.23	0.65	1.31	2.96	0.32
	T mean (Sv)	0.06	0.00	0.03	0.24	0.70	1.49	3.65	0.36

[a] A neutron RBE of 20 is assumed.
[b] $N = 20 \times$ neutron dose, $G = \gamma$ dose, $T =$ total $(\gamma + 20 \times$ neutron$)$ dose.

TABLE V
Increments or Decrements of Change in the Individual Frequencies of Congenital Malformation, Stillbirths, and Neonatal Deaths in the Original Cohort of Births per Sievert of Joint Parental Gonadal Dose Equivalent Based Upon an Assumed Neutron RBE of 20, DS86 Cohort Only

Variable	Regression coefficient	Standard error

Regression model: $P_i = \text{Constant} + \sum_{j=1}^{6} b_j x_{ij} (\text{Background}) + b_D \text{Dose}_i$

Malformation		
Joint parental exposure	0.00099	0.00184
Birth order of child	0.00087*	0.00042
Year of birth	0.00120**	0.00032

Stillbirths		
Joint parental exposure	0.00151	0.00199
Birth order of child	−0.00054	0.00042
Year of birth	−0.00048	0.00035

Neonatal deaths		
Joint parental exposure	0.00237	0.00233
Birth order of child	0.00011	0.00049
Year of birth	0.00094**	0.00041

Regression model:
$P_i = 1 - \exp - (\text{Constant} + \sum_{j=1}^{6} b_j x_{ij} (\text{Background}) + b_D \text{Dose}_i)$

Malformation		
Joint parental exposure	0.00106	0.00233
Birth order of child	0.00091***	0.00054
Year of birth	0.00116**	0.00042

Stillbirths		
Joint parental exposure	0.00138	0.00244
Birth order of child	−0.00064	0.00054
Year of birth	−0.00050	0.00044

Neonatal deaths		
Joint parental exposure	0.00233	0.00272
Birth order of child	0.00028	0.00059
Year of birth	0.00101	0.00049

Note. Significance levels: *($P < 0.05$), **($P < 0.01$), ***($P < 0.10$).

estimate of the total kerma for the subgroups where the DS86 dose is unknown.

Various methods of assigning an ad hoc dose come to mind. We proceeded on the principle that within the six dose intervals shown in Table II (0.01–0.19, ..., ≥1.0 Gy), the T65DR dose on an individual could be multiplied by the ratio of the DS86 to the T65DR mean total kerma to obtain an ad hoc kerma dose. Accordingly, within the DS86 known group, the mean total kerma and its γ and neutron components, for the six dose intervals, were computed for the two cities separately (data not shown). From these data, we have calculated the ratio of the means for the total DS86 and total T65DR kerma. These ratios we term the dose conversion factors. These factors are strikingly different between the two cities, as was anticipated because of the difference in the neutron component attributed to the two explosions in the T65DR doses. Furthermore, whereas the dose conversion factors are functionally dependent upon the dose interval in Hiroshima, there is a much lesser dependence, if one at all, in Nagasaki. Again, this is in keeping with the changes in kerma associated with the DS86 system [see (20)].

To obtain the ad hoc estimates of total kerma, we have multiplied each individual's T65 estimate of kerma within a dose interval in the unknown DS86 group (for the six intervals) by the dose conversion factor within that interval in the city where the parent was exposed. To partition the estimated total kerma into neutron and γ components, the ad hoc total kerma dose was multiplied by the proportion of the known DS86 dose within the interval ascribed to neutrons. Finally, the ad hoc total kerma was divided into neutron and γ doses using these proportions. Where the T65DR dose was stated to be zero, we have presumed the DS86 to be the same and assigned this value as the ad hoc estimate.

To obtain the putative gonadal doses, the ad hoc kerma estimates were multiplied by the DS86 average ovary and testis transmission factors for γ and neutron emanations based on the LSS cohort. These values are 0.74 and 0.16 for γ rays and neutrons, respectively, for the ovary and 0.78 and 0.32 for the testes. We have disregarded the neutron–gamma capture factors, for it seemed too tenuous to attempt to specify the appropriate fraction of the total kerma assignable to neutron-capture gamma rays. In keeping with the higher frequency of one or more symptoms of radiation exposure encountered in the DS86 unknown dose group within 2000 m at the time of bombing (ATB), which was 25% compared to 19% for the DS86 known group, we find that the system described above assigns to them an average kerma dose of 0.75 Gy in contrast to the dose of 0.53 Gy computed for the DS86 known group within 2000 m.

The approximate nature of this procedure is clear. Possibly the most important assumption on which it rests is the tacit one that the distribution of types of shielding is the same, or approximately so, in the DS86 known and unknown groups since the DS86 doses used to obtain the crude adjustment factors are estimates of shielded kerma. As shown earlier, this assumption seems reasonably well met by the data on shielding as categorically defined. In this connection, we are impressed at how well the biological dosimetry (frequency of epilation, chromosomal damage) reinforces the physical.

Table III, using only those pregnancy terminations for which there are DS86 doses, or one or both parents are known not to have been in Hiroshima or Nagasaki ATB, summarizes the number of pregnancies ascertained, the mean combined (summed) parental absorbed gonadal dose, and the dose equivalent, assuming an RBE for neutrons of 20, by city for six dose categories: 0, 0.01–0.09, 0.10–0.49, 0.50–0.99, 1.00–2.49, and ≥2.50 Gy.

Table IV gives the comparable values when the parents for whom ad hoc doses were calculated are included. The lowest combined dose group, that is, those individuals exposed to less than 0.01 Gy, includes not only those persons present in the city and exposed to less than the stated dose but also any spouses who were not present in the city at the time of the bombing [the so-called not-in-city group (NIC)]. As will be noted from a comparison of the average gonadal absorbed doses with the average dose equivalents given in Table III (see also Table IV), other values for the neutron RBE would not materially alter the results to be described subsequently, for under the DS86 system of dosimetry the neutron contribution is small, and the RBE assigned to neutrons is of minor importance in determining the slope of the dose–response relationship.

Statistical Methods

In the analysis of untoward pregnancy outcomes, the effects of radiation have been estimated using two overlapping samples of observations. The

TABLE VI

The Distribution of Untoward Pregnancy Outcomes (UPO) and Total Subjects by Parental Gonadal Dose Equivalents in Sieverts (DS86) Based on an Assumed Neutron RBE of 20, Sexes and Cities Combined, DS86 Cohort Only

Mother's gonadal dose equivalent (Sv)	<0.01 Subj	<0.01 UPO	0.01–0.09 Subj	0.01–0.09 UPO	0.10–0.49 Subj	0.10–0.49 UPO	0.50–0.99 Subj	0.50–0.99 UPO	1.0+ Subj	1.0+ UPO	Mean dose
1.00+	388	19	21	2	9	0	9	1	15	1	1.66
0.50–0.99	651	44	19	0	24	1	47	4	17	1	0.70
0.10–0.49	1655	81	124	9	209	13	41	2	20	0	0.24
0.01–0.09	3790	179	700	27	138	5	27	2	45	2	0.04
<0.01	45234	2257	1104	60	510	21	238	12	268	17	0.00
Mean dose	0.00		0.03		0.25		0.71		2.08		

first sample ($n = 55,303$) is restricted to those pregnancy terminations for which there are DS86 doses for both parents or it is known they were not in the city ATB. The second sample ($n = 69,706$) consists of the first sample augmented by those pregnancy terminations where an ad hoc dose could be computed on the basis of a previous T65DR dose estimate. Data were collected on 76,626 infants, but after exclusions because the pregnancy had not been previously registered or the biomedical or exposure data were deficient in some respect [see (*1*)], the number available for analysis was 69,706. In the original analysis (*1*), the offspring of consanguineous parents were excluded because of the possible heterogeneity they would introduce to the sample, but they are included in the present analysis since a subsequent analysis (*22*) did not indicate that they responded differently to radiation.

Information regarding the diagnosis of a major congenital defect, the occurrence of a stillbirth, or death in the first 14 days following birth is available on all these infants. The decision to restrict this treatment of death to the first 14 days rests on the following considerations: In the years from 1948–1954 over 90% of all pregnancy terminations occurred in the home rather than a hospital or clinic, and most mothers were reluctant to take their newly born child out until it was several weeks old. Of necessity, therefore, the initial examination of the infant occurred in the home, generally within the first 2 weeks following delivery. Thus the information available to us on mortality from this study can be presumed to be complete for only about the first 14 days of life. (Subsequent deaths became known through another program, the findings of which will be described elsewhere.)

Although a variety of dose–response models have been fitted to the data, the results of only two will be described in detail. The first involves fitting a linear dose–response model to the occurrence of the various indicators of radiation-related damage treated as binary response variables. Data were available on six background or concomitant variables, in addition to summed parental dose, which might influence the occurrence of an event of interest: city (Hiroshima, Nagasaki; Hiroshima was assigned the value 1 and Nagasaki 0), sex (male, female; males were assigned the value 1, females 0), maternal age (years), paternal age (years), year of birth following the bombing, and birth rank. Specifically, we have fitted a model of the following form:

$$P_i = \text{Constant} + \sum_{j=1}^{6} b_j x_{ij}(\text{Background}) + b_\text{D} \text{Dose}_i$$

where P_i is the expected frequency of the event of interest in the ith individual ($i = 1, 2, \cdots n$; i.e., the total number of subjects) having background characteristics x_{ij} ($j = 1, 2, \ldots, 6$), and dose i. The constants, b_j, and b_D are the parameters to be estimated. The effects associated with the background characteristics are useful in providing standards of comparison with the risk seen with exposure. When these latter variables are treated as risk factors to be estimated rather than as part of the background, the model indicated above takes a slightly different form, indicated in Table V.

The second dose–response model, employed in an earlier analysis of the data (*2*), involves fitting an exponential curve of the kind described as a "one-hit" model, of the following form:

$$P_i = 1 - \exp-(\text{Constant} + \sum_{j=1}^{6} b_j x_{ij}(\text{Background}) + b_\text{D}\text{Dose}_i)$$

where P_i, x_{ij}, and Dose$_i$ are as previously defined. It is important to note in this connection that in the earlier analysis employing this model (*2*) 25 dose categories were used (five maternal and five paternal groupings), and that the background variables were treated somewhat differently. Also, the simple linear model described above is a first approximation to the exponential, and the results of fitting the two models would not be expected to differ greatly. We offer the results of the exponential model only for continuity with earlier analyses which have used this model. In both instances, the parameters of the model have been estimated by the method described elsewhere (*2*).

RESULTS

Analysis employing DS86 doses only. Of the 55,303 pregnancy terminations analyzed, 2760 culminated in an untoward pregnancy outcome (Table VI). Among these latter terminations were 770 infants with a major congenital defect, 894 who were stillborn, and 1230 who, although live-born, died before the fifteenth day after birth. These individually designated categories are overlapping, and 134 of the 2760 events occurred in conjunction with another. As shown in Table VI, which presents the relationship between parental dose category and untoward pregnancy outcome, the background or control rate was 4.99%.

Table VII sets forth the results of the analysis of these data when the concomitant sources of variation which can be identified in these data are taken into account. We note no

TABLE VII

Increments or Decrements of Change in the Frequency of Untoward Pregnancy Outcomes in the Original Cohort of Births per Sievert of Joint Parental Gonadal Dose Equivalent Based upon an Assumed Neutron RBE of 20, DS86 Cohort Only

Variable	Regression coefficient	Standard error

Regression model: $P_i = \text{Constant} + \sum_{j=1}^{6} b_j x_{ij} (\text{Background}) + b_D \text{Dose}_i$

Background effects

Constant	0.04282	0.00662
City	0.00110	0.00189
Sex	0.00262	0.00184
Mean age of father	−0.00039*	0.00022
Mean age of mother	0.00034	0.00031
Birth order of child	0.00039	0.00075
Year of birth	0.00173**	0.00061

Excess risk

| Joint parental exposure | 0.00422 | 0.00343 |

Cov(Constant, Dose) = -0.2980×10^{-9}
Corr(Constant, Dose) = -0.0131

Regression model:
$P_i = 1 - \exp - (\text{Constant} + \sum_{j=1}^{6} b_j x_{ij} (\text{Background}) + b_D \text{Dose}_i)$

Background effects

Constant	0.04324	0.00714
City	0.00110	0.00203
Sex	0.00262	0.00198
Mean age of father	−0.00040*	0.00024
Mean age of mother	0.00034	0.00034
Birth order of child	0.00046	0.00080
Year of birth	0.00172**	0.00066

Excess risk

| Joint parental exposure | 0.00412 | 0.00364 |

Cov(Constant, Dose) = -0.3515×10^{-9}
Corr(Constant, Dose) = -0.0135

Note. Significance levels: *($P < 0.10$), **($P < 0.01$).

significant effects of city, sex, or ages of mothers or fathers. A somewhat larger number of untoward pregnancies occurred in Hiroshima than in Nagasaki, and among the progeny of irradiated males. Adverse outcomes decreased slightly with paternal age and increased with maternal age. These findings are consistent with our previous studies of this cohort (1, 2, 12, 23). The regression on parental radiation exposure is positive as expected on the hypothesis that radiation produces mutations but well below the level of statistical significance. The results obtained by fitting the one-hit model differ negligibly from those just described (see Table VII).

In Table V we present separate analyses for malformations, stillbirths, and neonatal deaths using both statistical approaches, to determine whether there might be a stronger suggestion of a radiation effect upon some one of the three indicators "hidden" within the gross regression coefficient. In fact, within the statistical limits imposed by the data, all three indicators behave in a similar manner.

The role of possible sources of extraneous variability is potentially troublesome. We have examined, albeit crudely, their effects on the estimate of the genetic risk by ignoring all of the concomitant variables and fitting a simple linear model to the data given in Table VII. We assigned as the dose in each of the 25 cells the sum of the individual mean parental doses given in the marginal entries (data not shown). A further purpose of this analysis was to determine the effect of grouping of doses on the estimates of intercept and slope. The estimates of the latter two parameters in the restricted sample (with an RBE of 20) obtained by the method of maximum likelihood are 0.0497 and 0.00354 (±0.00343), respectively. As can be seen from a comparison of these estimates with those given in Table VII, grouping and ignoring the extraneous sources of variation does not appear to have a profound effect on either estimate. The estimates change by about 16% when compared with the linear model including the background factors, suggesting that the extraneous sources of variation are more or less randomly distributed with respect to dose.

Analysis employing DS86 and ad hoc doses. The use of ad hoc doses results in the addition of 14,403 to the sample. Of the 69,706 pregnancy terminations included in this extended sample, 3498 culminated in an untoward pregnancy outcome (Table VIII). Among these untoward outcomes were 950 infants with major congenital defect, 1148 who were stillborn, and 1565 liveborn infants who died within 14 days of birth. When abnormal terminations were scored in this manner, 165 of the 3613 events occurred in conjunction with one other, resulting in the 3498 abnormal infants. As shown in Table VIII, which presents the distribution of untoward pregnancy outcomes with respect to parental dose, the background rate was 5.02%.

Table IX presents the results of the analysis of the extended cohort. All of the findings observed in the more restricted subset of data are repeated. Again the regression of untoward pregnancy outcome upon parental radiation exposure is positive, which would be expected if the exposure had induced deleterious dominant mutations, but the regression is well below statistical significance, and insignificantly lower than that obtained with the DS86 only cohort. Furthermore, none of the components of the untoward pregnancy outcomes gives any hint of responding differentially to parental radiation (Table X). There is no suggestion of significant heterogeneity between the two data sets. As

TABLE VIII
The Distribution of Untoward Pregnancy Outcomes (UPO), and Total Subjects by Parental Gonadal Dose Equivalent in Sieverts (DS86) Based on an Assumed Neutron RBE of 20, Sexes and Cities Combined, Extended (DS86 + ad hoc Dose) Cohort

Mother's gonadal dose equivalent (Sv)	<0.01 Subj	<0.01 UPO	0.01–0.09 Subj	0.01–0.09 UPO	0.10–0.49 Subj	0.10–0.49 UPO	0.50–0.99 Subj	0.50–0.99 UPO	1.0+ Subj	1.0+ UPO	Mean dose
1.00+	528	24	36	3	13	0	12	1	19	1	1.78
0.50–0.99	834	58	25	0	27	2	53	4	24	1	0.69
0.10–0.49	1984	95	151	10	235	15	53	2	30	1	0.23
0.01–0.09	4428	207	755	28	179	7	54	2	71	3	0.03
<0.01	57322	2880	1428	79	726	36	339	16	380	23	0.00
Mean dose	0.00		0.03		0.25		0.70		2.12		

expected, the standard errors of the various regression terms are appropriately decreased in the larger data set.

Comparison with previous analysis. Our previous analysis of these data with the one-hit model, employing the T65DR doses and a somewhat different computational technique, yielded a regression on summed parental doses per sievert of 0.001824 ± 0.003232 (2). The regression most nearly comparable to this in the present analysis (one-hit model, full data set, RBE of 20) is 0.00264 ± 0.00277. This increase in the regression under the current analysis reflects the reduction in the gonad dose estimated with the DS86 schedule and the use of slightly different statistical procedures. The reduction in the estimated gonadal dose is complex, depending, inter alia, on a major reduction in the neutron component attributed to the Hiroshima bomb which entailed a change in the neutron RBE from 5 to 20 and a decrease in the transmission of the radiation by Japanese-style buildings, but an increase in the estimate of tissue transmission of the radiation, especially for γ radiation. The exact reduction in gonadal dose varies with distance from the hypocenter, but for all parents receiving doses ≥ 0.01 Sv it is about 30%.

As noted, none of the models we have used reveals a statistically significant effect of combined parental exposure, although all show that the risk for an untoward outcome of pregnancy increases with increasing dose. Previous analyses have also failed to disclose a significant dose effect (1, 2). We note once again, however, that in associating errors with these regressions we do not imply a test of the null hypothesis. Radiation has resulted in an increased frequency of mutation in every well-studied organism, and it is inconceivable that humans are an exception. We will discuss below the appropriate uses of these data.

UNCERTAINTIES

As in most if not all epidemiological studies of similar scope, a number of uncertainties attend this one and the analyses we have presented. Two of these warrant particular comment, namely, errors in the estimation of the organ absorbed doses, and the completeness of the ascertainment of death and defect.

Errors in the Estimation of the Gonadal Absorbed Doses

All estimates of the doses to survivors of the A-bombing are subject to at least three sources of error that stem from the FIA dose curves themselves, the estimation of the attenuation of energy through tissues, materials, and the like, and the assertions of the survivors as to their locations. There is no particular reason, however, to believe that these errors are any larger or smaller in the present study than in any other study based on the survivors of the atomic bombing of these cities, where it has been estimated that individual doses could be in error by as much as 30% (24). Be this as it may, these errors can affect inferences on the overall shape of the dose–response curve as well as parameter values defining that shape (24–26).

Completeness of the Ascertainment of Death and Defect

As we have previously stated, ascertainment of the pregnancies whose outcomes we report occurred at or shortly after the fifth lunar month of gestation, and generally well before the pregnancy terminated. Thus the design was inherently prospective, and from a variety of lines of evidence, including examination of the births officially recorded in these cities, we estimate that more than 95% of qualifying pregnancies were identified. Many of those pregnancies which were not registered did subsequently come to our attention through the attending physician or midwife. Most involved children either conceived out of wedlock or born to women who had applied for rations elsewhere or were unaware of the provisions of the rationing system. There is no evidence that pregnancies persisting for 20 weeks or more and terminating untowardly went unregis-

TABLE IX
Increments or Decrements of Change in the Frequency of Untoward Pregnancy Outcomes in the Original Cohort of Births per Sievert of Joint Parental Gonadal Dose Equivalent Based upon an Assumed Neutron RBE of 20, Extended (DS86 + ad hoc) Cohort

Variable	Regression coefficient	Standard error

Regression model: $P_i = \text{Constant} + \sum_{j=1}^{6} b_j x_{ij} (\text{Background}) + b_D \text{Dose}_i$

Background effects

Constant	0.03856	0.00582
City	0.00100	0.00167
Sex	0.00238	0.00165
Mean age of father	−0.00023	0.00020
Mean age of mother	0.00034	0.00028
Birth order of child	0.00019	0.00066
Year of birth	0.00179**	0.00055

Excess risk

Joint parental exposure	0.00264	0.00277

Cov(Constant, Dose) = −0.2827 × 10⁻⁹
Corr(Constant, Dose) = −0.0175

Regression model:
$P_i = 1 - \exp - (\text{Constant} + \sum_{j=1}^{6} b_j x_{ij} (\text{Background}) + b_D \text{Dose}_i)$

Background effects

Constant	0.03868	0.00627
City	0.00101	0.00180
Sex	0.00237	0.00177
Mean age of father	−0.00024	0.00022
Mean age of mother	0.00035	0.00030
Birth order of child	0.00022	0.00071
Year of birth	0.00179**	0.00059

Excess risk

Joint parental exposure	0.00262	0.00294

Cov(Constant, Dose) = −0.3299 × 10⁻⁹
Corr(Constant, Dose) = −0.0179

Significance levels: **($P < 0.01$).

DISCUSSION

In principle, an increase in untoward outcomes in the pregnancies of women exposed to the atomic bombs could reflect an altered maternal physiology and/or depressed socioeconomic status and/or an increased mutation rate in the woman and/or her husband. The average child entering into this study was conceived about 5 years following the bombings, by which time any acute effects of the bombings had certainly disappeared. The intensive studies sponsored by RERF have revealed an increase in a variety of malignant tumors as a late sequela of the bombings [review in (27)], but aside from leukemia, these had not yet made their appearance at the time the present observations were made. We will therefore argue that an altered maternal physiology is unlikely to have affected the indicator, but would only inflate any observed effect. With respect to socioeconomic tered more frequently if one or both parents were exposed to the bombing.

TABLE X
Increments or Decrements of Change in the Individual Frequencies of Congenital Malformation, Stillbirths, and Neonatal Deaths in the Original Cohort of Births per Sievert of Joint Parental Gonadal Dose Equivalent Based Upon an Assumed Neutron RBE of 20, Extended (DS86 + ad hoc) Cohort

Variable	Regression coefficient	Standard error

Regression model: $P_i = \text{Constant} + \sum_{j=1}^{6} b_j x_{ij} (\text{Background}) + b_D \text{Dose}_i$

Malformation

Joint parental exposure	0.00101	0.00154
Birth order of child	0.00064*	0.00036
Year of birth	0.00131**	0.00028

Stillbirths

Joint parental exposure	0.00092	0.00163
Birth order of child	−0.00059	0.00038
Year of birth	−0.00028	0.00032

Neonatal deaths

Joint parental exposure	0.00128	0.00185
Birth order of child	0.00026	0.00043
Year of birth	0.00068*	0.00037

Regression model:
$P_i = 1 - \exp - (\text{Constant} + \sum_{j=1}^{6} b_j x_{ij} (\text{Background}) + b_D \text{Dose}_i)$

Malformation

Joint parental exposure	0.00120	0.00194
Birth order of child	0.00064	0.00047
Year of birth	0.00130**	0.00038

Stillbirths

Joint parental exposure	0.00091	0.00200
Birth order of child	−0.00075	0.00048
Year of birth	−0.00029	0.00040

Neonatal deaths

Joint parental exposure	0.00128	0.00218
Birth order of child	0.00047	0.00052
Year of birth	0.00074*	0.00044

Significance levels: *($P < 0.10$), **($P < 0.01$).

status, we have reported that the parents who were not exposed because they came to Hiroshima and Nagasaki following the bombings as released service men, repatriates, spouses, or immigrants were slightly younger and had a little more education and somewhat higher occupational ratings than the exposed parents (1-4). In principle, this might result in a higher indicator frequency in the children of exposed parents and also lead to an overestimate of radiation effects.

It is important to note that the estimate of genetic damage following exposure to ionizing radiation derived here rests on mutational events manifesting themselves as an untoward pregnancy outcome recognizable between 20 weeks after fertilization and 14 days following birth. Clearly this does not represent all of the mutational damage that could be expressed throughout life or even all of that from fertilization to 14 days postpartum. Pregnancies terminating prior to the 20th week were not ascertained in this study, and to the extent that such occurrences were dose-related, reflecting mutational damage, our data would underestimate the risk. The magnitude of this possible underestimation is uncertain, indeed impossible to estimate, since pregnancies terminating within 4 weeks of fertilization, before a menses is missed or other symptoms of pregnancy become apparent, are often unrecognized by the prospective mother. Although the proportion of pregnancies terminating this early in gestation is not precisely known, it appears relatively large. However, from the societal standpoint, these early losses are much less traumatic than the untoward pregnancy outcomes of this study.

Since radiation has caused genetic damage in every species properly studied in an experimental setting, we assume that some genetic damage resulted from the Hiroshima-Nagasaki experience. We must then also accept that the data from these children, despite their limitations, provide the best available basis for estimating the confidence limits to be placed on the computed genetic risk involved in exposure to ionizing radiation. Briefly, under a linear dose–response function, the excess relative risk for genetic effects can be defined as the ratio of the slope to the intercept. When concomitant sources of variability are taken into account (Tables VII and IX), this ratio at an RBE of 20 is 0.0986, based on the restricted sample. The 95% lower confidence limit on this doubling dose estimate is approximately 0.15 Sv. The comparable values for the extended sample are 0.0685, and 0.19 Sv, respectively. To utilize these estimates to derive a doubling dose applicable to the genetic component, however, it is necessary to postulate what fraction of the indicator may be attributed to spontaneous mutation in the preceding generation. We shall provide an estimate of this fraction based upon current information and combine the findings of this study with other genetic endpoints that have been measured in the children of survivors to derive estimates of both the minimal and most likely doubling dose in a future paper.[2]

ACKNOWLEDGMENTS

This research was conducted at the Radiation Effects Research Foundation, Hiroshima, Japan. The Radiation Effects Research Foundation (formerly ABCC) was established in April 1975 as a private non-profit-making Japanese foundation, supported equally by the Government of Japan through the Ministry of Health and Welfare, and the Government of the United States through the National Academy of Sciences under contract with the Department of Energy.

RECEIVED: June 7, 1989; ACCEPTED: October 30, 1989

REFERENCES

1. J. V. NEEL and W. J. SCHULL, *The Effect of Exposure to the Atomic Bombs on Pregnancy Termination in Hiroshima and Nagasaki.* Publ. No. 461, National Academy of Sciences–National Research Council, Washington, DC, 1956.
2. W. J. SCHULL, M. OTAKE, and J. V. NEEL, Genetic effects of the atomic bombs: A reappraisal. *Science* 213, 1220–1227 (1981).
3. J. V. NEEL, H. KATO, and W. J. SCHULL, Mortality in the children of atomic bomb survivors and controls. *Genetics* 76, 311–326 (1974).
4. H. KATO, W. J. SCHULL, and J. V. NEEL, A cohort-type study of survival in the children of parents exposed to atomic bombings. *Am. J. Hum. Genet.* 18, 339–373 (1966).
5. W. J. SCHULL and J. V. NEEL, Maternal radiation and mongolism. Letter to the Editor. *Lancet* I, 537–538 (1962).
6. W. J. SCHULL, J. V. NEEL, M. OTAKE, A. AWA, C. SATOH, and H. B. HAMILTON, Hiroshima and Nagasaki: Three and a half decades of genetic screening. In *Environmental Mutagens and Carcinogens* (T. Sugimura, S. Kondo, and S. Takebe, Eds.), pp. 687–700. University of Tokyo Press, Tokyo, 1982.
7. A. A. AWA, A. D. BLOOM, M. C. YOSHIDA, S. NERIISHI, and P. ARCHER, A cytogenetic survey of the offspring of atomic bomb survivors. *Nature* 218, 367–368 (1968).
8. A. A. AWA, T. HONDA, S. NERIISHI, T. SOFUNI, H. SHIMBA, K. OHTAKI, M. NAKANO, Y. KODAMA, M. ITO, and H. B. HAMILTON, Cytogenetic studies of the offspring of atomic bomb survivors. In *Cytogenetics: Basic and Applied Aspects* (B. Obe and A. Basler, Eds.), pp. 166–183. Springer-Verlag, Berlin, 1987.
9. C. SATOH, A. A. AWA, J. V. NEEL, W. J. SCHULL, H. KATO, H. B. HAMILTON, M. OTAKE, and K. GORIKI, Genetic effects of atomic bombs. In *Human Genetics Part A: The Unfolding Genome* (B. Bonne-Tamir, Ed.), pp. 267–276. A. R. Liss, New York, 1982.
10. J. V. NEEL, C. SATOH, K. GORIK, J. ASAKAWA, M. FUJITA, N. TAKAHASHI, T. KAGEOKA, and R. HAZAMA, Search for mutations altering protein charge and/or function in children of atomic bomb survivors: Final report. *Am. J. Hum. Genet.* 42, 663–676 (1988).
11. J. V. NEEL, A study of major congenital defects in Japanese infants. *Am. J. Hum. Genet.* 10, 398–445 (1958).
12. D. GRAHN, C. H. LEE, and B. F. FARRINGTON, Interpretation of cytogenetic damage induced in the germ line of male mice exposed for over 1 year to ^{239}Pu alpha particles, fission neutrons, or ^{60}Co gamma rays. *Radiat. Res.* 95, 566–583, (1983). [See also D. GRAHN, B. A. FARRINGTON, and C. H. LEE, Genetic injury in hybrid male mice exposed to low doses of ^{60}Co gamma-rays of fission neutrons. I. Response to single doses. *Mutat. Res.* 129, 215–229 (1984).]

13. J. A. AUXIER, *Ichiban—Radiation Dosimetry for the Survivors of the Bombings of Hiroshima and Nagasaki.* REDA Critical Review Series, TID-27080, NTIS 1977. [See also J. T. CHEKA, F. W. SANDERS, *et al.* Distribution of weapons radiation in Japanese residential structures. USAEC Report CEX-62, **11**, 1965.]

14. R. C. MILTON and T. SHOHOJI, *Tentative 1965 Radiation Dose Estimation for Atomic Bomb Survivors, Hiroshima and Nagasaki.* Technical Report 1-68, Atomic Bomb Casualty Commission, Hiroshima, 1968.

15. G. D. KERR, Organ dose estimates for Japanese atomic bomb survivors. *Health Phys.* **37**, 487–508 (1979).

16. P. P. WHALEN, Status of Los Alamos efforts related to Hiroshima and Nagasaki dose estimates. In *Reevaluation of Dosimetric Factors: Hiroshima and Nagasaki* (V. P. Bond and J. W. Thiessen, Eds.). Technical Information Center, U.S. Dept. of Commerce, Springfield, VA, 1982.

17. W. E. LOEWE and E. MENDELSOHN, Revised dose estimates at Hiroshima and Nagasaki. *Health Phys.* **41**, 663–665 (1981). [See also W. E. LOEWE and E. MENDELSOHN, Neutron and gamma doses at Hiroshima and Nagasaki. *Nucl. Sci. Eng.* **81**, (1982).]

18. G. D. KERR, Findings of a recent Oak Ridge National Laboratory review of dosimetry for Japanese Atomic Bomb survivors. In *Reevaluation of Dosimetric Factors: Hiroshima and Nagasaki* (V. P. Bond and J. W. Thiessen, Eds.). Technical Information Center, U.S. Dept. of Commerce, Springfield, VA, 1982.

19. G. D. KERR, T. HASHIZUME, and C. W. EDINGTON, Historical Review. In *Final Report of US–Japan Reassessment of Atomic Bomb Dosimetry in Hiroshima and Nagasaki* (W. C. Roesch, Ed.), pp. 1–13. Radiation Effects Research Foundation, Hiroshima, 1987.

20. W. C. ROESCH, Ed., *Final Report of US–Japan Reassessment of Atomic Bomb Radiation Dosimetry in Hiroshima and Nagasaki.* Radiation Effects Research Foundation, Hiroshima, 1987.

21. D. L. PRESTON and D. A. PIERCE, *The Effect of Changes in Dosimetry on Cancer Mortality Risks in the Atomic Bomb Survivors.* Technical Report 9-87, Radiation Effects Research Foundation, Hiroshima, 1987 [See also *Radiat. Res.* **114**, 437–466 (1988).]

22. W. J. SCHULL and J. V. NEEL, Atomic bomb exposure and the pregnancies of biologically related parents. *Am. J. Publ. Health* **49**, 1621–1629 (1959).

23. W. J. SCHULL, Empirical risks in consanguineous marriages: Sex ratio, malformation and viability. *Am. J. Hum. Genet.* **10**, 294–343 (1958).

24. S. JABLON, *Atomic Bomb Radiation Dose Estimation at ABCC.* Technical Report 23-71, Atomic Bomb Casualty Commission, Hiroshima, 1971.

25. E. S. GILBERT, *Some Effects of Random Dose Measurement Errors on the Analysis of Atomic Bomb Survivor Data.* Technical Report 12-82, Radiation Effects Research Foundation, Hiroshima, 1982.

26. E. S. GILBERT and J. L. OHARA, Analysis of atomic bomb radiation dose estimation at RERF using data on acute radiation symptoms. *Radiat. Res.* **100**, 124–138 (1984).

27. I. SHIGEMATSU and A. KAGAN, *Cancer in Atomic Bomb Survivors*, pp. 1–196. Plenum Press, New York/London, 1986.

THE CHILDREN OF ATOMIC BOMB SURVIVORS

Malignant Tumors during the First 2 Decades of Life in the Offspring of Atomic Bomb Survivors

Y. Yoshimoto,[*] J. V. Neel,[†] W. J. Schull,[‡] H. Kato,[*] M. Soda,[*] R. Eto,[*] and K. Mabuchi[*]

[*]Radiation Effects Research Foundation, Hiroshima and Nagasaki; [†]Department of Human Genetics, University of Michigan Medical School, Ann Arbor; and [‡]Center for Demographic and Population Genetics, University of Texas, Houston

Summary

The risk of cancer (incidence) prior to age 20 years has been determined for children born to atomic bomb survivors and to a suitable comparison group. Tumor ascertainment was through death certificates and the tumor registries maintained in Hiroshima and Nagasaki. The rationale for the study stemmed from the evidence that a significant proportion of such childhood tumors as retinoblastoma and Wilms tumor arise on the basis of a mutant gene inherited from one parent plus a second somatic cell mutation involving the allele of this gene. Gonadal radiation doses were calculated by the recently established DS86 system, supplemented by an ad hoc system for those children for one or both of whose parents a DS86 dose could not be computed but for whom an ad hoc dose could be developed on the basis of the available information. The total data set consisted of (1) a cohort of 31,150 live-born children one or both of whose parents received >0.01 Sv of radiation at the time of the atomic bombings (average conjoint gonad exposure 0.43 Sv) and (2) two suitable comparison groups totaling 41,066 children. Altogether, 43 malignant tumors were ascertained in the children of exposed parents, and 49 malignant tumors were ascertained in the two control groups. A multiple linear regression analysis revealed no increase in malignancy in the children of exposed parents. However, examination of the data suggested that only 3.0–5.0% of the tumors of childhood that were observed in the comparison groups are associated with an inherited genetic predisposition that would be expected to exhibit an altered frequency if the parental mutation rate were increased. There is thus far no confirmation of the positive findings that Nomura found in a mouse system.

Introduction

This paper will report on the occurrence of malignant tumors prior to age 20 years both in the children born between the years 1946 and 1982 to survivors of the atomic bombings and in a suitable control group. The justifications for having employed the finding as an indicator of the potential genetic effects of the atomic bombs are as follows: For many years certain of the tumors of childhood, of which retinoblastoma serves as the prototype, have been known to be occasionally familial—and, when familial, usually in the pattern of

Received September 21, 1989; revision received December 13, 1989.

dominant inheritance (reviewed in Vogel 1979). In 1971, Knudson suggested that children with retinoblastoma result from either of two mechanisms (Knudson 1971). On the one hand, the child may inherit from one of its parents an altered gene or a deletion at the retinoblastoma locus. A second somatic cell event in the child results in a change in or loss of the normal allele of this gene in a retinoblast, in consequence of which this cell is now homozygous (or hemizygous) for an abnormality at this locus, and a clonally derived retinoblastoma results (and, as was suggested later, there sometimes also result, by the same mechanism, osteosarcoma [Kitchin and Ellsworth 1974], other mesenchymal tumors [Friend et al. 1987], and breast cancer [Lee et al. 1988]). Multiple such somatic cell events result in multiple foci of retinoblastoma or sarcoma. Cytological and molecular evidence for the nature of both the first and second events has been forthcoming from sev-

eral laboratories (Cavenee et al. 1983; Friend et al. 1986; Lee et al. 1987; Bookstein et al. 1988). Technically, then, inherited retinoblastoma is a recessive trait. In some affected children (in fact, the majority), on the other hand, both of the necessary events occur in a somatic cell (e.g., retinoblast, osteocyte), as a consequence of which the child is now homozygous for an abnormality at this locus in this cell; but, although the child develops a malignant tumor, he or she does not transmit an altered allele to the next generation because the germ line is not affected. Parental radiation would not be expected to increase the frequency of this type of retinoblastoma.

Wilms tumor also rather clearly meets the specifications of this model (reviewed in Matsunaga 1981; Dao et al. 1987), as, probably, does neuroblastoma (Knudson and Meadows 1978; Bundey and Evans 1982). Although the evidence is less extensive, childhood gonadal dysgerminoma, pheochromocytoma, hepatoblastoma, and, more problematically, rhabdomyosarcoma and the central nervous system tumors may have a similar but less pronounced genetic basis (reviewed in Knudson and Meadows 1978; Koufos et al. 1985; Schimke 1985; Knudson 1986). A further argument for the genetic basis of some of these tumors is the reportedly more frequent occurrence of *other* malignancies in the parents of affected children (Hartley et al. 1986; Li et al. 1988).

The leukemias and the malignant lymphomas of childhood have not yet been shown to exhibit a pattern comparable to retinoblastoma or Wilms tumor (Videbaek 1947; Steinberg 1960), but, because of the poor survival of affected children in the past, the appropriate systematic genetic data to support an opinion concerning the proportion of the leukemias consistent with the above-described model are not yet available. There are, however, anecdotal data concerning familial clusterings of childhood leukemia (see Gunz et al. 1978), and the risk of leukemia in siblings of children with leukemia is increased about two- to fourfold (Miller 1971; Draper et al. 1977). Furthermore, concordance rates in identical twins are about 20% (MacMahon and Levy 1964). The observation that in more than half the concordant twin pairs described in the literature the leukemia was diagnosed during the first year of life, whereas overall the peak incidence is between 3 and 5 years of age (Zuelzer and Cox 1969), is consistent with (but does not demand) the hypothesis that the very-early-onset cases reflect a mutational event in one of the parents.

In recent years a variety of approaches has led to the recognition of a second class of genes concerned with carcinogenesis, generically termed proto-oncogenes, which as a result of a mutational event in a somatic cell become the basis for a clonally derived malignancy. A related phenomenon is the implication of precise chromosomal regions, including fragile sites, in the chromosomal breakpoint associated with various clonally derived leukemias, lymphomas, and other malignancies. The number of such proto-oncogenes is thought to be of the order of 50 (reviewed in Bishop 1987; Human Gene Mapping 9 1987), and the list continues to grow, although hard data relating many of those proto-oncogenes to the malignant transformation are still lacking, and there is evidence that, with respect to the childhood malignancies, the number could be relatively limited (Koufos et al. 1985). We are not aware of any clear evidence that a transmitted (germ-line) mutation in a proto-oncogene can be the basis for a childhood malignancy, but, if there are such mutations, in principle they should be no less responsive to radiation than are mutations in tumor-suppressor genes.

Finally, Batra and Sridharan (1964) reported an increase in leukemia persisting over 4 generations in the offspring of radiated mice, but Kohn et al. (1965) saw no increase in leukemia or other tumors under comparable experimental circumstances. More recently, Nomura (1982, 1983, 1986) has reported an increase in the frequency of a variety of adult-onset-type tumors in the offspring of irradiated male mice, an increased frequency which persisted for the several subsequent generations during which the strain was observed. This latter observation created an experimental precedent which in part motivated the present study. Moreover, Shiono et al. (1980) have reported a relative risk of childhood malignancy of 2.61 in the offspring of women who received preconception diagnostic X-ray exposures to the ovaries, the risk being computed from a comparison with matched controls ($P = .021$). An estimated mean ovarian exposure was not given for these women, but it may be presumed to be <0.01 gray (Gy).

Our hypothesis in investigating the possibility of an increase in malignancy in the children of survivors of the atomic bombings was that the proto-oncogenes and tumor-suppressor genes collectively constitute a sufficient target such that the frequency of cancer of relatively early onset might be a suitable indicator of the genetic effects of the atomic bombs, a thesis reinforced by the experimental findings just cited.

Design of Study

Definition of Sample and Sources of Data

The Radiation Effects Research Foundation (RERF), under whose aegis this study was conducted, maintains two sources of information concerning malignant disease in childhood. The first is a study of the survival of the children live born to the survivors of the bombings and to suitable control groups: the F_1 Mortality Study (see Kato et al. 1966; Neel et al. 1974; Schull et al. 1982). For the purposes of this study, three cohorts were identified in 1959, namely, (1) all children live born between 1946 and 1958 to survivors residing in Hiroshima or Nagasaki, one or both of whom were <2,000 m from the hypocenter at the time of the bombing (ATB), i.e., proximally exposed; (2) an equal number of age-, sex-, and city-matched children born either to survivors residing in either city ATB, both of whom were >2,500 m from the hypocenter, or to parents one of whom was >2,500 m (i.e., distally exposed) ATB and the other of whom was not in the city (NIC) ATB; and (3) an equal number of age-, sex-, and city-matched children born in either city to parents who ATB were NIC. Children born prior to May 1, 1946 (i.e., in utero ATB) were excluded from the cohorts. The radiation received by the more distally exposed ATB is generally assumed, on the basis of the current data on the atomic bomb explosions, to have been negligible. Children born to parents whose exposures occurred within the 2,001–2,500-m radius from the hypocenter have not been followed up because, although the gonadal doses are thought to be quite small at this distance, they are difficult to estimate and because the considerable number of children born to such parents would not be very informative with respect to the genetic effects of radiation. The original cohorts have been periodically enlarged by the inclusion of (a) all additional children born to proximally exposed parents between 1959 and 1984 and, again, (b) suitably matched controls.

The three cohorts combined consisted of 76,817 entries. However, for the purposes of this analysis, it was necessary to exclude 799 individuals either because they did not have Japanese citizenship (and so were not entered in the Japanese family record [koseki] which was the basis for follow-up) or, if Japanese, because their survival was uncertain. An additional 3,790 individuals were excluded because of inadequate data on parental exposure, and 12 individuals were excluded because they were born after the cutoff date of 1982 (see below). These deletions plus the repositioning of the hypocenter in Nagasaki, the introduction of new data on radiation exposure, and various other technical factors have considerably altered the originally envisioned 1:1:1 ratio of the three cohorts. The viability of the children in the cohorts is determined on an approximately 3-year cycle by a search of the koseki record to determine whether an individual is alive or dead; if the individual was listed as dead, the cause of death was transcribed from schedules based on death certificates kept in health centers throughout Japan. The last complete and verified cycle of checking death certificates terminated in 1985.

The second source of information is the tumor registries maintained (with assistance from RERF) by the Medical Association of Hiroshima and the Medical Association of Nagasaki, registries dating from 1957 and 1958, respectively (Ikeda et al. 1986; Monzen and Wakabayashi 1986). These registries overlap with the F_1 Mortality Study with respect to children who die of cancer, but the tumor registries are the only source of information on living children with malignancies. It is difficult to estimate the completeness of the tumor registries with respect to cancer morbidity, but, given the various procedures that have been adopted, we believe reporting of recognized malignant tumors in childhood is at least at the 90% level. Because of the cycle on which the registries are maintained, the data on morbidity are thought to be as complete as possible only through 1982. Accordingly, in order to bring the data from death certificates and the data from the tumor registry into temporal concurrence, we have included in this analysis only the registry cases ascertained through 1982 that occurred prior to age 20 years.

Diagnostic Standards

In both the F_1 Mortality Study and the tumor registries, diagnoses have been coded in keeping with successive publications of the World Health Organization's International Classification of Diseases (editions 7–9 plus the International Classification for Cancer of 1976). These coded entries were reviewed independently by two of the authors, and, for all possible or probable diagnoses of malignant tumor, the death certificates and the verifying medical information contained in the files were reexamined. Diagnoses of benign tumors, such as meningiomas, papillomas, lipomas, skeletal exostoses, polyps, or neurofibromas, were specifically excluded from consideration. For 73 individuals there was

histological verification of the diagnosis, but eight of the diagnoses in individuals still alive rest on clinical/operative findings without tissue studies, and another 11 diagnoses in deceased individuals rest only on clinical evidence/death certificates.

De Novo Origin of the Potentially Inherited Malignancies in the Study

Because of the extensive literature on inherited forms of retinoblastoma and Wilms tumor, we have conducted studies of the occurrence of cancer in the 19 nuclear families of individuals with these tumors, as well as with osteosarcoma, renal sarcoma, embryonal carcinoma of the testes, and neuroblastoma. Data are available on 38 parents, 57 siblings, and eight children. There have been seven diagnoses of cancer among the parents (one each of the urinary bladder, maxillary sinus, thyroid, pancreas, esophagus, colon, and liver). No cancer deaths were reported among siblings and children. The possibility of cancer occurring in the living members of these nuclear families was also explored through inquiries to the tumor registries of Hiroshima and Nagasaki, the Japanese National Retinoblastoma Registry, and the Japanese Childhood Malignancy Registry, with no additional reports of malignancies being encountered. Thus all the children in this series who have the above-enumerated malignancies can be considered to result either from mutation in the germ line of the preceding generation plus a somatic cell mutation or from a double somatic cell mutation in their generation.

Dosimetry

The procedures for assigning organ radiation doses to the individual survivors who are the parents of the children in the present study have recently been revised (Roesch 1987), resulting in the Dosimetry System 1986 (DS86). This replaces the previous tentative dosimetry system known as "T65D" and "T65DR" (Milton and Shohoji 1968). DS86 doses have been assigned to both parents of each of 67,574 of the children included in the study. For the parents of the remaining 4,642 children, the recently developed DS86 system cannot be used to assign gonadal doses, either because in the early years of the study the radiation histories were not sufficiently detailed to supply all the parameters necessary for the assignment of a DS86 dose or because of the present inability of the DS86 system to cope with the complex shielding ATB of some of the survivors. These individuals have been assigned gonadal doses on the basis of an empirical conversion from the previously assigned T65DR doses, the conversion factors being derived from persons for whom both T65DR and DS86 doses were available (for detailed description, see Otake et al. 1990). Individuals not in city ATB have, of course, been assigned zero doses. Resorting to this procedure added to the sample a total of 4,265 children whose parents were estimated to have received ≥0.01 Sv (a 15.9% increase in the sample of proximally exposed) but only 377 children (0.9%) whose parents sustained exposures less than this. The discrepancy is of course due to the greater ease in determining dose for distally exposed parents. We will use the term "extended data set" to refer to the sample of children whose parents were assigned either a DS86 dose or an ad hoc dose.

The unit of ionizing radiation is the gray (1 Gy = 100 rad), but, because of the mixed gamma-neutron exposure, gonadal doses will be expressed in sieverts (1 Sv = 100 rem). In calculating sieverts, the relatively small neutron component of the radiation exposure has been assigned an RBE of 20. In accordance with past RERF policy, surface doses (kerma) estimated to be >6 Gy were truncated at that value. Jablon (1971) has placed the total error involved in assigning individual doses based on the previous T65R schedule at ±30%; we assume a similar error for the DS86-assigned doses.

Statistical Procedures

The data were analyzed on the basis of a linear multiple regression model. We treat cancer in an individual as a binary observation, so $Y = 1$ if we observe a cancer in an individual and 0 otherwise. We assumed that Y had an expected function of the following form:

$$E(Y_{ij} \text{ given } x,d) = P_{ij}(x,d) = g_{ij}(x) + b_4 d,$$

where $P_{ij}(x,d)$ is the probability ($0 < P < 1$) of Y_{ij} in an individual with parental dose d and having background characteristics city i ATB, sex j, and x years between the bombing and the birth of the subject. This simple linear model is a first approximation to an exponential curve of the kind described as a "one-hit" model.

It is assumed that the three background characteristics (city, sex, and years between bombing and birth) act additively on the background rate $g_{ij}(x)$. $g_{ij}(x)$ can then be expressed by the function

$$g_{ij}(b_0, b_1, b_2, b_3, x) = b_0 + b_1 I(i) + b_2 I(j) + b_3(x-\bar{x}),$$

where $I(i) = 1$ if i is Hiroshima and $I(j) = 1$ if j is male but both are zero otherwise and where x is equal to the years between bombing and birth (\bar{x}) is the sample

mean of x, 10.5 years). The parameters of the model have been calculated by two different methods, namely, (1) the method of least squares and (2) the method of maximum likelihood for binary observations, where P ranges from 1 to 0 and where combined parental gonad dose is truncated at 7 Sv.

Results of Analysis

The estimated gonadal radiation exposures of the parents of the 67,574 children for whom parental DS86 doses are available are shown in table 1. Among the children of those parents receiving increased amounts of radiation (≥ 0.01 Gy), the conjoint parental gonad exposure for the cities combined averaged 0.002 Gy of neutron radiation and 0.358 Gy of gamma radiation, or 0.405 Sv; but there was marked skewing of the dose curve to the right. There were a total of 83 diagnoses of malignant tumors in the children. Because of the organization of the Japanese family-records system, all deaths from cancer in the children in the present study will have a very high probability of being identified, even if the family concerned has left Hiroshima or Nagasaki subsequent to enrollment in the study. This does not hold true for the registry data, since the cancer registries are limited to persons residing within the limits of the two cities at the time of diagnosis. If out-migration from the two cities during the period of this study were related to radiation exposure, this fact could introduce bias into the study. We have examined this possibility by a comparison of the parental exposures for the children ascertained through death certificates with the exposures when the child was ascertained through registry notification only, by using the radiation exposure categories of table 1 but combining the two highest dose categories (data not shown). There were 49 of the former children and 34 of the latter. The heterogeneity χ^2 with respect to parental gonad doses is 2.481, df = 4, and P = .648. Thus there is no suggestion of a biased ascertainment in children with malignancies known only through the tumor registries.

Table 2 lists the 83 tumors accepted as malignant, categorized by the exposure status of the parents and their city of birth. Among the children of the control (zero-dose) parents, the total incidence of all types of childhood cancer was 1.2/1,000 (6.6/100,000 person-years). Those malignancies that seem to conform most

Table 1

Distribution of Combined Parental Gonad Doses for the Subset of Children Both of Whose Parents Have Been Assigned DS86 Gonadal Doses

City and Parameter	Total	\multicolumn{6}{c}{Joint Parental Dose (Sv, RBE = 20)}					
		0	.01–.09	.10–.49	.50–.99	1.00–2.49	2.50+
Hiroshima:							
No. of children	43,181	25,920	6,597	7,143	1,847	1,273	401
Mean neutron	.001	.0	.000	.001	.004	.013	.049
Mean gamma	.132	.0	.041	.214	.604	1.283	2.847
Total (Sv)	.155	.0	.041	.233	.690	1.546	3.824
Nagasaki:							
No. of children	24,393	14,769	4,045	2,719	1,645	1,022	193
Mean neutron	.001	.0	.000	.000	.002	.005	.018
Mean gamma	.161	.0	.030	.243	.691	1.327	3.425
Total (Sv)	.172	.0	.030	.250	.729	1.423	3.793
Combined:							
No. of children	67,574	40,689	10,642	9,862	3,492	2,295	594
Mean neutron	.001	.0	.000	.001	.003	.009	.039
Mean gamma	.142	.0	.037	.222	.645	1.303	3.035
Total (Sv)	.161	.0	.037	.238	.708	1.491	3.814

NOTE.—We have assigned zero value as a parental dose for those parents who ATB were NIC. Parents whose joint gonadal exposure was estimated to be <4 mSv have been assigned to the zero-dose group. Neutron and gamma doses are expressed in grays, but total mean dose is expressed in sieverts, with the neutron component assigned an RBE of 20. An explanation of the distinction between this sample and the full sample can be found in the text.

Table 2

Number of Malignant Tumors prior to Age 20 Years in Hiroshima and Nagasaki in Children Born in 1946–82, in Relation to Gonadal Dose (DS86) Received by Parents (Sv, RBE = 20)

Site and Tumor	Hiroshima 0	.01–.09	.10–.49	.50+	Nagasaki 0	.01–.09	.10–.49	.50+	Total
Lip, oral cavity, and pharynx:									
Skin sarcoma, sublingual[a]					1				1
Carcinoma, lip					1				1
Basal cell carcinoma, skin	1								1
Digestive organs:									
Adenocarcinoma, stomach					1				1
Leiomyosarcoma, stomach[a]		1							1
Bone and connective tissue:									
Osteosarcoma[b]	1		1						2
Fibrosarcoma[a]	1			1					2
Hemangiopericytoma	1								1
Genitourinary organs:									
Wilms tumor[b]		1		1	1	1	1		5
Sarcoma, kidney[b]		1							1
Embryonal carcinoma, testis[b]		1			1				2
Lymphatic and hematopoietic tissue:									
Leukemia	13	6	3		4	1		4	31
Malignant lymphoma	3	1			1	1			6
Endocrine organs:									
Adenocarcinoma, thyroid	1								1
Nervous system:									
Retinoblastoma[b]	1				3			1	5
Neuroblastoma[b]	2				1				3
Brain									
Unspecified[a]	1								1
Ependymoma[a]	1								1
Meningiosarcoma[a]	1								1
Germinoma[a]					2				2
Medulloblastoma[a]			1				1		2
Astrocytoma[a]	2								2
Glioma[a]		1							1
Glioblastoma multiforme[a]			1						1
Miscellaneous:									
Metastatic malignancy, type and primary site unknown					1				1
Total	29	10	6	4	15	5	2	5	76
Total showing strong evidence for heritability (see text)	4	1	3	1	6	1	1	1	18
Total showing less evidence for heritability (see text)	6	2	0	3	1	2	1	0	15

[a] Less evidence for heritability.
[b] Strong evidence for heritability.

strongly to the retinoblastoma–Wilms tumor model are indicated by a superscript "b" and are termed "heritable"; a superscript "a" indicates those for which the evidence is less convincing. There is probably a continuum in the role to be assigned to inherited factors in the etiology of these various tumors; arbitrarily, however, for the purposes of certain analyses to come, we will, following the earlier discussion, apply the rubric "herita-

ble" only to those indicated by a superscript "b," a conservative procedure. The family studies mentioned earlier were restricted to the latter group.

Table 3 presents the findings based on the application of the multiple regression model to the 67,574 children both of whose parents have been assigned DS86 gonadal exposures; the parameters given are shown both as estimated by the method of least squares and as estimated by the method of maximum likelihood. The data were analyzed both as a whole and after subdivision into three categories, namely, leukemia, heritable (as defined earlier), and others. With neither procedure are there significant findings with respect to radiation dose in either the total or subdivided data. The results of the analyses of the three convenient subsets (leukemia, heritable, and other) are consistent except for the reversal of sign of the regression on dose of "other" in the two analyses, reflecting how very close the regressions are to zero. As table 2 shows, these negative regressions on dose were due primarily to the disproportionate clustering of tumors in children whose parents, although exposed, received less than the average amount of radiation experienced by all exposed parents.

The results of an analysis of the extended data set (see Dosimetry section above) are given in table 4. The tabulations on which this analysis is based are presented by Yoshimoto et al. (in press). In this analysis, although the number of children born to parents receiving no increased radiation remained almost unchanged, the extended set includes 4,265 additional children whose parents received significant amounts of radiation, an increase of 15.9% in the data base. There are nine additional tumors in the extended data set, bringing the total to 92. Now the conjoint parental exposure is 0.435 Sv, again with marked skewing of the dose to the right. Only the results based on the method of least squares are given; the method of maximum likelihood yielded similar results. The results of this analysis are so similar to the results of the earlier analysis that no specific discussion seems indicated; we note only the slightly greater precision of the analysis of the extended sample. For this extended sample, the mean age (if alive) of the members of the sample was 26.1 years; 79.7% of the children in the sample had completed their nineteenth year, the diagnostic cutoff point.

With regard to the other variables, the regression on

Table 3

Results of Linear Multiple Regression Analysis of the Incidence of Cancer Below the Age of 20 Years by Conjoint Parental Dose (DS86, Sv, RBE = 20), City, Sex, and Birth Year

Method and Category	Joint Dose	Hiroshima	Maleness	Years since Birth	Intercept[a]
Least squares:					
Leukemia	−.000033	.000126	.000301	−.000007	.000229
	(.000170)	(.000172)	(.000165)	(.000011)	(.000163)
Heritable[b]	−.000053	−.000153	−.000013	.000005	.000380
	(.000130)	(.000131)	(.000126)	(.000008)	(.000124)
Other	.000005	−.000052	.000270	−.000002	.000397
	(.000178)	(.000180)	(.000173)	(.000011)	(.000171)
All cases	−.000081	−.000079	.000558*	−.000003	.001006
	(.000278)	(.000282)	(.000270)	(.000017)	(.000267)
Maximum likelihood:[c]					
Leukemia[d]			No convergence		
Heritable[b,d]			No convergence		
Other	.000049	−.000021	.000266	−.000002	.000372
	(.000174)	(.000176)	(.000171)	(.000010)	(.000160)
All cases	−.000098	−.000004	.000538*	−.000001	.000971
	(.000230)	(.000276)	(.000268)	(.000017)	(.000251)

NOTE.—Values in parentheses are standard errors for the estimated parameter.
[a] Adjusted for the averaged years between bombing and birth.
[b] As defined in the text.
[c] Within P = 1–0 range (dose is 7 Sv truncated for convergence).
[d] Convergence cannot be obtained with this model with the use of maximum likelihood.
* .01 < p < .05.

Table 4

Results of Linear Multiple Regression Analysis on the Full Sample of the Incidence of Cancer Below the Age of 20 Years, by Conjoint Parental Dose (Sv, RBE = 20), City, Sex, and Birth Year

Category	Joint Dose	Hiroshima	Maleness	Years since Birth	Intercept[a]
Leukemia	−.0000003	−.000075	.000281	−.000007	.000266
	(.000151)	(.000165)	(.000159)	(.000010)	(.000156)
Heritable[b]	−.000073	−.000156	−.000040	.000007	.000395
	(.000114)	(.000125)	(.000121)	(.000008)	(.000118)
Other	−.000009	−.000165	.000361*	.000004	.000473
	(.000166)	(.000181)	(.000175)	(.000011)	(.000171)
All cases	−.000081	−.000247	.000602*	.000004	.001134
	(.000252)	(.000275)	(.000266)	(.000017)	(.000260)

NOTE.—The method of least squares was used. Values in parentheses are standard errors for the estimated parameter.
[a] Adjusted for the number of years between bombing and birth.
[b] As defined in the text.
* $.01 < p < .05$.

city was negative, in keeping with the slightly greater frequency of childhood tumors encountered in Nagasaki. The regression on sex was significantly positive (more males involved), in keeping with other Japanese data (Children's Cancer Association of Japan 1989). None of the analyses of subsets with respect to city, sex, or years between bombing and birth was noteworthy, with the possible exception of both the reversal of sign for the regression on maleness of heritable tumors and a similar reversal, for leukemias (leukemias of earlier onset), for the regression of years between bombing and birth.

Discussion

This failure to demonstrate a statistically significant increase in malignant tumors prior to age 20 years in children born to the survivors of the atomic bombings confirms an earlier, less detailed report by Schull et al. (1982), as well as a still earlier report, restricted to leukemia, by Ishimaru et al. (1981). The mechanism to follow these children into their adult years is now in place, and we anticipate in due time a follow-up to this report, directed at determining whether the pattern of adult-onset tumors is altered among the children of exposed parents.

Inasmuch as ionizing radiation has produced mutations in every properly studied plant and animal species, we accept the premise that there was genetic damage in the parents who were proximally exposed to the atomic bomb explosions. In our various studies of the children of survivors, our ultimate objective has been to take at face value the observed results, statistically significant or not, and to develop an estimate of the sensitivity of the human genome to radiation. An insignificant negative regression of indicator on dose, as in this study, is viewed as a random deviation from some small positive effect. In order to refine the present analysis, we must next deal with the fact that the frequency of only a fraction of the malignancies indicated as heritable should, according to current concepts, be altered by an increased mutation rate in the parents. This fraction provides an improved baseline against which to view the observed regressions.

To return to the retinoblastoma paradigm, we note that on a worldwide basis about 60% of such tumors are now thought to result from somatic cell events only (see Vogel 1979; Knudson 1986); Matsunaga and Minoda (1988) have recently endorsed this figure with reference to the Japanese population. The frequency of tumors due only to somatic cell mutations would not be altered in children by parental radiation. Furthermore, in as many as 5%–10% of the carriers of the appropriate germinal mutation the tumor is not diagnosed. For Wilms tumor the original estimate of the percent associated with a germ-line mutation was 38% (Knudson and Strong 1972), an estimate that was subsequently reduced to ≤10% by Matsunaga and Minoda (1988). However, Li et al. (1987) found no cases of either Wilms tumor or any other cancer among the 155

offspring surviving the neonatal period who were born to 99 patients who had been successfully treated for Wilms tumor, most of which patients had received abdominal radiotherapy in addition to surgery. Since these patients had unilateral disease, it may be presumed that, in the great preponderance, the malignancy was not associated with a germ-line mutation. The median age of the children was 6 years. Recently it has been suggested that two or three different genetic loci are involved in the Wilms tumor phenotype (Grundy et al. 1988; Huff et al. 1988). Data on genetic transmission are still quite scanty for the other heritable tumors. Familial cases of neuroblastoma are very rare, and only two have been reported from Japan (Kaneko and Sawaguchi 1988). Then, although some 20% of the tumors in this series may conform in the most general sense to the retinoblastoma model (our "b"-superscripted tumors in table 2), only a minority of those conforming to this model have a genetic basis that should reflect, in the next generation, any change in the parental mutation rate. That the transmissibility of the tumors indicated by a superscript "a" in table 2 is even less (<1%) is indicated both by the literature survey of Mulvihill and Byrne (1985) and by the multicenter study of Mulvihill et al. (1987).

Any attempt to generalize from these studies must, however, consider that these survivors may not be typical of the spectrum of the childhood malignancies. For instance, familial retinoblastoma and Wilms tumor are much more often bilateral (i.e., multifocal) than are sporadic cases and confer a greater risk of death (or failure to marry and reproduce) than do their nonfamilial counterparts. Nevertheless, although many of the children in the Mulvihill series were still quite young at the time of the reports, so that the ultimate estimate of cancer incidence below the age of 20 years will undoubtedly be greater, these empirical data suggest that a transmissible germinal mutation is involved in a fraction of the childhood tumors that is much smaller than the figures mentioned above for retinoblastoma and Wilms tumor. The sib recurrence risks lead to a similar conclusion (see Leck 1977).

Among the children of unexposed parents, the representation of tumors designated as heritable and indicated by a superscript "b" in table 2 is 21.6%. An additional 18.1% (indicated by a superscript "a") may ultimately be shown to conform to this model, but, as noted, the fraction of these that is associated with a germ-line mutation does not appear to be nearly as high as that for retinoblastoma and Wilms tumor. This is admittedly a small sample on which to base such a breakdown, but it is the only Japanese sample of "early-onset" cancers known to us that is based on both the morbidity and mortality data of defined cohorts. On the basis of the foregoing review, we conclude that approximately 20% of the tumors in this series conform to the retinoblastoma model, of which only 10%–20% are associated with a germ-line mutation in the parents. The corresponding percentage for the "less heritable" tumors cannot be >5%. This suggests that, in the control series, 3%–5% of all tumors are associated with a germ-line mutation. The approximate character of this calculation is clear, but it provides a rough perspective.

Returning now to the previous analysis, we can express the findings somewhat more concretely. The estimate of 3%–5% developed in the preceding paragraph suggests that, of the cancers observed in the control series in this study, roughly one or two are heritable in the sense of involving a germ-line mutation in a "tumor suppressor" gene and that, in the children of the exposed, the expectation in the absence of a radiation effect is very similar. Even if the additional 18.1% of tumors indicated by a superscript "a" in table 2 were clearly shown to conform to the retinoblastoma model, this demonstration would add only one or two tumors to the germ-line-associated baseline. Only two or three additional heritable tumors of the germ-line type in the children of proximally exposed would constitute a doubling. Given the role of the stochastic process in such small expectations, where the chance occurrence (placement) of one or two malignancies can alter the sign of the regression in either direction, the regression's closeness to zero is noteworthy. As noted earlier, the importance of *germ-line* mutations in proto-oncogenes in childhood cancer remains unclear, but, whatever that role, the impact that oncogene response to radiation has on live-born children was also implicitly examined in the foregoing analysis.

In view of these findings, it seems appropriate to return to a more detailed consideration of the murine data that were one of the compelling factors in this examination of tumor mortality and morbidity in these children. From Nomura's (1982) data, derived from acutely administered X-ray exposures of 36–504 rad, it can be estimated that, in the offspring of male mice receiving 150–200 rad to germ cells in the spermatogonial stage, the frequency of *all* tumors was approximately twofold greater than "normal." (Traditional radiation units are retained if they appear in the work being cited. In all current RERF reports, the International System of Units is employed.) From the results of enhancing the fre-

quency of tumors by postnatal treatment of the F_1 with urethane, Nomura (1983) has suggested that for this endpoint the "genetic doubling dose" of spermatogonial radiation should be placed at 50 rad. Different tumors appear to exhibit different doubling doses. These tumors in mice occurred over a period of 8 mo, a period which, in terms of life cycle, is proportionately longer than the restricted (human) age span in the present study. A precise comparison of the results of the two studies is therefore inappropriate, but it does appear that, if the totality of the human tumors encompassed by this study responded as the totality of the mouse tumors, we might already expect an effect greater than that observed, since the average conjoint gonadal exposure in the proximally exposed parents can be estimated to be 0.435 Sv. Given, however, that most of the tumors reported by Nomura are adult-onset tumors in humans, a definitive comparison of the present findings with the murine data will not be possible for some years to come.

There are, however, aspects of the Nomura data which render their relevance to the human situation moot. In his experiments with the ICR strain of mice, 87% of the tumors in the offspring of treated mice were scored as papillary adenomas of the lung. In the experiments with LT and N5 mice, 16.0% and 21.0%, respectively, of tumors were pulmonary, 25.3% and 22.8%, respectively, were ovarian tumors, and 5.3% and 3.9%, respectively, were leukemias. These endpoints had a relatively high frequency in the controls. Transplantation experiments involving 26 tumors suggested that 88% of the tumors were malignant (Nomura 1986). As Nomura points out, the predominance of a single or several tumor types in his data leads to the suspicion of a strain-specific effect which, given our previous analysis of the genetic basis of human childhood malignancies, in our opinion necessitates great caution in extrapolation from the mouse paradigm to the human situation.

Finally, we point out that the base for Nomura's doubling-dose calculations appears to be *all* the tumors of any specific type in the control animals. To the extent that some of the tumors in the mouse model are due to somatic mutation only, this biases the estimate of the doubling dose, i.e., if, as seems likely, some of the spontaneous tumors in the F_1 mice are entirely due to somatic cell events, their inclusion in a doubling-dose estimate tends to bias it to a higher absolute value. The correct calculation of the mouse doubling dose should yield a figure lower than that quoted above, thus further intensifying the discrepancy at this point between the mouse and the human data.

Our results are incompatible with the previously mentioned findings of Shiono et al. (1980) on 40 malignant tumors of childhood, especially when we consider that the relative risk of 2.61 that they derived applied to *all* tumors and not just to the fraction presumed to be related to a germ-line mutation in the mother. A correct treatment of their data would imply a genetic doubling dose for this indicator <0.01 Gy, far below any other genetic doubling-dose estimate based on either human or murine data.

Acknowledgments

We are grateful for the cooperation of Dr. Kensei Minoda, representing the Japanese National Retinoblastoma Registry, and of Dr. Nobaru Kobayashi, representing the Japanese Childhood Malignancy Registry, in the family studies. The investigation was supported by the Department of Energy.

References

Batra BK, Sridharan BN (1964) A study of the progeny of mice descended from x-irradiated females with special reference to the gonads. Acta Union Intern Contra Cancrum 20:1181–1186

Bishop JM (1987) The molecular genetics of cancer. Science 235:305–311

Bookstein R, Lee EY, To H, Young LJ, Sery TW, Hayes RC, Friedmann T, et al (1988) Human retinoblastoma susceptibility gene: genomic organization and analysis of heterozygous intragenic deletion mutants. Proc Natl Acad Sci USA 85:2210–2214

Bundey S, Evans K (1982) Survivors of neuroblastoma and ganglioneuroma and their families. J Med Genet 19:16–21

Cavenee WK, Dryja TP, Phillips RA, Benedict WF, Godbout R, Gallie BL, Murphree AL, et al (1983) Expression of recessive alleles by chromosomal mechanisms in retinoblastoma. Nature 305:779–784

Children's Cancer Association of Japan (1989) Japan Children's Cancer Registry, vol 3 (1979–1983) [in Japanese]. Children's Cancer Association of Japan, Tokyo

Dao DD, Schroeder WT, Chai L-Y, Kikuchi H, Strong LC, Riccardi VM, Pathak S, et al (1987) Genetic mechanisms of tumor-specific loss of 11p DNA sequences in Wilms tumor. Am J Hum Genet 41:202–217

Draper GJ, Heaf MM, Kinnear Wilson LM (1977) Occurrence of childhood cancers among sibs and estimation of familial risks. J Med Genet 14:81–90

Friend SH, Bernards R, Rogelj S, Weinberg RA, Rapaport M, Albert DM, Dryja TP (1986) A human DNA segment with properties of the gene that predisposes to retinoblastoma and osteosarcoma. Nature 323:643–646

Friend SH, Horowitz JM, Gerber MR, Wang XF, Bogenmann E, Li FP, Weinberg RA (1987) Deletions of a DNA sequence

in retino-blastoma and mesenchymal tumors: organization of the sequence and its encoded protein. Proc Natl Acad Sci USA 84:9059–9063

Grundy P, Koufos A, Morgan K, Li FP, Meadows AT, Cavenee WK (1988) Familial predisposition to Wilms' tumour does not map to the short arm of chromosome 11. Nature 336: 374–376

Gunz FW, Gunz JP, Vincent PC, Bergin M, Johnson FL, Bashir H, Kirk RL (1978) Thirteen cases of leukemia in a family. J Natl Cancer Inst 60:1243–1250

Hartley AL, Birch JM, Marsden HB, Harris M (1986) Breast cancer in mothers of children with osteosarcoma and chondrosarcoma. Br J Cancer 54:819–823

Huff V, Compton DA, Chao LY, Strong LC, Geiser CF, Saunders GF (1988) Lack of linkage of familial Wilms' tumour to chromosomal band 11 p 13. Nature 336:377–378

Human Gene Mapping 9 (1987) Cytogenet Cell Genet 46: 1–357

Ikeda T, Hayashi I, Matsuo T, Maeda T, Shimokawa I (1986) The cancer registry in Nagasaki City, with atomic bomb survivor data, 1973–1977. In: Shigematsu I, Kagan A (eds) Cancer in atomic bomb survivors. Gann Monogr Cancer Res 32. Plenum, New York, pp 41–52

Ishimaru T, Ichimaru M, Mikami M (1981) Leukemia incidence among individuals exposed in utero, children of atomic bomb survivors, and their controls: Hiroshima and Nagasaki, 1945–79. Tech rep 11–81. Radiation Effects Research Foundation, Hiroshima

Jablon S (1971) Atomic bomb radiation dose estimation at ABCC. Tech rep 23–71. Atomic Bomb Casualty Commission, Hiroshima

Kaneko M, Sawaguchi S (1988) Genetic aspect of neuroblastoma: epidemiology, familial neuroblastoma, and cytogenetic studies. In: Takabe H, Utsunomiya J (eds) Genetics of human tumors in Japan. Gann Monogr Cancer Res 35. Japan Scientific Press, Tokyo, pp 63–69

Kato H, Schull WJ, Neel JV (1966) A cohort-type study of survival in the children of parents exposed to atomic bombings. Am J Hum Genet 18:339–373

Kitchin FD, Ellsworth RM (1974) Pleiotropic effects of the gene for retinoblastoma. J Med Genet 11:244–246

Knudson AG (1971) Mutation and cancer: statistical study of retinoblastoma. Proc Natl Acad Sci USA 68:820–823

——— (1986) Genetics of human cancer. Annu Rev Genet 20:231–251

Knudson AG, Meadows AT (1978) Developmental genetics of neural tumors in man. In: Saunders GF (ed) Cell differentiation and neoplasia. Raven, New York, pp 83–92

Knudson AG, Strong LC (1972) Mutation and cancer: a model for Wilms' tumor of the kidney. J Natl Cancer Inst 48: 313–324

Kohn HL, Epling ML, Guttman PH, Bailey DW (1965) Effect of paternal (spermatogonial) X-ray exposure in the mouse: lifespan, X-ray tolerance, and tumor incidence of the progeny. Radiat Res 25:423–434

Koufos A, Hansen MF, Copeland NG, Jenkins NA, Lampkin BC, Cavenee WK (1985) Loss of heterozygosity in three embryonal tumors suggests a common pathogenetic mechanism. Nature 316:330–334

Leck I (1977) Congenital malformations and childhood neoplasms. J Med Genet 14:321–326

Lee EYPH, To H, Shew JY, Bookstein R, Scully P, Lee WH (1988) Inactivation of the retinoblastoma susceptibility gene in human breast cancers. Science 241:218–220

Lee WH, Bookstein R, Hong F, Young LJ, Shew JY, Lee E (1987) Human retinoblastoma susceptibility gene: cloning, identification, and sequence. Science 235:1394–1399

Li FP, Fraumeni JF, Mulvihill JJ, Blattner WA, Dreyfus MG, Tucker MA, Miller RW (1988) A cancer family syndrome in twenty-four kindreds. Cancer Res 48:5358–5362

Li EP, Gimbrere K, Gelber RD, Sallam SE, Flamant F, Green DM, Heyn RM, et al (1987) Outcome of pregnancy in survivors of Wilms' tumor. JAMA 257:216–219

MacMahon B, Levy MA (1964) Prenatal origin of childhood leukemia: evidence from twins. N Engl J Med 270:1082–1085

Matsunaga E (1981) Genetics of Wilms' tumor. Hum Genet 57:231–246

Matsunaga E, Minoda K (1988) Use of retinoblastoma and Wilms' tumor as sentinel phenotypes for population surveillance. In: Takabe H, Utsunomiya J (eds) Genetics of human tumors in Japan. Gann Monogr Cancer Res 35. Japan Scientific Societies Press, Tokyo, pp 127–133

Miller RW (1971) Deaths from childhood leukemia and solid tumors among twins and other sibs in the United States, 1960–1967. J Natl Cancer Inst 46:203–209

Milton RC, Shohoji T (1968) Tentative 1965 radiation dose (T65D) estimation for atomic bomb survivors, Hiroshima-Nagasaki. Tech rep 1–68. Atomic Bomb Casualty Commission, Hiroshima

Monzen T, Wakabayashi T (1986) Tumor and tissue registries in Hiroshima and Nagasaki. In: Shigematsu I, Kagan A (eds) Gann Monogr Cancer Res 32. Plenum, New York, pp 29–40

Mulvihill JJ, Byrne J (1985) Offspring of long-time survivors of childhood cancer. Clin Oncol 4:333–334

Mulvihill JJ, Connelly RR, Austin DF, Cook JW, Holmes FF, Krauss MR, Meigs JW, et al (1987) Cancer in offspring of long-term survivors of childhood and adolescent cancer. Lancet 2:813–817

Neel JV, Kato H, Schull WJ (1974) Mortality in the children of atomic bomb survivors and controls. Genetics 76:311–326

Nomura T (1982) Parental exposure to X-rays and chemicals induces heritable tumors and anomalies in mice. Nature 296:575–577

——— (1983) X-ray-induced germ-line mutation leading to tumours: its manifestation in mice given urethane postnatally. Mutat Res 121:59–65

——— (1986) Further studies on X-ray and chemically in-

duced germ-line alterations causing tumors and malformations in mice. In: Ramel C, Lambert B, Magnusson J (eds) Genetic toxicology of environmental chemicals. Part B: Genetic effects and applied mutagenesis. Alan R Liss, New York, pp 13-20

Otake M, Schull WJ, Neel JV (1990) Congenital malformations, stillbirths, and early mortality among the children of atomic bomb survivors: a reanalysis. Radiat Res 122:1-11

Roesch WC (ed) (1987) U.S.-Japan reassessment of atomic bomb radiation dosimetry in Hiroshima and Nagasaki. Vol 1. Radiation Effects Research Foundation, Hiroshima

Schimke RN (1985) Dominant inheritance in human cancer. In: Chaganti RSK, German JL (eds) Genetics in clinical oncology. Oxford University Press, New York, pp 103-121

Schull WJ, Neel JV, Otake M, Awa AA, Satoh C, Hamilton HB (1982) Hiroshima and Nagasaki: three and a half decades of genetic screening. In: Sugizura T, Kondo S, Takebe H (eds) Environmental mutagens and carcinogens. Alan R Liss, New York, pp 687-700

Shiono PH, Chung CS, Myrianthopoulos NC (1980) Preconception radiation, intrauterine diagnostic radiation, and childhood neoplasia. J Natl Cancer Inst 65:861-868

Steinberg A (1960) The genetics of acute leukemia in children. Cancer 13:985-999

Videbaek A (1947) Heredity in human leukemia and its relation to cancer. Opera ex Domo Biol Hereditariae Humanae Univ Hafmiensis. Vol 13. Nyt Nordisk, Arnold Busck, Copenhagen

Vogel F (1979) Genetics of retinoblastoma. Hum Genet 52: 1-54

Yoshimoto Y, Neel JV, Schull WJ, Kato H, Soda M, Eto R, Mabuchi K. The frequency of malignant tumors during the first two decades of life in the offspring (F_1) of atomic bomb survivors. Tech rep 4-90. Radiation Effects Research Foundation, Hiroshima (in press)

Zuelzer WW, Cox DE (1969) Genetic aspects of leukemia. Sem Hematol 6:228-249

RERF TR 1-91
Research Project 4-75

Technical Report Series

Mortality among the Offspring (F₁) of Atomic Bomb Survivors, 1946–85

Yasuhiko Yoshimoto, Ph.D.; William J. Schull, Ph.D.;
Hiroo Kato, M.D.; James V. Neel, Ph.D., M.D., Sc.D.

放影研 Radiation Effects Research Foundation
RERF A Cooperative Japan–United States Research Organization

Reproduced, with permission, from the *Technical Report Series*, Radiation Effects Research Foundation, Hiroshima and Nagasaki, Japan.

THE CHILDREN OF ATOMIC BOMB SURVIVORS

Technical Report Series

RERF TR 1-91
Research Project 4-75

Mortality among the Offspring (F₁) of Atomic Bomb Survivors, 1946–85[§]

Yasuhiko Yoshimoto, Ph.D.[1]; William J. Schull, Ph.D.[2]; Hiroo Kato, M.D.[3]; James V. Neel, Ph.D., M.D., Sc.D.[4]

[1]*RERF Department of Epidemiology;* [2]*RERF Permanent Director;* [3]*RERF Consultant and Director-General, National Institute for Minamata Disease;* [4]*RERF Consultant Emeritus and Department of Human Genetics, University of Michigan Medical School*

Summary

We compare deaths occurring in the years 1946–85 in a cohort of 31,159 children born to parents one or both of whom were exposed to the atomic bombing of Hiroshima or Nagasaki, and received a combined (i.e., joint) gonadal dose of 0.01 Sv or more, with deaths in a comparable control group, totaling 41,069 children. The average combined gonadal dose equivalent for the exposed parents was 0.435 Sv. Gonadal doses were calculated using the recently established DS86 system, supplemented by an ad hoc system for those children for whom a DS86 dose could not be computed for one or both parents. At the end of 1985, those members of the study groups born in 1946 had reached 39 years of age, whereas those born in the years 1966 through 1984 had not yet reached their 20th birthday. The mean age of living members of the cohorts was 28.8 years.

When a linear relative risk model is fitted to the data, no statistically significant increase in the risk of mortality attributable to diseases other than neoplasms is noted following parental exposure, the excess relative risk being 0.030 (±0.046) per sievert based on the subset of individuals with DS86 doses, assuming the RBE of neutrons to be 20. For fatal cancer, in confirmation of an earlier report on cancer incidence below the age of 20 in this same group, again no statistically significant effect was observed as parental radiation dose increased. Finally, although the present method of analysis using Poisson regression and person-years at risk of death seems more appropriate now, particularly as the cohort ages—since earlier analyses of mortality in the F1 cohort have been based on a simple linear regression of the frequency of death on parental dose—this model was also fitted to the data used in the relative risk estimate to provide some continuity with the past. The results give an intercept of 0.0420 (±0.0015) and a linear regression coefficient of 0.00169 (±0.00157) per sievert. This leads to the calculation of a (statistically nonsignificant) excess relative risk of 0.040, in good agreement with the excess

[§]The complete text of this report will not be available in Japanese.

Approved 28 January 1991 Printed February 1991

obtained by fitting the relative risk model. An analysis based on the full sample, using not only the DS86 dose group but also the ad hoc dose group, yields essentially the same result as the analysis restricted to the DS86 dose group.

Introduction

An extensive literature on the genetic effects of exposure to ionizing radiation in experimental situations supports the belief that exposure of human germ cells to ionizing radiation will result in deleterious mutations that may lead to disease, disability, or premature death.[1,2] The collection of the epidemiological evidence which will permit a precise estimate of the extent and nature of the genetic effects to be anticipated in human populations has proceeded much more slowly than the experimental studies. The principal sources of such evidence are the cohorts of children born to survivors of the atomic bombings (A-bombings) of Hiroshima and Nagasaki, who for some 40 years have been the subjects of an investigation into the possible genetic effects attributable to the exposure of human populations to ionizing radiation.

These investigations began in 1948 and in the succeeding six years some 70,000 pregnancies occurring in Hiroshima and Nagasaki were studied with respect to congenital malformations, stillbirths, and deaths in the first 14 days following parturition.[3,4] In addition to these early clinical studies, this series of investigations later assessed the frequency of biochemical mutations,[5,6] cytogenetic abnormalities,[7] and survival[8-10] among the children of A-bomb survivors and of controls, collectively referred to as the F_1. Recently the incidence of cancer with onset prior to age 20 among the children of survivors has also been investigated.[11] Each of these studies has failed to show a significant genetic effect of exposure to A-bomb radiation with the single exception of an early report[12] of changes in the sex ratio of the children of exposed survivors, a report not confirmed by later studies.[13] However, the interpretation of the biological significance of the early changes observed in the sex ratio among these children has been complicated by later discoveries of the Lyonization phenomenon (mosaic inactivation of one X-chromosome in females) and the occurrence of sex chromosomal aneuploidies.[13,14]

In 1990, Gardner et al.[15] reported an increased frequency of leukemia in children of fathers employed at the Sellafield nuclear reprocessing facility in England, an increase which had not been observed in an earlier Japanese study.[16] This report has prompted a renewed interest in the genetic effects of radiation, and in particular, in the risk of cancer among the F_1 children. In this paper, we focus on updating the earlier studies on mortality among liveborn infants, the data now encompassing the years 1946–1985. This is the first analysis of these data based on the revised estimates of the A-bomb survivors' radiation exposures which became available in 1986 (DS86). With particular reference to the Gardner report, this paper contains data on death from malignancies, including leukemia, not available in the earlier studies on A-bomb effects.[11,16]

Design of the Study

The study of the survival of children born alive to survivors of the A-bombings uses as the basis for death ascertainment the obligatory family registries (*koseki*) that exist in Japan. Some characteristics of these children and their parents obtained during the early clinical study and pertinent to the present study will now be reviewed.

The population

The Radiation Effects Research Foundation (previously the Atomic Bomb Casualty Commission [ABCC]) continually surveys the survival of a cohort of children born to the survivors of the A-bombings of Hiroshima and Nagasaki. The population to be examined in the present analysis consists of 72,228 children born alive from 1 May 1946 through December 1984, specifically 31,159 children one or both of whose parents received gonadal radiation ≥0.01 Sv and a control group of 41,069 children whose parents received <0.01 Gy gonadal radiation at the time of the bombing (ATB) (Table 1). Two criteria were used to select these individuals from the larger F_1 mortality sample, consisting of 76,817 persons. First, the current survival status of a sample member had to be established; this excluded 799 individuals because no *koseki* records existed (individuals of foreign nationality, such as Koreans, do not normally have such records), or because the survival status could not be determined. Second, sufficient information had to exist to assign a dose estimate for both parents of an individual; this excluded 3,790 subjects.

Originally, the F_1 mortality study consisted of three age-, sex-, and city-matched cohorts born between May 1946 and December 1958, namely, 1) all children born to survivors residing in Hiroshima or Nagasaki, one or both of whom were within 2,000 m of the hypocenter ATB (the so-called proximally exposed); 2) children born to parents from either city ATB both of whom had been >2,500 m from the hypocenter, or born to one parent who had been >2,500 m from the hypocenter with the other parent not in the city (NIC) ATB (the distally exposed); and 3) children born in either city to parents who were NIC ATB.[8] Collectively these three cohorts now consist of 50,529 individuals.

Through examining the pertinent family registers, the original study sample was subsequently enlarged to include births between 1959 and 1984 to parents in the three original groups—proximally exposed, distally exposed, or NIC—described above, and to parents who would have qualified for inclusion in these groups had they been old enough to have had children prior to 1959 (this addition is known as the F_1 mortality sample extension [F_1-Ext], and consists of 21,699 subjects, see Table 1). The exposed parents of these individuals were 0–34 years old ATB. Of these individuals, 10,811 had parents one or both of whom were exposed ATB to a total kerma dose of 0.10 Gy or more based on the previously employed T65D[17] dose schedule. The remainder (10,888 individuals) were born to parents who were exposed to doses of 0.09 Gy or less based on T65D total kerma dose; no children of

parents both of whom were NIC ATB were included in this F_1-Ext sample. Births to A-bomb survivors have decreased markedly with time and in 1984 only three such births occurred; thus, the original and extended samples embrace most, if not all, of the children to be born in these cities to the survivors proximally exposed to the A-bombings.

Dosimetry

The estimated parental doses used in the present study are based on the revised dosimetry system known as the DS86.[17] At present, combined (i.e., joint) parental gonadal doses based on this system have been assigned for 67,586 (94%) of the 72,228 children (Table 1). Among the remaining 6%, the principal reasons for not directly assigning a parental DS86 gonadal dose are either a lack of detailed information on the individual's shielding ATB or the present inability of the DS86 system to cope with the complex shielding of some of the survivors. These individuals had, however, been assigned a gonadal dose under an earlier system known as T65D. Since a disproportionate number of these children were born in the early years of the study, it was felt to be important to include them in the analysis. Accordingly, for the present analysis they were assigned a gonadal dose on the basis of an empirical conversion factor derived from persons for whom both T65D and DS86 doses were available (the details are described elsewhere).[4] Individuals NIC ATB have, of course, been automatically assigned zero dose. All analyses reported in this paper are based on the combined (joint) gonadal doses of father and mother. Since, as will be shown, none of the findings approach statistical significance, separate analyses for maternal and paternal effects were not undertaken.

Table 1. Number of subjects by city, sex, sample class, and parental gonadal dose equivalent categories. The neutron component of the radiation has been assigned an RBE of 20.

Group	F_1 mortality sample	
	Sample of DS86 dose estimates	Full sample
	Total 67586	72228
Parental gonadal dose	0 Sv = 40692	41069
	≥0.01 Sv = 26894	31159
City	Hiroshima = 43188	44852
	Nagasaki = 24398	27376
Sex	Male = 34684	37044
	Female = 32902	35184
Major sample	Original = 47330	50529
	F_1-Ext = 20256	21699

The unit of absorbed ionizing radiation is the gray, but because of the mixed gamma–neutron exposure, gonadal doses are expressed in sieverts. In calculating sieverts, the relatively small neutron component of the radiation exposure has been assigned an RBE of 20. The International Commission on Radiation Units and Measurements in its assessment of maximum RBE values at low doses, for fission neutrons versus gamma rays, has found a range of 10–45 for genetic endpoints in mammalian systems.[18] Likewise, the Committee on the Biological Effects of Ionizing Radiation (BEIR-V) has suggested a value of about 5 for high dose rate comparisons and a range of 15–50 for low dose rate studies.[2] We believe an RBE of 20 to be reasonable considering the uncertainty in current estimates. As we will subsequently show, the exact RBE chosen exerts only a minor influence on the results of our analysis, because the DS86 gonadal neutron dose is very small.

The mean combined parental gonadal dose equivalents are shown in Table 2. In the ≥0.01 Sv dose group, the combined parental gonadal dose equivalent averages 0.405 Sv (0.047 Sv of neutron and 0.358 Sv of gamma) for the subset of children both of whose parents have been assigned DS86 doses. Of the 4,642 children whose parents lacked DS86 doses, there were 4,265 (91.9%) for whom parental combined gonadal doses were ≥0.01 Sv under the ad hoc dosage system. The average combined gonadal dose for the full sample (DS86 plus ad hoc doses) with doses of ≥0.01 Sv is 0.435 Sv, slightly higher than for the subset on whom formal DS86 doses exist.

Characteristics of the study population
1. Year of birth, parental age, and the sex of the sample subjects. Year of birth is available on all of the sample subjects, since a known birth date was necessary for the inclusion of an individual in the study group. As a result of the early clinical study, maternal and paternal ages are available for 67.6% (48,838) and 54.8% (39,603) of the 72,228 children, respectively. Mean maternal age at the birth of these children was 28.2 years (range: 15–53), and mean paternal age was 32.6 years (range: 17–71). Mean parental age ATB in the F_1-Ext sample of course differs from that in the F_1 original sample. In the F_1 original sample, the average ages ATB are 28.3 (range: 7–65) for the father and 22.1 years (range: 5–46) for the mother if he or she received 0.01 Sv or more. However, since the extension sample involves a disproportionate number of survivors who were too young to have had a child prior to 1959, the average ages ATB in this group are 10.2 years (range: 0–42) for the father and 7.7 (range: 0–31) for the mother.

Given the importance of the sex of an individual in determining mortality rates, particularly in the early years of life, we have examined the ratio of male births to female births among the F_1 children to determine whether a sex bias might exist as a result of the addition of the F_1-Ext sample and the use of the DS86 doses.

RERF TR 1-91

Table 2. Distribution of combined parental gonadal dose equivalents and mean doses in dose categories.

City	Item	Total A[b]	Total B[b]	0[a] A	0[a] B	0.01–0.09 A	0.01–0.09 B	0.10–0.49 A	0.10–0.49 B	0.50–0.99 A	0.50–0.99 B	1.00–2.49 A	1.00–2.49 B	≥2.50 A	≥2.50 B	≥0.01 A	≥0.01 B
Hiroshima	Number	43188	44852	25922	26007	6601	7001	7143	7835	1848	2002	1273	1467	401	540	17266	18845
	Neutron	0.001	0.001	0.0	0.0	0.000	0.000	0.001	0.001	0.004	0.004	0.013	0.013	0.049	0.050	0.003	0.003
	Gamma	0.132	0.147	0.0	0.0	0.041	0.041	0.214	0.212	0.604	0.605	1.283	1.288	2.847	2.871	0.329	0.350
	Total dose equivalent	0.155	0.175	0.0	0.0	0.041	0.041	0.233	0.231	0.690	0.692	1.546	1.554	3.824	3.866	0.388	0.417
Nagasaki	Number	24398	27376	14770	15062	4047	4641	2720	3701	1646	2362	1022	1300	193	310	9628	12314
	Neutron	0.001	0.001	0.0	0.0	0.000	0.000	0.0004	0.0004	0.002	0.002	0.005	0.005	0.018	0.016	0.001	0.001
	Gamma	0.161	0.196	0.0	0.0	0.030	0.031	0.243	0.249	0.691	0.678	1.327	1.320	3.425	3.181	0.409	0.436
	Total dose equivalent	0.172	0.209	0.0	0.0	0.030	0.031	0.250	0.257	0.729	0.714	1.423	1.416	3.793	3.510	0.435	0.464
Combined	Number	67586	72228	40692	41069	10648	11642	9863	11536	3494	4364	2295	2767	594	850	26894	31159
	Neutron	0.001	0.001	0.0	0.0	0.000	0.000	0.001	0.001	0.003	0.003	0.009	0.009	0.039	0.038	0.002	0.003
	Gamma	0.142	0.166	0.0	0.0	0.037	0.037	0.222	0.224	0.645	0.644	1.303	1.303	3.035	2.984	0.358	0.384
	Total dose equivalent	0.161	0.188	0.0	0.0	0.037	0.037	0.238	0.240	0.708	0.704	1.491	1.489	3.814	3.736	0.405	0.435

NOTE: Neutron and gamma doses are expressed in gray, but total mean dose is expressed in sieverts with the neutron component assigned an RBE of 20.
[a] We have assigned zero value as a parental dose for those parents who were not in the city at the time of the bombing. Parents whose joint gonadal doses were estimated to be less than 4 mSv have been assigned to the zero dose group.
[b] A is the subset of children both of whose parents have been assigned DS86 doses. B is the full sample.

Because of the procedure for the sample selection, for children born during 1946–58 the proportion of males among the F_1 children has to be examined in the children of proximally exposed parents whereas for the F_1-Ext sample the data are for the group of children born to parents exposed to a dose of 0.10 Gy or more total kerma as calculated by the T65D dose schedule. In these samples there is no significant change in the proportion of males among the F_1 children in relation to parental dose (Table 3), and thus no evidence that the mortality surveillance sample is biased toward one sex or the other.

Parenthetically, it should be noted that the analysis presented in Table 3 using the DS86 dose schedule confirms the results of an earlier analysis based on the T65D dose schedule.[13] The least ambiguous sex ratio data are based on maternal exposure. In this case, the induction of sex-linked recessive mutations should decrease the proportion of male births. We note, however, that for the total data the regression term is (insignificantly) positive, i.e., opposite in direction from the negative value expected theoretically.

2. Other factors. Information is available on parity, duration of pregnancy, and birth weight for those 41,649 of the 72,228 (57.7%) children who were born between 1948 and 1958, since these data were routinely obtained at the time of the early clinical study.[3] Because of the correlation between duration of pregnancy and birth weight ($r = 0.342$), only the relatively uncorrelated distributions of birth weight and parity ($r = 0.117$) are shown in Figure 1. The chi-square statistics for tests of the homogeneity of the distributions of these variables among parental dose groups are 27.07 ($df = 15$, $p = .03$) for birth weight and 66.56 ($df = 20$, $p < .01$) for parity. Although there is no simple relationship between these two variables and parental dose, the statistical significance apparently results primarily from a higher frequency of immature infants (with birth weights of <2.5 kg) in the ≥1.00 Sv dose group than in the 0 Sv dose group.

A qualitative economic assessment of the parental home was made by the Japanese physician on a specified subset of pregnancy terminations at the time of the home visit examination in the years of the clinical program (1948–1954);[3] this assessment is available on approximately 6.0% (4,348) of the total of 72,228 individuals. The proportion of children with socioeconomic scores of "very poor" or "poor" in the ≥0.01 Sv dose categories is 12.7% (133/1,045), slightly higher than the 10.9% (359/3,303) noted in the 0 Sv dose group, but not significantly so. In 1964, a questionnaire was sent to the parents of about 10% of the living children in the original F_1 mortality sample to obtain socioeconomic information. Based on this survey, Kato et al.[9] noted that "In general, these differences in distribution of extraneous socioeconomic variables, ... lead to an inflation of mortality rates in the children of the heavily exposed group. ..."

Table 3. Sex ratio by year of birth for the subset of children one or both of whose parents were exposed within 2 km from the hypocenter. Parental gonadal dose equivalents are based on DS86. The neutron component of the radiation has been assigned an RBE of 20.

Year of birth		Father only exposed (Sv)					Mother only exposed (Sv)				
		0.01–0.09	0.10–0.49	0.50–0.99	1.00–2.49	≥2.50	0.01–0.09	0.10–0.49	0.50–0.99	1.00–2.49	≥2.50
Original F₁[a]											
1946–47	Male/total	62/125	61/125	27/53	20/39	2/4	183/356	203/386	61/120	35/65	4/7
	% of Males	49.6	48.8	50.9	51.3	50.0	51.4	52.6	50.8	53.8	57.1
			$P = 0.4942 + 0.0068\,d^{c}$					$P = 0.5181 + 0.0068\,d$			
			(0.0444)					(0.0346)			
1948–53	Male/total	180/363	280/545	143/264	109/225	41/69	621/1152	850/1726	340/680	187/363	33/67
	% of Males	49.0	51.4	54.2	48.4	59.4	53.9	49.2	50.0	51.5	49.3
			$P = 0.5047 + 0.0138\,d$					$P = 0.5134 − 0.0096\,d$			
			(0.0142)					(0.0136)			
1954–58	Male/total	176/344	226/425	102/213	67/162	34/63	313/629	443/844	183/358	117/220	24/48
	% of Males	51.4	53.2	47.9	41.4	54.0	49.8	52.5	51.1	53.2	50.0
			$P = 0.5032 − 0.0032\,d$					$P = 0.5139 + 0.0013\,d$			
			(0.0157)					(0.0159)			
F₁ extension[b]											
1959–69	Male/Total	343/657	764/1517	265/506	172/320	54/117	451/883	916/1780	270/517	164/310	36/58
	% of Males	52.2	50.4	52.4	53.8	46.2	51.1	51.5	52.2	52.9	62.1
			$P = 0.5171 − 0.0082\,d$					$P = 0.5141 + 0.0085\,d$			
			(0.0104)					(0.0142)			
1970–84	Male/Total	157/323	306/604	96/171	39/94	13/32	199/394	350/726	100/198	55/103	16/27
	% of Males	48.6	50.7	56.1	41.5	40.6	50.5	48.2	50.5	53.4	59.3
			$P = 0.5078 − 0.0198\,d$					$P = 0.4941 + 0.0075\,d$			
			(0.0197)					(0.0193)			
Total	Male/Total	918/1812	1637/3216	633/1207	407/840	144/285	1767/3414	2762/5462	954/1873	558/1061	113/207
	% of Males	50.7	50.9	52.4	48.5	50.5	51.8	50.6	50.9	52.6	54.6
			$P = 0.5096 − 0.0029\,d$					$P = 0.5116 + 0.0011\,d$			
			(0.0068)					(0.0075)			

[a] Either parent exposed to A-bombing within 2 km from the hypocenter.
[b] A linear model, $P = \beta_0 + \beta_1 d$, was applied in which the unit is the individual child, P is the probability of a male child being born alive, and d is a conjoint parental dose. The estimations were calculated by the maximum likelihood method for binary observations. Values within parentheses are standard errors for β_1.
[c] Either parent exposed to ≥0.10 Gy of T65D total dose (kerma).

Figure 1. Distribution of birth weight and parity for the subset of children both of whose parents have been assigned DS86 doses (38,903 individuals born during 1948–58). The neutron component of the radiation has been assigned a relative biological effectiveness of 20. For the homogeneity test of birth weight among parental gonadal dose groups, $\chi^2 = 27.07$, $df = 15$, $p = .03$. For the homogeneity test of parity among parental gonadal dose groups, $\chi^2 = 66.56$, $df = 20$, $p < .01$.

Method of analysis

As previously stated,[10] deaths during the years 1946–85 of children in the cohort were ascertained by searching the family *koseki* in which the child's life status is recorded, and the cause of death is based on the death certificate information maintained in the health centers throughout Japan. Confirmation of the fact of death is more than 99% complete.

Using the actuarial life table method, we have calculated survival rates for all causes or all diseases at each year of age in four groups of birth years, namely, 1946–50, 1951–58, 1959–69, and 1970–84. Mantel-Haenszel statistics,[19] comparing mortality in the ≥0.01 Sv group with that in the 0 Sv dose group based on 2 × 2 tables for each year of age, were employed. For convenience, survival rates at ages 1, 5, 10, 20, and 30 years are shown in Table 4. Except where specifically stated to the contrary, deaths from accidents or suicides are excluded from the analyses.

To provide a basis for the comparison of mortality in these cohorts of children with that among children in Japan generally, we have calculated the expected number of deaths at the attained ages (0, 1–4, 5–9, 10–19, 20–29, 30–39 years) based on the national vital statistics[20] for the same year.

Mortality rates (all ages and age <20 years) for all diseases except neoplasms and cancer by six parental dose categories were based on the person-years and were indirectly standardized to adjust for city, sex, year of birth, and attained age using the total sample as a standard.

Excess relative risk per sievert of combined parental gonadal dose was evaluated using an additive relative risk model in the Poisson regression analysis.[21] Deaths and person-years observed in the period 1946–85 were stratified by s strata of city (Hiroshima or Nagasaki), sex (male or female), year of birth (1946–50, 1951–58, 1959–69, 1970–84), k strata of combined parental dose (0, 0.01–0.09, 0.10–0.49, 0.50–0.99, 1.00–2.49, ≥2.5 Sv), and t strata of attained age (0, 1–4, 5–9, 10–19, 20–29, 30–39 years). Under the Poisson model, it is postulated that $Y_{t,s,k}$, the number of deaths from the causes of interest, follows the Poisson distribution with mean $E(Y_{t,s,k})$, where:

$$E(Y_{t,s,k}) = R_{t,s,k}\, g_{t,s}(1 + \beta D_{t,s,k}) \ .$$

Here $R_{t,s,k}$ is the person-year, $g_{t,s}$ is the background mortality rate per person-year, and $D_{t,s,k}$ is the mean combined parental dose (in sieverts) of the dose group k for the stratum s at attained age t. The parameter β is the excess relative risk per sievert of combined parental gonadal dose, i.e., the relative risk in the dose group k is $1 + \beta D_{t,s,k}$, as contrasted with that for the 0 Sv dose group. To examine the influence of city, sex, and year of birth, the background mortality rate, $g_{t,s}$, can be parameterized as follows:

Table 4. Survival rate [$S(t)$] at age t (1, 5, 10, 20, or 30 years), due to all causes of death or all diseases (May 1946–December 1985) by combined parental dose and year of birth for the subset of children both of whose parents have been assigned DS86 doses. The neutron component of the radiation has been assigned an RBE of 20.

Year of birth	$S(t)$	0 Sv			≥0.01 Sv			Test statistics[a]		
		No. of births	All deaths	All diseases	No. of births	All deaths	All diseases		All deaths	All diseases
1946–1950		15,099	1458	1113	6541	585	470	MH χ^2	2.61	0.24
	S(1)		0.9546(0.0017)	0.9629(0.0015)		0.9577(0.0025)	0.9652(0.0023)		(df=1, p=.106)	(df=1, p>.5)
	S(5)		0.9266(0.0021)	0.9385(0.0020)		0.9320(0.0031)	0.9411(0.0029)	RR	0.93[0.85-1.03]	0.99[0.88-1.10]
	S(10)		0.9220(0.0022)	0.9354(0.0020)		0.9263(0.0032)	0.9371(0.0030)	Homo χ^2	26.01	24.90
	S(20)		0.9174(0.0022)	0.9328(0.0020)		0.9222(0.0033)	0.9346(0.0031)		(df=38, p>.5)	(df=38, p>.5)
	S(30)		0.9089(0.0023)	0.9288(0.0021)		0.9156(0.0034)	0.9309(0.0031)			
1951–1958		18,262	993	708	7428	404	289	MH χ^2	<0.01	<0.01
	S(1)		0.9705(0.0013)	0.9765(0.0011)		0.9690(0.0020)	0.9765(0.0018)		(df=1, p>.5)	(df=1, p>.5)
	S(5)		0.9596(0.0015)	0.9681(0.0013)		0.9585(0.0023)	0.9685(0.0020)	RR	1.01[0.90-1.15]	1.02[0.89-1.18]
	S(10)		0.9565(0.0015)	0.9663(0.0013)		0.9550(0.0024)	0.9659(0.0021)	Homo χ^2	17.42	18.86
	S(20)		0.9521(0.0016)	0.9641(0.0014)		0.9510(0.0025)	0.9633(0.0022)		(df=34, p>.5)	(df=34, p>.5)
	S(30)		0.9466(0.0017)	0.9618(0.0014)		0.9464(0.0026)	0.9613(0.0022)			

(Continued)

Table 4. Continued

Year of birth	S(t)	0 Sv			≥0.01 Sv			Test statistics[a]		
		No. of births	All deaths	All diseases	No. of births	All deaths	All diseases		All deaths	All diseases
1959–1969		5469	121	96	9239	209	164			
	S(1)		0.9857(0.0016)	0.9866(0.0016)		0.9861(0.0012)	0.9871(0.0012)	MH χ^2	0.03	<0.01
	S(5)		0.9812(0.0018)	0.9837(0.0017)		0.9826(0.0014)	0.9842(0.0013)		(df=1, p>.5)	(df=1, p>.5)
	S(10)		0.9799(0.0019)	0.9832(0.0017)		0.9814(0.0014)	0.9835(0.0013)	RR	0.98[0.78-1.12]	0.96[0.75-1.25]
	S(20)		0.9785(0.0020)	0.9824(0.0018)		0.9785(0.0015)	0.9828(0.0014)	Homo χ^2	19.81	13.39
	S(30)		—	—		—	—		(df=25, p>.5)	(df=25, p>.5)
1970–1984		1862	27	24	3686	55	45			
	S(1)		0.9893(0.0024)	0.9903(0.0023)		0.9894(0.0017)	0.9897(0.0017)	MH χ^2	<0.01	<0.01
	S(5)		0.9876(0.0026)	0.9893(0.0024)		0.9870(0.0019)	0.9889(0.0017)		(df=25, p>.5)	(df=25,p>.5)
	S(10)		0.9860(0.0027)	0.9876(0.0026)		0.9853(0.0020)	0.9880(0.0018)	RR	0.94[0.60-1.47]	0.87[0.54-1.41]
	S(20)		—	—		—	—	Homo χ^2	2.99	3.81
	S(30)		—	—		—	—		(df=10, p>.5)	(df=7, p>.5)

NOTE: MH χ^2 = Mantel-Haenszel χ^2 statistics. RR = Weighted relative risk; the values in brackets are the 95% confidence limits. Homo χ^2 = Homogeneity χ^2 statistics for relative risk in each year interval of age at death. Values within parentheses in the "All deaths" and "All diseases" columns are standard errors.

[a]Test for relative risk of ≥0.01 Sv dose group compared with 0 Sv dose group.

$$g_{t,s} = g_t \exp(\beta_s Z_{t,s}) ,$$

where g_t is the background rate at attained age t, $Z_{t,s}$ is a covariate vector of variables for city, sex, and the mean number of years between the bombings and birth, and β_s is the corresponding regression vector. Hypothesis tests for $H_0: \beta = 0$ were based on likelihood ratio statistics. A chi-square statistic, using the Freeman-Tukey residual,[22] was employed to test the goodness-of-fit of a model.

Finally, although the present method of analysis using Poisson regression and person-years at risk of death seems more appropriate, particularly as the cohort ages—since earlier analyses of mortality in the F_1 cohort have been based on a simple linear regression of the frequency of death on parental dose[13]—we have undertaken a similar analysis here to provide continuity with the past and insight into the dependence of the results on the method of analysis chosen.

Results

All causes of death and survival rates

In Table 4, we have shown the survival rates at 1, 5, 10, 20, and 30 years of age by different birth years: 1946–50, 1951–58, 1959–69, and 1970–84. The survival rates were calculated for all causes and all diseases. Clearly, mortality has diminished in the more recent birth years. For example, in the 0 Sv dose group, the survival rates for all causes at age 10 years are 0.9220, 0.9565, 0.9799, and 0.9860 for the birth years 1946–50, 1951–58, 1959–69, and 1970–84, respectively. Within the four birth year groups, there is no statistically significant difference in mortality from all causes or all diseases between the 0 Sv and the ≥0.01 Sv dose groups. It should be noted here that in the ≥0.01 Sv dose group, 27.0% (7,255 individuals in the birth years 1966–84) had not as yet reached 20 years of age prior to 1985.

In the ≥0.01 Sv dose group from 1946–85, 1,253 deaths were observed in the subset of children both of whose parents have been assigned DS86 doses (Table 5). Among these deaths, 4.3% (40 cancers and 14 unspecified neoplasms) were attributed to neoplasms, 72.9% (914 deaths) to all diseases except neoplasms, 14.3% (179 deaths) to accidents or suicide, and the remainder, 8.5% (106 deaths), to unknown causes. Among 2,599 deaths in the 0 Sv dose group, these proportions were about the same (3.4% or 75 cancers and 14 unspecified neoplasms, 71.3% or 1,852 deaths from all diseases except neoplasms, 15.6% or 405 deaths from accidents or suicide, and 9.7% or 253 deaths from unknown causes). Since there is a difference in the structure of the denominators between the original F_1 sample and the F_1-Ext sample, the proportional mortality in the ≥0.01 Sv dose group should be compared with that in the 0 Sv dose group within the two samples. The chi-square statistics for tests of the equivalence of the proportional mortality in the two samples are 4.56 ($df = 3, p = .21$) for the F_1 original sample and 1.62 ($df = 3, p = .66$) for the F_1-Ext sample.

RERF TR 1-91

Table 5. Number of deaths (May 1946–December 1985) by causes and parental gonadal dose equivalent categories for the subset of children both of whose parents have been assigned DS86 doses. The neutron component of the radiation has been assigned an RBE of 20.

Cause of death	Original F₁ births, 1946–58 Combined parental dose (Sv) 0	≥0.01	Total	F₁ extension births, 1959–84 Combined parental dose (Sv) 0	≥0.01	Total	Total births, 1946–84 Combined parental dose (Sv) 0	≥0.01	Total
No. at risk	33,361	13,969	47,330	7331	12,925	20,256	40,692	26,894	67,586
Neoplasms									
Leukemia	25	12	37	2	5	7	27	17	44
Other cancers	47	18	65	1	5	6	48	23	71
Subtotal	72	30	102	3	10	13	75	40	115
U.N. neoplasms[a]	11	10	21	3	4	7	14	14	28
All diseases except neoplasms[b]									
Infectious	319	120	439	6	14	20	325	134	459
Respiratory	316	132	448	30	37	67	346	169	515
Digestive	235	114	349	0	10	10	235	124	359
Newborn	398	149	547	53	85	138	451	234	685
Others	369	161	530	23	48	71	392	209	601
Ill-defined	101	43	144	2	1	3	103	44	147
Subtotal	1738	719	2457	114	195	309	1852	914	2766
Injury and poisoning									
Accidents	279	102	381	26	47	73	305	149	454
Suicide	98	24	122	2	6	8	100	30	130
Subtotal	377	126	503	28	53	81	405	179	584
Unknown	253	104	357	0	2	2	253	106	359
	(223)[c]	(98)	(321)	(0)	(1)	(1)	(223)	(99)	(322)
All deaths	2451	989	3440	148	264	412	2599	1253	3852

[a] U.N. neoplasms = benign and unspecified tumors.
[b] All diseases except neoplasms are infectious = 9th ICD 001-139, respiratory = 9th ICD 460-519, digestive = 9th ICD 520-579, newborn = 9th ICD 760-779, or others = other diseases except ill-defined diseases.
[c] The values within parentheses are the number of deaths occurring at ages of less than 1 year among those deaths due to unknown causes.

As shown in Table 5, most of the 359 deaths (89.7%, or 322 deaths) from unknown causes occurred before one year of age. Among the 106 deaths from unknown causes in the ≥0.01 Sv dose group, 70.8% (75 deaths) were identified in the course of the early clinical study as neonatal deaths occurring within the first 14 days after birth. Of 253 deaths from unknown causes in the 0 Sv dose group, 66.8% (149 deaths) were neonatal deaths occurring at <14 days after birth. The difference observed between these two groups is not statistically significant (Yates's corrected chi-square is 0.37 [$df = 1$], $p = .54$).

The observed numbers of all deaths and fatal cancers, by age at death, among the children of the A-bomb survivors and the expected numbers based on the national vital statistics for the same years are shown in Table 6. Of all of the 3,852 deaths observed in 1946–85, 76.2% occurred at ages between 0 and 4 years, and thus the data presented here largely overlap with those presented in reports issued prior to the advent of the revised (DS86) dose schedule.[8-10] The ratios of the numbers of observed to expected deaths are about 0.7 for all deaths and 0.8 for fatal cancers. The mortality observed among the children of A-bomb survivors is never higher than that expected on the basis of the Japanese population as a whole in the same years.

Excess relative risk for mortality

Table 7 shows the mortality rates observed in the subset of children whose parents have been assigned DS86 doses for all diseases except neoplasms, and for cancer. The average number of person-years from 1946 to 1985 in the ≥0.01 Sv dose group is 25.5. In this dose group, 95.1% (869 deaths) of the 914 deaths from all diseases except neoplasms and 47.5% (19 deaths) of the 40 deaths from cancer occurred before the age of 20. The relative risk of death from all diseases except neoplasms is close to or less than unity in all of the dose groups except the ≥2.5 Sv dose group. The relative risk of death from cancer in the ≥0.01 Sv dose group varies around unity. The relative risk of 1.51 for all diseases except neoplasms at age <20 years in the ≥2.5 Sv dose group is barely statistically significant ($p = .046$). However, as noted in the description of study population characteristics, this group of 308 infants (Figure 1) is characterized by higher maternal parity and lower birthweight, and the statistical significance of this higher mortality disappears when birth weight and parity, in addition to combined parental dose, are included in the risk model (the data are not shown but see Table 8). Furthermore, mortality in the ≥2.50 Sv dose group is never higher than that expected on the basis of national vital statistics for the same year (the expected number of deaths from all diseases except neoplasms is 31.6 whereas the observed number is 29 in the ≥2.5 Sv dose group). Based on an additive relative risk model in the Poisson regression analysis, the excess relative risks per sievert of combined parental gonadal dose for all diseases except neoplasms and for cancer are shown in Table 9. The excess relative risk for mortality from all diseases except neoplasms is positive but not significant (0.030 ± 0.046 per sievert, assuming a neutron RBE of 20 and using the subsample

Table 6. Comparison of the observed number of all deaths (Ad) and cancer deaths (Ca) for the subset of children both of whose parents have been assigned DS86 dose equivalents with the expected number based on all Japan. The relative risk (RR) was calculated from the ratio of O/E for doses \geq 0.01 Sv to the O/E for the 0 Sv dose group, the O/E's being derived from a comparison with Japanese national statistics.

DS86 parental dose group (Sv)		0 Ad	0 Ca	1–4 Ad	1–4 Ca	5–9 Ad	5–9 Ca	Age at death 10–19 Ad	10–19 Ca	20–29 Ad	20–29 Ca	30–39 Ad	30–39 Ca	Total Ad	Total Ca
Total	Obs.	2011	2	923	16	215	13	240	18	320	36	143	30	3852	115
	Exp.	3162.3	3.3	1100.3	16.7	257.6	13.8	304.2	30.5	360.4	41.5	147.9	36.3	5328.3	142.2
	O/E	0.64	0.61	0.84	0.96	0.83	0.94	0.79	0.59	0.89	0.87	0.97	0.83	0.72	0.81
0	Obs.	1333	0	639	12	135	8	158	10	231	21	103	24	2599	75
	Exp.	2017.8	1.9	739.8	9.8	167.7	8.1	197.0	19.1	247.4	28.6	103.1	25.4	3472.8	92.9
	O/E	0.66	0.00	0.86	1.22	0.81	0.99	0.80	0.52	0.93	0.73	1.00	0.94	0.75	0.81
≥0.01	Obs.	678	2	284	4	80	5	82	8	89	15	40	6	1253	40
	Exp.	1037.3	1.4	360.5	7.0	89.9	5.7	107.2	11.4	112.9	12.9	44.8	11.0	1752.6	49.4
	O/E	0.65	1.43	0.79	0.57	0.89	0.88	0.77	0.70	0.79	1.16	0.89	0.55	0.71	0.81
	RR	1.00	—	0.91	0.47 Sug.	1.10	0.89	0.95	1.34	0.84	1.58	0.89	0.58	0.96	1.00

NOTE: Obs. = the observed number of deaths; Exp. = the expected number of deaths based on all Japan. Sug. = .05 < p < .10, test for H_0: RR = 1.

Table 7. Mortality rate from cancer and all diseases except neoplasms by combined parental dose equivalent categories for the subset of children both of whose parents have been assigned DS86 doses. The neutron component of the radiation has been assigned an RBE of 20.

		Combined parental gonadal dose categories (Sv)						
Item	0	0.01–0.09	0.10–0.49	0.50–0.99	1.00–2.49	≥2.50	≥0.01	Total
Mean dose	0	0.037	0.238	0.708	1.491	3.814	0.405	0.161
No. at risk	40692	10648	9863	3494	2295	594	26894	67586
PY[a] All ages 1199866	265811	250227	92850	62177	15251	686317	1886182	
	[29.5][b]	[25.0]	[25.4]	[26.6]	[27.1]	[25.7]	[25.5]	[27.9]
Age < 20 756717	190444	177643	63807	42234	10705	484833	1241550	
	[18.6]	[17.9]	[18.0]	[18.3]	[18.4]	[18.0]	[18.0]	[18.4]
Deaths from all diseases except neoplasms								
All ages Obs.	1852	349	335	120	81	29	914	2766
Exp.	1840.2	351.1	335.8	130.4	88.4	20.1	925.8	2766.0
Age < 20 Obs.	1740	332	318	114	76	29	869	2609
Exp.	1731.4	332.7	318.4	123.7	83.8	19.1	877.6	2609.6
Adjusted rate[c] and relative risk								
All ages Rate	147.6	145.8	146.3	135.0	134.4	211.1	144.8	146.7
RR	1.00	0.99	0.99	0.91	0.92	1.43	0.98	
Age < 20 Rate	211.2	209.7	209.9	193.7	190.7	318.7	208.1	210.1
RR	1.00	0.99	0.99	0.92	0.90	1.51*	0.99	

(Continued)

Table 7. Continued

Item		Combined parental gonadal dose categories (Sv)							
		0	0.01–0.09	0.10–0.49	0.50–0.99	1.00–2.49	≥2.50	≥0.01	Total

Item		0	0.01–0.09	0.10–0.49	0.50–0.99	1.00–2.49	≥2.50	≥0.01	Total
Cancer deaths									
All ages	Obs.	75(27)[d]	17(7)	14(5)	5(4)	4(1)	0(0)	40(17)	115(44)
	Exp.	76.4	15.0	13.6	5.4	3.6	0.9	38.6	115.0
Age < 20	Obs.	30(17)	10(6)	5(3)	3(3)	1(1)	0(0)	19(13)	49(30)
	Exp.	30.8	7.4	6.3	2.5	1.6	0.4	18.2	49.0
Adjusted rate and relative risk									
All ages	Rate	6.0	6.9	6.3	5.6	6.7	0.0	6.3	6.1
	RR	1.00	1.15	1.05	0.94	1.12	0.00	1.06	
Age < 20	Rate	3.8	5.3	3.1	4.8	2.4	0.0	4.1	3.9
	RR	1.00	1.39	0.82	1.26	0.64	0.00	1.07	

NOTE: Obs. = the observed number of deaths; Exp. = the expected number of deaths based on the internal comparison with the assumption that there is no parental dose effect on mortality.

[a]Person-years.
[b]The values within brackets are averages of the person-years.
[c]The adjusted rate is calculated by the indirect method, using the total mortality rate as a standard per 100,000 person-years.
[d]The values within parentheses are the numbers of leukemia cases.

*$p = .046$ for the test of H_0: RR = 1.0 compared with the 0 Sv dose group.

Table 8. Comparison of results of a Poisson regression analysis for mortality (1948–85) from all diseases except neoplasms which includes birth weight and parity in the model with the results of a model which excludes these two factors. The analysis is based on 1,872 deaths among the 38,903 individuals live born between 1948–58 for whom birth weight and parity are available, DS86 doses only.

Type of analysis	Multiplicative factors in background rate				Birth weight (kg)		ERR (Sv) for conjoint parental dose equivalent	Deviance
	Birth year (years since bombing)	Hiroshima	Male	Parity	Linear term	Quadratic term		
Background parameters only	-0.091* (0.008)	-0.210* (0.049)	0.128* (0.047)					2778.94 [df=4439]
Background parameters and dose effect	-0.091* (0.008)	-0.210* (0.049)	0.127* (0.047)				0.048 (0.058)	2778.13 [df=4438]
Background parameters, parity, birthweight, and dose effect	-0.086* (0.009)	-0.190* (0.049)	0.197* (0.048)	0.064* (0.015)	-0.843* (0.054)		0.009 (0.050)	2499.34 [df=4436]
As above, plus quadratic component	-0.088* (0.009)	-0.201* (0.049)	0.230* (0.048)	0.082* (0.019)	-2.035* (0.152)	2.679* (0.254)	-0.010 (0.064)	2241.66 [df=4435]

NOTE: ERR is excess relative risk per sievert. A model for mortality, λ_A, $\exp(\gamma'_0 Z_0)(1 + \beta d + \gamma'_1 Z_1)$, is used. λ_A = Mortality at attained age; Z_0 is a covariate vector for birthyear, city, and sex; Z_1 is a covariate vector of parity and birthweight; γ_0, γ_1 are vectors of parameters corresponding to Z_0 and Z_1, respectively; d is conjoint dose; and β is excess relative risk per sievert. Estimates are calculated after subtraction of the sample mean from birthyear, parity, and birthweight. The sample means are 6.8 years and 2.6 and 3.06 kg, respectively. Values within parentheses are standard errors.

*$p < .01$.

Table 9. Results of a Poisson regression analysis of the mortality (1946–85) from all diseases except neoplasms using stratification for attained age, birth years, city, and sex. The neutron component of the radiation has been assigned an RBE of 20.

Cause of death	No. of deaths	ERR (Sv) for combined parental gonadal dose	Test p value
All diseases except neoplasms			
All ages	2,766	0.030 (0.046)	.485
Age < 20	2,609	0.038 (0.048)	.379
Infectious, 9th ICD 001-139			
All ages	459	-0.066 (0.100)	.470
Age < 20	451	-0.062 (0.102)	.504
Respiratory, 9th ICD 460-519			
All ages	515	0.153 (0.121)	.119
Age < 20	502	0.166 (0.124)	.098
Digestive, 9th ICD 520-579			
All ages	359	0.152 (0.152)	.294
Age < 20	344	0.167 (0.158)	.266
Newborn, 9th ICD 760-779			
All ages	685	0.030 (0.090)	.711
Age < 20	685	0.030 (0.090)	.711
Others except ill-defined diseases			
All ages	601	-0.019 (0.092)	.841
Age < 20	491	0.006 (0.105)	.956
All cancer			
All ages	115	-0.195 [~ 0.271]	.432
Age < 20	49	-0.233 [~ 0.328]	.341
All cancer except leukemia			
All ages	71	-0.220 [~ 0.340]	.441
Age < 20	19	-0.224 [~ 0.310]	.281
Leukemia			
All ages	44	-0.117 [~ 0.934]	.801
Age < 20	30	0.021 [~ 1.516]	.945

DS86 doses only (67,586 children)

NOTE: The test p-value is calculated by likelihood ratio statistics for H_0: ERR = 0, where ERR is excess relative risk per sievert. The values within parentheses are standard errors; the values within brackets are 95% upper bounds.

with assigned DS86 doses). However, the excess relative risk for cancer tends to be negative. The 95% upper bound of the excess relative risk for cancer is shown in Table 9 instead of the standard error, since the distribution of the parameter estimates is skewed. There is no significant increase in the relative risk of mortality from all diseases except neoplasms or from cancer as combined parental gonadal dose increases. Parenthetically, fitting the simple linear model to which we have previously alluded to these data yields an intercept of 0.0420 (±0.0015) and a linear regression coefficient of 0.00169 (±0.00157) per sievert. These values imply a slightly higher excess relative risk, specifically 0.040, than resulted from the previous analysis, but the two values are really in very good agreement.

Excess relative risk was also evaluated for four major disease categories (based on the International Classification of Diseases [9th ICD] rubrics[23]): infectious and parasitic diseases (9th ICD 001-139), diseases of the respiratory system (9th ICD 460-519), diseases of the digestive system (9th ICD 520-579), and certain conditions originating in the perinatal period (9th ICD 760-779). There is no significant increase in the relative risk of mortality in any of these four categories as combined parental gonadal dose increases. At face value, the estimate of the excess relative risk varies slightly depending upon the cut point for attained ages, but no significant increase was observed with any of the alternative cut points.

In Table 10, the estimated parameters for birth year (years between bombing and birth), city, and sex are shown, to demonstrate the influence of these factors on the background mortality rate. The latter rate, for all diseases except neoplasms, in children born in Hiroshima is lower than that in Nagasaki; the Hiroshima value is about 0.86 times (exp[−0.15]) that in Nagasaki. The background mortality for all diseases except neoplasms for male children is slightly higher than that for female children by about 1.1 times (exp[0.09]), and the background mortality rate for all diseases in children born in 1946 is higher than that in children born in 1955 by about 2.2 times (exp[−0.088 × −9]). However, only the difference between the sexes in the background mortality rate from cancer at ages of <20 years is statistically significant.

The findings in Table 10 allow us to compare the results of a Poisson regression analysis assuming the RBE of neutrons to be 20 with an analysis assuming an RBE of 1. At face value, the excess relative risk for all diseases except neoplasms in the former instance is slightly higher than that in the latter. Yet it is clear that the choice of any RBE between 1 and 20 will have little effect on the estimated excess relative risk itself.

Table 10. Comparison of estimated excess relative risk (ERR) per sievert and coefficients for birth year, city, and sex on the Poisson regression analysis assuming an RBE of 20 with those assuming an RBE of 1 (67,586 children, DS86 dose only).

RBE	Cause of death	Age at death	Multiplicative factors in background rate[a] — Birth year (years since bombings)	Multiplicative factors in background rate[a] — Hiroshima	Multiplicative factors in background rate[a] — Male	ERR (Sv) for combined parental gonadal dose equivalents
RBE = 1	All diseases except neoplasms	All ages	-0.088** (0.004)	-0.148** (0.040)	0.090* (0.038)	0.026 (0.053)
		Age < 20	-0.088** (0.004)	-0.164** (0.041)	0.077* (0.039)	0.039 (0.055)
	All cancer	All ages	-0.011 (0.019)	-0.103 (0.196)	0.263 (0.189)	-0.164 [~ 0.366]
		Age < 20	-0.001 (0.020)	-0.184 (0.293)	0.619* (0.302)	-0.242 [~ 0.437]
RBE = 20	All diseases except neoplasms	All ages	-0.088** (0.004)	-0.146** (0.039)	0.091* (0.038)	0.036 (0.047)
		Age < 20	-0.088** (0.004)	-0.167** (0.041)	0.075 (0.039)	0.044 (0.048)
	All cancer	All ages	-0.010 (0.019)	-0.102 (0.196)	0.262 (0.189)	-0.206 [~ 0.250]
		Age < 20	-0.002 (0.020)	-0.168 (0.294)	0.634* (0.302)	-0.224 [~ 0.284]

[a] A model for mortality, $\lambda_A \cdot \exp(\gamma' Z)(1 + \beta d)$, is used, where: λ_A = mortality at attained age; Z is a covariate vector for birth year, city, and sex; d is conjoint dose; γ is a vector parameter corresponding to Z, and β is excess relative risk per sievert. Estimates are calculated after subtraction of the sample mean from the birth year (years since the bombings). The sample mean = 10.5 years. The values within parentheses are standard errors; the values within brackets are the 95% upper bounds.
* $.01 < p < .05$; ** $p < .01$.

As earlier stated, 91.9% (4,265) of the 4,642 children one or both of whose parents have been assigned an ad hoc dose among the full sample of 72,228 children are believed to have been exposed to ≥0.01 Sv of conjoint parental gonadal dose. In Table 11 we compare the results of a Poisson regression analysis for the subset of children assigned only DS86 doses with the results (DS86 plus ad hoc doses) for the full sample. The excess relative risk for all diseases except neoplasms in the full sample is a little higher, and the standard error a bit smaller, than in the subset of children whose parents have assigned DS86 doses, but these differences are not significant.

Table 11. Comparison of results of the Poisson regression analysis using stratifications for attained age, birth year, city, and sex of the mortality (1946–85) of all diseases except neoplasms in the children assigned only DS86 doses with those for the full sample. The neutron component of the radiation has been assigned an RBE of 20.

Cause of death	DS86 doses only		Full sample	
	Number	ERR (Sv) for combined parental gonadal dose	Number	ERR (Sv) for combined parental gonadal dose
At risk	67,586		72,228	
All diseases except neoplasms				
All ages	2766	0.030 (0.046)	2958	0.046 (0.041)
Age < 20	2609	0.038 (0.048)	2789	0.043 (0.042)
All cancer				
All ages	115	-0.195 [~ 0.271]	120	-0.223 [~ 0.106]
Age < 20	49	-0.223 [~ 0.328]	53	-0.223 [~ 0.293]

NOTE: ERR = excess relative risk. No coefficients of ERR for conjoint parental dose are statistically significant. The values in parentheses are standard errors; the values in brackets are 95% upper bounds.

Discussion

Continued surveillance of mortality among the liveborn children of A-bomb survivors has not revealed a significant increase in the relative risk of mortality from all diseases except neoplasms, nor from neoplasms, following parental exposure to A-bomb radiation. Although a significantly increased mortality cannot be shown, this does not permit one to assume that deleterious, indeed lethal, mutations did not occur. Accordingly, we accept the data at their face value, and on the assumption that the genetic effect due to A-bomb radiation increases linearly with dose, estimate the excess relative risk of death from all causes except neoplasms to be 0.030 (±0.046) per sievert, based on the DS86 gonadal doses using a neutron RBE of 20 (Table 9 or 11).

Survival rates (due to deaths from all causes or all diseases at age 1, 5, 10, 20, and 30 years for the different birth year cohorts have been computed to provide a simple but appropriate way to observe mortality at a glance (Table 4). As these rates show, mortality in the earlier years of life has declined markedly with time in this sample. This is in keeping with a Japan-wide trend: For males, mortality from all causes at an age of <1 year decreased between 1947 and 1955 from 8.8% to 4.3%; for females during the same period the decrease was from 7.7% to 3.7%.

The proportions of deaths due to neoplasms (cancer as well as unspecified neoplasms), all diseases except neoplasms, accidents or suicide, and deaths from unknown causes among all deaths do not differ significantly between the 0 Sv and the ≥0.01 Sv dose groups. Most of the 359 deaths from unknown causes (68%, or 244 deaths) exhibited in Table 5 involved neonatal deaths in the first 2 weeks after birth identified by the early clinical study.[3,4] We note that the present analysis of cancer differs from that reported elsewhere[11] in that it includes only cancer deaths, not incident cases, and that death could have occurred at any age during the surveillance period, not only before the age of 20.

Generally, the number of deaths observed among the children of the A-bomb survivors is never higher than the expected number based on the Japanese national vital statistics for the same years (Table 6). The ratio of observed to expected deaths is about 0.7 for all deaths and 0.8 for fatal cancer. A variety of explanations can be advanced for this discrepancy from an expected ratio of 1. Arguably the most important of these centers on the appropriateness of the national statistics as the basis for determining the expectations. These statistics are derived from all of Japan including rural areas, which generally have different death rates than those seen in urban areas, and the pooling of information obscures the regional variability in mortality common to all countries.

The small increase in mortality observed in the children of survivors, even if it were statistically significant, could not be assigned unequivocally to parental exposure to A-bomb radiation, since it is difficult, if not impossible, to account adequately for the risk due to environmental factors. Specifically, socioeconomic factors during the first 20 years of life could have affected survivorship among these children, since it is not unreasonable to assume that the aftermath of the bombings worsened the socioeconomic status of many of the survivors. To assess the role of factors other than parental exposure, we have examined the influence of birth weight and parity on mortality (see Table 8). These two factors may be related to socioeconomic status or the age of A-bomb survivors, and an infant's birth weight could be a more important factor in early mortality than parental gonadal dose. This possibility was borne out by the analysis of Table 8: The apparent radiation effect disappeared when parity of mother and infants' birth weights were included in the analysis.

Ideally an assessment of the relationship between A-bomb exposure and cancer occurrence should be based on tumor registry data involving histological verification and medical information. This would minimize the number of nonfatal cancers lost to follow-up. In the first RERF study of cancer in the F_1, 36 leukemia incidence cases (all ages) were identified during 1946–79 by Ishimaru et al.[16] No significant excess risk for leukemia among the children of exposed parents was observed. In subsequent reports we have described the risk of all cancer (1946–82) including leukemia in the first two decades of life, based on cancer incidence and the use of tumor registry data, and no statistically significant increase in cancer risk was observed as parental gonadal dose increased.[11] In a cancer incidence study it is impossible to systematically identify cancers occurring in people who have left the tumor registry reporting areas. Outmigration does not, however, influence the mortality study since the fact of death can be determined wherever in Japan an individual may have resided prior to death, but the death certificate is a less reliable means of identifying the occurrence of cancer than the tumor registries. At present, as shown in Tables 7, 9, and 10, there is no clear evidence of an increase in fatal cancer (either before the age of 20 or for all ages) with parental dose among the children of A-bomb survivors.

Generally, postspermatogonial cells (spermatocytes and spermatids) are more sensitive to radiation exposure than spermatogonia.[2] From the beginning of puberty, proliferation of spermatogonia continues throughout the lifetime of the human male, and the full cycle of spermatogenesis is normally about 64 days.[24] Gardner and his coworkers[15] suggest that the increased risk of leukemia observed in the Sellafield study is primarily due to radiation-related mutations in postspermatogonial cells. The RERF data pertinent to this issue are limited; the number of leukemia cases and of children born from May 1946 to December 1946 is small (i.e., children presumed to have been conceived during the 6 months after the bombing amount to about 2% of the total sample). The RERF studies are more pertinent, therefore, to the possible genetic effects due to radiation-induced mutations in the spermatogonia or oocytes of the A-bomb survivors. Since, however, mutations induced in spermatogonia may accumulate and be copied many-fold in descendant sperm cells—in fact, the bulk of the radiation experienced in the Sellafield study was to spermatogonia—it would require sensitivities of postspermatogonial cells far in excess of those established experimentally to bridge the gap between the present observations and the Sellafield findings (see also Little[25]).

Radiation is not selective in its spectrum of germinal mutational damage. The components of the spectrum of mutation types from spontaneous germinal mutation may show differing frequencies and each component may have a different radiosensitivity. Because we have failed to show a significant genetic effect among the children of the A-bomb survivors, we cannot present further speculation at this point. However, one would expect in Sellafield an increase in other mutational diseases that could scarcely have gone unnoticed if the excess risk of leukemia was due to radiation-induced germinal mutations.

As mentioned in the Introduction, a variety of other genetic studies have been performed on the children of A-bomb survivors, and recently all these data, combined with the data of the present study, have been used to generate an estimate of the genetic doubling dose of radiation for humans, on the basis of the DS86 dose schedule.[26] To prevent confusion, we note that the regression term for F_1 mortality employed in that doubling dose-ultimate is somewhat smaller than that developed in the present paper. This results from the fact that to avoid overlap between a previous study on stillbirths, congenital defects, and neonatal deaths[4] and the present study of all deaths among liveborn children, neonatal deaths scored in the course of the previous study have been subtracted from the present data sets and a new regression term derived for the purpose of obtaining a regression to be used in a doubling dose estimate.

References

1. United Nations Scientific Committee on the Effects of Atomic Radiation: *Sources, Effects and Risks of Ionizing Radiation*. New York, United Nations, 1988.

2. National Academy of Sciences-National Research Council: *The Effects on Populations of Exposure to Low Levels of Ionizing Radiation*. Washington, D.C., National Academy Press, 1990.

3. Neel JV, Schull WJ: *The Effect of Exposure to the Atomic Bombs on Pregnancy Termination in Hiroshima and Nagasaki*. Washington, D.C., National Academy of Sciences-National Research Council, Pub. No. 461, 1956.

4. Otake M, Schull WJ, Neel JV: Congenital malformations, stillbirths and early mortality among the children of atomic bomb survivors: A reanalysis. Radiat Res 122:1–11, 1990.

5. Satoh C, Neel JV: Biochemical mutations in the children of atomic bomb survivors. Gann Monogr Cancer Res 35:191–208, 1988.

6. Neel JV, Satoh C, Goriki K, Asakawa J, Fujita M, Takahashi N, Kageoka T: Search for mutations altering protein charge and/or function in children of atomic bomb survivors: Final report. Am J Hum Genet 42:663–76, 1988.

7. Awa AA, Honda T, Neriishi S, Sofuni T, Shimba H, Ohtaki K, Nakano M, Kodama Y, Itoh M, Hamilton HB: Cytogenetic study of the offspring of atomic bomb survivors, Hiroshima and Nagasaki. In *Cytogenetics*. Ed. by G. Obe and A. Basler. Berlin, Springer-Verlag, 1987. pp. 166–83.

8. Kato H, Schull WJ: Joint JNIH–ABCC life-span study of children born to atomic bomb survivors, research plan. ABCC TR 4-60

9. Kato H, Schull WJ, Neel JV: A cohort-type study of survival in the children of parents exposed to atomic bombings. Am J Hum Genet 1:339–73, 1966.

10. Neel JV, Kato H, Schull WJ: Mortality in the children of atomic bomb survivors and controls. Genetics 76:311–26, 1974.

11. Yoshimoto Y, Neel JV, Schull WJ, Kato H, Soda M, Eto R, Mabuchi K: The frequency of malignant tumors during the first two decades of life in the offspring of atomic bomb survivors. Am J Hum Genet 46:1041–52, 1990.

12. Schull WJ, Neel JV: Radiation and the sex ratio in man. Science 128:343–8, 1958.

13. Schull WJ, Neel JV, Hashizume A: Some further observations on the sex ratio among infants born to survivors of the atomic bombings of Hiroshima and Nagasaki. Am J Hum Genet 18:328–38, 1966.

14. Schull WJ, Otake M, Neel JV: Genetic effects of the atomic bombs: A reappraisal. Science 213:1220–7, 1981.

15. Gardner MJ, Snee MP, Hall AJ, Powell CA, Downes S, Terrell JD: Results of case-control study of leukaemia and lymphoma among young people near Sellafield nuclear plant in West Cumbria. Br Med J 300:423–9, 1990.

16. Ishimaru T, Ichimaru M, Mikami M: Leukemia incidence among individuals exposed in utero, children of atomic bomb survivors, and their controls. Hiroshima and Nagasaki, 1945–79. RERF TR 11-81, 1981.

17. Roesch WC (ed.): *US–Japan Joint Reassessment of Atomic Bomb Radiation Dosimetry in Hiroshima and Nagasaki*. Hiroshima, Radiation Effects Research Foundation, 1987

18. International Commission on Radiation Units and Measurements: *The Quality Factor in Radiation Protection* (Report 40). Bethesda, Md., ICRUM, 1986.

19. Kleinbaum DG, Kupper LL, Morgenstern H: *Epidemiologic Research: Principles and Quantitative Methods*. Belmont, Calif., Lifetime Learning Publications, 1982.

20. Ministry of Health and Welfare: *Vital Statistics Japan,* 1947 (part II) and 1951–1985 (vol. I). Tokyo, Division of Health and Welfare Statistics, Welfare Minister's Secretariat.

21. Preston DL, Kato H, Kopecky KJ, Fujita S: Studies of the mortality of A-bomb survivors. 8. Cancer mortality, 1950–82. Radiat Res 111:151–78, 1987.

22. Frome EL: The analysis of rates using Poisson regression models. Biometrics 39:665–74, 1983.

23. World Health Organization: *International Classification of Diseases* (1975 revision, vol. 1). Geneva, WHO, 1977.

24. McKusick VA: *Human Genetics*. Englewood Cliffs, N.J., Prentice-Hall, 1964.

25. Little, MP: A comparison between the risks of childhood leukemia from parental exposure to radiation in the Sellafield work force and those displayed among the Japanese bomb survivors. J Radiol Prot 10:185–98, 1990.

26. Neel JV, Schull WJ, Awa AA, Satoh C, Kato H, Otake M, Yoshimoto Y: The children of parents exposed to atomic bombs: Estimates of the genetic doubling dose of radiation for humans. Am J Hum Genet 46:1053–72, 1990

The Children of Parents Exposed to Atomic Bombs: Estimates of the Genetic Doubling Dose of Radiation for Humans

J. V. Neel,* W. J. Schull,† A. A. Awa,‡ C. Satoh,‡ H. Kato,‡ M. Otake,‡ and Y. Yoshimoto‡

*Department of Human Genetics, University of Michigan Medical School, Ann Arbor; †Center for Demographic and Population Genetics, University of Texas, Houston; and ‡Radiation Effects Research Foundation, Hiroshima

Summary

The data collected in Hiroshima and Nagasaki during the past 40 years on the children of survivors of the atomic bombings and on the children of a suitable control population are analyzed on the basis of the newly revised estimates of radiation doses. No statistically significant effects emerge with respect to eight different indicators. Since, however, it may confidently be assumed some mutations were induced, we have taken the data at face value and calculated the minimal gametic doubling doses of acute radiation for the individual indicators at various probability levels. An effort has also been made to calculate the most probable doubling dose for the indicators combined. The latter value is between 1.7 and 2.2 Sv. It is suggested the appropriate figure for chronic radiation would be between 3.4 and 4.5 Sv. These estimates suggest humans are less sensitive to the genetic effects of radiation than has been assumed on the basis of past extrapolations from experiments with mice.

Introduction

Recently our colleagues and ourselves have reported in detail on the current status of the various studies aimed at understanding the genetic effects of the atomic bombs detonated over Hiroshima and Nagasaki (Awa et al. 1987; Neel et al. 1988; Yoshimoto et al. 1990, and submitted; Otake et al., in press). The purpose of the present communication is to summarize briefly all these studies, in a fashion that will provide a perspective on the genetic effects of ionizing radiation for humans, by utilizing the latest developments in atomic bomb dosimetry. Then, in the closing sections of the present paper, we will examine the magnitude of the genetic effects which, by these studies, can be excluded at specified probability levels, and we will also attempt to generate an estimate of the amount of acute radiation which will have a mutational impact equivalent to the role of spontaneous mutation each generation.

Received October 27, 1989; revision received December 20, 1989.
Address for correspondence and reprints: James V. Neel, Department of Human Genetics, University of Michigan Medical School, M4708 Medical Sciences II, Ann Arbor, MI 48109-0618.

This is the convenient but somewhat elusive measure commonly referred to as the genetic "doubling dose."

As will become apparent, for none of the indicators of a genetic effect of radiation is there a statistically significant difference between the children of the parents receiving increased amounts of radiation at the time of the bombings (ATB) and the children of parents who are not thought to have received significantly increased radiation exposures. Since these studies without doubt constitute the single most expensive and extensive genetic undertaking on record, it is important to recognize that the possibility of such an "inconclusive" outcome was clearly foreseen from the outset (Genetics Conference 1947). However, the occasion is so singular in human history that the various advisory groups considering such a study felt there was no alternative but to proceed. What could not be anticipated when the study was initiated in 1947 was the advent of cytological and biochemical technologies that would greatly increase the rigor of the undertaking and, together with new insights into the morphological data collected during the early years of the investigation, would permit a more quantitative assessment of the outcome of the study than originally seemed possible.

1053

Reproduced, with permission, from the *American Journal of Human Genetics*, vol. 46, © 1990 by the American Society of Human Genetics.

Background

The details of the Genetics Program which evolved in Japan have been presented on several occasions (Neel and Schull 1956; Kato et al. 1966; Schull et al. 1981; Awa et al. 1987; Neel et al. 1988). In the immediate postwar period in Japan, the economic stringencies were such that a rationing program which had been initiated during the war years to benefit pregnant women was continued. This involved registration of the fact of pregnancy at the completion of the fifth lunar month of gestation. By incorporating the registrants into a genetic study, it was possible to mount a *prospective* investigation of pregnancy outcome in Hiroshima and Nagasaki that included reference to sex ratio, congenital defect, viability at birth, birth weight, and survival of child during the neonatal period. After approximately a year of preparation, this study was initiated in February 1948. The questionnaires and examination procedures have been presented in detail elsewhere (Neel and Schull 1956). This more clinical program was terminated in February of 1954, but the collection of data on the sex ratio and survival of live-born infants continued, with ascertainment of additional births now done through city records. In addition, from the birth registrations in these two cities, the sample for the consideration of these latter two indicators was extended backward in time, to include births between May 1946 and January 1948. During the clinical program, in Hiroshima, some 62% of all infants who were stillborn or died in the first 6 d of life were subjected to autopsy. A subset of some 30% of the total 76,617 infants born during this period were reexamined at approximately 9 mo of age, the examination including measurements of growth and development.

Relatively few survivors beyond 2,000 m from the hypocenter were exposed to radiation from the bombs; the mortality within the 2,000-m zone was high. Accordingly, in the original sample of 76,617 children plus subsequent additions, many more had been born to parents who had been >2,000 m from the hypocenter ATB—i.e., parents who seldom received increased radiation from the explosions—than had been born to parents who had been within the 2,000-m radius—i.e., parents whose gonadal doses were relatively large. In 1959, to increase the efficiency of the study of survival, three cohorts were defined from the children born in the two cities since the bombings. One cohort was comprised of all children born between May 1, 1946, and December 31, 1958, to parents one or both of whom had been <2,000 m from the hypocenter ATB (i.e., who were proximally exposed). The second cohort was composed of age-, sex-, and city-matched births to parents one or both of whom had been distally exposed to the bombs (>2,500 m from the hypocenter), and the third cohort was composed of age-, sex-, and city-matched children born to parents then residing in Hiroshima and Nagasaki but neither of whom had been in the cities ATB. Such parents are referred to as NIC (not in city). The latter two cohorts are thus subsets of a considerably larger pool. Children born to parents who had been within the 2,000–2,500-m radius of distance from the hypocenter ATB are omitted from consideration, both because of the exceptional difficulties in estimating the (low) exposures at this distance and because of the work involved in following children who would contribute relatively little information regarding radiation effects. Setting the earliest date for admission into the cohort at May 1, 1946, effectively eliminates from the sample all children in utero ATB.

These cohorts have been periodically expanded to include matched numbers of recent births, the limiting factor in the expansion being the number of births to the proximally exposed parents. In 1984, some 39 years after the birth of the first children conceived following the bombings, there were only three births to proximally exposed parents, and it was decided to close the cohorts, effective January 1985. At this writing the cohort of children ever to be born in Hiroshima and Nagasaki to proximally exposed survivors of the bombings is essentially complete. The original projection was for equality of numbers in all the cohorts, but various developments detailed by Yoshimoto et al. (1990) have altered these proportions. The cohort of children born through 1982 (the cutoff date for analyses) to parents receiving >0.01 Gy of radiation now equals 31,150; there are 41,066 "control" children, whose parents either are not known to have received any radiation ATB or received <0.01 Gy. In addition to the study of the survival of these children, in the 1960s and 1970s special studies were mounted with respect to the physical development of a subsample of these children during their school years. In 1968 a search for cytogenetic abnormalities in those children in the cohorts >12 years of age was initiated, and in 1975 a search in these same cohorts for mutation altering the electrophoretic mobility and/or function of a select battery of proteins was undertaken. A study of cancer incidence prior to age 20 years was initiated in 1985. For all these studies, the necessary controls were drawn from the children of parents not receiving increased radiation ATB.

An unusual feature of the Genetics Program is that

all of its various components have been based on a well-defined population subsampled in various (overlapping) ways. The result is a multifaceted appraisal of the bombs' genetic impact on what is essentially a single cohort. We note that, although these studies lack the elegance that can be achieved in an experimental setting, they have the virtue of being based on a single complex population exhibiting the full range of human heterogeneity rather than on laboratory animals (mostly mice) that are the product of previous genetic selection intended to reduce the genetic variance associated with traits of interest. Furthermore, the subjects of the study have been born over a sufficiently extended time interval that we begin to approximate the effects on an entire generation rather than on a "window" of access to a very limited period in the reproductive histories of the parents.

The Estimation of the Amount of Radiation Reaching the Gonads

Even after more than 40 years of intermittent evaluation, the issue of the type and amount of radiation received by survivors remains subject to considerable debate. This results primarily from continuing discussion of the yield and radiation spectrum of the Hiroshima bomb, the (in)ability of some survivors to reconstruct precisely their shielding from the effects of the bombs, difficulties in evaluating radiation attenuation by complex shielding, the possible existence of inherent and ineradicable biases in the dosage estimation procedures, and continuing uncertainty over the LD_{50} of whole-body radiation for humans. A recent thorough reconsideration of the "dose problem" (Roesch 1987) has resulted in a system for calculating organ doses which has been termed "Dosimetry System 1986" (DS86) which, while state of the art in its treatment of physical dosimetry, cannot overcome the uncertainties inherent in a complex human disaster such as this. This replaces a system known as "Tentative 1965 Estimate, Dose Revised" (T65DR). In expressing DS86 dose estimates, we shall follow the current practice of employing grays (1 Gy = 100 rad) and sieverts (1 Sv = 100 rem).

The DS86 dose estimate is based on (1) the spectrum and amount of radiation released by the bombs, (2) its attenuation with distance and physical shielding, (3) the free-in-air kerma reaching the person, (4) body position ATB, and, given position, (5) the further attenuation, by the intervening body tissues, of the radiation reaching the gonads. About 88%–96% of the parents entering into the various studies place themselves in ATB situations for which the calculation of the DS86 dose has been relatively straightforward, the exact percentage depending on the study in question. The remaining 4%–12% either (a) report being in ATB situations for which the calculation of a gonadal dose with the parameters of the DS86 system is not yet possible or (b) provide exposure data not sufficiently detailed for the calculation of a DS86 dose. This latter is especially true for some of the data collected in the early years of the study, before the currently employed exposure history was instituted. Fortunately, almost all of these individuals have in the past been assigned the (less demanding) T65DR doses. Inasmuch as gonadal doses must be available for both parents in a genetic study, and since the "DS86 dose unassigned group" tends to involve predominantly the proximally (more heavily) exposed parents, failure to assign a revised dose to this group would result in a sacrifice of, for example, 23% of the children in the first study we will summarize, i.e., the study on untoward pregnancy outcomes. We have accordingly developed an ad hoc procedure for assigning doses to these parents, a procedure which, with other details regarding dosimetry, is fully described by Otake et al. (1990). In brief, an empirical conversion factor has been developed which permits assigning a DS86-like gonadal exposure on the basis of a previously assigned T65DR gonadal exposure. In the two groups the pattern of occurrence of the critical symptoms of substantial radiation exposure (epilation, subcutaneous bleeding, and oropharyngeal lesions) has been used to provide an appropriate verification of the accuracy of the procedure. In the analyses to be presented we will restrict ourselves to those based on DS86 doses only, except where the sacrifice of data seems too great; the nature of the data will be indicated for each analysis. The term "conjoint" as applied to parental exposure indicates the sum of the parental gonad doses.

The radiation released by the two atomic bombs, although very predominantly gamma, included a small neutron component. The integration of these two types of radiation into a single figure requires that dose be expressed in Sv. For the generally low neutron doses now thought to characterize these atomic bomb exposures, we will, in making the transition from Gy to Sv, employ a value of 20 as the Relative Biological Effectiveness (RBE) for transmitted genetic effects (see Grahn et al. 1983, 1984; International Commission on Radiation Units and Measurement 1986). In our previous publications we have presented average estimated

gamma and neutron absorbed doses, to permit those who are so inclined to apply other RBE.

The Data

The data which have been accumulated over the years will be briefly presented under the following eight headings:

1. Untoward Pregnancy Outcomes

We define an untoward pregnancy outcome (UPO) as an infant stillborn and/or exhibiting major congenital malformation and/or dying within the first 2 wk of life expectancy. Between 1948 and 1954, data were collected on these outcomes for a total of 76,617 newborn infants in Hiroshima and Nagasaki, on 69,706 of whom the data were sufficiently complete in all respects to permit an analysis with the biomedical and dosage constraints of 1989. Of these, 12,409 were born to proximally exposed parents whose conjoint gonadal exposure, on the basis of use of both DS86 and our ad hoc gonad dose estimates, was 0.36 Sv. Shortly after birth all infants were examined, with reference to congenital defect, by staff physicians; an enumeration of the types of defect scored as "major" and of their frequencies will be found in the work of Neel and Schull (1956) and Neel (1958). (In view of the current interest in "sentinel phenotypes" in monitoring for mutation, it should be noted that, to the extent these were identified, they are included among the UPOs, but subsequent experience [Mulvihill and Czeizel 1983; Czeizel and Kis-Varga 1987] confirms the decision reached at the onset of the study—i.e., that cohort size would not be large enough to restrict this aspect of the study to those phenotypes only.)

Over the years these data have been analyzed in various ways, with no major differences in the conclusions reached. For this presentation, we will for this and subsequent indicators present only the results derived from fitting a linear dose-response model to the occurrence of the various indicators of radiation-related damage, these indicators being treated as binary-response variable (i.e., 1 if the event occurred in a given individual and 0 if it did not). In addition to conjoint parental dose, data were available on as many as six background or concomitant variables which could be assumed to influence the occurrence of the event of interest—namely, city (Hiroshima and Nagasaki; Hiroshima was assigned the value 1, and Nagasaki was assigned the value 0), sex (male and female; males were assigned the value 1, and females were assigned the value 0), maternal age (years), parental age (years), year of birth, and birth rank. Specifically, we have fitted a model of the following form:

$$P_i = \text{constant} + \sum_{j=1}^{6} b_j x_{ij} (\text{background}) + b_D \text{dose}_i,$$

where P_i is the expected frequency of the event of interest in the *ith* individual ($i = 1,2,\ldots,n$; i.e., the total number of subjects) having background characteristics x_{ij} ($j = 1,2,\ldots,6$ and dose i). The constant, the b_j, and the b_D are, of course, the parameters to be estimated.

The findings regarding UPOs, based on the extended cohort (DS86 plus ad hoc doses) are shown in table 1 (also see Otake et al. 1990). There were 3,498 UPOs; their frequency in the children of unexposed parents was 5.02%. The general thesis undergirding the search for an increase in UPOs as an indicator of the genetic effects of the atomic bombs is that some portion of this outcome is due to newly arisen, dominantly expressed mutation in the parental generation. We will discuss the probable magnitude of this fraction later. Meantime we note that the regression is small, far from the significance level, but positive, i.e., in the direction anticipated from an increase in dominantly expressed mutations resulting in such an endpoint. To conserve space, the analyses of the roles of concomitant variables—which introduce a certain perspective into the findings—are not presented but can be found in the basic publication; this is also true for the following two indicators:

2. Deaths among Live-born Infants (exclusive of those resulting from a malignant tumor)

As noted earlier, the mortality study embraces three

Table 1

Regression of Frequency of Indicator Trait on Conjoint Parental Gonadal Radiation Exposure (Sv)

Trait	Regression
Untoward pregnancy outcome	.00264 ± .00277
Mortality exclusive of cancer	.00076 ± .00154
Incidence of cancer:	
All cases	−.00008 ± .00028
Most genetic subset[a]	−.00005 ± .00013

NOTE.—Calculation of Sv is based on an assumed neutron RBE of 20. The papers from which these regressions were extracted contain additional regressions which pertain to the effect of concomitant variables and which are useful with respect to prospective.
[a] See text.

cohorts of infants live born in the two cities of interest between May 1946 and December 1985. The mortality data are updated on a 4-year schedule, the last update having been completed in 1985. Cumulative mortality in the live-born children of unexposed parents was 4.58%. At that time the average time elapsed since the birth of a "child" in the study was 26.2 years. The study thus provides coverage of a very high percentage of prereproductive mortality. To avoid overlap with the preceding indicator, the mortality during the first 2 wk of life of those children in the study of UPOs is not included in this mortality analysis. Deaths from childhood cancer were analyzed separately. Detailed descriptions of the organization of the study will be found in the work of Kato et al. (1966); Neel et al. (1974); and Yoshimoto et al. (submitted).

Thus far there have been, after the exclusions just mentioned, 2,584 deaths among the 67,202 members of the three cohorts for whom all the requisite information and DS86 doses are available. Of course, death during childhood has many determinants, only one of which would in this situation be radiation-induced, dominantly expressed mutations in the parental generation. (We include in this term the heterozygous effects of nominal recessives.) In a later section we will deal with what fraction of this indicator is to be assigned to mutation in the preceding generation. All the accumulated data have recently been analyzed in multiple ways (Yoshimoto et al., submitted). Only the results of the analysis based on the multiple linear-regression model applied to the DS86 dose cohort are presented in table 1. In this particular study, for the 26,963 children one or both of whose parents received ≥0.01 Sv of radiation ATB, the average conjoint parental gonadal exposure was 0.40 Sv.

3. Malignancies in the F_1

The rationale for regarding malignancies in the F_1 as a suitable and separate indicator of the genetic effects of the bombs is developed in detail in an accompanying paper in this issue of the *Journal*. In brief, with respect to the children described in the preceding section, studies were undertaken of the mortality and morbidity from malignancies with onset before the age of 20 years, in relation to parental radiation exposure (Yoshimoto et al. 1990). These studies were greatly facilitated by the existence of tumor registries in both Hiroshima and Nagasaki. At the close-out date for the study, 79.7% of the sample, if alive, would have completed 19 years. Thus far there have been 43 cancers in the 31,150 children born to exposed parents, versus 49 in the 41,066 children of unexposed parents. The incidence of cancer in the children of unexposed parents was 1.2/1,000 persons. The results of an analysis employing a linear-regression model and restricted to the DS86 dose cohort are shown in table 1. The regression of all malignancies (fatal and nonfatal) on parental exposure is a slightly negative term, which in fact is virtually zero. The regression was almost the same when the analysis was confined to those tumors for which the evidence that a parental germ-line mutation is a predisposing factor is strongest (i.e., Wilms tumor, the retinoblastoma-sarcoma complex, neuroblastoma, and embryonal carcinoma of the testes). The average conjoint parental gonadal dose for the parents receiving ≥0.01 Sv is, of course, the same as for the preceding data set.

4. Frequency of Balanced Structural Rearrangements of Chromosomes

In the late 1960s, as techniques for the accurate evaluation of the human karyotype became available, a program designed to detect an increase in chromosome abnormalities in the children of proximally exposed survivors was initiated, directed at children already enrolled in the F_1 Mortality Study (see Awa et al. 1968, 1973; Awa 1975). Two age- and sex-matched samples have been established, one drawn from the children of the proximally exposed and the other drawn from the children of the distally exposed. For practical reasons, no effort was made to obtain the necessary venous blood samples prior to the children's thirteenth birthdays. The sample will accordingly only provide reliable data on the two types of cytogenetic abnormalities which confer relatively little threat to survival among live-born infants—namely, sex-chromosome aneuploids (to be discussed in the next section) and balanced structural rearrangements of chromosomes, such rearrangements being defined as reciprocal translocations, pericentric inversions, and Robertsonian translocations.

The most recent summary of the findings by Awa et al. (1987) with respect to balanced structural rearrangements in the two samples is shown in table 2. These types of abnormalities may either have been inherited from a parent or have arisen de novo in the germ line. Thus far it has been possible to conduct for somewhat more than half of the children so diagnosed the family studies necessary to distinguish between these two possibilities. In each cohort a single mutant child has been identified. Awa et al. (1987), using an adjustment for the fact that family studies are not complete for all propositi, estimate the mutation rate as 1.0×10^{-4}/gamete/generation among both the exposed and the

Table 2

Frequency and Parent of Origin of Balanced Structural Rearrangements in the Cytogenetic Study of Awa et al. (1987)

	HIROSHIMA		NAGASAKI		TOTAL	
	Exposed	Control	Exposed	Control	Exposed	Control
Studied for parent of origin:						
De novo	1	1	0	0	1	1
Inherited:						
Father	2	4	2	4	4	8
Mother	0	1	0	1	0	2
Undetermined	6	3	0	2	6	5
Subtotal	9	9	2	7	11	16
Not studied for parent of origin	5	7	2	2	7	9
Grand total	14	16	4	9	18	25
No. of children studied	4,716	5,112	3,606	2,864	8,322	7,976

NOTE.—Rearrangements include reciprocal and Robertsonian translocations plus pericentric inversions.

controls—with, for both estimates, the large 95% confidence interval to be expected under these circumstances, namely, between 0.3×10^{-5} and 5.5×10^{-4}. At the time of the publication of their results, DS86 doses were not available. We have now calculated that, on the basis of an RBE of 20 for neutrons, the mean conjoint parental gonad dose for the parents receiving ≥ 0.01 Sv is 0.60. Unlike the results to be presented for the other studies, we have not calculated the regression of indicator on parental exposure because of the paucity of proven mutations.

5. Frequency of Sex-Chromosome Aneuploids

The types and frequencies of sex-chromosome aneuploidy in the individuals comprised by these same two samples have been presented by Awa et al. (1987). On the same population base as given in table 2, there were in the sample a total of 43 individuals with sex-chromosome aneuploidy—19 were children of proximally exposed parents and 24 were children of distally exposed parents. Individuals with some of these abnormalities are sterile (XXY), and with others may exhibit decreased fertility (XXX); furthermore, XYY and XXX individuals only rarely have similarly affected children (Evans 1977). Thus a very high proportion of these persons may be presumed to result from primary nondisjunction (i.e., mutation) in a parent, but specific family studies on these individuals were not undertaken. Sex-chromosome aneuploidy is increased in the offspring of older parents (Court Brown et al. 1969). Accordingly, an increase in such findings in the children of proximally exposed survivors could result from either a direct or an indirect effect of radiation, the latter being mediated by an acceleration of aspects of the aging process.

For the estimates of doubling dose that follow, it is necessary to express the findings as a linear regression on dose. We have accordingly extended the analysis of Awa et al. (1987), assigning parental gonad doses on the basis of the DS86 schedule. Table 3 presents the data as tabulated for the derivation of a linear regression, fitted by the method of maximum likelihood. The regression of indicator on dose (Sv) is .00044 ± .00069, with an intercept of .00252 ± .00044. To the extent that this abnormality may have been inherited in some cases, the mutation rate for the trait would be overestimated, but there should be no bias with respect to a radiation effect. The estimate of average conjoint parental gonad dose is, of course, that given in the preceding section.

The finding during the clinical phase of the Genetics Program that diagnoses of Down syndrome (trisomy 21) were no more common among the offspring of parents exposed to the bombs than they were among the offspring of unexposed parents (Schull and Neel 1962) is further evidence that the ATB exposure did not produce an increase in aneuploidy in this population.

6. Frequency of Mutation Altering Protein Charge or Function

During the past 40 years there have been frequent conjectures concerning hidden (i.e., recessive) genetic effects of the bombs, which effects would only become evident some generations after the bombings. The ad-

Table 3

Data for a Least-Squares-Fit Linear Regression of Sex-Chromosome Aneuploidy on Conjoint Parental Gonadal Exposure, Based on DS86 Doses

Conjoint Parental Gonad Exposure (Sv)	No. of Children	Mean Dose (Sv)	No. of Aneuploids
.0	8,225	0	24
.001–.050	1,346	.024	0
.051–.100	951	.073	2
.101–.500	2,693	.263	9
.501–1.000	1,531	.719	3
1.001–1.500	686	1.227	2
1.501–2.000	295	1.716	1
2.001–2.500	157	2.228	0
≥2.501	331	3.674	2
Unknown	83		0
	16,298		43

NOTE.—The data have been distributed so as to optimize the distribution for regression analysis.

vent of convenient techniques for the identification of variant proteins in the 1950s and 1960s provided an approach to this question. In 1972 a pilot study of the feasibility of electrophoretic examinations of what ultimately became a battery of 30 serum and erythrocyte proteins was initiated, followed in 1975 by a large-scale investigation which was only recently terminated. In 1979 a smaller-scale search for mutations that are characterized by loss of activity of a series of 11 carefully selected enzymes was initiated (Satoh et al. 1983); this program has also now terminated. The data from both programs have recently been presented in the *Journal* (Neel et al. 1988). The mutations detected by both these programs would be encountered in the heterozygous state and would have no discernible effect on the gross phenotype; that is, they conform to the classic definition of a recessive.

The program centered on the detection of rare (i.e., nonpolymorphic) protein variants, which were then subjected to family studies, to determine whether they had been inherited or were the result of mutation in the preceding generation. Only when both parents of a child with a variant have been studied is it possible to be certain concerning the occurrence of a mutation. Because both parents of a child who is found to possess a rare variant were not always available for study, an adjustment in the data, to "equivalent locus tests" (described in detail elsewhere [Neel et al. 1980]), is necessary. A child exhibiting a variant not present in either parent was termed "exceptional." Extensive tests were then performed to detect discrepancies between legal and biological parentage, resulting in some children being eliminated from further consideration. We have calculated that the a priori probability that a finding accepted as being due to mutation is in fact due to an undetected paternity discrepancy is approximately 4×10^{-8}. There were a total of seven apparent mutations in 1,256,555 locus tests.

The estimated mutation rate for loss-of-activity and electrophoretic variants was 0.6×10^{-5}/locus/generation, in both the proximally and the distally exposed parents, with the 95% confidence interval being $0.2-1.5 \times 10^{-5}$ for the former and $0.1-1.9 \times 10^{-5}$ for the latter. (Because disparate numbers of the two types of observations are involved, this is not an estimate of the total locus rate.) The estimated gonad doses in our recent description of this study were recognized as interim estimates, made while the DS86 dose system was still being phased in. We have now reestimated, with the standard procedure, these conjoint gonad doses for parents whose dose was ≥0.01 Sv, by using the total sample and value of 20 as the RBE for neutrons. The estimated conjoint dose for the proximally exposed 13,544 sets of parents is 0.41 Sv. Although the numbers are minimal for regression-type analysis, we have, in order to utilize these data in the calculations to come, proceeded with such an analysis, organizing the data as shown in table 4. Because the preceding tables are all on the basis of zygotes, we have in table 4 converted the locus determinations presented by Neel et al. (1988) to their zygotic equivalent by dividing the numbers of determinations in each exposure category by 2. The maximum-likelihood estimate of the slope of the regression is $-.00001 \pm .00001$ with intercept of $.00001 \pm$

Table 4

Data for a Least-Squares Fit, Linear Regression of Mutations Altering Protein Structure and/or Function on Conjoint Parental Gonadal Exposure, Based on DS86 Doses

Conjoint Parental Gonad Exposure (Sv)	No. of Tests[a]	Mean Dose	No. of Mutations
.0	264,310	0	4
.001–.050	109,776	.017	2
.051–.500	165,108	.201	0
.501–1.000	48,264	.708	1
≥1.001	40,820	1.988	0

NOTE.—The data have been distributed so as to optimize the distribution for regression analysis.

[a] The number of locus products examined for mutation will be twice the number of system tests.

.00001. As for the cytogenetic studies, since there were so few positive outcomes no effort was made to evaluate the effect of concomitant variables.

7. Sex Ratio among Children of Exposed Mothers

It was on the basis of a sex-ratio decrease among the offspring of irradiated female *Drosophila* that Muller (1927) first demonstrated the genetic effects of radiation, and in the early years of this Genetics Program the sex ratio—i.e., the ratio of male births to female births among the offspring of proximally and distally exposed survivors—received considerable attention (Neel and Schull 1956; Schull et al. 1966). This was based on the hypothesis that, because of the nature of sex-linked inheritance, any deleterious sex-linked mutations induced in irradiated mothers should reduce the sex ratio whereas similar mutations in irradiated fathers should increase the sex ratio. With the recognition that one X chromosome is inactive in the somatic cells of women (the Lyonization phenomenon), it became clear that sex-linked mutations induced in males were unlikely to have a dominant lethal effect in females. Furthermore, the discovery of the occurrence and frequency of sex-chromosome aneuploids in the late 1950s introduced a new variable into the study of the sex ratio, a variable the direction of whose impact on the sex ratio could not be precisely specified at that time.

The cytogenetic findings now reveal that in this sample there is no evidence that radiation significantly alters the frequency of sex-chromosome aneuploids. This potential complication in the use of the sex ratio as an indicator of the production of sex-linked lethal mutations by the atomic bombs is thus disposed of. Whether and to what extent, with Lyonization, there are sex-linked dominant "point" mutations in women remains a somewhat moot question. We suggest that the least equivocal aspect of the sex-ratio data involves the effects of maternal irradiation. In this event, male conceptuses of exposed mothers should be susceptible to the effects of both sex-linked dominant and recessive induced lethal mutations, whereas female conceptuses should not be expected to exhibit reduced survival from any induced "recessives"; and, in addition, Lyonization would temper to some unknown extent the manifestation of sex-linked "dominant" mutations, as is clear from the review of Wettke-Schafer and Kantner (1983). Thus maternal exposure should, in principle, reduce the sex ratio.

The last presentation of the sex-ratio data (i.e., that of Schull et al. 1966), which included some 70% of births ever to occur to proximally exposed parents, revealed that for children born to proximally exposed mothers (fathers unexposed), the regression of offspring maleness on mother's radiation dose was .0027 ± .0040/Sv. The observation is thus insignificantly counterhypothesis. Reanalysis might alter the magnitude but would not change the sign of the regression, but since we see no way to incorporate such an observation into a truly quantitative estimate of either the minimal or probable doubling dose (see below), we have not reanalyzed these data on the basis of the DS86 dose estimates.

8. Growth and Development of the F₁

The variables considered in the preceding seven sections are discrete or qualitative attributes, in whose etiology the role of single genetic events such as mutations can be estimated and the analysis of which is relatively straightforward. A body of data concerning the effect of parental exposure on certain quantitative traits is also available. For these traits, the production

of deleterious mutations in the parents in consequence of exposure to the bombs might be expected to depress the mean and to increase the variance—but in no simple manner amenable to a doubling-dose calculation (see below), unless rather restrictive assumptions are made.

During the period when the study population was under surveillance for untoward pregnancy outcomes (see section item 1 above), birthweights were obtained on all 76,617 infants who entered the study. An analysis restricted to the birthweights of live-born infants failed to reveal any suggestive effect of parental exposure (Neel and Schull 1956). A reexamination—with respect to body length, head and chest circumference, and body weight—of 18,498 of these infants between ages 8 and 10 mo also failed to reveal any between-infant differences related to parental exposure (Neel and Schull 1956). Furusho and Otake (1978a, 1978b, 1979, 1980, 1985) have extended this analysis by relating the annual measurements (stature, weight, and chest circumference) of Hiroshima school children to the irradiation history of their parents. Parental exposure had no certain or even suggestive effect on the means or variances of these measurements. These analyses employed T65DR doses; it has not seemed relevant to reanalyze the material by DS86 dose assignments. Since these data do not lend themselves to the types of statistical analysis that follow, we will not refer to them again.

Implications of the Data

We propose to develop a perspective on the present findings by three approaches. In all three of these approaches, we take it as a given that the exposure resulted in mutations in some survivors of the atomic bombings, inasmuch as, without exception, under controlled laboratory conditions ionizing radiation has produced mutations in every properly studied plant and animal species.

Approach 1

The simplest approach is to content ourselves with the statement that the children of what is almost certainly the most highly irradiated population in the world's history provide, with the present indicators, no statistically significant evidence that mutations were produced in their parents. The indicators employed in this study run the gamut of genetic change, from the effects of nucleotide substitution to the effects of gain or loss of an entire chromosome. The absence of statistically significant findings does not deny the possibility that the exposed survivors sustained an increased germinal mutation rate, but it characterizes this rate as being too small to be detected unambiguously in this sample by the approaches thus far employed. On a more positive note, these studies have produced an extensive body of data against which to evaluate empirically both past and future surmises concerning the genetic consequences of exposure to ionizing radiation. In particular, the studies should prove reassuring to that considerable group of exposed Japanese and their children, without whose magnificent cooperation these studies would have been impossible and who have over the years been subjected to a barrage of exaggerations concerning the genetic risks involved. To stop at this point in considering the implications of these findings would in our opinion be an abrogation of genetic responsibility.

Approach 2

The next level of sophistication in treating the data attempts to determine what the immediate mutational contribution to each indicator is and then, given this contribution, to determine what genetic "doubling dose" can be excluded for each indicator at various probability levels. Where endpoints such as UPOs and early death are employed, the estimates are to some extent time-place specific, depending on the severity of selection against deleterious genetic traits. For this and the next approach we must estimate the contribution of spontaneous mutation in the parents to each of the endpoints of this study. These critical estimates are developed in the Appendix.

In a linear model, the *excess* relative risk (which can by definition be negative), say RR, is (β/α), where α and β are the background frequencies of events of interest, i.e., the constant in the model and the slope of the regression of the frequency of events on dose, respectively. The variance (large sample) of this ratio is

$$V\left(\frac{\beta}{\alpha}\right) = \left(\frac{\beta}{\alpha}\right)^2 \left[\frac{V(\alpha)}{\alpha^2} + \frac{V(\beta)}{\beta^2} - 2\frac{\text{Cov}(\alpha,\beta)}{\alpha\beta}\right].$$

In any specific situation, α and β are, of course, replaced by their sample estimates. Now, by definition, the "doubling dose," D_d, is $\beta D_d = \alpha$ or $D_d = (\alpha/\beta)$, where it will be noted (α/β) is merely the reciprocal of the excess risk per unit dose. Thus, an estimate of the "doubling dose" is $D_d = (1/RR)$. If, now, the background rate, α, is assumed to consist of both a genetic component, C_G, which is augmented by parental exposure to ionizing radiation, and a nongenetic which does not respond to radiation, then the doubling-dose estimate

given above must be multiplied by the proportion of the background events presumed to be genetic.

In our situation, we reject the possibility of a negative risk implied by the negative regressions on parental exposure given earlier for malignancies and protein mutations, regarding these as random departures from some "true," small positive term. We can, however, even with a negative regression, use the variance term given above to calculate what lower value to the D_d can be excluded at stated probability levels for each of these indicators. Thus,

$$D_{d(min)} = \frac{1}{[(\beta/\alpha) + Z_\alpha \sqrt{V(\beta/\alpha)}]},$$

where Z_α is the normal deviate at the desired probability level.

Table 5 presents the results of the appropriate calculations. As noted, the data on balanced reciprocal translocations, sex ratio, and physical development were not felt to be adequate for this type of calculation. The first three columns in table 5 develop the proportion of the indicator deemed due to mutation in the parental generation, from the Appendix. There follow the values of β and α as derived from the appropriate regression, as presented in earlier sections and in previously published papers. Finally, the lower values of the doubling dose at three probability levels are presented. The minimal doubling doses for the various endpoints at the 95% probability level range from 0.05 to 2.27 Sv.

In previous calculations of the doubling dose, having reached this point in the calculation, we have been persuaded to divide the result by 2, to convert from a zygotic doubling dose to the gametic doubling dose in which radiation effects are more usually presented (see Schull et al. 1981; more recently Neel et al. 1989a, 1989b). We now believe this correction was incorrect. Consider the UPO data set. The estimated contribution of spontaneous parental mutation to the indicator is the contribution of both sexes. Assume the contributions are equal, so that $\alpha_{GF} = \alpha_{GM}$ and the total contribution is 2α. The regression of the indicator on dose (Sv) will be $\beta_F + \beta_M$, and that on the assumption of equal effects in the sexes will be 2β. Consider a hypothetical situation where $\alpha_{GF} = \alpha_{GM} = .001$ and where $\beta_F = \beta_M = .001/\text{Sv}$, i.e., where 1.0 Sv is the doubling dose. As we now approach this problem, $\Sigma\alpha/\Sigma\beta = .002/.002 = 1.0$, and we have the correct estimate of the D_d without reduction by a factor of 2. Since we cannot separate male and female effects, this estimate is the average of the sexes and in the strict sense is valid only when, as for the Hiroshima and Nagasaki situations, approximately equal numbers of the two sexes were exposed. It cannot be applied to either sex alone. Departures from the assumptions of the equality of spontaneous and induced mutation rates in the two sexes will not alter the principle. For instance, were the female to exhibit both spontaneous and induced rates of zero, the ratio of $\Sigma\alpha/\Sigma\beta$ would still be the *average* gametic doubling dose. To restate the now obvious, while the estimate of the impact that spontaneous mutation in the parental generation has on the zygote

Table 5

Estimate of Gametic Doubling Doses That Can Be Excluded at Specified Confidence Levels by These Data

Trait	Observed Total Background (a)	Mutational Contribution Background (b)[a]	Mutational Component (%[b/a])	Regression β_{Sv}	α	Sv at Lower Confidence Limit of 99%	95%	90%
UPO	.0502	.0017–.0027	3.4–5.4	.00264 ± .00277	.03856 ± .00582	.14–.23	.18–.29	.21–.33
F_1 mortality	.0458	.0016–.0026	3.5–5.7	.00076 ± .00154	.06346 ± .00181	.51–.83	.68–1.10	.81–1.32
F_1 cancer	.0012	.00002–.00005	2.0–4.0	–.00008 ± .00028	.00104 ± .00033	.04–.07	.05–.11	.07–.15
Sex-chromosome aneuploids	.0030[b]	.0030	100	.00044 ± .00069	.00252 ± .00043	1.23	1.60	1.91
Loci encoding proteins	.000013[b]	.000013	100	–.00001 ± .00001	.00001 ± .00001	.99	2.27	7.41

NOTE.—See text for further explanation.
[a] Per diploid locus.
[b] Observed zygotic mutation rates.

reflects the contribution of two gametes, the regression term also reflects the contribution of two haploid sets of chromosomes, and the factors of 2 cancel out.

Estimates such as these can only be combined to yield an average estimate on the assumption there is no heterogeneity among the individual estimates, i.e., on the assumption they have been drawn from a common pool. An extensive literature suggests that the true doubling dose for radiation-induced sex-linked chromosomal aneuploids and protein variants (resulting from nondisjunction and nucleotide substitutions, respectively) is probably higher than the doubling dose for the other three indicators (resulting predominantly from insertion/deletion/rearrangement events and unrepaired chromosomal breaks). We have therefore obtained separate estimates for these two types of indicators, using the formula

$$D_d \text{ pooled} = \frac{1}{[(\beta/\alpha) + Z_\alpha \sqrt{V(\beta/\alpha)}]},$$

where Z_α is again the desired probability level and where (β/α) is now the average of the sum of information-weighted (reciprocal-of-variance) individual estimates and where $V(\beta/\alpha)$ is the variance of this average. The minimal doubling dose at the 95% probability level which results from pooling the estimates for UPOs, F_1 cancer, and F_1 mortality is 0.63–1.04 Sv. The similar estimate for the sex-chromosome aneuploids and protein mutations is 2.71 Sv. These estimates must not be misinterpreted. They are the minimal (95%-probability-level) doubling doses derived from each of these five indicators or some combination thereof but are *not* a lower bound to the estimate to be derived in the next section.

Approach 3

Finally, we come to the third and ultimate level in treating these data, an effort to tease out of these findings a reasonable and preliminary estimate of the most likely gametic doubling dose of radiation for these indicators combined. *In the societal context, the doubling-dose estimate most needed is of the amount of parental radiation that in the first generation will result in the same impact of genetic mortality and morbidity as would result from spontaneous mutation in these same parents.* Ideally, the genetic doubling dose is calculated as that amount of radiation which increases the *total* impact of spontaneous mutation by 100%. It covers the spectrum from the impact of nucleotide substitutions to gains or losses of entire chromosomes. Since the components of this spectrum may present differing frequencies, the doubling dose should be an integrated treatment of both the *frequency* and the *radiosensitivity* of each component.

Because the lack of statistically significant findings raises the possibility that we are simply manipulating the "noise" in the study, some would advocate that the analysis be suspended at this point. Again we reiterate the point that unless humans differ from every other properly studied animal, as well as from plants, mutations must have been produced by this exposure; we wish cautiously to explore what the further genetic message in these data might be. At the mutagenic exposures which any substantial proportion of a human population will survive, there may never be a statistically significant demonstration of the genetic effects of the radiation of humans—and certainly there will be none within the next decade, pending the development of molecular strategies. However, the methodological problems presented by the attempt to generate a doubling dose from these data are unique. We present here one treatment, which in time may be superseded by others.

Each of the eight estimators of a genetic effect of the bombs which we have been able to generate is associated with a relatively large error term, and individually these estimators are fragile reeds on which to lean in building a case. Collectively, however, they amount to a major corpus of data. The estimate to be developed, like all such estimates, is of course *specific for these indicators*, which, however, represent our best effort, under the circumstances, to cover the morbidity-mortality resulting from the total spectrum of mutation. It is also, we repeat, time specific and place specific; were infant and childhood mortality greatly reduced (and selection against genetic imperfections presumably less stringent), the doubling-dose estimate might be different.

In an epidemiological study such as the present one there is a problem which the laboratory-based investigator is spared. Whereas in the laboratory one can by suitable doses and sample sizes usually contrive to demonstrate a significant positive response of the genetic endpoint to radiation, epidemiological studies involve a nonmanipulative situation. In the present study, we are confronted with several small negative regressions, with no current prospect of extending the data. Since we accept the proposition that exposure to the bombs produced genetic damage, these findings must be viewed as random fluctuations around some positive value. Inasmuch as these spuriously negative effects are presumably counterbalanced by some spuriously high positive effects, the former cannot be disregarded and must be

folded in to the estimate. There is no precedent for attempting to develop a doubling-dose estimate under these circumstances.

Calculation of the Doubling Dose

We suggest that the population samples for the various studies are sufficiently overlapping representatives of the same cohort that the various regressions can be additively combined to yield a composite picture. The ability to pursue this approach is a unique feature of this study not previously feasible for other studies. In particular, we contrast this approach to that of the murine studies, where estimates derived from different inbred strains have been combined in various fashions to obtain an average doubling dose (see Lüning and Searle 1971).

To implement this approach we need to estimate the total impact on the morbidity-mortality indicators of this study of spontaneous mutation in the parental generation. We will adopt the position that both the impact of mutation resulting in unbalanced chromosomal exchanges (which we presume to be similar numerically to the impact of the balanced chromosomal exchanges we have measured) and the impact of the heterozygous effects of recessive mutations as studied through protein variants are subsumed within the UPOs and prereproductive mortality in the live born (exclusive of cancer). As table 6 shows, a summation of the appropriate estimators of the impact of spontaneous parental mutation given in table 5 yields a figure of between 0.632% and 0.835%. (The total is, of course, not accurate to the third decimal place but results from simply summing the various indicators.) To obtain the most reasonable estimate of the genetic impact of radiation on the totality of the endpoints, we propose simply to combine the regressions of table 5 additively. We justify such cumulation on the grounds that each regression involves the relationship to radiation exposure, in the same cohort, of specific, independently determined genetic events. As table 6 shows, the value is +.00375. Note, however, that this involves the manipulation of negative (counterhypothesis) terms.

The lower estimate of the doubling dose which results from this procedure is .00632/.00375, or 1.69 Sv; the upper estimate is .00835/.00375, or 2.23 Sv. Note that these limits reflect biological uncertainties about the parameters but do not take into consideration the additional error inherent in the estimation procedure, an error which must be relatively large. Recall (1) that, because of the extreme paucity of numbers, we elected *not* to try to incorporate into this regression approach the data on mutation resulting in reciprocal translocations but (2) that the rates were the same in the two populations. Recall also our inability, because of technical reasons, to incorporate into the calculations the sex-ratio results (which are counterhypothesis) and the data on growth and development (which reveal no hint of a radiation effect). Any use of these data in these calculations could only drive the estimate of the doubling dose upward.

We find it impossible at this time to place true confidence bounds on this estimate. In addition to the errors in estimating both the true value of the parameters and their regression on dose, there is the additional factor that our estimates are geared to specific indicators and to a specific time and place. We believe, however, that throughout we have been conservative in our assumptions. Furthermore, we have reported that the parents who were unexposed to the effects of the atomic bombings (who came to Hiroshima and Nagasaki after the bombings, as released servicemen, repatriates,

Table 6

Summary of Regression of Various Indicators on Parental Radiation Exposure and of Impact of Spontaneous Mutation on Indicator

Trait	Regression/Combined Parental Sv	Contribution of Spontaneous Mutation
UPO	+.00264	
F_1 mortality	+.00076	.0033–.0053
Protein mutations	−.00001	
Sex-chromosome aneuploids	+.00044	.0030
F_1 cancer	−.00008	.00002–.00005
	.00375	.00632–.00835

spouses, or immigrants) were slightly younger and had a little more education and somewhat higher occupational ratings than did the exposed (Neel and Schull 1956; Kato et al. 1966; Neel et al. 1974). In addition, the analysis has assumed complete recovery of the mother from her radiation exposure, i.e., has assumed no maternal effect in this regard. To the extent socioeconomic and maternal effects exist, they can only inflate the apparent effects of the bombs. The true value for the genetic doubling dose of radiation under these circumstances is thus apt to be higher rather than lower than our estimate.

Discussion

We have summarized an effort extending over >40 years to quantify the genetic effects of the atomic bombs detonated over Hiroshima and Nagasaki. No statistically significant findings have emerged. Since, however, we can scarcely doubt that some genetic damage occurred in consequence of that exposure, we have taken the observations at face value and have explored the inferences which can be drawn from them. On the one hand, we have generated estimates of the minimal doubling dose, at various probability levels, consistent with each of the five indicators which lend themselves to this calculation. On the assumption that three of the individual estimates of minimal doubling doses (UPO, F_1 mortality, and F_1 cancer) were all fluctuations about some common mean, we have derived an estimate of the average 95% exclusionary power of these indicators, which is a gametic doubling dose of 0.63–1.04 Sv. For protein mutations and sex-chromosome aneuploids, which from the genetic literature may be assumed to exhibit higher doubling doses, the average is 2.71 Sv. Even the lower of these two figures is larger than the value commonly cited as the doubling dose on the basis of the mouse data (see below). On the other hand, by a different approach, we have calculated a most probable gametic doubling-dose estimate, which is 1.69–2.23 Sv. This estimate is somewhat higher than that given in several recent, preliminary reports (Neel et al. 1989a, 1989b), primarily because of the introduction of the factor of 2 mentioned earlier. In 1981, on the basis of a smaller body of data, and employing somewhat different analytic methods and the now superceded T65DR dose schedule, we suggested that the most likely estimate of the gametic doubling dose of acute radiation was 1.56 Sv (Schull et al. 1981). This estimate was in error by the factor of 2; the value should have been 3.12 Sv. Given (1) that the gonad radiation exposures (Sv) estimated with the DS86 dose schedule are on average about 70% as large as those estimated with the T65DR schedule, (2) the additional data which have become available, and (3) the somewhat more conservative nature of the present analysis, the agreement between the two estimates is satisfactory.

We emphasize once again that these estimates are specific to the mix of gonadal exposures sustained by approximately equal numbers of mothers and fathers, the socioeconomic conditions in Japan during this period, and the collection of endpoints under scrutiny—but that we believe that, with respect to the latter, they are both representative and societally important. We also recognize the obvious—that genetic effects resulting in fetal loss prior to the fifth lunar month of pregnancy would not be detected by this study but that early fetal losses would have both a relatively small emotional impact on the parents and a relatively small cost to society.

The Extrapolation to the Effects of Chronic Radiation

Short of a nuclear disaster of some type, most human radiation exposures will be a mixture of chronic but very-low-level ionizing radiation punctuated by small, low-level bursts of occupational or medical exposure. In the specific locus test system, the fractionated or chronic irradiation of mice yielded, for a given total (high) dose, about one-third to one-quarter as many point mutations in the spermatogonia of males as did acute, relatively high-dose X-ray irradiation (Russell et al. 1958). For point mutations in female mice, and for such other indicators as chromosomal damage, the dose-rate effect is even greater (see review in Searle 1989). The radiation doses received from the atomic bombings were very high energy and very brief—the estimated individual gonadal doses among those receiving increased radiation were 0.01–3.0 Sv, the average being about 0.25 Sv.

The complex issues involved in extrapolating from such a situation to the effect of the low-level, chronic or intermittent exposures which are the usual lot of human populations are well discussed in report 64 of the National Council on Radiation Protection (1980). There are, unfortunately, no experimental data on the spectrum of radiation exposures observed in Hiroshima and Nagasaki. We must therefore employ an indirect approach. We will assume a linear-quadratic relationship for acute radiation dose and genetic effects. The precise relationship varies for the endpoint under study; we will, to be conservative, use the relationship derived for "point" mutations in the mouse, which has a smaller

quadratic component than do other endpoints, such as chromosomal damage. Employing both the gonadal dose distribution observed in the parents entering into this study and the parameters (developed by Abrahamson and Wolff [1976]) for the linear and quadratic components of the radiation effect for specific locus mutations in the mouse, we suggest the dose-rate factor to be applied to this spectrum of radiation and endpoints is about 2. (Technically, the neutron component does not need adjustment for a dose-rate factor, but the neutron component in the dose is so small it can be included in the adjustment with small error.) A major consideration in this suggestion—which is really quite critical to the argument—is the fact that about 40% of the total gonadal dose in the population of parents under study results from estimated individual exposures of ⩾0.50 Sv, at which point the quadratic component begins to become significant. In our view, given the higher dose-rate factors for many of the other endpoints included in this study, adjustment by a factor of 2 is quite conservative. This leads to a gametic doubling-dose estimate of 3.38–4.46 Sv for chronic or low-level intermittent low LET radiation.

The data have been presented in such a fashion that those who wish either to (a) make other assumptions about the contribution that spontaneous mutation in the parental generation makes to the indicator, (b) weight the various indicators in some other fashion, (c) apply a different RBE to the neutron values, or (d) develop alternative statistical approaches are in a position to do so. A question of this magnitude deserves multiple approaches. Two key aspects of this treatment are (1) the assumptions concerning the contribution that mutation in the parental generation makes to the endpoints and (2) the dose-rate factor appropriate to data of this type. The future will undoubtedly bring refinements to these assumptions.

The Discrepancy between This Estimate of a Genetic Doubling Dose and That Projected from Studies on Mice

Thus far the principal human surrogate in efforts to set genetically permissible limits to radiation exposures has been the domesticated house mouse. On the basis of extrapolation from studies on mice (primarily male), various national and international committees have set the doubling dose of acute gonial radiation for humans at 0.3–0.4 Sv, with limits of 0.1 and 1.0 Sv (reviewed in Committee on the Biological Effects of Ionizing Radiations, National Research Council 1980; United Nations Scientific Committee on the Effects of Atomic Radiation 1986). Because of the relatively very large gonadal doses employed in the mouse experiments, doses unrealistically large from the human perspective, the estimate of the effects of the *acute* exposures employed in mice cannot be compared directly with the present estimate of the doubling dose for the atomic bomb survivors, but the extrapolations to chronic or low-level pulse exposures can be compared. For this type of exposure, the United Nations Scientific Committee on the Effects of Atomic Radiation has for some years projected a doubling-dose value of 1.0 Gy (or Sv) (United Nations Scientific Committee on the Effects of Atomic Radiation 1986), whereas the Committee on the Biological Effects of Ionizing Radiations in 1980 suggested a range of 0.5–2.5 Sv (Committee on the Biological Effects of Ionizing Radiations, National Research Council 1980). Our estimate thus points to higher values than do past modal extrapolations from the mouse. The uncertainties in both the mouse extrapolation and human estimate are such that we do not see our way to applying statistical tests to the significance of this apparent difference, but it seems unlikely that at the 5% probability level there is overlap between the upper range of the mouse projection and the lower range of the present estimate.

There are many differences between the derivation of the present estimate and the past extrapolations from the murine data to humans. Elsewhere we are undertaking an extensive review and reanalysis of the murine data, to determine to what extent the difference results from the methodologies employed and to what extent it results from true species differences (Neel and Lewis accepted). Here we note only four salient points: (1) The sex-chromosome aneuploids, whose frequency of occurrence is relatively uninfluenced by radiation, play a much larger role in the present data than they have played in the extrapolations from mice. We argue that these are a very prominent component of the mutational burden in humans and that they must be treated accordingly. Even when these data are excluded, however, the human doubling-dose estimate for acute radiation would be some three times higher than the estimate based on the mouse data. (2) We suggest, from both direct and circumstantial evidence, that the Russell (1951) seven-locus test system, which has quite properly served as the cornerstone in most extrapolations, may have inadvertently utilized an unusually mutable set of loci (see Favor 1989; Neel et al. 1989b). (3) The extended period over which these observations have been assembled allows for any attenuation of genetic effects

associated with cell proliferation phenomena. (4) Data on exposed females contribute to the present findings to about the same extent as do data from exposed males; were the human female to exhibit the same paucity of induced mutation in offspring conceived many years after the bombings as is observed in the late litters of irradiated female mice (Russell 1965), this phenomenon would be incorporated into the estimate.

The Future of These Studies

The implications of these findings with respect to the vulnerability of the human species to extraneous genetic insult are such that every effort should be made to confirm or modify the conclusions of this study. Although different mutagens undoubtedly have different modes of action, and although one should not generalize from these studies to the consequences of other mutagenic exposures, the situation in Hiroshima and Nagasaki will provide, for the foreseeable future, humankind's best opportunity to understand the homeostatic properties of the human genome. The Genetics Program at RERF is currently establishing constellations of lymphocytoid cell lines, each constellation derived from a father, a mother, and one or more children. It is planned to develop ≥500 such constellations from proximally exposed survivors and their offspring and to develop another 500 such constellations from parents receiving no increased radiation and from their children. It is contemplated that, as the necessary techniques achieve the requisite level of efficiency, these lines will be examined with reference to mutations involving selected proteins visualized on two-dimensional polyacrylamide gels (see Neel et al. 1984, 1989b) and also at the DNA level with reference to selected DNA probes (see Delehanty et al. 1986; U.S. Congress, Office of Technology Assessment 1986). These studies should substantially extend our understanding of what are usually termed point mutations. In these studies, we strongly advocate the use of a battery of *representative* probes derived from genes of known function, since the relevance to human affairs of mutation in DNA of unknown function, as studied with so-called anonymous probes, is dubious; in view of the evidence for locus differences in mutability in humans (Chakraborty and Neel 1989), these probes must be selected with extreme care. Even with some of the technical improvements in the offing, these studies will be almost as laborious and time-consuming as those that have preceded them. There are no quick and easy answers in this field of investigation.

Acknowledgments

We are greatly indebted to Drs. Donald Pierce and Ranajit Chakraborty for many helpful discussions regarding the nuances of the minimal doubling-dose approach. This program was supported by the Radiation Effects Research Foundation, a United States–Japan binational foundation; J.V.N. also acknowledges the support of U.S. Department of Energy grant ER60089.

Appendix

The Mutational Component in the Indicators

The indicators to be used in the calculation of a minimal and most probable doubling dose are of two types. With respect to protein variants, sex-chromosome aneuploids, and balanced structural rearrangements of chromosomes, the individuals resulting from mutation in the parental generation can be directly identified by the appropriate studies on the parents, and we have already indicated in the text the base-line mutation rates observed in the offspring of the distally exposed. For three indicators—untoward pregnancy outcomes, mortality among live-born infants, and F_1 cancer—we must proceed more circumspectly. We can only rarely among the children exhibiting untoward pregnancy outcomes, early death, or childhood cancer clearly specify which children exhibit this endpoint because of parental mutation; they are usually anonymous in a much larger pool of similar outcomes because of a variety of other causes, some known and some unknown. For these endpoints, a different approach must be pursued. We must attempt to estimate the fraction of these children that results from spontaneous mutation in their parents. The genetic events contributing to these endpoints in the case of untoward pregnancy outcomes and early death can be subsumed under three headings:

1. *Autosomal aneuploidy.* The most important example of this is trisomy 21 (Down syndrome), concerning whose responsiveness to maternal irradiation there is a considerable and controversial literature (reviewed in Cohen et al. 1977; Uchida 1977). The studies on untoward pregnancy outcomes were conducted prior to the advent of modern cytogenetics, but on clinical grounds we have estimated the frequency of Down syndrome to be 0.13% in a series of 9,440 8–10-month-old infants born to unexposed Japanese mothers and fathers (Schull and Neel 1962), a figure in satisfactory agreement with the results of later cytogenetic studies on newborn Japanese infants (0.08%; Maeda 1979;

Kuroki 1983). An effort on our part to locate, some 20 years later, the children found to exhibit Down syndrome in the course of the earlier study revealed a high mortality rate. The other autosomal trisomies which would be encountered in a study such as this (i.e., trisomy 13 and trisomy 18) are much less common (together summing to about 0.04% in Caucasian [de Grouchy and Turleau 1983; Goldstein and Nielsen 1988] and Japanese [Matsunobu et al. 1989] newborn infants), and the affected children, if liveborn, die within the first week. We will take the position that during the period of this study 0.10% of the children born to unexposed parents died in childhood because of autosomal trisomy, almost always because of mutation in a parental gamete. This high death rate reflects the relatively harsh postwar conditions during which the bulk of our cohorts were assembled, and it would not prevail in Japan today.

2. *Unbalanced structural rearrangements of chromosomes.* Children with major deletions or duplications of the autosomes usually exhibit gross abnormality and, if not stillborn, seldom survive for long. Cytogenetic techniques were not available during the years the clinical data were being collected. Presumably, the frequency in this series would be very similar to that observed in the surveys of Maeda (1979) and Kuroki (1983) on Japanese newborn infants—namely, 0.10%. The appropriate studies on Caucasian infants reveal that, among live-born infants, the rate with which live-born infants exhibit such chromosomal abnormalities because of mutation in the preceding generation is 1.6×10^{-4} (Jacobs 1981). The frequency with which abortuses exhibit such findings is 50-fold greater (Machin 1974; Alberman and Creasy 1977; Evans 1977). Accordingly, we surmise that in our series of live- and stillborn pregnancy terminations after 20 wk of gestation, the observed mutation rate would be at least twice that in live-born term infants—namely, 3.2×10^{-4} (0.032%). This is a minimal figure because it is based on "unbanded" chromosomes (see Hook et al. 1989), but a precise correction for nondetection is not possible at this time.

Children with balanced chromosomal rearrangements occurred in our series with a frequency of 0.22%, in good agreement with the findings of Maeda (1979), Kuroki (1983), and the data on Caucasians (Jacobs 1981). Our family studies on the children of unexposed parents revealed a spontaneous mutation rate for this endpoint (based on a single event) of 1.0×10^{-4}/gamete/generation. This estimate is consistent with the results of studies on live-born Caucasian infants (1.9×10^{-4}/gamete/generation; Jacobs 1981). Although this rate should be factored into our thinking, since it sets the stage for children who will in later generations die of the consequences of unbalanced chromosomal abnormality, in many respects this type of mutation presents the same problem in "forward accounting" as do the "recessive" mutations, and it will be further discussed when they are considered. Neglecting this cause of abnormality, we will assume from the data of this and the preceding paragraph that some 0.03% of all newborns will exhibit gross defect and/or die early because of mutation resulting in a chromosomally unbalanced gamete.

3. *Point mutations.* We come now to the very difficult task of estimating the contribution that spontaneous, so-called point mutation in parents makes to death and congenital defect in their children of the immediately following generation. Under the term "point" we include small chromosomal deletions and duplications. It is customary to distinguish between so-called dominant and recessive mutation—with the dominants visualized as exerting immediate impact and with the effect of the recessives for the most part becoming apparent only generations later—but in fact there is no clear dividing line. Dominant deleterious traits arising through mutation may persist for some generations before ultimate elimination by negative selection, but, conversely, "recessive" mutations may have an immediate impact through a deleterious effect on survival in heterozygotes, estimated in *Drosophila* to amount to 2%–4% in the case of newly arisen spontaneous and radiation-induced mutations (Stern et al. 1952; Hiraizumi and Crow 1960; Mukai et al. 1972). The various attempts to identify an average heterozygous effect of induced recessive lethals in mice have yielded conflicting results; Grahn et al. (1972) concluded that "heterozygote disadvantage is probably no more than 5% for mammalian genetic systems." More recently, Russell and Rinchik (1987) have presented additional evidence that some specific deletion-type "point" mutations as heterozygotes have deleterious effects on survival and weaning weight. An estimate of 2% disadvantage would seem to be conservative. It is arbitrary whether those heterozygous effects of "recessives" should be included in the impact of the dominant or recessive component of mutation; we include them with the latter because, in principle, their impact is in general spread over so many more generations than is the impact of the dominants.

In this situation, there are two "benchmarks" with respect to humans: (1) the numerous studies on the rate

with which spontaneous mutation results in a variety of dominantly inherited phenotypes and (2) the more recent data on the rate with which spontaneous mutation results in "recessively" inherited protein abnormalities. An absolutely lower bound to the impact of "point" mutation is provided by studies of the cumulative rate with which mutation results in the genetic syndromes exhibiting dominant or sex-linked inheritance. Thus, considering the results of a variety of specific studies as well as population surveys, the successive Committees on the Biological Effects of Ionizing Radiation of the U.S. National Academy of Sciences (see Committee on the Biological Effects of Ionizing Radiations, National Research Council 1980) and the various United Nations Scientific Committees on the Effects of Atomic Radiation (see the 1986 Report) have in their various publications tended to converge on the estimate that, in each generation, about 0.2% of all newborns will to some extent be handicapped (and sometimes will die prematurely) from spontaneous dominant or sex-linked mutation in their parents that results in identifiable diseases/syndromes.

This figure does not appear to include some of the growing list of syndromes due to chromosomal microdeletions (reviewed in Dallapiccola and Forabosco 1987), syndromes for which early death or sterility prevent a demonstration of genetic transmission. This figure also does not appear to include either any fraction of the cytogenetically normal children with major congenital defect not corresponding to a known genetic entity or the dysplastic children with impaired survival and mental retardation (with whom all pediatricians are familiar and who are so often encountered in institutions), some proportion of which must be due to dominant mutation, chromosomal or "point." The figure also does not include children with "failure to thrive," a category comparable to that of mouse "runts," whose frequency is increased by radiation (Searle and Beechey 1986).

With respect to autosomal recessively inherited disease, we note our own recent estimates, arrived at by electrophoretic studies (see Neel et al. 1986), that spontaneous mutations characterized by amino acid substitutions in the proteins under study occur with a frequency of about 1×10^{-5}/gene/generation, or 1×10^{-8}/nucleotide/generation. If this rate can be extrapolated to the entire genome, then with 3×10^9 nucleotides/gamete, or perhaps 50,000 functional genes, the implication is that each zygote might receive on average one mutation of this type in the introns of a functional gene, or some 60 nucleotide substitution–type mutations scattered throughout the genome.

The ultimate potential for phenotypic ill effects of some of these mutations is best documented by the extensive studies of the hemoglobin loci, with the demonstration that nucleotide substitutions may result in abnormal hemoglobins (e.g., S or C) or in thalassemia.

In addition, enzyme variants characterized by total loss of activity (the basis ranging, at the DNA level, from key nucleotide substitutions or abnormally situated stop codons to insertion/deletion/inversion events) have about the same frequency in populations as do the rare protein variants presumably maintained by the pressure of mutation resulting in nucleotide substitutions (see Satoh et al. 1983). This similarity suggests that these may arise through the mutation of functional genes with at least the same rate (1×10^{-5}/locus/generation) as nucleotide substitutions. The ill effects of these null variants is documented by the many recessively inherited inborn errors of metabolism. If, again, there were 50,000 genes in the haploid genome, the average zygote might carry one newly arisen mutation resulting in either loss of the activity or nonsynthesis of a gene product. We suggest that the high level of child care exercised by humans reduces dramatically the heterozygote effects in our species prior to age 20 years— to perhaps 10%–20% of those observed in experimental animals. It follows that the heterozygous effects of spontaneous point mutation might result in the pre-reproductive death of approximately 0.2%–0.4% of all children.

In the light of all these considerations concerning point mutations, we will suggest that in this series *at least* 0.20% of the infants reaching the twentieth week of gestation who do not exhibit gross chromosomal abnormality will be characterized by an untoward pregnancy outcome/prereproductive death because of so-called point mutation of all types in the parents—and the total may well be 0.40%. In view of the total frequency of congenital defect/genetic disease documented for Japanese and other populations (Schull and Neel 1965; Baird et al. 1988; Czeizel et al. 1988), and in view of the data on heterozygote disadvantage, this seems a *conservative figure—conservative in the sense that, the lower this estimate, the lower the estimates of the minimal and probable doubling doses.*

When the results of the considerations of these three sections are totaled, we arrive at the following estimate of the impact of spontaneous mutation on untoward pregnancy outcomes/early deaths: approximately 0.10% + 0.03% + (0.20%–0.40%), or 0.33%–0.53%. We assign half of this to UPOs and half to F_1 mortality exclusive of cancer (see table 5). *Inasmuch as UPOs*

among the children of the unexposed had a 5.02% frequency, and inasmuch as mortality among live borns was 4.58%, this estimate attributes 3.44%–5.52% of these events to mutation in the parental generation. (We emphasize this point because of Sankaranarayanan's [1988] recent criticism of our previous doubling-dose calculation [Schull et al. 1981], a criticism based on his failure to understand that only the estimated mutational fraction entered into the calculation [see Neel 1988].) Again we draw attention to the fact that these estimates are for a period of Japanese history (1946–85) during the first third of which approximately 10% of children either were stillborn or died prior to the age of reproduction—i.e., in a period of Japanese history when selection against deleterious genetic traits may be assumed to have been higher than at present.

It is important to recognize that the estimate of 0.33%–0.53% is very specific—it is for the impact, in the *first* postbomb generation, of spontaneous mutation on the endpoints we have measured. Given the frequency we have attributed to "recessive" mutation, the cumulative impact of the recessive component of these newly arisen mutations, in both the heterozygous and homozygous state, will be greater than that in the first generation. How much greater the perceived impact is depends on how selection operates on these recessives. Since, unlike many of the committee reports referenced above, we are not estimating the total impact of an increased mutation rate but a doubling dose for specified endpoints, we do not have to deal with this very difficult issue of the ultimate impact of recessive effects.

With respect to childhood cancer, we observed 49 cases among 41,066 children of parents not exposed to the effects of the bombs, a normative frequency of 0.12%. However, our detailed analysis of these data and the literature on the genetics of childhood cancer suggested that only 3.0%–5.0% of these malignancies (i.e., a frequency of 0.002%–0.005% of all children) had the type of genetic basis (a germ-line mutation) such that their frequency might be increased by parental irradiation (Yoshimoto et al. 1990); this is the frequency to be used for the doubling-dose estimate.

References

Abrahamson S, Wolff S (1976) Reanalysis of radiation induced specific locus mutations in the mouse. Nature 264:715–719

Alberman ED, Creasy MR (1977) Frequency of chromosomal abnormalities in miscarriages and perinatal deaths. J Med Genet 14:313–315

Awa AA (1975) Genetic effects: cytogenetic study. J Radiat Res (Tokyo) 16 (Suppl): 78–81

Awa AA, Bloom AD, Yoshida MC, Neriishi S, Archer P (1968) A cytogenetic survey of the offspring of atomic bomb survivors. Nature 218:367–368

Awa AA, Honda T, Neriishi S, Sufuni T, Shimba H, Ohtaki K, Nakano NM, et al (1987) Cytogenetic studies of the offspring of atomic bomb survivors. In: Obe G, Basler A (eds) Cytogenetics: basic and applied aspects. Springer, Berlin, pp 166–183

Awa AA, Neriishi S, Honda T, Sufuni T, Hamilton HB (1973) A cytogenetic survey of the offspring of A-bomb survivors in Hiroshima and Nagasaki, Japan. Proceedings of the Fourth International Conference on Birth Defects, Excerpta Medica International Congress Series 297:92. Excerpta Medical, Amsterdam

Baird PA, Anderson TW, Newcombe HB, Lowry RB (1988) Genetic disorders in children and young adults: a population study. Am J Hum Genet 42:677–693

Chakraborty R, Neel JV (1989) Description and validation of a method for simultaneous estimation of effective population size and mutation rate from human populations data. Proc Natl Acad Sci USA 86:9407–9411

Cohen BH, Lilienfeld AM, Kramer S, Hyman CC (1977) Parental factors in Down's syndrome—results of the second Baltimore case-control study. In: Hook EB, Porter IH (eds) Population cytogenetics. Academic Press, New York, pp 301–352

Committee on the Biological Effects of Ionizing Radiations, National Research Council (1980) The effects on populations of exposure to low levels of ionizing radiation: 1980. National Academy Press, Washington, DC

Court Brown WM, Law P, Smith PG (1969) Sex chromosome aneuploidy and parental age. Ann Hum Genet 33:1–14

Czeizel A, Kis-Varga A (1987) Mutation surveillance of sentinel anomalies in Hungary, 1980–1984. Mutat Res 186: 73–79

Czeizel A, Kovács M, Kiss P, Méhes K, Szabo L, Oláh É, Kosztolányi G, et al (1988) A nationwide evaluation of multiple congenital abnormalities in Hungary. Genet Epidemiol 5:183–202

Dallapiccola B, Forabosco A (1987) Human microcytogenetics. Acta Med Auxol 19:5–33

de Grouchy J, Turleau C (1983) Autosomal disorders. In: Emery AEH, Rimoin DL (eds) Principles and practice of medical genetics. Vol 1. Churchill Livingstone, Edinburgh, pp 170–192

Delehanty J, White RL, Mendelsohn ML (1986) Approaches to determining mutation rates in DNA. Mutat Res 167: 215–232

Evans HJ (1977) Chromosome abnormalities among livebirths. J Med Genet 14:309–312

Favor J (1989) Risk estimation based on germ cell mutations in animals. Proceedings of the Sixteenth International Congress of Genetics. Genome 31:844–852

Furusho T, Otake M (1978a) A search for genetic effects of atomic bomb radiation on the growth and development of the F_1 generation. I. Stature of 15- and 17-year-old senior high school students in Hiroshima. RERF Tech rep 4-78. Radiation Effects Research Foundation, Hiroshima

——— (1978b) A search for genetic effects of atomic bomb radiation on the growth and development of the F_1 generation. II. Body weight, sitting height, and chest circumference of 15- to 17-year-old senior high school students in Hiroshima. RERF Tech rep 5-78. Radiation Effects Research Foundation, Hiroshima

——— (1979) A search for genetic effects of atomic bomb radiation on the growth and development of the F_1 generation. III. Stature of 12- to 14-year-old junior high school students in Hiroshima. RERF Tech rep 14-79. Radiation Effects Research Foundation, Hiroshima

——— (1980) A search for genetic effects of atomic bomb radiation on the growth and development of the F_1 generation. IX. Body weight, sitting height, and chest circumference of 12- to 14-year-old junior high school students in Hiroshima. RERF Tech rep 1-80. Radiation Effects Research Foundation, Hiroshima

——— (1985) A search for the genetic effects of atomic bomb radiation on the growth and development of the F_1 generation. V. Stature of 6- to 11-year-old elementary school pupils in Hiroshima. RERF Tech rep 9-85. Radiation Effects Research Foundation, Hiroshima

Genetics Conference (1947) Genetic effects of the atomic bombs in Hiroshima and Nagasaki. Science 106:331-333

Goldstein H, Nielsen K (1988) Rates and survival of individuals with trisomy 13 and 18. Clin Genet 34:366-372

Grahn D, Carnes BA, Farrington BH, Lee CH (1984) Genetic injury in hybrid male mice exposed to low doses of ^{60}Co γ-rays or fission neutrons. I. Response to single doses. Mutat Res 129:215-229

Grahn D, Lee CH, Farrington BF (1983) Interpretation of cytogenetic damage induced in the germ line of male mice exposed for over 1 year to ^{239}Pu alpha particles, fission neutrons, in ^{60}Co gamma rays. Radiat Res 95:566-583

Grahn D, Leslie WP, Verkey FA, Lea RA (1972) Determination of the radiation-induced mutation rate for sex-linked lethals and detrimentals in the mouse. Mutat Res 15:331-347

Hiraizumi Y, Crow JF (1960) Heterozygous effects on viability, fertility, rate of development and longevity of Drosophila chromosomes that are lethal when homozygous. Genetics 45:1071-1083

Hook EB, Healy NP, Willey AM (1989) How much difference does chromosome banding make? Ann Hum Genet 53:237-242

International Commission on Radiation Units and Measurements (ICRUM) (1986) The quality factor in radiation protection. Rep 40. ICRUM, Bethesda, pp 1-32

Jacobs PA (1981) Mutation rates of structural chromosome rearrangements in man. Am J Hum Genet 33:44-54

Kato H, Schull WJ, Neel JV (1966) A cohort-type study of survival in the children of parents exposed to atomic bombings. Am J Hum Genet 16:214-230

Kuroki Y (1983) A cytogenetic survey of 5,026 newborn infants. J Yonagao Med Assoc 34:1-7

Lüning KG, Searle AG (1971) Estimates of the genetic risks from ionizing radiation. Mutat Res 12:291-304

Machin GA (1974) Chromosome abnormality and perinatal death. Lancet 1:549-551

Maeda T (1979) Screening of chromosomal abnormalities in newborns. Shusanki Igaku 9:1455-1462

Matsunobu A, Ohno M, Funato T, Yoshihara K, Yabe N, Maeda T (1989) A cytogenetic survey of consecutive newborn infants Jpn J Hum Genet 34:52

Mukai T, Chigusa SI, Mettler LE, Crow JF (1972) Mutations rate and dominance of genes affecting viability in *Drosophila melanogaster*. Genetics 72:335-355

Muller HJ (1927) Artificial transmutation of the gene. Science 46:84-87

Mulvihill JJ, Czeizel A (1983) A 1983 view of sentinel phenotypes. Mutat Res 123:345-361

National Council on Radiation Protection (1980) Influence of dose and its distribution in time on dose-response relationships for low-LET radiation. National Council on Radiation Protection, Washington, DC

Neel JV (1958) A study of major congenital defects in Japanese infants. Am J Hum Genet 10:398-445

——— (1988) Should editorials also be peer-reviewed? Am J Hum Genet 43:981-982

Neel JV, Kato H, Schull WJ (1974) Mortality in the children of atomic bomb survivors and controls. Genetics 76:311-326

Neel JV, Lewis S. A comparison of the genetic doubling dose of radiation for mice and humans. Annu Rev Genet (accepted)

Neel JV, Rosenblum BB, Sing CF, Skolnick MM, Hanash SM, Sternberg S (1984) Adapting two-dimensional gel electrophoresis to the study of human germ-line gel electrophoresis of proteins. In: Celis JE, Bravo R (eds) Two-dimensional gel electrophoresis of proteins. Academic Press, New York, pp 259-306

Neel JV, Satoh C, Goriki K, Asakawa J, Fujita M, Takahashi N, Kageoka T, et al (1988) Search for mutations altering protein charge and/or function in children of atomic bomb survivors: final report. Am J Hum Genet 42:663-676

Neel JV, Satoh C, Goriki K, Fujita M, Takahashi N, Asakawa J, Hazama R (1986) The rate with which spontaneous mutation alters the electrophoretic mobility of polypeptides. Proc Natl Acad Sci USA 83:389-393

Neel JV, Satoh C, Hamilton HB, Otake M, Goriki K, Kageoka T, Fujita M, et al (1980) Search for mutations affecting protein structure in children of atomic bomb survivors: preliminary report. Proc Natl Acad Sci USA 77:4221-4225

Neel JV, Schull WJ (1956) The effect of exposure to the atomic bombs on pregnancy termination in Hiroshima and

Nagasaki. Pub 461. National Academy of Sciences–National Research Council, Washington, DC

Neel JV, Schull WJ, Awa AA, Satoh C, Otake M, Kato H, Yoshimoto Y (1989a) The genetic effects of the atomic bombs: problems in extrapolating from somatic cell findings to risk for children. In: Baverstock K, Stather JW (eds) Low dose radiation: biological bases of risk assessment. Taylor & Francis, London, pp 42–53

——— (1989b) Implications of the Hiroshima-Nagasaki Genetic Studies for the estimation of the human "doubling dose" of radiation. Proceedings of the Sixteenth International Congress of Genetics. Genome 31:853–859

Neel JV, Strahler JR, Hanash SM, Kuick R, Chu EHY (1989c) 2-D PAGE and genetic monitoring: progress, prospects, and problems. In: Endler AT, Hanash S (eds) Two-dimensional electrophoresis. VCH, Weinheim, Federal Republic of Germany, pp 79–89

Otake M, Schull WJ, Neel JV (1990) The effects of parental exposure to the atomic bombings of Hiroshima and Nagasaki on congenital malformations, stillbirths and early mortality among their children: a reanalysis. Radiat Res 122:1–11

Roesch WC (ed) (1987) U.S.-Japan joint reassessment of atomic bomb radiation dosimetry in Hiroshima and Nagasaki. Vol 1. Radiation Effects Research Foundation, Hiroshima

Russell LB, Rinchik EM (1987) Genetic and molecular characterization of genomic regions surrounding specific loci of the mouse. Banbury Rep 18:109–121

Russell WL (1951) X-ray induced mutations in mice. Cold Spring Harbor Symp Quant Biol 16:327–336

——— (1965) Effect of the interval between irradiation and conception on mutation frequency in female mice. Proc Natl Acad Sci USA 54:1552–1557

Russell WL, Russell LB, Kelly EM (1958) Radiation dose rate and mutation frequency. Science 128:1546–1550

Sankaranarayanan K (1988) Invited review: prevalence of genetic and partially genetic diseases in man and the estimation of genetic risks of exposure to ionizing radiation. Am J Hum Genet 42:651–662

Satoh C, Neel JV, Yamashita A, Goriki K, Fujita M, Hamilton HB (1983) The frequency among Japanese of heterozygotes for deficiency variants of 11 enzymes. Am J Hum Genet 35:656–674

Schull WJ, Neel JV (1962) Maternal radiation and mongolism. Lancet 1:537–538

——— (1965) The effects of inbreeding on Japanese children. Harper & Row, New York

Schull WJ, Neel JV, Hashizume A (1966) Some further observations on the sex ratio among infants born to the survivors of the atomic bombings of Hiroshima and Nagasaki. Am J Hum Genet 18:328–338

Schull WJ, Otake M, Neel JV (1981) Genetic effects of the atomic bombs: a reappraisal. Science 213:1220–1227

Searle, AG (1989) Evidence from mammalian studies on genetic effects of low level irradiation. In: Baverstock KF, Stather JW (eds) Low dose radiation: biological bases of risk assessment. Taylor & Francis, London, pp 123–138

Searle AG, Beechey C (1986) The role of dominant visibles in mutagenicity testing. In: Ramal C, Lambert B, Magnusson J (eds) Genetic toxicology of environmental chemicals. Part B: Genetic effects and applied mutagenesis. Alan R. Liss, New York, pp 511–518

Stern C, Carson G, Kinst M, Novitski E, Uphott D (1952) The viability of heterozygotes for lethals. Genetics 37:413–449

Uchida IA (1977) Maternal radiation and trisomy 21. In: Hook EB, Porter IH (eds) Population cytogenetics. Academic Press, New York, pp 285–299

United Nations Scientific Committee on the Effects of Atomic Radiation (1986) Genetic and somatic effects of ionizing radiation. United Nations General Assembly Official Records, Forty-first Session, suppl 16A/41/16. United Nations, New York

U.S. Congress, Office of Technology Assessment (1986) Technologies for detecting heritable mutations in human beings. OTA-H-298. U.S. Government Printing Office, Washington, DC, pp xii, 144

Wettke-Schafer R, Kantner G (1983) X-linked dominant inherited diseases with lethality in hemizygous males. Hum Genet 64:1–23

Yoshimoto Y, Neel JV, Schull WJ, Kato H, Makuchi K, Soda M, Eto R, et al (1990) Malignant tumors during the first two decades of life in the offspring of atomic bomb survivors. Am J Hum Genet 46:000–000

Yoshimoto Y, Schull WJ, Neel JV, Kato H. Mortality among the offspring (F_1) of atomic bomb survivors, 1946–1985. RERF Tech rep. Radiation Effects Research Foundation, Hiroshima (submitted)

THE COMPARATIVE RADIATION GENETICS OF HUMANS AND MICE

James V. Neel

Department of Human Genetics, University of Michigan Medical School, Ann Arbor, Michigan 48109–0618

Susan E. Lewis

Research Triangle Institute, P. O. Box 12194, Research Triangle Park, North Carolina 27709

KEY WORDS: mutation rates/induced/mouse, mutation rates/induced/man, genetic doubling dose/radiation/man, genetic doubling dose/radiation/mouse, low-level radiation/genetic effect

CONTENTS

INTRODUCTION	328
A PRÉCIS OF THE HUMAN DATA ON THE GENETIC EFFECTS OF ACUTE RADIATION	328
THE GENERATION OF AN ESTIMATE OF THE DOUBLING DOSE FROM THE HUMAN DATA	335
SOME UNUSUAL FEATURES OF THE ESTIMATE OF THE HUMAN DOUBLING DOSE	337
A PRÉCIS OF THE MOUSE DATA ON THE GENETIC EFFECTS OF ACUTE RADIATION	338
THE GENERATION OF AN ESTIMATE OF THE DOUBLING DOSE FROM THE MOUSE DATA	352
SOME UNUSUAL FEATURES OF THE ESTIMATE OF THE MOUSE DOUBLING DOSE	355
CONCLUSIONS	356

327

Reproduced, with permission, from the *Annual Review of Genetics*, vol. 24, © 1990 by Annual Reviews, Inc.

INTRODUCTION

The attempt by geneticists to predict the genetic consequences for humans of exposure to ionizing radiation has arguably been one of the most serious social responsibilities they have faced in the past half century. Important for its own sake, this issue also serves as a prototype for the effort to evaluate the ultimate genetic impact on ourselves of other human perturbations of the environment in which our species functions. Recently we (67–69) have been developing the thesis that according to the results of studies on the children of survivors of the atomic bombings, humans may not be as sensitive to the genetic effects of radiation as has been projected by various committees on the basis of data from the most commonly employed paradigm, the laboratory mouse (cf 12, 104, 105). In this review we attempt as detailed a comparison as space permits of the findings on humans and mice, presenting the data in a fashion that will enable those who at certain critical points in the argument wish to make other assumptions, to do so. We argue that a reconsideration that includes *all* the data now available on mice brings the estimate of the doubling dose for mice into satisfactory agreement with the higher estimate based on humans.

Since the concept of a doubling dose is critical in these discussions, a clear definition is needed at the outset. The genetic doubling dose of radiation is the amount of acute or chronic radiation that will produce the same mutational impact on a population as occurs spontaneously each generation. To be societally relevant, that estimate is best expressed in terms of morbidity and mortality. For both spontaneous and induced mutation, the reference is customarily first generation effects. This definition in principle encompasses the impact of the gamut of all possible alterations in DNA, from single nucleotide substitutions and very small insertions and deletions to gain or loss of entire chromosomes. The definition is as simple as the concept is useful. Expressing genetic damage as a doubling dose provides a convenient societal frame of reference.

Unfortunately, for neither mice nor humans is the impact of spontaneous mutation each generation as yet satisfactorily defined. Thus, in the calculation of a doubling dose, the errors inherent in estimating the magnitude of a radiation effect may be further compounded by imprecision in the definition of baseline values. We present radiation in the units employed by the investigator, but for our own studies and analyses present the results in gray (1 Gy = 100 rad) and sieverts (1 Sv = 100 rem). In the experimental settings reviewed herein, one r may be equated to one rad (rem).

A PRÉCIS OF THE HUMAN DATA ON THE GENETIC EFFECTS OF ACUTE RADIATION

This treatment of the human data is confined to the information accumulated in Hiroshima and Nagasaki over the past 40 years, because of their high

information content compared to other sets of human data, the relative accuracy of radiation-exposure estimates, and a study design that provides essentially nonoverlapping indicators of radiation effects. The Genetic Program that evolved in Japan has been described in detail on several occasions (4, 39, 64–67, 93), Briefly, in the immediate postwar period in Japan, the economic stringencies were such that a rationing program initiated during the war years to benefit pregnant women was continued. This involved registration of the fact of pregnancy at the completion of the fifth lunar month of gestation. By incorporating the registrants into a genetic study, a *prospective* investigation of pregnancy outcome in Hiroshima and Nagasaki was initiated in February, 1948, that included reference to sex ratio, congenital defect, viability at birth, birth weight, and survival of child during the neonatal period. This more clinical program, which included a physical examination of all newborn infants by a physician, was terminated in February, 1954, but the collection of data on the sex ratio and survival of live-born infants continued, with ascertainment of additional births now through city records. In addition, from the birth registrations in these two cities, the sample for the consideration of these latter two indicators was extended backward in time, to include births between May, 1946, and January, 1948. During the clinical program in Hiroshima, some 62% of all infants who were stillborn or died in the first six days of life were subjected to autopsy. A subset of some 26% of the infants born during this period was reexamined at approximately 9 months of age.

Relatively few survivors beyond 2000 meters from the hypocenter were exposed to radiation from the bombs; the mortality within the 2000 meter zone was high. Accordingly, the original sample of 76,617 children plus subsequent additions contained many more born to parents at distances from the hypocenter > 2000 meters ATB, who seldom received increased radiation from the explosions, than to parents within the 2000 meter radius, whose gonadal doses were relatively large. In 1959, to increase the efficiency of the study of survival, three cohorts were defined from the children born in the two cities since the bombings: The first comprised all children born between May 1, 1946, and December 31, 1958, to parents one or both of whom were within 2000 meters of the hypocenter ATB (proximally exposed); the second was composed of age, sex, and city-matched births to parents one or both of whom were distally exposed to the bombs (> 2500 meters from the hypocenter); and the third, of age, sex, and city-matched children born to parents now residing in Hiroshima and Nagasaki, neither of whom were in the cities ATB. Children born to parents within the 2000–2500 meter radius of distance from the hypocenter are omitted from consideration, because of the exceptional difficulties in estimating the (low) exposures at this distance and the work involved in following children who would contribute relatively little information regarding radiation effects. Setting the earliest date for admission into the

cohort at May 1, 1946, effectively eliminates from the sample all children in utero ATB.

These cohorts have been periodically expanded to include matched numbers of recent births, the limiting factor in the expansion being the number of births to the proximally exposed parents. In 1984, some 39 years after the birth of the first children conceived following the bombings, there were only three births to proximally exposed parents and the cohorts were closed, effective January, 1985. The cohort of children born through 1982 (the cut-off date for analyses) to parents receiving > 0.01 Gy of radiation now equals 31,150. In addition to the study of the survival of these and the "control" children, special studies have been mounted on the physical development of these children during their school years and on the incidence of cancer. In 1968, a search for cytogenetic abnormalities in those children in the cohorts over 12 years of age was initiated, and, in 1975, a search was undertaken in these same cohorts for mutation altering the electrophoretic mobility and/or function of a select battery of proteins. For both these studies, the necessary controls were drawn from the children of parents not receiving increased radiation ATB.

The approach to assigning gonadal doses to the survivors underwent a drastic revision during the early 1980s, resulting in the system known as Dose Schedule 1986 (DS86) (81). All the genetic data have now been analyzed on the basis of the new dosage schedules, supplemented, where DS86 dose are not available, by an ad hoc procedure (cf 74). The radiation resulting from the explosions was predominantly gamma, but included a small neutron component. These two components have been separately estimated, and for these exposure levels, the neutron component assigned a Relative Biological Efficiency (RBE) of 20 (cf 37). This permits assigning the total dose in sieverts. The average conjoint parental gonadal exposures of course vary from study to study, as the composition of the parents of the children examined varies, but in general has been between 0.32 and 0.60 Sv; the distribution of conjoint doses is markedly skewed to the right. Given the chronology of this study, the findings must apply almost exclusively to the results of the radiation of spermatogonia and early, immature oocyte.

We turn now to a consideration of the data.

Untoward pregnancy outcome Because of the interrelatedness of major congenital defect, stillbirth, and neonatal death, we have treated these endpoints as an entity, termed Untoward Pregnancy Outcome (UPO), defined as an outcome resulting in a child with major congenital defect and/or stillborn and/or dying within the first two weeks of life. Between 1948 and 1954, data were collected on these outcomes for a total of 76,617 newborn infants in Hiroshima and Nagasaki, on 69,706 of whom the data were sufficiently

complete in all respects to permit inclusion in the analysis (cf 66). A total of 3,498 children were classified as UPOs. The congenital defects considered major are listed in Neel & Schull (66) and Neel (62). Although the data have been analyzed in many ways, we will for this and the subsequent endpoints to be considered present only the results derived from fitting a simple linear dose-response model to the occurrence of the various indicators of radiation-related damage. This model yields a regression coefficient per Sv (β), and an intercept term, α. Table 1 presents the results of the most recent analysis, based on the use of both DS86 and the ad hoc doses (74). The regression of indicator on dose is positive, and so consistent with the hypothesis that genetic damage was produced, but well below the level of significance.

"Prereproductive" deaths among live-born infants (exclusive of those resulting from a malignant tumor) The mortality data on the three cohorts of live-born children described earlier extend through 1985, when the mean age of the members of the cohorts was 26.2 years. The data are thus relatively complete for death during the prereproductive years. Details of the organization of the study have been presented elsewhere (39, 64). After exclusion of the neonatal deaths included in the analysis of the foregoing section, and cancer deaths (to be discussed in the next section), the three cohorts together include 67,202 individuals, among whom there have been 2,584 deaths. The most recent estimate of the regression of death on parental exposure, employing DS86 doses, is presented in Table 1 (63). Again, the regression is positive, as might be expected if deleterious mutations had been produced, but less than its standard error.

Malignancies in the F_1 A minority of certain cancers of childhood, such as retinoblastoma and Wilms' tumor, result from the presence of one defective gene inherited from a parent plus a second defective allele of this gene resulting from a somatic mutation in the child. Most cancers of childhood, however, seem to be entirely due to somatic cell events occurring in the child. Only the frequency of the former type might be expected to respond to parental radiation. The normal alleles at the loci associated with these "genetic" cancers are usually referred to as "tumor-suppressing" genes. In addition, the occurrence of proto-oncogenes is well established; in principal, radiation could introduce mutations in proto-oncogenes, although thus far there are no clear examples in humans of a genetic (inherited) predisposition to cancer (or any other trait) due to a mutant proto-oncogene. On the thesis that the frequency of tumor-suppressor and proto-oncogenes was probably such that collectively they constituted a significant target for radiation, the incidence of cancer in the populations under consideration was studied, using both death

Table 1 An estimate of the genetic doubling doses that can be excluded at specified confidence levels by the human data. Further explanation in text.

Trait	Regression βSV	Regression α	Observed total background (a)	Mutational contribution background* (b)	Mutational component (%, b ÷ a)	Lower confidence limit (SV) 99%	95%	90%
UPO	0.00264 ±0.00277	0.03856 ±0.00582	0.0502	0.0017 – 0.0027	3.4–5.4	0.14 – 0.23	0.18 – 0.29	0.21 – 0.33
F$_1$ Mortality	0.00076 ±0.00154	0.06346 ±0.00181	0.0458	0.0016 – 0.0026	3.5–5.7	0.51 – 0.83	0.68 – 1.10	0.81 – 1.32
F$_1$ Cancer	−0.00008 ±0.00028	0.00104 ±0.00033	0.0012	0.00002 – 0.00005	2.0–4.0	0.04 – 0.07	0.05 – 0.11	0.07 – 0.15
Sex-chromosome aneuploids	0.00044 ±0.00069	0.00252 ±0.00043	0.0030	0.0030	100	1.23	1.60	1.91
Protein variants	−0.00001 ±0.00001	0.00001 ±0.00001	0.000013	0.000013	100	0.99	2.27	7.41

*per diploid locus

certificates and the cancer registries maintained in Hiroshima and Nagasaki (109).

There were 43 cancers in the 31,150 children born to exposed parents and 49 in the 41,066 children of unexposed parents, for an incidence in the latter of 1.2/1000 persons. As shown in Table 1, the regression on exposure was very small and negative, i.e. counter hypothesis. This was true both for all cancers and for those thought most likely to fit the retinoblastoma model, i.e. retinoblastoma, Wilms' tumor, neuroblastoma, embryonal carcinoma of the testis, and sarcomas (considered an alternative expression of the retinoblastoma allele). None of these latter tumors was in this series associated with a positive family history.

Frequency of balanced structural rearrangements of chromosomes and of sex-chromosome aneuploidy Cytogenetic studies on the children of survivors were initiated in the 1960s, involving members of the F_1 Mortality Study described earlier (2–4). Two age- and sex-matched samples were established, one drawn from the children of proximally exposed and the other from the children of the distally exposed. To increase participation, no effort was made to obtain the venous blood samples necessary to the study from children under age 13. The study will therefore not supply reliable data on chromosomal abnormalities associated with high childhood mortality (including Down syndrome under the conditions of postwar Japan) but is thought to provide appropriate data on the occurrence of sex-chromosome aneuploidy and balanced structural rearrangements of chromosomes, the latter defined as reciprocal translocations, pericentric inversions, and Robertsonian translocations.

Awa et al (4) detected 19 individuals with sex-chromosome aneuploidy in the 8322 children of proximally exposed parents who were examined, and 25 such individuals in the 7976 children of controls. Individuals with some of these abnormalities are sterile (XXY) and with others may exhibit decreased fertility (XXX); furthermore, XYY and XXX individuals only rarely have similarly affected children (19). Thus a very high proportion of these persons may be presumed to result from primary nondisjunction (i.e. mutation) in a parent, but specific family studies were not undertaken on these individuals. DS86 doses were not available to Awa et al (4); we have now distributed these children according to joint parental gonad exposure and derived the appropriate regressions (67). The regression of incidence on dose (Sv) is 0.00044 ± 0.00069, with an intercept of 0.00252 ± 0.00044 (Table 1).

Awa et al (4) also encountered 18 balanced structural rearrangements in the 8322 children of proximally exposed parents and 25 in the 7976 children of controls. Family studies could be performed on about 60% of the children, and one de novo mutational event was detected in the children of the exposed

and one in the controls. With adjustment for the children not subjected to family studies, Awa et al (4) estimate a mutation rate for both groups of 1.0×10^{-4}/gamete/generation, with a 95% confidence interval between 0.2 and 4.5×10^{-4}. The number of mutations was not felt sufficient to warrant a regression analysis.

"Point" mutations affecting protein charge and/or function As used throughout this treatment, the term "point" mutation of necessity encompasses not only specific nucleotide changes but also small insertions/deletions/rearrangements. Between 1972 and 1985, a battery of 30 serum transport and erythrocyte enzyme proteins were screened for the occurrence of mutations altering electrophoretic mobility, and a subset of 11 of these proteins (all enzymes) were screened for mutations resulting in loss of function (65). Three mutations altering electrophoretic mobility were encountered in the equivalent of 667,404 locus product tests on children of proximally exposed persons, and the same number (3) in the equivalent of 466,881 tests on the children of distally exposed parents. These mutations were teased out of family studies on some 964 rare protein variants that had the potential of resulting from mutation in the parental generation.

The examination of a subset of 60,529 locus products for loss of enzyme activity in the children of proximally exposed parents yielded 26 rare variants, one of which proved to be a newly arisen mutation; no mutations were encountered in the 21 variants identified in the 61,741 determinations on the children of the comparison group. When the data on the mobility and loss-of-function mutations were combined, the mutation rate in the children of proximally exposed was 0.60×10^{-5}/locus/generation, with 95% confidence intervals between 0.2 and 1.5×10^{-5}/locus/generation. In the comparison group the rate was 0.64×10^{-5}/locus/generation, with 95% intervals between 0.1 and 1.9×10^{-5}/locus/generation. The average conjoint parental exposure for the proximally exposed was 0.41 Sv and the regression of mutation occurrence on conjoint parental gonadal exposure was -0.00001 ± 0.00001/Sv with an intercept of 0.00001 ± 0.00001 (Table 1) (67).

Sex ratio The relation of sex ratio to parental radiation history in the Japanese experience has been presented on several occasions. With the discovery of sex-chromosome aneuploids in 1959 and recognition of the Lyonization phenomenon in 1961, it became clear that the interpretation of any observed change in the phenotypic sex-ratio at birth would not be simple, and no further analyses have been conducted on sex-ratio since our 1966 publication. The demonstration of no significant difference in the occurrence of sex-chromosome aneuploidy in the two groups, as described earlier, now indicates this phenomenon is not a complication in our findings regarding

sex-ratio. Because of Lyonization, the least complicated data on sex-ratio concern the offspring of exposed mother:unexposed father marriages, where the expectation is for a decrease in the frequency of male offspring if the mutation rate had been increased. In our last analysis (92), the regression, based on the previous (T65DR) dose schedule, was $0.0027 \pm 0.0040/\text{Sv}$, i.e. insignificantly counter hypothesis. Since we see no way to bring these data into the framework of a doubling-dose estimate (see below), these data have not been reanalyzed or extended with the advent of the DS86 dose system.

Anthropometric studies Birthweights were obtained on the 69,706 children examined during the clinical phase of the program, and weight, body length, head circumference, and chest circumference obtained on a subset of 18,498 of these infants reexamined between ages 8 and 10 months (66). In addition, the annual school measurements (height, weight, sitting height, and chest circumference) have for a subset of Hiroshima school children been analyzed in relation to parental radiation exposure, using the T65DR schedule (22–26). No significant or suggestive differences existed between the children of proximally exposed and the control series.

THE GENERATION OF AN ESTIMATE OF THE DOUBLING DOSE FROM THE HUMAN DATA

The attempt to derive both the lower limits to the genetic doubling dose(s) compatible with these data and the most probable doubling dose consistent with all the data has presented a number of unique problems recently described in detail (67). On the other hand, an unusual and favorable feature of this study is that all of its various components have been based on a well-defined population subsampled in various, essentially nonoverlapping ways. The result is a multifaceted appraisal of the genetic impact of the bombs on what is essentially a single cohort. We will enumerate below some of the ways in which these studies lack the elegance that can be achieved in an experimental setting. They are, however, based on a single complex population exhibiting the full range of human heterogeneity. Furthermore, the subjects of the study have been born over a sufficiently extended time interval that we begin to approximate the effects on an entire generation rather than on a "window" of access to a very limited period in the reproductive histories of the parents.

We take the position that in the studies in Hiroshima and Nagasaki we are not testing the hypothesis that radiation is mutagenic. It has been so for every properly studied species of prokaryote and eukaryote, and *Homo sapiens* cannot be an exception. The present data are thus the best approximation to the genetic effect of the bombs to be available for some years. To explore

their precise implications requires specification for each indicator of the contribution of mutation in the parental generation to the frequency of the trait. For sex-chromosome aneuploids, reciprocal chromosome exchanges, and protein variants, this frequency can and has been directly established by the appropriate family studies. For UPOs, and death and cancer among live-born infants, this is not the case; elsewhere we have attempted to generate the estimate of the contribution of spontaneous mutation in the parental generation to these events that is necessary to derive a doubling dose (67).

Table 1 summarizes the results of this effort. Column 4 presents the incidence of the events to be used in subsequent calculations. For the first three items, it is total incidence; for the last two, where it can be specified, it is mutational incidence. Because of the paucity of balanced chromosomal rearrangement that can be attributed to mutation in the parental generation, this indicator is not incorporated in the calculations to come. The fifth column presents the estimated incidence of the trait as a result of spontaneous parental mutation; for the last two indicators it is of course identical with the entry in column 4. The range reflects the uncertainty in the estimate. A detailed justification of these estimates is presented elsewhere (67). Column 6 attempts to provide perspective by indicating the percent of the entry in column 4 to be attributed to spontaneous parental mutation. Note—contrary to Sankaranarayanan's (90) unfortunate erroneous perception of how the doubling dose has been estimated—that only a small fraction of these three indicators, ranging from 2–5%, is attributed to mutation in the parental generation. Finally, by a standard statistical procedure presented elsewhere (67), we have calculated (cols 7–9) for the various indicators lower confidence limits for the doubling dose at the specified probabilities. The range given for the first three indicators results from the range in the estimated mutational component for each.

Such estimates as these five can be combined only on the assumption that they are drawn from a common pool of estimates, i.e. that the doubling doses for the individual phenomenon are essentially the same. An extensive literature suggests that the true doubling dose for radiation-induced sex-linked chromosome aneuploids and protein variants (resulting from nondisjunction and nucleotide substitutions, respectively) is probably higher than the doubling dose for the other three indicators (resulting predominantly from insertion/deletion/rearrangement events and unrepaired chromosome breaks). We have therefore pooled the first three and last two estimates of Table 1, using a statistical procedure described elsewhere (67). The resulting lower confidence limits, at the 95% confidence limit, are between 0.63–1.04 Sv for the first three indicators pooled and 2.71 Sv for the last two.

The data of Table 1 can also be used to derive the most probable overall doubling dose suggested by these data. For this we assume the effect on the F_1

of any mutations resulting in protein variants (i.e. point mutations) is subsumed within the UPOs and F_1 mortality. Since we are basically working with independent observations on the same cohort, we have suggested that the doubling dose can be derived by dividing the sum of the first four entries in column 4 (0.00632–0.00835) by the sum of the five regression terms (0.00375), resulting in an estimate of 1.69–2.23 Sv. There is satisfactory agreement between this estimate and the minimal 95% probability estimates.

The most probable estimate cannot be compared directly with any estimate based on acutely irradiated mice because with humans, both sexes enter into the estimate, the average conjoint gonadal exposure being 0.3–0.6 Sv, depending on the precise study. Moreover, the dosage curve is highly skewed to the right, some couples having received an estimated joint gonadal dose as high as 6.0 Sv. The mouse data, on the other hand, are largely derived from gonadal doses to one sex, the male, of 3.0 or 6.0 Gy (=Sv). The difference in radiation exposures can in principle be met by converting both the human and mouse data to equivalent results from low-dose rate, low-LET radiation (LET = linear energy transfer). The many difficulties in extrapolating from acute, relatively high-level radiation effects to the effects of low-level, chronic, or intermittent exposure are detailed in Report 64 of the National Council on Radiation Protection (61), and yet the effort must be made, both to render the man-mouse comparison more valid and to supply guidelines for human exposures. Russell et al (89) have demonstrated with the 7-locus system that high-level, acute radiation exposures are *at least* three times more mutagenic than low-dose pulsed or chronic radiation. Other indicators, notably those on translocation frequencies, have yielded higher dose rate factors (revs. in 98, 106). Following the linear quadratic treatment of the mouse data by Abrahamson & Wolff (1), we suggest applying a dose rate factor of 2 to convert the Hiroshima-Nagasaki data into their low-dose rate, low-LET radiation equivalent. The resulting figure is 3.38–4.46 Sv. This is a conservative estimate, since it does not incorporate three sets of data that showed no hint of an exposure effect, namely, the data on the balanced chromosomal exchanges, the sex-ratio, and physical development. We find it impossible to set limits in the usual statistical sense on this estimate, because of statistical sampling errors and the uncertainty we have described in establishing the contribution of parental spontaneous mutation to the indicators in the control population.

SOME UNUSUAL FEATURES OF THE ESTIMATE OF THE HUMAN DOUBLING DOSE

As we turn to a similar treatment of the accumulated mouse data, the following features of the human data must be kept in mind:
1. None of the studies has yielded a statistically significant difference be-

tween the children of proximally exposed and controls, and the "numerators" for the individual studies are indeed small. On the other hand, collectively these numerators sum to a substantial body of data, based, for the largest study, on approximately 13,395 Sv of parental gonadal exposure.

2. Exposed males and females contribute about equally to the gonadal exposures of the human data. In the absence of significant radiation effects, there is no prospect of effectively distinguishing the contributions of the two sexes. The human estimate is specific to this situation.

3. The human data cover the entire postexposure reproductive spans of the exposed males and females, whether they were aged 40 or aged 1 ATB, and not, as is customary in experimental studies, the period of maximum fertility following the treatment of young adults.

4. The estimates of the mutational contribution to UPOs and prereproductive mortality are time-place specific, in any strict sense valid only for this particular postwar population. In a more favorable environment, the relative and absolute frequency of children dying because of a mutation in the parental generation should be less, and a somewhat different estimate of the doubling dose would be obtained. This fact, plus the nature of the assumptions necessary to obtaining the estimate of the human doubling dose, impart to this estimate a greater error than is typical of an experimental situation.

5. The parents who were not exposed to the explosions of the bombs (i.e. who came to Hiroshima or Nagasaki following the bombings, as released service men, repatriates, or spouses) were slightly younger and had a little more education and somewhat higher occupational ratings than the exposed (39, 64, 66). It is also assumed that the mother had completely recovered from her exposure when she reproduced. To the extent socioeconomic and maternal effects exist, they will inflate the estimate of the genetic effects of the bombs.

PRÉCIS OF THE MOUSE DATA ON THE GENETIC EFFECTS OF ACUTE RADIATION

The data on mice are summarized in the same sequence as the human data. With rare exceptions, only those murine data are cited that involve transmitted genetic effects following single or double acute exposures to radiation, the effects studied in late fetuses or full-term offspring. The justification is that the primary concern in the societal context is with genetic damage that impacts on the population, whereas damage that results in nonfunctional sperm and eggs, or even the early death of zygotes (i.e. early-acting dominant lethals), although of basic biological importance, does not result in the

transmitted effects of most concern to society. We emphasize the data most relevant to the totality of human reproduction following exposure to ionizing radiation, namely, those based on the irradiation of spermatogonia and early oocytes. Since we wish to have maximum comparability with the treatment of the human data, we will, insofar as possible, present the findings in terms of doubling dose rather than rate per unit radiation. One dramatic discovery in radiation genetics was the demonstration by W. L. Russell in 1965 (88) that in his 7-locus system no mutations were recovered in the "late" litters of irradiated females. Unfortunately, because female mice rapidly lose fertility after radiation, for most indicators there is relatively little early oocyte data for mice, a deficiency that greatly complicates the calculation of a mouse doubling dose.

Untoward pregnancy outcome Because parturient mice tend to devour stillborn offspring as well as those dying shortly after birth, the acquisition of reliable data on the frequency of congenital defect, stillbirth, or early postnatal death comparable to the human data is difficult. The pioneer study of Charles et al (11), involving what must be termed on the average high-level chronic radiation (daily, lifetime exposure schedules of either 0.1, 0.5, 1.0, or 10.0 r), scored the offspring of irradiated DBA males × unirradiated C57BL females for defects shortly after birth and also on the basis of an autopsy at maturity. Deceased offspring with major defects present at birth had probably been eaten by the mother by the time of scoring, so that the at-birth data are of limited value. With respect to the autopsy results, many of the malformations listed in their report would have been considered as minor (and not included in the analysis) by the standards of the Japanese studies. The control frequency of rare anomalies at autopsy was $0.73 \pm 0.11\%$. Averaged over all experiments, the regression of scored defect on dose was reported to be 2.4×10^{-5} per r. The data appear to indicate a dose-rate effect; it would be conservative to consider this an experiment with chronic radiation.

The data on congenital malformation most nearly comparable to the human data were acquired by scoring fetuses obtained from sacrificing pregnant mice at various stages of gestation. Lyon et al (58) exposed the male offspring of a C3H/HeH × 101/H cross to 600 + 600 r of acute radiation and then mated them, at an interval that ensured the results would reflect spermatogonial radiation, with CBA/H females. The females were sacrificed at about 14 days of gestation and scored for corpora lutea and small and large moles, dead embryos, malformed live embryos and normal live embryos. We will not consider the data on preimplantation losses or on numbers of moles (i.e. early postimplantation losses), with respect to which an excess would result from early-acting dominant lethals and would have no counterpart in the human data, and which in the human societal context have relatively little signifi-

cance. Of the 1109 control fetuses, 6 were malformed (0.54 ± 0.22%) and 23 dead (2.07 ± 0.43%). Of 955 fetuses in the irradiated series, 3 (0.31 ± 0.18% were malformed and 25 dead (2.62 ± 0.52%). There was thus no significant difference between the two series, but the very high doses used in this study are at the level where genetic radiation effects "fall off" in many studies, for complex reasons.

In further experiments, Kirk & Lyon (40, 41) sacrificed at 18 days gestation the offspring of male or female F_1 hybrid C3H/HeH × 101/H mice treated in one experiment with 0, 108, 216, 360, and 504 cGy and in another, with 0 and 500 cGy and scored for moles and living fetuses with gross malformations and dead fetuses (with or without malformation). Since the data on exposed females are based on matings within 28 days of radiation, they presumably reflect the results of radiation of mature or semimature oocytes. Over 50% of the fetuses scored as malformed were "dwarfs" or "runts" (body weight less than 75% of mean of litter mates). Malformations observed in dead fetuses were not scored. There was an increase in defective fetuses with radiation, from values of 1.2 ± 0.2% in the controls to values of 1.8 ± 0.3% in the offspring of males receiving an average of 390 r and 3.4 ± 0.4% in the offspring of females receiving an average of 278 r, the latter increase significant. To meet the possibility that maternal radiation per se was a teratogenic influence, West et al (107, 108) undertook a further series of experiments involving the transplantation of preimplantation embryos after irradiation of C3H/HeH × 101/H females with 3.6 Gy X-rays. These experiments demonstrated that the effect of radiation was truly genetic, as did further experiments with 3.3 or 3.7 Gy of localized (versus whole-body) radiation, which revealed that irradiation that did not involve the ovaries had no effect on the frequency of abnormal embryos.

In other experiments, Nomura (70–73) irradiated male and female ICR mice at doses of 36, 216, and 504 rad, and then scored for anomalies at 19 days gestation and 7 days postpartum. There was a significant effect of radiation in both series (all data combined). Thus, among living fetuses scored on day 19 of gestation, the control value was 0.39 ± 0.19%, whereas following male (spermatogonial) radiation it was 2.20 ± 0.33% and following female radiation 2.65 ± 0.52%, both these latter two frequencies being significantly different from controls. Likewise, among offspring surviving more than 7 days, the control frequency was 0.12 ± 0.12%, whereas following male radiation the frequency was 0.86 ± 0.19% and following female radiation, 1.65 ± 0.31%. Again the differences are significant. The average radiation dose for males was about 250 rad and for females, 200 rad (S. Kondo, personal communication). The female data all appear to have been derived from oocytes treated in the follicular stage.

These experiments establish that radiation increases the frequency of con-

genital defect in the F_1 of irradiated fathers and mothers, and provide empiric justification for employing this indicator in the human studies. Defective and dead fetuses correspond to the first two components of the triad we have termed an UPO. There are, however, several reasons why care must be exercised in comparing observations of this type made in mice with those on humans (see also 18).

To begin with, the mouse fetus at term differs in many respects from the term human fetus. Otis & Brent (75) have suggested a rough equivalence of a term mouse to about a 100-day old human fetus, but there are many asymmetries in the comparison, and a sweeping generalization is impossible. The equivalence would be somewhat less than 100 days in the above-quoted experiments. In the program in Japan, pregnant women were eligible to register at the completion of the fifth lunar month of pregnancy—140 days of gestation—but there was no penalty for nonregistration and many delayed registration a few weeks or even longer after their eligible date. The data on newborn mice thus correspond in some but not all respects to the first half of the second trimester stage of human development, when, for example, although the bulk of the malformed human fetuses with cytogenetic abnormalities have been eliminated, there is still a higher frequency of defect due to karyotypic abnormality than is observed at term (cf 35), and it is now documented that a significant proportion of these will be lost before term (36). Thus, malformed offspring that would have been aborted before the pregnancy came to attention in the study in Japan may still be represented in the mouse series under discussion.

The findings are further rendered noncomparable to the data on humans by what we will term the litter-size effect, of much less importance in the human data. Postnatal mortality is higher in large litters (11) as well as in very small litters (58). This effect for large litters is also present in the prenatal period; in the controls to the experiments just cited, the intrauterine fetal death rate was higher in large than in small litters. In these irradiation experiments, because of the induction of early-acting dominant lethals (identified in the autopsies as "moles"), the average number of fetuses was decreased in those litters where one or the other parent had been irradiated (40, 41, 58, 71). There is thus a major bias in studying the effect of parental radiation on late fetal death, recognized by Lyon et al (58). But the situation is complicated even further because defective but dead fetuses were not scored as defective in these experiments; since defective fetuses may be expected to compete less well *in utero*, the elimination of defective offspring may be higher in the larger control litters than in the experimental material.

Finally, there is the question of the comparability of diagnostic standards for the small mouse fetus and the human infant at term. For instance, even severe club foot or hypospadias might not be recognized in a mouse fetus at

18 or 19 days of gestation. Thus, while these experiments certainly establish the principle that ionizing radiation increases the frequency of congenital defect in the F_1 of treated male and female mice, and provide empiric justification for the use of this indicator in the studies on humans, precise comparisons of the findings in mice and humans are not possible.

"Prereproductive" death among live-born offspring The murine data most comparable to the human data on mortality prior to the age of reproduction are those on preweaning mortality. Given the complications in interpretation introduced through inbreeding when radiation experiments are continued over many generations, in this section we consider only the results of studies on the F_1. As noted above, a serious problem with the use of this indicator in the polytocous mouse is the negative correlation of survival with litter size, and the smaller mouse litter size in the offspring of irradiated mice. As recognized by the investigators, this fact vitiates many of the observations on parental radiation and offspring survival (79). The various experimental strains used by investigators also differ in mean litter size. Finally, the relative immaturity of the mouse at birth (and differences in maternal care) may result in different postnatal environment-genotype interactions in mice and humans that render precise comparisons difficult.

A representative set of data (S. E. Lewis, unpublished data) is given in Table 2. We note first the variation in control survival rates in the different experiments, a fact that underscores the importance of concurrent controls in studies of this type. Survival is better in the treated than in the control series in four of six experiments. We note further, however, that litter size averages about one animal more in the controls. To test to what extent this is responsible for the *better* survival of the offspring of treated animals observed in four of the experiments, we have conducted a preliminary analysis of the last data set in Table 2, in which each control litter greater than size 5 but less than 11 was randomly matched with a litter in the experimental series, the range 6–10 chosen to optimize the likelihood of survival. This resulted in two subsets of 480 births. Survival in the controls was $97.5 \pm 0.7\%$ and in the treated series, $91.7 \pm 1.3\%$. A much more extensive analysis along these lines will be presented elsewhere; for now it is sufficient to recognize the need for this type of adjustment, either empirically (as here) or statistically, and the possibility it creates for comparable analyses of human and mouse data.

Malignancies in the F_1 Batra & Sridharan (8) reported an increase in leukemia in the offspring of irradiated male mice, but this was not confirmed by Kohn et al (44). More recently, Nomura (70–73) has reported a significant increase in malignant tumors in 8-month-old offspring of both male and female ICR mice following acute radiation exposures of 0, 36, 108, 216, 360,

Table 2 Preweaning mortality in mice in relation to spermatogonial radiation exposure and litter size.[1] Note the impaired survival in the *controls* in four of six experiments.

	No. litters	No. progeny at birth	Birth average litter	No. progeny at weaning	% Survival
Female 300R					
Treated	488	3,265	6.7	2,706	82.9 ± 0.7
Control	51	390	7.6	319	81.2 ± 2.0
Male 300R Spermatogonia					
Treated	536	3,506	6.5	2,701	77.0 ± 0.7
Control	87	584	6.7	413	70.7 ± 1.9
Male 300R Presterile					
Treated	454	2,329	5.1	2,010	86.3 ± 0.7
Control	87	643	7.4	521	81.0 ± 1.6
Male 600R Spermatogonia					
Treated	861	5,569	7.5	4,597	82.5 ± 0.5
Control	175	1,036	5.9	901	87.0 ± 1.0
Male 600R Presterile					
Treated	337	1,291	3.8	761	58.9 ± 1.4
Control	115	784	6.8	529	67.5 ± 1.7
Male 300R × 2 Spermatogonia					
Treated	547	3,755	6.9	3,478	92.6 ± 0.4
Control	102	776	7.6	715	92.1 ± 1.0

[1] S. E. Lewis, unpublished data.

and 504 rad (average dose of 250 rad for males and 200 rad for females). The control data yielded a tumor frequency of 5.29 ± 0.96% whereas the corresponding (consolidated) figure from male exposure was 10.01 ± 0.77%, and from female exposure, 8.74 ± 0.83%. The findings were repeated on a smaller scale in the LT and N5 strains.

The transfer value of these data to the human studies is moot (cf 109). Mouse strains differ widely in their tumor profiles, and Nomura (71) recognized that "the results may be a special property of the mouse strain used." In the experiments with the ICR strain, 87% of the tumors in the offspring of treated mice were scored as papillary adenomas of the lung. In the experiments with LT and N5 mice, 16.0% and 21.0% of the tumors were pulmonary, 25.3% and 22.8% were ovarian tumors, and 5.3% and 3.9% were leukemias. These endpoints had a relatively high frequency in the appropriate controls. These were not tumors observed with any frequency in the human series, but 8 months covers a relatively larger part of a mouse's life span than the fraction of the human span represented in the human series. These differences in the pattern of malignancy in the two species—which will persist even with further follow-up of the human material—render any detailed comparison of the results from the two species suspect.

Furthermore, none of the human tumors based on the genetic mechanism

that presumably should be most responsive to radiation (retinoblastoma, Wilms' tumor, sarcomas of various types, and embryonal carcinoma of the testis) were observed in the mouse series. There is no basis for an estimate of the proportions of the control malignancies that result from mutation in the parental generation, an estimate essential to the derivation of a doubling dose. We conclude that these data represent a special circumstance in mouse strains many of which were at one time selected for high frequencies of malignant tumors, and do not believe the data can be used in the calculation of a mouse doubling dose to be compared with the human doubling dose (for further discussion, see 109).

Frequency of structural rearrangements and of sex-chromosome aneuploidy The subject of sex-chromosome aneuploidy has been comprehensively reviewed by L. Russell (82). Whereas in humans the most commonly encountered abnormalities are the XXY, XYY, and XXX state, in mice the most frequent abnormality is the XO state, and unlike humans with this constitution, these are fertile individuals. Furthermore, XXX mice are extremely rare (17). There are no large-scale cytogenetic surveys of newborn mice comparable to those on humans; the frequencies of sex-chromosome abnormalities have been determined by genetic techniques whereby individuals exhibiting the various types of abnormalities have distinctive phenotypes. To a first approximation the frequency of primary XO and XXY individuals in mice is about 3/1000, very similar to the human total for sex-chromosome aneuploidy, but predominantly due to XOs to which the male parent has not contributed a sex chromosome.

Acute gamma radiation resulted in a clear increase in sex-chromosome anomalies in offspring from males in whom the germ cells were exposed as spermatocytes, spermatids, and spermatozoa, of the order of 2×10^{-5}/r, but this effect was not observed in offspring resulting from irradiated spermatogonia (Table 7 in 82). The result of the acute radiation of dictyate oocytes as reflected in matings up to 49 days post treatment, based on the occurrence of OX^P offspring, appears more variable but is of the order of 1×10^{-5}/r (Table 8 in 82). The collection of data on the effect of *acute* radiation on immature oocytes has been hampered by the extreme sensitivity of these cells to radiation. Two studies in which small doses of *acute* radiation were employed revealed no increased incidence of trisomy in midterm fetuses (60, 103). On the other hand, two studies in which larger doses were delivered as *chronic* radiation revealed either a nonsignificant increase in X-chromosome loss in term mice (82) or a significant increase in hyperhaploidy in metaphase II oocytes (33). These latter findings, while interesting, are of dubious value in establishing a reference point for the situation in Japan.

Spontaneous or induced mutations characterized by structural rearrange-

ments of chromosomes will result in zygotes with either a balanced or unbalanced chromosomal complement. In mice, as in humans, the latter are usually eliminated early in pregnancy but some few survive until the later stages of pregnancy. The cytogenetic studies on the children of survivors in Japan were not initiated until age 13 and so would only very rarely detect children resulting from chromosomally unbalanced zygotes. Such of the latter type of children as survived beyond the fifth gestational month presumably were detected as grossly abnormal newborn infants and their impact was scored in the analysis of malformations. Likewise, in the above-referenced murine studies of Lyon and colleagues, the more-or-less comparable impact of the mutations resulting in chromosomally unbalanced gametes should be reflected in the malformations scored on the 18th gestational day although, as we noted, the differences between the human and mouse fetus at term or near-term renders a direct comparison with the human data problematic.

On the basis of the studies of high, acute doses of X-rays to the C3H/HeH × 101/H strain (21, 58, 94), Lüning & Searle (50) suggested a spermatogonial doubling dose for reciprocal translocations of about 31 r. They reject from this calculation the data of Reddi (78), which showed a much lesser effect of radiation on the CBA strain. The later data of Generoso et al (27), employing male (101 × C3H)F_1 mice, with a spontaneous rate for heritable reciprocal translocation in spermatogonia of 1.8×10^{-4}/gamete/generation and an induced rate of 3.9×10^{-5}/r/gamete, yields a doubling-dose estimate of 4.6 r; this estimate rests on a single mutation in the controls. As noted, the sparse human data were deemed insufficient for the calculation of a doubling dose for mutations resulting in reciprocal translocations, but since the findings were nearly identical in the children of exposed and controls, the human data appear to be inconsistent with the mouse doubling dose reported by Generoso et al (27).

"Point" mutations Perhaps in recognition of the various complications in the study of the indicator traits thus far reviewed, most studies of the genetic effects of radiation on mice, especially those of recent years, have involved some version of a specific-locus/specific-phenotype test. Again we use the traditional term "point" mutation but recognize that, in addition to radiation-induced specific nucleotide changes [as suggested by the reversibility of radiation-induced mutations in Neurospora (14, 59)] and intragenic insertions/deletions/rearrangements, some of the radiation-induced mutations encompass several loci (83), in which case the term "point" is a misnomer, but we use it for convenience. In this section we draw heavily on the excellent reviews of Searle (96) and Favor (20). The estimates obtained with eight different test systems are summarized in Table 3. *All entries are based primarily on the results of the acute, relatively high-dose (usually 300 or*

Table 3 A summary of the gametic doubling doses for acute, "high-dose" radiation of spermatogonia yielded by the various specific-locus/specific-phenotype systems developed in the laboratory mouse.

System	Data summarized in:	Doubling dose (Gy)	Calculated by:	Origin of treated males
1. Russell 7-locus	16, 96	.44	This paper	101 × C3H
2. Dominant visibles	50	.16	50	various
3. Dominant cataract	20	1.57	This paper	101/E1 × C3H/E1
4. Skeletal malformations	15	.26	50	101
5. Histocompatibility loci	6	>2.60	6	C57BL/6JN
6. Recessive lethals	102	.51 ⎫	50	DBA
	52, 58	.80 ⎬ 1.77	6	C3H/HeH × 101/H
	52, 95	4.00 ⎭	6	CBA, C3H
7. Loci encoding for proteins	This paper	.11	This paper	various
8. Recessive visibles	58	<u>3.89</u>	This paper	C3H/HeH × 101/H
		Av. 1.35		

600r) irradiation of spermatogonia. In assembling this table, we have insofar as possible drawn on doubling-dose estimates already in the literature. Each entry requires some brief discussion.

(a) The multiple, specific-loci test system The most widely used murine test system has been that developed by W. L. Russell (84). This entails the cross of an irradiated mouse homozygous for the normal alleles at 7 test loci with a test strain homozygous with respect to mutant alleles at the same 7 loci. A mutation is immediately scorable in the F_1. The loci are *a, b, c, p, d, se,* and *s.* Because of its efficiency, the method has been widely used and has led to many important insights in radiation genetics. When the results of various laboratories with single or multiple, acute high doses of spermatogonial radiation are combined, (cf summaries in 16, 96), the mutation rate/locus/r is 2.2×10^{-7} (2,346,687 locus tests), with a control rate of 9.0×10^{-6}/locus/generation (6,417,082 locus tests). This leads to a doubling-dose estimate of 41 r for acute, high-dose spermatogonial radiation.

Female mice treated with doses readily tolerated by males produce a few posttreatment litters but then rapidly become sterile due to the extreme sensitivity of early oocytes to ionizing radiation. Even so, because of the importance of obtaining data on the results of female radiation, a particular effort has been made to obtain such data with this system. The first few litters conceived by female mice following single, acute radiation doses of 50 r (involving mature dictyate oocytes) revealed an induced rate/r of from 2.1 to 5.5×10^{-7}, depending on the dose (1,628,277 locus tests), as contrasted with a spontaneous rate/locus/generation of 4.9×10^{-6} (202,812 locus tests). This yields a doubling dose of 9 to 23 r. However, when breeding of females who

had received 50 r continued, no mutations were recovered in the "late" litters of these same females (0 mutations in 547,337 locus tests) (88).

There are major differences between loci in induced rates following acute spermatogonial radiation (96; Table 14). Induced mutations are approximately 18 times more frequent at the *s* than at the *a* locus. Unfortunately, only fragmentary data seem to have been published on locus differences in spontaneous mutation rates, insufficient for a proper statistical analysis. However, we note (96; Table 3) that the 3 (*b, d, s*) loci (among 7) showing the highest spontaneous rates are those with the highest induced rates, and the 3 with the lowest induced rates are those with 0 spontaneous mutations. We return to the great significance of this later.

An alternative 6-locus system was developed by Lyon & Morris (54, 55), consisting of the *a, bp, fz, ln, pa,* and *pe* loci. The rate of mutation induction following acute, high-dose spermatogonial radiation was 7.6×10^{-8}/locus/r (149,004 locus tests), approximately one third of the rate in the Russell 7-locus system at the same exposure level. Unfortunately, no mutations were encountered in the 124,614 control locus tests, so that these observations cannot be put into a doubling-dose context.

(b) The "gross" dominant visible test The observations on the induction of dominant mutations by acute spermatogonial radiation have been well reviewed (96, 99). Since these data are for the most part an incidental by-product of other (specific-locus) studies, they must be placed on a per gamete rather than per locus basis. Searle's analysis (96) reveals a spontaneous rate of 0.8×10^{-5}/gamete and an induced rate of 4.7×10^{-7}/gamete/rad, resulting in a doubling-dose estimate of 17 rad. Although we include this estimate in our summary, the opportunistic use of certain phenotypes suggests that the estimate is based on the more spontaneously mutable loci, and we have just seen that these loci tend to show the higher induced rates (see below also). Otherwise stated, loci with low spontaneous and/or induced rates that failed to yield any mutations would not enter into this estimate.

(c) The dominant cataract system This approach involves scoring the offspring of male mutagenized mice for lens opacities with the aid of a slit lamp after dilation of the pupils with atropine (47). Animals with such opacities were bred, and only transmitted defects scored. In these studies locus rates are calculated on the assumption that 30 loci may contribute to the phenotype. In fact, the number of loci at which mutation may result in this phenotype is unknown, for mice or humans, and for our purposes it is sufficient to express doubling dose in terms of phenotype. In a recent summary of the data from five different experiments, Favor (20) reported that the phenotypic doubling dose ranged from 0.96 to 2.14 Gy. The unweighted average of these estimates

is 1.57 Gy. That this estimate of the doubling dose is based on a single mutation in the control series imparts to it a wide error.

(d) The skeletal malformation system Because of the obvious relevance to human disease and because of the potential immediate expression in the F_1 generation, skeletal defects are, like cataract, an attractive mutational endpoint. That dominant mutations resulting in skeletal defects can be identified in the mouse is well documented in M. Green's catalog of known mouse mutations (32). However, dominant skeletal mutations detected through a standard "physical examination" of the mouse in general appear to be relatively rare both as spontaneous (91) and induced (96) occurrences.

The screening of F_1 generation, cleared and stained mouse skeletons have provided a means for intensifying and expanding the search for skeletal mutations (15, 100). The use of these methods has indeed identified skeletal defects that could reasonably be ascribed to treatment of parents with mutagenic agents (15, 100, 101). However, the number of minor defects found with this method, possibly due to developmental accidents, produces a large background noise level that confounds the identification of truly heritable events.

Heritability tests of presumed skeletal defects are especially cumbersome because all F_1 animals to be tested must be bred to assess transmission of these traits, inasmuch as with this method presumptive mutants can only be identified after death. To offset the need for breeding tests, various criteria have been suggested by Selby & Selby (101) to identify skeletal defects due to mutation. The accuracy of these criteria is yet to be rigorously validated. Selby & Selby (101) apparently did not include controls in their studies. We will therefore accept for this indicator the doubling-dose estimate of 26 r by Lüning & Searle (50), which employed the data of Ehling (15). However, since this estimate is based on a single mutation in the controls, it has a wide error.

(e) The histocompatability system Bailey & Usama (7; see also 5) developed a system of testing for mutations at loci determining histocompatability, by orthoptic tail-skin transplantation involving graft exchange in a 'reciprocal circle' system that should detect both gain and loss mutations. It was suggested that the system would detect mutation at approximately 30 different loci, but as for the other 'phenotype tests,' we shall only present a phenotypic doubling dose. Bailey & Kohn (6) report that in experiments in which the male parent received 522 rads acutely, spermatogonial mutant frequency was higher among the control group than among the offspring of treated, although the difference was not significant. This finding precludes the calculation of a doubling dose but does permit assigning it a lower limit. They calculate that the spontaneous rate per gamete is $<2.6 \times 10^{-5}$, from which the doubling

dose is estimated to be >260 rads. Similar data published by Godfrey & Searle (28) and later data published by Kohn et al (46) are consistent with this calculation.

(f) The prenatal recessive lethal system The recessive lethals scored in murine experiments are mostly very early acting, manifesting as moles. The majority of such early-acting lethals could not be recognized in human studies, and their societal impact would come through their heterozygous effects. The murine data are extremely heterogeneous and involved, and cannot be presented in detail in these confines. Very competent investigators have viewed the data in different lights. A basic problem has been to estimate the baseline total gametic rate of mutation to lethals. In general this has been measured indirectly, using assumptions concerning heterozygote disadvantage and the results of inbreeding. Lüning & Searle (50), using the data of Sheridan & Wårdell (102) on 14 generations of male irradiation at 276 r/generation, developed a doubling-dose estimate for this indicator of 51 r (see also 49). On the other hand, Bailey & Kohn (6), using the control data of Lyon (52) and Searle (95) and the radiation data of Lyon et al (58), developed doubling-dose estimates of 80 and 400 rads, respectively, depending on which control figure is used.

Roderick (80), treating DBA/2J males, has described a direct approach to the detection of spontaneous and induced lethals in a chromosomal inversion covering approximately 3% of the mouse genome, but thus far the system does not seem to have been applied to spermatogonial mutation rates. For postspermatogonial cells, however, he adduces, in a small series, a lethal rate per locus of 0.35×10^{-8}/r, based on an estimated 1750 loci in the segment of chromosome 1 being screened. The mutation rate for the same cell stage in the 7-locus system was 45.32×10^{-8} (84); approximately 80% of these mutations were homozygous lethal. As Roderick points out, this is a 100-fold difference, albeit with a large (indeterminate) error. Unfortunately, there were no lethal mutations in Roderick's control series so that a doubling dose cannot be calculated.

(g) The electrophoretic system In recent years considerable information has accumulated on the frequency in mice of spontaneous and induced mutation that can be detected by the electrophoresis of proteins, either as mobility or absence-of-gene-product variants, or by enzyme activity assays (10, 38, 76, 77). In mice, test crosses can often be designed so that the absence-of-gene-product can be recognized by the absence of an electrophoretic band rather than through the more laborious enzyme assays necessary to the detection of similar variants in humans. In principle, these studies should provide one of the strongest links with the human studies, since there is a substantial overlap in the proteins employed in the two sets of studies. Table 4 summarizes the

Table 4 The frequency of X-ray induced mutations in mouse spermatogonia involving the electrophoretic mobility or loss of physiological function in proteins.

Attribute	Investigator	Strain	Treatment	No. loci	No. observations	Mutations
Protein Charge	W. Pretsch (See also 20)	101/E1 × C3H/E1	0	23	133,676	0
	S. E. Lewis, unpubl.	DBA/2J × C57BL/6J	0	32	1,289,500	1
	J. Peters, unpubl.	C3H/HeH	0	4	1,452	0
	W. Pretsch (See also 20)	101/E1 × C3H/E1	3 + 3 Gy	23	23,069	0
	S. E. Lewis, unpubl.	DBA/2J × C57BL/6J	3	32	91,105	0
			3 + 3	32	115,395	0
			6	32	142,385	0
	J. Peters, unpubl.	C3H/HeH	3 + 3 Gy	4	40,004	0
Enzyme Activity	D. J. Charles & W. Pretsch	101/E1 × C3H/E1	0	12	86,640	0
	S. E. Lewis, unpubl.	DBA/2J × C57BL/6J	0	32	1,289,500	0[a]
	J. Peters, unpubl.	C3H/HeH	0	4	1,452	0
	D. J. Charles & W. Pretsch	101/E1 × C3H/E1	3 + 3 Gy	12	40,656	1[b]
		101/E1 × C3H/E1	5.1 + 5.1 Gy	12	38,244	5[c]
	S. E. Lewis, unpubl.	DBA/2J × C57BL/6J	3	32	40,515	2[a,d]
			3 + 3	32	49,555	2
			6	32	66,784	1
	J. Peters, unpubl.	C3H/HeH	3 + 3 Gy	4	40,004	3

[a] The cross results in heterozygosity at 16 of the loci
[b] This mutation significantly increased specific activity
[c] Of these, 2 increased and 3 decreased specific activity
[d] One of these mutations was detected in a non-heterozygous genotype

data currently available. (We omit reference to publications in which there are no control data.)

Inasmuch as absence-of-gene-product can be reliably detected at only a subset of loci, the data presented in the table are not really descriptive of total mutation rates for these types of variants at the loci concerned. Since, however, the proportion of loci scored for nulls is roughly comparable in the experimental and control series, the data on electrophoretic and null frequencies can be combined in this preliminary evaluation of the doubling doses suggested by this system. For the denominator in the calculation, we use the total number of gene products scored for electrophoretic mutations. The mutation frequency is 0.7×10^{-6}/locus/generation in the controls whereas in the experimental series, the probability of mutation/.01 Gy is 6.3×10^{-8}. The doubling dose (based on the occurrence of a single mutation in the controls) is .11 Gy. We note, by comparison with the data on humans (65), a relatively low spontaneous mutation rate for electromorphs; this observation plays a major role in the low doubling dose. It is perhaps pertinent that some 60%–70% of the control data is derived from female mice, whereas the data on induced rates are derived from spermatogonial radiation. Lyon et al (56) have suggested from specific-locus systems data a control spontaneous mutation rate for females of 2×10^{-6}/locus/generation, approximately one fourth the male rate. A rough correction for the use of female controls would increase this doubling dose estimate by at least a factor of 2.

(h) Recessive visible mutations In addition to the test systems designed to detect recessive (and dominant) mutations at specific loci, several experiments have been designed to permit the detection of recessive visible mutations at all loci at which such mutations can occur. In a comprehensive experiment involving the male offspring of a C3H/HeH \times 101/H cross, Lyon et al (58) recovered one recessive mutation in the equivalent of 142.07 control gametes and six in the equivalent of 142.02 gametes from animals treated with 600 + 600 r acute spermatogonial radiation. They calculate an overall gamete mutation rate to recessive visibles of 7.0×10^{-3} in the controls and 28.2×10^{-3} after radiation. After correction for control frequency, the rate of induction of recessive visible mutations was 1.8×10^{-5}/gamete/r. This leads to a doubling-dose estimate of 3.9 Gy. This estimate is based on a single mutation in the controls and, of course, again implies a large error.

The simple unweighted average of the results of these eight systems is 1.35 Gy. A weighting system and/or the calculation of an error is impossible because the number of loci involved in systems 2, 3, 4, 5, 6, and 8 is unknown. We note, however, that because of the paucity of mutations detected with some of the approaches, the error term should be Poisson in type, i.e. asymmetrical, the upper range at any given probability level exceeding the lower range.

Sex-ratio In the mouse, the sex-chromosome is about 1/20th of the total genome. Thus, in principle there could be of the order of 1000–2000 genes on the X susceptible to radiation damage, perhaps half of which could yield lethal mutations. The published data on the effect of radiation on the mouse sex-ratio are confusing (reviewed in 29, 30, 57). Most early studies found no effect of radiation of males on the sex-ratio of the F_1 or F_2; in those studies that report findings in the F_2 of irradiated males consistent with the induction of sex-linked lethals, efforts to isolate these presumed lethals have failed (cf 29), leading Lüning & Sheridan (51) to conclude that changes in the sex-ratio are an "unacceptable" way to search for sex-linked recessive lethals. Lyon et al (57) met many of the early problems by developing a strain with an X-chromosome inversion encompassing about 85% of the X-chromosome. Treated (500 + 500 rad) males of the F_1 C3H/HeH × 101/H stock were mated to females heterozygous for the inversion. Two of 536 irradiated and 0/529 control X-chromosome carried a confirmed lethal, corresponding to a rate for recessive lethals of 1.9×10^{-6}/rad/X-chromosome. The authors suggest that this points to a locus rate much below that of the 7-locus system, but in the absence of control mutations, no doubling dose can be calculated.

The murine data most comparable to the least ambiguous aspects of the human data would be the proportion of males in the offspring of irradiated females. We have found no published murine data of this type. Accordingly, we have tabulated the (tertiary) sex-ratio observed at weaning in several relatively small experiments involving female radiation performed in the laboratory of one of the authors (S. E. L.) (Table 5). Note that these results are based on the radiation of mature oocytes. Two observations are noteworthy. First, in the controls (and irradiated) there is an insignificant excess of females ($\chi^2 = 1.658$, d.f. = 1, $0.10 < p < 0.20$), whereas in humans at a corresponding stage in development, there is a slight excess of males. Second, the difference between the offspring of control and treated animals is (nonsignificantly) counter-hypothesis in sign ($\chi^2 = 0.124$, d.f. = 1, $0.70 < p < 0.80$). Once again, loci on the X-chromosome appear resistant to the induction of lethal mutations.

Table 5 The sex ratio at weaning in the offspring of females treated with 300r and mated during the week following treatment.

	No. litters	No. progeny at birth	Average birth litter size	Females	Males	Total	M/F Ratio	% Survival
Treated	488	3,265	6.7	1,475	1,331	2,806	.902	85.9 ± 0.7
Concurrent control	51	390	7.6	171	148	319	.865	81.8 ± 2.2

Anthropometrics We have found no analysis of data on mouse body weight or other measurements in the *first* generation following parental radiation that allows for the litter-size effect. Accordingly, there do not seem to be any data suitable for comparison with the measurements on humans.

THE GENERATION OF A DOUBLING DOSE FROM THE MOUSE DATA

Unlike the human situation, in the various experiments with mice it has usually been possible to drive the system under study with sufficient radiation to obtain significant differences between the experimental and control material. This should increase the precision in estimation procedures and provide us in principle with two alternatives in calculating a doubling dose. On the one hand, we can simply combine the results of the various specific-locus/ specific-phenotype studies. This provides a very definite and clear-cut endpoint, which, however, does not readily translate into a morbidity-mortality figure comparable to the basis for the estimation of the human doubling dose. On the other hand, we can attempt to proceed as with the human data. This requires estimating the contribution of mutation in the parental generation to the incidence of malformed and dead fetuses, preweaning mortality, and sex-chromosome aneuploids, with the specific-locus findings represented in the calculation through their heterozygous effects on malformations and fetal deaths and preweaning mortality. While this procedure produces a more relevant figure, it requires assumptions that are probably even softer for mice than for humans.

In the derivations of a mouse gametic doubling dose to be extrapolated to humans, it is essentially the first approach that has been employed in the past. On this basis, various national and international committees have set the doubling dose of acute low-LET gonadal radiation for humans at 30 to 40 r, with limits of 10 to 100 r (reviewed in 12, 105). That estimate appears to be based primarily on the data summarized pictorially by Lüning & Searle in 1971 (50; see also 97), namely, data on semisterility (= reciprocal translocations), the 7-locus system, dominant visible mutations, dominant skeletal mutations, and recessive lethals. The average of these values at that time was 31 r. Our Table 3 in essence extends the approach of Lüning & Searle (50), adding new data but, for consistency in the use of specific-locus/specific-phenotype results, omitting the findings on reciprocal translocations (treated elsewhere in this review). The average of the eight estimates of Table 3 is 1.35 Gy.

As with the human data, these data sets are of variable information content, but none can justifiably be disregarded, and collectively they comprise a formidable body of data. Three of these estimates, however, are based on the

occurrence of a single mutation in the controls (cataract, skeletal defects, protein variants). One additional mutation in the controls would double the estimate of the doubling dose for these three outcomes. Thus, this estimate of 1.35 Gy carries a considerable potential for error. For the high doses of radiation employed in these studies, the appropriate dose-rate factor in converting to the effects of low-dose, chronic, or intermittent radiation is no less than 3.0 (86, 87). This would lead to an estimate of the doubling dose for chronic radiation of approximately 4.0 Gy. The data in Table 3 are based on spermatogonial radiation, and are thus a male-oriented figure. When the failure to recover specific-locus mutations in the late litters of irradiated females (88) is considered, that doubling-dose estimate must be increased, to an indeterminate degree (but less than doubled).

The "information content" of the 7-locus test in particular outweighs that of any of the seven other specific-locus/specific-phenotype systems that have been employed, and some will argue a weighting system should be employed in estimating the doubling dose. However, both empirical and theoretical reasons considerations suggest that the component loci of the 7-locus system may be unusually responsive to the mutagenic effects of radiation. In the context of this discussion, it is important to bear in mind that W. L. Russell in his earliest presentations of the results of the system he had developed expressed concern over the "representativeness" of the loci he had selected (84). Since then significant progress has been made on the subject of "representativeness."

To begin with, although the existence of mutable loci has been recognized for many years, it is now clear that loci in general show a wide range in spontaneous and induced mutability. In the mouse, five coat-color loci differed in their spontaneous mutability by a factor of 7 (but the individual estimates carry large errors) (31, 91). Favor (20; Table 5) has emphasized the differing responses to radiation of the 7-locus, the cataract, and the enzyme-activity systems; data to that effect have also been presented in this review. With humans, recently we have suggested a tenfold difference in the mutation rates of loci encoding for serum transport proteins and erythrocyte enzymes, these loci having been selected for study only because of the clarity of resolution of their products with one-dimensional electrophoresis (9). We also suggest that the loci responsible for the cellular proteins visualized with two-dimensional polyacrylamide gel electrophoresis and silver staining, proteins sometimes referred to as 'structural', are as a group less mutable than the loci encoding for erythrocyte and serum transport proteins (63).

Finally, in a recent mutagenization experiment using the TK6 line of lymphocytoid cells and employing ethylnitrosourea as mutagen, the products of 263 loci were scored for the occurrence of mutation resulting in electrophoretic variants, using the two-dimensional gel technique (34). Ten of

these 263 loci were known from family-oriented studies on peripheral lymphocytes to be associated with genetic polymorphisms. The induced mutation rate at these 10 loci was 3.6 times greater than at the monomorphic loci, an observation with a probability <0.004. Now, a specific-locus test system such as developed by Russell can only use loci associated with genetic variants. The possibility of a bias towards mutability in selecting the components of the specific-locus system is obvious. On the other hand, there may also be bias towards lesser mutability in some of the other systems; this will presumably become apparent with the passage of time. The least prejudicial procedure at present is simply to average the results of all valid approaches.

The alternative approach to developing an estimate of the doubling dose for the mouse is to attempt to analyze the mouse data in the context of radiation-induced morbidity and mortality, in keeping with the basis for the human estimate. At the outset of this review, it was indeed our intention to proceed in this direction, taking into due consideration the early studies on F_1 characteristics following paternal radiation so well summarized by E. Green (30), as well as the later studies on nondisjunction, congenital malformations, and cancer. Unfortunately, as we have progressed, it has become clear that this is simply not feasible. The three dominant components of the human estimate are UPOs, prereproductive mortality among live-born offspring, and sex-chromosome aneuploids. As we have pointed out, each of the murine equivalents of these indicators differs in such important respects from the human data that a valid comparison does not seem possible. Even if this were not the case, however, there is a further reason for the inability to use the first two of these indicators. For the human data, we can estimate, albeit with considerable error, the contribution of mutation in the parental generation to the indicator; this is the necessary basis for an estimate of a doubling dose. For the mouse, no such estimates exist. Indeed, given the complex backgrounds of the various strains of experimental mice used in radiation experiments, it is not clear a normative figure comparable to the estimates for humans could be developed. Finally, relatively few of the published observations on litter size and survival in the older literature are on F_1. Under the circumstances, then, we must compare a specific-locus/specific-phenotype estimate from the mouse with an estimate for humans based on the presumed genetic component in neonatal and childhood morbidity-mortality and sex-chromosome aneuploids.

SOME UNUSUAL FEATURES OF THE ESTIMATE OF THE MOUSE DOUBLING DOSE (FROM THE STANDPOINT OF EXTRAPOLATION TO HUMANS)

As with the human data, several features must be born in mind concerning the murine data:

1. In the several decades following the advent of the atomic bombs, when major programs on mammalian radiation genetics were funded to provide guidance for human risks, the effort was about evenly distributed between programs oriented towards specific-locus tests and towards such strain characteristics as litter size, preweaning and lifetime mortality, body weight at various stages, and sex-ratio. When the latter programs failed to reveal the expected strain deterioration even after repeated generations of massive radiation (cf 30), whereas the specific-locus test was revealing rates greatly in excess of those encountered in Drosophila (cf 85), a consensus developed that the "strain deterioration" approach—which should reflect the impact of mutation at *all* loci—was too blunt an instrument for the problem (cf 30, 48). At the same time, however, investigators repeatedly suggested that the loci of the specific-locus test of Russell (84) might be unusually sensitive to the genetic effects of radiation (e.g. 42, 45, 48, 80, 84), and also recognized the many difficulties in extrapolating from mice to humans (see 43, 48, 53). The present analysis has attempted to meet those concerns by using all the specific-locus/specific-phenotype data available.
2. The exclusive use of specific-locus/specific-phenotype data in reaching a murine doubling dose entails fewer assumptions than in the human estimate but also reflects a smaller sampling of the total genome.
3. A problem with the mouse data that may never be rectified is lack of data on a variety of genetic endpoints in the late litters of irradiated females. If, for whatever reason, all other specific-locus/specific-phenotype indicators behave as do those in the 7-locus test, then the results of female radiation can be only a fraction of the results of male, and the doubling dose of chronic radiation for a normal (two-sexed) mouse population will be substantially higher than the 4.0 Gy we have estimated, and would exceed the estimate for humans, although, given the looseness of these estimates, probably not "significantly" so.

CONCLUSIONS

Studies of eight indicators in the children of atomic bomb survivors and suitable controls have led us to suggest that the gametic doubling dose for the spectrum of acute gonadal radiation experienced by survivors of the atomic bombings is in the neighborhood of 2.0 Sv. Inasmuch as statistically significant effects were not encountered with any of the indicators, this estimate carries a large but indeterminate error. Additional error may be introduced by the assumptions made concerning the extent to which mutation in the parental generation contributes to some of these indicators. For an extrapolation to the effects of chronic radiation, we employ for these specific exposures a dose-rate factor of 2, resulting in an estimate of 4.0 Sv.

A review of all of the pertinent murine data was undertaken, organized to be as comparable as possible to the human data. Because of the differences in the maturity of mice and humans at birth, plus radiation-induced litter-size effects that, as we have shown, strongly influence pre- and postnatal survival, much of the mouse data cannot be compared directly with the human data. The strongest aspect of the mouse data—which is the weakest aspect of the human data—are the specific-locus, specific-phenotype studies. Averaged, these lead to a doubling-dose estimate for acute spermatogonial radiation of 1.35 Gy. Because of the higher acute doses employed in the mouse experiments than obtain for the human experience, we suggest a dose-rate factor of not less than 3, leading to a doubling-dose estimate for chronic radiation of approximately 4.0 Gy. Although the situation requires that we express the human and mouse estimates in different units, in any practical sense the human and mouse doubling dose estimates for chronic radiation are very similar.

The current general agreement between these estimates, reached by relatively independent approaches, is noteworthy, albeit, as we have repeatedly emphasized, the error in both estimates is large but essentially indeterminate. There is, of course, no reason to expect identical doubling doses in two such different species. Together, however, these estimates suggest that the genetic risk of low-dose rate, low-LET radiation for humans is less than has generally been assumed during the past 30 years.

As noted in the Introduction, the attempt to develop a rounded appraisal of the genetic effects of radiation on either the mouse or human genome represents one of the most complex undertakings of modern genetics. The forthrightness with which we have dealt with certain problems with both the human and murine data should not divert attention from the great progress that has been made. Inasmuch as the general issue of genetic damage from environmental exposures will be with society for some time to come, we hope that this discussion of the problems inherent in the present data, both human and mouse, will be useful in planning the experiments and observations of the future. Space constraints unfortunately do not permit discussion of possible future developments and especially the exciting prospects inherent in the parallel study in the two species of radiation-induced mutation in DNA.

This review was well along when the most recent report of the Committee on the Biological Effects of Ionizing Radiation (13) appeared. With respect to the human data, the Committee treatment was appropriate to the time, but, unfortunately, the Committee did not have access to the most recent analyses-syntheses of the Japanese data, based on the DS86 dose schedule, which, we suggest, place the human estimates on somewhat firmer footings than previously. With respect to the mouse data, where we have analyzed essentially the same findings, our approach in an effort to use the mouse data most appropriate to the human situation leads to a somewhat higher estimate of the doubling dose of chronic radiation for the mouse than the Committee's, and

lessens the apparent conflict between the two sets of data that the Committee identifies in its report.

ACKNOWLEDGMENTS

James V. Neel gratefully acknowledges the support of Department of Energy Grant FG02-8760533 and Susan E. Lewis similarly acknowledges the support of NIH Contract No. 1-ES-55078. We are greatly indebted to Dr. Josephine Peters for critiquing the manuscript and making available unpublished data on mutation rates in an electrophoretic system, and to Drs. Michael Shelby and Robert P. Erickson for a critical reading of the manuscript.

Literature Cited

1. Abrahamson, S., Wolff, S. 1976. Reanalysis of radiation-induced specific locus mutations in the mouse. *Nature* 264:715–19
2. Awa, A. A. 1975. Cytogenetic study. *J. Radiat. Res.* 16:75–81 (Suppl.)
3. Awa, A. A., Bloom, A. D., Yoshida, M. C., Neriishi, S., Archer, P. 1968. A cytogenetic survey of the offspring of atomic bomb survivors. *Nature* 218:367–68
4. Awa, A. A., Honda, T., Neriishi, S., Sufuni, T., Shimba, H., et al. 1987. Cytogenetic study of the offspring of atomic bomb survivors, Hiroshima and Nagasaki. In *Cytogenetics: Basic and Applied Aspects*, ed. G. Obe, A. Basler, pp. 166–83. Heidelberg: Springer-Verlag
5. Bailey, D. W. 1963. Histoincompatability associated with the X chromosome in mice. *Transplantation* 1:70–74
6. Bailey, D. W., Kohn, H. I. 1965. Inherited histocompatability changes in the progeny of irradiated and unirradiated inbred mice. *Genet. Res.* 6:330–40
7. Bailey, D. W., Usama, B. 1960. A rapid method of grafting skin on tails of mice. *Transplant. Bull.* 7:424–25
8. Batra, B. K., Stridharan, B. N. 1964. A study of the progeny of mice descended from X-irradiated females with special reference to the gonads. *Acta Unio Int. Contra Cancrum* 20:1181–86
8a. Baverstock, K. F., Stather, J. W., eds. 1989. *Low Dose Radiation: Biological Bases of Risk Assessment*. London: Taylor & Francis
9. Chakraborty, R., Neel, J. V. 1989. Description and validation of a method for simultaneous estimation of effective population size and mutation rate from human population data. *Proc. Natl. Acad. Sci. USA* 86:9407–11
10. Charles, D. J., Pretsch, W. 1986. Enzyme-activity mutations detected in mice after paternal fractionated irradiation. *Mutat. Res.* 160:243–48
11. Charles, D. R., Tihen, J. A., Otis, E. M., Grobman, A. B. 1960. Genetic effects of chronic x-irradiation exposure in mice. *AEC Res. Devel. Rep. UR-565.* Off. Tech. Serv., Dept. Commerce. 354 pp.
12. Comm. Biol. Effects of Ionizing Radiat. 1980. *The Effects of Populations of Exposure to Low Levels of Ionizing Radiation (BEIR III)*, pp. xv & 524 Washington, DC: Natl. Acad. Press
13. Comm. Biol. Effects of Radiat. 1990. *Health Effects of Exposure to Low Levels of Ionizing Radiation (BEIR V)*, pp. xiii & 421 Washington, DC: Natl. Acad. Press
14. de Serres, F. J. 1958. Studies with purple adenine mutants in *Neurospora crassa*. III. Reversion of X-ray-induced mutants. *Genetics* 43:187–206
14a. de Serres, F. J., Sheridan, W., eds. 1983. *Utilization of Mammalian Specific Locus Studies in Hazard Evaluation and Estimation of Genetic Risk*. New York: Plenum
15. Ehling, U. H. 1966. Dominant mutations affecting the skeleton in offspring of X-irradiated male mice. *Genetics* 54:1381–89
16. Ehling, U. H., Charles, D. J., Favor, J., Graw, J., Kratochvilova, J., et al. 1985. Induction of gene mutations in mice: The multiple endpoint approach. *Mutat. Res.* 150:393–401
17. Endo, A., Watanabe, T. 1989. A case of X-trisomy in the mouse. *Cytogenet. Cell Genet.* 52:98–99
18. Erickson, R. P. 1989. Why isn't a mouse more like a man? *Trends Genet.* 5:1–3

19. Evans, H. S. 1977. Chromosome abnormalities among live-births. *J. Med. Genet.* 14:309–12
20. Favor, J. 1989. Risk estimation based on germ cell mutations in animals. *Proc. 16th Int. Congr. Genet., Genome* 31: 844–52
21. Ford, C. E., Searle, A. G., Evans, E. P., West, B. J. 1969. Differential transmission of translocations induced in spermatogonia of mice by irradiation. *Cytogenetics* 8:447–70
22. Furusho, T., Otake, M. 1978. A search for genetic effects of atomic bomb radiation on the growth and development of the F_1 generation. 1. Stature of 15- to 17-year-old senior high school students in Hiroshima. Hiroshima: *RERF Tech. Rep. 4–78.* 33 pp.
23. Furusho, T., Otake, M. 1978. A search for genetic effects of atomic bomb radiation on the growth and development of the F_1 generation. 2. Body weight, sitting height, and chest circumferences of 15- to 17-year-old senior high school students in Hiroshima. Hiroshima: *RERF Tech. Rep. 5–78.* 16 pp.
24. Furusho, T., Otake, M. 1979. A search for genetic effects of atomic bomb radiation on the growth and development of the F_1 generation. 3. Stature of 12- to 14-year-old junior high school students in Hiroshima. Hiroshima: *RERF Tech. Rep. 14–79.* 26 pp.
25. Furusho, T., Otake, M. 1980. A search for genetic effects of atomic bomb radiation on the growth and development of the F_1 generation. 4. Body weight, sitting height, and chest circumference of 12- to 14-year-old junior high school students in Hiroshima. Hiroshima: *RERF Tech. Rep. 1–80.* 18 pp.
26. Furusho, T., Otake, M. 1985. A search for the genetic effects of atomic bomb radiation on the growth and development of the F_1 generation. 5. Stature of 6- to 11-year-old elementary school pupils in Hiroshima. Hiroshima: *RERF Tech. Rep. 9–85.* 33 pp.
27. Generoso, W. M., Cain, K. T., Cacheiro, N. C. A., Cornett, C. V. 1984. Response of mouse spermatogonial cells to X-ray induction of heritable reciprocal translocations. *Mutat. Res.* 126:177–87
28. Godfrey, J., Searle, A. G. 1963. A search for histocompatibility differences between irradiated sublines of inbred mice. *Genet. Res.* 4:21–29
29. Grahn, D., Leslie, W. P., Verkey, F. A., Lea, R. A. 1972. Determination of the radiation-induced rate for sex-linked lethals and detrimentals in the mouse. *Mutat. Res.* 15:331–47
30. Green, E. L. 1968. Genetic effects of radiation on mammalian populations. *Annu. Rev. Genet.* 2:87–120
31. Green, E. L., Schlager, G., Dickie, M. M. 1965. Natural mutation rates in the house mouse: plan of study and preliminary estimates. *Mutat. Res.* 2:457–65
32. Green, M. C., ed. 1981. *Genetic Variants and Strains of the Laboratory Mouse,* pp. xvi & 476. New York: Fischer
33. Griffin, C. S., Tease, C. 1988. γ-ray induced numerical and structural chromosome anomalies in mouse immature oocytes. *Mutat. Res.* 202:209–13
34. Hanash, S. M., Boehnke, M., Chu, E. H. Y., Neel, J. V., Kuick, R. D. 1988. Nonrandom distribution of structural mutants in ethylnitrosourea-treated cultured human lymphoblastoid cells. *Proc. Natl. Acad. Sci. USA* 85:165–69
35. Hook, E. B. 1985. The impact of aneuploidy upon public health: mortality and morbidity associated with human chromosome abnormalities. In *Aneuploidy: Etiology and Mechanisms,* ed. V. L. Dellarco, P. E. Voytek, A. Hollaender, pp. 7–33. New York: Plenum
36. Hook, E. B., Topol, B. B., Cross, P. K. 1989. The natural history of cytogenetically abnormal fetuses detected at midtrimester amniocentesis which are not terminated electively: New data and estimates of the excess and relative risk of late fetal death associated with 47, & 21 and some other abnormal karyotypes. *Am. J. Hum. Genet.* 45:855–61
37. Int. Comm. Radiat. Units Meas. 1986. *The Quality Factor in Radiation Protection. Rep. 40.* Bethesda: ICRUM. 32 pp.
38. Johnson, F. M., Lewis, S. E. 1981. Electrophoretically detected germinal mutations induced in the mouse by ethylnitrosourea. *Proc. Natl. Acad. Sci. USA* 78:3138–41
39. Kato, H., Schull, W. J., Neel, J. V. 1966. A cohort-type study of survival in the children of parents exposed to atomic bombings. *Am. J. Hum. Genet.* 18:339–73
40. Kirk, K. M., Lyon, M. F. 1982. Induction of congenital anomalies in offspring of female mice exposed to varying doses of x-rays. *Mutat. Res.* 106:73–83
41. Kirk, K. M., Lyon, M. F. 1984. Induction of congenital malformations in the offspring of male mice treated with x-rays at pre-meiotic and post-meiotic stages. *Mutat. Res.* 125:75–85
42. Kohn, H. I. 1979. X-ray mutagenesis:

Results of the H-test compared with others and the importance of selection and/or repair. *Genetics* 9 2(Suppl. 1 : 1):203–9
43. Kohn, H. I. 1983. Radiation genetics: The mouse's view. *Radiat. Res.* 94:1–9
44. Kohn, H. I., Epling, M. L., Guttman, P. H., Bailey, D. W. 1965. Effect of paternal (spermatogonial) X-ray exposure in the mouse: Lifespan, X-ray tolerance, and tumor incidence of the progeny. *Radiat. Res.* 25:423–34
45. Kohn, H. I., Melvold, R. W. 1976. Divergent x-ray-induced mutation rates in the mouse for H and "7-locus" groups of loci. *Nature* 259:209–10
46. Kohn, H. I., Melvold, R. W., Dunn, G. R. 1976. Failure of x-rays to mutate Class II histocompatibility loci in BALB/c mouse spermatogonia. *Mutat. Res.* 37:237–44
47. Kratochvilova, J. 1981. Dominant cataract mutations detected in offspring of gamma-irradiated male mice. *J. Hered.* 72:302–7
48. Lüning, K. G. 1979. Some problems in the assessment of risks. *Genetics* 92(Suppl. 1 : 1):121–26
49. Lüning, K. G., Eiche, A. 1975. X-ray induced recessive lethal mutations in the mouse. *Mutat. Res.* 34:63–174
50. Lüning, K. G., Searle, A. G. 1971. Estimates of the genetic risks from ionizing radiation. *Mutat. Res.* 12:291–304
51. Lüning, K. G., Sheridan, W. 1971. Changes in sex-proportion: an unacceptable way to estimate sex-linked recessive lethals. *Mutat. Res.* 13:77–83
52. Lyon, M. F. 1959. Some evidence concerning the 'mutational load' in inbred strains of mice. *Heredity* 13:341–52
53. Lyon, M. F. 1983. Problems in extrapolation of animal data to humans. See Ref. 14a, pp. 289–305
54. Lyon, M. F., Morris, T. 1966. Mutation rates at a new set of specific loci in the mouse. *Genet. Res.* 7:12–17
55. Lyon, M. F., Morris, T. 1969. Gene and chromosome mutation after large fractionated or unfractionated radiation doses to mouse spermatogonia. *Mutat. Res.* 8:191–98
56. Lyon, M. F., Phillips, R. J. S., Fisher, G. 1979. Dose-response curves for radiation-induced gene mutations in mouse oocytes and their interpretation. *Mutat. Res.* 63:161–73
57. Lyon, M. F., Phillips, R. J. S., Fisher, G. 1982. Use of an inversion to test for induced X-linked lethals in mice. *Mutat. Res.* 92:217–28
58. Lyon, M. F., Phillips, R. J. S., Searle, A. G. 1964. The overall rates of dominant and recessive lethal and visible mutations induced by spermatogonial x-irradiation of mice. *Genet. Res.* 5:448–67
59. Malling, H. V., de Serres, F. J. 1973. Genetic alterations at the molecular level in X-ray induced ad-3B mutants of *Neurospora crassa*. *Radiat. Res.* 53:77–87
60. Max, C. 1977. Cytological investigation of embryos in low-dose X-irradiated young and old inbred mice. *Hereditas* 85:199–206
61. Natl. Counc. Radiat. Prot. Meas. 1980. Influence of dose and its distribution in time on dose-response relationships for low-LET radiations. Washington, DC: *NCRP Rep. 64*:vi & 216
62. Neel, J. V. 1958. A study of major congenital defects in Japanese infants. *Am. J. Hum. Genet.* 10:398–445
63. Neel, J. V. 1990. Average locus differences in mutability related to protein "class": A hypothesis. *Proc. Natl. Acad. Sci. USA* 86:9407–11
64. Neel, J. V., Kato, H., Schull, W. J. 1974. Mortality in the children of atomic bomb survivors and controls. *Genetics* 76:311–26
65. Neel, J. V., Satoh, C., Goriki, K., Asakawa, J-I., Fujita, M., et al. 1988. Search for mutations altering protein charge and/or function in children of atomic bomb survivors: Final report. *Am. J. Hum. Genet.* 42:663–76
66. Neel, J. V., Schull, W. J. 1956. The effect of exposure to the atomic bombs on pregnancy termination in Hiroshima and Nagasaki. *Natl. Acad. Sci. Natl. Res. Counc. Publ. 461*. Washington, DC. pp. xvi & 241
67. Neel, J. V., Schull, W. J., Awa, A. A., Satoh, C., Kato, H., et al. 1990. Children of parents exposed to atomic bombs: Estimates of the genetic doubling dose of radiation for humans. *Am. J. Hum. Genet.* 46:1053–72
68. Neel, J. V., Schull, W. J., Awa, A. A., Satoh, C., Otake, M., et al 1989. The genetic effects of the atomic bombs: Problems in extrapolating from somatic cell findings to risk for children. See Ref. 8a, pp. 42–53
69. Neel, J. V., Schull, W. J., Awa, A. A., Satoh, C., Otake, M., et al. 1989. Implications of the Hiroshima-Nagasaki genetic studies for the estimation of the human "doubling dose" of radiation. *Proc. 16th Int. Congr. Genet., Genome* 31:853–59
70. Nomura, T. 1978. Changed urethan and radiation response of the mouse germ

cell to tumor induction. In *Tumours of Early Life in Man and Animals*, ed. L. Severi, pp. 873–91. Monteluce, Italy: Perugia Quadrenn. Int. Conf. Cancer
71. Nomura, T. 1982. Parental exposure to X-rays and chemicals induces heritable tumours and anomalies in mice. *Nature* 296:575–77
72. Nomura, T. 1983. X-ray-induced germ-line mutation leading to tumors. Its manifestation in mice given urethane postnatally. *Mutat. Res.* 121:59–65
73. Nomura, T. 1986. Further studies on X-ray and chemically induced germ-line alterations causing tumors and malformations in mice. See Ref. 77a, pp. 13–20
74. Otake, M., Schull, W. J., Neel, J. V. 1990. The effects of exposure to the atomic bombing of Hiroshima and Nagasaki on congenital malformations, stillbirths and early mortality among the children of atomic bomb survivors: A reanalysis. *Radiat. Res.* 122:1–11
75. Otis, E. M., Brent, R. 1952. Equivalent ages in mouse and human embryos. *Anat. Res.* 120:33–64
76. Peters, J., Ball, S. T., Andrews, S. J. 1986. The datechain of gene mutations by electrophoresis and their analyses. *Prog. Clin. Biol. Res.* 209B:367–74
77. Pretsch, W. 1986. Protein-charge mutations in mice. *Prog. Clin. Biol. Res.* 209B:383–89
77a. Ramel, C., Lambert, B., Magnussen, J., eds. 1986. *Genetic Toxicology of Environmetal Chemicals, Part B: Genetic Effects and Applied* Mutagenesis. New York: Liss
78. Reddi, D. S. 1965. Radiation induced translocations in mouse spermatogonia. *Mutat. Res.* 2:95
79. Roderick, T. H., ed. 1964. The effects of radiation on the hereditary fitness of mammalian populations. *Genetics* 50:1213–17
80. Roderick, T. H. 1983. Using inversions to detect and study recessive lethals and detrimentals in mice. See Ref. 14a, pp. 135–67
81. Roesch, W. C., ed. 1987. *US-Japan Reassessment of Atomic Bomb Radiation Dosimetry in Hiroshima and Nagasaki* 1:434. Hiroshima: RERF
82. Russell, L. B. 1976. Numerical sex-chromosome anomalies in mammals: Their spontaneous occurrence and use in mutagenesis studies. In *Chemical Mutagens. Principles and Methods for Their Detection*, ed. A. Hollaender, 4:55–91. New York: Plenum
83. Russell, L. B., Rinchik, E. M. 1987. Genetic and molecular characterization of genomic regions surrounding specific loci of the mouse. *Banbury Rep.* 28:109–21
84. Russell, W. L. 1951. X-ray induced mutations in mice. *Cold Spring Harbor Symp. Quant. Biol.* 16:327–36
85. Russell, W. L. 1956. Comparison of x-ray-induced mutation rates in Drosophila and mice. *Am. Nat.* 90:69–80
86. Russell, W. L. 1963. The effect of radiation dose rate and fractionation on mutation in mice. In *Repair from Genetic Radiation*, ed. F. H. Sobels, pp. 205–17. London: Pergamon
87. Russell, W. L. 1965. The nature of the dose-rate effect of radiation on mutation in mice. *Proc. Conf. Mech. Dose Rate Effect Radiat. Jpn. J. Genet.* 40:128–40 (Suppl.)
88. Russell, W. L. 1965. Effect of the interval between irradiation and conception on mutation frequency in female mice. *Proc. Natl. Acad. Sci. USA* 54:1552–57
89. Russell, W. L., Russell, L. B., Kelly, E. M. 1958. Radiation dose rate and mutation frequency. *Science* 128:1546–50
90. Sankaranarayanan, K. 1988. Prevalence of genetic and partially genetic diseases in man and the estimation of genetic risks of exposure to ionizing radiation. *Am. J. Hum. Genet.* 42:651–62
91. Schlager, G., Dickie, M. M. 1967. Spontaneous mutations and mutation rates in the house mouse. *Genetics* 57:319–30
92. Schull, W. J., Neel, J. V., Hashizume, A. 1966. Some further observations on the sex ratio among infants born to survivors of the atomic bombings of Hiroshima and Nagasaki. *Am. J. Hum. Genet.* 18:328–38
93. Schull, W. J., Otake, M., Neel, J. V. 1981. Genetic effects of the atomic bombs: A reappraisal. *Science* 213:1220–27
94. Searle, A. G. 1964. Effects of low-level irradiation on fitness and skeletal variation in an inbred mouse strain. *Genetics* 50:1159–78
95. Searle, A. G. 1964. Genetic effects of spermatogonial X-irradiation on productivity of F_1 female mice. *Mutat. Res.* 1:99–108
96. Searle, A. G. 1974. Mutation induction in mice. In *Advances in Radiation Biology*, ed. J. T. Lett, H. I. Adler, A. Zelle, 4:131–207. New York: Academic
97. Searle, A. G. 1977. Radiological protection and assessment of genetic risk. *J. Med. Gen.* 14:307–8
98. Searle, A. G. 1989. Evidence from

mammalian studies on genetic effects of low-level irradiation. See Ref. 8a, pp. 123–38

99. Searle, A. G., Beechey, C. 1986. The role of dominant visibles in mutageneity testing. See Ref. 77a, pp. 511–18

100. Selby, P. B., Selby, P. R. 1977. Gamma-ray induced dominant mutations that cause skeletal abnormalities in mice. I. Plan, summary of results, and discussion. *Mutat. Res.* 43:357–75

101. Selby, P. B., Selby, P. R. 1978. Gamma-ray-induced dominant mutations that cause skeletal abnormalities in mice. II. Description of proved mutations. *Mutat. Res.* 51:199–236

102. Sheridan, W., Wårdell, I. 1968. The frequency of recessive lethals in an irradiated mouse population. *Mutat. Res.* 5:313–21

103. Speed, R. M., Chandley, A. C. 1981. The response of germ cells of the mouse to the induction of non-disjunction by X-rays. *Mutat. Res.* 84:409–18

104. UN Sci. Comm. Effects of At. Radiat. 1977. *Sources and Effects of Ionizing Radiation.* New York: United Nations. 725 pp.

105. UN Sci. Comm. Effects At. Radiat. 1986. *Genetic and Somatic Effects of Ionizing Radiation.* New York: United Nations. 366 pp.

106. Van Buul, P. P. W. 1983. Induction of chromosomal aberrations by ionizing radiation in stem cell spermatogonia of mammals. In *Radiation-induced Chromosome Damage in Man,* ed. T. Ishihara, M. S. Sasaki, pp. 369–400. New York: Liss

107. West, J. D., Kirk, K. M., Goyder, Y., Lyon, M. F. 1985. Discrimination between the effects of X-ray irradiation of the mouse oocyte and uterus on the induction of dominant lethals and congenital anomalies. I. Embryo transfer experiments. *Mutat. Res.* 149:221–30

108. West, J. D., Kirk, K. M., Goyder, Y., Lyon, M. F. 1985. II. Localized irradiation experiments. *Mutat. Res.* 149:231–38

109. Yoshimoto, Y., Neel, J. V., Schull, W. J., Kato, H., Mabuchi, K., et al. 1990. Malignant tumors during the first two decades of life in the offspring of atomic bomb survivors. *Am. J. Hum. Genet.* 46:1041–52

The Future of These Studies

We have emphasized in the preceding two chapters that there is considerable error to be attached to the estimate that the doubling dose of chronic or intermittent, low level, low-LET ionizing radiation for both humans and mice is in the neighborhood of 4.0 Sv. Given the continuing concern with the genetic effects of exposures to ionizing radiation, further data bearing on both the estimate and its limits are highly desirable.

CURRENT RESEARCH PROJECTIONS

The additional studies of the F_1 that seem to be indicated are of two types. On the one hand, an extension of some of the present studies, plus a reanalysis in due time of some of the data already at hand, should prove fruitful. On the other hand, new genetic technologies that in principle have the potential to yield additional insights are becoming available.

With respect to an extension of present programs, the studies of cancer incidence and mortality among the F_1 continue. Furthermore, developments in clinical genetics will surely result in a refinement of the estimate of the contribution of spontaneous mutation to congenital defects and pre-reproductive-age deaths, and hence in an improvement in that aspect of the current doubling dose estimate which depends on these indicators. In this connection, we emphasize again that the estimate that we have generated of the contribution of spontaneous mutation in the parental generation to congenital defects and pre-reproductive death, namely, between 0.33% and 0.53%, is more-or-less specific to a Japanese population in which during the time span of the investigation some 10% of all pregnancies completing the fifth lunar month of gestation terminated in one of the end points concerned. For instance, unlike the current situation, in post-war Japan Down's

syndrome carried a relatively high mortality. Thus, this figure of 0.33%–0.53% cannot be automatically applied to other populations. Continuing advances, in Japan and elsewhere, in understanding the genetic basis of congenital defect should result in a substantial refinement of any estimate of this type. In particular, progress may be expected in defining additional chromosomal microdeletion syndromes, comparable to such already recognized entities as the Langer-Gideon, Prader-Willi-Angelman, Miller-Dieker, thalassemia/mental retardation, choroideremia/deafness/mental retardation, and DiGeorge Syndromes.

With respect to new technologies that can be brought to bear upon the question, the most obvious direction in which to proceed involves an attempt to extend the genetic studies to the DNA level. For a development of this type, as for any other study of mutation, *both* parents as well as offspring must be available. At this writing, the modal age of a living parent whose child is in the various study cohorts is 60-70, i.e., the parents are entering a period of high mortality. Recognizing the problem the passage of time was posing for genetic investigations, in 1985 the RERF began to develop as a resource for future genetic studies, a bank of transformed B-lymphocyte cell lines "immortalized" by infection with the Epstein-Barr virus. The current projection is to establish approximately 500 constellations comprised of one or more children and their father and mother for which one or both parents received relatively high radiation exposures ATB, and a matching number of similar constellations not involving parental exposure. An aliquot of untransformed cells is held in reserve for each individual. The donors for these cell lines are being selected for birth years which reflect proportionately the distribution of births to survivors since 1946. This major undertaking should be completed in 1993.

This decision in part dictates the DNA strategy to be pursued. With respect to any particular "gene" (however that be defined), there will be approximately 1,000 copies to scrutinize for mutational events in the children of exposed, and another 1,000 in the children of unexposed. Elsewhere, we have presented calculations to the effect that the spontaneous nucleotide germinal mutation rate is of the order of 1 per 10^8 nucleotides per generation; we assume a similar rate for relatively small insertion/deletion/rearrangement events (Neel et al., 1986). Given these baselines and the estimate of the human genetic doubling dose employed at that time, and accepting a type I error of 0.05 and a type II error of 0.20, we further estimated that a demonstration of a significant difference between the children of proximally exposed and the children of non-exposed parents would require two samples of approximately 7×10^9 nucleotides each (Satoh et al., 1990). This was a very approximate calculation, but it did serve to indicate the order of magnitude of any study expected to yield the decisiveness of a statistically significant difference between the two groups of children. With current technology and staffing, these are formidable numbers. It is important to recognize, however, that any substantial body of new data, whether or not it yields statistically significant differences between the two sets of children, can be folded in with the existing data to refine further the understanding of the genetic effects of the atomic bombs, both with respect to the minimal doubling dose at specified probability levels and the most probable dose.

The foregoing calculation suggests the need to scrutinize a relatively large number of nucleotide sequences from each of the 1,000 children represented in the cell line panels if the DNA studies are to make a meaningful contribution to the question of genetic effects. The choice of exactly which sequences should be examined from among the estimated 6×10^9 nucleotides of a (diploid) human cell line is a critical decision. Two considerations seem important in this selection. First, since it is highly desirable to be able to translate the results of any new study into the criteria by which the public judges genetic risk—

morbidity and mortality —the genetic probes should predominantly involve functional genes, preferably genes whose normal products can be related to human well-being, although genes associated with presently anonymous polypeptides, such as are visualized by two-dimensional polyacrylamide gel electrophoresis of cellular contents, are acceptable, inasmuch as in time the functions of these polypeptides will become known. Data on mutations in pseudogenes, highly repetitive DNA sequences, or nonsense sequences are of relatively little value in the context of evaluating the phenotypic impact of radiation-induced genetic damage.

Second, since it is now apparent that the genes of higher eukaryotes, including man, vary widely in spontaneous and induced mutation rates (e.g., Searle, 1974; Chakraborty and Neel, 1989; Neel, 1990; Neel and Lewis, 1990), an effort must be made to ensure that the genes whose spontaneous and induced mutation rates are being studied are "representative." On the assumption that the number of different alleles associated with any given locus is a rough index of locus mutability, "representativeness" translates into selecting probes exhibiting the range of genetic variability of probes in general. The need for many different probes imposed by the study sample sizes discussed earlier will to some extent protect against over-reliance on a few, perhaps atypical probes, but even so, as knowledge of the occurrence of variation at the genome level unfolds, it will be important to develop the battery of probes to be used in any study with an appreciation of probe variability in mutation rate.

With these numerical demands and the limitations on the personnel to be committed to the project, it seems likely that the generation of a substantial data set at RERF will require an effort extending over at least a decade. The DNA technologies are evolving so rapidly that any experimental design adopted today will probably be superseded by a more efficient procedure in the course of the next 3 or 4 years. Nevertheless, it has seemed desirable that a start be made, to provide a standard against which to judge future developments. At this writing the three most efficient techniques for the detection of nucleotide substitutions and/or chromosomal insertions/deletions/rearrangements (I/D/R) appear to be those described by Fischer and Lerman (1979, 1983), by Cotton et al. (1988), and by Mohrenweiser et al. (1989).

The experience of the personnel of RERF engaged in applying the DNA methodologies suggests that with appropriate modifications, the technique of Fischer and Lerman (1979, 1983) should permit the screening of approximately 2.7×10^8 nucleotide equivalents per technician year (Satoh et al., 1990; Takahashi et al., 1990; Satoh, personal communication). This technique is especially suitable for the detection of nucleotide substitutions but will also detect small I/D/R events (Takahashi et al., in press). On the other hand, an extension of the restriction-enzyme-site-mapping approach of Mohrenweiser et al. (1989), especially oriented toward the detection of larger I/D/R events (but detecting some nucleotide substitutions), should permit the screening of some 10^9 nucleotide equivalents per technician year (Mohrenweiser, personal communication). Thus on the basis of the foregoing, present technology is such that screening 10^8 nucleotide equivalents for mutational events with these two techniques requires 5.6 technician months, an estimate which, rough though it has to be at this stage in the development of the technologies, can be used to plan the approximate magnitude of a study designed to reach a given endpoint with current technologies. This estimate, however, does not include the investment of time that was required to collect and immortalize the samples, nor the time to verify variants or to characterize and perform family studies on putative mutations, and to manage the data.

THE LIMITATIONS OF THE NEW TECHNOLOGIES

As currently practiced, the DNA technologies of course will not usually detect the chromosomal aneuploidies that constitute about half of the mutational burden each generation as we have defined it. Furthermore, since as no effort is made at RERF to collect blood samples from the offspring of survivors prior to age 13, many of the larger chromosomal I/D/R types of mutation associated with impaired survival will presumably have been lost from the population. Thus the principal finding in a DNA-oriented program directed at surviving children will be nucleotide substitutions and relatively small I/D/R events. These DNA data, while they have the potential to extend the understanding of the genetic effects of radiation, cannot replace the earlier observations. In this connection, an interesting possibility created by the polymerase chain reaction technique is a return to preserved samples from the 717 infants autopsied in Hiroshima during the period 1948-1953, to search for transmitted major DNA damage.

It seems important to emphasize that although investigations at the DNA level bring greater precision into the studies of genetic effects than previously, there may be real difficulties in translating the results of such studies into a phenotypic equivalent. Many of the mutations that may be detected in DNA may have no biomedical significance—even mutations in genes encoding for structural, regulatory, or enzymatic proteins. Given the current evidence for genetic regulation of mutation rates, it is entirely possible that noncoding sequences have higher spontaneous and induced mutation rates than the coding sequences; the relevance to human disease of the results of studies directed at the former will be particularly difficult to interpret. While all mutations are of interest to the geneticist, mutations with no fundamental impact are of little concern to a society worried about the deleterious effects of radiation exposure. In our view, then, the full potential of DNA studies will not be realized without a major effort to understand the functional impact of the mutations detected by this methodology. Developing this understanding promises to be difficult.

PARALLEL ANIMAL STUDIES

The issue of whether there should be a study involving an animal model, presumably the mouse, to parallel the human DNA studies, invites extensive discussion. In one sense, the focus has now shifted to human studies, and it is no longer necessary to extrapolate to humans from an animal model. On the other hand, the ability to manipulate the animal model should result in additional insights that cannot be obtained from the human material. At the DNA level, the use of homologous probes in parallel studies on mice could at long last provide truly comparable indicators of a radiation effect for the two species. It would be highly desirable for any study on mice to employ at least one dose rate schedule as nearly comparable as possible to the Nagasaki-Hiroshima experience. Unfortunately, the use of such low doses, low relative to the 3.0, 6.0, or 10.0 Gy gonadal exposures employed in so many of the earlier murine experiments, will require either sample sizes or end-points per mouse far beyond the experiments of the past.

STAYING THE COURSE

It is difficult in the 45th year of this study not to stray into philosophical considerations. The reprints which constitute the bulk of this volume reflect the most major effort to date directed at a single question in the history of human genetics. At what point can it be said, 'mission accomplished'? Are further extensive studies of any type justified?

The proposed DNA studies will bring a precision into the genetic enterprise that many experimental geneticists will find more satisfying intellectually than the studies of the past. But DNA studies are not the only desirable aspect of future genetic studies. Continuing surveillance of the morbidity and mortality patterns of the F_1 are also in order. These latter studies are important not only in their own right, but as a frame of reference through which to view the results of epidemiological studies on radiation effects in other populations.

The most recent example of this latter role for the RERF genetic data is provided by the report by Gardner et al. (1990) that the parents of children developing leukemia prior to age 25 in the West Cumbria district of the United Kingdom are characterized by occupational exposures to radiation in the Sellafield nuclear plant located in this district which are not observed in the parents of matched controls. The sample is small; for instance, the much discussed relative risk of leukemia to their children of 6.4 when the fathers had received an average total preconceptual occupational (chronic) exposure to ionizing radiation of 100 mSv or more, is based upon a total of four fathers. The investigators suggest a genetic explanation for the finding.

This report finds no support in our studies on malignancies in the F_1 summarized in Chapter 11, although we recognize that the circumstances of exposure differ in the two studies. In this connection, we again note that, commentary to the contrary notwithstanding (Roberts, 1990), the study on malignancy in the F_1 includes the first children conceived following the bombings. The suggestion that the studies in Japan may have "missed" a sensitive period in gametogenesis has no basis. The birth dates of the children comprising the cohorts studied for malignancy extended from May 1946 through 1982, i.e., include the first children conceived following the bombings. During 1946, 263 children were born to parents whose combined exposures were \geq0.01 Sv, and 1189 to parents receiving <0.01 Sv. In the former group, the mean paternal gonad dose was 0.26 Sv and the mean maternal dose, 0.12 Sv. No cases of leukemia were detected in this group, but one case is on record among the children whose parents received <0.01 Sv. Although we find it difficult to compare precisely the total gonadal dose for the Sellafield cohort of fathers with that of the Japanese cohort, it is clear that the total dose of *acute* radiation is definitely much higher for the latter than the dose of *chronic* radiation for the former. Little (1990), equating film badge exposures to gonadal doses, concludes that the Sellafield data imply genetic sensitivities 50 to 80 times higher than the Japanese data. He further concludes that the apparent Sellafield sensitivities are statistically incompatible with the Japanese findings. We believe the gonadal doses are so poorly defined for Sellafield that it is inappropriate to apply precise statistical tests to the situation, but certainly accept the thrust of his analysis. As pointed out by Abrahamson (1990), the Sellafield findings also imply much higher genetic sensitivities than the studies on mice reviewed in Chapter 14.

But the role of the RERF data in providing a continuing perspective to such reports as that of Gardner et al. (1990) goes beyond the specific disease entities that may be involved. These data provide a broad framework of reference within which to view the results of any *ad hoc* epidemiological study. Radiation is not specific in its genetic effects: the damage that it produces is distributed throughout the genome. The childhood leukemias, unlike Wilms' tumor and retinoblastoma, have not in the past appeared to be malignancies in whose etiology mutation in a parent plays a prominent etiological role (rev. in Chapter 11). Genetic damage sufficient to result in this increase in leukemia should thus have had other effects, reflected in morbidity and mortality, of a magnitude that could not have gone unnoticed in the West Cumbria health district. The totality of the Japanese data can and should be brought to bear in the evaluation of the results of disease-oriented epidemiological studies.

It seems likely at this writing that in the wake of the report by Gardner et al. (1990) there will be a series of epidemiological studies on the potential genetic effects of low-level radiation on humans. While our own experience leads us to entertain many reservations concerning the scientific value of such low-level effects studies, we recognize the political and sociological imperatives that may dictate such undertakings. These additional studies, should they appear to differ in outcome from the investigations summarized in this book, will be perceived by some as challenging the conclusions stemming from the studies in Japan. Accordingly, a clear statement *a priori* of certain elementary principles might be helpful.

Let us assume that a future study is directed at certain categories of congenital defect or malignancy in children. We suggest that if the broad category (congenital defect or malignancy) reveals no radiation effect, little purpose is served by extending the analysis to subtypes (and this is the principle followed in the analysis of the Japanese material). The emergence of an apparently "significant" finding with respect to a subtype under these circumstances implies a compensatory decrease (below normative expectation) in some other subtype or subtypes, but the latter phenomenon is seldom noted. Furthermore, the likelihood of false positives must always be borne in mind. Forty years ago, Tippett (1952; see also Wakeford et al., 1989) pointed out that if the significance level is set to α for each of n independent tests, and if the test statistics have continuous distributions, then under the null hypothesis the probability (τ) of encountering at least one outcome that is "significant" at the α level is:

$$\tau = 1 - (1 - \alpha)^n.$$

Thus, if 20 tests are run on a set of data for which α has been set at 0.05, then in addition to the detection of "true" associations there is also a 0.64 probability of finding one or more "significant" results by chance alone. If a series of independent epidemiological studies were to be undertaken, this expectation applies to each study.

An example of this principle is presented by the study of Sever et al. (1988) on congenital defects in two counties in the southeastern portion of the State of Washington, where the Hanford nuclear facility has been a major employer. There was no significant association of occurrence of defect with either paternal employment at Hanford or the magnitude of the estimated (very low) radiation exposures incident to that employment. However, the analyses were extended to 12 subtypes of defect. With a one-tailed probability test, two of these showed a statistically significant association with employment of parents at Hanford, and one showed an association with the (very low) estimated paternal exposure levels. Different malformations were involved in the three associations. These two analyses cannot be regarded as statistically independent. From Tippett's formula we calculate a two-tailed probability of one or more "significant" findings as 0.46 in either series, so that the observations are well within null hypothesis expectations. Nevertheless, the authors discuss at some length why, in view of the studies in Japan, the associations must be regarded as spurious. It is true that in a question of this importance, one does not wish summarily to discard possible leads (see recent discussion in Rothman, 1990). We suggest that particular consideration in evaluating the outcome of the multiple studies of the future should be given to the consistency of any findings. Specifically, if several independent studies do yield findings of borderline significance, it would lend credence to the "findings" if they involved the same category of malformation or cancer. These same

considerations apply to any epidemiological study of the genetic effects of a chemical mutagen.

Periodically the question surfaces of an extension of these genetic studies to the next generation (F_2), and even subsequent generations, on the premise that the recessively inherited effects of radiation would not be manifest in the first filial generation. The biochemical studies described in Chapter 9 were a first approach to the question of recessive damage. The variants detected by electrophoresis and enzyme activity studies, which would not be detected by the kinds of gross phenotypic classification that originally led to the terms "dominant" and "recessive," fit the classical definitions of recessively inherited traits. The DNA studies would now be the definitive approach to the question of induced "recessive" damage. Carried out on a sufficient scale, the results of such studies, combined with the other data, should render unnecessary any studies on subsequent generations of Japanese. By extrapolation, the results of these studies should provide the data with which to deal in a factual fashion with concerns over the hidden genetic effects of lesser exposures in occupational and other settings.

REFERENCES

Abrahamson, S. 1990. Childhood leukemia at Sellafield. *Rad. Res.* 123:237-238.

Chakraborty, R. and Neel, J.V. 1989. Description and validation of a method for simultaneous estimation of effective population size and mutation rate from human population data. *Proc. Natl. Acad. Sci. USA* 86:9407-9411.

Cotton, R.G.H., Rodrigues, N.R., and Campbell, R.D. 1988. Reactivity of cytosine and thymine in single-base-pair mismatches with hydroxylamine and osmium tetroxide and its application to the study of mutations. *Proc. Natl. Acad. Sci. USA* 85:4397-4401.

Fischer, S.G. and Lerman, L. 1979. Length-independent separation of DNA restriction fragments in two-dimensional gel electrophoresis. *Cell* 16:191-200.

Fischer, S.G. and Lerman, L.S. 1983. DNA fragments differing by single base-pair substitutions are separated in denaturing gradient gels: Correspondence with melting theory. *Proc. Natl. Acad. Sci. USA* 80:1579-1583.

Gardner, M.J., Snee, M.P., Hall, A.J., Powell, C.A., Downes, S., and Terrell, J.D. 1990. Results of case-control study of leukemia and lymphoma among young people near Sellafield nuclear plant in West Cumbria. *Brit. Med. J.* 300:423-429.

Little, M.P. 1990. A comparison between the risks of childhood leukemia from parental exposure to radiation in the Sellafield work force and those displayed among the Japanese bomb survivors. *J. Radiol. Prot.* 10:185-198.

Mohrenweiser, H.W., Larsen, R.D., and Neel, J.V. 1989. Development of molecular approaches to estimating germinal mutation rates. I. Detection of insertion/deletion/rearrangement variants in the human genome. *Mut. Res.* 212:241-252.

Neel, J.V. 1990. Average locus differences in mutability related to protein "class": a hypothesis. *Proc. Natl. Acad. Sci. USA* 87:2062-2066.

Neel, J.V. and Lewis, S.E. 1990. The comparative radiation genetics of humans and mice. *Annu. Rev. Genet.* 24:327-362.

Neel, J.V., Satoh, C., Goriki, K., Fujita, M., Takahashi, N., Asakawa, J., and Hazama, R. 1986. The rate with which spontaneous mutation alters the electrophoretic mobility of polypeptides. *Proc. Natl. Acad. Sci. USA* 83:389-393.

Roberts, L. 1990. British radiation study throws experts into tizzy. *Science* 248:24-25.

Rothman, K. 1990. No adjustments are needed for multiple comparisons. *Epidemiology* 1:43-46.

Satoh, C., Hiyama, K., Takahashi, N., Kodaira, M., and Neel, J.V. 1990. Approaches to DNA methods for the detection of heritable mutations in humans. In: M. L. Mendelsohn and R. J. Albertini (eds.), *Mutation and the Environment, Part C* (Proc. Fifth Intn. Congr. on Environ. Mut.). New York: Wiley-Liss, Inc. pp. 197-206.

Searle, A.G. 1974. Mutation induction in mice. *Advances in Radiation Biology* 4:131-207. New York: Academic Press.

Sever, L.E., Gilbert, E.S., Hessol, N.A., and McIntyre, J.M. 1988. A case-control study of congenital malformations and occupational exposure to low-level ionizing radiation. *Am. J. Epidem.* 127:226-242.

Takahashi, N., Hiyama, K., Kodaira, M., and Satoh, C. 1990. An improved method for the detection of genetic variations in DNA with denaturing gradient gel electrophoresis. *Mut. Res.* 234:61-70.

Takahashi, N., Hiyama, K., Kodaira, M., and Satoh, C. Detection of the length polymorphism in the 5¹ flanking region of the human b-globin gene with denaturing gradient gel electrophoresis in a Japanese population. *Hum. Genet.*, in press.

Tippett, L.H.C. 1952. *The Methods of Statistics, 4th Edition.* New York: Wiley. pp. 395.

Wakeford, R., Binks, K., and Wilkie, D. 1989. Childhood leukemia and nuclear installations. *J. Roy. Stat. Soc.*, Series A, 152:61-86.

Yoshimoto, Y., Neel, J.V., Schull, W.J., Kato, H., Mabuchi, K., Soda, M., and Eto, R. 1990. Malignant tumors during the first 2 decades of life in the offspring of atomic bomb survivors. *Am. J. Hum. Genet.* 46:1041-1052.

Biographical Sketches

JAMES V. NEEL is Lee R. Dice Distinguished University Professor of Human Genetics, Emeritus, and Professor of Internal Medicine, Emeritus, at The University of Michigan Medical School, where for 25 years he was chairman of the Department of Human Genetics, the first such department in the country. He received his Ph.D. and M.D. degrees at the University of Rochester. Elections to learned societies include the National Academy of Sciences, the Institute of Medicine, the American Philosophical Society, and the American Academy of Arts and Sciences. He has been President of the American Society of Human Genetics and of the Sixth International Congress of Human Genetics, and has served on numerous national and international advisory committees. Honors include the Allan Award of the American Society of Human Genetics, the Lasker Award of the American Public Health Association, the Presidential Medal of Science, and the Medal of the Smithsonian Institution. His association with the investigations reported in this book began in 1946, when he went to Japan as a member of a five-man team charged to advise the National Academy on the feasibility of studies on the delayed effects of atomic bombs.

WILLIAM J. SCHULL is Director of the Genetics Center of the Graduate School of Biomedical Sciences of the University of Texas Health Science Center in Houston, and Ashbel Smith Professor of Academic Medicine. Soon after the completion of his doctoral degree at Ohio State University, he went to Japan (in 1949) to participate in the genetics studies initiated by the Atomic Bomb Casualty Commission. In the 40 subsequent years, he has served the Commission, and its successor, the Radiation Effects Research Foundation, in a variety of positions—as head of the Department of Statistics and Epidemiology, Chief of Research, Vice-Chairman, and is currently one of the Foundation's Permanent Direc-

tors. Prior to 1972, when he assumed his current position in Texas, he was a member of the Department of Human Genetics at the University of Michigan. Dr. Schull is a member of the International Commission on Radiological Protection, the U.S. National Council on Radiation Protection and Measurements, and a consultant to the United Nations Scientific Committee on the Effects of Atomic Radiation.

Name Index

A

ABCC Semi-annual Report (1954), 57
Abrahamson, S., 444, 448, 461, 482, 491, 493
Adachi, H., 357
Alberman, E.D., 446, 448
Albert, D.M., 398
Albertini, R.J., 365, 374, 493
Alexander, M.L., 236, 239, 258
Amano, T., 356
Anderson, N.G., 339, 341
Anderson, R.C., 13, 81, 262
Anderson, T.W., 448
Andrews, S.J., 485
Antonarkis, S.E., 365, 374, 376
Antonov, A.N., 89, 138, 159, 258
Arakawa, E.T., 318, 323, 328, 340
Archer, P.G., 357, 386, 448, 482
Armitage, P., 48
Aronis, S., 376
Asakawa, J., 363, 375, 376, 428, 449, 484, 493

Ashman, C.R., 365
Austin, D.F., 399
Auxier, J., 283, 322, 329, 340 368, 374, 387
Awa, A.A., 344, 345, 357, 358, 376, 378, 386, 400, 428, 429, 431, 432, 435, 436, 448, 450, 457, 482, 484

B

Baba, K., 374, 376
Bachelor, A.L., 340
Bailey, D.W., 399, 472, 473, 482, 484
Bailey, H.J., 340, 341
Baird, P.A., 447, 448
Bakker, E., 376
Balfour, M.I., 87, 258
Ball, D.W., 358
Ball, S.T., 485
Barnett, H.L., 262, 341
Barr, D., 357
Bartlett, M.S., 203, 207, 258
Bashir, H., 399

Basler, G., 482
Batra, B.K., 390, 398, 466, 482
Baverstock, K.F., 450, 482
Beadle, G.W., 9, 29, 30, 47
Beardsley, R., 49, 258
Beebe, G.W., 281, 290, 318, 323
Beechy, L., 447, 450, 486
Bell, M., 259
Bellows, M.T., 261
Benedict, W.F., 398
Bergin, M., 399
Bernards, R., 398
Bernstein, M., 117, 258
Bertelson, C.J., 375
Binks, K., 494
Birch, J.M., 399
Bishop, J.M., 390, 398
Blattner, W.A., 399
Block, M., 9, 29
Bloom, A.D., 357, 386, 448, 482
Boehnke, M., 483
Bogenmann, E., 398
Bond, V.P., 78, 258
Bonnier, G., 236, 258
Bookstein, R., 390, 398, 399
Borges, W., 13, 61, 97, 259
Bowers, J.Z., 64, 263
Box, G.E.P., 112, 207, 258
Boxer, C.R., 52, 258
Boyd, W.C., 56, 258
Bradshaw, E.S., 83, 262
Brandt, A.E., 48, 110
Brent, R., 465, 485
Brewer, R.C., 13, 81, 262
Brinkley, F., 49, 258
British Mission to Japan, 56, 258
Bronk, D.W., 9, 29
Brown, A., 259
Brues, A., 9, 29
Buckton, K.E., 354, 356, 357
Bundey, S., 390, 398
Burke, B.S., 87, 258
Butterfly, Madame 63
Byrne, J., 397, 399

C

Cacheiro, N.C.A., 483
Cain, K.T., 483

Call, K.M., 375
Campbell, R.D., 493
Canning, N., 357
Carnes, B.A., 375, 449
Carson, G., 450
Carter, C., 83, 133, 258
Caskey, C.T., 365, 374, 376
Caspersson, T., 345, 357
Cavenee, W.K., 390, 398, 399
Chai, L.-Y., 398
Chakraborty, R., 373, 445, 448, 482, 489, 493
Chandley, A.C., 486
Chao, L.Y., 399
Chapman, A.B., 319, 323, 336, 340
Charles, D.J., 474, 482
Charles, D.R., 9, 30, 47, 229, 230, 258, 463, 482
Cheka, J.S., 340, 387
Chen, A.T.L., 358
Chigusa, S.I., 449
Children's Cancer Association of Japan, 396, 398
Chu, E.H.Y., 450, 483
Chung, C.S., 400
Ciocco, A., 83, 117, 258
Cochran, W.G., 110, 258, 283, 290
Coffey, V.P., 133, 258
Cogan, D.G., 61, 97, 258
Cohen, B.H., 445, 448
Colletti-Feener, C., 375
Collins, V.P., 77, 258
Compton, D.A., 399
Connell, F.H., 47
Connelly, R.R., 399
Cook, J.W., 399
Copeland, N.G., 376, 399
Cornett, C.V., 483
Cotton, R.G.H., 489, 493
Court Brown, W.M., 436, 448
Cox, D.E., 390, 400
Cox, D.F., 319, 320, 323, 336, 340
Cox, D.W., 288, 290, 336, 340
Cox, G., 114
Craig, C.C., 47, 48
Creasy, M.R., 446, 448
Cronkite, E.P., 78, 258
Cross, P.K., 483

Crow, J.F., 231, 259, 320, 323, 336, 338, 340, 446, 449
Cupp, M.B., 322, 324, 340, 341
Czeizel, A., 434, 447, 448, 449

D

Dallapiccola, B., 447, 448
Danforth, C.H., 9, 30
Dao, D.D., 390, 398
Davidson, B.L., 376
Davidson, F., 48
Davidson, R.L., 365, 374
Dean, R.F.A., 87, 259
de Boer, J.G., 375
de Grouchy, J., 446, 448
de Jong, P.J., 375
Delehanty, J., 374, 445, 448
Dellarco, V.L., 483
de Serres, F., 364, 375, 482, 483
Dickie, M.M., 485
Dobzhansky, T., 237, 259
Downes, S., 429, 493
Draper, G.J., 390, 398
Draper, N.R., 335, 340
Dreyfus, M.G., 399
Drobetsky, E.A., 375
Dryja, T.P., 398
Dunham, C.L., 78, 258
Dunn, G.R., 484
Dwyer, P.S., 48

E

Ebbs, J.H., 87, 259
Edington, C.W., 356, 387
Ehling, U.H., 472, 482
Eiche, A., 484
Eisenhart, C., 110, 259, 316, 323
Ellsworth, R.M., 389, 399
Endler, A.T., 450
Endo, A., 482
Epling, M.L., 399, 483
Erickson, R.P., 482
Eto, R., 389, 400, 428, 450, 494
Evans, E.P., 330, 341, 483
Evans, H.J., 357, 436, 446, 448, 483
Evans, K., 390, 398
Evans, R.D., 238, 239, 240, 259

F

Faaborg-Andersen, J., 358
Falk, R., 234, 259
Falls, H.F., 241, 262
Farrington, B.H., 375, 386, 449
Favor, J., 444, 448, 469, 471, 478, 482, 483
Ferrell, R.E., 368, 374, 375, 375
Fillmore, P.G., 97, 259
Fischer, S.G., 489, 493
Fisher, G., 484
Fisher, R.A., 112, 259, 316, 323
Folley, J.H., 61, 97, 259
Flamant, F., 399
Forabosco, A., 447, 448
Ford, C.E., 483
Forrestal, J.T., 9, 29
Fraumeni, J.F., 399
Friedmann, T., 398
Friedrich, U., 354, 357
Friend, S.H., 389, 390, 398
Frölen, H., 319, 323, 336, 340
Frome, E.L., 429
Fujita, K., 340, 356
Fujita, M., 363, 375, 376, 428, 449, 450, 484, 493
Fujita, S., 429
Funato, T., 449
Furuhata, T., 259
Furusho, T., 439, 449, 483

G

Gallie, B.L., 398
Gardner, M.J., 404, 427, 429, 491, 492, 493
Garriott, M.L., 369, 374
Gedeon, M.M., 357
Geiser, C.F., 399
Gelber, R.D., 399
Generoso, W.M., 469, 483
Genetics Conference (1947), 9, 30, 431, 449
Gerald, P.S., 354, 358
Gerber, M.R., 398
Gershowitz, H., 375
Gibbs, R.A., 365, 374
Gilbert, E.S., 368, 369, 374, 387, 494

Gimbrere, K., 399
Glass, H.B., 236
Glickman, B.W., 375
Godbout, R., 398
Goldstein, H., 446, 449
Godfrey, J., 483
Gordon, J.E., 83, 261
Goriki, K., 363, 375, 376, 386, 428, 449, 450, 484, 493
Goyder, Y., 486
Grahn, D., 369, 374, 386, 433, 446, 449, 483
Graw, J., 482
Green, D.M., 399
Green, E.L., 48, 479, 483
Green, M., 472, 483
Griffin, C.S., 483
Griffith, P., 357
Grobman, A.B., 482
Grosovsky, A.J., 365, 375
Gruenwald, P., 133, 259
Grundy, P., 397, 399
Gunz, F.W., 390, 399
Gunz, J.P., 399
Guttman, P.H., 399, 483

H

Hadorn, E., 230, 259
Haldane, J.B.S., 2, 8, 237, 239, 245, 259
Haliandros, M., 376
Hall, A.J., 429, 493
Hamerton, J.L., 353, 354, 357
Hamilton, H.B., 344, 372, 374, 375, 376, 386, 400, 428, 448, 449, 450
Hammond, E.C., 262, 341
Hanash, S.M., 375, 449, 450, 483
Hansen, J.L., 319, 323, 340
Hansen, K.B., 358
Hansen, M.F., 399
Hansteen, I.-L., 354, 357
Harris, H., 113, 259
Harris, M., 399
Harris, T., 50
Hartley, A.L., 390, 399
Hasebe, K., 199, 259
Hashizume, A., 280, 322, 324, 429, 450, 485

Hashizume, T., 368, 375, 387
Hastay, M.W., 316, 323
Havenstein, J.L., 319, 323, 340
Hayashi, I., 24, 212, 216, 217, 218, 219, 259, 399
Hayes, R.C., 398
Haywood, F.F., 340
Hazama, R., 356, 363, 375, 386, 449, 493
Heaf, M.M., 398
Healy, N.P., 449
Hechter, H., 78
Hegnauer, H., 83, 259
Heintz, N.H., 374
Hempelmann, L.H., 64, 73, 77, 259
Henshaw, P., 9, 29
Hertwig, P., 228, 230, 231, 259
Hessol, N.A., 494
Heyn, R.M., 399
Hill, B., 48
Hiraizumi, Y., 446, 449
Hiyama, K., 493, 494
Hoffman, J.G., 64, 73, 77, 259
Hofker, M.H., 376
Hollaender, A., 483
Holmes, F.F., 399
Holmes, R., 126
Honda, T., 344, 357, 386, 428, 448, 482
Hong, F., 399
Hook, E.B., 352, 357, 446, 448, 449, 450, 483
Horowitz, J.M., 398
House, V.L., 241, 259, 260
Hsiao, B., 48
Huff, V., 397, 399
Hulse, F., 199, 260
Human Gene Mapping 9, 390, 399
Hurst, G.S., 318, 324, 328, 341
Hvidman, L., 358
Hyman, C.C., 448

I

Ichimaru, M., 399, 429
Iida, S., 356
Ikeda, T., 391, 399
Inamizu, T., 374
Ingalls, T.H., 130, 262

International Commission on Radiation Units and Measurements, 429, 433, 449, 483
ISCN, 345, 357
Ishida, M., 281, 290, 318, 323
Ishimaru, T., 340, 396, 399, 429
Itoh, H., 373, 375, 428
Itoh, M., 344, 357, 386
Ives, P.T., 236, 239, 260
Iwamoto, K., 376
Izumi, T., 51, 260

J

Jablon, S., 281, 290, 318, 323, 329, 340, 368, 375, 392, 399
Jacobs, P.A., 354, 356, 357, 446, 449
Jacobson, K.B., 376
James, A.P., 319, 324
Japan Science Council, 62, 260
Japanese Empire Statistical Annual, 51, 261
Jenkins, N.A., 376, 399
Jessop, W.J.E., 133, 258
Jobo, T., 357
Johnson, F.L., 399, 483
Johnson, F.M., 364, 375
Johnson, H., 48
Jones, T.O., 340

K

Kaempfer, E., 50, 52, 199, 260
Kagan, A., 387, 399
Kageoka, T., 363, 375, 386, 428, 449
Kalmus, H., 228, 260
Kaneko, M., 397, 399
Kantner, G., 438, 450
Kaplan, I.I., 320, 324, 336, 340
Karn, M.N., 83, 159, 260
Kastenbaum, M., 48, 97, 261, 262, 275, 279
Kato, H., 290, 291, 327, 328, 329, 332, 333, 336, 340, 357, 366, 375, 376, 378, 386, 389, 391, 399, 400, 401, 403, 409, 428, 429, 431, 432, 435, 443, 449, 450, 483, 484, 486, 494
Kato, Y., 357
Kazazian, H.H., 374, 376

Keay, A.J., 357
Kelleher, P.C., 374
Kelley, E.M., 319, 322, 324, 339, 341, 450, 485
Kelley, H.C., 48
Kelley, W.N., 376
Kendall, M.G., 114, 260
Keosian, J., 244, 260
Kerkis, J.J., 234, 260
Kerr, G.D., 368, 375, 387
Kikuchi, H., 398
Kimball, A.W., 289, 290
Kimura, S., 61, 97, 258
Kinch, R.A.H., 358
Kinnear Wilson, L.M., 398
Kinst, M., 450
Kirk, K.M., 464, 483, 486
Kirk, N.T., 9, 29
Kirk, R.L., 399
Kirwood, J.B., 130, 262
Kishimoto, S., 376
Kis-Varga, A., 434, 448
Kiss, P., 448
Kitamura, S., 13, 262
Kitchin, F.D., 389, 399
Kiyono, K., 49, 260
Klag-Olsen, B., 358
Kleinbaum, D.G., 429
Knudson, A.G., 389, 390, 396, 399
Kobayashi, N., 398
Kobayashi, R., 48
Kodaira, M., 493, 494
Kodama, Y., 344, 386, 428
Kodani, M., 13, 46, 81, 262
Kohn, H.I., 319, 324, 336, 340, 390, 399, 472, 473, 482, 483
Komai, T., 48
Kondo, S., 464
Kopecky, K.J., 429
Kosztolányi, G., 448
Koutos, A., 390, 399
Kovács, M., 448
Koya, Y., 47, 260
Kramer, S., 448
Krasteff, T., 375
Kratochvilova, J., 482, 484
Krauss, M.R., 399
Krooth, R., 48, 103, 133, 213, 242, 260

Krzesicki, R., 375
Kubat, M., 288, 290
Kuick, R., 450, 483
Kuji, V., 129
Kunkel, L.M., 375, 376
Kupper, L.L., 429
Kurasaki, H., 97, 260
Kuroki, Y., 446, 449
Kutlaca, R., 376

L

Lambert, B., 450, 485
Lamphiear, D.E., 48
Lampkin, B.C., 399
Landtman, B., 83, 133, 260
Langard, S., 357
Lange, R.D., 61, 97, 260, 261
Larsen, R.D., 493
Law, P., 448
Lawrence, P.S., 228, 230, 231, 260, 320, 324, 336, 341
Lea, R.A., 449, 483
Leck, I., 397, 399
Lee, C.H., 375, 386, 449
Lee, E.Y.P.H., 389, 398, 399
Lee, W.H., 390, 399
Lefevre, G., 236, 260
Leong, P.H., 375
Lerman, L., 489, 493
Lerner, I.M., 234, 260
LeRoy, G.V., 75, 260, 262, 341
Leslie, W.P., 449, 483
Levy, M.A., 390, 399
Lewis, S.E., 364, 375, 444, 449, 451, 466, 474, 482, 483, 489, 493
Li, E.P., 396, 399
Li, F.P., 390, 398, 399
Liber, H.L., 364, 365, 375
Lieben, A.A., 341
Liebow, A.A., 262
Lilienfeld, A.M., 448
Lin, C.C., 354, 357
Lindahl, T., 364, 375
Lisco, H., 64, 73, 77, 259
Little, J.B., 375, 376
Little, M.P., 427, 429, 491, 493
Loeffler, R.K., 77, 258
Loewe, W.E., 387

Lomakka, G., 357
Long, M., 375
Lorenz, E., 97, 260
Los Alamos Scientific Laboratory, 56, 64, 260, 328, 340
Lowe, C.R., 83, 117, 260
Lowry, R.B., 448
Lubs, H.A., 354, 357
Lucas, H.C., 48, 102
Lüning, K.G., 236, 258, 339, 340, 442, 449, 469, 472, 473, 476, 477, 484
Lyon, G.M., 9, 29
Lyon, M.F., 340, 341, 463, 464, 465, 471, 473, 475, 476, 483, 484, 486

M

Mabuchi, K., 389, 400, 428, 450, 486, 494
Machin, G.A., 446, 449
Macht, S.H., 228, 230, 231, 260, 320, 324, 336, 341
MacMahon, B., 83, 261, 390, 399
Maeda, T., 354, 357, 399, 445, 446, 449
Magnusson, J., 450, 485
Maki, H., 48
Malling, H., 364, 365, 375, 483
Marsalek, J., 288, 290
Marsden, H.B., 399
Martin, S.F., 61, 97, 258
Maruyama, T., 368
Masunari, N., 376
Matsubayashi, I., 31
Matsumoto, H., 375
Matsumoto, S., 32, 261
Matsumura, A., 199, 261
Matsunaga, E., 390, 396, 399
Matsunaga, H., 97, 260
Matsunobu, A., 446, 449
Matsuoka, S., 97, 260
Matsuo, T., 399
Max, C., 484
McBeath, S., 357
McCarthy, D., 133, 258
McDonald, D.J., 13, 262
McGregor, J.F., 319, 324
McIntosh, R., 132, 261
McIntyre, J.M., 494
McKeown, T., 83, 117, 260, 262
McKusick, V.A., 429

Meadows, A.T., 390, 399
Meerdervoort, P., 50
Méhes, K., 448
Meigs, J.W., 399
Melville, M., 357
Melvold, R.W., 484
Mendelsohn, E., 387
Mendelsohn, M.L., 374, 448, 493
Merritt, K.K., 261
Metrakos, J.D., 228, 260, 263
Mettler, L.E., 449
Mikami, M., 399, 429
Miller, R.W., 7, 8, 390, 399
Millis, J., 159, 261
Milton, R.C., 329, 341, 344, 357, 368, 375, 387, 392, 399
Ministry of Health and Welfare (1947), 429
Minoda, K., 396, 398, 399
Mitani, S., 129
Mitchell, M., 357
Mohrenweiser, H.W., 366, 375, 489, 493
Moloney, W.C., 61, 97, 260, 261
Monaco, A.P., 365, 375, 376
Monzen, T., 391, 399
Morgan, K., 399
Morgenstern, H., 429
Morley, A.A., 376
Morris, T., 471, 484
Morton, J., 47
Morton, N.E., 13, 159, 261, 262, 279, 319, 323, 340
Moulvad, I., 358
Mourant, A.E., 56, 261
Moyle, W.J., 259
Mukai, T., 446, 449
Muller, C., 288, 290
Muller, H.J., 9, 30, 231, 234, 235, 236, 238, 239, 240, 241, 243, 244, 261, 321, 324, 438, 449
Mulvihill, J.S., 397, 399, 434, 449
Munro, N.G., 49, 216
Murdoch, J., 49, 52, 261
Murphree, A.L., 398
Murphy, D.P., 230, 261
Musil, J., 288, 290
Myers, R.J., 83, 261
Myrianthopoulos, N.C., 400

N

Nachtsheim, H., 237, 261
Nagai, I., 48
Nagasaki Since the Restoration 51, 261
Nakano, M., 344, 386, 428, 448
Naruge, T., 293, 324
National Academy of Sciences (1972), 337, 340
National Academy of Sciences (1956), 321, 324
National Academy of Sciences (1980), 444, 447, 448, 482
National Academy of Sciences (1990), 428
National Council on Radiation Protection and Measurements, 443, 449, 461, 484
Neel, J.V., 4, 8, 9, 13, 29, 30, 47, 73, 81, 83, 124, 236, 237, 238, 241, 261, 262, 271, 273, 279, 280, 281, 282, 283, 288, 289, 290, 291, 293, 297, 298, 299, 301, 318, 321, 322, 324, 327, 328, 329, 330, 332, 333, 336, 339, 340, 341, 355, 357, 358, 363, 364, 366, 368, 371, 374, 375, 376, 377, 378, 386, 387, 389, 391, 399, 400, 401, 403, 428, 429, 431, 432, 434, 435, 436, 437, 438, 439, 440, 443, 444, 445, 447, 448, 449, 450, 451, 455, 482, 483, 484, 485, 486, 488, 489, 493, 494
Neriishi, S., 344, 357, 375, 386, 428, 448, 482
Newcombe, H.B., 297, 319, 324, 448
Newton, D.R., 357
Newton, M.S., 357
Nicklass, J.A., 374
Nickson, J.J., 77, 262
Nielsen, J., 354, 357, 358
Nielsen, K., 446, 449
Nishida, M., 357
Nixon, W.C.W., 87, 262
Nomura, T., 389, 390, 397, 398, 399, 464, 466, 467, 484, 485
Novitski, E., 83, 117, 262, 289, 290, 450

O

Obe, G., 482
Ohara, J.L., 387

Ohkura, K., 293, 320, 324, 336, 342
Ohno, N., 376, 449
Ohtaki, K., 344, 386, 428, 448
Oláh, E., 448
Olkin, I., 48
O'Neill, J.P., 374
Ono, M., 357
O'Riordan, M.L., 357
Orkin, S.H., 374
Otake, M., 8, 356, 357, 375, 376, 377, 378, 386, 400, 428, 429, 431, 433, 434, 439, 449, 450, 483, 484, 485
Otis, E.M., 465, 482, 485
Oughterson, A.W., 9, 29, 61, 261, 329, 341

P

Pai, G.S., 345, 358
Palella, T.D., 376
Parkes, A.S., 228, 262
Pathak, S., 398
Paulson, E., 286, 290
Pearson, P.L., 376
Penrose, L.S., 48, 83, 159, 243, 260, 262
Perry, Commodore 50, 53
Peters, J., 474, 482, 485
Phelps, L.V., 35, 48
Phillips, D.G., 376
Phillips, R.A., 398
Phillips, R.J., 330, 340, 341, 484
Pierce, D.A., 378, 387, 445
Plough, H., 47
Popp, R.A., 376
Porter, I.H., 448, 450
Powell, C.A., 429, 493
Preston, D.L., 378, 387, 429
Pretsch, W., 474, 482, 485
Prindle, R.A., 130, 262

R

Radiation Effects Research Foundation (1987), 364, 376
Ramal, C., 450, 485
Rao, C.R., 48, 110, 114, 165, 196, 203, 262
Rapaport, M., 398
Ratcliffe, S., 357

Ray, M., 357
Ray, P., 376
Record, R.G., 83, 262
Reddi, D.S., 469, 485
Reed, T.E., 237
RERF Research Protocol (1975), 344, 358
Rhoads, C.P., 9, 29
Riccardi, V.M., 398
Rice, R.G., 83, 264
Richards, M.R., 261
Rinchik, E.M., 365, 376, 446, 450, 485
Ritchie, R.H., 318, 324, 328, 341
Ritterhoff, R.K., 236
Rivers, T.M., 9, 29
Roberts, L., 3, 8, 491, 493
Robson, E., 159, 262
Roderick, T.H., 320, 324, 336, 341, 473, 485
Rodrigues, N.R., 493
Roesch, W.C., 378, 387, 392, 400, 429, 433, 450, 485
Rogelj, S., 398
Rosenbaum, J.D., 262, 341
Rosenblum, B., 375, 449
Rothman, E., 375
Rothman, K., 492, 493
Roy, S.N., 106, 262, 275, 279
Ruddle, F.H., 354, 357
Russell, L.B., 319, 322, 324, 336, 339, 340, 341, 365, 376, 446, 450, 468, 485
Russell, W.L., 227, 228, 229, 230, 238, 239, 262, 289, 290, 319, 320, 322, 324, 330, 336, 339, 340, 341, 365, 376, 443, 444, 445, 450, 461, 463, 470, 478, 480, 483, 485

S

Salber, E.J., 83, 262
Sallam, S.E., 399
Sams, C.F., 30, 48
Samuels, M.H., 261
Sanders, F.W., 387
San Jerson, B.J., 376
Sandler, L., 289, 290
Sankaranarayanan, K., 448, 450, 460, 485
Sansom, G.B., 49, 52, 262
Sasano, A., 51, 262

Satoh, C., 363, 366, 368, 369, 370, 371, 374, 375, 376, 378, 386, 400, 428, 429, 431, 447, 449, 450, 484, 488, 489, 493, 494
Saunders, G.F., 399
Sawaguchi, S., 397, 399
Sawyer, J.R., 358
Schinke, R.N., 390, 400
Schneider, B.A., 262, 341
Schneidewind, J., 48
Schroeder, W.T., 398
Schlager, G., 485
Schull, W.J., 4, 7, 8, 13, 47, 82, 133, 236, 237, 238, 262, 271, 273, 279, 280, 281, 282, 283, 288, 290, 291, 293, 297, 298, 299, 301, 318, 321, 322, 324, 327, 328, 329, 330, 332, 333, 336, 340, 341, 355, 357, 358, 363, 364, 366, 374, 375, 376, 377, 378, 386, 387, 389, 391, 396, 399, 400, 401, 403, 428, 429, 431, 432, 434, 436, 438, 439, 440, 443, 445, 447, 448, 449, 450, 455, 483, 484, 485, 486, 494
Schultz, J., 235, 262
Scott, W.A., 259
Scully, P., 399
Seabright, M., 345, 358
Searle, A.G., 330, 339, 340, 341, 442, 443, 447, 450, 469, 471, 472, 473, 477, 483, 484, 485, 486, 489, 493
Seijas, B., 48
Selby, P.B., 472, 486
Selby, P.R., 472, 486
Seng, Y.P., 160, 261
Sergovich, F.R., 354, 358
Sery, T.W., 398
Sever, L.E., 492, 494
Sevitt, S., 24, 212, 216, 219, 263
Sewell, L.M., 357
Shapiro, H.L., 200, 263
Shaw, M., 374
Shelby, M., 482
Sheridan, W., 473, 476, 482, 484, 486
Shew, J.Y., 399
Shigematsu, I., 387, 399
Shimba, H., 344, 386, 428, 448, 482
Shimizu, Y., 356

Shimokawa, I., 399
Shiono, P.H., 390, 397, 400
Shohoji, T., 329, 341, 344, 357, 368, 375, 387, 392, 399
Short, M., 357
Sillesen, I., 354, 358
Silverberg, M., 228, 260
Sinclair, W.K., 369, 376
Sing, C.F., 375, 449
Sinskey, R.M., 61, 97, 263
Skolnick, M.M., 374, 375, 376, 449
Slatis, H.M., 238, 239, 240, 263
Slight, J., 357
Smink, W.K., 357
Smith, C.A., 87, 138, 159, 263
Smith, C.A.B., 113, 259
Smith, H., 335, 340
Smith, H.F., 48
Smith, P.G., 448
Smith, S., 357
Smout, M.S., 358
Snedecor, G.W., 114, 263
Snee, M.P., 429, 493
Snyder, L.H., 9, 30, 47
Soda, M., 356, 357, 389, 400, 428, 450 494
Spassky, B., 237, 259
Spassky, N., 237, 259
Speed, R.M., 486
Spencer, W.P., 231, 235, 263
Spowart, G., 357
Spuhler, J.N., 240, 263
Sridharan, B.N., 390, 398, 466, 482
Stather, J.W., 450, 482
Steen-Johnsen, J., 357
Steinberg, A.G., 390, 400
Stern, C., 47, 48, 231, 235, 263, 446, 450
Sternberg, S., 375, 449
Stevenson, S.S., 83, 258, 264
Stout, J.T., 365, 376
Strahler, J.R., 450
Strandskov, H.H., 117, 230, 231, 263
Strong, L.C., 398, 399
Stuart, H.C., 258
Sturtevant, A.H., 239, 263
Sufuni, T., 344, 357, 358, 386, 428, 448, 482
Sugahara, T., 319, 324, 336, 342

Sumner, A.T., 345, 358
Sutherland, I., 83, 263
Suzuki, M., 13, 262
Syme, J., 357
Szabo, L., 448

T

Taguchi, A., 357
Takabe, H., 399
Takada, M., 357
Takahashi, N., 363, 375, 376, 386, 428, 449, 489, 493, 494
Takeshima, K., 13, 262
Tanabe, K., 356, 358
Tanaka, K., 320, 324, 336, 342
Tanis, R.J., 368, 374, 375, 376
Tavendale, O.G., 297, 324
Taylor, G., 47
Tease, C., 483
Terrell, J.D., 429, 493
Terry, V.H., 375
Tessmer, C.F., 47
Thomas, G.H., 345, 358
Thorngate, J.H., 340
Thunberg, K.P., 53, 263
Tietze, C., 116, 263
Tiffany, T.O., 339, 341
Tihen, J.A., 482
Timoféeff-Ressovsky, N.W., 231, 234, 263
Tippett, L.H.C., 492, 494
Tisdall, F.F., 259
To, H., 398, 399
Tomonaga, M., 97, 260
Topol, B.B., 483
Trasler, D.G., 228, 263
Tsiftis, G., 376
Tsunoda, R., 297, 325
Tsuzuki, M., 48
Tucker, M.A., 399
Turleau, C., 446, 448
Turner, D.R., 365, 376

U

Uchida, I.A., 355, 358, 445, 450
Ueda, N., 368, 374, 376
Ullman, J., 48
Ullrich, F., 9, 29
United Nations Scientific Committee on the Effects of Atomic Radiation (1958), 321, 325
United Nations Scientific Committee on the Effects of Atomic Radiation (1972), 339, 342
United Nations Scientific Committee on the Effects of Atomic Radiation (1977), 486
United Nations Scientific Committee on the Effects of Atomic Radiation (1986), 444, 447, 450
United Nations Scientific Committee on the Effects of Atomic Radiation (1988), 428
United States Strategic Bombing Survey, 56, 263
Uphoff, D., 450
U.S. Congress, Office of Technology Assessment (1986), 374, 376, 445, 450
U.S.-Japan Joint Workshop on Reassessment of Atomic Bomb Radiation Dosimetry, 344, 358
Urakawa, Y., 356
Usama, B., 472, 482
Utsunomiya, J., 399

V

Valcovic, L.R., 365, 375
Valencia, J.I., 236
Valencia, R.M., 236
Valentine, G.H., 358
Van Buul, P.P.W., 486
van Ommen, G.J.B., 365, 376
Varslot, K., 357
Vaughan, C., 376
Verkerk, J.M.H., 376
Verkey, F.A., 449, 483
Videbaek, A., 390, 400
Videbech, P., 358
Vincent, P.C., 399
Vogel, F., 389, 396, 400
Vor der Bruegge, C.F., 64, 263
Voytek, P.E., 483

W

Wakabayashi, T., 391, 399
Wakeford, R., 492, 494
Wallace, B., 244, 263
Wallis, W.A., 101, 263, 286, 290, 316, 323
Walzer, S., 354, 358
Wang, X.F., 398
Wårdell, I., 473, 486
Warkany, J., 133, 138, 263
Warren, S.L., 9, 29, 62, 263
Warren, Sh., 9, 10, 29, 62, 64, 263
Watanabe, T., 482
Wedemeyer, W.J., 13
Weed, L.H., 9, 29
Weinberg, R.A., 398
West, B.J., 330, 341, 483
West, J.D., 464, 486
Westergaard, M., 240, 263
Wettke-Schafer, R., 438, 450
Whalen, P.P., 387
Whipple, G.H., 9, 29
White, R.L., 374, 448
Wieringa, B., 376
Wilkie, D., 494
Wilkie, L., 357
Wilks, S.S., 114, 207, 263
Willey, A.M., 449
Willham, R.L., 319, 323, 336, 340
Wilson, J.M., 365
Wilson, R.R., 62, 74, 264
Wishart, J., 165, 264
Wohlert, M., 358
Wolff, S., 444, 448, 461, 482
Wood, J.W., 13, 262
Woodbury, L., 71, 126, 264
Woolley, W., 52, 264
Worcester, J., 83, 258, 264
World Health Organization (1977), 429
Worton, R., 376
Wright, S., 238, 239, 241, 260, 264
Wright, S.W., 13, 97, 260, 262, 264
Wurzinger, K.H., 375

Y

Yabe, N., 449
Yamashita, A., 376, 450
Yamawaki, T., 61, 97, 259, 260
Yamazaki, J., 13, 262
Yanase, T., 293, 325
Yandell, D.W., 365, 376
Yasuda, N., 375
Yates, F., 114, 264, 316, 323
Yerushalmy, J., 83, 181, 264, 297, 325
Yesley, G., 48
Yoshida, M.C., 357, 386, 448, 482
Yoshihara, K., 449
Yoshimaru, H., 8
Yoshimoto, Y., 378, 389, 395, 400, 401, 403, 428, 429, 431, 435, 448, 450, 486, 494
Young, L.J., 398, 399
Youssoufian, H., 365, 370
Yunis, J.J., 345, 358

Z

Zech, L., 357
Zelle, M., 47
Zirkle, R.E., 9, 29
Zuelzer, W.W., 390, 400

Subject Index

A

Abortions, 10
 induced, 89, 91, 92–95, 101, 104
 irradiation subclasses, 87, 89, 91, 92–95, 101, 104, 121
 social/legal issues, 3
 study methodology, 37, 46, 47
Accumulation factor, recessive genes, 240–241
Acetabulum, dysplasia, 43, 130, 132
Age, parental, 117, 180, 273, 297–298, 300
 autopsied infants, 213–214, 216
 birthweight, 159, 160–161, 165, 173, 176, 178
 child deaths, 307, 308, 309, 311, 407
 ethnicity and, 54–55
 irradiation subclasses, 83–85, 97, 101, 104, 133, 134–135, 138, 142, 143
 malformation, 133, 134–138, 141, 142, 143, 383
 maternal, 83, 101, 104, 117, 133, 134–138, 141, 142, 143, 147,148–152, 158, 159, 160–161, 173, 176, 180, 217
 neonatal death, 180–181, 182–185, 191, 383
 sex ratios and, 117, 289
 stillbirths, 147, 148–152, 158, 383
Ainu, 49
Americans, in Nagasaki, 51
Analysis of variance, 100, 104, 161, 163, 165, 166, 168, 171, 178, 185, 197, 203, 207
 see also Covariance analysis
Analytic methods, *see* Methodology; Statistical approaches and programs
Anamnestic data, 37, 42
Animal studies, 11, 30, 64, 97, 194, 231, 291, 319–320, 339, 490
 guinea pigs, 230–231, 241
 see also Drosophila; Murine studies
Anthropometry, 43, 100, 110–111, 130, 192–211, 224, 457, 477
 chest circumference, 130, 192, 196, 439, 459

consanguineous parents, 275
head circumference, 130, 192, 195, 439, 459
length of child, 192, 194, 199–200, 439, 459
sex differences, 138, 152, 163, 193–196, 198, 202, 204–207, 211, 220
weight of child, 192, 199, 439
see also Birthweight
Armed Forces Institute of Pathology, 75
Army-Navy Joint Commission, 29, 61, 62, 75, 77
Army of Occupation, 31
Atomic Bomb Casualty Commission, 1, 8, 30, 32, 272, 281
dosimetry, 5
Atomic Energy Commission, 1, 4, 29, 31, 47, 115
Attitudes
parents, 2, 91, 97
public, 8
Attribute data, 106–110, 190, 224
Austrians, in Nagasaki, 51
Autopsy findings, 4, 37, 42, 97, 147, 212–219, 230
congenital malformation, 215, 217
parental age and, 213–214, 216
sampling, 212–213, 217, 219

B

Background radiation, 99, 220
Bartlett's test, 155, 203, 207
Bias, sampling, 99, 106, 219, 228–229
Biochemistry, 378
methodology, 5, 8, 367–368
protein electrophoresis, 6, 363–374, 436–438, 441, 442, 447, 454, 458, 473–475
Birth injury, 147, 180
Birth rates, 47
Birth/pregnancy registrations, 10, 32, 33–37, 91, 96, 101, 105, 272, 281–282, 292–293, 332, 378, 391, 432, 453
Birthweight, 97, 100, 101, 110–111, 159–178, 192, 193, 224, 231, 241, 273, 299, 311–312, 426, 439
duration of pregnancy, 409
induced pregnancy termination, 105–106

midwife records, 159
parental age and, 159, 160–161, 165, 173, 176, 178
parity, 159, 160–161, 165, 173, 176, 178, 193, 409, 411, 421, 426
sex differences, 164–178, 220
socioeconomic status, 159, 176, 177–178, 426
"Black market," 32, 160
Blood
hemogloblin, 365
petechiae, 63, 64, 65, 380
types, 53, 56, 368
British Medical Research Council, 271
Burn injuries, 61, 69

C

Cancer, 389–398, 404, 415, 417, 418, 419–420, 422, 423, 424, 425, 426, 427, 434, 435, 448, 455, 457, 466–468, 487, 491
leukemia, 61, 97, 98, 390, 395–396, 404, 422, 427, 491
murine, 390, 398
neuroblastomas, 397, 457
retinoblastomas, 389–390, 392, 394, 396, 397, 457, 468
study methodology, 37, 46, 47
Wilms tumor, 390, 392, 394, 396–397, 457, 468
Cataracts, 61, 97, 471–472
Chest circumference, 130, 192, 196, 439, 459
Chi square, 83, 91, 93, 96, 101, 110, 133, 142, 155, 157, 185, 189, 409
Chickuzen type, of Japanese, 199
Child deaths, 6, 291–323, 327–340, 455, 462, 466, 479, 487
parental age and, 307, 308, 309, 311, 407
parity effects, 307, 310, 409, 411, 421, 426
sex differences, 306–307, 396, 407–408, 423, 424, 426
socioeconomic status, 299–300, 301, 335, 409, 426
study methodology, 293–295, 328–333, 405–415

SUBJECT INDEX

study samples, 292–293, 328–332, 405–406, 407, 409
see also Neonatal death
China, 49
Chinese, in Nagasaki, 50, 56
Christianity
 ban against, 50, 52
 consanguineous marriages, 83
 Franciscans, 50
 Jesuits, 49
Chromosomes, 5, 6, 116, 231–232, 321, 337, 338, 344–361, 364, 378, 443–444, 488
 autosomal trisomics, 352–353, 353, 354, 355
 carcinogenesis, 390
 sex chromosomes, abnormalities, 348, 359–361
 sex-specific, 116, 232, 280, 288, 289, 319, 344, 347, 349, 435, 436, 438, 441, 444, 445–446, 457, 458–459, 468–469, 475, 476, 479
 structural rearrangements, 349–352, 354, 356, 435–436, 446, 457–458, 468–469
 X-chromosome, 116, 438, 468, 475
 Y-chromosome, 116, 232
Codes for data, 56, 246, 249, 253
Committee for the Investigation of the Effects of the Atomic Bombs, 30, 48
Committee on Atomic Casualties, 9, 29, 47, 115
Comparability of irradiation subclasses, 81, 83, 87, 89, 91, 97, 99, 101, 220, 301, 308–311, 313
Concomitant variation, 101, 103–104
Congenital defects, *see* Malformation, congenital
Consanguineous parents, 4, 271–279, 281
 anthropometric findings, in children, 275
 Christianity and, 83
 irradiation subclasses, 81–83, 97, 101, 104, 106, 272, 275
 malformation, congenital, in children, 275, 276
 rural areas, 83
 sex ratios, in children, 275, 281
 stillbirths, in children, 275, 277, 278
Contingency analysis, 332, 333, 334

Control groups, 46–47, 114–115, 121, 129–130, 220, 225, 228, 336–337, 345, 389, 391, 405, 453, 454
Covariance analysis, 103, 111, 112, 151, 165, 170, 172, 175, 199, 203
Cultural factors
 Christianity, 49, 50, 52, 83
 Western contacts, 49–56
Cytogenetics, 5, 8, 46, 56, 344–361, 435–436, 438, 445, 447, 454, 457–458, 469, 478–479, 488

D

Danes, in Nagasaki, 51
Deshima, 50, 52
Development, *see* Growth and development
Diagnoses
 malformation, 127–132
 see also Autopsy findings
Diarrhea, 57, 63, 64, 65, 180, 185
"Disaster effect," versus "radiation effect," 61, 178, 220
Distance-dosage relation, 62, 64, 70, 74, 75, 78, 79–80, 121, 217, 273, 308, 318, 328–329, 330, 373, 379, 432, 453
Distance-symptom relation, 62–72, 75, 76, 77, 78, 98, 121, 123, 273, 329–330, 379, 380, 453
DNA, 7, 445, 447, 452, 481, 488–491, 493
Dose and dosimetry, 61–80, 317–319, 321–322, 328–331, 378–381, 389, 392, 403, 408
 distance-dosage relation, 62, 64, 70, 74, 75, 78, 79–80, 121, 217, 273, 308, 318, 328–329, 330, 373, 379, 432, 453
 doubling dose, 431–448, 452, 456, 459–462, 471, 472–482, 487
 DS86, 5, 355, 369, 373, 377–386, 389, 392, 403, 404, 406, 407, 409, 411, 413–414, 416, 419–425, 428, 433, 435, 439, 443, 454, 455
 exposure categories, 72–99, 81–211, 273, 275, 281–282, 308–315, 318, 328–329, 331, 379–380

gamma radiation, 368, 407, 433–434, 468
lethal irradiation, 10
neutron, 329–330, 368, 380, 381, 393, 403, 407, 410, 413–414, 416, 419–420, 422, 433–434, 436, 444, 454
protein electrophoresis study, 368–369, 373, 436–438, 441, 442, 447, 454, 458, 473–475
residual radiation, 51, 61–66, 79, 127, 217–218
roentgen units, 10, 62, 73–74, 78, 330
shielding and, 64–66, 67, 68, 69–78, 273, 328, 329, 368, 379, 380, 381, 392, 433
T57, 318, 328
T65DR, 5, 329, 330, 331, 345, 368, 369, 377, 379, 380–381, 382, 384, 392, 405, 406, 409, 433, 439, 443, 459
Down syndrome, 436, 445–446
Drosophila, 31, 227, 231–232, 234, 236, 237, 239, 240, 241, 244, 321, 438, 446, 480
DS86, 5, 355, 369, 373, 377–386, 389, 392, 403, 404, 406, 407, 409, 411, 413–414, 416, 419–425, 428, 433, 435, 439, 443, 454, 455
Dutch, in Nagasaki, 50, 51, 52–53

E

Economic factors
 "Black market," 32, 160
 post-War Japan, 32, 138, 160, 378, 443, 453
 research funding, 29
 see also Socioeconomic status
Education
 attainment, 300, 301, 443, 462
 midwives, 32
 pregnancy registration pamphlet, 8
Electrophoresis, see Protein electrophoresis
English, in Nagasaki, 50, 51
Epilation, 57, 63, 64, 65–69, 75, 273, 329–330, 380
Error of measurement, 313, 316
 doubling dose, 384–385, 480, 487

malformation data, 91, 97, 130
murine studies, 7
neonatal studies, 43, 91
protein electrophoresis studies, 368–369, 374
sex ratio analyses, 286–287
stillbirths, 333
Ethnicity, see Racial and ethnic factors
Exposure to radiation, 3, 4–5, 73, 192, 273, 321–323
 anthropometric findings, 43, 100, 110–111, 130, 192–211, 224, 457, 477
 autopsy findings, 212–219
 birthweight, 159–178
 cancer and, 389, 393–398
 categories, 72–99, 81–211, 273, 275, 281–282, 308–315, 318, 328–329, 331, 379–380
 death, neonatal, 179–191, 277–278
 mothers versus fathers, 73, 82–96, 116–126, 134, 139–140, 152–155, 162, 166, 186–191, 193–198, 202, 213–218, 223, 226, 276, 277, 338, 347, 348, 350, 410
 sequelae, 61, 97, 99
 sex ratio, 116–126
 stillbirths, 146–158
 see also Dose and dosimetry; Shielding

F

Finland, 49
Foreigners, genetic influence, 49, 51–56
Formosa, 49
Franciscans, in Nagasaki, 50
French, in Nagasaki, 51
Funnel chest, 130

G

Gamma radiation, 78, 79, 80, 217, 330, 363, 368, 381, 407, 433–434, 468
Gender factors, see Sex differences; Sex ratio
Genes, 233, 240, 245
 accumulation factor, 240–241
 additive effects, 241
 DNA, 7, 445, 447, 452, 481, 488–491, 493

drosophila, 240
genomes, 2, 364–365, 396, 447, 473, 480, 481
Japanese inhabitants, genetic history, 49, 51–56, 199
mRNA, 365
see also Chromosomes
Genetic damage, 9–11, 30, 132–138, 217, 365–366, 396, 452–482
consanguineous parents, 271–279
"disaster effect" versus "radiation effect," 61, 178, 220
Down syndrome, 436, 445–446
indicators, 31, 33, 100–101, 102–104, 224–225, 445–448
see also Malformation, congenital; Mutation
Genetics Conferences, 30, 57, 245, 259
Genetics Long Form, 37, 42, 89, 121, 249–257, 272–273
Genetics 9-Month Follow-up, 43, 253
Genetics Short Form, 33, 37, 52, 81, 89, 179, 246–248, 272
Genomes, 2, 364–365, 396, 447, 473, 480, 481
Gingivitis, 63, 65–69, 75, 273, 330
Goa, 49
Government role, Japanese, 1, 10, 30–31
National Institute of Health, 30, 33, 48, 272, 281
National Research Council, 30
see also specific departments and agencies
Growth and development
animal studies, 231
socioeconomic status, 299
Guinea pigs, 230–231, 241

H

Hair, *see* Epilation
Hawaii, 56, 201
Head circumference, 130, 192, 195, 439, 459
Heart disease, 53, 130
Height, *see* Length of child
Hemorrhages, 63, 64, 65, 147, 380
Hernia, 53, 130
Heterogeneity, exposure cells, 113, 161, 165, 167, 175, 176, 196, 198, 203, 208
Histocompatibility, 472–473
Historical perspectives
genetics study, 2–5, 7, 9–11, 29–47, 57, 60, 432–433, 453
Japanese inhabitants, genetic determinants, 49, 51–56, 199
methodologies, 2–5, 7, 9–11, 29–32, 291–293
patient histories, 37, 42, 44
Western contacts, 49–56
see also Radiation histories
Honshu, 21
Hydrocephaly, 130

I

Indonesia, 49
Infants, 42–46
classification, irradiation, 81–99
sex differences, 87, 138, 139–140, 142, 152–155, 186–191, 193–198, 202, 204–207, 213, 214, 220, 226, 301
see also Birthweight; Neonatal deaths; Stillbirths
International Classification of Diseases, 391, 423
Irradiation, lethal dose, 10
Ishikawa type, of Japanese, 199

J

Japanese, origin of, 49, 51–56, 199
Japan Science Council, 62, 217
Jesuits, 49

K

Korea, 49, 201
"Korean colony," in Hiroshima, 56
Kyushu, 49

L

Legal issues, abortion, 3
Length of child, 192, 194, 199–200, 439, 459
Leucopenia, 62, 74–75, 76, 77, 78

Leukemia, 61, 97, 98, 390, 395–396, 422, 427, 491
Likelihood ratio, 176, 198
Litter size, 319–320, 339, 465, 477, 481
L statistic, 176, 198
Lymphomas, 390

M

Macao, 49, 52
Malaysia, 49
Malformation, congenital, 2, 3, 4, 6, 37, 97, 98, 99, 100, 101, 104, 127–145, 222, 223–224, 227, 229–230, 339, 378, 381, 382, 434, 454–456, 463–466
 autopsy findings, 215, 217
 concealment, 91, 97
 consanguineous parents, 275, 276
 diagnosis, 127–132
 error of measurement, 91, 97, 130
 heart disease, 53, 130
 midwives records, 127–128
 parental age, 133, 134–138, 141, 142, 143, 383
 parity effects, 133
 skeletal, 43, 130, 132, 472
 see also specific types
Manchuria, 49, 200
Marshall Islands, 49, 77–78
Medical Association of Hiroshima, 391
Medical Association of Nagasaki, 391
Meiji era, 50
Mental defects, 130, 132
 Down syndrome, 436, 445–446
Methodology, 2–8, 33–48, 293–295, 487–493
 abortion data, 37, 46, 47
 administrative framework, 29–30, 57, 60
 anamnestic data, 37, 42
 attribute data, 106–110, 190, 224
 biochemistry, 5, 8, 367–368
 cancer study, 391–393, 427
 childhood deaths studies, 293–295, 328–333, 405–415
 codes for data, 56, 246, 249, 253
 cytogenetic study, 345
 historical perspective, 2–5, 7, 9–11, 29–32, 291–293

 International Classification of Diseases, 391, 423
 pregnancy registration, 10, 32, 33–37, 91, 96, 101, 105, 272–273, 281–282, 292–293, 332, 378, 391, 432, 453
 protein electrophoresis, 364–369, 436–437, 441, 447, 454, 458, 473–475
 sex ratio studies, 3, 4–5, 7, 31, 101, 112, 229
 see also Diagnosis; Dose and dosimetry; Sampling; Statistical approaches and programs
Midwives, 3, 32, 33, 35, 37, 91
 birthweight records, 159
 malformation records, 127–128
Midwives Association, 33
Ministry of Health and Welfare, Japanese, 1, 30
Miscarriages, 10, 37
mRNA, 365
Multiple births, 274, 294
 litter size, mice, 319–320, 339, 465, 477, 481
Multivariate means, 192, 193, 196, 198, 199–200, 203, 204–205, 207, 211, 222
Murine studies, 7, 228, 229–230, 231, 232, 236, 237, 239, 319, 321, 339–340, 443, 444–445, 452, 462–482, 490
 cancer, 390, 398
 nutrition, 138
 protein electrophoresis, 364, 365
Mutation, 9, 10, 328, 378, 386, 431–448, 451–482
 detrimental, 192, 234–235, 291
 doubling dose, 431–448, 452, 456, 459–462, 471, 472–482
 invisible, 42, 159, 234, 235
 lethal, 116, 146, 232, 233, 234, 235, 236, 237, 275, 280, 288, 289, 317, 321–322, 327–328, 364, 438, 473, 476
 versus natural selection, 241–244, 271, 276
 protein electrophoresis, 6, 363–374, 436–438, 441, 442, 447, 454, 458, 473–475
 radiation-induced, 5–6, 31, 228, 238–240, 243–245, 321, 431, 440–441, 444–445, 446, 461–482

SUBJECT INDEX

ratio of types, 234, 235
sex differences, 116, 231, 237, 238–239, 337, 436, 438, 447
spontaneous, 31, 233–238, 236, 237, 271, 321, 337, 366, 431, 440–441, 442, 445, 446–447, 452, 460, 468–469, 471, 473, 478
tumors, 389–398, 327
visible, 9, 10, 43, 132, 232, 233, 234, 236, 237, 475–477
X-radiation effected, 319–320, 364–366, 390, 397, 443, 469, 473–475

N

Nagasaki Medical School, 57, 130, 212, 216, 217
National Academy of Sciences-National Research Council (U.S.), 2, 4, 29, 31, 47, 48, 271
 see also Committee on Atomic Casualties
National Institute of Health, Japanese, 30, 33, 48, 272, 281
National Research Council, Japanese, 30
Natural selection, 241–244, 271, 276
Neonatal deaths, 3, 6, 31, 43, 91, 97, 100, 101, 105, 180, 179–191, 222, 224, 226, 227, 231, 275, 277–278, 381, 382, 417, 434–435, 455, 460
 parental age and, 180–181, 182–185, 191, 383
 sex differences, 185–187, 189, 220, 300, 301
Neutron radiation, 74, 79, 80, 329, 330, 337, 338, 363, 368, 380, 381, 393, 403, 407, 410, 413–414, 416, 419–420, 422, 433–434, 436, 444, 454
Nine-months program, 42–46, 110, 127, 130–132, 143–145, 179–191, 193
Non-orthogonality, 114, 185
Neuroblastomas, 397, 457
Nutrition, 61, 133, 138, 147–148, 159–160, 180, 200, 300, 301

O

Okayama type, Japanese, 171
Omnibus tests, 112

Operating characteristic curves, 126, 316–317
Oropharyngeal lesions, 57, 63, 64, 65, 330, 380

P

Parents and parenting, general
 cooperation with study, 91, 97
 irradiation categories, 72–99, 81–211, 273, 275, 281–282, 308–315, 318, 328–329, 331, 379–380
 see also Age, parental; Consanguineous parents; Exposure
Parity effects, 83, 85–87, 104, 271, 297–298
 birthweight, 159, 160–161, 165, 173, 176, 178, 193, 409, 411, 421, 426
 child deaths, 307, 310, 409, 411, 421, 426
 differences in irradiation subclasses, 83–85, 97, 101, 104, 133, 136–137, 138, 143, 148–152
 growth and development, 193
 malformation frequency, general, 133
 neonatal death, 180–181, 182–185, 188, 190
 stillbirths, 148, 156, 157, 181
Petechiae, 66–69, 75, 273, 329–330
Physicians
 birth, attended by, 32
 malformation diagnosis, 127, 128
 pregnancy/birth registration, 35, 37, 281
Plants, irradiation effects, 30
Plutonium, 46, 56, 57
Political factors
 Christianity banned, 50, 52
 missionaries, 50
 research influences, 4, 492
Polynesia, 49
Portmanteau tests, 112
Portuguese, in Japan, 49–50, 52
Power curves, 225, 226
Pregnancy and pregnant women
 duration of pregnancy, 298–299, 409
 miscarriages, 37
 registration, 10, 32, 33–37, 91, 96, 101, 105, 272, 281, 292–293, 332, 378, 391, 432, 453

termination effects, 13–269, 434
 see also Abortion
Protein electrophoresis, 6, 363–374, 436–438, 441, 442, 447, 454, 458, 473–475
Purpura, 57, 63, 64, 65

Q

Questionnaires, forms
 births, 34–35, 38–41
 Genetics Long Form, 37, 42, 89, 121, 249–257, 272–273
 Genetics Short Form, 33, 37, 52, 81, 89, 179, 246, 272
 pediatric follow-up, 44–45
 response bias, 229

R

Racial and ethnic factors, 117, 225
 age and, 54–55
 Japanese inhabitants, genetic history, 49, 51–56, 199
Radiation, indicators of genetic effects, 31, 33, 100–101, 102–104, 271
 anthropometric findings, 192–211
 autopsy findings, 212–219
 birthweight, 159–178
 consanguineous parentage, 271–279, 281
 death, neonatal, 179–191, 227, 277–278
 malformation, 127–145
 sex ratio, 116–126, 280–290
 stillbirth, 146–158
Radiation, residual, 51, 61–66, 79, 127, 217–218
Radiation categories, 72–99, 81–211, 273, 275, 281–282, 308–315, 318, 328–329, 331, 379–380
 defined, 72, 117, 273
 dosage in, 73, 78
Radiation Census, 57, 72
Radiation Effects Research Foundation, 1, 8, 344, 391, 397, 405, 427, 445, 488, 489, 490, 491
Radiation histories, 33, 37, 66, 72, 281, 330
 Genetics Long Form, 37, 42, 89, 121, 249, 272–273
 Genetics Short Form, 33, 37, 52, 81, 89, 179, 246, 272
Radiation sickness, symptoms of, 57
 distance and, 62–72, 75, 76, 77, 78, 98, 121, 123, 273, 329–330, 379, 380, 453
 see also Diarrhea; Epilation; Gingivitis; Hemorrhages; Oropharyngeal lesions; Petechiae; Purpura
Regression analyses, 100, 160, 161, 165, 167, 169, 170, 172, 173, 176, 274, 275–276, 278, 282–287, 333, 335, 336, 338, 381, 384, 389, 392–393, 395–396, 403, 421, 423, 425, 428, 434, 435, 436, 439–443, 456, 457, 461
Retinoblastoma, 389–390, 392, 394, 396, 397, 457, 468
Rockefeller Foundation, 48
Roentgen units, 10, 62, 73–74, 78, 330
Roentgen equivalent physical, 73, 74, 79, 231
Rural areas
 birthplace, 99, 117, 172
 consanguineous marriages, 83
 residence, 146, 294
Russia, 49
Russians, in Nagasaki, 51, 53
Ryukyu Islands, 49

S

Sampling, 99, 292–299, 313, 316–317, 336, 378, 379, 381–382, 432, 453–454
 autopsies, 212–213, 217, 219
 balanced, 103–104
 bias, 99, 106, 219, 228–229
 cancer study, 391
 childhood death studies, 292–293, 328–332, 405–406, 407, 409
 control groups, 46–47, 114–115, 121, 129–130, 220, 225, 228, 336–337, 345, 389, 391, 405, 453, 454
 cytogenetic study, 345, 346
 protein electrophoresis, 366–367
 "Ten-percent sample," 37, 299–300, 335, 345
Satsuma type, of Japanese, 199

Secondary infections, 61
Sex chromosomes, 116, 232, 280, 288, 289, 319, 344, 347, 349, 435, 436, 438, 441, 444, 445–446, 457, 458–459, 468–469, 475, 476, 479
Sex differences, 109, 462
　anthropometric, 138, 152, 163, 193–196, 198, 202, 204–207, 211, 220
　autopsy findings, 214
　birthweight, 164–178, 220
　cancer, 396, 423, 424
　childhood deaths, 306–307, 396, 407–408, 423, 424, 426
　chromosomal abnormalities, 348, 359–361
　infants, 87, 138, 139–140, 142, 152–155, 186–191, 193–198, 202, 204–207, 213, 214, 220, 226, 301
　mothers versus fathers, exposure, 73, 82–96, 116–126, 134, 139–140, 152–155, 162, 166, 186–191, 193–198, 202, 213–218, 223, 226, 276, 277, 338, 347, 348, 350, 410
　mutation, 116, 231, 237, 238–239, 337, 436, 438, 447
　neonatal death, 185–187, 189, 220, 300, 301
　protein electrophoresis, 369–371, 441
　socioeconomic status, 301
　wives versus husbands, data on, 33, 66–69
Sex ratios, 85, 100, 116–126, 222–223, 226, 227, 228–229, 232, 280–290, 293, 302–305, 322, 410, 438, 458–459, 476
　consanguineous parentage, 275, 281
　error of measurement, 286–287
　exposure to radiation, 116–126
　parental age, 117, 289
　stillbirths, 152–155
　study methodology, 3, 4–5, 7, 31, 101, 112, 229
Shielding, 64–66, 67, 68, 69–78, 273, 328, 329, 368, 379, 380, 381, 392, 433
Skeletal system
　acetabulum, dysplasia, 43, 130, 132
　malformation, general, 472
Siberia, 49
Sinus, pilonidal, 43, 132

Social factors, 1, 10–11, 32
　abortion laws, 3
　cultural, 2, 49–56, 83
　midwives, 32, 91
　parental cooperation with study, 91, 97
Socioeconomic status, 42, 117, 273, 385–386, 443, 462
　autopsied infants, 213, 214
　birthweight, 159, 176, 177–178, 426
　child deaths, 299–300, 301, 335, 409, 426
　educational attainment, 300, 301, 443, 462
　growth, 299
　irradiation subclasses, 87–89, 99, 101, 104
　nutrition and, 138, 148
　sex differences, 301
Southern blotting, 365, 366
Spanish, 50, 70
Statistical approaches and programs, 2–8, 100–115, 331, 412–415
　analysis of dispersion, 111, 114, 196, 197, 198, 201, 202, 208–209
　attribute data, 106–110, 190, 224
　Bartlett's test, 155, 203, 207
　cancer study, 392–393
　chi square, 83, 91, 93, 96, 101, 110, 133, 142, 155, 157, 185, 189, 409
　concomitant variation, 101, 103–104
　confidence limits, 225
　contingency analysis, 332, 333, 334
　control groups, 46–47, 114–115, 121, 129–130, 220, 225, 228, 336–337, 345, 389, 391, 405, 453, 454
　covariance analysis, 103, 111, 112, 151, 165, 170, 172, 175, 199, 203
　determinants and matrices, 196
　F test, 179
　heterogeneity, within-cell, 113, 114, 161, 165, 167, 175, 176, 196, 198, 203, 208
　historic inhabitants of Japan, 50–51
　index of absolute difference, 133, 213
　likelihood statistic, 176, 198
　measurement data, 111, 112
　means, multivariate, 192, 193, 196, 198, 199–200, 203, 204–205, 207, 211, 222

omnibus tests, 112
portmanteau tests, 112
power curves, 225, 226
pyramidal handling of data, 101
Radiation Census, 57, 72
radiation data, 57, 62, 66–69, 72
regression analyses, 100, 160, 161, 165, 167, 169, 170, 172, 173, 176, 274, 275–276, 278, 282–287, 333, 335, 336, 338, 381, 384, 389, 392–393, 395–396, 403, 421, 423, 425, 428, 434, 435, 436, 439–443, 456, 457, 461
significance test, one-tailed, 97, 222, 226, 339, 492
variance, generalized, 85, 87, 111–112, 113, 114, 176, 192, 196, 201, 203, 207
variance analysis, 100, 104, 161, 163, 165, 166, 168, 171, 178, 185, 197, 203, 207
vital statistics, 10, 56–57, 121, 281, 295, 391, 426
Wilks' test, 196
see also Error of measurement; Questionnaires, forms; Sampling
Stillbirths, 3, 6, 9, 31, 37, 100, 101, 146–158, 152, 179, 181, 224, 226, 227, 230–231, 272, 333, 381, 382, 434, 447, 448, 454–456, 463–466
concealment of, 91, 97
consanguine parents, 275, 277, 278
error of measurement, 333
induced termination and, 105
parental age, 147, 148–152, 158, 383
parity effects, 148, 156, 157, 181
sex ratios, 152–155
Sweden, 49
Syphilis, 37, 42, 89, 90, 101, 104, 147, 157, 180, 213, 215

T

T57, radiation dose, 318, 328
T65DR, radiation dose, 5, 329, 330, 331, 345, 368, 369, 377, 379, 380–381, 382, 384, 392, 405, 406, 409, 433, 439, 443, 459

"Ten-percent sample," 37, 299–300, 335, 345
Thalassemia, 365
Togugawa Shogunate, 50, 52
Tumors, malignant, see Cancer

U

United Kingdom, 491
English, in Nagasaki, 50, 51
United Nations, 271, 444
"Unregistered Series," 35, 212, 272
Uranium, 46, 56
Urban areas
origin, 117
residence, 99, 146, 294
Uterus, dilation and curettage, 89, 91, 95, 104

V

Variance, generalized, 85, 87, 111–112, 113, 114, 176, 192, 196, 201, 203, 207
see also Analysis of variance; Covariance analysis
Variances, test for equality of, 112
Viral diseases, 133
Vital statistics, 10, 56–57, 121, 281, 295, 391, 426

W

Weight of child, 192, 199, 439
see also Birthweight
Wilks' test, 196
Wilms tumor, 390, 392, 394, 396–397, 457, 468
World Health Organization, 271, 391

X

X-chromosome, 116, 438, 468, 475
X-ray irradiation, 319–320, 364–366, 390, 397, 443, 469, 473–475

Y

Y-chromosome, 116, 232

A00002213B392